BASIC
ANIMAL
NUTRITION
AND
FEEDING

BASIC ANIMAL NUTRITION AND FEEDING

FIFTH EDITION

W. G. Pond, Ph.D.
Courtesy Professor
Department of Animal Science
Cornell University
Ithaca, New York

D. C. Church, Ph.D.
Professor Emeritus
Department of Animal Science
Oregon State University
Corvallis, Oregon

K. R. Pond, Ph.D.
Professor and Chair
Department of Animal and Food Science
Texas Tech University
Lubbock, Texas

P. A. Schoknecht, Ph.D.
Information Services
Center for Teaching, Learning, and Technology
University of Richmond
Richmond, Virginia

WILEY

ACQUISITIONS EDITOR	Rebecca Hope
EDITORIAL PROGRAM ASSISTANT	Dana Kasowitz
PRODUCTION MANAGER	Pamela Kennedy
PRODUCTION EDITORS	Cassandra Raphael and Sarah Wolfman-Robichaud
MARKETING MANAGER	Clay Stone
SENIOR DESIGNER	Madelyn Lesure
ILLUSTRATION COORDINATOR	Gene Aiello
SENIOR PHOTO EDITOR	Lisa Gee

This book was set in 10/12 pt Century Old Style by Matrix Publishing and printed and bound by Malloy Lithographers. The cover was printed by Phoenix Color.

This book is printed on acid free paper. ∞

To order books or for customer service, please call 1-800-CALL WILEY (225-5945).

ISBN: 0-471-21539-2
WIE ISBN: 0-471-65893-6

Printed in the United States of America

10 9 8 7 6 5 4 3 2 1

Preface

THIS FIFTH EDITION of *Basic Animal Nutrition and Feeding,* like the first four editions, is intended for students interested in the basic principles of animal nutrition and the application of these principles, the role of animal nutrition in modern agriculture and society, the current knowledge of nutrient metabolism, the formulation of diets from an array of available feedstuffs, and the life-cycle feeding of animals.

The fifth edition retains the general style and format of previous editions and the thorough coverage of concepts and principles of animal nutrition, while introducing important new knowledge about nutrient metabolism, nutrient requirements, and current life-cycle feeding systems. We have added two new chapters to address developing areas. One new chapter, Chapter 16 (Regulation of Nutrient Partitioning), contributed by Michael J. Vande-Haar, provides a timely discussion of emerging technologies in modifying and increasing the efficiency of nutrient metabolism and the composition of animal source food. The second new chapter, Chapter 13 (Natural Toxicities in the Food Chain), addresses the role of agricultural production and animal nutrition in protecting the environment from excessive levels of minerals and other nutrients entering the food chain. Heightened attention to the environment has driven recent advances in animal nutrition and feeding practices.

An exciting new dimension of the book is the use for the first time (to our knowledge), in a basic animal nutrition textbook, of electronic images and animations depicting various processes in nutrient digestion and metabolism, photographs of signs of specific nutrient deficiencies in animals, and other powerful learning tools. We believe these features will add appreciably to the value of this textbook and will provide a solid foundation of knowledge for students of animal nutrition and feeding.

The section on life-cycle feeding of individual animal species includes chapters contributed by authorities in their respective fields of animal nutrition. Contributors of these chapters are Calvin L. Ferrell (Beef Cattle, Chapter 22), Michael J. VandeHaar (Dairy Cattle, Chapter 23), Richard E. Austic (Poultry, Chapter 26) Harold F. Hintz (Horses, Chapter 27) and Duane E. Ullrey (Dogs and Cats, Chapter 28; Fish, Chapter 29; and Zoo Animals, Chapter 30).

The fifth edition is unique in that, unlike most current animal nutrition textbooks, it covers fundamental nutrition and applied animal feeding all in the same book. If the instructor prefers to emphasize only one of these two areas, however, the teaching and reading assignments can be concentrated there. The authors are enthused about the book's potential for serving as an enhanced learning resource for undergraduate students of animal nutrition, especially through the supplementary information provided in electronic form.

Finally, we thank Dr. Alan Bell, Chair of the Department of Animal Science, Cornell University, for providing office space and other generous physical resources for one of us (WGP) during the revision process, Dr. Harry J. Mersmann for his valuable contributions to the discussion of lipids (Chapter 8), Dr. Hugo Varela-Alverez for his contribution to Diet Formulation (Chapter 21) through authorship and coauthorship of that chapter in the Third and Fourth Editions, respectively, Sue McGinnis and Betsy Miles of the University of Richmond for their technical expertise, and the editorial and production of staffs (Rebecca Hope, Dana Kasowitz, Cassandra Raphael, Sarah Wolfman-Robichaud, and Betty Pessagno) of John Wiley & Sons, Inc., for their collaboration and efficiency during the preparation of the Fifth Edition.

Wilson G. Pond
David C. Church
Kevin R. Pond
Patricia A. Schoknecht

Table of Contents

Preface   v

PART 1—INTRODUCTION TO ANIMAL NUTRITION

1. Concepts of Nutrition   **1**

2. Animal Nutrition: Its Role in Modern Agriculture and Society   **5**

3. Common Methods of Analysis for Nutrients and Feedstuffs   **15**

4. The Gastrointestinal Tract and Nutrition   **25**

5. Measurement of Feed and Nutrient Utilization   **49**

PART 2—NUTRIENT METABOLISM

6. Water   **61**

7. Carbohydrates   **73**

8. Lipids   **91**

9. Proteins and Amino Acids   **113**

10. Energy Metabolism   **145**

11. Macromineral Elements   **163**

12. Micro- (Trace) Mineral Elements   **185**

13. Mineral Toxicities and Organic Toxins in the Food Chain   **217**

14. Fat-Soluble Vitamins   **229**

15. Water-Soluble Vitamins   **251**

16. Regulation of Nutrient Partitioning (by Michael J. VandeHaar)   **275**

PART 3—APPLIED ANIMAL NUTRITION AND FEEDING

17. Factors Affecting Feed Consumption — 291

18. Feeding Standards and Productivity — 307

19. Feedstuffs — 321

20. Feed Preparation and Processing — 369

21. Diet Formulation — 381

22. Beef Cattle by Calvin L. Ferrell — 395

23. Dairy Cattle by Michael J. VandeHaar — 413

24. Sheep and Goats — 439

25. Swine — 461

26. Poultry by Richard E. Austic — 479

27. Horses by Harold F. Hintz — 501

28. Dogs and Cats by Duane E. Ullrey — 515

29. Fish by Duane E. Ullrey — 531

30. Zoo Animals by Duane E. Ullrey — 549

1

Concepts of
Nutrition

IN ANIMAL AGRICULTURE, ADEQUATE feeding of animals for production of meat, milk, eggs, or fiber is an essential component of the total enterprise. In animals as well as in humans, nutrition clearly may affect health and welfare, emotions, physical capabilities, susceptibility to and recovery from disease, and incidence and severity of chronic metabolic diseases of aging.

What is nutrition? A dictionary may define nutrition as the interrelated steps by which a living organism assimilates food and uses it for growth, tissue repair and replacement, or elaboration of products. Such a broad definition encompasses all forms of life, including plants and animals. On a global scale, the balance of nutrients and all energy transformations must be in equilibrium. Janick et al. (1976, page 75) described the nutritional aspects of this phenomenon succinctly:

The diversion of the flow of nutrients through the food cycle is the aim of all agricultural technologies. The distinction of modern agriculture is that it has augmented the food supply by increasing the rate at which nutrients flow through the cycle. This has been accomplished by several methods, but by far the most common and one of the most important consists of speeding the return of nurients to the soil, where they can be reabsorbed. Hence, in order to feed the human population we must ensure the nutrition of an assortment of plants, animals, and microorganisms.

In this book, we focus on animal nutrition. Most plants require only inorganic elements, nitrogen in the form of nitrate or ammonia, water, carbon dioxide, and solar energy captured by the chlorophyll in the plant through photosynthesis. In this way plants provide the essential link between soil and animal life. The organic molecules (proteins,

1

carbohydrates, lipids, vitamins) produced by plants provide the nutrients for animals, for certain plants, and for microorganisms. The science of nutrition requires the application of chemistry, physics, and mathematics as well as the integration of advances in soil science, plant science, animal science, biochemistry, engineering, genetics, production systems, and other disciplines. The competent practicing nutritionist must not only be knowledgeable about the nutrients and their chemistry, functions, occurrence, metabolism, and metabolic interrelationships, but must also be familiar with animal behavior, physiology, genetics, and husbandry and with microbiology and other disciplines.

What is a nutrient? A *nutrient* may be defined as any chemical element or compound in the diet that supports normal reproduction, growth, lactation, or maintenance of life processes. The six classes of nutrients are water, proteins and amino acids, carbohydrates, lipids, vitamins, and inorganic elements. Energy, which is required in the diet of all animals, can be provided by fat, by carbohydrate, and by the carbon skeleton of amino acids after removal of nitrogen. The nutrients provided by each of the nutrient classes are described later in the book within separate chapters. The nutrients support cellular needs for water, fuel, structural constituents (skin, muscle, bone, nerves, fat), and metabolic regulation. Nutrients that are required in the diet because they cannot be synthesized by the body in sufficient amounts to satisfy metabolic needs are termed "essential" or "indispensable" nutrients.

❑ DEVELOPMENT OF NUTRITIONAL SCIENCE

The science of nutrition is partly the outgrowth of observations by farmers and livestock feeders over many centuries and, more recently, advances in science and technology developed by animal scientists and those in other disciplines. Consulting nutritionists and veterinarians also provide input, particularly on aspects of applied nutrition. The quantitative aspects of nutrition, the ability to describe nutrient requirements for different species in a variety of situations, the definition of deficiency signs and symptoms, and the accumulation of knowledge about the metabolism of nutrients are the result of research done by scientists throughout the world. This knowledge has been developed primarily with domestic and laboratory animals, but also with animal tissues and cells or with microorganisms. For example, the laboratory rat has made a tremendous contribution to our knowledge of vitamins, amino adds, minerals, and toxicants. The dog played an important part in the discovery of insulin and metabolism of carbohydrates and in the discovery of the role of nicotinic acid in the prevention and cure of pellegra. The guinea pig was the animal model used to elucidate the cause and prevention of scurvy. Hamsters, pigs, monkeys, chickens, quail, and other mammals and birds have all played a role in expanding knowledge of nutrition.

Nutritional science has progressed rapidly because of the availability of different types of models (e.g., tissue cultures, cell cultures, bacteria, laboratory animals) to substitute for various domestic species or humans. Such procedures have been used to develop information that might otherwise have been impossible to obtain with the target species because of the constraints of cost, ethics, or time.

Nutrition, like other biological sciences, is less precise than the physical sciences, primarily because biological organisms are variable. In higher animals, not even identical twins are exactly alike; the environments of any two animals are nearly always different. Therefore, nutrient needs are apt to be different between and among animals. Animal nutrition, as practiced today, requires that the nutritionist be able to formulate diets or supplemental feeds that are sufficiently palatable to ensure an intake adequate (but not necessarily maximal) for the purposes desired. Formulated diets must nearly always supply adequate and balanced levels of nutrients at reasonable cost for the desired level of animal performance. The diet formula may include growth promotants, medicants, or other nonnutritive additives for specific purposes. In addition, the diets so formulated must have adequate milling, mixing, handling, and storage properties. A nutritionist may be called on to know or to do numerous other things, but these functions would seem to be the minimum that should be expected.

The rapid growth in the body of scientific nutritional information has resulted in the specialization of animal nutritionists. Most animal nutritionists specialize in either nonruminant (monogastric or simple-stomached) animals or ruminant (multicompartmented stomach) animals. Those dealing with monogastric species may be involved with poultry (e.g., chickens, turkeys, ducks, geese), swine, horses, pets (dogs, cats), fish and other aquatic species, laboratory species (e.g., rats, mice, guinea pigs, monkeys), or captive animals in zoos. Ruminant nutritionists may specialize in beef cattle, dairy cattle, sheep, goats, deer, and other species. Beef cattle nutrition is often divided into cow-calf and feedlot specialities. Specialization allows the nutritionist to become more familiar with new literature and commercial practices than would be possible attempting to keep abreast of the total field.

In addition to specialization related to animal species, advances in biochemistry, molecular biology, and mathematical modeling of cellular and whole-animal metabolism have stimulated emphasis on approaches to nutrition involving research at the subcellular and molecular levels. This trend has allowed development of a more complete understanding of metabolism but has diverted research attention away from "whole-animal" nutrition. A blend of molecular and whole-animal nutrition research, and the spectrum of activities between, promises to yield greater progress in improving animal nutrition and productivity than strict emphasis on either approach alone. In practice, teams of individuals with different specialties often produce more complete and imaginative research and greater progress than the same people working individually.

The French chemist, Antoine Lavoisier (1743–1794), is often called the founder of the science of nutrition. Following early studies in the eighteenth century, progress was relatively slow in the nineteenth century. The need for protein, fat, and carbohydrates was recognized, and most of the research emphasis was on these nutrients or energy utilization, along with a gradual development of data for some of the mineral elements. During the latter half of the twentieth century, there was greater accumulation of knowledge of vitamins, amino acids, fatty acids, inorganic elements, energy metabolism, protein metabolism, and nutrient requirements and interrelationships. In the first decade of the twenty-first century, rapid advances in molecular nutrition, computer technology, and the broad area of biotechnology promise to enhance knowledge of the complex field of animal nutrition, all to the benefit of society.

Today, we recognize more than 40 nutrients needed in the animal diet (the number depends on the animal species) in contrast to the three classes of nutrients (carbohydrate, fat, and protein) that were first recognized. Some nutritionists believe that additional food constituents will be identified as nutrients for some species under some conditions. Certainly, the unknown nutrients must not be critical to life (or are needed in very small amounts), since domestic animals can be maintained through a complete reproduction cycle when fed purified diets containing all of the known required nutrients.

Developments in nutrition have been facilitated greatly by improved analytical techniques (see Chapter 3) and a vastly greater knowledge of chemistry and biochemistry, animal physiology, pathology, genetics, genomics, proteomics, and other related sciences. Information on mineral element nutrition has been expanded through instrumentation that can now detect some of these elements in the parts per billion range. In addition, advances in mass spectrometry permit the use of stable isotopes to study the metabolism of carbohydrates, lipids, amino acids, proteins, and mineral elements in feedstuffs intrinsically labeled with the isotopes of interest (carbon, nitrogen, oxygen, and other elements). Such analytical capabilities provide a new threshold for achieving a more detailed understanding of nutrient metabolism and dietary requirements for all stages of the life cycle of animals and humans.

These advances in science have resulted in steady improvements in diet formulation and more efficient animal production. Genetic and animal management improvements also have been achieved, resulting in improved efficiency of production for all food animal species.

❑ VARIATIONS IN KNOWLEDGE OF NUTRIENT REQUIREMENTS AMONG ANIMAL SPECIES

Nutrient needs of chickens have been defined more precisely than those of other domestic species, primarily because chickens are grown under more uniform environmental conditions than other domestic animals and are less genetically diverse. In addition, the age and weight of broilers when they are marketed is relatively uniform, and they are usually grown under similar conditions and fed diets of similar composition regardless of geographic location. All of these factors make it less difficult to determine the quantitative nutrient requirements of chickens compared to those of other species. Unique features of the anatomy and functions of the digestive system of various species of animals (e.g., the complex stomach of ruminants compared with the simple stomach of nonruminants) are associated with differences in nutrient requirements among domestic animal species. Furthermore, it is now recognized that selection for specific traits (e.g., growth rate, body composition) within a given species produce genetic changes in quantitative nutrient requirements.

❑ CHALLENGES IN HUMAN NUTRITION

Even with our relatively advanced knowledge about animal nutrition, however, human malnutrition remains a very serious concern in the developing countries and among the poor in other countries. Surveys conducted by world health organizations indicate that the infants and children in these countries face nutritional marasmus (deficiency of energy, protein, and possibly other nutrients), kwasiorkior (deficiency of protein quality and/or quantity), and vitamin A deficiency. For all ages, iodine deficiency (goiter) and nutritional anemias (iron, vitamin B_{12}, folic acid, and other nutrients) remain severe problems. The number of chronically undernourished people in the world is estimated at more than 815 million, of which 95.3% live in developing countries, 3.3% in transition countries, and 1.4% in the industrialized countries (FAO, United Nations, 2001). According to this information, there has been a clear overall decline in the number of undernourished people in the developing regions of the world. Some countries (e.g., China, Peru, Indonesia, Nigeria, Thailand, Vietnam, and Brazil) have made good progress, yet many others have suffered reverses. In Chapter 2, we address the role of animals in providing improved human nutrition in a world whose population is predicted to increase from the current 6 billion to about 9 billion in 2035, a remarkable 50% increase in 30 years (one generation). This problem—providing adequate nutrition to all people—represents one of the most critical challenges facing society at the beginning of the twenty-first century.

❑ SUMMARY

- A nutrient is any chemical element or compound in the diet that is required for normal reproduction, growth, lactation, or maintenance of life processes.

- Animals depend on plants as a source of nutrients. Without plants, animal life could not be sustained.

- Plants require only inorganic elements, nitrogen, and solar energy captured by chlorophyll through photosynthesis to produce proteins, carbohydrates, lipids, and vitamins required by animals.

- More than 40 nutrients are required in the diet of most animal species for normal life processes.

- Advances in nutritional science have resulted in continuous improvements in efficiency of animal production and human welfare. Achieving adequate nutrition for the human population remains a major challenge to society and to animal agriculture.

❑ REFERENCES

FAO (Food and Agriculture Organization) of the United Nations. 2001. The state of food insecurity in the world 2001. www.fao.org/docrep/003/Y1500e03.htm

Janick, J., el al. 1976. The cycles of plant and animal nutrition. *Sci. Am.* **235:** 74–86.

2

Animal Nutrition: Its Role in Modern Agriculture and Society

THE RAPIDLY GROWING HUMAN population in a world with finite dimensions and resources presents a major challenge for agriculture and all of society. Adequate food supplies depend on the continuing advances in new scientific knowledge, the success of agricultural research and related disciplines, and the application of science and technology to the challenges related to the production of nutritious, safe, and wholesome food. Animal products have been constituents of human diets for centuries. Their consumption historically increases as economic status improves, except in the richest countries, where per capita consumption of animal products tends to plateau as family incomes rise. Food products from animals enhance the opportunity for adequate intake of nutrients in forms of high biological availability. Exclusion of animal products from the diet is not incompatible with human existence. However, current evidence indicates that major reductions in animal agriculture would disturb the ecological balance and create economic instability. In this chapter, we consider the nutrient requirements of plants and animals, describe the chemical and physical composition of feeds, and provide some information on components common to both plant and animal tissues.

❑ DEFINITION OF TERMS

A **nutrient** is any chemical element or compound in the diet that is required for normal reproduction, growth, lactation, or maintenance of life processes. It is difficult to give a completely accurate short definition of a nutrient. For example, some compounds, such as starch, are readily utilized by most species as a source of energy (and thus provide nourishment), yet starch is not specifically required by an animal as a source of energy or for any other function.

Food is an edible material that provides nutrients.

Feed (noun) refers to food but more commonly is used to designate animal food.

Foodstuff or **Feedstuff** is any material made into or used as food or feed, respectively.

Diet is a mixture of feedstuffs used to supply nutrients to an animal.

Ration is a daily allocation of food (or feed).

❑ NUTRIENTS REQUIRED BY PLANTS AND ANIMALS

Plants

In contrast to animals, the nutrient requirements of plants are simple. In general, plants take up nitrogen (N) in the form of nitrate or ammonia, and they synthesize complex proteins by incorporating these forms of N into amino acids and other intermediate products. Plants require a large number of inorganic elements. The qualitative mineral element requirements for plants are similar to those of animals. Plants may also require aluminum (Al), bromine (Br), cesium (Cs), and strontium (Sr). Thus, the primary nutrient needs of plants are the required mineral elements and N, which are normally obtained from soil through the roots. Through the process of photosynthesis, the plant takes in atmospheric CO_2, releases O_2, and synthesizes glucose, the fundamental biochemical required for plant growth. Using these basic components, the plant is capable of synthesizing all of the complex biochemicals that it requires for completing its life cycle.

Animals

Depending on animal age and species, animals require a source of N in the form of essential amino acids, fat in the form of essential fatty acids, essential mineral elements, fat-soluble and water-soluble vitamins, and a source of energy that may vary from primarily fat and protein for carnivorous animals to coarse fibrous plant tissue for some herbivorous animals. The amounts and proportions of nutrients required are influenced by the type of gastrointestinal tract, the age of the animal, its level and type of productivity (i.e., maintenance of body tissues, work, growth, milk, eggs, pregnancy), the dietary constituents available, and other factors. Because animals require about 40 nutrients, meeting dietary requirements may be difficult, depending on the availability of appropriate feedstuffs. Humans require the same nutrients as animals, although the quantitative needs of each nutrient for each productive function may differ. The complete list of nutrients known to be required by plant, animals, and humans is shown in Table 2.1.

❑ COMPOSITION OF ANIMAL FEED

The feed an animal consumes may vary from simple compounds such as salt (NaCl) or glucose to the complex mixtures provided by some plant and most animal products. Not all components provide usable nutrients. Indeed, some of the material consumed may be insoluble and/or indigestible, and some may be toxic under some conditions.

Table 2.2 contains an abbreviated list of some elements and compounds (including classes of nutrients) that may be present in feeds and foods. The names and terminology are introduced here to provide a background for discussion of nutrients in greater detail in later chapters. Quantitative data are not included in Table 2.2 because feed composition varies so much that to do so would be meaningless. Of course, not all feed formulas would contain all of the compounds listed in the table. Even for a single feedstuff, it is difficult to find published analytical values for all constituents of that feedstuff. This is because tables of feed composition are normally derived from compilations of individual data sets originating from many different laboratories.

Water, a major constituent of many feedstuffs, is not listed in Table 2.2. The other ingredients of a feedstuff constitute the dry matter of the feed,

TABLE 2.1 Nutrients required by plants, animals, and humans.

NUTRIENT	PLANT	REQUIRED ANIMAL	HUMAN	NUTRIENT	PLANT	REQUIRED ANIMAL	HUMAN
Water	x	x	x	Inorganic elements *(Continued)*			
Energy	x	x	x	Potassium	x	x	x
Carbohydrate		?[a]	?	Selenium	x	x	x
Fat		x	x	Silicon	x	x	x
Linoleic acid		x	x	Sodium	x	x	x
Linolenic acid		x	x	Zinc	x	x	x
Protein		?	x	Aluminum	x	?	?
Nitrogen	x			Bromine	x	?	?
Amino acids		x	x	Cesium	x		
Arginine		x	x	Strontium	x		
Histidine		x	x	Cadmium		?	?
Isoleucine		x	x	Arsenic		?	?
Leucine		x	x	Lithium		?	?
Lysine		x	x	Lead		?	?
Methionine		x	x	Nickel	x	?	?
Phenylalanine		x	x	Tin		?	?
Proline		x	?	Vanadium		?	?
Threonine		x	x	Vitamins			
Tryptophan		x	x	Vitamin A		x	x
Valine		x	x	Vitamin C		x	x
Inorganic elements		x	x	Vitamin D		x	x
Boron	x	x	x	Vitamin E		x	x
Calcium	x	x	x	Vitamin K		x	x
Cobalt	x	x	x	Vitamin B12		x	x
Copper	x	x	x	Biotin		x	x
Chromium	x	x	x	Choline		x	x
Chlorine	x	x	x	Folacin		x	x
Fluorine	x	x	x	Niacin		x	x
Iron	x	x	x	Pantothenic acid		x	x
Iodine	x	x	x	Pyridoxine		x	x
Magnesium	x	x	x	Riboflavin		x	x
Molybdenum	x	x	x	Myoinositol		?	?
Phosphorus	x	x	x	PABA		?	?

[a]A question mark (?) means that there is no unequivocal evidence for a dietary requirement; in some cases, one or more animal species have shown a requirement but others have not.

composed of organic compounds and inorganic elements, as shown in the table.

Organic compounds include N-containing compounds (proteins containing a range of amino acids, and an array of nonprotein N-containing compounds); lipids (generally fat soluble), carbohydrates (generally water soluble); and vitamins.

Nearly all animal feeds contain proteins, which are complex molecules containing various amino acids and other nonprotein components. Both animal and plant proteins may be very complex and may vary in the content, sequence, and configuration of their constituent amino acids, resulting in differences in molecular size, solubility, and digestibility of the protein. In addition, plants contain many amino acids not present in proteins, as well as other nitrogenous compounds, including nitrates and nucleic acids.

Lipids of many types are found in both animal and plant tissues. Lipids known to be required by animals are the fatty acids, linoleic and linolenic acid, and the fat-soluble vitamins A, D, E, and K.

TABLE 2.2 A simplified list of elements and compounds that may be present in food.

Organic compounds
  Nitrogenous
    Protein
      Amino acids
    Nonprotein (partial list)
      Peptides
      Amines and amides
      Nucleic acids
      Nitrates
      Urea
  Lipids
    Fatty acids
    Phospholipids (e.g., lecithin, sphingomyelin)
    Triacylglycerol (triglycerides)
    Sterols (e.g., hormones, cholesterol, vitamin D)
    Terpenoids (e.g., carotene, xanthophyll)
    Waxes (e.g., cutin)
  Carbohydrates
    Monosaccharides (e.g., glucose, xylose)
    Disaccharides (e.g., sucrose, lactose)
    Oligosaccharides
    Polysaccharides, fibrous (e.g., hemicellulose, cellulose, xylans)
    Polysaccharides, nonfibrous (e.g., starches, dextrins, pectins
  Vitamins
    Fat soluble
    Water soluble
  Other
    Polyphenols (e.g., lignin)
    Organic acids
    Compounds contributing to color, flavor, odor; toxins; inhibitor
Inorganic elements
  Essential
  Nonessential
  Possibly essential
  Toxic

Carbohydrates make up the major fraction of most plant tissues, but they represent less than 1% of the weight of animal tissues. Animals require no specific dietary carbohydrate. Glucose, a monosaccharide, synthesized by plants from carbon dioxide and water through photosynthesis in the presence of chlorophyll, is the simple sugar that is required by all animal cells as an energy source.

Vitamins account for only a minute fraction of the weight of feedstuffs. Feed sources vary widely in vitamin content, owing partly to inherent differ-

ences in tissue vitamin concentrations and partly to vitamin degradation by exposure to light, heat, and environmental variables.

Inorganic (mineral) elements present in feeds include the macroelements, and micro- (trace) elements, which are required in much lower amounts in the diet. Several mineral elements present in plants have no known function in animals. Eventually, it may be shown that some animal species require some of these trace elements for specific functions. Both macro- and microelements may be present in some plant tissues under some conditions at toxic concentrations. Any mineral element (in fact, any nutrient) can be toxic when ingested in excessive amounts.

❑ COMPOSITION OF THE PLANT

The chemical composition of whole plants is exceedingly diverse, being affected greatly by stage of growth and plant species. Data on a few animal feeds are shown in Table 2.3. Note that the water content of pasture grass and of the whole corn plant is much higher than that of other feeds listed. However, if one expresses the percentage composition on a water-free (dry matter) basis, pasture grass and alfalfa hay are similar in protein content (5.0 divided by dry matter content, or 0.321 = 15.57%). This illustrates a common practice to be used when comparing feedstuffs (i.e., express nutrient content on a water-free (usually referred to as air-dried) basis.

With regard to other comparisons shown in Table 2.3, note that the protein content of alfalfa hay and pasture grass is relatively high compared with that of other plant materials listed. The whole corn and wheat plants are lower, whereas wheat straw is much lower. On the other hand, soybean meal and meat meal are concentrated sources of protein. Except for meat meal, none of these feeds contains much fat. With regard to carbohydrate content, meat meal is low, and that present is largely an artifact of the method of analysis used. Total carbohydrate includes readily digested carbohydrates plus relatively unavailable fibrous components (listed as crude fiber); therefore, total carbohydrate values in Table 2.3 are not very meaningful. Feeding value is generally negatively related to fiber content.

Mineral content of feedstuffs is variable. Generally, legume forages are relatively high in calcium

TABLE 2.3 Percentage composition of selected animal feeds.

FEED	WATER	PROTEIN	FAT	CRUDE FIBER	TOTAL CARBO-HYDRATE	ASH	CALCIUM	PHOSPHORUS
Pasture grass (young, leafy)	67.9	5.0	1.1	7.9	23.1	2.8	0.12	0.06
Corn plant, whole	75.7	2.0	0.6	5.8	20.4	1.3	0.07	0.05
Wheat straw	12.2	3.2	1.4	38.3	76.9	6.3	0.14	0.07
Alfalfa hay	8.6	15.5	1.7	28.0	65.1	9.0	1.29	0.21
Corn grain	14.6	8.9	3.9	2.1	71.3	1.3	0.02	0.27
Soybean meal	10.9	46.7	1.2	5.2	35.3	5.9	0.30	0.65
Meat meal	5.8	54.9	9.4	2.5	5.0	24.9	8.49	4.18

Data taken from various NRC publications.
Source: Adapted from National Research Council (1982), United States-Canadian Tables of Feed Composition and Fonnesbeck et al. (1984), International Feeds Institute Tables of Feed Composition.

(Ca), whereas grasses and the seeds of cereal grains and corn are low. Meat meal is moderately high and variable in Ca and phosphorus (P), depending on the amount of Ca–P-rich bone present in the meal. The seeds of cereal grains, corn, and soybeans contain moderate levels of P, but 30 to 50% or more is present as phytate P, which may be biologically unavailable to the animal. These examples illustrate the many ways in which plants differ in composition and nutritive value for animals. Many other examples of differences in plant composition and nutrient utilization by animals are given in other chapters and in the Appendix tables.

❏ COMPOSITION OF THE ANIMAL BODY

In contrast to plants, animals contain very low concentrations of carbohydrate and generally higher concentrations of fat, on a dry matter basis. The chemical composition of animals tends to be more uniform, across species, than that of plants. Typical body composition of an adult mammal is about 60% water, 16% protein, 20% fat, and 4% mineral matter. Carbohydrates represent less than 1% of body tissue weight (primarily as blood glucose and muscle and liver glycogen). Differences among species in body composition are not as large as one might infer from the values shown in Table 2.4.

Indeed, the proportions of water, protein, and ash (inorganic minerals) in the fat-free bodies of

animals are remarkably constant. A survey (Clawson et al., 1991) of nearly 200 published research reports involving several thousand animals (mammals, birds, fish) over a wide age range revealed that the water, protein, and ash content of the fat-free body is in a ratio of about 19:5:1 (74 to 76% water, 20 to 22% protein, and 3 to 5% ash) in cattle, goats, mice, rats, sheep, swine, chickens, quail, turkeys, and fish. In the limited data shown in Table 2.4, age (and changing fat content) is responsible for more variation than species. Humans tend to vary more in fat content than most domestic or wild species. Athletes may have less than 15% body fat, whereas sedate obese adults may have 40% or more fat. Wild terrestrial species,

TABLE 2.4 Composition of the animal body.[a]

SPECIES	COMPONENTS, %			
	WATER	PROTEIN	FAT	MINERALS
Calf, newborn	74	19	3	4
Steer, thin	64	19	12	5
Pig, 100 kg	49	12	36	2–3
Hen	57	21	19	3
Horse	60	18	18	4
Rabbit	69	18	8	5
Human	60	18	18	4

[a]Values do not include contents of the gastrointestinal tract.

TABLE 2.5 Relationship between body weight, body protein, and body fat in pigs of different ages.

BODY WEIGHT (KG)	BODY PROTEIN (KG)	BODY FAT (KG)	PROTEIN/FAT
1.3	0.15	0.06	2.5
5.9	0.83	1.28	0.6
22	3.9	3.3	1.2
38	6.2	7.5	0.6
65	10.6	15.4	0.7
85	13.0	27.0	0.5
110	16.6	45.0	0.3

Source: Reeds et al., 1993, with permission from CABI Publishing, Wallingford, UK.

TABLE 2.6 Change in body composition of cattle with increasing body size.

ITEM[a]	EMPTY BODY WEIGHT, KG 300	400	500
Protein	163	157	152
Fat	299	431	573
Energy	15.6	20.6	26.1
Calcium	14.9	14.2	13.7
Phosphorus	8.1	7.8	7.5

[a]Expressed as g/kg or, for energy, as Mj/kg.
Source: Nutrient Requirements of Ruminant Livestock. Agricultural Research Council (1980).

except for those that hybernate, do not accumulate nearly so much fat as do domestic species. Aquatic species (such as seals and whales) that accumulate large amounts of subcutaneous fat do so to improve body insulation as an aid in maintaining their body temperature above that of the environment.

The changes in body protein and fat content with increasing body weight in pigs from birth to slaughter weight are illustrated in Table 2.5 (from Reeds et al., 1993). Note that both protein and fat accretion continue over the entire growth period, but fat accumulates at a faster rate, resulting in a substantial decrease in the protein:fat ratio as the animal approaches mature size.

Changes in empty body composition of Holstein cattle with increases in body weight from 300 to 500 kg are shown in Table 2.6. Protein content, expressed as g/kg body weight, decreased slightly, as did Ca and P content. Fat increased by 192%, and the energy value of the tissues increased by 167%. It is clear that the two major variables in animal body composition are the concentrations of water and fat, and that these two components vary inversely.

The changes in protein, fat, water, and ash content of the carcass and offal of male sheep from birth to maturity are shown in Table 2.7. Note that the offal contributes a significant part of the total nutrient accretion.

TABLE 2.7 Protein, water, fat and ash (inorganic elements) accretion in edible carcass and of fat of male sheep from age one day to four years.

	BODY WT., KG	PELT-FREE[a] EMPTY BODY WT., KG	PROTEIN CARCASS, G	OFFAL,[b] G	FAT CARCASS, G	OFFAL,[b] G	WATER CARCASS, G	OFFAL,[b] G	ASH CARCASS, G	OFFAL,[b] G
1 day	5.4	4.2	428	163	100	100	2,145	1,079	136	45
13 wks	31.9	19.6	2,382	825	2,400	800	7,984	4,220	806	219
6 mo	52.1	37.1	4,410	1,243	8,400	2,200	12,932	6,016	1,421	346
1 yr	74.9	56.1	6,330	1,540	13,000	5,500	20,689	6,955	1,960	461
2 yr	103.4	64.0	8,464	2,117	10,700	3,800	26,154	9,034	2,921	717
4 yr	110.9	81.8	10,606	2,224	18,000	7,400	30,899	8,805	3,457	763

[a]Live body weight minus head, feet, skin, wool, and gastrointestinal tract contents.
[b]Includes inedible parts of the body, e.g., lungs, heart, liver, intestinal tract, reproductive organs.
Source: Jenkins and Leymaster (1993).

Mineral composition of the whole body varies with age and, to a lesser extent, with species. Bone mineral content increases as the animal grows and cartilage is replaced by bone tissue in the skeleton. For cattle, average values for the whole body are (%): Ca, 1.33; P, 0.74; potassium (K), 1.19; sodium (Na), 0.16; sulphur (S), 0.15; chloride (Cl), 0.11; and magnesium (Mg), 0.04. The ratio of protein to ash (total minerals) in the fat-free dry matter of the body of a wide array of mammals, birds, and aquatic animals tends to be similar, but nutritional factors, level of feed intake, and such factors as age, sex, and genetic background affect the ratio (Clawson et al., 1991). This variation may have important physiological and economic implications in the nutrition of food animals for efficient production.

❏ CROP PRODUCTION AND ANIMAL FEEDING

Marked and continuous improvements have been made in crop production for many decades. The almost universal use of hybrid corn and other crops in all areas of intensive crop production, the success in modifying important crops such as corn and soybeans so that they are adaptable to a wider range of environments, and the "green revolution" (which resulted in high-yielding varieties of rice and wheat and their use worldwide) have allowed substantial increases in food-grain production. Developments in molecular biology have created genetically modified (GM) crops for disease and pest resistance and improved nutrient composition to accommodate desired changes in the nutrient content of human food and animal feed. World crop production increased from 1.450 billion metric tons in 1980 to more than 2.0 billion metric tons in 2002, while human population increased from about 4.4 billion to about 6 billion in the same time period. Grain and total food production must continue to increase to feed a rapidly growing human population.

Although percentage increases in production of some cereal grains are probably higher than those for animals, animal productivity has been improved considerably in the recent past and continues to improve. Production of milk, meat, and eggs is markedly higher on a per animal basis, resulting in more efficient use of feed, labor, land, and capital. Inputs of energy and protein for production of milk, beef, swine, and poultry products

TABLE 2.8 Inputs and returns of animal production[a]

| PRODUCT | TOTAL ENERGY AND PROTEIN | | | |
| | ENERGY | | PROTEIN | |
	INPUT, MCAL	RETURN, %	INPUT, KG	RETURN, %
Milk	19,960	23.1	702	28.8
Beef	20,560	5.2	823	5.3
Swine	1,471	23.2	66	37.8
Poultry	23.2	15.0	1.2	30.0

[a]Inputs are calculated as digestible energy and digestible protein and include cost of maintaining breeding herds and flocks. *Source:* Data from Bywater and Baldwin (1980).

and the percentage of return for each commodity are summarized in Table 2.8 (Bywater and Baldwin, 1980). Aquaculture is becoming more important in a number of areas; where water is available, production of fish is more efficient than meats from our typical warm-blooded animals.

Even with the many marked improvements in crop and animal production, there is much concern that population growth may outrun the world's capacity to produce food and feed because of limited arable land, usable water, and energy. Water for irrigation is in short supply in many areas. In some instances where ground water has been used for irrigation, the water level has been dropping, resulting in greater costs to get it to the surface. Increased energy costs, unless compensated for by comparable increases in product prices, will make it less feasible to use ground water for irrigation. Similarly, increased prices for natural gas and other petroleum products directly affect fertilizer costs because some manufacturing processes use natural gas as a primary ingredient.

The rising demand for food and feed can likely be met by continual technological developments, improvements in marketing, and reduction in wastage. Recent estimates (Bradford et al., 1999) indicate that satisfying the projected global demand for foods of animal origin in 2020 will require an increase of more than 50% for meat, milk, and eggs. To accomplish this goal, increases will probably be needed both in feed supply and in efficiency of conversion of feed to food by animals. Table 2.9 summarizes projected trends in the use of cereals as feed between 1983 and 2020. Note

TABLE 2.9   Projected trends in use of cereal as feed.

REGION	TOTAL CEREAL USE AS FEED (MILLION TONS)			PER CAPITA CEREAL USE (KILOGRAMS)	
	1983	1993	2020	1993	2020
Developing countries	126	194	390	45	62
Developed countries	453	443	536	346	386
United States	132	159	199	603	622
World	579	637	927	115	120

Source: From Bradford et al. (1999).

that both total and per capita cereal use as feed are projected to increase substantially in both the economically developing and developed countries. These projections may be altered by many unforeseen factors both extrinsic and intrinsic to agriculture, which are beyond the scope of our discussion here. One important factor is the predicted critical shortage of water for use in agriculture which in itself may prove to be the most severe constraint on food production.

Animal products almost certainly will have a major role in meeting increased demands in the future. Many of the ingredients in farm animal diets can originate from materials that are not edible for humans. In fact, in most areas of the world, the milk and meat produced by cattle, sheep, buffaloes, and goats is derived directly from grazing land that is not in cultivation and from crop residues, milling byproducts, or wastes that normally never get into the food chain. On the other hand, a considerable amount of feed fed to animals in some countries is directly competitive for human use. It has been estimated that feed fed to pets in the United States could feed some 40 million people, although some pet food ingredients (animal offal of various types such as lungs and condemned livers) generally are not considered as edible in the United States.

Future developments in animal and human nutrition are not highly predictable because many modulating factors cannot be foreseen. However, we can be certain that animals will be important to humans for the foreseeable future. It is therefore important to continue to expand our knowledge of all phases of animal nutrition so that we will be able to respond constructively to human population growth, environmental and ecological needs, new agricultural production technologies, and factors external to agriculture. As we progress through this book, it will be evident that many gaps in our knowledge remain. Given time and resources, the deficiencies will be remedied gradually by the scientists, livestock producers, and other agricultural professionals of tomorrow who are students today.

❑ SUMMARY

- A major challenge for society and for animal agriculture for the twenty-first century is to provide sufficient food nutrients for a burgeoning population

- Animal diets can be supplied by a broad array of feedstuffs of plant and animal origin, of which a major portion can be supplied from sources not edible for humans.

- The nutrients required in the diet of most animal species are supplied from mixtures of feedstuffs containing organic compounds (amino acids, lipids, carbohydrates, vitamins), inorganic elements (macro- and trace minerals), and water.

- Quantitative nutrient requirements of animals are affected by age (stage of the life cycle), species, productive function (gestation, lactation, growth, maintenance), genetics, and a host of environmental factors.

- Marked improvements have occurred and are occurring in crop production through improved genetics, methods of pest control, and advances in soil management and conservation, fertilizer use, reduced postharvest losses, and other and new technologies, all of which enhance the efficiency of animal production.

- Trends in demand for animal products indicate that animal production worldwide will continue to increase for the foreseeable future, particularly in the developing countries with increasing purchasing power.

- The projected continued growth of the human population must be accompanied by continued advances in knowledge and application of animal nutrition as well as agricultural technology, and attention to environmental and ecological stability to ensure an adequate food supply. This is an exciting challenge for animal nutritionists of the future.

❏ REFERENCES

ARC. 1980. *The Nutrient Requirements of Ruminant Livestock.* Agricultural Research Council and the Commonwealth Agricultural Bureaux, England.

Bradford, E., et al. 1999. *Animal Agriculture and Global Food Supply.* Council for Agricultural Science and Technology (CAST), Task Force Report No. 135, pp. 1–92. CAST, Ames, IA.

Bywater, A. C., and R. L. Baldwin. 1980. In *Animals, Feed, Food, and People. Alternative Strategies in Food Animal Production,* pp. 1–29, R. L. Baldwin (ed.). Westview Press, Boulder, CO.

Clawson, A. J., et al. 1991. *J. Anim. Sci.* **71:** 2952.

Fonnesbeck, P. V. et al. 1984. IFI Tables of Feed Composition. International Feedstuff Institute, Utah State University, Logan, UT. 607 pp.

Jenkins, T. G., and K. A. Leymaster. 1993. *J. Anim. Sci.* **71:** 2982.

National Research Council. 1982. United States-Canadian Tables of Feed Composition: Nutritional Data for United States and Canadian Feeds. Third edition. Washington D.C.: National Academy Press. 772 pp.

Reeds, P. J., et al. 1993. In *Growth of the Pig,* pp. 1–33, G.R. Hollis (ed.). CAB International, Wallingford, Oxon, UK.

3

Common Methods of Analysis for Nutrients and Feedstuffs

PROGRESS IN NUTRITIONAL SCIENCE depends largely on the availability of accurate analytical methods to measure the chemical composition of feedstuffs and animal tissues. Expanded knowledge and an improved understanding of nutrient requirements and metabolism are possible, in part, because methods that quantitatively evaluate the nutrient content of foods, feeds, and animal tisues are continually being developed and improved. Thus, to have a good understanding of nutrition, the student must have some familiarity with laboratory analyses and equipment and at least some background in chemistry. Although detailed knowledge is not required, some understanding of analytical procedures will permit easier reading, and some familiarity with organic structures will facilitate learning about specific nutrients.

❏ SAMPLING FOR ANALYSIS

Modern chemical methods are geared to procedures that require small amounts of material that must be collected and prepared to allow the best reasonable estimate of the total batch. For example, if we are interested in the protein content of hay produced from a field, where do we begin? We certainly cannot grind up all of the hay produced; even one bale would tax the facilities of most laboratories. Consequently, we resort to the use of pooled core samples taken from many different bales. Perhaps as many as 25 to 50 core samples may be taken from one stack of bales, which represents the hay from a field many acres in area. The assumption is that each core will correspond reasonably well to the total composition of the bale from which it came and that, if we sample enough bales, our composite sample will be representative of the hay crop.

The core samples are brought to the laboratory, ground, and mixed well, and then small subsamples are taken for analyses. The most important consideration in this procedure is that the sample must have the same proportion of particle types and sizes as the original feed so that it is representative of what the animals are actually fed. A common mistake made when subsampling a hay sample is to shake the sample before transferring it. Shaking causes loss of small particles and enriched stem content, resulting in a higher than actual fiber content of the hay. Quartering and sampling is a good way to mix and subsample. In this method, a feed sample is placed on a plastic sheet, and then mixed by lifting one corner of the sheet after another so that the sample mixes without sorting of particle sizes. Then the pile is quartered into four pie slices, and two opposite slices are removed from the sheet; the process is repeated until the sample size is appropriate for the analysis needed.

For the common Kjeldahl N analysis that is used for crude protein (see the section on proximate analyses), a typical sample size is 2 g of material. A micro-Kjeldahl procedure now in use allows the use of a sample that contains about 1 mg of N, or a sample of about 100 mg of the hay in question. Thus, because we may base our estimate of protein content of the total field on a very small amount of material, the sample being analyzed must be representative to produce meaningful results.

Similar procedures are used for other commodities. One small sample of grain may be used to evaluate a carload. Liquids are assumed to be more homogeneous than solids, but this is not always true and errors may creep in if care is not taken in sampling. Poor sampling procedures of any material may be worse than none.

The same concerns about accurate sampling techniques apply for estimates of body composition in animals. This subject is addressed briefly in Chapter 5 in relation to nutrient utilization and storage in animals.

❏ ANALYTICAL METHODS

Most of the analytical methods in common use today depend on various chemical procedures that are specific for a given element, compound, or group of compounds. Quantitative data may be obtained by gravimetric procedures, but more often they are obtained by other methods that involve the use of acid or base titration, colorimetry, chromatography, two-dimensional gel electrophoresis, mass spectrometry, or other sophisticated procedures. Chemical methods often involve drastic degradation of feeds with reagents such as concentrated solvents or other treatments that are biologically harsh. As a result, a serious problem in nutrient analysis is that a given procedure may provide accurate quantitative measurement of a particular nutrient or compound in the feed, but no information related to the animal's degree of utilization of the nutrient. For example, a forage might be analyzed for its Ca content, but the data provide no information on how much of the Ca is available to the animal.

Because chemical methods often leave unanswered questions regarding the availability of nutrients from feeds, biological procedures are sometimes used. Although biological methods are usually more tedious and expensive, they may give a much more accurate estimate of the bioavailability of a specific nutrient in a feedstuff for animals. Chicks or rats are often used in tests of this type. If, for example, one desires to determine the effect of heating on utilization of proteins from soybean meal, one or more groups of chicks could be fed unheated meal along with other necessary nutrients, while other groups might be fed meal that had been heated to different temperatures or for different lengths of time. Such comparisons provide a reasonably good biological estimate of the effect of heat treatments on soy proteins.

Microbiological methods may be used much as the biological procedures just described for chicks. Bacteria have been isolated that have specific requirements for one or more of the nutrients required by animals (e.g., a particular essential amino acid or a specific water-soluble vitamin). These microorganisms can be used to determine how much of a given amino acid or vitamin is available in a given product or mixture. The information may or may not be applicable to rats, chicks, or swine, but it is more likely to be applicable than that from a chemical method.

❏ SPECIFIC METHODS OF ANALYSIS

Dry Matter

The determination of dry matter is probably the most common procedure carried out in nutrition laboratories because plant feedstuffs, animal tissues, or other samples of interest may be quite variable in water content, and one must know the amount of water present to permit comparisons of different feeds. When grain is purchased or fed, obviously the value of grain with 20% moisture will not be the same as the value of grain with 10% moisture. After analysis, nutrient composition can be expressed on a dry basis or a normal "as fed" basis, which is about 90% dry matter for most grains.

The simplest way to determine dry matter is to place the test material in an oven and leave it until all of the free water has evaporated. The temperatures used are usually 100 to 105°C. Moisture also can be estimated with moisture meters, devices that give immediate results by means of a probe inserted into the test material (Fig. 3.1a). Some of these devices depend on electrical conductivity; they are very useful for quick answers, such as when buying grain or hay, but results are

Figure 3.1 *Examples of moisture-determining equipment that can give rapid information on water content. (a) A simple meter depending on electrical conductivity. Courtesy of Epic, Inc. (b) A more elaborate instrument, useful in the laboratory, that depends on microwave energy for moisture analysis. Courtesy of Photovolt Co.*

not as precise as those obtained by actually drying the test material in an oven. Other equipment includes microwave ovens adapted for obtaining weights before and after drying (Fig. 3.1*b*).

Determining dry matter, as with most procedures, is not always as simple as our discussion implies. This statement applies to any material that has a relatively high content of volatile compounds. Most fresh plant tissues contain volatile compounds, but the amount is low enough that the volatiles usually can be ignored with little error. However, some plants contain large amounts of essential oils, terpenes, and other volatile substances that may be lost in drying and, thus, use of the usual procedures will give an erroneous answer. Of the common feedstuffs, silages or other fermented products may contain large amounts of easily vaporized compounds such as volatile fatty acids (acetic, propionic, butyric) and ammonia. Furthermore, some sugars may decompose, and many proteins may become partially insoluble at temperatures above 70°C. Several means are available to avoid excessive losses of volatiles. Drying in vacuum ovens, freeze drying, oven drying at 70°C or less, and distillation with toluene have been used. The effect of these different procedures on the estimated dry matter content of silage was demonstrated by Brahmakshatriya and Donker (1971). Substantial losses may occur if silage is oven-dried at 100°C.

Proximate Analysis

Proximate analysis, a combination of analytical procedures developed in Germany well over a century ago, is intended for the routine description of feedstuffs. Although from a nutritional point of view it has many faults, it is still used, though for forage analysis it is being supplanted by the detergent extraction methods of Van Soest (1982) described later in this chapter. In some instances, the use of proximate analysis of feeds has been encouraged and prolonged because of laws that require the listing of minimum and maximum amounts of components that may be present in commercial feed mixtures. The different fractions that result from the proximate analysis include water, crude protein, ether extract, ash, crude fiber, and nitrogen-free extract. We have already discussed water (or dry matter); more detailed discussion of the other components follows.

Crude Protein.   The procedure used to determine crude protein is known as the Kjeldahl

method. The average protein contains about 16% N. The Kjeldahl analysis depends on the measurement of N in the test material. To convert the measured N content of a feed to crude protein, the value of N is multiplied by 6.25. For example, the calculated crude protein content of a feed containing 8% N would be 50% (8% × 6.25 = 50%). Material to be analyzed first is digested in concentrated H_2SO_4, which converts the N to $(NH_4)_2SO_4$. This mixture is then cooled, diluted with water, and neutralized with NaOH, which changes the N into the form of ionized ammonium. The sample is then distilled, and the distillate containing the ammonium is titrated with acid. This analysis is accurate and repeatable, but it is relatively time consuming and involves the use of hazardous chemicals. A micro apparatus is shown in Fig. 3.2. Spectrophotometic methods of N analysis that do not require distillation following sample digestion are more commonly used.

From a nutritional point of view, crude protein data are applicable to ruminant species that can efficiently utilize almost all forms of N, but the information may be of little value for nonruminant species (such as swine or poultry). Nonruminants, as well as some ruminant species under certain physiological states, have specific requirements for various amino acids (see Chapter 9). Nonruminants do not efficiently utilize non-protein N compounds such as amides, ammonium salts, and urea. The crude protein analysis does not distin-

Figure 3.2   *A micro-Kjeldahl apparatus that is designed for nitrogen analysis of small samples. The digestion apparatus is shown on the left and the distillation equipment immediately to the right.*

guish one form of N from another, so it is not possible to determine whether a feed mixture has urea or a high quality of protein such as casein (from milk). In addition, nitrate N is not converted to ammonium salts by this procedure, so N in this form is not included.

Ether Extract. This procedure requires that ground samples be extracted with diethyl ether for a period of 4 hours or more. Ether-soluble materials include quite a variety of organic compounds (see Chapter 8), only some of which have high nutritional significance. Those of quantitative importance include the true fats and fatty acid esters, some of the compound lipids, and fat-soluble vitamins or provitamins such as the carotenoids. The primary reason for obtaining ether extract data is an attempt to isolate the fraction of a feedstuff that has a high caloric value. Provided the ether extract is made up primarily of fats and fatty acid esters, this may be a valid approach. If the extract contains large percentages of plant waxes, essential oils, resins, or similar compounds, it has little meaning because compounds such as these are of little value to animals.

Ash. Ash is the residue remaining after all the combustible material has been burned off (oxidized completely) in a furnace heated to 500 to 600° C. Nutritionally, ash values have little importance, although excessively high values may indicate contamination with soil or dilution of feedstuffs with such substances as salt and limestone. In the proximate analysis, data on ash are required to obtain other values. It should be noted that some mineral elements, such as iodine and selenium, may be volatile and are lost on ashing. Normally, these elements represent only very small percentages of the total, so little error is involved.

Crude Fiber. Crude fiber is determined by using an ether-extracted sample, boiling in dilute acid, boiling in dilute base, filtering, drying, and burning in a furnace. The difference in weight before and after burning is the crude fiber fraction. This is a tedious laboratory procedure that is not highly repeatable. It is an attempt to simulate digestion that occurs first in the gastric stomach and then in the small intestine of animals. Crude fiber is made up primarily of plant structural carbohydrates such as cellulose and hemicellulose (see Chapter 7) but it also contains some lignin, a highly indigestible material associated with the fibrous

portion of plant tissues. For the nonruminant animal, crude fiber is of a variable but low value; for ruminants, it is of variable value, but it is much more highly utilized than it is by nonruminants.

Nitrogen-Free Extract [NFE]. This term is a misnomer in that no extract is involved. NFE is determined by difference; that is, it is the difference between the original sample weight and the sum of weights of water, ether extract, crude protein, crude fiber, and ash. It is called N-free because ordinarily it would contain no nitrogen. NFE is made up primarily of readily available carbohydrates, such as the sugars and starches (see Chapter 8), but it may also contain some hemicellulose and lignin, particularly in such feedstuffs as forages. A more appropriate analysis would be one specifically for readily available carbohydrates—one in which starches are hydrolyzed to sugars and the mixture is analyzed for all sugars present. Nutritionally, the NFE fraction of grains is utilized to a high degree by nearly all species, but NFE from forages and other roughages is less well utilized.

A diagram of the proximate analysis scheme illustrating the sequence of procedures as well as the major fractions that are isolated is shown in Fig. 3.3.

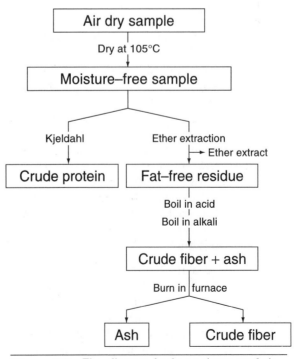

Figure 3.3 *Flow diagram for the proximate analysis.*

Detergent Extraction Methods

Analytical methods, intended primarily for forages, have been developed by Van Soest (1982), his co-workers, and other scientists. Micro methods also have been developed. Use of these methods allows plant components to be divided as follows:

Neutral Detergent Extraction.   Samples are boiled for 1 h in a solution containing primarily sodium laurel sulfate. This detergent extracts lipids, sugars, organic acids, and other water-soluble material, pectin (usually classified as a fibrous carbohydrate), nonprotein N compounds, soluble protein, and some of the silica and tannin. The nonsoluble material is referred to as neutral detergent fiber (NDF). The NDF contains the major cell wall components such as cellulose, hemicellulose, and lignin. It may also contain minor cell wall components, including some protein, bound N, minerals, and cutin. The soluble material, often referred to as cell wall contents (CWC), is highly digestible by all species, with the possible exception of the pectins and any silica and tannin. The NDF is only partially digestible by any species, but it can be used to a greater extent by animals such as ruminants, which depend on microbial digestion for utilization of most fibrous plant components.

Acid Detergent Extraction.   Samples are boiled for 1 h in a solution containing cetyl trimethylammonium bromide in H_2SO_4. Components soluble in acid detergent include primarily hemicelluloses cell wall proteins, and the residue includes cellulose, lignin, and lignified N (indigestible N), cutin, silica, and some pectins. It is usually referred to as acid detergent fiber (ADF).

These detergent methods are often used alone but may be used together, or the ADF method may be substituted for the crude fiber method partly because it is more repeatable and faster. The ADF fraction can be further extracted with sulfuric acid to isolate lignin. The pros and cons of the methods have been discussed in detail by Van Soest (1982).

pH of Feedstuffs

The pH of feedstuffs is rarely used to evaluate materials except for fermented products such as silage, cannery residues, or other similar mixtures. It should be pointed out that pH of mineral supplements may be of importance with respect to palatability or metabolism by the animal. With respect to silage, pH may be determined by mixing 100 g of silage with 100 ml of water, expressing the juice, and measuring with a pH meter. Good quality silages should have a pH between 3.8 and 5.0.

❏ SPECIALIZED ANALYTICAL METHODS

A wide variety of analytical methods find some use in nutrition. Such methods may be used for feedstuffs, mixed diets, animal tissue, or for urine and fecal samples. The list of such methods is far too long to discuss in a book of this type; however, several methods involving specialized equipment are used extensively and deserve some discussion.

Bomb Calorimetry

The oxygen bomb calorimeter (Fig. 3.4) is used to determine the energy values of solids, liquids, or gases. The energy value of a given sample is determined by burning it in a pressurized oxygen atmosphere. When the sample is burned, the heat produced raises the temperature of water surrounding the container in which the sample is enclosed, and the temperature increase provides the basis for calculating the energy value. Bomb calorimetry finds extensive use for evaluating

Figure 3.4   An oxygen bomb calorimeter used for energy determinations of feed and other combustible materials.

fuels such as natural gas and coal. Its most useful application in nutrition is in determining the digestible energy of feedstuffs or rations. The gross energy value (the value obtained by burning) of feedstuffs has little or no direct application, for it is almost impossible to distinguish between constituents that are well utilized by animals and those that are poorly utilized (see Chapter 10).

Amino Acid Analysis

Chemical methods for amino acid analysis have been available for many years, but now semiautomated equipment, such as that shown in Fig. 3.5, are available. This type of equipment is capable of fractionating protein preparations that have been hydrolyzed into the constituent amino acids. The preparations are placed on chromatographic columns, and various solutions are passed through the columns, resulting in separation and measurement of the individual amino acids in a relatively short time (minutes or hours). This methodology has greatly facilitated collection of data on amino acid composition of foods and feeds as well as on metabolism and requirements of amino acids.

Atomic Absorption Spectrophotometry

Atomic absorption spectrophotometric instruments (Fig. 3.6) have greatly facilitated analyses

Figure 3.6 *A modern atomic absorption spectrophotometer. Courtesy of the Instrumentation Laboratory.*

for most mineral elements (cations). In the operation of these instruments, liquid or solid materials are ashed and resuspended in liquid, which may be put directly into the instrument. Body fluids such as blood plasma and urine may be used directly. The solution passes through a flame that serves to disperse the molecules into individual atoms. Radiation from a cathode lamp is passed through the flame, and the atoms absorb some of this radiation at specific wavelengths. With instru-

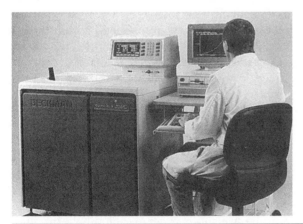

Figure 3.5   *(left) Analytical Ultracentrifuge for separation of cellular constituents and body fluid constituents differing in specific gravity. Beckman Instruments, Inc., Fullerton, CA. (right) A versatile system combining a high-performance liquid chromatography (HPLC) instrument with prepackaged reagents, HPLC column, and eluents provides an accurate analysis of hydrolysate amino acids. This Waters Amino Acid Analysis System provides fully automated analysis of up to 96 samples with no operator intervention in femto-molar limits of detection for micorvolume quantities of test material. Courtesy of Waters Chromatography Division. Millipore Corporation, Milford, MA.*

ments such as this, vast numbers of samples can be analyzed in a short time.

Gas–Liquid Chromatography

The forerunner of the gas–liquid chromatograph (GLC) was developed in the early 1950s to analyze rumen volatile fatty acids. Since that time, tremendous improvements have been made in this technique and in the available instrumentation (Fig. 3.7). Such instruments are capable of handling almost any compound that exists in gaseous form or can be vaporized.

The sample to be analyzed is placed in the instrument and is moved through a heated chromatographic column by means of gas. This process allows the quantitative separation of closely related chemical compounds (such as acetic and propionic acid) quite rapidly. This process requires only very small samples. In nutrition, GLCs are particularly useful for fatty acid analyses but are capable of handling many other organic compounds.

Automated Analytical Equipment

The gradually increasing cost of labor has stimulated the development of instrumentation designed to perform simultaneous repetitive analyses (Fig. 3.8). Such equipment is used widely in the medical field in particular, but it has application as well in the nutrition laboratory. For example, it is possible to obtain simultaneous data on blood serum for glucose, total lipids, cholesterol, Ca, P, Mg, urea, and total protein as well as other compounds. This is just one example of the type of information that may be obtained on one tissue. Increased availability of more complete data on animals would greatly improve our knowledge of nutrient metabolism of healthy as well as sick animals. Such

Figure 3.7   *High-Performance Liquid Chromatography (HPLC) gradient analytical system with an autosampler. Rainin Instrument Company, Inc., Emeryville, CA.*

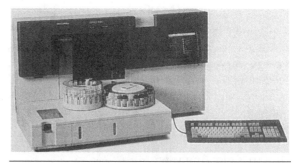

Figure 3.8   *A fully automated clinical chemical analyzer capable of measuring the concentrations of up to 48 different metabolites and ions in blood plasma and other body fluids at the rate of up to 180 tests per hour. Courtesy of Ciba-Corning Diagnostics Corporation, Norwood, MA.*

rapid and highly automated analysis has greatly increased the volume of information that may be obtained at a given cost, even though the equipment itself is expensive.

Infrared.   Use of infrared light rays for feed analyses is of relatively recent origin. Analyses are obtained by placing a sample in a receptacle and impinging infrared light on the sample. The reflected light goes back into the instrument, and the changes caused by the sample can be detected and related to composition of the sample by a built-in computer. Analyses are usually restricted to lipids, protein, fiber, and moisture, although some instruments have been used for Ca, P, salt, and occasionally for other ingredients. These instruments were initially developed for use with grains, but now are used with other feedstuffs, including mixed feeds and ground forages. The major advantage of infrared analysis is time: it takes only about 20 seconds per sample. Obviously, in the grain or feed trade, this would be a tremendous asset as compared to most analytical methods, which may require one to several days to get data back from the laboratory. By that time the feed may be gone or may have been consumed. One disadvantage of infrared analysis is that a range of samples must be available in order to calibrate the machine. Consequently, samples with at least as much variation as the test samples must be available for this purpose. In addition, calibration samples must be available for every type of feed to be used.

Other Instrumentation.   Other types of instrumentation may be used for nutritionally related research. Some of the instruments or methods available include automated instruments for measuring blood flow, blood cell counters, high-pressure liquid chromatography, nuclear magnetic resonance (NMR), mass specrometry (used in stable isotope measurement in nutritional physiological research), gamma and beta counters used in radioimmune assays (RIA) for hormones, and other metabolites, plate readers for enzyme-linked immunosorbent assay (ELISA) tests of antibody titers, DNA synthesizers, inductively coupled plasma emission spectrophotometers, and flow cytometers.

Several new instruments have been produced for identification, isolation, fractionation, and characterization of the myriad proteins present in the animal body. It has been estimated (Bunk, 2003) that about one million different protein molecules exist in the human body (termed the *proteome*). Comparable numbers of different proteins are likely present in the proteomes of many farm animal species. An automated, two-dimensional chromatographic system that fractionates high- and low-abundance proteins from cell lysates (Constan, 2003) is shown in Fig. 3.9.

Another instrument used for automated high resolution protein characterization is a capillary electrophoresis-based system. When available, such instrumentation and methods allow the collection of much more data than would otherwise be possible and, in some instances, the collection of data that could not otherwise be obtained.

Figure 3.9 *Beckman Coulter's ProteomeLab PF 2D Protein Fractionation System. Photo courtesy of Beckman Coulter, Inc.*

On-Farm Methods for Feed Analysis

Most of the analytical methods and instrumentation described in the preceding section are more appropriate for laboratory use than for use on commercial farms. Two common on-farm procedures currently in use to evaluate and characterize feedstuffs and mixed diets are (1) microwave measurement of dry matter and (2) particle size analysis by a sifting procedure. Such a procedure, developed at Pennsylvania State University, uses a series of boxes with acreens of progressively decreasing size. Feed is placed in the top box, the unit is shaken, and particles drop through the screens until they are too large to drop to the next screen. The amount in each box is then weighed to provide a profile of the range of particle sizes of the feed.

❑ SUMMARY

- Analytical methods used for nutrients, feeds, animal tissues allow very rapid accumulation of biological and nutritional data.

- When such instrumentation and methodology are used as designed, they enhance our knowledge of animal biology and nutrition.

- The rapid generation and acquisition of information obtained from elaborate instrumentation is not a guarantee that these reams of data are accurate or applicable to a particular biological question; this requirement must be given consideration when using such equipment.

❑ REFERENCES

Brahmakshatriya, R. D., and J. D. Donker. 1971. *J. Dairy Sci.* **54:** 1470.

Bunk, S. 2003. Proteomic players pick plasma. *The Scientist,* March 10, pp. 18–19.

Constan, A. 2003. The proteomic toolbox. *The Scientist,* March 10, p. 37.

Van Soest, P. J. 1982. *Nutritional Ecology of the Ruminant.* 0 & B Books, Corvallis, OR.

4

The Gastrointestinal Tract and Nutrition

SOME KNOWLEDGE OF THE gastrointestinal tract (GI tract) is important to those who study nutrition because of its influence on the utilization of food and nutrients. The organs, glands, and specialized structures of the gastrointestinal tract are concerned with procuring, chewing, and swallowing food; with digesting and absorbing nutrients; and with performing secretory and excretory functions.

Digestion and *absorption* are terms that are used frequently in this chapter. **Digestion** may be defined simply as the preparation of food for absorption. In the broad sense, it may include mechanical forces (chewing or mastication; muscular contractions of the GI tract), chemical action (hydrochloric acid in the stomach; bile in the small intestine), or hydrolysis of ingesta by enzymes produced in the GI tract or from microorganisms in various sites in the tract. The overall function of the various digestive processes is to reduce food to a molecular size or solubility that allows absorption and cellular utilization of the individual nutrients released in the process.

Absorption consists of the processes that result in the passage of small molecules from the lumen of the gastrointestinal tract through the mucosal cells lining the surface of the lumen and into the blood or lymph systems.

25

❏ ANATOMY AND FUNCTION OF THE GASTROINTESTINAL TRACT

The GI (gastrointestinal) tract of mammals includes the mouth and associated structures and glands, the esophagus, the stomach, the small intestine, and the large intestine (including a cecum in some species). Associated organs that are intimately involved with digestion and absorption are the liver (secretes bile into the small intestine) and pancreas (secretes digestive enzymes into the small intestine). The tract itself is essentially a modified tubular structure used for ingestion and digestion of food and the elimination of some of the wastes of metabolic activity produced by the animal body. Its ultimate purpose is to provide for the efficient assimilation of nutrients and to reject dietary constituents unnecessary for or potentially harmful to the animal.

Types of GI Tracts

Among the many species of mammals and birds there are wide variations in the structure and functions of individual components of the GI tract. The animal does not, of course, exist in a sterile environment, and in many instances the GI tract has been modified to take advantage of symbiotic relationships with various microorganisms. Concentrated microbial populations are found in the large intestine of all species; some species have developed modifications of the upper GI tract that allow microbes to thrive and generate products that are beneficial to the animal in the process of partially

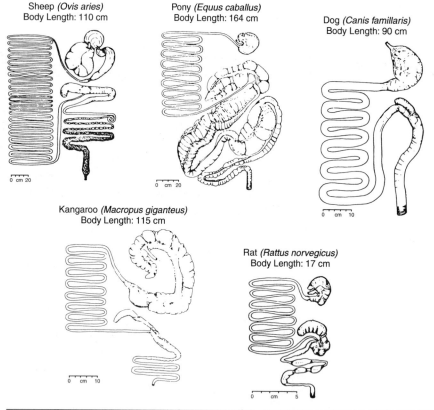

Sheep *(Ovis aries)*
Body Length: 110 cm

Pony *(Equus caballus)*
Body Length: 164 cm

Dog *(Canis famillaris)*
Body Length: 90 cm

Kangaroo *(Macropus giganteus)*
Body Length: 115 cm

Rat *(Rattus norvegicus)*
Body Length: 17 cm

Figure 4.1 *Some examples of mammalian digestive tracts drawn to illustrate major anatomical differences. The sheep and kangaroo are examples of species with pregastric digestion. Note that there is some similarity between the sacculated stomach of the kangaroo and the cecum of the pony and, to a lesser extent, the pig (see Fig. 4.2), both hind gut fermentors. Sacculation is a distinguishing characteristic of herbivores whether in the stomach, cecum, or large intestine. The cecum is small in the dog and absent in such carnivores as the mink (not shown). Redrawn from an illustration in Duke's Physiology of Domestic Animals, 9th ed. (Duke, 1977).*

TABLE 4.1 Classification of some animal species according to areas in the gastrointestinal tract where extensive rermentation occurs.

CLASS	SPECIES	DIETARY HABIT
Hind gut fermentors		
Cecal fermentors	*Various rodents*	
	Capybara	Grazer
	Rabbit	Selective herbivore
	Rat	Omnivore
Colonic digesters with		
Sacculated colon	Horse, donkey, zebra	Grazer
	New World monkey	Selective herbivore
	Pig, man	Omnivores
Unsacculated colon	Dog, cat	Carnivores
	Fruit-eating bats	Herbivores
Pregastric fermentors		
Ruminants and		
pseudoruminants	All species	Herbivores
Nonruminants	Colobine monkey	Selective herbivore
	Hamster, vole	Selective herbivores
	Kangaroo, wallaby, quokka	Herbivores
	Hippopotamus and possibly	
	other suisformes	Herbivores
	Three-toed tree sloth	Herbivore

digesting some of the ingested food. Mammals with an uncomplicated stomach are sometimes referred to as monogastrics* or as nonruminants.

Existing animal species have evolved many variations in their digestive tracts that allow them to utilize diets of varying composition or quality ranging from nectar (hummingbird and other nectar-eating birds) to coarse fibrous plant material (elephant and other large herbivorous wild species, horse, some ruminants). Some of these differences are characterized in schematic drawings shown in Fig. 4.1 and in photographs (Fig. 4.2 and 4.6) and 4.7 or listed in Table 4.1.

Carnivores are animals whose diet is composed primarily of nonplant material, for example, meat, fish, and insects. They are represented in Fig. 4.1 by the dog. In general, the diet of a carnivore is relatively concentrated and highly digestible—except for hair, feathers, and other types of resistant

proteins. The GI tract of carnivores is represented by a gastric stomach and a relatively short and uncomplicated intestine. The large intestine is uncomplicated in that it is not sacculated. Carnivores are classified as hind gut fermentors; they are further subdivided into cecal or colonic digesters (see Table 4.1). They have the capability of digesting limited amounts of fibrous feeds, but fermentation of fiber is quite limited compared to that in other species listed in Table 4.1. It is believed that the fiber consumed by most omnivores (plant and animal eaters) and by all carnivores has little importance as a source of nutrients, although the fiber may serve a useful function in providing bulk in the GI tract. Although dogs and cats have little or no cecal capacity and an unsacculated colon (not typical of fiber digesters), they are known to eat grass, presumably because they have some specific need for it.

Omnivores and herbivores (plant eaters) generally have more complicated GI tracts that have been modified in some manner to improve utilization of plant tissues. The pig and rat are two examples of omnivorous species. Note that the pig has a long but simple small intestine, a moderately

*Some writers prefer to use terms such as *ruminant* and *nonruminant* to designate the type of GI tract. Monogastric is a term that has been used for many years. Strictly speaking, all mammals have only one stomach; ruminant stomachs have several compartments but only one glandular stomach compartment.

large cecum, and a sacculated large intestine. In Table 4.1, the pig (and humans) are classified as colonic digesters. In comparison, the rat has a shorter but simple small intestine, an enlarged cecum, and an unsacculated large intestine (classed as a cecal fermentor). Both of these species depend on hind gut fermentation to varying degrees; the pig has fermentation in both the cecum and colon, whereas most of the fermentation in the rabbit GI tract occurs in the cecum.

The sheep, pony, and rabbit are three examples of herbivorous species with quite different adaptations to handle fibrous diets. The sheep, because it is a ruminant, has a complex large stomach with extensive fermentation followed by a long but simple small intestine, a relatively large cecum, and a rather short large intestine—allowing both pregastric and hind gut fermentation (although not listed in this manner in Table 4.1). The horse has a small, simple stomach with a relatively short small intestine, a large cecum, and a very large sacculated hind gut (large intestine). On the other hand, the rabbit (Fig. 4.2) has a medium-sized stomach, a relatively short and simple small intestine, a large sacculated cecum, and a medium-sized unsacculated large intestine. Both the horse and the rabbit have a substantial amount of hind gut fermentation.

Within these groups the ruminants represent a highly specialized class because of their ability to digest fiber and other carbohydrates more completely than the other groups. Some pregastric fermentation occurs in other species of nonruminants, such as kangaroos, that have sacculated stomachs (Fig. 4.1 and Table 4.1), but it has not been described as well in the literature as that of ruminants. Among the hind gut fermentors, the large herbivores (horse, rhinoceros, elephant) depend on fermentation of fiber in the large intestine. Their diet may be as fibrous as that of some ruminants, but such animals tend to eat more per unit of metabolic size (body weight$^{0.75}$) while digesting less of low-quality feeds. Another example is the Chinese panda (*Ailuropoda melanoleuca*). This bearlike animal exists almost exclusively on bamboo shoots. It is said that captive pandas weighing less than 90 kg may consume almost 20%

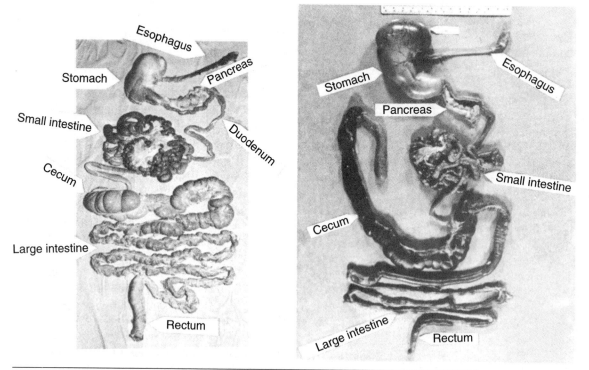

Figure 4.2   *Photographs of dissected tissues showing the principal parts of the gastrointestinal tract (GI tract) of the pig (left) and the rabbit (right). Note the differences in relative size of the various parts and the much more apparent sacculation in the large intestine of the pig. Approximate weights of the animals were pig, 100 kg; rabbit, 2 kg. Photos by D. C. Church.*

of their body weight of bamboo leaves per day (Schaffer, 1981). In humans, the cecum is reduced to a point of little if any function, but the colon is sacculated and fermentation occurs there.

When animals depend heavily on cecal fermentation, there is often an association with coprophagy (feces eating), as in rabbits. Such adaptations often result in two kinds of feces. Only the finer material is selected for recycling through coprophagy. This practice allows these small herbivores to consume fibrous diets that would otherwise be inadequate in some of the essential nutrients such as essential amino acids or vitamins. Microbial activity in the GI tract provides a more complete supply of some vitamins and amino acids which, when coprophagy is practiced, is beneficial to the animal. Further detail on nonruminant and ruminant GI-tract structure and function is provided in later chapters of this book and in other publications such as Stevens and Sellers (1977), Stevens (1977), Hoffmann (1973), Church (1987), and Moran (1982). More specific information follows on nonruminant, avian, and ruminant species.

Nonruminant Species

The mouth and associated structures—tongue, lips, teeth—are used for grasping and masticating food; however, the degree of use of any organ depends on the species of animal and the nature of its food. In omnivorous species, such as humans or swine, the incisor teeth are used primarily to bite off pieces of the food, and the molar teeth are adapted to mastication of nonfibrous materials. The tongue is used relatively little. In carnivorous species the canine teeth are adapted to tearing and rending, whereas the molars are pointed and adapted only to partial mastication and the crushing of bones. Herbivorous species, such as the horse, have incisor teeth adapted to nipping off plant material, and the molars have relatively flat surfaces that are used to grind plant fibers. The jaws are used in both vertical and lateral movements, which shred plant fibers efficiently. Rodents have incisor teeth that continue to grow during the animal's lifetime, allowing the animal to use its teeth extensively for gnawing on hard material such as nut shells. Their incisor teeth would not withstand such rugged use without the continual growth and would be worn down greatly.

In the process of mastication, saliva is added, primarily from three bilateral pairs of glands—submaxillary, at the base of the tongue; sublingual, underneath the tongue; and the parotids, below the ear. Other small salivary glands are present in some species. Saliva aids in forming food into a bolus, which may be swallowed easily, and has other functions such as keeping the mouth moist, aiding in the taste mechanisms, and providing a source of enzymes (see a later section) for initiating enzymatic digestive processes.

Photographs of portions of the GI tract of the rabbit and pig are shown in Fig. 4.2. These pictures illustrate the relative differences in size of the stomach compared to the intestines. The pig, for example, has a relatively large stomach, with the capacity in the adult of about 6 to 8 L. The shape of the stomach of different species varies, as does the relative size. Mucosal tissues lining the interior of the stomach are divided into different areas that are supplied with different types of glands as illustrated in Fig. 4.3. In the cardiac region the cells produce pri-

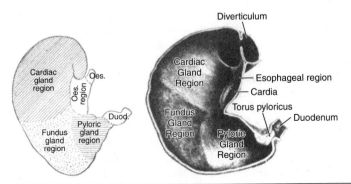

Figure 4.3 *Diagrams of the stomach of the pig illustrating the various zones and types of mucosal areas found in the stomach lining. The fundus area is the site of production of acid and digestive enzymes.*

marily mucus, which helps to protect the stomach lining from gastric secretions. In the peptic gland region or fundus, the lining is covered with gastric pits (Fig. 4.4), which open into gastric glands. These produce a mixed secretion of acid, enzymes, and mucus. The glands consist of two main types of cells, the body chief or peptic cells, which produce proteolytic enzymes, and the parietal or oxyntic cells, which secrete hydrochloric acid. Mucus-producing cells are also found in the pyrolic region. The enzymes produced and their functions are discussed in the section, The Role of GI Tract Secretions in Digestion, later in this chapter.

The relative length of the small intestine of animal species varies greatly. In the pig it is comparatively long (15–20 m), but in the dog it is short (about 4 m). The duodenum, the first (anterior) section, is the site of production of various diges-

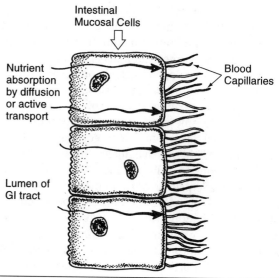

Figure 4.5   *Diagram of arrangement of intestinal mucosal cells.*

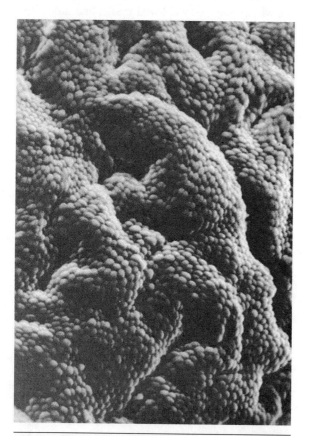

Figure 4.4   *Surface of human stomach's inner lining (glandular mucosa) is seen enlarged some 260 diameters in this scanning electron micrograph. The view shows the tops of epithelial cells, the gastric pits, and characteristic folds of a normal stomach. Courtesy of Jeanne M. Riddle, Wayne State University School of Medicine.*

tive secretions. Furthermore, a variety of digestive secretions from the pancreas as well as bile come into the duodenum within a short distance of the pylorus of the stomach.

Ducts from the liver and pancreas join to form a common bile duct that empties into the duodenum in some species; other species have separate ducts. An appreciable amount of absorption may occur in the duodenum (see later section on absorption), but most nutrients are absorbed primarily from the jejunum (middle section) and upper ileum (last or posterior section) of the small intestine. The small intestine, in general, accounts for most of the absorption in the GI tract. It is lined with a series of fingerlike projections, the villi (Fig. 4.5), which serve to increase enormously the absorptive surface area of the small intestine. Each villus contains an arteriole and a venule, together with a drainage tube of the lymphatic system, a lacteal. The venules drain ultimately into the portal blood system, which goes directly to the liver; the lymph system empties via the thoracic duct into the vena cava, a large vein.

The large intestine is made up of the cecum, colon, and rectum. Relative sizes of these organs vary considerably in different animal species. In the adult pig, the large intestine is about 4 to 4.5 m in length and is appreciably larger in diameter than the small intestine. The length and diameter of the cecum vary considerably, generally being much larger in herbivorous species than in omnivorous or carnivorous species. Note the very large size (Fig. 4.2) of the cecum of the rabbit compared to that of the pig. The horse also has a very large cecum and large intestine relative to other segments of its GI tract. In general, the large intestine acts as an area for absorption of water and secretion of inorganic elements. An appreciable amount of bacterial fermentation takes place in the cecum and colon. Data on horses indicate that the volatile fatty acids (acetic, propionic, butyric) may be absorbed from the cecum, as are some amino acids and other small molecules. The microbial populations in the cecum may be vital for synthesis of some of the water-soluble vitamins and amino acids in species such as the horse and rabbit. Because proteins and other large molecules originating in the cecum and colon are not subject to action of the digestive juices, they are assumed to be of little use to the host. Further data are required to understand the overall function of the cecum and large intestine.

The pancreas and liver are vital to digestive processes because of the digestive secretions they produce. Bile, from the liver, functions in emulsification of lipid and micelle formation for improved absorption of lipids. The liver is an extremely active site of synthesis and detoxification of metabolites and is an important storage site for many vitamins and trace elements.

Avian Species

The GI tract of avian species (Fig. 4.6) differs considerably in anatomy from typical nonruminant species. Birds have no teeth, for example, although some prehistoric forms did have teeth. Thus, the beak and/or claws serve to partially reduce food to a size that may be swallowed. Although some insect-eating birds have no crops, other types have crops of variable sizes. Ingested feed goes directly to the crop, where fermentation probably occurs in some species. The proventriculus of birds is the site of production of gastric juices, and the gizzard serves some of the functions of teeth in mammalian species, acting to physically reduce the particle size of food. Current data indicate that little proteolytic digestion occurs in either the proventriculus or the gizzard and that removal of the gizzard has little effect on digestion if the food is ground. Most of the enzymes found in mammals are present in the avian small intestine, with the

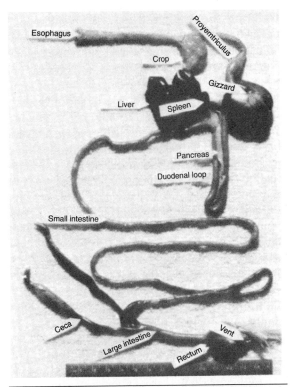

Figure 4.6 *Digestive tract of the chicken. Note the rather uncomplicated and unsacculated small and large intestines and the presence of two ceca. Photo by Don Heifer, Oregon State University.*

exception of lactase. The pH of the small intestine is slightly acid, and protein digestion is assumed to result from a combination of the common proteolytic enzymes. Data on absorption indicate that it is similar to that in mammals. The ceca and large intestine are sites for water resorption. Some fiber digestion occurs in the ceca because of bacterial fermentation, but at much lower levels than in most mammals. Removal has very little effect on digestion. Total digestibility of nonfibrous diets by birds is similar to that in mammals. Further information on the GI tract structure and function in birds may be found in Moran (1982).

Ruminants

The mouths of ruminants differ from those of other mammals in that they have no upper incisor teeth, and only a few species have canine teeth. Thus, they depend on an upper dental pad and lower incisors in conjunction with lips and tongue for prehension of food. Ruminant species may be divided into roughage eaters, selective eaters, and transitional types (Hoffmann, 1973), and these various types utilize differences in tongue mobility and lip structure, particularly, to facilitate selection of plants being consumed. With respect to mastication, ruminants have molar teeth so shaped and spaced that the animal can chew on only one side of the jaw at one time. Lateral jaw movements aid in shredding tough plant fibers.

Saliva production in ruminants is copious, reaching amounts of 150 or more liters per day in adult cows and 10 L or more in sheep. Production of saliva is relatively continuous, although greater quantities are produced when eating and ruminating than when resting. Saliva produces a source of N (urea and mucoproteins), P, and Na, which are utilized by rumen microorganisms. It is also highly buffered and aids in maintaining an appropriate pH in the rumen, in addition to other functions common to nonruminant species.

The stomach of the ruminant (Fig. 4.7) is divided into four compartments—**reticulum, rumen, omasum,** and **abomasum.** The reticulum and rumen are only partially separated but have different functional purposes. The reticulum moves ingested food into the rumen or into the omasum and functions in regurgitation of ingesta during rumination. The rumen acts as a larger fermentation vat and has a very high population of microorganisms. The function of the omasum is not clearly understood, although it appears to aid in reducing particle size of ingested food, and it obvi-

ously has some effect on controlling passage of ingesta into the lower tract. Some absorption occurs in the omasum. The abomasum is believed to be comparable in function to the glandular stomach of nonruminants.

The stomach of ruminants makes up a greater percentage of the total GI-tract capacity than is the case for other species. In adults the stomach may contain 65 to 80% of the digesta in the entire GI tract. The intestinal tract is relatively long; typical values for cattle and sheep, respectively, are small intestine, 40 and 24–25 m; cecum, 0.7 and 0.25 m; and colon, 10 and 4–5 m.

In the young ruminant the reticulum, rumen, and omasum are relatively underdeveloped because the suckling animal depends primarily on the abomasum and intestine for digestive functions. As soon as the animal starts to consume solid food, the other compartments develop rapidly, reaching relative mature size by about 8 weeks in lambs and goats, 3 to 4 months in black-tailed deer, and 6 to 9 months or more in domestic cattle.

Another anatomical peculiarity of ruminants is that they have a structure called the **reticular groove** (sometimes erroneously called the esophageal groove). This structure begins at the lower end of the esophagus and, when closed, forms a tube from the esophagus into the omasum. Its function is to allow milk from the suckling animal to bypass the reticulorumen and escape bacterial fermentation. Closure of this groove is stimulated by the normal sucking reflexes of the suckling animal, by certain ions, and by solids in suspension in liquid. It does not appear to remain functional in older animals unless they continue to suckle liquid diets.

❏ COMPARATIVE CAPACITY OF THE GASTROINTESTINAL TRACT

The relative capacities of the GI tract from different species vary greatly, as shown in Table 4.2. Assuming that these values are truly representative, it is evident that there are tremendous differences among species. Humans have a very small GI tract compared to any of the other species. The pig has a very large stomach capacity for a nonruminant animal, and the horse shows the adaptation of herbivorous animals to handle the large amounts of roughage naturally consumed. Ruminants have, by far, the largest stomachs. It should be recognized

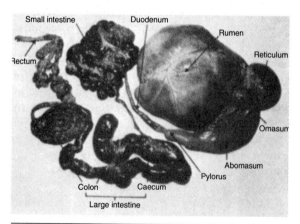

Figure 4.7   *The stomach and intestines of the sheep dissected out to illustrate the relative volumes of the various parts. Note the very large size of the stomach as compared to that of nonruminants. Courtesy of CSIRO.*

TABLE 4.2   Comparative capacity of the gastrointestinal tract of different species.[a]

| ANIMAL | RELATIVE CAPACITY, % | | | | RATIOS | |
	STOMACH	SMALL INTESTINE	CECUM	COLON AND RECTUM	INTESTINAL TO BODY LENGTH	GASTROINTESTINAL SURFACE TO BODY SURFACE AREA
Cattle	71	18	3	8	20:1	3.0:1
Sheep, goat	67	21	2	10	27:1	
Horse	9	30	16	45	12:1	2.2:1
Pig	29	33	6	32	14:1	
Dog	63	23	1	13	6:1	0.6:1
Cat	69	15		16	4:1	0.6:1
Man	17	67		17		

[a]Values are based on old European data obtained on organs after autopsy. It should be noted that the volume of an organ after removal from the body may be quite different from the relative capacity in the live, functioning animal. Nevertheless, these data probably provide a reasonable comparison for differences in capacities of the various parts of the GI tract.

that volume is subject to change, depending upon the bulkiness of the diet and with time after feeding, so these values are not fixed.

Note, also, that the intestinal length (expressed as a ratio of intestinal to body length) is much greater in the ruminant species, and that the values for the horse and pig are much greater than those for the dog and cat. This gives a crude estimate of the relative difference in absorptive surface in the gut of these different species. Also, the ratio of intestinal surface to body surface is much greater for the cow and horse than for the dog and cat.

❏ RUMEN METABOLISM

Rumen Microorganisms

In the GI tract of the ruminant animal, in contrast to that of other types of animals that have no pregastric fermentation, ingested feed is exposed to very extensive microbial fermentation before it reaches the glandular stomach (abomasum) or intestinal juices. The reticulorumen provides a very favorable environment for survival and activity of anaerobic microbes; it is moist and warm, and there is an irregular introduction of new digesta and a generally continuous removal of fermented digesta and end products of digestion so that metabolites such as acids do not build up to inhibitory levels. Most of the bacteria are obligate anaerobes, meaning that such organisms grow best in an atmosphere that has very low oxygen levels. Other types of microorganisms may be present at times, but they find it difficult to compete with the usual rumen population. A vast number of rumen bacteria may be found in the rumen, typical counts approaching 25 to 50 billion/ml. Table 4.3 shows the names of some of the species that have been isolated and cultured in purified media. In the 11 different groups shown here, the bacteria either grow on a specific ingredient or produce a specific end product. Note that a number of species, such as *Butyrivibrio fibrisolvens* or *Bacteroides ruminicola*, are listed in several different groups. The ability to metabolize different energy or nitrogen sources allows the host animal to utilize a wider variety of feed. The characteristics of these and other bacterial species have been studied in many different laboratories. More than 22 genera and over 60 species have been described. Their characteristics are too detailed to discuss here, except to note that a wide variety exists in cell size, shape, and structure in addition to differences in their metabolism.

In addition to bacteria, about 35 species of ciliate protozoa have been identified from the rumens of animals in different situations, although the variety that may be found in any one animal is considerably less. Protozoal counts vary widely, but typical values to be expected are on the order of 20,000 to 500,000/ml. Six different genera and 14 species are shown in Table 4.4. Most, if not all, protozoal species ingest rumen bacteria, and many

TABLE 4.3   Major groups of rumen bacteria that ferment important feed ingredients in the rumen.

Major Cellulolytic Species
Bacteroides succinogenes
Ruminococcus flavefaciens
Ruminococcus albus
Butyrivibrio fibrisolvens

Major Hemicellulolytaic Species
Butyrivibrio fibrisolvens
Bacteroides ruminicola
Ruminococcus sp.

Major Pectinolytic Species
Butyrivibrio fibrisolvens
Bacteroides ruminicola
Lachnospira multiparus
Succinivibrio dextrinosolvens
Treponema bryantii
Streptococcus bovis

Major Amylolytic Species
Bacteroides amylophilus
Streptococcus bovis
Succinimonas amylolytica
Bacteroides ruminicola

Major Sugar-Utilizing Species
Treponema bryantii
Lactobacillus vitulinus
Lactobacillus ruminus

Major Ammonia-Producing Species
Bacteroides ruminicola
Megasphera elsdenii
Sellenomonas ruminantium

Major Ureolytic Species
Succinivibrio dextrinosolvens
Selenomonas sp.
Bacteroides ruminicola
Ruminococcus bromii
Butyrivibrio sp.
Treponema sp.

Major Proteolytic Species
Bacteroides amylophilus
Bacteroides ruminicola
Butyrivibrio fibrisolvens
Streptococcus bovis

Major Acid-Utilizing Species
Megasphaera elsdenii
Selenomonas ruminantium

Major Methane-Producing Species
Methanobrevibacter ruminaantium
Methanobacterium formicicum
Methanomicrobium mobile

Major Lipid-Utilizing Species
Anaerovibrio lipolytica
Butyrivibrio fibrisolvens
Treponema bryantii
Eubacterium sp.
Fusocillus sp.
Micrococcus sp.

also ingest food particles such as grains of starch or particles of feed. Note in Table 4.4 that most of the species listed ferment starch and smaller carbohydrate molecules such as glucose. All the *Diplodinium* species shown ferment cellulose, as do the examples shown for *Epidinium* and *Ophryoscolex* genera. Qualitatively, the end products produced by ciliate protozoa are very similar to bacterial species.

Other microorganisms are also found in the rumen, but they are not as well identified and have not been studied nearly as much as the bacteria and ciliate protozoa. For example, considerably lower concentrations of flagellated protozoa are sometimes found, particularly in young animals. In addition, relatively large counts of phages (bacterial viruses) have been observed from time to time. Anaerobic fungi that can digest cellulose have been observed, and at least two species of my-

coplasmas have been identified (Yokoyama and Johnson, 1987). Yeasts are sometimes found in relatively high numbers. For that matter, almost any microorganism found on food or in water may be identified in the rumen, but the conditions do not favor survival of nonrumen microbes. The fate of rumen microorganisms is that, eventually, they pass into the abomasum and intestines where they are digested by enzymes produced by the host animal and provide nutrients for the animal. Studies with rats fed dried rumen microbes suggest that protozoa are more highly digested than bacteria.

Rumen Fermentation

Carbohydrates.   Carbohydrates make up the major portion of the diet of herbivorous animals. There may be many different types (see Chapter 7), but for our discussion here they can be classed

TABLE 4.4 Feed ingredients fermented and end products produced by rumen ciliate protozoa.[a]

GENERA	INGREDIENT FERMENTED	END PRODUCTS
Isotricha		
Intestinalis	Starch, sucrose, glucose, pectin	A, p, B, L, H, Li
Prostoma	Starch, sucrose, glucose, pectin	A, p, B, L, H, Li, C
Dasytricha		
Ruminantium	Starch, maltose, cellobiose, glucose	A, B, L, H, C
Entondinium		
Bursa	Starch, hemicellulose	—
Caudatum	Starch, cellobiose, glucose, maltose, Sucrose	A, P, B, L, H, Li, C
Fuca bilobum	—	Li
Simplex	Starch	Li
Diplodinium		
Polyplastron	Cellulose, glucose, starch, sucrose	A, P, B, L, H, C
Diploplastron	Cellulose, hemicellulose, starch	—
Eudiplodinium	Cellulose, hemicellulose, starch	A, P, B, L, H, F, C
Ostracodinium	Cellulose, hemicellulose, starch	—
Eremoplastron	Cellulose, hemicellulose, starch	—
Epidinium		
Ecaudatum caudatum	Cellulose, hemicellulose, starch sucrose, maltose	A, p, B, I, H, f, Li
Ophryoscolex		
Caudatus	Cellulose, hemicellulose, starch	A, p, B, H

[a]Note that the protozoa were grown on purified substrates. Relative amounts of end products probably would be modified if typical mixtures of ingredients, such as found in normal feedstuffs, were used. Capital letter indicates major end product; lower case indicates trace. Acetate (a), proprionate (P or p), butyrate (B), lactate (L or l), hydrogen (H), formate (F or f), lipids (Li), CO2 (C).

as fibrous (hemicellulose, cellulose, xylans) or readily available (primarily sugars, starches). In the rumen of an animal fed at a low level (i.e., a maintenance diet), essentially 100% of the readily available carbohydrates (RAC) will be fermented by the rumen microorganisms. The principal end products of fermentation are the volatile fatty acids (primarily acetic, propionic, and butyric), carbon dioxide, methane, and heat. The animal, in turn, uses the volatile fatty acids as a source of energy for its life processes. This is in contrast to the situation in most animal species, in which the major product of carbohydrate digestion is glucose. Glucose is the principal energy source for cell metabolism in animals. The fibrous carbohydrates also are fermented by rumen microorganisms and the end products are the same, although less propionic acid is produced normally than from RAC. Because animal tissues do not produce cellulase or hemicellulase, microbial fermentation is the only means by which animals can indirectly use these complex carbohydrates. The ruminant animal is unusual in that it is capable of utilizing a larger proportion of fibrous carbohydrates than other herbivorous animals, such as the horse or rabbit, which depend on microbial digestion in the cecum and large intestine for the digestion that occurs.

When ruminant animals are fed large amounts of grain or pelleted roughage, moderate amounts of RAC or cellulose may pass out of the rumen before fermentation can be completed. In this case, the RAC is exposed to pancreatic and intestinal enzymes that may digest some part of it. However, the ability of the animal to digest large amounts of sugars (especially sucrose) or starches is limited because of limited production of the appropriate enzymes. In such cases, microbial digestion of RAC also may occur in the lower gut.

Feeding unadapted animals large amounts of cereal grains such as wheat, may result in abnormal rumen fermentation, which is associated with high production of lactic acid and a very acid rumen. This condition may cause death if severe enough.

Proteins.   Although some rumen bacteria may require amino acids or peptides, and most ciliate protozoa probably must ingest bacteria for survival, the overall mixtures of rumen microorganisms generally do not respond in this manner. There are many proteolytic organisms that attack dietary protein, with the result that a considerable amount of it is degraded to ammonia and organic acids. These in turn, may be utilized by other microbial species to synthesize amino acids and bacterial proteins. Most cellulose-digesting bacteria require ammonia. The net result of feeding plant (or animal) protein is that a considerable portion of it is degraded and resynthesized into different microbial proteins.

Rumen microorganisms also are capable of utilizing simple N sources such as urea (a normal mammalian excretory product), amino acids, nitrates, biuret, and amines. Thus, urea or other economically competitive nonprotein nitrogen sources may be substituted for protein in ruminant diets in many instances. The net effect of normal rumen metabolism of protein is that it allows the animal to exist on a wide variety of diets, but overall efficiency of protein utilization is low because it is biologically inefficient to degrade and resynthesize complex molecules such as proteins.

Recent research data suggest that solubility and degradability of protein sources are important to ruminants expected to perform at a high level. There is some evidence that high-producing dairy cows require larger amounts of some amino acids, particularly methionine, lysine, and leucine, than they can obtain from microbial protein. If a protein is fed that is either not soluble or only partially degraded in the rumen but that can be digested in the gut, then it is sometimes possible to provide the animal with the needed amino acids. This type of protein, called rumen bypass protein, is present to some degree in many feed sources. Information on the feed content of proteins of this type can be found in NRC bulletins.

Lipids.   Lipid consumption by herbivorous animals is low because most forage contains only limited amounts. Of that present, a considerable portion is in the form of galactosyl diglycerides, compounds in which galactose (a simple sugar) has replaced one of the fatty acids normally found in a common triglyceride. Another characteristic of herbage lipids is that a high percentage of the fatty acids are unsaturated (see Chapter 8).

In the rumen, the microbes do not greatly alter the fat fraction, although some lipids may be synthesized. When fed low to moderate levels of dietary fat, the rumen microorganisms hydrolyze fat, producing free fatty acids and glycerol. Rumen microbes modify the unsaturated fatty acids by either saturating them or causing changes in the location of double bonds or altering the normal cis bond to a trans double bond. These are two of the characteristics that are common to ruminant milk or body fat. In addition, some of the microbial fat synthesis results in the production of fatty acids with an odd number of carbon atoms in the chain and with branched chains in the molecule. For example, butterfat may have fatty acids with 9, 11, 13, 15, 17, and 19 carbon atoms, numbers not at all characteristic of plant fatty acids, which usually have even numbers of carbon atoms.

The rumen microbial population is intolerant to high dietary levels of fat. When fat is added to the diet, it is fed normally at no more than 5 to 7% of the total diet. Higher levels are apt to result in abnormal rumen fermentation unless the fat is protected from the organisms by coating fat droplets with casein, and then treating the complex with formaldehyde.

Gas Production.   Anaerobic fermentations such as those that occur in the rumen result in the production of copious amounts of gases. A fairly typical composition of the gases would be: 65% CO_2, 25 to 27% CH_4, 7% N, and trace amounts of O_2, H_2, and H_2S. Calorimetry data indicate that cattle may produce up to 600 L of gases per day (Hoffmann, 1973). Methane, which has a high heat equivalent, represents a direct loss of energy to the animal. It is produced in anaerobic fermentation as a means of getting rid of excess hydrogen. If eructation (see section on anatomy and function) is inhibited, bloat may result; bloat can be chronic or acute, and the acute form can cause death very rapidly.

Vitamin Synthesis.   Rumen microorganisms have the capability of synthesizing essentially all of the B-complex vitamins (see Chapter 11) required by the host animal. Although some synthesis may occur in the large intestine or cecum of other species, the amount synthesized in the rumen is probably greater than that in the lower GI tract. Normally, deficiencies of only two B-complex vitamins ever cause any problem for ruminants. Cobalamin (B_{12}) may be deficient at times because of a dietary shortage of cobalt, an essential component of the vitamin. This deficit can be remedied by supplying more cobalt in the diet. Thiamin (B_1) deficiency also may occur. Data suggest that synthesis is reduced in the rumens of animals fed high-RAC diets. There is also some evidence that insufficient niacin may be involved in ketosis of lactating dairy cows. Some evidence also shows that rumen activity may result in a decrease in vitamin A or carotene concentrations of ingesta, even though the diet contains sufficient amounts. However, the consumption of carotene is usually at sufficiently high levels that this effect of the rumen is not a problem for the animal.

Rumination and Eructation

There is a well-developed pattern of rhythmic contractions of the various stomach compartments that act to circulate ingesta into and throughout the rumen, into and through the omasum, and on to the abomasum. Of special importance also are contractions that aid in regurgitation during **rumination.** This is a phenomenon peculiar to ruminants. In effect, it is a controlled form of vomiting, allowing semiliquid materials to be regurgitated up the esophagus, swallowing of the liquids, and a deliberate remastication of and swallowing of the bolus formed in the process. Ruminants may spend 8 h/day or more in rumination, depending on the nature of their diet. Coarse, fibrous diets stimulate longer rumination time. Inhibition of rumination results in a marked reduction in feed consumption. The evolutionary origin of ruminating is not clear; perhaps it was evolved to allow animals to consume feed hastily and then retire in relative safety from predators to rechew their food at leisure.

Eructation (belching of gas) is another mechanisms important to ruminants. Microbial fermentation in the rumen results in production of large amounts of gases that must be eliminated. This is accomplished by contractions of the upper sacs of the rumen which force the gas forward and down; if the esophagus is not obstructed, it then dilates and allows the gas to escape. During this process much of the gas penetrates into the trachea and lungs. A common problem in ruminants is bloat, a condition that develops mostly as a result of formation of froth in the rumen. If froth accumulates in the area where the esophagus enters the rumen,

it inhibits eructation, a safety mechanism preventing inhalation of froth into the lungs. Bloat may be produced by mechanical injury to various parts of the GI tract, but the most common cause is production of froth from consumption of lush young legumes such as alfalfa. Dry alfalfa may also stimulate froth when included in dry rations such as those used in feedlots, particularly when the particle size of the ration ingredients is quite fine. Froth formation can largely be prevented *if* the animal regularly consumes a froth-inhibiting compound.

❏ THE ROLE OF GI-TRACT SECRETIONS IN DIGESTION

Digestive secretions have very important roles in the overall digestive processes. In nonruminant mammals and avian species, the digestive enzymes attack the food before it is subjected to microbial action in the cecum and large gut. In ruminants the digestive secretions are supplementary to the digestion that occurs first in the rumen as a result of microbial fermentation.

Digestive enzymes are found in saliva in small amounts. The glandular stomach (or abomasum or proventriculus) is a major source of proteolytic enzymes and hydrochloric acid. The pancreas is an important source of enzymes that act on proteins, starches, and fats, and glands in the wall of the duodenum produce a variety of enzymes that act on sugars, protein fragments, or lipids. The enzymes produced in the glandular stomach of the animal are made in the peptic gland region and are released into the lumen of the organ. Pancreatic secretions enter the duodenum directly via a pancreatic duct or, in some species, via a duct that joins the bile duct from the gall bladder (or liver in species that do not have a gall bladder).

The principal digestive enzymes secreted by the GI tract and pancreas are listed in Table 4.5, along with information on origin, target substrate, and end products. The enzymes are listed according to the general type of compound hydrolyzed—amylolytic (carbohydrates), lipolytic (lipids), or proteolytic (proteins). A number of animal species produce small amounts of salivary amylase, but it is not believed to be of major importance because of the relatively short time it may act in the mouth and because the usual stomach pH is too low for continued activity. Numerous saccharidases are produced by the pancreas and intestinal mucosa.

TABLE 4.5   Principal digestive enzymes secreted by the gastrointestinal tract, substrates attacked, and end products produced.

TYPE, NAME	ORIGIN	SUBSTRATE, ACTION	END PRODUCTS	COMMENTS
Amylolytic				
Salivary amylase	Saliva	Starch, dextrins	Dextrins, maltose	None in ruminants; of minor importance in other species
Pancreatic amylase	Pancreas	Starch, dextrins	Maltose, isomaltose	Low in ruminants
Maltase, isomaltase	Sm. intestine	Maltose, isomaltose	Glucose	Low in ruminants
Lactase	Sm. intestine	Lactose	Glucose, galactose	High in young mammals
Sucrase	Sm. intestine	Sucrose	Glucose, fructose	None in ruminants
Oligoglucosidase	Sm. intestine	Oligosaccharides	Misc. monosaccharides	
Lipolytic				
Salivary lipase	Saliva	Triglycerides	Diglyceride + 1 fatty acid (FA)	Of minor importance in young mammals
Pancreatic lipase	Pancreas	Triglycerides	Monoglyceride + 2 FA	
Intestinal lipase	Sm. intestine	Triglycerides	Glycerol + 3 FA	
Lecithinase	Pancreas, sm. intestine	Lecithin	Lysolecithin, free FA	
Proteolytic				
Pepsin[a]	Gastric juice	Native protein	Proteoses, peptones, polypeptides	Clots milk; hydrolyzes native proteins in acid pH
Rennin[a]	Abomasum	Clots milk (casein)	Ca caseinate	Important in young mammals
Trypsin[a]	Pancreas		Peptides with terminal arginine or lysine group	
Chymotrypsin	Pancreas	Native proteins or products of pepsin and rennin digestion	Peptides with terminal aromatic amino acid	
Elastase	Pancreas		Peptide with terminal alphatic ammo acid	
Carboxypeptidase A	Pancreas	Peptides with aromatic or alphatic amino acid	Small peptides, neutral amino acids, acidic amino acids	
Carboxypeptidase B	Pancreas	Peptides with terminal arginine or lysine	Basic amino acids	
Aminopeptidases	Sm. intestine	Peptides	Amino acids	
Dipeptidases	Sm. intestine	Dipeptides	Amino acids	
Nucleases (several types)	Pancreas, Sm. intestine	Nucleic acids	Nucleotides	
Nucleotidases	Sm. intestine	Nucleotides	Purine and pyrimidine bases, phosphoric acid, pentose sugars	

[a]These enzymes are given off in inactive forms, probably as a means of protecting the body tissues. Pepsinogen is activated by HCl to the active form, pepsin; rennin is activated by HCl; trypsinogen by an intestinal enzyme, enterokinase, and by trypsin. The proforms of chymotrypsin, carboxypeptidases, and amino peptidases are activated by trypsin.

Pancreatic amylase is of major importance in hydrolyzing starch, glycogen, or dextrins to maltose.

The various enzymes that act on oligosaccharides or disaccharides are produced by the small intestine mucosa. These include maltase, isomaltase, lactase, and sucrase. However, not all enzymes are found in all animals. For example, little if any lactase is found in species other than mammals. This is logical, of course, because lactose is found only in milk. In young mammals, lactase levels are high while they are consuming milk, but production of the enzyme tapers off after milk consumption ceases. Similarly, sucrase is not found in ruminant (and some other) species. With a natural diet, sucrose normally entering the stomach of a ruminant animal would be fermented in the rumen and would not reach the small intestine in large amounts.

If young animals are fed sucrose in milk replacers, any digestion that takes place occurs in the lower gut as a result of fermentation associated with microbial activity. Diarrhea may occur if much sucrose is fed. Also, in ruminants the pancreatic production of amylase is low, presumably because natural diets do not contain much starch and because most of that entering the stomach would be fermented there. Some starchlike compounds (microbial storage polysaccharides) do enter the lower GI tract in these situations, but much less than would occur in animals fed high-grain diets. The overall result is that ruminant animals have a more limited ability than nonruminants to digest starch.

Animals do not produce enzymes that hydrolyze complex carbohydrates such as xylan, cellulose, and hemicellulose. Consequently, any hydrolysis of such compounds is the result of microbial enzymatic activity. This appears to be an illogical evolutionary development for herbivorous species, which consume relatively large amounts of these carbohydrates. Perhaps, with increasing knowledge of digestive physiology, we will one day have a logical explanation for the absence of such enzymes.

With respect to the proteolytic enzymes, the proenzymes are converted to active enzymes by the activation of trypsinogen by enterokinase, an enzyme produced by the intestinal mucosa. Tyrpsin, in turn, activates chymotrypsinogen to chymotrypsin and procarboxypeptidase to carboxypeptidase. Pepsin is relatively inactive except at a rather low pH (below pH 3.5). Thus, in the young suckling animal, in which stomach pH is apt to be on the order of 4–4.5, pepsin is relatively inactive. In fact, some information indicates very little pepsin secretion in calves until they start to consume solid food. Rennin is an important enzyme in young suckling mammals. Its major action is to coagulate milk into a clot (also produced by HCl and pepsin), which allows a continuous and prolonged flow of milk into the small intestine, thus avoiding an excessive overload of the duodenum immediately following a meal. Rennin production is reduced considerably after milk feeding ceases.

The proteolytic enzymes, as well as other digestive enzymes, have relatively high substrate specificity. Pepsin, for example, tends to attack peptide bonds involving an aromatic amino acid (phenylalanine, tryptophan, or tyrosine), and it also has significant action on peptide bonds involving leucine and acidic residues. However, it liberates only a few free amino acids. Trypsin acts on peptide linkages involving the carboxyl group of arginine and lysine, and chymotrypsin is most active on peptide bonds involving phenylalanine, tyrosine, and tryptophan. The action of trypsin and chymotrypsin is additive, resulting in more complete degradation of proteins to small peptides. These enzymes are termed endopeptidases because they hydrolyze interior bonds. Those that act on terminal amino acids are called exopeptidases.

Carboxypeptidase A rapidly liberates carboxyl terminal amino acids, but carboxypeptidase B acts only on peptides with terminal arginine or lysine residues.

Young mammals are capable of absorbing relatively large protein molecules from colostrum (legally, colostrum is not milk, but fluid produced by the mammary glands during the first 2 to 3 days after parturition), which provides them with a source of antibodies (immunoglobulins). These proteins apparently avoid digestion because of some delay in enzyme production and because, in some species, of the presence of trypsin inhibitors that protect the immunoglobulins without completely inhibiting the digestion of other milk proteins.

Hydrochloric acid, produced by specialized cells in the gastric mucosa, has an important effect on gastric digestion. It activates both pepsin and rennin and provides a pH that is near the optimum for pepsin activity. HCl, along with pepsin and rennin, also has the property of coagulating milk proteins and has some hydrolytic activity.

Bile from the liver enters the small intestine via the bile duct and has an important role in digestion.

Note that not all species have a gall bladder; for example, rats, horses, deer, elk, moose, and camels do not. In other species, including humans, swine, chickens, cattle, and sheep, the gall bladder serves as a reservoir for temporary storage of bile. Bile contains salts of various bile acids (glycocholates and taurocholates) which are important in providing an alkaline pH in the small intestine and in preparing fats for absorption by the formation of micelles (aggregates of molecules of lipids and bile acids) in the lumen of the intestine. The bile salts are absorbed easily in the small intestine and are recirculated back to the liver. Bile pigments are responsible for its color and, ultimately, for most of the color in feces and urine. Bilirubin or its oxidation products account for much of the pigment. These pigments are waste products, being derived from the porphyrin nucleus of hemoglobin, which is metabolized in the liver. Bile also serves as a route of excretion for many different metallic elements, inactivated hormones, and various harmful substances.

Fat digestion is a result of the emulsifying action of bile salts and the enzymatic action of pancreatic and intestinal lipases. Pancreatic lipase preferentially hydrolyzes fatty acids in the 1 and 3 positions of triglycerides to produce two free fatty acids and a 2-monoglyceride. The bile salts and hydrolyzed fats form micelles, which greatly increase the surface area of fat droplets (up to a 10,000-fold increase). The micelles are absorbed by the intestinal epithelium, and a major portion passes into the lymph system, which empties into the vena cava.

❏ CONTROL OF GASTROINTESTINAL ACTIVITY

The gastrointestinal system of an animal is a very complex system controlled by a combination of nerve stimulation and inhibition and by a number of gastrointestinal peptide hormones. Cranial parasympathetic neurons provide innervation via the vagus nerve to the esophagus, stomach, small intestine, and proximal portions of the large intestine. The sacral segment of the spinal cord provides parasympathetic innervation to the distal portion of the colon, the rectum, and the internal anal sphincter. Controls are also provided by gastrointestinal peptide hormones acting alone or in conjunction with neural controls. A gastrointesti-

nal hormone may be defined as a peptide that is synthesized primarily in cells of the digestive tract and that influences in an endocrine, paracrine, neurocrine, or neuroendocrine manner the physiological activity of some organ or organs of digestion. More simply, the actions of gastrointestinal peptide hormones and hormonelike peptides include stimulation and reduction of secretions, motility, and absorption. Some may have other functions (Titchen, 1986).

Note that most of the major GI tract peptide hormones are produced in the stomach or in duodenal tissues (Table 4.6). The exceptions are pancreatic polypeptide, which is produced in the pancreas, and vasoactive intestinal peptide (VIP), which is produced primarily by nervous tissues in many locations in the body. The main point is that the neural and neuroendocrine controls result in a coordinated series of secretory and motor activities that provide for an orderly sequence of action. The coordinated muscle activity serves to mix and transport digesta through the total GI tract. In addition, the intestinal hormones stimulate or inhibit secretions of various glands necessary for digestion and absorption of nutrients. Other hormones may also be involved; that is, various adrenal hormones and parathyroid hormone have marked effects on absorption and excretion of some inorganic elements. Several hormones—insulin and glucagon from the pancreas, adrenocortical hormones, thyroid hormone, an array of pituitary hormones, and others—have effects on nutrient utilization.

❏ ROLE OF THE GI TRACT IN TRANSPORT OF NUTRIENTS

Most of the absorption of nutrients takes place in the upper intestinal tract, including the duodenum and the jejunum and, to a lesser extent, the ileum and large intestine. The passage of ingesta through the entire length of the digestive tract takes only a few hours in most species, so it is clear that opportunity for the processing and absorption of nutrients is limited. The degree of absorption of nutrients from the intestinal tract is increased enormously by an increase in the absorptive surface. For example, in the human adult, if the surface area of the intestine were in the form of a simple cylinder, it would have a functional area of approximately 3300 cm^2. However, its area in-

TABLE 4.6 Gastrointestinal peptide hormones and functions related to activity of the gastrointestinal tract.

HORMONE	ORIGIN	RELEASING MECHANISM	FUNCTION
Gastrin	Pyloric antrum of the stomach or abomasum of ruminants	Vagal nerve stimulation; food in the stomach; stomach distension	Stimulation of acid secretion by gastric glands
Gastric-Inhibitory Polypeptide (GIP)	Gastric antrum, duodenum, jejunum	Fats and fatty acids plus bile in the duodenum	Inhibition of gastric secretion and mobility
Secretin	Duodenal mucosa	Acidification of the duodenum, peptones in duodenum	Stimulation of volume and bicarbonate outputs of pancreatic secretion and, in some species, of bile
Cholecystokinin (CCK)	Duodenal mucosa, brain	Long-chain fatty acids, amino acids, peptones	Contraction of gallbladder and pancreas; stimulates synthesis of pancreatic enzymes; inhibits gastric acid secretion; enhances insulin release; may induce satiety
Somatostatin	Abomasal antrum and duodenum, nerve cells in GI tract	Vagal stimulation and changes in the composition of intestinal chyme	Inhibition of release of gastrin, secretin, and CCK; inhibition of ion transport in intestines
Pancreatic Polypeptide	Pancreas	Nerve stimulation; entry of food into the duodenum; insulin hypoglycemia	Reduction or inhibition of pancreatic secretions
Vasoactive Intestinal Peptide (VIP)	Many neural tissues throughout body	Neural stimulation, prolonged exercise, fasting	Stimulation of pancreatic exocrine secretions and potent vasodilator

creased to about 2 million cm^2 by virtue of folds, villi, and microvilli, each of which increases surface area by several times. Figure 4.8 presents photographs of a jejunal intestinal, epithelial cell from newborn and 4-day-old pigs. The microvilli and subcellular components of a single cell are visible at a magnification of 1800 times.

The rate of metabolism of the intestinal mucosa (the intestinal epithelium) is one of the fastest of any tissue in the body. In the human adult, there is a turnover of about 250 g of dry matter per day, an amount that contributes significantly to the maintenance requirements of the individual.

The passage of individual nutrients from the intestinal lumen into the intestinal epithelial cell and then into the blood or lymph may occur by passive diffusion (the pore size of cells in the jejunum is 7.5 angstroms, or 0.0000075 mm), by active trans-

port, or by pinocytosis (phagocytosis). Pinocytosis involves engulfment of large particles or large ions in a manner similar to the way an amoeba surrounds its food. This occurs in newborn animals to allow absorption of immune globulins and other proteins and peptides from colostrum.

The passage of a nutrient across the intestinal mucosal cell membrane and into the blood or lymph, whether by diffusion or by active transport, involves (a) penetration of the microvillus and of plasma membrane, which encapsulates the epithelial cell, (b) migration through the cell interior, (c) possible metabolism within the cell, (d) extrusion from lateral and basal aspects of the cell, (e) passage through the basement membrane, and (f) penetration through the vascular or lymphatic epithelium into blood or lymph. The exact means of transfer of each individual nutrient is discussed in subsequent chapters on individual nutrient requirements and metabolism.

❏ ROLE OF BLOOD AND LYMPH IN NUTRIENT TRANSPORT

A broad concept of nutrient flow through the body is illustrated in Fig. 4.9 (lymph system not shown). Briefly, a nutrient passes across the epithelial cell and enters either blood capillaries or the lymph system and is carried through the portal vein to the liver or, when materials enter the lymph, through the thoracic duct to the heart. The liver acts as a central organ in metabolism because many complex and vital reactions occur there. After traversing the liver, venous blood reaches the right atrium of the heart and passes into the right ventricle, from which it is pumped through the lung for oxygenation. From the lung it returns to the left atrium of the heart and passes into the left ventricle, the largest and most muscular chamber of the heart. From the left ventricle it is pumped through the aorta to enter all tissues of the body, carrying oxygenated blood and nutrients from the

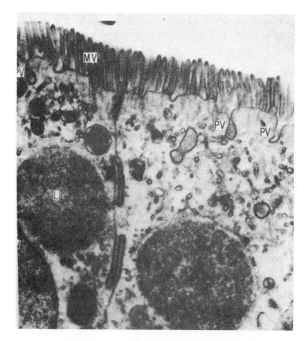

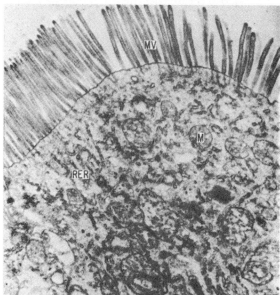

Figure 4.8   Top: Electromicrographs of piglet jejunal intestinal epithelium 1.5 h after feeding cow's colostrum (open). Bottom: same except specimen came from a 4-day-old piglet (closed). Note indentations and vesiculation of plasmalemma in open piglet and lack of activity in closed piglet. PV = pinocytotic vesicle; MV = microvilli; L = lysosome; N = nucleus; RER = rough endoplasmic reticulum; M = mitochondria. From Broughton and Lecce (1970). Courtesy of J. C. Lecce.

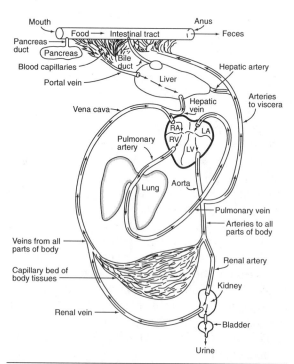

Figure 4.9   Diagram of routes of nutrient absorption, flow of nutrients and fluids in the body, and routes of excretion. The lymph system is not shown.

GI tract and those synthesized in other tissues, including the liver. The nutrients enter the capillaries in all tissues of the body and, in this way, nourish every cell. The same capillary system carries waste products from the cell into veins for transport to sites of further metabolism or for excretion.

The kidney plays an important role in acting as a filter to dispose of waste materials and as an organ for catabolism of metabolites for excretion or use elsewhere in the body. In addition to waste product excretion through the kidney, other waste products can be disposed of by reexcretion back into the intestinal lumen. This takes place by accumulation in the liver and excretion via the bile. In addition, excretory products pass through the blood capillaries into the intestinal epithelium. Thus, two-way traffic moves across the epithelial cells—both absorption of nutrients from the intestinal lumen and excretion of metabolites and waste materials across the intestinal wall—but in opposite directions. In addition to excretory products, there is considerable loss from the body of nutrients that have already served a productive function. For example, the loss

via feces of cells sloughed off from intestinal mucosa represents an obligatory loss of protein, minerals, and other nutrients, even though they have been utilized in normal body function.

The GI tract provides the most readily accessible route for nongaseous substances to enter the body. Consequently, it must carefully select those substances through a variety of mechanisms, including rapid rejection of toxic compounds (vomiting or increased rate of passage) and gastric digestion, before they reach the more permeable intestinal tract. In addition, selective permeability of the GI tract epithelium is vital to minimize the absorption of toxic compounds or excessive absorption of natural nutrients—particularly some of the mineral elements—or reabsorption of toxic metabolites such as ammonia.

Summary of Major Activity in the GI Tract

For the benefit of students, the major activities in the GI tract have been tabulated in Table 4.7. This material is self-explanatory and does not require further discussion in the text.

TABLE 4.7 A summary of the major actions in the GI tract.

Mouth/Esophagus

In species that chew their food, the activities are well defined. The food is normally grasped by teeth and/or tongue and masticated and ensalivated enough to swallow. Some species don't chew their food at all. For example, owls and hawks which prey on small rodents and most snakes just swallow their prey intact. This may require the ability to greatly enlarge the throat and esophagus to swallow some of their prey.

Crop

Except for insectivore species, most birds have a crop for temporary storage of ingested food. A few birds, such as the pigeon, have some glandular-type cells in the crop that secrete "pigeon milk," a product used to nourish the very young hatchlings. This "milk" is available only for a few days after the young birds have hatched out.

Reticulorumen

Rumen microorganisms ferment to some degree all but the most insoluble dietary components. The intensity and amount of fermentation are greatly affected by the time spent in the rumen, which, in turn, is affected by particle size of the digesta, the level of feeding (a high level pushes food through faster), and solubility of the diet. The rumination process, where food is regurgitated and rechewed, results in a somewhat more uniform rate of fermentation. The eructation process gets rid of massive amounts of gases produced by microbial fermentation. Fermentation of carbohydrates results in the production of volatile fatty acids, carbon dioxide, and methane. Ammonia and volatile fatty acids result from fermentation of soluble proteins and amino acids. N is recycled into the rumen from saliva and through the rumen wall. In the rumen N is utilized for synthesis of bacterial or protozoal protein. Unsaturated fatty acids in the diet are mostly saturated or modified into *trans*-fatty acids by microorganisms. Odd-length or branched-chain volatile fatty acids or amino acids may be synthesized into longer odd-length and branched-chain fatty acids. Most of the water-soluble vitamins are synthesized in great abundance in the rumen. Rumen contractions are coordinated so as to move freshly ingested food backward into the central and posterior portions of the rumen so that it will be mixed in with previously ingested food. Rhythmic contractions suffice to bathe the ingesta with

continued

TABLE 4.7   A summary of the major actions in the GI tract—*Continued.*

liquids and to move smaller particles into the reticulum so that they can be regurgitated in the rumination process. Also, the reticulo-omasal orifice opens from time to time when the reticulum is contracting, thus allowing smaller particles to pass into the omasum or on into the abomasum.

Omasum

The combination of the reticulo-omasal orifice and the omasum obviously act as filters to prevent excessively large particles of ingesta from passing into the abomasum and small intestine. Forage or other food particles found between the leaves of the omasum are usually rather small in size. Some absorption of components such as volatile fatty acids occurs in the omasum.

Stomach/Abomasum/Proventriculus

The function of different types of glandular stomach in various species is quite similar. Chyme entering from proximal organs is acidified by the addition of HCL, and initial digestion of proteins by pepsin occurs. In young mammals, milk is clotted by the acid and enzymes, thus prolonging its stay in this part of the stomach. Some fat digestion may occur in young mammals.

Gizzard

In birds, the proventriculus flows into the gizzard where, with the aid of grit picked up by the bird, the digesta is ground to a fineness suitable for better utilization in the intestines.

Small Intestine

Bile from the liver enters the intestine producing an alkaline medium that is conducive to digestion by the pancreatic and intestinal enzymes which enter with the bile via the common bile duct in most species. Some absorption occurs in the duodenum, but most occurs in the jejunum and ileum. The bile salts are absorbed (and recycled to the liver), lowering the pH, thus allowing a gradual increase in microbial activity.

Cecum

The cecum is of little importance in carnivores and in some omnivores such as man. In other species digesta from the ileum enters the cecum where a considerable amount of microbial activity occurs. Among other things, the microorganisms may produce microbial protein of much better quality than that in the diet. In addition, vitamin synthesis occurs. For species that practice coprophagy, the fecal material consumed thus allows them to exist on a diet of lower quality than otherwise might be the case. Note also that some interspecies benefits of intestinal production of vitamins and proteins may occur. Dogs, as well as other species such as swine, often eat some feces of other species such as horses or cattle, and it is most probable that they derive some nutritional benefit from this practice.

Large Intestine

Continued microbial fermentation occurs in the large intestine, although therre is uncertainty about the quantity of absorption of amino acids, peptide, or vitamins from the large gut. Excretion of a number of mineral elements occurs in the large gut. Large amounts of water may be absorbed, especially in species adapted to arid climates. Many species adapted to dry climates produce fecal pellets, examples being kangaroo, rats, rabbits, sheep, goats, deer, and antelope of various types.

❏ BLOOD AND NUTRITION

The literature contains a great body of information relating normal blood components to pathology, both animal and human, but considerably less information on mild (subclinical) nutritional problems with animals. Regardless of this deficiency, some discussion on this topic is justified in a book on animal nutrition. Further and more detailed information can be found in many reference books on the topic.

Very briefly, blood is composed of the formed elements—erythrocytes or red blood cells, leukocytes (neutrophils, eosinophils, basophils, lymphocytes, and monocytes), and platelets; water; gases (O_2, CO_2, N_2); electrolytes (e.g., ions of Na, K, Cl, carbonate, and other inorganic compounds); and various substances dissolved or suspended in the blood.

If blood is allowed to clot, the fluid remaining after the cells are removed is called serum. If unclotted blood is centrifuged, the fluid is called plasma.

The difference between serum and plasma is that fibrin, the protein that forms the clot, is precipitated when blood is allowed to coagulate and remains with the cells on centrifugation, whereas when clotting is prevented by an anticoagulant, fibrinogen, the precursor of fibrin, remains in solution in the plasma.

Typical values for some blood components for some of the domestic animal species and humans are shown in Table 4.8. In addition to the cellular and extracellular components listed above, many other constituents are present in low concentrations. Blood is the principal vehicle for transport of nutrients and metabolites among organs, tissues, and cells of the body. Nutrient transport is facilitated in many cases by carrier or binding proteins; for example, ferritin is a transport protein for iron, ceruloplasmin for copper. Some lipids are transported as lipoproteins of varying composition, for example, high-density lipoproteins (HDL) and low-density lipoproteins (LDL), which contain a high or low percentage of protein, respectively. In addition, many different enzymes, hormones, and other metabolites are found in blood, some of which have useful diagnostic value in nutrition.

Blood Evaluation in Nutrition

Anemia is a fairly common problem in animals. It may develop as a result of deficiencies of Fe, Cu, Co (as a component of vitamin B_{12}), vitamin B_{12} itself, folic acid, or protein. Anemia can be confirmed by the presence of a low-packed cell volume (percentage of space of blood occupied by red cells after centrifugation), by a low hemoglobin content in blood, or by a microscopic examination of red cells. Red cells may be smaller than normal (microcytic), normal (normocytic), or larger than normal (macrocytic) and have normal amounts (normochromic), more (hyperchromic), or less (hypochromic) hemoglobin in the cells.

Groups of blood chemistries (profiles) now are done routinely on humans and animals by the use

TABLE 4.8 Typical amounts or ranges of some blood components in domestic animals.

	ANIMAL SPECIES					
BLOOD CHEMICAL	CATTLE	SHEEP	SWINE	HORSE	CHICKEN, LAYER	HUMANS[a]
Serum proteins, g/dL						
Total serum protein	7.7	5.4	6.3	6.5	5.4	6.0–8.4
Albumin	3.6	3.1	2.0	3.25	2.5	3.5–5.0
Globulin	3.9	2.3	4.3	3.25	2.9	2.3–3.5
Other whole blood components, mg/dL						
Total nonprotein N	20–40	20–38	20–45	20–40	20–35	
Urea N	6–27	8–20	8–25	10–20	0.4–1	8–25
Uric acid		0.05–2		0.05–1	0.8–5.0	3–7
Amino N	4–8	5–8	6–8	5–7	4–9	3.0–5.5
Lactic acid	5–20	9–12		10–16	20–100	5–10
Glucose	40–70	30–50	80–120	60–110	130–290	70–115
Serum components						
Total cholesterol, mg/dL	50–230	100–150	100–250	75–150	125–200	120–220
Calcium, mg/dL	8–12	8–12	8–12	8–12	10–30	8.5–10.5
Phosphorus, inorganic, mg/dL	4–8	4–8	4–8	3–6	3–8	3.0–4.5
Magnesium, mg/dL	1.8–2.3	1.8–2.3	2–5		3–4	1.5–2.0
Potassium, meq/L	3.9–5.8	3.9–5.8	3.5–5.5		3.8–6.0	3.5–6.0
Chloride, meq/L	130–150	150–160	140–160	145–150		100–106

[a]Adapted from P. A. Potter and A. G. Perry. 1993. *Fundamentals of Nursing.* Mosby Yearbook, St. Louis, MO, pp. 1697–1704.

of automated clinical instruments such as shown earlier in Fig. 3.8. Some of the blood values often included in profiles are packed cell volume; hemoglobin; serum glucose; urea; inorganic P, Ca, Mg, Na, K; total protein, albumin, and globulin; total cholesterol; HDL–cholesterol, triglycerides, and often, thyroxin and other hormones and a series of liver and heart enzymes reflective of cellular disruption. Appraisal of groups of animals in a population allows the evaluation of the overall nutritional status. Recognizing that there may be substantial differences caused by breeds, seasons (diet and environment), age, stage and level of production, and other factors, this approach has a useful place in evaluating the nutritional status of individual herds. Also, sampling of clusters of herds in specific geographical areas offers the potential for identifying nutritionally related problems associated with feeds grown on soils in those areas. Blood profiles on individual animals may aid in identifying nutritional deficiencies or imbalances or nutritionally related metabolic disorders on an individual basis.

❏ DIGESTIBILITY AND PARTIAL DIGESTION

Digestion, as indicated previously, may be defined as the preparation of food for absorption into the body from the GI tract. However, in common usage, digestibility is taken to mean disappearance of food from the GI tract. This broader definition includes absorption as well as digestion.

Digestibility data are used extensively in animal nutrition to evaluate feedstuffs or study nutrient utilization (see Chapter 5 for more detail). It is important to recognize that digestibility is variable; that is, the same feedstuff given to the same animal is not always digested to the same extent. Several factors may alter the extent of digestion; these include level of feed intake, digestive disturbances, frequency of feeding, feed processing, and associative and interactive effects of feedstuffs (nonadditive effects of combining different feedstuffs). Marked differences also exist in the ability of different animal species to digest a particular feedstuff, particularly roughages. For example, when seven different animal species were fed the same alfalfa hay, there were appreciable differences in digestion of the different fractions listed (Table 4.9). This information reflects the inherent differences among animal species in their ability to utilize fibrous feeds, clearly indicating an advantage of ruminants and horses over swine, rabbits, and guinea pigs.

The term **partial digestion** implies that total digestion in the GI tract can be subdivided into fractions that are digested in different parts of the tract. There is a considerable amount of information on this subject in ruminants; less information is available on other domestic animals. Data on sheep and cattle indicate that about two-thirds of digestible organic matter and energy disappear from the forestomach (reticulorumen, omasum), the remainder being digested in the intestines. The more fibrous portions (crude fiber, cell walls), as well as readily available carbohydrates, are di-

TABLE 4.9   Digestibility of alfalfa hay by different species.[a]

	DIGESTION COEFFICIENTS, %					
SPECIES	ORGANIC MATTER	CRUDE PROTEIN	ETHER EXTRACT	CRUDE FIBER	NFE	TDN
Cattle	61	70	35	44	71	48.3
Sheep	61	72	31	45	69	48.1
Goat	59	74	19	41	69	46.9
Horse	59	75	10	41	68	46.0
Pig	37	47	14	22	49	30.5
Rabbit	39	57	21	14	51	30.9
Guinea pig	52	58	12	33	65	40.5

[a]From Maynard and Loosli (1969). The alfalfa hay in question contained 86.1% dry matter, 16.2% crude protein, 1.6% ether extract, and 26.9% crude fiber.

gested to a somewhat greater extent (65 to 90% of total digestion) in the forestomach. Digestibility of N by ruminants is quite variable but is relatively more important in the small intestine; this reflects the passage of large amounts of microbial proteins into the small intestine.

❑ FECAL AND URINARY EXCRETION

Fecal material excreted by animals is comprised of undigested residues of food material; residues of gastric secretions, bile, pancreatic, and enteric secretions; cellular debris from the mucosa of the intestine; catabolic products excreted into the lumen of the GI tract; and cellular debris and metabolites of microorganisms that inhabit the large intestine, or, in the case of ruminants, the forestomach (rumen). The amounts of undigested food residues are largely dependent on the type of food consumed and the type of GI tract; thus, in roughage eaters, undigested residues will usually account for more of the total than would be the case with nonruminants consuming a diet lower in fiber. Sloughed cellular debris from the GI tract mucosa also may contribute substantially to the amount of feces. Based on data from rats, it has been estimated that the dairy cow may slough about 2500 g of cells daily from the interior of the gut wall. The color of feces comes from plant pigments and stercobilinogen produced by bacterial reduction of bile pigments. The odor is from aromatic substances, primarily indole and skatole, that are derived from deamination and decarboxylation of tryptophan in the large intestine. Species adapted to arid climates generally excrete fecal pellets.

Urine represents the main route of excretion of nitrogenous and sulfurous metabolites of body tissues. In addition, it is usually the principal route for excreting some of the inorganic elements and electrolytes, particularly Cl, K, Na, and of P. Urine is an aqueous solution of these components, urea, and other nitrogenous metabolites, and small quantities of vitamins and their metabolites, trace elements, minor amounts of pigments and other compounds, and sloughed cells from the urogenital tract.

The color of urine is due primarily to urochrome, a metabolite of bile pigments complexed with a peptide. The urine from ruminants (other than suckling animals and those on high-grain feed) usually is basic, being in the pH range of 7.4 to 8.4. The basic reaction is characteristic of herbivorous animals because of the relatively large amounts of Na and K ions found in vegetation. Urine usually is in the acidic range in monogastric species. In mammals, urea is the principal N-containing compound excreted, with lesser amounts of other compounds such as ammonia, allantoin, creatine, and creatinine. In birds the major N-containing compound in urine is uric acid.

Only traces of protein are normally found in urine. The presence of proteins such as albumin and globulins in quantity is indicative of kidney disease in adults, although it may occur in young animals under normal conditions, particularly within a few days after normal consumption of colostrum. Carbohydrates such as glucose or fructose sometimes may be found in mammalian urine following ingestion of a meal high in soluble carbohydrates. However, carbohydrates in the urine are usually indicative of disease. Ketones are found at times, particularly in animals suffering from starvation or ketosis. Consumption of very low levels of carbohydrates, such as with the Atkins program, will increase ketone excretion via the kidneys and lungs (Atkins, 1999).

❑ SUMMARY

- The gastrointestinal tract of higher animals is a very complicated system with complex nervous and hormonal controls. Although the anatomy of the GI tract varies among species or groups such as ruminants and nonruminants or between mammals and birds, the essential digestive functions are similar; that is, the GI tract will break the feed down to a molecular size and solubility that can be absorbed and utilized. Animals eat foods that their GI tract can accommodate in terms of bulk and nutrient content and indigestible components. Domestic animals, however, do not have such specific food requirements as many wild species, some of which may exist primarily on nectar or only one plant species.
- Nutrient requirements for ruminants are simplified by rumen synthesis of microbial proteins and B vitamins. Similarly, vitamin requirements for species that have extensive fermentation in pregastric compartments or in the large intestine are also simplified. In addition, amino acid

requirements are less complex for species that practice coprophagy as a result of microbial protein synthesis in the large intestine.

❏ REFERENCES

Atkins, R. C. 1999. *Dr. Atkins New Diet Revolution.* Avon Books, New York.

Broughton, C. W., and J. C. Lecce. 1970. *J. Nutr.* **100:** 445.

Church, D. C. (ed.). 1987. *The Ruminant Animal.* Prentice-Hall, Englewood Cliffs, NJ.

Duke, G. E. 1977. In *Duke's Physiology of Domestic Animals* (9th ed.), Chapter 25, M. J. Swenson (ed.). Cornell University Press, Ithaca, NY.

Hoffmann, R. R. 1973. *The Ruminant Stomach.* East African Literature Bureau, Nairobi, Kenya.

Maynard, L. A., and J. K. Loosli. 1969. *Animal Nutrition* (6th ed.). McGraw-Hill, New York.

Moran, E. T., Jr. 1982. *Comparative Nutrition of Fowl and Swine: The Gastrointestinal Systems.* Office of Educational Practice, University of Guelph, Guelph, Ontario, Canada.

Potter, P. A., and A. G. Perry. 1993. *Fundamentals of Nursing.* Mosby Yearbook, St. Louis, MO, pp. 1697–1704.

Schaffer, G. B. 1981. *National Geograpic* **160**(6): 735.

Stevens, C. E. 1977. In *Duke's Physiology of Domestic Animals* (9th ed.), Chapter 18, pp. 216–232. Cornell University Press, Ithaca, NY.

Stevens, C. E., and A. F. Sellers. 1977. In *Duke's Physiology of Domestic Animals* (9th ed.), Chapter 17, pp. 210–215. Cornell University Press, Ithaca, NY.

Titchen, D. A. 1986. In *Control of Digestion and Metabolism of Ruminants,* p. 227. Prentice-Hall, Englewood Cliffs, NJ.

Yokoyama, M. T., and K. A. Johnson. 1987. In *The Ruminant Animal,* Chapter 7, pp. 125–144. Prentice-Hall, Englewood Cliffs, NJ.

5

Measurement of Feed and Nutrient Utilization

KNOWLEDGE AND UNDERSTANDING OF nutrient utilization are necessary in evaluating feedstuffs and in defining nutrient requirements to facilitate the development of feeding standards for animals. Nutrient utilization from a given feedstuff is affected by many factors, including the animal's species, age, physiological state, and type of GI tract; level of consumption (plane of nutrition); physical form in which the nutrient is provided, that is, pelleted, ground, or otherwise processed; presence of infectious diseases and parasites; and balance of nutrients within the feedstuff itself and in associated diet constituents. Consequently, if we calculate, for example, the utilization of protein from alfalfa for a young, growing pig, the value will be different from that for a mature sow or a ruminant.

Methods used for evaluating nutrient utilization are, in general, similar for all animal species. In this chapter we discuss some of the methods most widely used to measure feed utilization. Specific information on individual nutrients is presented in more detail in chapters on those nutrients.

❏ METHODS

Growth Trials

Growth, as defined by Brody (1945), is "the constructive or assimilatory synthesis of one substance at the expense of another (nutrient) that undergoes dissimilation." In the broadest sense, growth of an animal consists of an increase in body weight resulting from assimilation by body tissues of ingested nutrients. Growth of the whole animal is usually accompanied by an increase in weight and in height or other measures of skeletal size. The increase in weight is composed of the sum of the increases in weight of individual components making up the body, namely, water, fat, protein, carbohydrate, and minerals (ash). Growth of the whole animal consists of a highly complex array of activities in cells making up the specialized organs and tissues in all parts of the body. One organ may begin rapid growth at a time when the growth rate of other organs is slower. The maximum rate of growth of the brain and nervous system occurs earlier in development than the skeleton, followed by muscle, and finally by fat. Thus, the nutrient requirements for growing animals change throughout the growing period in direct response to the changing needs of the individual organs and organ systems making up the whole animal. Growth can be expressed as the increase in absolute weight (or another parameter) in a given period of time or as the increase in relative weight (or another parameter) (usually expressed as a percentage). Growth experiments usually include the measurement of absolute gain in body weight during a period of feeding a test diet. Rate of gain is then expressed as average daily or weekly gain (absolute gain) or in terms of final weight as a percentage of initial weight (relative gain).

Animals used in a growth experiment usually are fed the test diet concurrently with similar animals fed a standard (or basal) diet of known nutritive quality that allows normal growth. In this way direct comparisons can be made among various feed or nutrient sources, and the diets can be ranked in order of their ability to promote weight gain or efficiency of feed utilization. The term *growth* is often used interchangeably with weight gain; strictly speaking, however, the two terms are different because an equal increase in body weight between animals does not necessarily indicate equal growth of body tissues. One animal may deposit more lean muscle mass and its penmate may deposit more fat, which has a much higher energy content. Thus, rate of weight gain does not provide a precise estimate of diet utilization.

Normally, a growth trial involves ad libitum feeding of the experimental diet. Knowing the rate of gain and the total feed consumption, feed efficiency (feed required per unit of weight gain or weight gain per unit of feed) can be computed. Feed efficiency also is a very useful estimate of nutrient adequacy of a test diet. Diets that promote a high rate of gain will usually result in a greater efficiency than diets that do not allow rapid gain. This is illustrated in Table 5.1. The simple explanation for this response is that the rapidly gaining animal utilizes less of the total feed consumption for maintenance and more is available for gain.

Factors such as physical characteristics of the diet, its nutrient content, or its palatability may affect total voluntary feed intake. To rule out this type of variation, paired feeding experiments can be done. In such studies, test diets are fed to animals of comparable size and at equalized intake based on the voluntary consumption by the member of the pair eating the least. Paired feeding eliminates differences in animal performance related to voluntary intake of the feed, but it penalizes the animal consuming the more adequate diet and tends to reduce the magnitude of difference in growth of animals fed the test diets. The wide biological variation in the voluntary feed intake of animals makes it desirable to have data on individual animals rather than on groups. Individual feeding allows measurement of the feed consumed by each ani-

TABLE 5.1   Inverse relationship between daily gain and feed required per unit of gain.

DIET	DAILY GAIN	FEED TO GAIN RATIO
Pigs[a]		
Diet 1	717 g	3.19
Diet 2	702 g	3.49
Diet 3	695 g	3.57
Diet 4	689 g	3.74
Cattle[b]		
Diet 1	1.54 kg	6.76
Diet 2	1.31 kg	7.14

[a]Taken from Lindemann et al. (1986).
[b]Taken from Zobell (1986).

Figure 5.1 *An electronic feeder used for obtaining data on feed consumption for individual pigs maintained in pens. Courtesy of American Calan, Inc.*

mal in the experiment rather than having to rely on the group average. In research with large animals, the measurement of individual feed intake not only provides more powerful tests of real differences between diet groups, but also requires less feed and other costs to obtain the data required. Electronically controlled feeding devices such as that illustrates in Fig. 5.1 are available to measure individual feed intake of animals penned in groups. Electronic feeding devices are available for pigs, cattle, sheep, and other animals. The animal wears a "key" in its ear (or around its neck) that triggers the feeder to open; only this animal can gain access to the feed. Feeders of this type allow the animal to eat at any time it desires. In addition, animals can be penned in groups where their performance will be measured under conditions more nearly resembling those found in commercial animal enterprises.

Growth trials have the following advantages: (1) They allow the accumulation of relatively large amounts of data at reasonable costs; (2) the animals usually can be maintained under conditions that are similar to normal environmental situations; (3) the measurements are easily obtained; (4) the results are a reflection of a fundamental biological response, yet can be applied directly to commercial production systems. A disadvantage is that growth is a more variable biological parameter than many other measures of metabolic or physiological response. If we feed the same diet to 10 similar animals, it is unlikely that all 10 will gain at the same rate. Their rate of gain (and feed effi-

ciency) will be affected by so many different factors for which we cannot easily account that large numbers of animals are required for a valid estimate of a dietary effect. Such variation frequently requires 12 to 15 or more animals per treatment to detect statistical differences between dietary treatments. Some of the biological variation can be removed by using full- or half-sibs (such as littermate pigs) or animals of similar genetic background. Most of the genetic variation in chicks and laboratory animals such as mice and rats has been removed by many generations of inbreeding or other selection programs. Genetically identical twins from natural matings and genetically identical cattle produced by gene-cloning techniques have been used in nutrition experiments as a means of reducing variation in response to dietary variables.

Another disadvantage of the growth trial is the simple matter of getting an accurate weight. Live body weight includes not only body tissue mass, but also water and ingesta within the GI tract. The amount of this "fill" varies, depending on factors such as variation in ingesta mass resulting from a change in feed consumption, time of defecation or urination, and the interval after ingestion of feed or water. These weighing errors are more serious in ruminants, whose GI tract capacity is higher than that of nonruminants. For this reason it is common to use the average of two or three consecutive daily weights to improve the accuracy of weight gain data in cattle and sheep. The net result of these variations in GI tract fill is that measuring body weight is not as simple as it first seems, and any errors that result make it more difficult to evaluate differences in animal performance.

Growth trials may be focused on determining changes in body composition during growth. Such information may be useful as an indicator of different rates of deposition of fat, protein and bone in growing animals as influenced by diet or age. Chemical composition (fat, protein, water, minerals) may be obtained by analyzing the whole body or a particular anatomical location that is known to be representative of the composition of the whole body. A reliable estimate of the chemical composition of the entire body (or its parts) is obtained only by adequate grinding, mixing, and sampling procedures as described for feeds and other substances in Chapter 3. The added information available from chemical analysis of a sample of the whole body or a representative part, compared to obtaining only body weight gain as in the conventional growth trial, is more costly to obtain, but may be biologically more informative. In beef cattle, rather than grinding the entire carcass, the 9-10-11 rib cut has been shown to give a relatively accurate estimate of the total carcass for fat, protein, water and ash (minerals) content. As a result, we can obtain this cut from one side of the carcass, remove the bone or leave it in place, and analyze it for the constituents of interest.

Other Types of Feeding Trials

Many types of feeding trials do not involve young, growing animals. For example, a feeding trial during lactation can be conducted with any mammal in which it is desired to measure the quantity and/or quality of milk produced in response to diet composition or intake. In some instances, milk production is estimated indirectly by measuring gain of the young or weighing them before and after suckling (sometimes referred to as "weigh-suckle-weigh"). This is a common practice for laboratory animals, pigs, and beef cattle. In birds, feeding experiments are commonly used to measure the effects of diet or feedstuffs on egg production. Feeding trials are also done with horses to study the effect of diet on endurance or speed.

In the types of experiments just discussed, there is measurement of a quantitative trait for animals producing at a high level. However, in many instances we may want to assess nutrient utilization or requirements of animals at a maintenance level of feed intake. Alternatively, we may wish to determine the effect of the level of a specific nutrient in the diet on the response of the animal to a toxic substance in the diet. Or we may want to determine the effect of an iodine deficiency on goiter incidence in young lambs, the likelihood of a urea toxicity when it is fed to wintering cows eating poor quality hay, or the vitamin A requirements of male roosters for adequate sperm production. One could cite many other examples of feeding trials that may be used in which some common measure of production (e.g., body weight gain, milk production, egg production) is not the criterion for evaluation.

Digestion Trials

Conventional Methods. Digestion trials are used to determine the proportion of the nutrients in a feed or diet that are absorbed from the GI tract. Animals are fed a diet of known composition over a time period of several days, during which the feces are collected and later analyzed for the components of interest. Maintaining a constant daily feed intake is advisable to minimize day-to-day variation in fecal excretion. Time required for feed residues to traverse the GI tract is 1 to 2 days or less for most nonruminants and longer for ruminants. Therefore, a preliminary period of 3 to 10 days is needed to void the GI tract of residues of pretest feed and to allow adaptation of the animal to the diet. A collection period of 4 to 10 days follows the preliminary adjustment period. Diets and animals are controlled more closely than for a typical feeding trial to increase accuracy of the measurements by reducing uncontrolled variables. Between-animal variability tends to be lower than for growth. Therefore, usually four to six animals per treatment will be sufficient to detect differences among diets.

Values can be obtained for apparent digestibility of any desired nutrient, but data may be meaningless for nutrients such as vitamins and some minerals that are present in extremely small amounts, synthesized by GI tract microorganisms, or recirculated into the intestinal lumen in significant amounts by digestive secretions and sloughing of intestinal mucosal cells. A fraction of nitrogen, fat, carbohydrates, and inorganic elements appearing in the feces is from endogenous sources, that is, sloughed intestinal mucosal cells, digestive secretions. Thus, the feces contain unabsorbed feed residues plus endogenous sources of nitrogen, carbon, and inorganic elements. The term **apparent digestibility** takes into account both the unabsorbed feed residues and the components of the feces that are of endogenous origin. Digestibility is calculated as shown:

Apparent digestibility (%)
$$= \frac{\text{Nutrient intake} - \text{Nutrient in feces}}{\text{Nutrient intake}} \times 100$$

Indicator Methods. A second method, involving the use of nonabsorbed indicators, is often the one of choice when it is impossible or inconvenient to measure total feed intake or to collect total feces. This method depends on the use of a reference substance, which should be indigestible, nonabsorbable, nontoxic, and easily analyzed in feed and feces. **Internal indicators** or markers are those such as lignin that are present in the feed but that are digested to a negligible degree, if at all. Use of lignin is plagued with problems of incomplete recovery and difficulty in analysis, but even so it has been used extensively with herbivorous species of range animals. Silica has been used to a limited extent, but recovery is a problem, presumably because of contamination with soil. Acid-insoluble ash (ash in the feed insoluble in boiling HCl) has been used with a reasonable degree of success. **External indicators** or markers are chemicals, such as chromic oxide and rare earth elements, that are either added to the feed or given to the animals orally in capsules or by drenching or by introduction into the rumen or through a cannula. External markers used with success in various situations when added to the feed include particles stained with color-fast insoluble, nontoxic dyes; chromic oxide; rare earth elements such as lanthanum, cerium, samarium, ytterbium, or dysprosium; chromium affixed to fiber in a "mordanting" process; and various water-soluble markers. Rare earth markers are used to estimate digestion kinetics, rate of passage, and retention times of different feed nutrients, ingredients and particles (Pond et al., 1986; Moore et al., 1992). Chromic oxide has been used extensively, but it cannot be administered readily to range animals, because it must be given orally via a capsule or mixed in a supplemental feed. Irregular excretion and incomplete recovery are common problems. When stained particles are used, it is necessary to recover them from the feces by the tedious process of washing and sieving. Some of the rare earths show promise for less effort in analysis using neutron activation procedures. Water-soluble markers are usually used to measure flow of fluids in the GI tract. A common marker is polyethylene glycol. Further information on the use of these markers has been presented by Merchen (1987) and Adeola (2001).

A formula for calculating the digestibility of a nutrient using the indicator method is as follows.

Apparent digestibility (%)
$$= 100 - \left(100 \times \frac{\% \ ind._{fd}}{\% \ ind_{fc}} \times \frac{\% \ nutr._{fc}}{\% \ nutr._{fd}} \right)$$

where *ind.* stands for "indicator," *fd* for "in food," *fc* for "in feces," and *nutr.* for "nutrients."

This method provides an estimate of digestibility of any or all nutrients without need to know either the total feed consumption or the total excretion of feces. Indicators lend themselves to use with group-fed animals in pens by "grab sampling" of feces from the rectum (or pen floor) and in pasture and range studies where measurement of both

feed consumption and fecal output may be difficult, if not impossible. Consumption of feed on pasture may be estimated by the following equation, where dry matter (DM) is used as an illustration:

DM_{fc} (units/day)
$$= \frac{\times \ (\text{amt. } ind. \text{ per unit dry feces})}{ind._{fd}/\text{unit } DM_{fd}}$$

In this situation total fecal collection is required. A bag attached to the animal (Fig. 5.2) can be used for this purpose.

In pasture studies it is difficult to estimate actual consumption accurately because grazing animals feed selectively, especially when there is a wide variety of plant species available. Apparent digestibility and total feed intake from pasture can be estimated with the simultaneous use of two indicators. The external indicator (such as chromic oxide) can be administered to the animal in known amounts, and the internal indicator (such as lignin), since it is essentially undigestible, can be measured as a percentage of the diet and used as described previously to calculate apparent digestibility.

Digestibility by Difference

In many instances, it is desirable to evaluate the digestibility of a feedstuff when fed in a mixture with one or more other feeds. Examples include protein supplements or single feedstuffs that are normally never used as a complete diet by themselves. In this situation it is necessary to determine digestibility by difference. With this method a basal diet is fed, and the basal diet plus the test feed are fed at one or more levels. If time and numbers of animals permit, more valid estimates can be obtained if all animals are fed alternately the basal and basal + test feeds, although this is often not the practice. After digestibility of the complete diets has been determined, the digestibility of the test feed can be calculated as shown:

$$D_T(\%) = \frac{D_{T+B}(\%) - D_B(\%) \, (N_{B+T})}{N_T \times N_{B+T}}$$

where D = "digestibility," T stands for "test feed," N is "fraction of nutrient," and B is "basal feed."

An example of data from a trial with lambs is shown in Table 5.2. When using this formula, values for a specific nutrient must be used in each instance.

Figure 5.2   *One example of a harness and plastic bag used for fecal collections with sheep. Courtesy of G. Fishwick, Glasgow University Vet School.*

TABLE 5.2   An illustration of digestion coefficients calculated for the total diet and by difference and of an associative effect of feeds.

| | | DIGESTIBILITY, % | | |
| | | EXPERIMENTAL N SOURCES | | |
ITEM	BASAL DIET	1	2	3
Complete diet				
Crude protein	57.3	71.5	72.8	72.9
Organic matter	76.9	78.7	79.5	78.5
By difference				
Crude protein[a]	—	85.8	88.2	89.0
Organic matter	—	87.3	104.7	95.1

[a]Use of the formula (see text) requires fewer computations. The formula for N source 1 is:

$$\frac{71.5 - (57.3 \times 0.5_a)}{0.5_b} = 85.8,$$

where 0.5_a is the fraction of crude protein in the diet containing basal feed + test feed and 0.5_b is the fraction of crude protein in the test feed in the same diet.

the mixture. This response is referred to as a nonadditive or an **associative effect.** This is illustrated in Table 5.2 with data on organic matter digestibility. Note that digestibility, calculated by difference, of the organic matter for N source 2 is greater than 100%. Obviously, this is not possible. What has happened is that the addition of the N source to the diet stimulated digestion of the basal diet so that the mixture was more digestible than when the basal diet was fed alone. This type of response is often observed, particularly for herbivorous animals that depend to a great extent on microbial fermentation. Components of the test diet that stimulate fermentative activity are usually responsible for the increased digestibility.

Apparent vs. True Digestibility

Except for fibrous carbohydrates, major classes of nutrients (protein, fat, carbohydrate) are excreted in the feces from endogenous sources. Sloughed intestinal cells and digestive secretions make a substantial contribution to these endogenous losses. Apparent digestibility of a nutrient represents the difference between the amount ingested and the amount appearing in the feces. The total amount in the feces includes not only undigested

Associative Effects

A common phenomenon observed with digestibility data is that mixtures of feedstuffs do not always give results that would be predicted from digestibility values of the individual components of feed residues but also endogenous sources of the same nutrient. This endogenous fraction cannot be distinguished from the undigested portion of the ingested nutrients.

The true digestibility of a nutrient is the proportion of the dietary intake that is absorbed from the GI tract, excluding any contributions from body (endogenous) sources. Fecal N derived directly from ingested food is called **exogenous N** (not from body tissues); that derived from body tissues is termed **fecal metabolic N** (endogenous). Thus, for protein, true digestibility can be estimated by subtracting the amount of N appearing in the feces of an animal fed a low-protein diet from the amount of N appearing in the feces of animal fed a test diet as illustrated in Table 5.3.

The apparent digestibility of protein in a feed is influenced by the level of protein in the feed. This is so because the amount of endogenous protein tends to be constant, so at a high-protein intake the endogenous fraction represents a smaller percentage of the total nitrogen excretion. This is illustrated in Table 5.3. Note that true digestibility remains the same for animals fed both a low and high level of protein.

TABLE 5.3 Steps in calculating true digestibility of a protein.

LINE NO.[a]	ITEM	PROTEIN INTAKE HIGH	PROTEIN INTAKE LOW
1	Daily N intake, g	20	10
2	Daily fecal N, g	5	3
3	Apparent N absorption, g (line 1 − line 2)	15	7
4	Apparent dig., % (3 ÷ 1 × 100)	75	70
5	Metabolic fecal N, g	1	1
6	Unabsorbed dietary N, g (line 2 − line 5)	4	2
7	True N absorption, g (line 1 − line 6) (1-6)	16	8
8	True N dig., % (7 ÷ 1 × 100)	80	80

[a]To determine true digestibility of a protein, proceed sequentially through the 8 steps as indicated in the column labeled "Line No." Note that true digestibility is not changed by level of protein intake, even though apparent digestibility is increased by a high protein intake.

With some nonruminant and avian species that have a very short ingesta retention time, it may be feasible to feed a protein-free diet for a short time. Briefly, the methods used to estimate endogenous excretion of nitrogen are: (1) feeding a nitrogen-free diet and determining the amount of nitrogen in the feces; (2) feeding several levels of the nutrient (nitrogen) and calculating the fecal level by regression analysis to a zero intake of the nutrient as described below; (3) feeding a completely digestible protein. In the pig, protein and amino acids may be synthesized or degraded in the large intestine by microbial action. Thus, the availability of dietary amino acids to the animal may be assessed more accurately by measuring digestibility at the terminal end of the small intestine rather than in the feces. Surgical cannulation of the lower ileum allows collection and analysis of digesta before its exposure to microbial action by large intestine bacteria and may provide a more accurate value for digestibility of dietary amino acids (Young et al., 1991; Gabert et al., 2001; Nyachoti and Stein, 2005). Another surgical procedure available to estimate protein and amino acid digestibility involves cannulation of the duodenum and introduction of the test feed sample placed in a nylon bag. The bag is allowed to traverse the intestinal tract and is retrieved in the feces. The disappearance of nitrogen from the nylon bag during its transit is measured to determine digestibility.

In contrast to nonruminants, ruminant animals depend to a larger extent on microorganisms residing in the GI tract for digestion of some of their nutrients. If rumen bacteria do not have a source of N, they do not sustain normal function for long. The usual method for determining true digestibility of nitrogen in ruminants, therefore, is to feed graded levels of protein, determine excretion, and then construct a regression line and extrapolate the regression line back to a zero protein intake and calculate the fecal excretion (see Fig. 5.3).

Balance Trials

Balance trials are similar to digestion trials but provide information on utilization of nutrients after absorption from the GI tract. The intent is to get an accurate measure of total intake and total excretion in order to determine whether there is a net retention (positive balance) or loss (negative balance) of the nutrient in question. This type of study usually is applicable to N, energy, and some of the

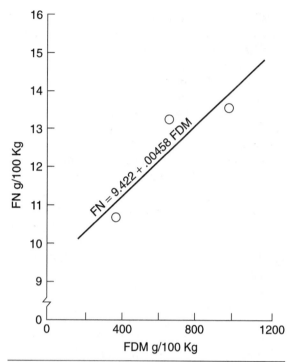

Figure 5.3   *This graph illustrates the type of a regression line that can be calculated to estimate metabolic fecal N (FN) in relation to fecal dry matter (FDM). The same type of line could be constructed in relationship to N consumption. From Strozinski and Chandler (1972).*

major minerals. The major routes of excretion are feces, urine, and expired air. To determine nitrogen balance, it is essential to collect feces and urine. Other losses include those from sloughed skin, shed hair (or feathers), sweat, expired gases (CO_2), and energy lost as heat via respiration and sweating.

Therefore, for precise accounting of nutrient balance one must also collect shed body hair (or wool or feathers), sloughed cells from skin, and sweat. For balance data in a lactating animal, one also measures milk production and composition. One type of metabolism crate useful for such trials with cattle is shown in Fig. 5.4.

Although balance trials are somewhat more tedious to carry out than digestion trials, the added data (particularly for N, energy, and some mineral elements) provide more complete information on animal nutrient requirements or nutrient utilization from feedstuffs. This is true particularly if we are dealing with a young, growing animal or an animal producing at a relatively high level (milk,

eggs, work). It is important to note that short-term studies as in conventional balance trials can lead to exaggerated retention values (either positive or negative) that would not be typical of animal response over a period of weeks or months.

Purified Diets

Much of the current knowledge of specific nutrient requirements of animals has been derived from experiments in which pure or semipurified sources of nutrients were used to formulate diets. Such diets allow one to make quantitative changes in the concentration of a single nutrient without greatly affecting the concentrations of other nutrients in the diet. In this way, graded levels of an individual nutrient can be fed to determine the quantitative requirement of that nutrient for growth or other productive functions in a particular animal species. The use of purified diets also minimizes the presence of unknown or undesired constituents present in natural feedstuffs that might affect the utilization of or requirement for the nutrients under study. Purified diets are also referred to as synthetic or semipurified. Depending on the species of animal in-

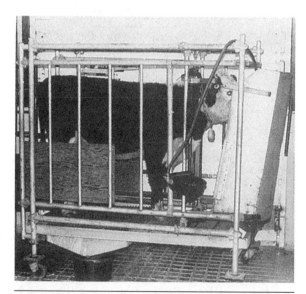

Figure 5.4   *One type of metabolism crate for cattle designed for animal comfort and for easy collection of urine and feces for the determination of digestibility and metabolism of feed. Water is provided free-choice, and bulky feeds such as hay and silage can be accommodated. Sometimes eating behavior is also determined by measuring jaw movements with specialized computers and halters. Photo courtesy of North Carolina State University.*

volved, its age, and other factors, energy sources for such diets usually are based on ingredients with a sugar source such as glucose, some starch, and various amounts of cellulose. Nitrogen may be supplied by a purified protein source such as casein (from milk) or isolated soybean protein or by crystalline amino acids. Inorganic element requirements can be met by using various mineral salts that are soluble and digestible. Vitamins may be supplied by a rich source such as yeast or liver concentrates, or, more commonly, as synthetic crystals. Essential fatty adds can be supplied as a component of purified corn oil, lard, or other purified fat source. In some instances only a portion of the ingredients may be highly purified.

The use of purified diets has allowed us to understand more completely the requirements for most of the essential nutrients in an array of ruminant and nonruminant animals. It is noteworthy that it still may be difficult to determine whether some elements are required when the amount needed is extremely small. Even highly purified inorganic element sources may contain trace amounts of contaminating elements. For this reason, there is uncertainty about the dietary essentiality of some trace mineral elements.

❏ RUMEN DIGESTION TECHNIQUES

The high cost of digestion trials, especially with cattle, has prompted the use of in vitro techniques that allow simulation of rumen fermentation under controlled conditions. A variety of methods has been developed. Some are more suitable than others for a specific purpose. Some of the common methods are described briefly.

Batch trials: A small amount of rumen fluid is obtained from a rumen-fistulated animal (Fig. 5.5). After most of the feed particles have been removed, the fluid is placed in a container along with some buffer (to simulate saliva) and the test sample. The combination then is fermented at rumen temperature (39°C) for a period of time, usually 24 to 48 h. Where the object is to predict (or correlate) live animal digestion from in vitro digestion, adequate methods have been developed that give a more valid estimate of animal digestion of roughage and forage than can be obtained by using chemical analyses. These methods are not as useful for evaluating animal utilization of grains and other concentrates.

Figure 5.5 *A dairy cow fitted with a rumen cannula allowing direct access to the rumen. This animal serves as a donor of rumen fluid to other animals and aids in determining the digestibility of various feedstuffs. Samples of feed, in small bags, can be placed in the rumen for various times to determine disappearance and thereby digestibility.*

Continuous fermenters (Merchan, 1987): If these are designed and maintained properly, it is possible to simulate feed input and rumen outflow (and/or absorption) for a period of days or weeks. It is therefore possible to simulate rumen conditions more closely than can be done with batch cultures where conditions would be more representative of feeding an animal once a day.

Rumen fermentation procedures have proved useful for screening feedstuffs or generally characterizing them for use by ruminants, especially when only small samples are available (as, for example, from a plant breeding trial in the greenhouse). These methods also are useful for studying rumen function and metabolism of specific compounds, for example, determination of what types of nonprotein N can be used by rumen microorganisms, production of volatile fatty acids from a diet processed in different ways.

Nylon bag technique: In this procedure the feedstuff in question is placed in a nylon (or other undigestible cloth) bag suspended in the rumen of an animal with a rumen fistula. The bags then are removed after a predetermined time, and the loss of material (from fermentation) in the bag is determined. Such methods are useful in evaluating relative differences between feedstuffs but do not give values similar to those from live animal digestion trials. This method is perhaps more useful for studying rumen digestion of concentrates (grains) than the in vitro method in which relative digestion is of interest.

❏ BIOLOGICAL VARIATION IN DIGESTION AND ABSORPTION OF NUTRIENTS

Biological availability of nutrients is not fixed at a constant value, but is affected by a number of factors. Bioavailability of energy (Ferrell, 2005), amino acids (Nyachoti and Stein, 2005), mineral elements (Lei, 2005), and vitamins (Baker, 2005a, 2005b) has been recently summarized. Increased feed intake above maintenance tends to depress digestibility, especially in ruminants (Grovum, 1987). Part of this variation results from more rapid passage through the GI tract, and so there is less time for microbial or enzymatic activity on the digesta. Studies with sheep have demonstrated, even in animals of the same breed, sex, and age, that there may be a considerable difference in stomach capacity, which in turn affects eating rate, amount eaten, and passage rate through the GI tract. Factors such as mastication of forages affect digestion of diet constituents and total feed intake (Olubobokun et al., 1990). Undoubtedly, other unrecognized factors alter digestibility, but it is clear that all animals do not digest a given diet to the same extent. An example with sheep is shown in Table 5.4. Data are shown on digestibility of crude protein, energy, and dry matter. In this instance, digestibility of dry matter was less variable than the others, but there was still a range of 6.4 units.

TABLE 5.4 An example of typical variation in digestibility encountered with sheep.

ANIMAL NO.	DIGESTIBILITY, %		
	DRY MATTER	CRUDE PROTEIN	ENERGY
1	76.9	77.1	80.8
2	73.1	73.0	77.3
3	72.3	77.2	80.1
4	78.7	77.3	73.1
5	76.0	67.4	74.6
Overall mean	76.5	73.6	77.2

The ranges shown in this table are not extreme by any means.

Aside from differences related to source of feedstuffs, variation exists in absorptive capacity based on the specificity of a wide variety of specialized transport systems present in the intestinal epithelium. This specialization is illustrated by comparing the maximum absorptive capacity of humans for vitamin B_{12}, which has been estimated at 1 mcg/day, with that for glucose, which is 3600 g/day. The recognition of the existence of such specific absorption pathways with widely differing capacities has allowed development of a number of research techniques for absorption studies. Techniques include various surgical preparations, cannulation of the portal vein or thoracic duct, whole-body counting of radioactivity after administration of radioisotopes, use of plants intrinsically labeled with stable isotopes whose enrichment in various metabolites within the body indicates absorption and metabolism, and others, many of which are discussed in more detail in later chapters in conjunction with the metabolism of individual nutrients.

❏ SURGICAL PROCEDURES FOR STUDYING NUTRIENT ABSORPTION AND UTILIZATION

A wide variety of surgical techniques has been developed to aid in the study of nutrient absorption. Yen (2001) summarized this subject in swine. Fistulation of the rumen of cattle, sheep, or goats allows sampling of contents and infusion of known quantities of substances at known rates directly into the rumen. Stomach fistulation of nonruminants and abomasal cannulations of ruminants (Fig. 5.6) are common procedures. Cannulas can be inserted in various parts of the intestinal tract. Most of these procedures have been done on sheep or cattle (less commonly in cattle than in sheep because of the expense). However, it is now a more usual practice to establish cecal and ileal cannulas in swine or horses.

In addition to these methods, several procedures have been perfected that allow catheters to be implanted in specific veins or arteries. Thus, blood going into or out of specific organs can be sampled or chemicals can be injected at the same site. With calves, the catheters are frequently exteriorized in the lumbar area, where they are ac-

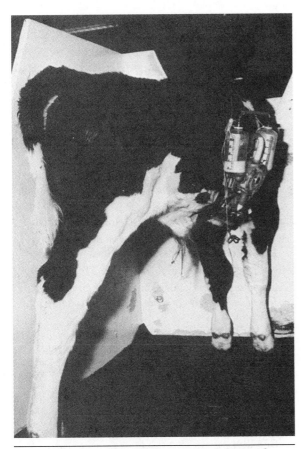

Figure 5.6 *An animal with one type of abomasal–intestinal cannula which allows volumetric measurements to be made of digesta. Courtesy of C. Noel, University of Laval, Quebec, Canada.*

cessible to the researcher but not damaged easily by the animal.

Rates of absorption of individual radioisotopes can be measured effectively by surgical techniques that consist of tying off a segment of the intestinal tract and measuring disappearance over a short time span of known amounts of a substance injected directly into the lumen of the gut between the two ligatures. This is done in the live (anesthetized) animal. An extension of this in vivo procedure involves a related in vitro technique wherein a segment of intestinal tract is removed, it is everted so that the mucosal side is exterior, and the ends are ligated as above. The substance whose transport across the intestinal lining is being studied is then injected into the lumen of the everted intestinal segment which has been placed in a beaker containing physiological saline solu-

tion. The appearance of the injected substance in the beaker contents or its disappearance from the everted intestinal segment is then used to estimate its absorption. Such in vivo and in vitro techniques based on surgical procedures add another dimension to the study of nutrient absorption and utilization and, although they do not substitute for more conventional methods, frequently they can add significant knowledge. Guidelines and recommendations related to surgery and care of animals have been cited (Yen, 2001; National Research Council, 1996).

❑ LABORATORY ANIMALS AS MODELS FOR FARM ANIMAL NUTRITION

Laboratory animals contribute to knowledge of nutritional requirements and interrelationships in farm animals. The numbers of animals and housing facilities needed to obtain meaningful information on some nutritional questions in farm animals are often so large as to become economically prohibitive. Also, the shorter life cycle of small animal species compared with farm animals offers the opportunity to obtain data covering several generations per year. Thousands of laboratory mice, rats, guinea pigs and other mammals, as well as birds such as Japanese quail, and a variety of reptiles and fishes, are used in nutritional research. Of course, the results provide information directly applicable to the species used, but often the information can be extrapolated to farm animals or to humans. Among the farm animals, the pig is especially useful as a model for human nutrition because its digestive system is similar anatomically and functionally to that of humans. The National Research Council (NRC) publishes a continuous series on the nutritive requirements of laboratory and farm animals. Appendix Table 1 lists the NRC publications on the nutrient requirements of laboratory and farm animals and of pets. The science of nutrition knows no limits in terms of the range of animal life selected for obtaining new information on the feeding and nutrition of animals.

❑ SUMMARY

- Commonly used methods for either evaluating feedstuffs or measuring nutrient utilization by animals are discussed.

- Growth trials and other feeding experiments are designed to quantify some particular entity related to feed or nutrient utilization.

- Nutrient digestion and balance experiments normally are carried out with fewer animals under more controlled conditions. Data from digestion experiments are used widely to evaluate feedstuffs; subsequently, they are used for diet formulation or for refining information on nutrient requirements of animals.

- Various in vitro rumen procedures that simulate rumen fermentation are useful in characterizing feedstuffs for use by ruminants.

- Surgical procedures, including rumen fistulation in ruminants, ileal cannulas in swine, and implantation of catheters in blood vessels serving specific organs, are used to study nutrient digestion and absorption from the GI tract and nutrient metabolism in animals.

❏ REFERENCES

Adeola, O. 2001. In *Swine Nutrition* (2nd ed.), Chapter 40, pp. 903–916, A. J. Lewis and L. L. Southern (eds.). CRC Press, Boca Raton, FL.

Baker, D. H. 2005a. Bioavailability of fat-soluble vitamins. In *Encyclopedia of Animal Science,* W. G. Pond and A. W. Bell (eds.). Marcel Dekker, New York (In Press).

Baker, D. H. 2005b. Bioavailability of water-soluble vitamins. In *Encyclopedia of Animal Science,* W. G. Pond and A. W. Bell (eds.). Marcel Dekker, New York (In Press).

Brody, S. 1945. *Bioenergetics and Growth.* Reinhold Publishing Corp., New York.

Ferrell, C. L. 2005. Bioavailability: energy. In *Encyclopedia of Animal Science,* W. G. Pond and A. W. Bell (eds.). Marcel Dekker, New York (In Press).

Gabert, V. M., H. Jorgensen, and C. M. Nyachoti. 2001. In *Swine Nutrition,* (2nd ed.), Chapter 9, pp. 151–186, A. J. Lewis and L. L. Southern (eds.). CRC Press, Boca Raton, FL.

Grovum, W. L. 1987. In *The Ruminant Animal,* p. 202, D. C. Church (ed.). Prentice Hall, Englewood Cliffs, NJ.

Lei, X. G. 2005. Bioavailability of mineral elements. In *Encyclopedia of Animal Science,* W. G. Pond and A. W. Bell (eds.). Marcel Dekker, New York (In Press).

Lindemann, M. D., et al. 1986. *J. Anim. Sci.* **62**: 412.

Merchen, N. R. 1987. In *The Ruminant Animal,* p. 172, D. C. Church (ed.). Prentice-Hall, Englewood Cliffs, NJ.

Moore, J. A., et al. 1992. *J. Anim. Sci.* **70**: 3528.

National Research Council. 1996. *Guide for the Care and Use of Laboratory Animals.* National Academy Press, Washington, DC.

Nyachoti, C. M., and H. H. Stein. 2005. Bioavailability: Amino Acids. In *Encyclopedia of Animal Science,* W. G. Pond and A. W. Bell (eds.). Marcel Dekker, New York (In Press).

Olubokum, J. A., et al. 1990. *J. Anim. Sci.* **68**: 3371.

Pond, W. G., et al. 1986. **63**: 1140.

Strozinski, L. L., and P. T. Chandler. 1972. *J. Dairy Sci.* **55**: 1281.

Yen, J. T. 2001. In *Swine Nutrition* (2nd ed.), Chapter 42, pp. 961–983, A. J. Lewis and L. L. Southern (eds.). CRC Press, Boca Raton, FL.

Young, L. G., A. G. Low, and W. H. Close. 1991. In *Swine Nutrition,* Chapter 39, pp. 623–630, E. R. Miller, D. E. Ullrey, and A. J. Lewis (eds.). Butterworth-Heinemann, Boston, MA.

Zobell, D. R. 1986. Proc. Western Section, *Amer. Soc. Anim. Sci.* **37**: 96.

6

Water

WATER IS NOT NORMALLY thought of as a nutrient even though it clearly meets all criteria for definition as one. Life could not be sustained without water. It makes up about one-half to two-thirds of the body mass of adult animals and up to 90% of that of newborn animals, and it accounts for more than 99% of the molecules in the body. (The latter statistic is possible because water molecules are smaller than most others.) The importance of an adequate supply of potable water for livestock is well recognized and currently is receiving more emphasis in the quest to clean up polluted environments by improving the quality and dependability of water supplies.

❑ FUNCTIONS

Water serves two basic functions for all terrestrial animals: (1) as a major component in body metabolism and (2) as a major factor in body temperature control. These functions result from the unique properties of water (summarized by Oster, 2003), and are discussed in the following sections.

Water and Body Metabolism

From a functional viewpoint, water is essential for life. The death of plants and animals occurs quickly when water is insufficient, as contrasted to relatively long-term life when the supply of other nutrients is restricted. All of the biochemical reactions that take place in an animal require water.

Water as a Solvent.   Many of the biological functions of water are dependent on its property of acting as a solvent for a wide variety of compounds; many compounds ionize readily in water. Solvent properties are extremely important because most protoplasm is a mixture of colloids and crystalloids in water.

Water as a Transport Medium.   In addition, water serves as a medium for transportation of semisolid digesta in the GI tract, for various solutes in blood, tissue fluids, cells, and secretions, and in excretions such as urine and sweat.

Water as a Dilutent.   Water provides for dilution of cell contents and body fluids so that relatively free movement of chemicals may occur within the cells and in the fluids and GI tract. Thus, water serves to transport absorbed substances, conveying them to and from their metabolism sites.

Water in Hydrolysis and Oxidation.   Water is involved in many chemical reactions. In hydrolysis, water is a substrate in the reaction, and in oxidation, water is a product of the reaction.

Metabolic water, or water of oxidation, results from the oxidation of organic components in the cells of the body. Oxidation of 1 mol of glucose requires 6 mol of O_2 and produces 6 mol of CO_2 and 6 mol of water. The amount of O_2 required for oxidation of starch, fat, and protein is shown in Table 6.1. Note that much more O_2 is required to oxidize fat, 2.02 L/g of fat versus 0.83 L/g of starch or 0.97 L/g of protein. If expressed as O_2/g of water formed, protein requires 2.44 L of O_2. The metabolic water produced per gram of food is much higher for fat (1.07) than for protein or carbohydrates.

Ingestion and metabolism of fat, carbohydrate, and protein result in increased respiration and heat dissipation and, for protein, increased urinary excretion of urea, the principal excretory product of N metabolism in mammals. Large amounts of water are required for dilution and excretion via the kidney of the urea, and the amount of water derived from oxidation is not sufficient to meet the increased respiratory and excretory demands.

It has been calculated that in a hot, dry environment (ambient temperature 26°C, relative humidity 10%) an animal loses 23.5 g of respiratory water in the process of producing 12.3 g of metabolic water. The heat generated amounts to about 100 Cal. Part of the heat (13.6%) is offset by the heat of vaporization of the expired water. If the remainder (86 Cal) had to be dissipated by sweating, it would cost 149 mL of water. Because of the added needs for excretion when protein is ingested, it has a negative effect on conservation of water and should be avoided in periods of short-term water deprivation or deficit. With respect to fats, Schmidt-Nielsen (1964) showed that, in dry climates at least, ingestion of fat results in less net metabolic water than ingestion of carbohydrates (because of the added respiration required). The overall result is that carbohydrates supply more net metabolic water than either proteins or fats.

TABLE 6.1   Metabolic water production from nutrients.

NUTRIENT	OXYGEN (L) PER GRAM OF		METABOLIC WATER PRODUCED PER GRAM OF FOOD, G
	FOOD	WATER FORMED	
Starch	0.83	1.49	0.56
Fat	2.02	1.88	1.07
Protein	0.97	2.44	0.40

In hibernating animals, metabolic and preformed water (that associated with body tissues catabolized during a period of negative energy balance) is apparently adequate to supply total body needs to sustain normal functions.

Water and Body Temperature Regulation

Water has several properties that allow it to have a marked effect on temperature regulation. Its high specific heat, high thermal conductivity, and high latent heat of vaporization allow accumulation of heat, ready transfer of heat, and loss of large amounts of heat on vaporization. These physical properties of water are enhanced by physiological characteristics of animals. The fluidity of the blood and the rapidity with which it is circulated in the body, the large evaporative surfaces in the lungs and the body surface for sweating (in most species) or panting, the ability to constrict blood flow away from body surfaces during cold stress, as well as other factors allow animals to control body temperatures within desired ranges in most instances.

The specific heat of water is considerably higher than that of any other liquid (or solid). Many animals rely on the cooling capacity of water as it gives up its latent heat during evaporation by panting or sweating. As 1 g of water changes from liquid to vapor, whether by panting or sweating, it takes up about 580 Cal of heat. In terms of heat exchange, this is a very efficient use of water when it is realized that to heat 1 g of water from freezing to boiling requires only 117 Cal. Because of this great capacity to store heat, any sudden change in body temperature is avoided. Water has greater thermal conductivity than any other liquid, and this property is important for the dissipation of heat from deeply situated regions in the body. Many animals dissipate internal and absorbed heat by evaporation of body water. For example, in one study sweating accounted for 21% conduction and convection from the skin 16%, and respiratory evaporation 5% of the net loss of heat in a *Bos indicus* steer.

❑ WATER ABSORPTION

Water is readily absorbed from most sections of the GI tract. In ruminants, usually there is a net absorption from the rumen and omasum. In the abomasum of ruminants or the glandular stomach in other species, usually there is a marked net outflow of fluids (which accompany gastric secretions). The same is true for the duodenum where fluids from the pancreas, bile, and intestinal glands cause a net inflow of water. In all species, there is a net absorption from the ileum, jejunum, cecum, and large gut, but the amount absorbed (and moisture in the feces) varies considerably from species to species and from diet to diet within a species.

Osmotic relationships within the particular organ have a marked effect on absorption. Following a meal, normally there will be more solutes in the digesta; this will increase the osmotic pressure, which may result in an inflow of water into the organ (rumen, small gut), depending on the amount of fluid consumed prior to, during, and after the meal. This mechanism allows the body to maintain optimum consistency of the digesta throughout the GI tract. If fluids are ingested without food, absorption is more rapid and complete because of the osmotic relationships described.

Other factors also affect absorption. For example, polysaccharides such as pectin tend to form gels in the GI tract. Such gels hold water and reduce absorption from the GI tract and, as a result, are usually laxative. For some species the ingestion of indigestible fiber also tends to reduce water absorption. In addition, any factor promoting diarrhea, whether from the diet ingested, microbial toxins, altered osmotic relationships, or other physiological reaction (e.g., stressors that may stimulate intestinal motility), will result in reduced water absorption from the GI tract.

❑ BODY WATER

Water content of the animal body varies considerably; over the long term it is influenced by the age of the animal and the amount of fat in the tissues. Water content is highest in fetuses and in newborn animals, declines rapidly at first, and then slowly declines to adult levels. When body water is expressed on the basis of the fat-free body, the water content is relatively constant for many different animal species, including cattle, sheep, swine, mice, rats, chickens, and fish. The range is from 70 to 75% of fat-free weight, with an average of about 73%. As a result of this relatively constant relationship, the composition of the animal body can be estimated with reasonable accuracy if either the fat or water content is known. Body water can be estimated with various dyes or isotopes of hydrogen (deuterium, as deuterium oxide, or tritium) by administering them intravenously and determining the amount of dilution of the dye or isotope; fat content of the tissues may be calculated by the formula:

$$\text{Percent fat} = 100 - \frac{\text{percent water}}{0.732}$$

The greatest amount of water in the body tissues is present as intracellular fluids, which may account for 40% or more of total body weight. Most of the intracellular water is found in muscle with lesser amounts in other tissues. Extracellular water is found in interstitial fluids, which occupy spaces between cells, blood plasma, and other fluids such as lymph, synovial, and cerebrospinal fluids. Extracellular water accounts for the second largest water "compartment," or roughly one-third of the total body water, of which about 6% is blood plasma water.

Most of the remaining body water (not strictly body water, as it is outside of body tissues) is found in the contents of the GI tract and urinary tract. The amount present in the GI tract is quite variable, even within species, and is greatly affected by type and amount of feed consumed. As indicated, body water tends to decrease with age and has an inverse relationship with body fat. Body water is apt to be higher in lactating cows than in dry cows, and extracellular water is greater in young male calves than in female calves of the same age.

Water easily passes through most cell membranes and from one fluid compartment to another. Passage from compartment to compartment is controlled primarily by differences in osmotic or hydrostatic pressure gradients. It is a passive diffusion in that energy for this movement is not required (or only in extremely small amounts).

Water absorbed from the GI tract enters the extracellular fluid in the blood and lymph. Blood volume is largely regulated by body Na, the major cation in blood plasma. The volume and osmotic pressure of the extracellular fluid are also regulated by thirst and by antidiuretic hormone produced by the pituitary, and other endocrine factors, under control of the adrenal gland and kidney, control tubular reabsorption of water by the kidney, thereby regulating water loss. Thus, variations in water intake and excretion largely control osmoconcentration. Flow of water between the extracellular and intracellular fluids also is important in maintaining osmoconcentration.

Physiological abnormalities or diseases (fever, diarrhea) may result in body dehydration or in retention of excess water in the tissues (edema) because of faulty blood circulation or adrenal gland activity. Therefore, values for body water may not be within normal limits in cases of edema or dehydration.

Water Turnover

The term *water turnover* is used to express the rate at which body water is excreted and replaced in the tissues. Tritium-labeled water has been used to estimate turnover time in a variety of different species. In cattle, a typical half-life value (time for half of tritium to be lost) is about 3.5 days. Nonruminant species probably have a more rapid turnover because they have less water in the GI tract. Those species that tolerate greater water restriction (camel, some breeds of sheep) also have lower turnover rates than species that are less tolerant to water restriction (horse, European cattle). Water turnover is greatly affected by climatic factors such as temperature and humidity or by ingestion of compounds such as sodium chloride (common salt) that increase urinary or fecal excretion.

❏ WATER SOURCES

Water available to an animal's tissues is derived from (a) drinking water, (b) water contained in or on feed, (c) metabolic water, (d) water liberated from metabolic reactions such as condensation of amino acids to peptides, and (e) preformed water associated with body tissues catabolized during a period of negative energy balance. The importance of these different sources differs among animal species, depending on diet, habitat, and ability to conserve body water. Some species of desert rodents and antelopes are said not to require drinking water except in rare situations. This condition does not apply to domestic farm animals, all of which require copious amounts of water when producing at a high level, particularly when they are heat-stressed (illustrated later under Water Consumption and Requirements).

The water content of feedstuffs consumed by animals is highly variable. For example, in forage it may range from a low of 5 to 7% in mature plants and hays to as high as 90% or more in lush young vegetation. Likewise, precipitation and dew on ingested feed may, at times, account for a very substantial amount of water consumption.

One example of water consumption is given in Table 6.2. In this case, sheep were maintained indoors at a stable temperature and were fed diets made up of the dried ingesta collected from steers fitted with esophageal fistulas (surgical preparation enabling the steer to graze normally and with a col-

TABLE 6.2 Water metabolism of sheep maintained indoors at 20–26°C.[a]

| ITEM | MONTH OF COLLECTION OF RATIONS | |
	JUNE	SEPT.
Feed Consumption		
Dry matter consumed, g/day	795	789
Crude protein, g/day	122	50
Metabolizable energy, Mcal/day	2.00	1.39
Water input		
Water drunk, g/day	2093	1613
% of total	87.8	88.1
Water in feed, g/day	51	50
% of total	2.1	2.7
Metabolic water,[b] g/day	240	167
% of total	10.1	9.1
Total input, g/day	2384	1830
Water output		
Fecal water, g/day	328	440
% of total	13.8	24.0
Urinary water, g/day	788	551
% of total	33.0	30.1
Vaporized water,[c] g/day	1268	839
% of total	53.2	45.9
Total output	2384	1830

[a]Diets were dried samples collected during June or September from steers with esophageal fistulas.
[b]Calculated to be produced at rate of 120 g/Meal of Metabolizable Energy (ME).
[c]Equal to insensible water.
Source: Wallace et al. (1972).

TABLE 6.3 Relationship between water drunk and moisture content of forage.

WATER INTAKE L/KG OF DM	MOISTURE CONTENT OF FORAGE, %
3.7	10
3.6	20
3.3	30
3.1	40
2.9	50
2.3	60
2.0	65
1.5	70
0.9	75

Source: From Hyder et al. (1968).

lection vial attached to the fistula to retrieve newly ingested forage). The samples were collected at various times in the growing season and were dried after collection. Although the amount of diet fed was rather constant, the diet varied in protein content and in metabolizable energy content. In this example water in the feed amounted to only about 50 g/day, and the sheep drank about 88% of their total water. Metabolic water (calculated values) contributed 9 to 10% of the total. Total water input amounted to 2.95 g/g of feed dry matter for the June diet and 2.31 g/g of dry matter for the September diet.

The amount of water provided by green forage can be very substantial. Data in Table 6.3 show the relationship between the moisture content of the forage and the amount of free water consumed by sheep. Observations generally indicate that sheep will seldom drink water when the forage moisture content is 65 to 70% or higher.

❏ WATER LOSSES

Loss of water from the animal body occurs by way of urine, feces, insensible water (that lost via vaporization from the lungs and dissipation through the skin), and sweat from the sweat glands in the skin during warm or hot weather. Loss from the lungs, skin, and kidney occurs continuously, although at variable rates. Loss via urine and feces also occurs continuously, although this water is voided only during defecation and urination.

Water excreted via urine acts as a solvent for products excreted from the kidney. Some species have greater ability to concentrate urine than others. In some cases, urine concentration is related to the type of compound excreted. For example, birds excrete primarily uric acid rather than urea as an end product of protein metabolism. These species excrete urine in semisolid form with only small amounts of water. However, mammals cannot concentrate urine to nearly such an extent. Birds also have another slight advantage in that production of uric acid results in production of more metabolic water than does urea.

The kidney of most species has great flexibility in the amount of water that may be excreted. Minimal amounts required for excretion (called obligatory water) usually are exceeded greatly except when water intake is restricted. Consumption of excess water during periods of heat stress or consumption of diuretics (e.g., caffeine and alcohol in humans) may increase kidney excretion of water considerably. Within a given species the amount of kidney concentration of urine depends on the type of compound being excreted. The amount required to excrete chlorides and carbonates is similar and appears to be additive. That required for urea is somewhat less than for chlorides and carbonates, but the sum of that for urea and these salts is not additive.

Examples of water loss via the urine are shown in Tables 6.2 and 6.6. When sheep were on dry feed and not heat-stressed, urine accounted for 30 to 33% of water losses (Table 6.2). When dairy cows were given ad lib or restricted water and some were heat-stressed, urine volume ranged from 10 to 30 L/day and from 24 to 43% of water excreted (Table 6.6).

Fecal losses of water in humans are usually about 7 to 10% of urinary water. In ruminants such as cattle, fecal water loss usually exceeds urinary losses, although not during heat stress. Other species tend to be intermediate. Animals that consume fibrous diets usually excrete a higher percentage of total water via feces, and those that form fecal pellets (e.g., sheep, goats, deer) usually excrete drier feces and, presumably, are more adapted to drier climates and more severe water restriction than is the case for species that do not form fecal pellets. Note in Table 6.2 that fecal water represented 13 to 24% of total excretion with sheep, but with normal dairy cows (Table 6.6) it was 30 to 32% of total and 17 to 20% when the cows were heat-stressed.

Insensible water losses account for a relatively large amount of total loss, particularly at temperatures that do not induce sweating or in species that do not sweat. For example, insensible water losses are 45 to 55% of total losses for sheep in a metabolism chamber compared with about 30 to 35% for humans under similar conditions.

Air inhaled into the lungs may be very dry but is about 90% saturated with water when exhaled. During periods of hyperventilation that occur with hot temperatures, loss from the lungs is increased greatly and will increase to 50% or more of the total excretion in sweating cows that are heat-stressed. Insensible water loss through the skin is relatively low.

Loss of water via sweat may be very large in species such as humans and horses whose sweat glands are distributed over a large portion of the body surface. Sweating is used as a means of dissipating body heat and is said to have an efficiency of about 400% compared to respiratory heat loss. Heat-tolerant species generally have well-developed sweat glands. This helps explain why *Bos indicus* cattle are more heat tolerant than *Bos taurus* breeds. Species that have poorly developed sweat glands must keep cool either by panting (e.g., dog, chicken) or by finding shade or water (e.g., swine) to cool the body.

❑ REGULATION OF DRINKING

The regulation of drinking is a highly complex physiological process. It is induced as a result of dehydration of body tissues. However, drinking may also occur when there is no apparent need to rehydrate tissues. When an animal is thirsty, salivary flow usually is reduced and dryness of the mouth and throat may stimulate drinking—a relationship that may, indirectly, be traced to a decrease in plasma volume. Other information indicates that salivary flow is not a critical factor in initiation of drinking. Oral sensations apparently are involved that may be influenced by osmotic receptors in the mouth. For example, dogs with esophageal fistulas (which prevent orally ingested water from entering the stomach) will stop drinking after sham-drinking a normal amount of water. However, the sham-drinking will be repeated again in a few minutes. There is ample evidence that passage of water through the mouth is required for a feeling of satiety, because placing water in the stomach by a tube leaves animals restless and unsatisfied.

Most domestic animals generally drink during or soon after eating periods if water is close at hand. Frequency of drinking is increased in hot weather, as is the amount consumed. Under herded conditions in some of the dry areas of Africa and India, cattle, sheep, and goats may be watered on a 3-day cycle. This frequency is undoubtedly inadequate for maximum production, but maximum production is not the objective under such adverse conditions.

❑ WATER CONSUMPTION AND REQUIREMENTS

Water consumption by livestock and poultry has been reviewed (Parker and Brown, 2003). Water requirements for any class or species of healthy animals are difficult to delineate except in very specific situations. This is so because numerous dietary and environmental factors affect water absorption and excretion and because water is so important in regulation of body temperature. Other factors, such as ability to conserve water or differences in activity and in physiological state (i.e., growth, gestation, and lactation) compound the problem when different classes or species of animals are compared. As a result of these factors, relatively little effort has been made to quantify water needs, except in a few specific situations.

It is well recognized that water consumption is related to heat production and sometimes to energy consumption. Also, water requirements can be related to body surface area in nonstressing situations. At environmental temperatures that do not result in heat stress, a good linear relationship exists between dry matter consumption and water consumption. However, when the temperature reaches stressing levels, feed consumption is apt to decrease and water consumption is increased greatly. One example is shown in Fig. 6.1. Note in the figure that water consumption per unit of feed consumption by cattle goes from about 0.35 gal/lb (2.9 L/kg) of dry matter consumed at 40°F (15.3°C) to about 1.9 gal/lb (18 L/kg) at 100°F (38°C) for *Bos taurus* cattle. When expressed on the basis of consumption per unit body weight, non-heat-stressed, nonlactating cattle may drink 5 to 6% of body weight per day. Water consumption may increase to 12% or more of body weight per day during heat stress. Seasonal differences in wa-

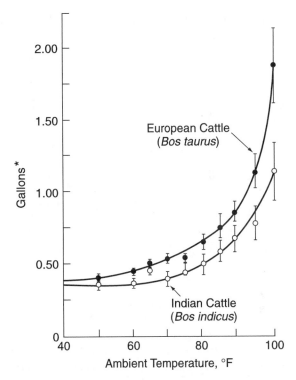

Figure 6.1 *Water requirements of European and Indian cattle as affected by increasing temperatures. From Winchester and Morris (1956).*

ter intake of confined cattle primarily reflect temperature stress. Data on feedlot steers (Hoffman and Self, 1972) show average daily water consumption in the winter to be 19 L of water as opposed to 31 L in the summer.

Dietary Factors

Dry matter intake is highly correlated with water intake at moderate temperatures. The water content of the feed consumed also affects total water intake. This results in consumption of more water than needed when forage is very lush with a high water content. High levels of protein also increase water intake because of greater urinary excretion of excess protein as urea. When urea has been used as the major source of nitrogen (N) in ruminant diets, some evidence shows that it results in greater urine volume than an equivalent amount of N in the form of purified soy protein. Young mammals consuming only milk need additional water, especially in hot weather. Although milk of most

TABLE 6.4   Effect of rations and level of feeding on water intake of holstein heifers.

| ITEM[a] | ROUGHAGE FED AND LEVEL OF FEEDING | | | |
| | HAY | | SILAGE | |
	AD LIB.	MAINTENANCE	AD LIB.	MAINTENANCE
Dry matter intake, kg/100 kg BW	2.06	1.24	1.70	1.15
Feed water, kg/kg feed DM	0.11	0.12	3.38	3.38
Waterdrunk, kg/kg feed DM	3.36	3.66	1.55	1.38
Total water, kg/kg feed DM	3.48	3.79	4.93	4.76
Urine, kg/kg feed DM	0.93	1.14	1.85	1.68

[a]BW = body weight; DM = dry matter.
Source: Waldo et al. (1965).

species is about 80 to 88% water, the high content of protein results in high obligatory urinary water loss, and without additional water performance is likely to be reduced.

An increased intake of fats also may increase water intake. Consumption of feeds such as silages tends to increase intake and urinary excretion as illustrated in Table 6.4. In this instance, note that total water consumed per kilogram of dry matter was appreciably higher in animals fed silage. Perhaps excess water was consumed because the silage-fed cattle also excreted more urine.

There is ample evidence that consumption of NaCl or other salts increases consumption and excretion of water greatly by different animal species. Some salts may cause diarrhea and greater fecal excretion of water, but those, such as NaCl, that are absorbed almost completely, result in much greater urinary excretion; tissue dehydration occurs if water is not available.

Environmental Factors

High temperature, as mentioned previously, is the major factor causing increased consumption of water. Associated with heat stress is high humidity, which also increases the need for water because heat losses resulting from evaporation of water from the body surface and lungs are reduced with high humidity.

In confined animals, design and accessibility of watering containers influences intake, as does cleanness of the containers. In range animals the distance that must be traveled between water and forage affects the frequency of drinking and the amount consumed; that is, the greater the distance

the less frequently the animal drinks and the less it consumes in a 24-hour period.

Volume of Water Required

In very general terms, animals will consume 2 to 5 kg of water for every kg of dry feed consumed when they are not heat-stressed. Those species that have the ability to conserve water require less, and those adapted to a wet environment consume more. For example, cattle generally consume at the 4:1 ratio, but sheep are more apt to be between 2.5:1 and 3:1. Cattle require more water than sheep because they are not as efficient at conserving water. Birds generally require less than mammalian species. Young animals generally require more water per unit of body weight than adult animals. Activity increases requirements; nervous animals also will likely consume more water than less active animals. Physiological, dietary, and environmental factors affecting absorption and excretion alter water requirements.

Swine generally consume 2 to 2.5 kg of water/ kg of dry feed at moderate temperatures, and horses and poultry consume 2 to 3 kg/kg of dry feed (National Research Council, 1974). Cattle appear to need 3 to 5 kg of water/kg of dry feed, and young calves appear to need 6 to 8 kg/kg of feed. In addition to these general values, water consumption will be increased by milk production; the additional requirement for cattle has been estimated at 1 to 1.8 kg of water/kg of feed over that required for a comparable animal not in lactation. Pregnancy also increases the need for water, and ewes bearing multiple fetuses require more water than those bearing singles. The expected water consumption of adult farm animals and poultry in

TABLE 6.5   Expected water consumption of livestock and poultry.

CLASS	WATER INTAKE
	Liters/animal/day
Growing heifers and steers, 180 kg	15–22
Dry cows, pregnant	26–49
Lactating cows, 22.7 kg milk/day	91–102
45.4 kg milk/day	182–197
Swine, 11 kg body wt.	1.9
90 kg body wt.	9.5
pregnant	17–21
lactating	22–23
Horses, 500 kg body wt., maintenance	23–57
500 kg body wt., moderate work	45–68
weanling, 300 kg body wt.	23–30
Sheep, nonlactating ewes	7.6
lactating ewes	11.3
lambs, 2–9 kg body wt	0.4–1.1
Poultry, Broiler chickens	Liters/bird/week
Age, 4 weeks	0.10
Age, 8 weeks	0.20
Laying hens 16–20 weeks old	1.20–1.60
Turkeys	
Age, 1 week	0.38
Age, 10 weeks	4.4–5.4
Age, 16 weeks	4.7–6.9

Source: Adapted from Parker and Brown (2003).

temperate environments as recommended by Parker and Brown (2003) has been summarized in Table 6.5.

Water Restriction

In many areas in the world, water supplies are more restrictive than feed supplies, either because of a lack of surface water or well water or because the available water is brackish and not suitable for consumption in adequate amounts to sustain the animals. Consequently, a relatively high percentage of animals, both domestic and wild, are faced with water shortages at some time during the year.

Fluid intake by animals is intermittent, even more so than food intake for most species, but the loss of water from the body is continuous, though variable. Thus, the body must be able to compensate in order to maintain its physiological functions. The most noticeable effect of moderate water restriction is reduced feed intake and reduced productivity. Urine and fecal water excretions drop markedly. With more severe restriction, weight loss is rapid as the body dehydrates. These changes are illustrated in Table 6.6. The dehydration is accompanied by increased renal excretion of N and electrolytes such as Na^+ and K^+. Water restriction causes more severe or quicker responses at high environmental temperatures. In addition to the dehydration that occurs, pulse rate and rectal temperatures usually increase, especially at hot temperatures, because the animal no longer has sufficient water to evaporate to maintain normal body temperatures. Respiration rate also increases in these conditions. Eventually, there is a relatively marked increase in blood concentration, loss of intra- and extracellular and total body water, nausea, difficulty in muscular movements, and (in humans) loss of emotional stability. Animals tend to become highly irritable and bad tempered. Eventually, prostration and death will follow if severe water deprivation persists.

Animals vary greatly in the amount of dehydration they can withstand; the camel is an example

TABLE 6.6   Effect of 50% water restriction at 18° or 32°C in dairy cows.

	18°C		32°C	
MEASUREMENT	AD LIB.	50% AD LIB.	AD LIB.	50% AD LIB.
Body weight, kg	641	623	622	596
Feed consumption, kg/day	36.3	24.9	25.2	19.1
Urine vol., L/day	17.5	10.1	30.3	9.9
Fecal water, kg/day	21.3	10.5	11.7	8.2
Total vaporization, g/h	1133	583	1174	958
Total body water, %	64.5	50.9	67.9	52.6
Extravascular fluids, %	59.0	45.5	61.5	46.9
Plasma volume, %	3.9	3.9	4.4	3.9
Metabolism, Kcal/day	798	694	672	557
Metabolic water, kg/day	2.5	2.0	2.1	1.9
Rectal temperature, °C	38.5	38.5	39.2	39.5

Source: Data from Self et al. (1973).

of one that can withstand weight loss of about 30% or more. Most mammals cannot survive such severe dehydration as this. With moderate restriction most species show some adaptation and can compensate partially for it by reducing excretion.

Water intoxication may occur in some species (humans, calves) as a result of sudden ingestion of large amounts after a short time period of deprivation. In calves, death may result because of the kidney's slow adaptation to the sudden high-water load.

❑ WATER QUALITY

Generally, it has been assumed that water safe for human consumption may be used safely by stock, but it appears that animals can tolerate higher salinity than humans. Thus, it is probable that tolerances for other substances may also be different. An NRC (1974) report on this topic is recommended for additional reading.

Water quality standards for livestock have been summarized (Sweeten et al., 2003). Water quality may affect feed consumption directly because low-quality water normally will result in reduced water consumption and, hence, lower feed consumption and production. Substances that may reduce the palatability of water include various salts. At high rates of consumption, these salts may be toxic. Substances that may be toxic without much effect on palatability include nitrates, fluoride, and salts

of some heavy metals. Other materials that may affect palatability or be toxic include pathogenic microorganisms (bacteria, protozoa, fungi), algae, hydrocarbons and other oily substances, pesticides, and many industrial chemicals that sometimes pollute water supplies.

The Environmental Protection Agency of the United States government has set upper acceptable limits of an array of pesticides in the drinking water for humans. These limits continue to be adjusted as new information becomes available. Although comparable standards have not been set for animals, the guidelines for humans can be generally applied to animals. Thulin and Brumm (1991) and Thacker (2001) summarized the water requirements for swine and published work related to water quality. Safe upper limits of potentially toxic inorganic elements for livestock are summarized in Table 6.7.

Mineral salts apt to be present in high amounts include carbonates and bicarbonates, sulfates and chlorides of Ca, Mg, Na, and K. At a given concentration the sulfates generally are more detrimental to the animal than the carbonates or chlorides. Ray (1989) reported that saline drinking water containing 6000 mg/L dissolved chloride, sulfate and carbonate salts of calcium, and sodium reduced the weight gain of feedlot steers by 10% compared with that of steers given water containing 1300 mg/L of dissolved salts (considered normal water). Jaster (1978) noted that dairy cows drinking water containing 2500 mg/L had in-

TABLE 6.7 Safe concentrations of nitrate, nitrite, sulfate, total dissolved solids (TDS),[a] and potentially toxic inorganic elements[b] in water for domestic animals.

SUBSTANCE	SAFE UPPER LIMIT, MG/LITER
Nitrate, NO_3	<440
Nitrate, NO_3-N	<100
Nitrite, NO_2	<33
Nitrite, NO_2-N	<10
Sulfate, SO_4	<300
Total dissolved solids, TDS	<3000
Arsenic, As	0.2–0.5
Boron, B	10.0
Cadmium, Cd	0.05–0.5
Chromium, Cr	1.0–1.5
Cobalt, Co	1.0
Copper, Cu	0.5
Fluoride, F	2.0–3.0
Lead, Pb	0.1
Mercury, Hg	0.01
Nickel, Ni	1.0
Selenium, Se	0.1
Vanadium, V	0.1–1.0
Zinc, Zn	25.0

[a]Adapted from Sweeten et al. (2003).
[b]Adapted from National Research Council (1974) and CAST (1974).

creased water consumption and a slight reduction in feed consumption and milk production. Higher levels of NaCl would produce more drastic effects.

Most domestic animals can tolerate a total dissolved solids (TDS) concentration of 15,000 to 17,000 mg/L, but production is apt to be reduced at this concentration. Water classified as good should have less than 2500 mg/L of TDS. Water containing 1000 to 5000 mg/L TDS is safe but may cause temporary mild diarrhea in animals not accustomed to it. Concentrations between 3000 and 5000 mg/L may cause watery feces and increased mortality in poultry. Water with 5000 to 7000 mg/L TDS is considered acceptable to all domestic animals except poultry. Concentrations greater than 7000 mg/L are unfit for poultry and swine and should not be used for pregnant or lactating cattle, horses, or sheep or for young growing animals of any species. Water containing more than 10,000 mg/L (1%) of soluble salts is unfit for use as drinking water under any conditions.

All of the dietary essential mineral elements are usually found in most water supplies, particularly in surface waters such as lakes. A substantial portion of the Na, Ca, and S requirement may be supplied in this manner. Consequently, water supplies should not be overlooked as a source of some of the needed mineral elements.

Nitrates and nitrites are widely dispersed in the environment and often find their way into drinking water. Although the concentrations of nitrate (NO_3) normally present in drinking water are well tolerated by animals, nitrite (NO_2), the reduced form of nitrate, is readily absorbed from the GI tract and can reach toxic levels. Nitrate concentrations in drinking water as high as 1320 mg/L may be well tolerated by animals (CAST, 1974), but nitrite levels above 33 mg/L are toxic (National Research Council, 1974; CAST, 1974). At toxic concentrations, nitrite in the blood oxidizes the iron in hemoglobin to methemoglobin and reduces the oxygen-carrying capacity of the blood. High levels of nitrate in water supplies may signal bacterial contamination. Bacteria can convert nitrate to nitrite, and this contaminated water may create a health hazard for animals and humans drinking the water.

❏ SUMMARY

- Water is required by animals in larger amounts than any other material ingested, and it has many different functions. It is a solvent, a transport medium, and a diluent, and it acts in hydrolysis and oxidation of nutrients and metabolites.

- Body tissues, on a fat-free basis, are about 73% water. The concentration gradually decreases from birth to maturity and is inversely related to body fat content.

- Water is absorbed rapidly from the stomach and intestinal tract and passes freely into most tissues and organs, depending on osmotic pressure gradients. The turnover of water is rapid; the half-life is generally about 3 days or less.

- Water is supplied primarily by drinking and by that present in feed, but metabolic water oxidation and other reactions may supply a substantial amount for some species adapted to dry environments.

- The ability to reduce water losses via feces, kidneys, lungs, or body surface is related to species adaptation to dry climates.

- Water requirements are directly related to dry feed consumption. In general, about two to five units of water are consumed per unit of dry feed, but this is affected by animal species, by type of diet, and particularly by high environmental temperatures.

- Water restriction reduces feed consumption and may result in dehydration and death.

- Water quality is of considerable importance to animals. Water supplies in many parts of the world are saline or have excess salts of other kinds. Many species can tolerate water containing 1% total dissolved solids, but the amount that can be tolerated depends on the specific chemical compounds present.

❏ REFERENCES

Council for Agricultural Science and Technology (CAST). 1974. *Quality of Water for Livestock,* Report No. 26. CAST, Ames, IA.

Hoffman, M. P., and H. L. Self. 1972. *J. Anim. Sci.* **35:** 871.

Hyder, D. N., et al. 1968. *J. Range Mgmt.* **21:** 392.

Jaster, E. H., et al. 1978. *J. Dairy Sci.* **61:** 66.

National Research Council. 1974. *Nutrients and Toxic Substances in Water for Livestock and Poultry.* National Academy of Sciences, Washington, DC.

Oster, J. D. 2003. In *Encyclopedia of Water Science,* pp. 1017–1019, B. A. Stewart and T. A. Howell (eds.). Marcel Dekker, New York.

Parker, D. B., and M. S. Brown. 2003. In *Encyclopedia of Water Science,* pp. 588–592, B. A. Stewart and T. A. Howell (eds.). Marcel Dekker, New York.

Ray, D. E. 1989. *J. Anim. Sci.* **67:** 357–363.

Schmidt-Neilsen, K. 1964. *Desert Animals, Physiological Problems of Heat and Water.* Oxford University Press, Oxford, England.

Self, S. M., et al. 1973. *J. Dairy Sci.* **56:** 581.

Sweeten, J. M., et al. 2003. In *Encyclopedia of Water Science,* pp. 596–600, B. A. Stewart and T. A. Howell (eds.). Marcel Dekker, New York.

Thacker, P. A. 2001. In *Swine Nutrition* (2nd ed.), Chapter 17, pp. 381–398, A. J. Lewis and L. L. Southern (eds.). CRC Press, Boca Raton, FL.

Thulin, A. J., and M. C. Brumm. 1991. In *Swine Nutrition,* Chapter 18, pp. 315–324, E. R. Miller, D. E. Ullrey, and A. J. Lewis (eds.). Butterworth-Heinemann, Boston, MA.

Waldo, D. R., et al. 1972. *J. Dairy Sci.* **48:** 1473.

Wallace, J. D., et al. 1972. *Am. J. Vet. Res.* **33:** 921.

Winchester, C. F. and M. J. Morris. 1956. *J. Anim. Sci.* **15:** 722.

7

Carbohydrates

CARBOHYDRATES ARE THE MAJOR components in plant tissues. They comprise up to 70% or more of the dry matter of forages. Higher concentrations (up to 85%) may be found in some seeds, especially cereal grains. The chloroplasts in plant leaves synthesize their carbohydrates, using solar energy, carbon dioxide, and water, and give off oxygen. This is a vital process for animals because they could not exist without this transformation of energy and the free oxygen produced as a byproduct of the photosynthetic reaction.

Carbohydrates, consisting mainly of glucose and glycogen, make up less than 1% of the weight of animals.

❏ CLASSIFICATION AND STRUCTURE

One method of classifying carbohydrates is shown in Table 7.1. Classification, as done in this manner, is strictly on the basis of the number of carbon atoms per molecule of carbohydrate and on the basis of the number of molecules of sugar in the compound. Thus, a monosaccharide has only one molecule of sugar, a disaccharide has two molecules, an oligosac-

charide (not shown) may have 3 to 10 sugar units, and a polysaccharide more than 10 sugar units. Most plant tissues contain many different types of carbohydrates; only sources with relatively high concentrations are shown in Table 7.1. The chemistry and nutritional properties of complex carbohydrates have been summarized (Van Soest, 1994).

Many different carbohydrate compounds or carbohydrate-containing compounds are found in

TABLE 7.1   Classification of carbohydrates and their occurrences.

COMPOUND	MONOSACCHARIDE CONTENT	OCCURRENCE
Monosaccharides (simple sugars)		
Pentoses (5-C sugars) ($C_5H_{10}O_5$)		
Arabinose		Pectin; polysaccharide, araban
Xylose	Corn cobs, wood; polysaccharides	
Ribose	Nucleic acids	
Hexoses (6-C sugars) ($C_6H_{12}O_6$)		
Glucose	Disaccharides; polysaccharides	
Fructose	Disaccharides (sucrose)	
Galactose	Milk (lactose)	
Mannose	Polysaccharides	
Disaccharides ($C_{12}H_{22}O_{11}$)		
Sucrose	Glucose–fructose	Sugar cane, sugar beets
Maltose	Glucose–glucose (Glucose-4-α-glucoside)	Starchy plants and roots
Lactose	Glucose–galactose	Milk
Cellobiose	Glucose–glucose (glucose-4-β-glucoside)	Fibrous portion of plants
Trisaccharides ($C_{18}H_{32}O_{16}$)		
Raffinose	Glucose–fructose–galactose	Certain varieties of eucalyptus, cottonseed, sugar beets
Polysaccharides		
Pentosans ($C_5H_8O_4$)$_n$		
Araban	Ababinose	Pectins
Xylan	Xylose	Corn cobs, wood
Hexosans ($C_6H_{10}O_5$)$_n$		
Starch (a polyglucose glucoside)	Glucose	Grains, seeds, tubers
Dextrin	Glucose	Partial hydrolytic product of starch
Cellulose	Glucose (Glucose-4-β-glucoside)	Cell wall of plants
Glycogen	Glucose	Liver and muscle of animals
Inulin (a polyfructose fructoside)	Fructose	Potatoes, tubers, artichokes
Mixed polysaccharides		
Hemicellulose	Mixtures of pentoses and hexoses	Fibrous plants
Pectins	Pentoses and hexoses mixed with salts of complex acids	Citrus fruits, apples
Gums (partly oxidized to acids)	Pentoses and hexoses	Acacia trees and certain plants

plant and animal tissues. However, from a nutritional point of view, the list of important compounds is rather short. Some compounds, such as glycerol, may be considered to be intermediate between a carbohydrate and a lipid. Structurally, glycerol is an alcohol and it is partially soluble in water. However, it is a component of triglycerides (fats), and it is also soluble in ether, a characteristic of most lipids. In addition, one or more molecules of different monosaccharides (simple sugars) may be found in complex compounds. For example, ribose, a simple five-carbon sugar, is an essential component of adenosine triphosphate (ATP), which is important in the transfer of energy at the cellular level.

Simple carbohydrates have the empirical structure of $C_nH_{2n}O_n$, whereas most complex compounds have a structure corresponding to $C_nH_{2n-2}O_{n-1}$. Carbon is combined with hydrogen and oxygen, which are in about the same ratio as in water. When two or more molecules of monosaccharides are joined together, the formula represents the loss of one or more molecules of water. Chemically, carbohydrates are polyhydroxyl aldehydes and ketones, with an aldehyde $\left(\begin{smallmatrix}-C=O\\|\\H\end{smallmatrix}\right)$ or ketone $\left(\begin{smallmatrix}C-C-C\\\|\\O\end{smallmatrix}\right)$ group in their structure. Structures of some of the common monosaccharides or disaccharides are illustrated in Fig. 7.1. Although the chain-type structure shown in Fig. 7.1 is a common way to depict sugar formulas, the biologically active structures of these compounds are more aptly shown in the two pyranose forms depicted at the top of page 76 (right side).

❏ FUNCTIONS

In animal nutrition carbohydrates serve primarily as a source of energy for normal life processes. However, in plants some of the simple sugars, especially glucose and ribose, are involved in energy transformations and tissue synthesis. Less soluble forms, such as starch, serve as energy reserves in roots, tubers, and seeds. The relatively insoluble fractions (cellulose, hemicellulose) are most important in providing structural support for the living plants.

A class of complex carbohydrates known as glycans, ubiquitous in nature, form an intricate sugar coat surrounding and occupying space between cells in virtually all organisms (Sasisekharan and

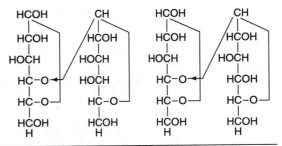

Figure 7.1 *Structure of monosaccharides and disaccharides.*

OPEN-CHAIN FORM

TWO WAYS OF DEPICTING THE PYRANOSE FORM

Myette, 2003). Glycans have an important role in transmitting biochemical signals into and between cells. They vary in shape and size, ranging from linear chains (polysaccharides) to highly branched molecules. Because they are synthesized in the body they are not nutrients, so they are not discussed further. The broad array of simple carbohydrates found in a variety of foods and feeds is illustrated in Table 7.1.

Soluble components of the plant cell wall include starches, hemicellulose, and beta-glucans, which are not linked to lignin and are available to fermentation. All have alpha 1-4 linkage, in contrast to cellulose, which has beta 1-4 linkage and is insoluble and resistant to breakdown except by cellulase produced by microbes in the rumen and lower intestinal tract of nonruminants. The structures of cellulose and pectins are shown in Figs. 7.2a and 7.2b, respectively.

Carbohydrates and lipids are the two major sources of energy for the animal body. The lipid content of most diets of food-producing animals is usually less than 5%. Since most of the energy stored in the animal body is in the form of fat, dietary carbohydrate not broken down immediately for energy is stored in the animal body as fat. In humans and in cats, dogs, mink, and several other animal species, a greater amount of food of animal origin is consumed. Therefore, the proportion of dietary energy coming from carbohydrate is considerably less in these animals and in humans than in most farm animals. The energy contribution from fat in human diets in the United States is usually greater than 30% of total dietary calories.

Although carbohydrates serve as a significant source of energy for body tissues, only limited data are available demonstrating a dietary requirement for any specific carbohydrate by higher animals and humans, even though certain insects may have a requirement (Dadd, 1963). Virtually all natural food sources contain some carbohydrate. Brambila and Hill (1966) showed that chicks can grow normally on carbohydrate-free diets if the calorie to protein ratio is optimum and if triglycerides are included in the diet. When fatty acids replaced triglycerides, growth depression resulted from the body's inability to produce glucose from amino acids in the absence of the glycerol from triglycerides. More recently, it has been shown (Koski and Hill, 1990; Koski et al., 1990) that maternal dietary carbohydrate is essential during late gestation for postnatal survival of rat pups and that optimum lactation requires a dietary carbohydrate source.

The ultimate source of energy for most animal cells is glucose. This basic unit is made available to animal cells either by ingestion of glucose or its precursors by the animal or by conversion from other metabolites. The carbon skeletons of these substances and of products of fat metabolism (Chapter 8) provide the energy for maintaining normal life processes.

❏ METABOLISM

Preparation for Absorption

Digestion in the Small Intestine.   Only monosaccharides can be absorbed from the GI tract, except in newborn animals capable of absorbing larger molecules. Thus, for absorption to occur, poly-, tri-, and disaccharides must be hydrolyzed by digestive

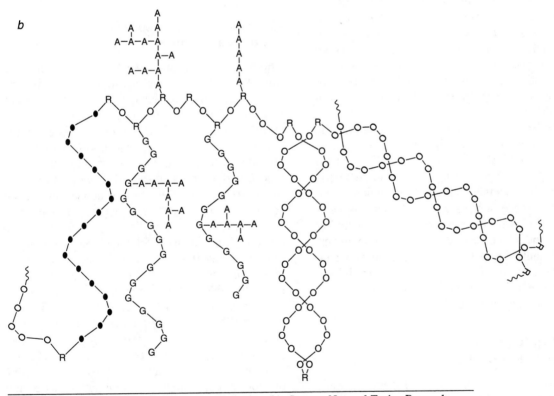

Glucose units
(Glucose-β(1-4)-Glucose)

Structure of a portion of a cellulose molecule in which n = repeating pair of glucose units.

Figure 7.2 *Structures of cellulose (a) and pectins (b). Source: Natural Toxins Research Center, Texas A&M University, Kingsville,TX. www.ntri.tamak.edu/cell//carbohydrates.html (a) (Cellulose) and Van Soest (1994) (b) (Pectins).*

enzymes elaborated by the host or by microflora inhabiting the GI tract of the host. Principal digestive enzymes elaborated by animals were discussed and listed in Chapter 4. The carbohydrate-splitting enzymes (carbohydrases) are effective in hydrolyzing most complex carbohydrates to monosac-charides except for those with a glucose-4-beta-glucoside linkage (see Table 7.1), such as in cellulose (see Fig. 7.2). Starch, which is quantitatively the most important soluble complex carbohydrate in plants, occurs in two important molecular configurations of glucose polymers, amylose and amylopectin. Both amylose and amylopectin occur as glucose polymers linked via a glucose-(1–4)-alpha-glucoside linkage. Amylose is a linear polymer of glucose (molecular weight of 150,000 to 1 million daltons). Amylopectin is a linear polymer with numerous branches (molecular weight 10 to 100 million daltons). The ratio of amylose to amylopectin varies among maize (corn), cereal grains, and other high-starch feeds. For example, standard maize contains 24% amylose and 76% amylopectin, whereas waxy maize contains 1% amylose and 99% amylopectin; rice and potatoes contain about 20% amylose and 80% amylopectin, whereas wheat generally contains an amylose to amylopectin ratio similar to that of standard maize. The hydrolysis of starch in amylose and amylopectin to glucose is similar. The milling and baking characteristics of different varieties of wheat are affected by the amylose to amylopectin ratio in the wheat; an important selection criterion for wheat geneticists is the starch composition of the wheat for specific uses. The digestive enzymes secreted by animals are generally able to hydrolyze both types of starch efficiently.

Microflora of the rumen of ruminants and the cecum and colon of some nonruminants, such as the horse and rabbit, produce cellulase, which is capable of hydrolyzing the glucose-4-beta-glucoside linkage (see Table 7.1) of cellulose. Consequently, these species can utilize large quantities of cellulose. Other nonruminants, including humans and swine, also utilize cellulose by anaerobic fermentation in the large intestine by virtue of the production of cellulase by some of the microorganisms residing in the lower intestinal tract, but not by mammalian cells.

Plant Cell Walls and Carbohydrate Digestion

Cellulose is a major component of plant cell walls and, together with hemicelluloses, often is present in combination with lignin, a highly insoluble and biologically unavailable mixture of polymers of phenolic acid. Lignification increases with plant age, and, in mature trees, lignin is the chief structural component. Vegetables and cereals are low in lignin, whereas grasses are intermediate and legumes are higher in lignin. The amount of lignin present affects the bioavailability of the cellulose and hemicellulose for microbial use and in this way affects the nutritive value of a plant material for animals.

Microbial Fermentation of Cellulose and Other Plant Fibers

In ruminants and other species with large microbial populations in the GI tract, anaerobic fermentation of carbohydrates results in the production of large quantities of volatile fatty acids (VFA), mainly acetic, propionic, and butyric acids, and provides a large proportion of the total energy supply. Even in pigs, whose ability to utilize cellulose is less than that of ruminants, some of the energy required for maintenance can be provided by VFA produced by microbial action on fiber in the large intestine. Pathways of carbohydrate metabolism in the rumen and in the lower intestinal tract of nonruminants are shown in Fig. 7.3. The identity and characteristics of microbes in the rumen are described in Table 7.2.

Absorption and Transport of Monosaccharides

The upper or proximal section of the small intestine, the duodenum and jejunum, has the greatest capacity to absorb monosaccharides. The lower or distal small intestine (lower ileum) absorbs less, and the stomach and large intestine absorb little, if any, sugars. Selective absorption of monosaccharides occurs from the GI tract of the rat. Galactose and glucose are absorbed very efficiently, but mannose, which differs from glucose only in the configuration of the hydroxyl group at carbon 2 (see Fig. 7.1), is absorbed at only about 20% of the efficiency of glucose. Wilson (1962) compiled a table showing the selective absorption of six different monosaccharides by a variety of animal species (Table 7.3). In general, glucose and galactose are absorbed at the highest rate and arabinose at the lowest rate among the six monosaccharides compared, regardless of the species tested. Active (energy-dependent) transport is established for glucose and galactose, but it probably does not operate for other sugars. Glucose and galactose appear unchanged in the portal vein after absorption. Glucose is transported across cell membranes by at least two transport families (Devaskar and Mueckler, 1992). The first family, sodium–glucose cotransporters, actively transport glucose across the intestinal mucosal cell during absorption from the lumen and across the kidney tubule. Glucose is transported against its concentration gradient by coupling its transport to sodium. The sodium is transported down its concentration gradient, which is maintained by the active trans-

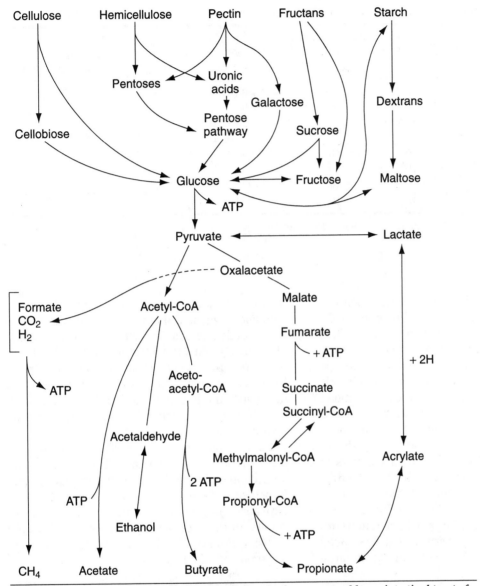

Figure 7.3 *Pathways of carbohydrate metabolism in the rumen and lower intestinal tract of nonruminants. Source: Van Soest (1994).*

port of sodium across the brush border (intestinal or kidney) cells by membrane-bound sodium–potassium ATPase (Na^+–K^+–ATpase). (This system contributes significantly to maintenance energy expenditure in animals). The second family, facilitative glucose transporters, transport glucose actively down a concentration gradient; they are present in nearly all mammalian (and probably nearly all vertebrate) cells.

The number of known glucose transporters has expanded to 6 Na-dependent glucose transporters (SGLT1 to SGLT-6) and 13 facilitative sugar transporters (GLUT1 to GLUT12 and H-coupled myo-inositol, HMIT) (Woods and Trayhum, 2003). Each of these transporters has different kinetic properties and substrate specificities and varied profiles of tissue expression, suggesting a very complex process for delivering glucose into cells. These two families of glucose transporters are proteins containing about 500 to 660 amino acid residues. The glucose transport proteins span the cell membrane in a particular configuration and

TABLE 7.2   Relative volumes and numbers of microbial organisms in the rumen.

GROUP	NUMBER PER ML	MEAN CELL VOLUME (μ^3)	NET MASS[a] (MG/100 ML)	GENERATION TIME	% OF TOTAL RUMEN MICROBIAL MASS
Small bacteria	1×10^{10}	1	1600	20 min	60–90
Selenomonads	1×10^8	30	300		
Oscillospira flagellates	1×10^6	250	25		
Ciliated protozoa					
Entodinia	3×10^5	1×10^4	300	8 h	
Dasytricha + Diplodinia	3×10^4	1×10^5	300		
Isotricha + Epidinia	1×10^4	1×10^6	1100	36 h	
Fungi	1×10^4	1×10^{5}		24 h	5–10

[a]Estimated cell weight per 100 ml rumen fluid, assuming a density of 1.0.
Source: Van Soest (1994).

facilitate glucose transit into the cell. In adipose and muscle cells, both insulin and glucose modulate glucose transporter activity (gene expression).

Conversion of some monosaccharides to glucose occurs within the intestinal mucosal cell. Conversion of fructose to glucose remains relatively constant over a wide range of fructose concentrations, but the rate of movement of fructose into the cell is roughly proportional to the luminal concentration.

Fructose is converted to lactic acid by the intestine of some animal species; the rat converts up to 50% of fructose to lactic acid, as measured by mesenteric vein cannulation, but the guinea pig produces very little. Fructose in the blood of mature animals is very low, but in fetal and newborn lambs and pigs it is high. Aherne et al. (1969) found that fructose was absorbed with little or no intestinal conversion to glucose in 3-, 6-, or 9-day-old pigs and suggested that fructokinase activity (enzyme needed for conversion of fructose to glucose) of the liver but not the intestine increases with age in the pig. Intravenous administration of sucrose, fructose, or lactose to hypoglycemic baby pigs fails to alleviate the hypoglycemia, and intravenous administration of disaccharides to animals generally results in their excretion in the urine.

Sugars apparently share a common pathway of transport across the intestinal mucosal cell. Therefore, competitive inhibition between glucose and galactose, as well as between glucose and several derivatives of glucose, is not a surprise. Further details on the mechanism of transport of sugars across the mucosal cell to the blood are beyond the scope of this discussion.

TABLE 7.3   Selective absorption of sugars by different animals.

ANIMAL	RATE OF SUGAR ABSORPTION (GLUCOSE TAKEN AS 100)					
	GALACTOSE	GLUCOSE	FRUCTOSE	MANNOSE	XYLOSE	ARABINOSE
Rat	109	100	42	21	20	12
Cat	90	100	35	—	—	—
Rabbit	82	100	—	—	—	60
Hamster	88	100	16	12	28	10
Man	122	100	67	—	—	—
Pigeon	115	100	55	33	37	16
Frog	107	100	46	51	29	
Fish	97	100	62	—	57	49

Several important factors affect the absorption of glucose. It is reduced by short-term (24- or 48-hour) fasting, but increased by chronically restricted food intake. The basis for this apparent difference in the functional capacity of the small intestine in fasting versus underfed animals is not well understood.

Digestive Enzymes and Utilization of Carbohydrates

Deficiency of specific disaccharidases (enzymes) in the GI tract results in serious gastrointestinal upsets. Young mammals fed large amounts of sucrose develop severe diarrhea, and death may occur from an insufficiency of sucrase during the first few weeks of life. Ruminant species apparently produce no sucrase. Feeding appreciable amounts of sucrose to liquid-fed young animals results in severe diarrhea. In addition, ruminants tend to have lower levels of starch-splitting enzymes than do nonruminants. Adult pigs fed lactose may develop diarrhea because of a deficiency of lactase; lactase deficiency is also prevalent in some human population groups and appears to have a genetic basis. Affected individuals are unable to tolerate milk products containing lactose.

Xylose is not transported actively against a concentration gradient and does not inhibit galactose transport appreciably. Yet evidence suggests a common carrier even for such widely different sugars as glucose and xylose.

Xylose feeding of young pigs results in depressed appetite and growth and causes eye cataracts. The mechanism whereby xylose causes these abnormalities in young animals is not known.

In the absence of diseases such as infectious diarrhea or of other pathological conditions affecting absorption, the absorption of soluble carbohydrates often exceeds 90%. Endogenous (fecal metabolic) losses result in a net (apparent) absorbability approximating 80% or more in nonruminant animals. Thus, except in cases of digestive enzyme insufficiency or impaired absorptive function, absorption of soluble carbohydrates is similar for a wide variety of carbohydrate sources.

The rate of digestion of starch in preparation for absorption is affected by many factors, including particle size, nature of the starch (amylase and amylopectin content), interactions of starch with protein and fat, and the presence of antinutrients such as phytate, tannins, saponins, and enzyme inhibitors (Jenkins et al., 1986).

Metabolic Conversions

Monosaccharides not converted to glucose in the intestinal mucosal cell during absorption may be converted to glucose by reactions in the liver. The animal body stores very little energy as carbohydrate, but some glucose is converted to glycogen, which is stored in liver and muscle tissues. Glycogen is a starchlike compound and can be hydrolyzed rapidly back to glucose. Thus, the level of blood glucose is maintained within a rather narrow range in normal animals by conversion of circulating blood glucose to glycogen (glycogenesis) and by reconversion to glucose by the process of glycogenolysis when the blood level declines. The blood glucose concentration increases after a meal but returns to the fasting level within a few hours. This homeostasis is under endocrine control, with insulin and glucagon from the pancreas playing an important role in maintaining the blood glucose concentration within normal limits for the species. No glucose appears in the urine of normal animals given a glucose load because the kidney functions to retain it in the body. Animals with kidney damage or diabetes, a metabolic disease associated with faulty glucose metabolism, lose glucose in the urine. Storage of glycogen after a carbohydrate meal prevents marked elevation in blood sugar (hyperglycemia), and release of glucose by breakdown of glycogen during fasting prevents low blood glucose (hypoglycemia).

The formation of glycogen from glucose (glycogenesis) requires two molecules of adenosine triphosphate (ATP) for every molecule of glucose. Glucose molecules are added, one unit at a time, to form the long chain of glycogen. Uridine triphosphate (UTP) also is involved in the conversion of glucose to glycogen. ATP is a high-energy compound supplied to the cell mainly by biological oxidations, involving the transfer of electrons from specific substrates to oxygen. Free energy produced with this transfer is captured in the form of the energy-rich ATP. Phosphorus uptake by adenosine diphosphate (ADP) to form ATP is called oxidative phosphorylation. The process is a driving force for many biochemical processes, including absorption of nutrients from the GI tract and synthesis of proteins, nucleic acids, fats, and carbohydrates. ATP-dependent reactions include those in which ATP provides the energy for enzyme-catalyzed processes such as nutrient synthesis (for example, glycogen from glucose), and those in which part of the ATP molecule (a high-energy phosphate) is transferred to an appropriate receptor such as glucose in the reaction:

$$\text{Glucose} + \text{ATP} \xrightarrow{\text{glucokinase}} \text{glucose-6-PO}_4 + \text{ADP}$$

The breakdown of glycogen to form glucose is, in essence, the reverse process (glycogenolysis).

Gluconeogenesis

Glucose can be formed by body tissues (primarily liver and, to a lesser extent, kidneys) from noncarbohydrate metabolites, including lipids and amino acids. This process is called **gluconeogenesis.** All of the nonessential amino acids along with several of the essential ones (arginine, methionine, cystine, histidine, threonine, tryptophan, and valine) are glucogenic. That is, when metabolized they can give rise to a net increase in glucose. Some (isoleucine, lysine, phenylalanine, and tyrosine) are both glucogenic and ke-

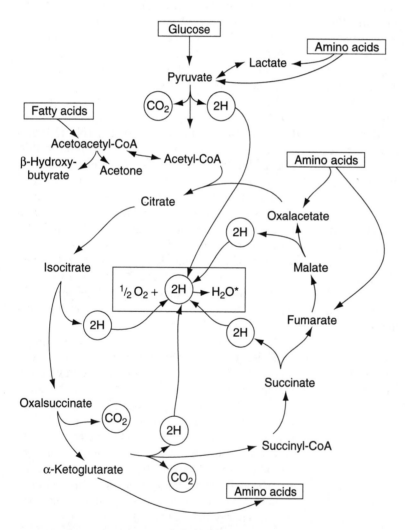

*Heat Production = 57,000 calories per molecule of water formed (represents wasted energy and is the amount of energy that must be ingested and absorbed for the animal to stay in energy balance).

Figure 7.4   *The citric acid (Krebs) cycle through which carbohydrates are oxidized to release energy.*

togenic; they can give rise to glucose and acetone or other ketones. Leucine is strictly ketogenic.

The amino acids used for gluconeogenesis or for energy enter the citric acid cycle (Fig. 7.4) as acetate, pyruvate, or α-ketoglutarate as listed in Table 7.4. Amino acids not used for protein synthesis enter the general pool of metabolites that provide energy for normal body maintenance and productive functions.

Important differences exist between ruminants and nonruminants in the importance of gluconeogenesis in energy balance. Ruminants, through fermentation of their digesta by microbes residing in the rumen, constantly depend on gluconeogenesis to meet their glucose needs. Ruminants depend on acetate instead of glucose as the major product of digestion, whereas nonruminants absorb glucose directly and do not need to rely on gluconeogenesis except in long-term meal deprivation. Lactate and pyruvate are also important sources of C atoms for glucose synthesis by gluconeogenesis. During anaerobic glycolysis in skeletal muscle, pyruvate is reduced to lactate by lactate dehydrogenase. The lactate produced is transported via the blood to the liver where it is converted to glucose.

Glycogen storage is limited. Therefore, when ingestion of carbohydrate exceeds current needs for glycogen formation, glucose is converted to fat. This conversion is accomplished by the breakdown of glucose to pyruvate, which then is available for fat synthesis (see Chapter 8).

The conversion of glycogen to glucose-6-phosphate and finally to pyruvate and to lactate under anaerobic conditions in muscle (glycolysis) occurs through a series of transformations (Fig. 7.5),

which depicts the interrelationships between glucose and lipid metabolism.

A fraction of the glucose-6-phosphate produced from glycogenolysis can enter the hexose monophosphate pathway (pentose shunt) under the influence of the coenzyme nicotine adenine dinucleotide phosphate (NADP), ultimately to form ribose-5-phosphate, a source of ribose for nucleic acids (DNA, RNA) and nucleotide coenzymes (adenosine tri-phosphate, ATP), and to form hexose and triose phosphates. However, most of the glucose-6-phosphate metabolism is by the glycolytic (or Embden-Meyerhof) pathway. Here, glucose-6-phosphate is rearranged to form fructose-6-phosphate and phosphorylated by ATP to fructose-1,6-diphosphate, which is, in turn, cleaved to glyceraldehyde-3-phosphate and dihydroxyacetone phosphate. Glyceraldehyde-3-phosphate is used for glyceride synthesis and undergoes a series of transformations ultimately to form pyruvate. Pyruvate is an important metabolite in glucose and lipid metabolism because it can be converted to acetyl-CoA in the mitochondria or it can be reduced to lactate in the cytoplasm by oxidation of NADH (reduced nicotine adenine dinucleotide). Mitochondrial pyruvate metabolism includes removal of one carbon (decarboxylation) by thiamin decarboxylase and the ultimate formation of acetyl-CoA. Further metabolism of acetyl-CoA occurs in the citric acid (Krebs) cycle (Fig. 7.4), the final common energy pathway for carbohydrate, fat, and carbon skeletons of amino acids. Lactate accumulates in the cytoplasm by oxidation of NADH to NAD and is related to the acidity that causes muscle fatigue following strenuous exercise. These two important metabolic cycles, the Krebs citric acid cycle and the Embden–Meyerhof glycolytic pathway, form

TABLE 7.4 Metabolism of the carbon chain of amino acids.

CITRIC ACID CYCLE			NICOTINIC ACID AND SEROTONIN
ACETATE	PYRUVATE	α-KETOGLUTARATE	
Isoleucine	Alanine	Arginine	Tryptophan
Leucine	Cystine	Glutamic acid	
Phenylalanine	Glycine	Hydroxyproline	
Treonine	Methionine	Histidine	
Tyrosine	Serine	Lysine	
Valine		Proline	

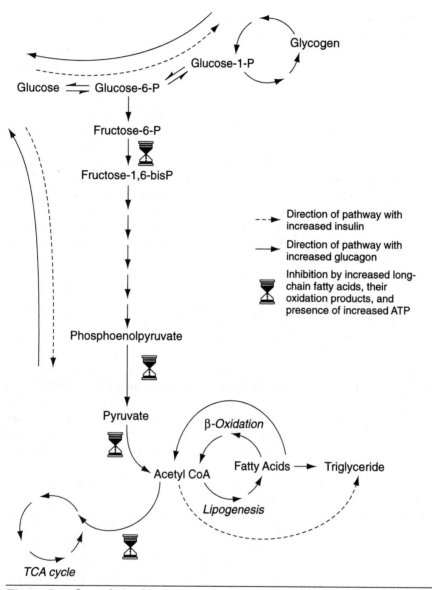

Figure 7.5 *Interrelationships between glucose and lipid metabolism. From Zeisel (1993). A more detailed account showing the pentose shunt is diagrammed in Figure 8.3 in Chapter 8.*

the basis for understanding energy metabolism in animals. The importance of vitamins, including thiamin, riboflavin, niacin, and pantothenic acid, as cofactors in catalyzing these reactions is discussed in more detail in Chapter 15. The complex, yet systematic, interplay of the nutrients in controlling and coordinating the living animal through cellular and subcellular activities is an exciting and fascinating picture, only the generalities of which are within the realm of discussion for this book.

Energetics of Glucose Catabolism

The total energy released in the conversion of glucose to CO_2 and H_2O is 673 Kcal/mole. This can be illustrated as follows:

$$C_6H_{12}O_6 \rightarrow 6CO_2 + 6H_2O + 673 \text{ Kcal}$$

The molecular weight of glucose is 180.2. Thus, the gross energy value of glucose is 673/180.2 = 3.74 Kcal/g. In the oxidation of metabolites via the citric acid cycle (Fig. 7.2), the 57 Kcal/mole of water formed (total of $6 \times 57 = 342$ Kcal) represents heat production and is wasted energy, equivalent to the amount of energy that must be ingested and absorbed for the animal to stay in energy balance.

Catabolism of 1 mole of glucose by the glycolytic pathway is associated with the following amounts of adenosine triphosphate (ATP) trapped at each stage of oxidation to CO_2 and H_2O:

Glycolytic pathway	8 mol of ATP
2 pyruvate to 2 acetyl CoA	6 mol of ATP
2 acetate to CO_2 and H_2O	24 mol of ATP
Total	38 mol of ATP

ATP serves as a major form of high-energy phosphate bonds. One mole of ATP has a value of about 8 Kcal/mole. That is,

$$ATP \rightarrow ADP + 8 \text{ Kcal/mole}$$

The conversion of the free energy of the oxidation of glucose has an efficiency of 40 to 65%, depending on the assumptions and calculations made.

❑ ABNORMAL CARBOHYDRATE METABOLISM

Of the many problems in abnormal metabolism that occur in animals, diabetes and ketosis are concerned primarily with faulty carbohydrate metabolism. Although diabetes does occur in domestic animals, adequate information is not at hand to quantify its overall importance. In humans it is an important disease, affecting people of all ages. Ketosis, on the other hand, appears to be more of a problem with domestic animals.

Ketosis

This syndrome involves an excess of ketones (acetone, acetoacetate, and β-hydroxybutyrate) accumulating in body tissues because of a disorder in carbohydrate or lipid metabolism. Increased concentration in blood is termed ketonemia or acetonemia; if levels are high enough to spill into the urine, the condition is called ketonuria. The disease is common in cattle at the peak of lactation and in sheep in late pregnancy. It is characterized by hypoglycemia, depleted liver glycogen, elevated mobilization of adipose tissue lipids, increased production of ketones, and lipemia. These

changes in energy metabolism resemble diabetes mellitus of humans and animals. The impaired utilization of energy results in increased breakdown of tissue proteins for energy, loss of body weight, decreased milk production in lactating animals, and abortion in pregnant animals. Water consumption is increased because of excessive loss of body fluids in the urine in response to ketonuria. The ketones in urine are accompanied by excessive losses of electrolytes (K and Na ions); this triggers tissue dehydration and induces increased water intake. In ruminants, propionic acid is the major volatile fatty acid (VFA) used for glucogenesis. The only pathway for fatty acids to be converted to glucose, except for propionic, is through acetyl-CoA and the tricarboxylic acid cycle, and there is no net increase in glucose synthesis by these reactions.

Because limited glucose is absorbed from the GI tract in ruminants, liver synthesis of glucose is a major source for maintenance of blood glucose and tissue glycogen levels. Thus, during periods of great physiological demand for glucose, such as in lactation or pregnancy, ketosis becomes a serious practical problem in ruminants. Metabolic vitamin B_{12} deficiency has been implicated in ketosis in dairy cattle; this possible relationship needs further study. There is also evidence that supplementary niacin may prevent or alleviate ketosis in cows. In pregnant ewes, ketosis is partly precipitated by reduced feed intake caused by reduced stomach capacity as a result of increased uterine size, especially when multiple fetuses are present.

Ketosis also occurs in swine and other nonruminants during starvation or chronic underfeeding, often at the onset of a sudden new energy demand such as at the beginning of lactation. Treatment of ketosis generally has centered on the restoration of normal blood glucose concentration. Thus, intravenous glucose and oral propylene glycol are common. Hormones, such as adenocorticotropic hormone (ACTH) and adrenal corticoid hormones, also have been used, but the mechanism of their action is outside the realm of this discussion.

❑ DIABETES MELLITUS

Diabetes mellitus is relatively common in humans but apparently occurs less frequently in other animals. The disease is diagnosed partly on the basis of higher than normal fasting levels of blood glucose. The disease is not a single entity; instead, it is a spectrum of clinical disorders of which excessive blood glucose (hyperglycemia) is a common denominator (Arky, 1983). The three basic types are insulin-dependent diabetes mellitus (IDDM), also called juvenile onset or Type I diabetes; non-insulin-dependent diabetes (NIDDM), also called adult onset or Type II diabetes; and gestational diabetes. The common metabolic defect in each of these three forms is improper production and/or utilization of the hormone insulin, which is produced by the pancreas. Insulin acts on cells throughout the body to promote glucose utilization; it also increases protein synthesis, decreases protein catabolism, increases the entry of amino acids into the cell, stimulates formation of triglycerides (lipogenesis), and inhibits triglyceride breakdown (lipolysis). All of these metabolic actions of insulin reflect its influence on cell membrane permeability. This results in increased entry of glucose, amino acids, and fatty acids into many tissues, including, particularly, muscle and adipose tissues.

Gestational diabetes is induced by increased tissue resistance to the action of insulin and involves an increase in blood glucose and blood ketones. Gestational diabetes is probably the most common form of the disease in domestic animals.

Because insulin is required for normal glucose utilization, its absence or cellular resistance to it results in hyperglycemia, urinary loss of glucose (glucosuria), and other changes as described for ketosis. Humans and animals with a tendency toward diabetes show an impaired glucose tolerance (an abnormally long time for clearance of oral or injected doses of glucose from the blood). Glucose tolerance curves in the normal, diabetic, and hyperinsulin human are shown in Fig. 7.6. A glucose tolerance test indicates not only the ability of the pancreas to secrete insulin, but also the ability of the liver to utilize glucose. Mild diabetes mellitus can be partially controlled by feeding low-carbohydrate, high-protein diets.

Diabetes can be induced artificially by administration of alloxan, a drug that selectively destroys pancreatic cells that produce insulin. Protein and amino acid metabolism as well as carbohydrate and lipid metabolism are affected by insulin deficiency. Insulin deficiency results in severe negative N balance, glucose loss in urine (glycosuria),

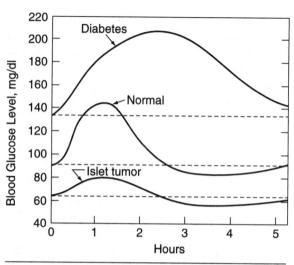

Figure 7.6 *Glucose tolerance curve in the normal person, in a diabetic person, and in a person with an islet tumor (hyperinsulinism). From Guyton (1968).*

excessive urine volume (polyuria), and electrolyte loss associated with hyperglycemia and ketonemia.

In any discussion of carbohydrate metabolism it is important to recognize that other hormones produced by the pancreas are also important. Glucagon, a polypeptide hormone produced by the alpha cells of the pancreas and in some cells of the stomach and duodenum, is a catabolic hormone that decreases carbohydrate and lipid energy stores and increases the amount of glucose and fatty acids available for oxidation. It stimulates secretion of insulin. Somatostatin is a peptide secreted by pancreatic delta cells, upper GI tract, and hypothalamus. Its endocrine effects include inhibition of secretion of insulin, glucagon, somatotropin (growth hormone), and several other hormones. It also inhibits gastric secretion, gastric emptying, gallbladder contraction, and pancreatic enzyme release and has a number of other actions outside the realm of our consideration.

Insulinlike Growth Factors

Before we leave the subject of the role of insulin in diabetes, we introduce briefly the concepts of its relationship to insulinlike growth factors present in blood serum and other body tissues. Insulinlike growth factors I and II (IGF-I and IGF-II, respectively), are small polypeptides that control the growth of several types of cells in animals and humans (Clemmons and Underwood, 1991). They have a structure very similar to that of insulin. Human IGF-I and IGF-II have been isolated, and their amino acid sequence has been determined (Rinderknecht and Humble, 1978). Porcine and human IGF-I are identical in structure. Bovine IGF-I and -II also have been isolated and characterized (Honegger and Humble, 1986). Several IGF-I and IGF-II binding proteins have been described; the role of these binding proteins in controlling the metabolic activity of IGF-I and -II is not completely understood. The exact role of IGF-I, IGF-II, and other peptide growth factors in animal growth and metabolism and how they interrelate with insulin, growth hormone, and other hormones in controlling cellular metabolism is unknown.

Details of this rapidly growing field of biology are beyond the scope of this discussion, but it seems certain that unfolding knowledge of growth will reveal important links between insulin, insulinlike growth factors, and carbohydrate, protein, and energy metabolism.

❑ PLANT FIBERS AS ENERGY SOURCES

Except in the case of certain oil-bearing seeds such as soybeans, the energy from most plants is available largely as carbohydrate, with only a small fraction of the calories provided as fat. The carbohydrates are present as mono-, di-, and polysaccharides, including starch, and a mixture of other complex carbohydrates, including cellulose and lignin, which resist hydrolysis by digestive enzymes elaborated by the host animal. In the ruminant animal, the large microbial population residing in the rumen contains species that elaborate enzymes capable of hydrolyzing cellulose present in plant cell walls, particularly those associated with the leaf and stem portion of the plant and the outer bran layer of seeds. Nonruminant animals, including the horse, pig, and human, are able to utilize some of these fibrous feeds by virtue of the presence of similar microbial populations residing in the colon and cecum. Symbiotic relationships between the host animal and the microbes residing in the gastrointestinal tract have been summarized (Varel, 2005). Van Soest (1989) classified dietary fibers according to type and source. Cellulose, hemicellulose (polymers of hexoses and pentoses, more digestible than cellulose), and lignin (a highly insoluble and biologically unavailable mixture of polymers of phenolic acids) are present in the outer bran layers of cereal grain seeds, in the stems of vegetative portions of grasses and legumes, and in the woody structure of trees and shrubs. In trees, lignin is the chief structural constituent. Termites are a notable example of organisms that elaborate enzymes capable of breaking down lignin. The structural components of plants also contain proteins (extensin), which are generally of low bioavailability owing to their inaccessibility to proteolytic enzymes. Other plant constituents that are often considered as plant fibers are the hemicelluloses; gums, including pectins of fruits and vegetables; and beta-glucans of cereal grains, particularly certain cultivars of barley and oats. They are present as cellular membranes and storage forms of energy in the plant. Tannins and tannin–protein complexes are present in the seed coats of some plants, notably in dark seeded cultivars of grain sorghum. Tannins are unpalatable to animals and birds, and their presence probably represents an adaptive protective mechanism for plants.

Van Soest (1967) described a method for partitioning plant tissues into two fractions based on nutritional availability: cell wall constituents and cell contents. Cell walls contain large amounts of lignin and cellulose, whereas cell contents consist of simple sugars and starches and are easily hydrolyzed to glucose by the enzymes secreted by the host. Fibrous portions of plants (stems, leaves, seed coats) are high in cellulose. Therefore, their energy is unlocked by anaerobic microbial fermentation in the rumen or cecum–colon to yield volatile fatty acids (VFA). Other measurements of fibrous material following enzymatic removal of starch can be used (Englyst and Cummings, 1984) along with the acid and neutral detergent methods of Van Soest (1963) and Van Soest and Wine (1967) to characterize energy sources. The Van Soest neutral detergent fiber (NDF) system of feed and forage evaluation has become standard as an index of dietary energy value of feedstuffs for animals consuming diets high in fiber.

Ruminants consuming all-forage diets obtain most of their energy as VFA produced by anaerobic microbial fermentation of fiber in the rumen. VFAs are readily absorbed from the rumen; the amounts absorbed from the large intestine of nonruminant are less certain, but may be as much as 30% of the digestible energy intake in swine (Rerat et al., 1985). A cellulolytic bacterium has been isolated from the large intestine of pigs and may offer promise as a means of increasing plant fiber utilization in swine (Varel and Pond, 1992).

Recombinant DNA (gene-splicing) techniques are being applied in carbohydrate nutrition by cloning cellulase genes from microorganisms. Changing the microbial activities in the gastrointestinal tract of animals may be applied in this way to enhance the breakdown of the holocellulose–lignin bond to increase the utilization of cellulose. Cloning procedures could also provide the basis for production in monoculture of superior cellulolytic microbes for degradation of fibrous feedstuffs. The extent to which such biotechnology will be applied in animal feeding will ultimately depend on economic incentive for developmental effort as well as on constraints imposed by the biological limits of the host animal and its gastrointestinal microbial ecosystem.

❏ CARBOHYDRATE COMPOSITION OF FOOD AND FEED

Detailed carbohydrate composition of a few common animal feeds and human foods is shown in Table 7.5. This table provides the basis for an appreciation of the large variation that exists among feed and food sources in carbohydrate composition. More detailed information on the carbohydrate composition of many more feeds and foods may be found in other books and food composition tables.

❏ SUMMARY

- Carbohydrates are the major component of plant tissues.

- Plants synthesize carbohydrate (glucose) using solar energy, carbon dioxide, and water, and they give off oxygen. This process, photosynthesis, is vital for animals; they could not exist without this transformation of energy and the free oxygen released.

- Carbohydrates contain C, H, and O in a ratio of approximately 1:2:1.

- The simplest carbohydrate is a monosaccharide such as glucose, which contains 6 C atoms.

- A disaccharide has two molecules of a simple sugar (for example, common table sugar is su-

crose, composed of glucose and fructose). An oligosaccharide may have 3 to 10 sugar units, and a polysaccharide over 10 sugar units.

- Animal tissues contain only trace amounts of carbohydrates, present as glucose and its polymer, glycogen, the storage form of carbohydrate energy.

- Plant tissues contain mono-, di-, and polysaccharides. Polysaccharides include starches, which are hydrolyzed by digestive enzymes of animals, and cellulose, which is not. Cellulose is a major constituent of plant cell walls, and together with hemicellulose, is often present in combination with lignin, a biologically unavailable mixture of polymers of phenolic acid.

- Lignification increases with plant age; it is a major structural component of mature plants and of trees. Vegetables and cereals are low, grasses are intermediate, and legumes high in lignin. The amount of lignin present affects cellulose and hemicellulose bioavailability for microbes inhabiting the animal digestive tract and in this way affects the nutritive value of plant materials.

- Other plant fibers include gums, beta glucans, and glycans.

- The Van Soest neutral detergent fibe methods are standard procedures for predicting the digestible energy content of feeds high in plant cell wall constituents.

- Fermentation of starch and cellulose by microbes in the rumen and in the lower digesetive tract of nonruminants produces large quantities of volatile fatty acids (acetic, propionic, and butyric), which are used to provide much of the total energy supply to the host animal.

- Soluble carbohydrates ingested by nonruminants are hydrolyzed in the small intestine and absorbed by active transport as monosaccharides.

- Glucose can be formed in body tissues from lipids and amino acids (gluconeogenesis); absorbed glucose and that resulting from gluconeogenesis, can be stored in limited amounts as liver and muscle glycogen or converted to body fat by breakdown first to pyruvate.

- The complex cascade of reactions by which glucose and its metabolites are oxidized to provide energy is called the Krebs (citric acid) cycle. Glucose catabolism to pyruvate and lactate for entrance into the citric acid cycle proceeds through the pentose shunt and glycolytic pathways.

TABLE 7.5 Carbohydrates in various foods and feeds, % as fed.[a,b]

FOOD OR FEED	MONOSACCHARIDES		DISACCHARIDES		POLYSACCHARIDES			LIGNIN	FIBER[c]
	GLUCOSE	FRUCTOSE	SUCROSE	LACTOSE	STARCH	DEXTRANS	CELLULOSE		
Alfalfa hay, early bloom	—[a]	—	—	—	—	—	24	6	19
Apple	1.7	5.0	3.1	0	0.6	0	0.4	—	1.8
Beans, navy	0	0	0	0	35.2	3.7	3.1	3	—
Barley, grain	—	—	—	—	—	—	—	—	5
Bermuda grass hay	—	—	—	0	—	—	25	8–12	25–30
Corn, sweet, fresh	0.5	0	0.3	0	14.5	0.1	0.6	—	—
Corn grain	—	—	—	0	62	—	—	—	2
Corn silage, dough stage	—	—	—	0	—	—	10	—	7
Corn silage, mature	—	0	0	0	—	—	26	—	7
Milk, cow, whole	0	0	0	4.9	0	0	0	0	0
Molasses	8.8	8.0	53.6	0	0	0	0	0	0
Oat, mill byproduct	—	—	—	0	—	—	—	13.5	54.5
Oats, grain	—	—	—	0	—	—	—	10–12	60–65
Oat hay, early bloom	—	—	—	0	—	—	23	24 (cell walls 49%)	43
Peas									
immature	0	0	5.5	0	4.1	0	1.1	—	—
mature, dry	0	0	6.7	0	4.1	0	5.0	—	—
Potatoes (white)	0.1	0.1	0.1	—	17.0	—	0.4	0.6	19
Rice, with hull	—	—	—	—	—	—	—	0.4	66
Rice, polished	2.0	—	0.4	—	72.9	0.9	0.3	0.4	80
Wheat flour	—	—	0.2	—	68.8	5.5	—	2–3	65–70
Wheat bran	—	—	—	—	—	—	—	10–11	50–55
Sunflower seeds, w/hull	—	—	—	—	—	—	—	29	19
Sunflower seeds, w/o hull	—	—	—	—	—	—	—	10–13	25
Timothy hay, early bloom	—	—	—	—	—	—	26–34	30–36 (cell walls, 66%)	40–43

[a]The values listed in this table represent only a small fraction of the total information available on carbohydrate composition of individual feedstuffs. They are presented to illustrate the wide variation that exists in carbohydrate composition of plants.

[b]Adapted from Hardinge et al. (1965) and National Research Council (1975).

[c]The analytical procedure for describing fiber components in terms of neutral detergent fiber (NDF) (cell walls) and acid detergent fiber (ADF) eventually will provide more meaningful information on biologically available energy of fibrous feedstuffs. The NDF, ADF, and lignin content of some unusual feedstuffs (plant residues, wood products, manures) has been reported by Van Soest and Robertson (1976).

[d]No value indicated.

- The total energy released in the conversion of glucose to carbon dioxide is 673 Kcal/mol (molecular weight of glucose is 180.2; therefore, the gross energy value of glucose is 673/180.2 = 3.74 Kcal/g).

- Abnormal carbohydrate metabolism in animals is associated with diabetes and ketosis.

- Diabetes is associated with elevated blood glucose resulting from an insufficiency of insulin from the pancreas or defective cellular sensitivity to insulin.

- Ketosis is the presence of excess ketones (acetone, acetoacetate, and B-hydroxybutyrate) in blood and body tissues. It is common in cattle at the peak of lactation and in sheep in late pregnancy, which represent periods of great physiological demand for glucose. Treatment of ketosis focuses on restoration of normal blood glucose concentration.

❏ REFERENCES

Aherne, F. X., et al. 1969. *J. Amin. Sci.* **29**: 444.

Arky, R. A. 1983. Prevention and therapy of diabetes mellitus. *Nutr. Rev.* **41**: 165.

Brambila, S., and F. W. Hill. 1966. *J. Nutr.* **88**: 84; 1967, **91**: 261.

Clemmons, D. R., and L. E. Underwood. 1991. *Annu. Rev. Nutr.* **11**: 393.

Dadd, R. H. 1963. *Adv. Insect Physiol.* **1**: 447, Academic Press, New York.

Devaskar, S. U., and M. M. Mueckler. 1992. *Pediatr. Res.* **31**: 1–13.

Englyst, H. N., and J. H. Cummings. 1984. *Analyst* **109**: 937.

Guyton, A. C. 1968. *Textbook of Medical Physiology* W. B. Saunders Co., Philadelphia, PA.

Hardinge, M. G., J. G. Swarnes, and H. Crooks. 1965. *J. Am. Dietet. Assoc.* **46**: 198.

Jenkins, D.J.A., A. Jenkings, T.M.S. Wolever, L. K. Thompson, and A. K. Rao. 1986. *Nutr. Rev.* **44**: 44.

Koski, K. G., and F. W. Hill. 1990. *J. Nutr.* **120**: 1016.

Koski, K. G., et al. 1990. *J. Nutr.* **120**: 1028.

Montencourt, B. S. 1983. *Biotechnology* **1**: 166.

National Research Council. 1975. *Atlas of Nutritional Data on U.S. and Canadian Feeds.* NAS-NRC and Agriculture Canada, Washington, DC.

Natural Toxins Research Center, Texas A&M University. 2003. Carbohydrate Structure and Function. 2003. www.nutr.tamuk.edu/cell/carbohydrates.html

Prior, R. L., and S. B. Smith. 1983. *J. Nutr.* **113**: 1016.

Rerat, A., M. Fiszlewies, P. Herpen, P. Vaugelade, and M. Durand. 1985. *C.R. Acad. Sci. Paris Ser. III* **300**: 467.

Rinderknecht, E., and R. E. Humbel. 1978. *J. Biol. Chem.* **253**: 2769.

Sasisekharan, R., and J. R. Myette. 2003. *Amer. Sci.* **91** (5): 432.

Van Soest, P. J. 1963. *J. Assoc. Off. Agric. Chem.* **46**: 829.

Van Soest, P. J. 1967. *J. Anim. Sci.* **26**: 119.

Van Soest, P. J. 1989. In *Proceedings Lectureships in Animal Science, 1986 through 1989,* pp. 3–67. University of Tennessee, Knoxville, TN.

Van Soest, P. J. 1994. *Nutritional Ecology of the Ruminant* (2nd ed.), pp. 1–476. Cornell University Press, Ithaca, NY.

Van Soest, P. J., and J. B. Robertson. 1976. *Proceedings Cornell Nutrition Conference,* p. 102. Cornell University, Ithaca, NY.

Van Soest, P. J., and R. H. Wine. 1967. *J. Assoc. Off. Agric. Chem.* **50**: 50.

Varel, V. H. 2005. GI-tract: animal/microbial symbiosis. In *Encyclopedia of Animal Science,* W. G. Pond and A. W. Bell (eds.). Marcel Dekker, New York (In Press).

Varel, V. H., and W. G. Pond. 1992. *Appl. Environ. Microbiol.* **58**: 1645.

Wilson, T. H. 1962. *Intestinal Absorption.* W. B. Saunders, Philadelphia, PA.

Woods, I. S., and P. Trayhum. 2003. *Brit. J. Nutr.* **89**: 3.

Zeisel, S. H. 1993. *J. Nutritional Biochem.* **4**: 549.

8

Lipids

LIPIDS ARE ORGANIC COMPOUNDS that are relatively insoluble in water but relatively soluble in organic solvents and serve important biochemical and physiological functions in plant and animal tissues. The lipids important in nutrition of humans and animals can be classified as follows:

Simple lipids are esters of fatty acids with various alcohols. Fats and oils and waxes are simple lipids. Fats and oils are esters of fatty acids with glycerol, and waxes are esters of fatty acids with alcohols other than glycerol.

Compound lipids are esters of fatty acids containing nonlipid substances such as phosphorus, carbohydrates, and proteins in addition to an alcohol and fatty acid. They include phospholipids, glycolipids, and lipoproteins. Phospholipids (phosphatides) are fats containing phosphoric acid and N. Glycolipids are fats containing carbohydrate and, often, N, and lipoproteins are lipids bound to proteins in blood and other tissues.

Derived lipids include substances derived from simple or compound lipids by hydrolysis, that is, fatty acids, glycerol, and other alcohols.

Sterols are lipids with complex phenanthrene-type ring structures.

Terpenes are compounds that usually have isoprene-type structures.

Fats and oils quantitatively make up the largest fraction of lipids in most food materials and are characterized by their high-energy value. One gram of a typical fat yields about 9.45 kcal of heat when completely combusted, compared with about 4.1 kcal for a typical carbohydrate.

❏ STRUCTURE

The most important lipid constituents in animal nutrition include fatty acids; glycerol; mono-, di-, and triacylglycerols (also called triglycerides); and phospholipids. Glycolipids, lipoproteins, and sterols are very important in metabolism but are present in the body in much lower amounts than triglycerides, the main storage form of energy in the animal body. Lipoproteins play a key role in transport via the blood of triglycerides, cholesterol, and other lipids from one organ system to another for metabolism and processing. Waxes and terpenes are quantitatively unimportant nutritionally or are poorly utilized.

Fatty Acids

The fatty acids (FA) consist of chains of carbon atoms ranging from 2 to 24 or more in length; there is a carboxyl group on the end of each chain. The general structure is RCOOH, where R is a carbon chain of variable length. Acetic acid, a major product of microbial fermentation of glucose in ruminants, has 2 carbons. Its formula is CH_3COOH.

Myristic acid, a constituent of milk fat, has 14 carbons. Its formula is $CH_3(CH_2)_{12}COOH$.

Acetic acid and myristic acid are saturated. That is, each carbon atom (C) in the chain (except the carboxyl group) has two hydrogen atoms attached to it (three hydrogens at the terminal C). Some fatty acids are unsaturated; one or more pairs of C atoms in their chain are attached by a double bond, and hydrogen (H) has been removed.

Linoleic acid, a constituent of corn oil and other plant oils high in polyunsaturated fatty acids, has 18 carbons and two double bonds. Its formula is:

$$CH_3(CH_2)_4CH = CHCH_2CH = CH(CH_2)_7COOH$$

LINOLEIC ACID

Most fatty acids commonly found in animal tissues are straight chained and contain an even number of carbons. Branched-chain fatty acids and those with an odd number of carbons are more common in microorganisms; however, tissues of ruminant animals, in particular, contain relatively large amounts of these fatty acids as a result of rumen fermentation and the subsequent absorption of these microbially derived acids.

Fatty acids containing double bonds can occur as the cis or the trans isomer, as illustrated in the next column:

$$\begin{array}{l} H-C-(CH_2)_7-CH_3 \\ \ \ \ \ \| \\ H-C-(CH_2)_7-COOH \end{array}$$

OLEIC ACID (cis)

$$\begin{array}{l} CH_3-(CH_2)_7-C-H \\ \qquad\qquad\quad \| \\ \ \ \ \ \ \ H-C-(CH_2)_7-COOH \end{array}$$

ELAIDIC ACID (trans)

Trans-fatty acids in the diet of nonruminant animals increase plasma low-density lipoprotein (LDL) cholesterol and triglycerides and decrease high-density lipoprotein (HDL) cholesterol. They are formed in ruminants as intermediates in hydrogenation of dietary unsaturated fatty acids by H produced during rumen microbial fermentation and by industrial hydrogenation of vegetable and fish oils (Katan and Zock, 1995).

The names, number of carbons, and number of double bonds for fatty acids most common in plant and animal tissues are given in Table 8.1.

Omega-3 and Omega-6 Fatty Acids

Depot fats (adipose tissue) of fish contain several long-chain unsaturated fatty acids not present in significant amounts in other animal fats. Evidence indicates a protective effect of fish oils against atherosclerosis in humans. The unique fatty acids in fish oils that set them apart from other fats containing high amounts of polyunsaturated fatty acids are the so-called omega-3 and omega-6 fatty acids, consisting of linolenic acid (18 carbons, 3 double bonds), eicosapentaenoic acid (20 carbons, 5 double bonds, and docosahexaenoic acid (DHA) (22 carbons, 6 double bonds). These acids contribute to the "fishy odor" of most seafish. It is generally believed that linoleic, linolenic, and arachidonic acids cannot be synthesized by animals, although some work (Babatunde et al., 1967) suggests the pig may synthesize some linoleic and arachidonic. The position of the double bond in the carbon chain is critical to biological activity. Table 8.2 shows the position of the double bonds in each of the common unsaturated fatty acids.

Conjugated Linoleic Acid

Several isomers of linoleic acid exist in animal fats. The structures of two isomers of conjugated linoleic acid (CLA) and linoleic acid (cis-9, cis-12, C18:2) are shown in Figure 8.1. Note that in CLA

TABLE 8.1 Fatty acids most common in plant and animal tissues.

ACID	NO. CARBONS	NO. DOUBLE BONDS	ABBREVIATED DESIGNATION
Butyric (butanoic)	4	0	C4:0
Caproic (hexanoic)	6	0	C6:0
Caprylic (octanoic)	8	0	C8:0
Capric (decanoic)	10	0	C10:0
Lauric (dodecanoic)	12	0	C12:0
Myristic (tetradecanoic)	14	0	C14:0
Palmitic (hexadecanoic)	16	0	C16:0
Palmitoleic (hexadecenoic)	16	1	C16:1
Stearic (octadecanoic)	18	0	C18:0
Oleic (octadecenoic)	18	1	C18:1
Linoleic (octadecadienoic)	18	2	C18:2
Linolenic (octadecadienoic)	18	3	C18:3
Arachidic (eicosanoic)	20	0	C20:0
Arachidonic (eicosatetraenoic)	20	4	C20:4
Lignoceric (tetracosanoic)	24	0	C24:0

the double bonds are conjugated or adjacent to each other and are cis,trans rather than cis,cis as in linoleic acid. CLA isomers have several beneficial biological effects on animals. They have been reported to be protective against cancer, diabetes, atherogenesis, and obesity, as well as to modulate immune function and bone growth (Pariza, 1999; Belury, 2002; Whigham et al., 2000). Biological activity varies among isomers of CLA: cis-9, trans-11 CLA is anticarcinogenic, and trans-10,cis-12 CLA inhibits milk fat synthesis in dairy cows (Baumgardt et al., 2001) and backfat depth in pigs (Ostrowska et al., 1999). Dairy products and red meat

provide about 70% and 25%, respectively, of human dietary sources of CLA (Pariza, 1999).

It was first assumed that CLA in milk fat originated from incomplete biohydrogenation of linoleic acid in the rumen. It was later revealed that cis-9, trans-11 CLA is only a transitory product in biohydrogenation of linoleic acid, while vaccinic acid, VA (trans-11 C18:1), is an intermediate of both linoleic and linolenic acids, and is the major intermediate that accumulates in the rumen and is the substrate used to form CLA in animal tissues. The cis-9, trans-11 CLA in milk fat originates from endogenous synthesis, primarily in the mammary gland, whereas that present in depot fat is synthesized primarily in the adipocyte (Bauman et al., 2003).

The origin of CLA in ruminant fat, factors affecting CLA content of foods, and biological effects of CLA have been reviewed (Bauman and Lock, 2005).

Glycerol

The formula for glycerol is:

$$
\begin{array}{l}
\text{HOCH}_2 \\
|\ \\
\text{HOCH} \\
|\ \\
\text{HOCH}_2
\end{array}
$$

GLYCEROL

TABLE 8.2 Position of double bonds in unsaturated fatty acids.

ACID	POSITION OF DOUBLE BONDS[a]	PRECURSOR
Palmitoleic	9	Palmitic
Oleic	9	Stearic
Linoleic	9, 12	None
Linolenic	9, 12, 15	None
Arachidonic	5, 8, 11, 14	Linoleic

[a]C atoms are numbered from the carboxyl end.

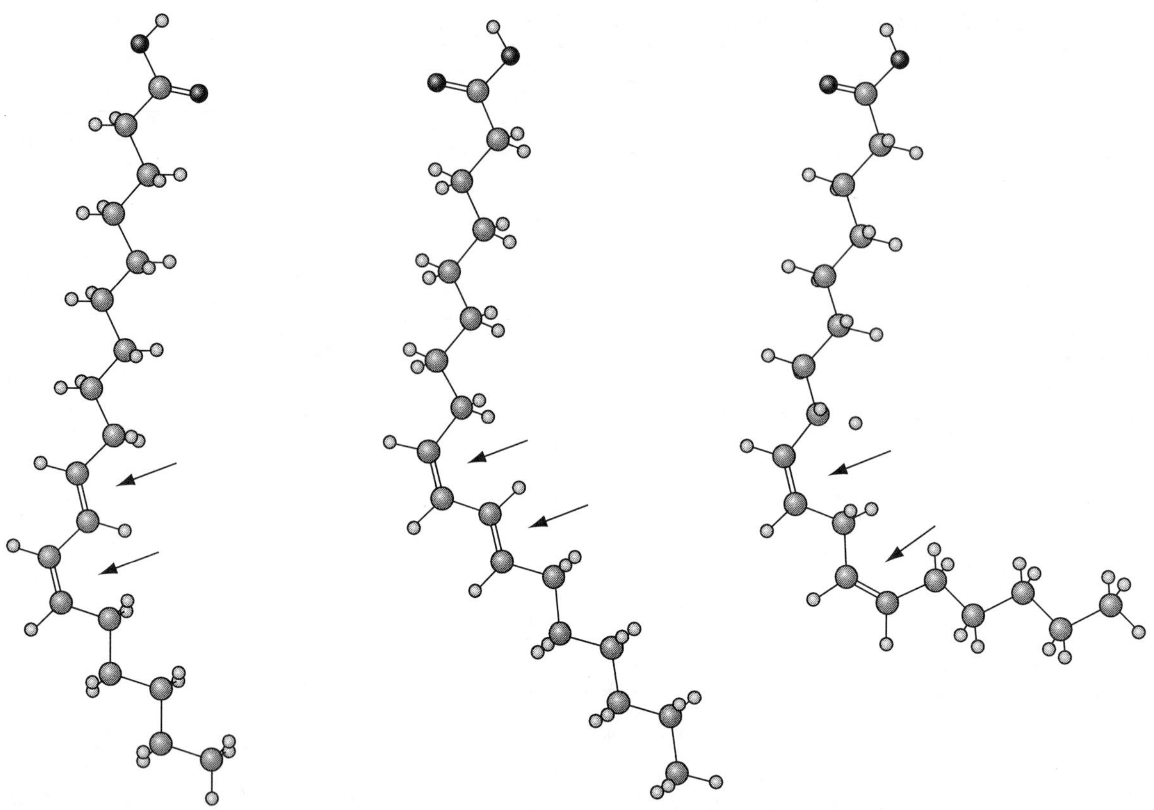

| trans-10, cis-12 CLA | cis-9, trans-11 CLA | linoleic acid |

Figure 8.1   *Structures of two isomers of conjugated linoleic acid (CLA) and linoleic acid (cis-9, cis-12 18:2). Arrows indicate double bond position. Source: Bauman and Lock (2005) with permission by Marcel Dekker, NY.*

Glycerol, the alcohol component of all triglycerides common in animal and plant tissues, is a component of the phosphatides—lecithin, cephalin, and sphingomyelin.

Mono-, Di-, and Triglycerides

Monoglycerides, diglycerides, and triglycerides are esters of glycerol and fatty acids. An ester is formed by reaction of an alcohol with an organic acid; the structure of an ester and the linkage between glycerol and fatty acids in glycerides are as follows:

$$
\underset{R}{R}-\overset{\overset{\displaystyle O}{\|}}{C}-OH + HOR' \rightleftharpoons R-\overset{\overset{\displaystyle O}{\|}}{C}-OR' + H_2O
$$

A monoglyceride, diglyceride, and triglyceride would have the following general structures, where R, R′, and R″ represent three different fatty acids:

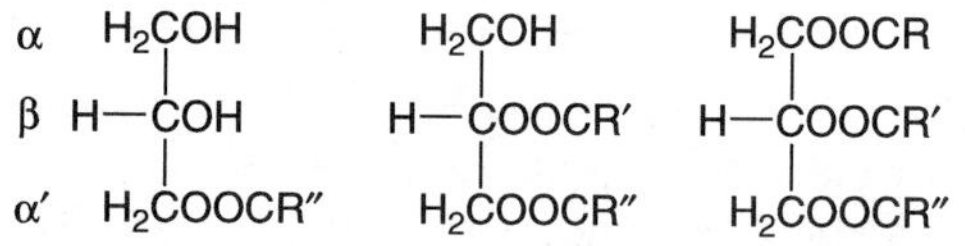

α	H₂COH	H₂COH	H₂COOCR
β	H—COH	H—COOCR′	H—COOCR′
α′	H₂COOCR″	H₂COOCR″	H₂COOCR″

The fatty acid composition of triglycerides (triacylglycerols) is variable. The same or different fatty acids may be in all three positions. For example, if stearic acid occupied all three positions, the compound would be termed tristearin (a simple triglyceride), whereas if butyric, lauric, and palmitic acid each occupied one position, the compound would be called butyrolauropalmitin (glyceryl butyrolauropalmitate), a mixed triglyceride.

The chain length and degree of unsaturation of the individual fatty acids making up the triglyceride (also called triacylglycerol, TAG) determine its physical and chemical properties. Simple triglycerides of saturated fatty acids containing 10 or more carbons are solid at room temperature, whereas those with fewer than 10 carbons usually are liquid. Triglycerides containing only long-chain saturated fatty acids are solids, whereas those containing a preponderance of unsaturated fatty acids are liquids.

Most of the information describing triaclyglycerol composition of fats provides only the overall fatty acid composition of the mixture. This type of data is important, but it does not reveal the position of the individual fatty acids on the glycerol moiety, a factor of significance in the absorption and utilization of fat (see the section on absorption). Many dietary fats have 10 or more different fatty acids. Thus, the number of potential individual triacylglycerols becomes enormous. For example, butterfat has short- and long-chain fatty acids as well as saturated and unsaturated ones, so the number of different triacylglycerols in milk is very large. The same is true of other fats. The major triacylglycerols of common animal and plant fats and their melting points are shown in Table

8.3 to illustrate the wide range in the composition of triglycerides (Small, 1991).

Several constants are commonly used to characterize the chemical properties of fats. Constants of some common fats are given in Table 8.4. Each of these has some application in nutrition.

Saponification number is the number of milligrams of KOH required for the saponification (hydrolysis) of 1 g of fat. The saponification number of a low molecular weight fat (short-chain fatty acids) is large and becomes smaller as the molecular weight of the fat increases. Thus, the saponification number gives a measure of the average chain length of the three fatty acids in the fat.

The **Reichert–Meissl (RM) number** is the number of ml of $0.1N$ KOH solution required to neutralize the volatile water-soluble fatty acids (short chain) obtained by hydrolysis of 5 g of fat. Beef tallow and other high molecular weight fats

TABLE 8.3 Composition and melting point of some common animal and plant triacylglycerols.[a]

FAT	M.P., °C	MAJOR TAG'S[b]		
Butterfat	37–38	PPB	PPC	POP
Lard	46–49	SPO	OPL	OPO
Beef tallow	40	PPO	POP	POS
Cocoa butter	28–36	POS	SOS	POP
Coconut oil	24–27	DDD	CDD	CDM
Palm kernel oil	24–29	DDD	MOD	ODO
Corn oil	−14	LLL	LOL	LLP
Cottonseed oil	5–11	PLL	POL	LLL
Linseed oil	−17	LnLnLn	LnLnL	LnLnO
Olive oil	−7	OOO	OOP	OLO
Palm oil	30–36	POP	POO	POL
Peanut oil	−8–12	OOL	POL	OLL
Rapeseed oil (low-Er)	5	OOO	LOO	OOLn
Rapeseed oil (high-Er)		ErOEr	ErLEr	ErLnEr
Safflower oil	−15	LLL	LLO	LLP
Soybean oil	−14	LLL	LLO	LLP
Sunflower oil	−17	LLL	OLL	LOO

[a]The terms *triacylglycerol* (*TAG*) and *triglyceride* are used interchangeably. Triacylglycerols consist of esters of glycerol and three fatty acids, one fatty acid at each hydroxyl (−OH) group (see structure of glycerol on p. 00). Adapted from Small, 1991, Table 1.

[b]Abbreviations for fattya cids in the triacylglycerols (indicated, left to right in each triplet, as positions 1, 2, 3, respectively, on glycerol skeleton): B, C4:0 (butyric); C, C10:0 (capric); D, C12:0 (lauric); M, C14:0 (myristic); P, C16:1 (palmitic); S, C18:0 (stearic); O, C18:1 (oleic); L, C18:2 (linoleic); Ln, C18:3 (linolenic); Er, C-22:1 (erucic).

TABLE 8.4   Constants of some common fats.[a]

FAT	SAPONIFICATION NO.	REICHERT-MEISSL NO.	IODINE NO.
Beef	196–200	1	35–40
Butter	210–130	17–35	26–38
Coconut	253–262	6–8	6–10
Corn	187–193	4–5	111–128
Cottonseed	194–196	1	103–111
Lard	195–203	1	47–67
Linseed	188–195	1	175–202
Peanut	186–194	1	88–98
Soybean	189–194	0–3	122–134
Sunflower	188–193	0–5	129–136

[a]The constants for animal fats may vary outisde the range of values listed because of unusual composition of dietary fats.

contain practically no volatile acids and therefore have RM numbers of near zero, but butter contains a higher proportion of volatile acids and has a RM number of 17 to 35. The **iodine number** is the number of grams of iodine that can be added to the unsaturated bonds in 100 g of fat. Iodine number is a measure of the degree of hydrogenation (saturation) of the fatty acids in the fat. A completely saturated fat such as tristearin has an iodine number of zero, whereas a liquid fat such as linseed oil has an iodine number of 175 to 202.

Phospholipids (phosphatides)

Phospholipids on hydrolysis yield fatty acids, phosphoric acid, and usually glycerol and a nitrogenous base. The general formula for lecithin is shown. Cephalins are similar to lecithins except that choline is replaced by hydroxyethyl amine in the molecule. Sphingomyelins do not contain glycerol, but contain fatty acids, choline, phosphoric acid, and the nitrogenous base, sphingosine.

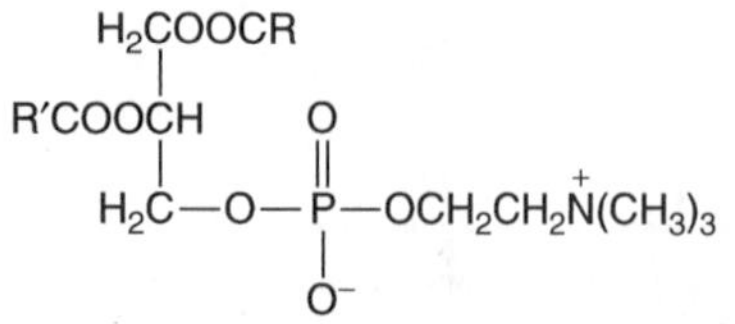

$$H_2COOCR$$
$$R'COOCH \qquad O$$
$$\qquad \qquad \parallel$$
$$H_2C-O-P-OCH_2CH_2\overset{+}{N}(CH_3)_3$$
$$\qquad \qquad \mid$$
$$\qquad \qquad O^-$$

L-α-LECITHIN

These formulas are general representations of each group of compounds. The exact composition varies as to fatty acid composition and in other ways. Phospholipids of animal tissues are higher in unsaturated fatty acids than are the triglycerides of adipose tissue; phospholipids are more widely dispersed in body fluids than are neutral fats and have emulsifying properties that allow them to serve important functions in lipid transport.

Sterols

The most abundant sterol in animal tissue is cholesterol, as follows:

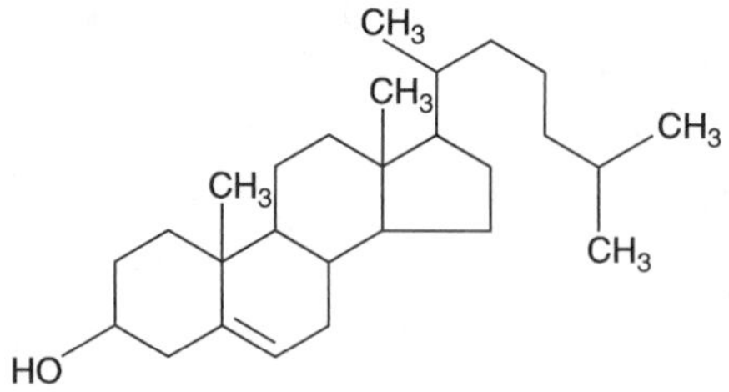

Other important sterols in animals are ergosterol (yields vitamin D_2 when irradiated), 7-dehydrocholesterol (yields vitamin D_3 when irradiated), bile acids, androgens (male sex hormones), and estrogens and progesterones (female sex hormones).

❏ FUNCTIONS

The functions of the lipids can be listed broadly as follows: (1) they supply energy for normal maintenance and productive functions; (2) they serve as a source of essential fatty acids; (3) they serve as a carrier of the fat-soluble vitamins; and (4) they serve as an integral constituent of cell membranes.

Energy Supply

The hydrolysis of triglycerides (triacylglycerols) yields glycerol and fatty acids, which serve as concentrated sources of energy. Most of the variation among fat sources in the amount of utilizable energy they contain is related to their digestibility, but except in abnormal or special conditions of malabsorption, the true digestibility of fats exceeds 80%. When the total lipid content of the diet is low (<10%), as often occurs when animals are fed all-plant diets, apparent digestibility may be much less than this because of the higher proportion of metabolic fecal lipids on a low-fat diet. The true digestibility, however, is not generally affected at moderate levels of fat intake. A high proportion of waxes or sterols in the diet tends to reduce absorbability of the lipid, as these components are usually poorly digested and absorbed.

All of the energy in the diet except that present in essential fatty acids (see next section) may be provided by carbohydrate. Strictly speaking, therefore, no dietary requirement exists for lipids, except for the essential fatty acids (EFA) they contain and their role as a solvent in fat-soluble vitamin absorption. Animals fed fat-free diets may develop fat-soluble vitamin deficiencies.

Essential Fatty Acids (EFA)

Linoleic acid (Cl8:2) and linolenic acid (Cl8:3) apparently cannot be synthesized by animal tissues, or at least not in sufficient amounts to prevent pathological changes, and so must be supplied in the diet. Arachidonic acid (C20:4) can be synthesized from C18:2 and, therefore, is required in the diet only if C18:2 is not available.

The exact mechanisms by which EFA function in maintaining normal body functions are not known, but two probable vital areas are: they are an integral part of the lipid–protein structure of cell membranes, and they appear to play an important part in the structure of several compounds called eicosanoids that play a role in the regulation of release of hypothalamic and pituitary hormones. The eicosanoids include thromboxane, leukotrienes, and prostaglandins (hormonelike compounds widely distributed in re-

productive organs and other tissues of humans and animals). Prostaglandins are biosynthesized from arachidonic acid and have a wide variety of metabolic effects, including the following: lower blood pressure, stimulate smooth muscle contraction, inhibit norepinephrine-induced release of fatty acids from adipose tissue, and modulate immune function. Prostaglandin E-2 has been shown to be immunosuppressive. This could explain the action by which fatty acids may suppress immune function (Hwang, 2000).

Dermatitis and other abnormalities have been traced to deficiencies of certain fatty acids in nonruminant species. Skin lesions develop in pigs fed a fat-free diet. The following effects of a deficiency of EFA have been reported: scaly skin and necrosis of the tail; growth failure; reproductive failure; elevation of the triene:tetraene ratio of tissue fatty acids; and edema, subcutaneous hemorrhage, and poor feathering in chicks.

Holman (1960) suggested that in the rat a ratio of trienoic acids to tetraenoic acids of more than 0.4 in tissue lipids indicates a deficiency of EFA and that a level of linoleic acid at or exceeding 1% of the calories in the diet is sufficient to maintain a ratio of less than 0.4. Babatunde et al. (1967) found increased triene to tetraene ratios in heart, liver, and adipose tissues of pigs fed diets containing no fat or 3% hydrogenated coconut oil compared to values obtained with 3% safflower oil, but found no skin lesions and no reduction in weight gain of pigs fed coconut oil, which aggravates EFA deficiency in the rat. Their work and that of Kass et al. (1975) suggested the possibility that some linoleic acid synthesis occurs in the pig. Young ruminants (calves, kids, lambs) apparently require EFA in their diet, but no reports of EFA deficiency have appeared in adult ruminants. This is somewhat of an anomaly, because rumen microflora hydrogenate most unsaturated fatty acids so that one would expect EFA deficiency to be more likely than in other animals. Apparently, sufficient amounts of unsaturated fatty acids escape hydrogenation in the rumen to provide the animal with needed EFA. Arachidonic acid has been found in high concentration in reproductive tissue of cattle and presumably is synthesized there as a precursor of prostaglandins. Additional studies clearly are needed to determine the degree of importance of linoleic, arachidonic, and other fatty acids such as omega-3 fatty acids in the diet of animals, including humans.

Carrier of the Fat-Soluble Vitamins

Absorption of the fat-soluble vitamins (A, D, E, and K) is a function of digestion and absorption of fats. Fat-soluble vitamins are dispersed in micelles similar or identical to those formed in the absorption of fatty acids. Mixed micelles containing monoglycerides and free fatty acids take up fat-soluble vitamins more efficiently than micelles not containing them. A bile acid sequestrant, cholestyramine, has been shown to reduce absorption of vitamin K when added to the diet, supporting the concept of an obligatory formation of bile salt-containing micelle formation. Because only a low level of dietary fat is needed for micelle formation, frank deficiencies of vitamin A, D, E, or K are unlikely to occur under normal dietary conditions.

❏ ABSORPTION

The digestion of lipids and their preparation for absorption and the absorption process itself have been described (Allen, 1977; Masoro, 1968; Small, 1991).

The upper small intestine is the site of the major processes of preparation for absorption. Dietary lipids, mainly triglycerides, are discharged slowly from the stomach and are mixed with bile and pancreatic and intestinal secretions. Emulsification occurs here through the detergent action of the bile salts and the churning action of the intestine; the lipid particle size is reduced to spheres of 500 to 1000 μm in diameter. This smaller particle size allows for greater surface exposure to pancreatic and intestinal lipases, which adsorb on the particle surface and attack fatty acids (FA) in the 1 and 3 (alpha) positions, resulting in hydrolysis of triglycerides to beta-monoglycerides (fatty acid in the C2 position) and free fatty acids (FFA). The beta-monoglycerides and FFA then combine with salt–phospholipid–cholesterol micelles (in about a 12.5:2.5:1 molar ratio) to form mixed micelles; these are essential for efficient absorption. Bile salts are detergent-like compounds that facilitate digestion and absorption of lipids. The presence of bile is necessary for efficient fat and fat-soluble vitamin absorption; in its absence, cholesterol absorption is reduced to near zero. Pancreatic lipase activity and resynthesis of triglycerides in the intestinal mucosal cells are promoted by bile salts.

The mixed micelles join with cholesterol and fat-soluble vitamins to form larger and more complex mixed micelles, each containing hundreds of molecules and having a diameter of 5 to 10 μm. These mixed micelles form microemulsions that then render the lipid ready for absorption as hydrolysis proceeds.

Bile salts are secreted in copious quantities (30 g/day in adult humans). Bile salts are reabsorbed readily from the GI tract in the lower jejunum and ileum and recycled to the liver. Thus, the amount of daily secretion exceeds by several-fold the amount present in the body at any one time and the amount of daily synthesis by the liver.

Several bile acids are common in animals, differing only in minor changes in the steroid portion of the molecule. Common bile acids are cholic, deoxycholic, taurocholic, and glycocholic acids. In the last two, the amino acids taurine and glycine, respectively, are part of the molecule. The structure of cholic acid is as shown:

CHOLIC ACID

The main site of absorption of lipids is the proximal (upper) jejunum, but some absorption occurs along the intestinal tract from the distal (lower) duodenum to the distal ileum. Glycerol and short-chain FA (C_2 through C_{10}) are absorbed by passive transport into the mesenteric blood and pass to the portal blood system. Monoglycerides and long-chain FA enter the brush border (microvilli) and the apical core of the absorptive intestinal mucosal cells by diffusion. To a limited extent, some triglycerides may be absorbed intact as a fine emulsion of particles averaging 500 Å in diameter.

Most of the phospholipids in the intestinal lumen are hydrolyzed partially by pancreatic and intestinal lipases to yield FFA. The remainder of the molecule (lysophospholipid) is absorbed intact along with a small proportion of unhydrolyzed phospholipid. Although free cholesterol is readily absorbed, most other dietary sterols except vitamin D are absorbed poorly. Cholesterol esters must be hydrolyzed by pancreatic and intestinal lipases to form free cholesterol for absorption. After entering the mucosal cell, free cholesterol is again esterified before transfer to the lymph system via the lacteals.

After entering the epithelial cell, long-chain FA are converted to derivatives of coenzyme A in the presence of ATP. This fatty acid–coenzyme A complex (termed fatty acyl coenzyme A) reacts with monoglyceride within the cell to form di- and then triglycerides. The triglycerides thus formed contain only FA of C_{12} or greater chain length because shorter chain FA are absorbed directly into the portal system.

Before leaving the mucosal cell, the mixed lipid droplets become coated with a thin layer of protein adsorbed to the surface. These protein-coated lipid droplets are called chylomicrons and consist mainly of triglycerides with small quantities of phospholipids, cholesterol esters, and protein. The chylomicrons leave the mucosal cell by reverse pinocytosis and enter the lacteals via the intercellular spaces. Lacteals lead to the lymphatic system, which carries the chylomicrons to the blood via the thoracic duct. A summary of the major conversions that occur in transport of long-chain FA, phospholipids, cholesterol, and monoglycerides by the intestinal mucosal cell is diagrammed in Fig. 8.2.

Although mammals absorb most of these long-chain FA into the lymphatic system, the chicken apparently absorbs its dietary lipids directly into the portal blood, which carries them directly to the liver. Nevertheless, the process of reesterification of FA to triglycerides in the mucosal cell is similar in birds and mammals.

❏ TRANSPORT AND DEPOSITION

Absorption of fat after a meal is associated with a large increase in lipid concentration of the blood, referred to as lipemia. Blood lipids consist of chylomicrons formed within the intestinal mucosal cell during absorption, as well as lipids arising from mobilized depot stores and from synthesis in body tissues, especially the liver and adipose tissues. Blood lipids are transported as lipoproteins ranging from very low density to high density. The density is increased as the proportion of protein in the complex increases and the lipid decreases. Density, composition, and electrophoretic mobility have been used to divide lipoproteins into four major classes: chylomicrons (lowest density; that is, high

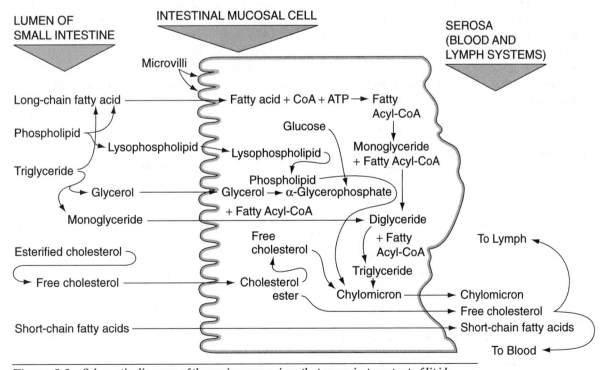

Figure 8.2 *Schematic diagram of the major conversions that occur in transport of lipids across the intestinal mucosal cell during absorption.*

ratio of lipids to protein), very low-density lipoproteins (VLDL), low-density lipoproteins (LDL), and high-density lipoproteins (HDL) (Schaefer and Levy, 1985). The composition of each of these four classes is summarized in Table 8.5.

Chylomicrons are synthesized in the small intestine from dietary fat; VLDL, intermediate-density lipoproteins (IDL), LDL, and HDL are synthesized in the liver and small intestine. Free fatty acids (FFA) are transported as a complex with albumin. Chylomicrons and other lipids are removed very rapidly from the blood by the liver, fat depots, and other tissues. For example, half of an injected dose of ^{14}C-labeled tripalmitin is removed from the blood plasma of dogs within 10 min.

The type and quantity of dietary lipid and the time after a meal are major determinants of the composition and concentration of lipids in blood at a particular time. In addition, such factors as species, age, and endocrine status of the individual have an influence. Levels of total cholesterol and fractions of LDL, HDL, and VLDL cholesterol in the blood serum are affected by diet as well as by endogenous synthesis in liver, intestine, and other tissues. Fasting levels of serum cholesterol are determined by genetic factors related to endogenous cholesterol synthesis and to trafficking of cholesterol among organs and tissues, and its catabolism and excretion. The ratio of free cholesterol to cholesterol esters and free cholesterol to phospholipid is rather constant in normal animals within a given species.

All tissues of the body store triglycerides. Adipose tissues (fat depots) are the most notable stor-age sites. Adipose tissue is capable of synthesizing fat from carbohydrate and of oxidation of fatty acids. Because stored triglycerides are a ready source of energy, continuous deposition and mobilization clearly occur in adipose tissue. Energy intake in excess of current needs results in a net deposition of triglycerides (fattening), and energy intake less than current needs (as in fasting) results in a net loss of triglycerides.

Triglycerides of depot fat tend to have a fatty acid composition characteristic for each animal species. In nonruminants, however, the fatty acid composition of the depot fat resembles that of the diet. This is illustrated in Table 8.6 for the pig fed semipurified diets containing various sources of fat. The depot fat of ruminant animals is less responsive to dietary fatty acid composition because of the action of the ruminant microflora in metabolizing dietary fatty acids, although minor changes can be produced by dietary changes. Depot fat of ruminants can be changed to resemble dietary fat if the action of the rumen microflora is circumvented. Ogilvie et al. (1961) accomplished this result by duodenal feeding of sheep, and others have been able to obtain softer depot fat in cattle and sheep by feeding very high levels of unsaturated FA, a portion of which presumably traverses the rumen without being metabolized by rumen microflora. Other workers showed that fatty acid composition of body fat can be modified through protection of dietary fat from rumen metabolism by coating with formaldehyde-treated casein. The depot fat composition of ruminants also can be changed by altering the proportion of readily

TABLE 8.5   Composition of lipoprotein isolated from normal subjects.

LIPOPROTEIN CLASS[a]	DENSITY RANGE (G/ML)	ELECTRO-PHORETIC MOBILITY	COMPOSITION (WEIGHT %)		CHOLESTEROL		
			PROTEIN	TRIGLYCERIDE	FREE	ESTER	PHOSPHOLIPID
Chylomicrons	<0.94	Origin	1–2	85–95	1–3	2–4	3–6
VLDL (beta lipoprotein)	0.94–1.006	Prebeta	6–10	50–65	4–8	16–22	15–20
LDL (beta lipoprotein)	1.006–1.063	Beta	18–22	4–8	6–8	45–50	18–24
HDL (alpha lipoprotein)	1.063–1.21	Alpha	45–55	2–7	3–5	15–20	26–32

[a]VLDL denotes very low-density lipoprotein, LDL low-density lipoprotein, and HDL high-density lipoprotein. From Schaefer and Levy (1985).

TABLE 8.6 Effect of dietary fatty acid composition on fatty acid composition of the depot fat of growing pigs.[a]

FATTY ACID DESIGNATION	DIETARY FAT		PIG DEPOT FAT	
	SAFFLOWER[b]	HYDROGENATED COCONUT[b]	SAFFLOWER[b]	HYDROGENATED COCONUT[b]
8:0		7.8		
10:0		5.7	0.2	0.2
12:0		44.5	0.2	0.6
14:0	trace	17.2	2.0	3.9
16:0	8.8	9.2	15.6	21.0
16:1			13.0	16.5
18:0	1.5	10.4	50.7	55.9
18:1	9.3	5.0		
18:2	80.4	0.2	17.2	0.9
18:3	trace		0.1	0.1

[a]From Babatunde et al. (1967).
[b]Percent of total fatty acids.

fermentable carbohydrates fed. Ruminant depot fats are also characterized by odd-length and branched-chain fatty acids, which are derived from volatile fatty acids, and by the presence of trans isomers, which result from microbial metabolism of dietary unsaturated fatty acids.

Body lipids clearly are in a dynamic state of metabolism. The turnover rate of triglycerides in adipose tissue is extremely rapid. For example, the half-life in mice is 5 days; in rats, it is 8 days. The turnover rate of phospholipids and cholesterol also is rapid and may vary from 1 day for liver in some species to 200 days for brain in other species.

❏ ONTOGENY OF ADIPOSE TISSUE

In most animals, depot fat (adipose tissue) accumulates steadily from late prenatal life to maturity. Early triacylglycerol (TAG) accretion, like that of lean tissue and bone, is a result of hyperplasia (increase in cell number) and hypertrophy (increase in cell size). Adipose tissue produces endocrine factors, including leptin and insulinlike growth factors (IGFs). Its extensive vascularity (Crandall et al., 1997) supports the concept of an endocrine function for adipose tissue. Detailed structural and functional ontogeny of adipose tissue is beyond the scope of this discussion; it has been reviewed by Hausman and Hausman (2005).

❏ FATTY ACID AND TRIGLYCERIDE (TRIACYLGLYCEROL) METABOLISM

Liver, mammary gland, and adipose tissue are the three major sites of biosynthesis of fatty acids and triglycerides. The liver is the central organ for lipid interconversion and metabolism, and its role can be summarized as follows: synthesis of fatty acids from carbohydrates; synthesis of fatty acids from lipogenic amino acids; synthesis of cholesterol from acetyl-CoA; synthesis of phospholipids; synthesis of lipoproteins; synthesis of ketone bodies (acetone, β-hydroxybutyric and acetoacetic acid); degradation of fatty adds; degradation of phospholipids; removal of phospholipids and cholesterol from blood; lengthening and shortening of fatty acids; saturating and desaturating of fatty acids; control of depot lipid storage; and storage of liver lipids.

Synthesis of fatty acids by liver and adipose tissue follows similar pathways, but the relative contribution made by each tissue differs greatly among the species studied. For example, in the mouse and rat about half of the synthesis occurs in the liver, but in the chicken and pigeon nearly all occurs in the liver; in the pig nearly all occurs in adipose tissue and in the cow and sheep both liver and adipose tissue are important, although the latter predominates.

Factors controlling fat synthesis in ruminant animals are not elucidated fully. A major portion of the energy absorbed from the GI tract for lipid synthesis is in the form of volatile FA (VFA) (acetic, propionic, and butyric acids). Although acetate is the primary substrate for fatty acid synthesis in ruminants, in some cases lactate may be an important substitute. Prior and Scott (1980) ranked relative potency of substrates in inducing lipgeneis in vitro in ruminant adipose tissue as follows: glucose > propionate > lactate > acetate. These in vitro data differ slightly from observations in intact cattle; the cow preserves glucose for other uses.

Fatty Acid Biosynthesis

Synthesis begins with acetyl-CoA (two carbons) derived from carbohydrates, from certain amino acids, or from degraded fats (see tricarboxylic acid cycle, Fig. 7.2). The fatty acid chain is assembled in two-carbon units by joining of the carboxyl head of one fragment to the methyl tail of another. Fatty acids build up from acetic acid units that are made reactive by combining first with coenzyme A to form acetyl-CoA and then with carbon dioxide to form the CoA ester of malonic acid. Malonyl-CoA and acetyl-CoA can condense into an intermediate compound. Further reactions then reduce the intermediate to the CoA ester of a four-carbon fatty acid, butyryl-CoA. This, like acetyl-CoA, can condense with a molecule of malonyl-CoA, ultimately giving the ester of the six-carbon fatty acid; the chain thus lengthens by two carbons at each successive step. Animal tissues synthesize carbon chains up to C_{16} in this way. Malonyl-CoA is, in effect, the source of the C_2 units. It combines with the even-numbered fatty acid–CoA esters.

Fatty acid chain length increases or decreases in two-carbon units under the control of two acetyl coenzyme A carboxylase (ACC) systems. ACC-alpha is the rate-limiting enzyme in the biosynthesis of long-chain fatty acids, while ACC-beta controls fatty acid oxidation (Kim, 1997). Fatty acid and triacylglycerol biosynthesis are tightly controlled by nutritional and hormonal conditions. Feeding fasted animals a high carbohydrate diet causes an induction of fatty acid synthase (FAS) and mitochondrial glycerol-3-phosphate acyltransferase (GPAT), which are enzymes involved in fatty acid and triacylglycerol (TAG) synthesis, respectively. The activity of these two enzymes and others involved in lipid synthesis is regulated by complex coordination among nutrients, including glucose, and hormones, including insulin, glucagon, glucocorticoids, and thyroid hormone (Sul and Wang,

1998; Coleman et al., 2000). Partitioning of fatty acids between degradation and synthetic pathways is regulated short term through modulation of mitochondrial membrane enzymes.

The details of fatty acid biosynthesis are illustrated in Fig. 8.3. The biosynthesis of fatty acids and of triglycerides occurs primarily in the endoplasmic reticulum of cells, but fatty acid oxidation occurs mainly in the mitochondria. Enzymes can use acetyl CoA to add two-carbon units to existing long-chain FA (C_{12}, C_{14}, C_{16}), but formation of FA longer than C_{18} in this way is restricted to conversion of C_{18} to C_{20} and C_{22}.

Desaturation of FA occurs in animal tissues, but the extent is limited, as evidenced by the inability of most animals to synthesize C18:1, C18:3, and C20:4 from saturated FA of the same chain lengths. Desaturation of C18:0 to C18:1 and of C16:0 to C16:1 occurs at the 9,10-position and subsequently moves three carbons toward the carboxyl end of the chain after elongation of the chain by two carbons at the carboxyl end. Animals elongate and desaturate the C_{18} dietary polyunsaturated fatty acids to obtain a range of 20- and 22-carbon highly unsaturated fatty acids. For example, linolenic acid (18:3) provides 20:5, 22:5, and 22:6, whereas linoleic acid (18:2) provides 20:3, 20:4, 22:4, and 22:5 acids. These highly unsaturated fatty acids are esterified into the lipids of cellular membranes, and they influence the rate of synthesis of prostaglandins and other physiologically important eicosanoids of similar composition (e.g., prostacyclin, thromboxane). These eicosanoids regulate the pulsatile release of pituitary hormones and have other effects on the endocrine system. The biosynthesis of prostaglandins and their role in essential fatty acid metabolism and nutrition has been reviewed (Lands, 1991; Hwang, 2000).

Triglyceride Biosynthesis

Synthesis occurs by fatty acid acyl-CoA reacting with alpha-glycerol phosphate to form a phospholipid that is then converted to a diacylglycerol (diglyceride) and thence to a triacylglycerol (triglyceride), or by fatty acyl-CoA reacting with a monoglyceride to form a diglyceride and thence a triglyceride. The details of triglyceride biosynthesis are illustrated in Fig. 8.4.

Triglyceride Catabolism

Adipose tissue, composed primarily of triacylglycerol (triglyceride), is in a dynamic state, with

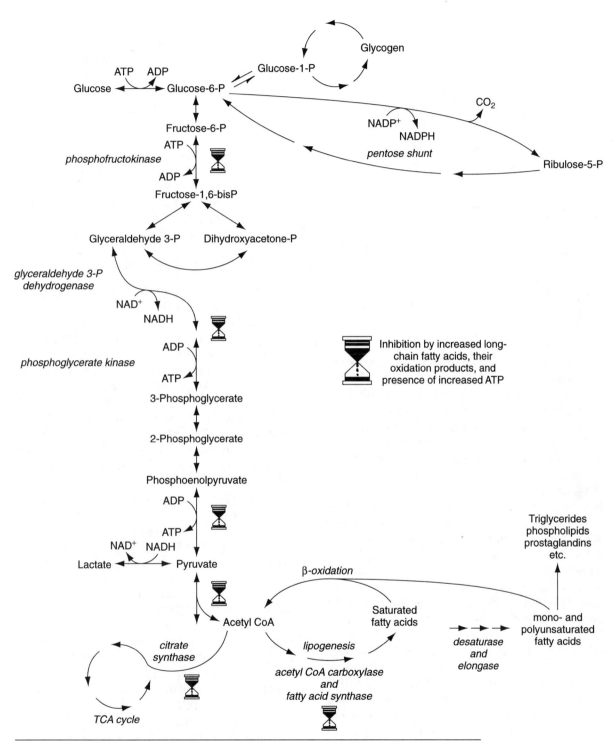

Figure 8.3 *Fatty acid biosynthesis. Note the enzymes involved at various stages. The interrelationships between glucose and lipid metabolism are also shown. From Zeisel (1993) J. Nutritional Biochemistry* **4***: 550 (1993).*

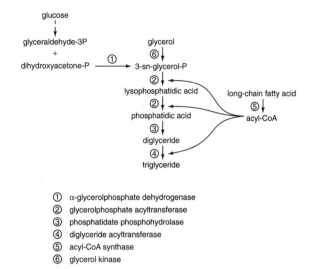

① α-glycerolphosphate dehydrogenase
② glycerolphosphate acyltransferase
③ phosphatidate phosphohydrolase
④ diglyceride acyltransferase
⑤ acyl-CoA synthase
⑥ glycerol kinase

Figure 8.4 *Triglyceride biosynthesis. Note the enzymes involved at various stages. From Mersmann (1986).*

triglycerides undergoing continuous synthesis and degradation (lipolysis). Lipoprotein lipase (LPL) is the major enzyme responsible for hydrolysis of circulating (blood plasma) triglycerides in chylomicrons and very low-density lipoproteins (Bensadoun, 1991). LPL is synthesized in many tissues of the body in most animals. It is secreted to the cell surface and transported to the capillary endothelial surface. The cativity of LPL in the capillary bed of a tissue is reflected in the movement of fatty acids to that tissue. The functional enzyme activity varies with the energy needs of the individual tissues. In animals in the fed state, the activity in adipose tissue is high; in adipose tissue of fasted animals, it is low. The intracellular lipolytic process is under endocrine control. Initiation of the process is controlled by formation of a complex between the hormone and its receptor located on the adipocyte (fat cell) surface, followed by intracellular generation of cyclic adenosine monophosphate (cAMP) and a cascade of other enzymatic steps to yield fatty acids and glycerol. Lipolysis is stimulated by a number of hormones, including epinephrine, and by synthetic beta-adrenergic agonists such as isoproterenol, adrenocorticotropin, and somatotropin (growth hormone). Lipolysis is inhibited in some mammals by receptors on the adipocyte surface called alpha adrenergic receptors. In swine up to market weight (about 5 months of age), no such receptors have been demonstrated (Mersmann, 1986). Inhibition of lipolysis also oc-

curs in the presence of adenosine and in response to prostaglandins and insulin in most species. It appears that each animal species has mechanisms of physiological control of lipolysis that are unique to that species. Measurements of glycerol release from adipose tissue in humans suggest that regional differences related to gender and degree of obesity exist in lipolysis of body fat stores (Jensen, 1997).

It seems clear that physiological factors that control net fat accretion in animals involve complex regulatory systems that include lipogenesis as well as lipolysis. The ultimate fat content of the animal body depends on the net balance of these two opposing processes at the cellular and subcellular level.

Fatty Acid Catabolism

Fatty acids released from hydrolysis of triglycerides are transported via the blood to various body tissues to be used as an oxidative energy source. Oxidation occurs in the mitochondria of skeletal muscle, liver, cardiac muscle, adipose tissue, and other tissues to yield CO_2 and ketones as products. Breakdown of long-chain FA proceeds by stepwise removal of two carbons at a time, beginning at the carboxyl end (β oxidation). The process is not the reverse of synthesis, although acetyl-CoA is the form in which the C_2 fragments are removed. Before oxidation begins, the fatty acid is activated by esterification with CoA to form acyl-CoA. At least three different enzymes are involved in oxidation of FA, one specific for short-chain (C_2 and C_3), one for medium-chain, and one for long-chain (C_{12} to C_{18}) fatty acids. These enzymes (dehydrogenases) remove hydrogens and C_2 units as acetyl-CoA, as illustrated in Fig. 8.5. The acetyl-CoA released in oxidation is available for resynthesis of fatty acids, for synthesis of steroids or ketones, or for entry into the tricarboxylic acid cycle. The total energy produced by complete degradation of long-chain fatty acids comes partly from the beta-oxidation sequence and partly from the oxidation of acetyl-CoA in the tricarboxylic acid (TCA) cycle.

Although even-numbered carbon FA are by far the most prevalent in animal tissues, some odd-numbered carbon FA are present that are oxidized by a similar route but with slight variations.

Steroid Metabolism

Cholesterol is the most abundant steroid in the diet and the precursor of most other animal steroids.

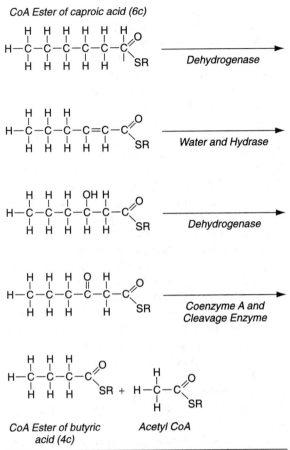

Figure 8.5   *Breakdown of a fatty acid is an oxidative process; that is, hydrogens are removed by the actions of enzymes (dehydrogenases). The fatty acid here (caproic acid, which has a six-carbon chain) is not broken down in its free form but is in the form of an ester of CoA (top). After oxidation and hydration (first three steps), two carbon units split off from the chain in the form of acetyl-CoA (last step). The remaining chain, still in the form of a CoA fatty acid ester can go through the whole process again. Thus, a fatty acid of any length can be disassembled two units at a time until it is all reduced to acetyl-CoA or, if an odd-length acid, to propionyl-CoA. Adapted from Green (1960).*

Biosynthesis is from acetyl CoA. Regulation of biosynthesis is partly by dietary intake; a high intake depresses the activity of HMG CoA reductase (the rate-limiting enzyme in cholesterol synthesis) in liver and ileum; low intake results in increased synthesis. This phenomenon has been shown in baby pigs fed diets high or low in cholesterol content (McWhinney et al., 1993). Whole-body cholesterol fractional synthesis rate can be determined by the deuterium enrichment of cholesterol iso-

lated from the red blood cells of animals or humans injected daily for several days with deuterium oxide (Wong et al., 1991, 1993). This procedure allows one to measure cholesterol synthesis in living animals. The liver of adult man contains about 3 to 5 g, and the blood pool is 10 to 12 g. Daily synthesis is 1 to 1.5 g, of which about half is converted to bile acids. Secretion of cholesterol and bile acids into the intestinal lumen via the bile duct approximates 2 and 20 to 30 g/day, respectively, but because of reabsorption by the enterohepatic circulation, less than 1 g of each is lost in the feces. Therefore, compounds that reduce absorption of cholesterol and bile acids may have a profound effect on the body pool of cholesterol and on its biosynthesis. One such compound is cholestyramine (a nonabsorbable resin that sequesters bile acids so that they may not be reabsorbed), which has been used to control hypercholesterolemia (Ballard et al., 1969).

In addition to excretion of cholesterol in the bile and its conversion to bile acids, cholesterol can be used for steroid hormone synthesis (progesterone, adrenal cortical hormones, testosterone, and estrogen) or stored as a component of pathological deposits in bile ducts (gall stones) and in arteries (atherosclerotic plaques). Conjugation of bile acids with taurine or glycine results in the excretion of these conjugated bile acids in the bile as taurocholic and glycocholic acid, respectively.

Research in baby pigs (reviewed by Pond, 2003) has shown that neonatal dietary cholesterol increases brain cholesterol content (presumably associated with increased myelination of brain cells) and improves learning behavior. This suggests the possibility that dietary cholesterol may be required during the suckling period to promote normal brain growth and function. Omega-3 and -6 fatty acids are known to improve early brain development and are added to human infant formulas. Cholesterol is not. The high concentration of cholesterol in milk of all mammals is suggestive of a dietary requirement for cholesterol in neonatal animals. Further research is needed to determine the validity of such a possibility.

Phospholipid Metabolism

The most abundant phospholipid in animal tissues is lecithin (phosphatidyl choline). It can be synthesized by two pathways, either by making use of choline directly or by methylation of phosphatidyl ethanolamine. The latter conversion occurs only in the liver, but the former can occur also in other tissues.

Degradation of phosholipids occurs in most mammalian tissues by hydrolysis of carboxyl esters and phosphate esters. Fatty acids must be oxidized to acetyl CoA before they are hydrolyzed to enter the tricarboxylic acid cycle. Glycerol can enter the glycolysis pathway or be used in triglyceride or phospholipid synthesis.

Ketones

Formation of ketones (ketogenesis) is a continuous process, but it may become excessive in certain disorders such as diabetes mellitus. The ketones are acetone, acetoacetic acid, and β-hydroxybutyric acid. The ketones are removed rapidly from the blood by skeletal muscle and other peripheral tissues and provide a substantial supply of energy for use by these tissues. Their synthesis originates with acetyl-CoA:

$$\text{Acetyl CoA} \longrightarrow \text{Acetoacetyl CoA}$$
$$\text{Acetoacetic acid} \begin{array}{l} \nearrow \text{Acetone} \\ \searrow \beta\text{-Hydroxybutyric acid} \end{array}$$

❏ EFFECTS OF FREQUENCY OF FEEDING ON METABOLISM

The composition of the diet and the level of dietary intake are important factors responsible for controlling body composition and metabolism of energy. Distribution of intake during a fixed period of time also affects lipogenesis and body composition in some species. Such terms as meal eater versus nibbler, single feeder versus multiple feeder, and others have been used to describe the phenomenon of feeding frequency. Bodies of meal-eating rats (those trained to consume the entire 24-h ration in 2 h) contain more fat and less protein and water than nibblers. Adipose tissue accounts for only 50 to 90% of total fatty acids synthesized in nibbling rats, but at least 95% in meal-fed rats. Meal eating alters the activities of enzymes involved in both carbohydrate and fat metabolism. Metabolic adaptations developed in the rat by meal-eating persist for several weeks after feed is provided ad libitum. Meal-fed chickens have been shown to have elevated plasma cholesterol and much higher incidence of atherosclerosis than ad libitum-fed

chickens (both groups on a high-cholesterol diet), even though daily feed intake was less in meal-fed chickens. Livers of meal-fed chickens incorporate more acetate into fatty acids than those of controls; chickens apparently are similar to the rat in their response to feeding frequency. In the pig, meal-eating also appears to influence weight gain and efficiency of feed utilization, but longer time of fasting (one 2-h feeding every 48 h) is needed than in the rat or bird to achieve increased adipose tissue lipogenesis associated with a longer postabsorptive state. The capacity of the GI tract also increases in meal-fed as compared with nibbling animals. Ruminants, whose rumen provides a reservoir of nutrients for absorption on a comparatively continuous basis, have not been shown to respond to frequency of feeding changes with appreciable changes in body composition.

Long-term fasting results first in depletion of liver and muscle glycogen and then oxidation of tissue lipids to meet energy requirements. Increased lipogenesis occurs in liver and to a smaller extent in adipose tissue after refeeding of fasted animals. The rates of fatty acid and cholesterol synthesis are affected by prefasting diet (high-fat diets inhibit the increase in lipogenesis), prefasting body weight (restricted-fed animals have greater fatty acid incorporation into depot fats than controls, but less turnover of cholesterol), and postfasting diet (high-fat diets inhibit lipogenesis).

❏ OBESITY IN HUMANS AND ANIMALS AS RELATED TO LIPID METABOLISM

The study of obesity is important in human nutrition because of the established relationship between obesity (it has been estimated that nearly 50% of adults in the United States are obese) and breast, colon, and prostate cancer; heart disease; and diabetes. It is important in animal nutrition because of consumer demands for low-fat edible animal products and the influence of body composition on energetic efficiency. A large number of animal models has been used for studies of obesity (Johnson et al., 1991). Genetically lean and obese pigs have contributed significantly to current knowledge of factors affecting fat metabolism (Mersmann, 1986; 1991; Pond and Mersmann, 1996). Mounting evidence shows that obesity has a strong genetic basis, involving differences between lean and obese individuals in activities of tissue enzymes associated with lipid synthesis and oxidation. Fat cells (adipocytes) from genetically obese rats convert more pyruvate or glucose to glyceride–glycerol than those from rats made obese by hypothalamic lesions, which cause excessive feed intake. Adipose tissue enzymes associated with lipogenesis are higher in obese than in lean strains of swine. The net deposition of fat in the body is a balance between lipogenesis and lipolysis processes that occur simultaneously at varying rates. The degree of saturation of body fat varies among and within animal species. Ruminants tend to have more highly saturated fat than nonruminants, and the body fat composition of ruminants is less responsive to diet. Pigs vary in degree of saturation of body fat according to genetic background via unknown mechanisms, possibly associated with their selection for altered lipid metabolism or for degree of obesity. Obese pigs tend to have more highly saturated body fat than do lean pigs.

Lipid deposition in the body is related both to number of adipose cells and to their individual size. There is a close spatial and temporal relationship between formation of adipocytes and the blood vessel formation required to supply nutrients to the adipocyte; it seems reasonable to assume that formation of blood vessels precedes the formation of adipocytes at a particular site. The histological origin of the adipocyte is still uncertain, but it now appears that preadipocytes (cells that ultimately accumulate lipids to become adipocytes) can proliferative postnatally, even during adulthood. This postnatal increase in adipocyte number probably varies among species. In swine, there is a different distribution of small and large adipocytes in obese and lean swine; obese swine have more small adipocytes (20 to 30 microns diameter) than observed in lean swine, although maximum cell diameter is greater in obese (190 microns) than in lean (140 microns) swine. The appearance of biphasic distribution of cell diameter both in lean and obese swine 1 year old has been interpreted as evidence that adipocyte hyperplasia continues into adulthood in swine.

Leptin, a hormone produced by adipocytes, has been shown to affect feed intake, tissue metabolism, body fatness, stress responses, and other physiological functions (Harris, 2000; Hausman and Hausman, 2005). It appears to function both directly and indirectly, via the brain, to coordinate metabolic changes in an array of tissues and organs, so as to modulate nutrient use in favor of energy expenditure over energy storage in the body (Baile et al., 2000).

Body fat deposition in humans and animals is a dynamic phenomenon, encompassing both anatomical (adipocyte size and number) and biochemical (lipogenesis and lipolysis) variables. The role of nutrition in affecting adipocyte size and number and in controlling lipid synthesis and oxidation is not understood fully. Nutritionists must be aware of and concerned about the importance of obesity in health and must continue to identify interactions between genetics and nutrition with respect to body composition and metabolism.

❏ EFFECT OF DIETARY CARBOHYDRATE SOURCE ON LIPID METABOLISM

Considerable recent interest has focused on the effects of dietary carbohydrate on lipid metabolism, especially in view of reports linking high-sucrose intake with atherosclerosis. Epidemiological studies with humans implicating such a relationship stimulated research activity in this area. Plasma triglyceride and cholesterol tend to be elevated in animals fed high levels of sucrose, and these plasma components have been implicated in human atherosclerosis. The effect of sucrose is thought to be related to the metabolism of the fructose moiety. It seems clear that refined glucose or starch do not induce increased plasma triglycerides. Much more research is needed before the exact role of dietary carbohydrate source on lipid metabolism is understood.

❏ ABNORMALITIES IN METABOLISM OF LIPIDS

Abnormal metabolism of lipids may occur in animals and humans as a result of genetic factors or in response to alterations in the environment, including diet. Persons whose genetic makeup results in high levels of lipoproteins circulating in the blood (familial hyperlipoproteinemia) have been categorized according to the type of lipoprotein patterns associated with their high blood lipid levels. Several types of familial hyperlipidemia have been identified.

Persons with familial Type II hyperlipidemia lack cell surface receptors for LDL (low-density lipoproteins) in the liver, a defect that blocks removal of these lipoproteins from the blood. These patients not only cannot remove LDL from blood efficiently, but they produce more LDL than normal. LDL are released from the liver in the form of VLDL (very low-density lipoproteins), which carry triglycerides as well as some cholesterol. The VLDL are carried to body tissues (e.g., muscle, adipose tissue, mammary glands) that remove the triglycerides and cleave them to fatty acids for oxidation or synthesis of complex lipids. In the normal liver, the LDL receptors bind the VLDL remnants and degrade them. In most species, only a small fraction escapes this degradation, whereas in humans almost half is converted to LDL. Drugs such as cholestyramine that lower blood cholesterol in Type II patients act by reducing cholesterol absorption from the intestine, thus depriving the liver of this source of cholesterol. Several cholesterol-lowering compounds are available for oral administration in humans. These "statin" compounds inhibit HMG (hydroxymethylgluta) CoA reductase activity (a key enzyme in cholesterol synthesis) and thereby reduce cholesterol biosynthesis; gemfibrozil reduces both serum triglycerides and cholesterol.

Fatty Livers

Because the liver is a key organ in metabolism, it is not surprising that factors affecting liver function have far-reaching effects on the overall well-being of the animal. A common manifestation of abnormal liver function is accumulation of lipids in the liver. Normally, fat constitutes about 5% of the wet weight of the liver, but the value may be 30% or more in pathological conditions. Fatty liver may arise from high-fat or high-cholesterol diet; increased liver lipogenesis caused by excessive carbohydrate or excessive intake of certain B vitamins (biotin, riboflavin, thiamin); increased mobilization of lipids from adipose tissue caused by diabetes mellitus, starvation, hypoglycemia, increased hormone output (growth hormone, adrenal corticotrophic hormone, adrenal corticosteroids); decreased transport of lipids from liver to other tissues caused by deficiencies of choline, pan-

tothenic acid, inositol, protein, or certain amino acids (methionine, threonine); and cellular damage to the liver (cirrhosis, necrosis) because of infections, vitamin E–Se deficiency, or liver poisons such as chloroform and carbon tetrachloride.

Atherosclerosis

Atherosclerosis is a disease characterized by progressive degenerative changes in the blood vessels and heart of humans and animals, and it ultimately is responsible for more deaths among humans in the United States than any other single cause. More than half (54%) of all deaths in the United States result from cardiovascular disease. About two-thirds of these are caused by atherosclerotic heart disease and one-eighth by cerebral accidents (strokes) resulting from atherosclerosis. Chapman and Goldstein (1976) reviewed the problem. Among the factors responsible for development of atherosclerosis, nutrition has received perhaps the most attention. The observation that serum cholesterol concentration seems to be correlated with the incidence of atherosclerosis has led to numerous studies of the influence of diet on serum cholesterol. The cholesterol of blood is transported in association with other lipids and with proteins. Each species of animals has its own peculiar serum lipoprotein composition, but the profile for normal humans provides a useful guide (Table 8.5).

The effect of saturated animal fats (such as butter, tallow, lard) and eggs (high in cholesterol) on serum cholesterol of humans has received much attention and publicity. Saturated fats tend to raise serum cholesterol, and, of course, consumption of products high in cholesterol such as eggs add to the body burden of cholesterol in addition to that synthesized by the liver.

Although the ultimate work needed to clarify the role of diet in atherosclerosis must be done in humans, experimental animal models provide much useful data. The pig is a superior model because the morphology and biochemistry of the atherosclerotic lesion resembles that observed in humans, and severe lesions can be induced by diet. There is a close resemblance between swine and human serum lipoprotein fractions (Birchbauer et al., 1993; Lee et al., 1990). The hypercholesterolemia induced in swine by feeding a high-cholesterol diet is distinguished by an increase in low-density lipoprotein concentration. Low-density lipoproteinemia is associated with increased atherogenesis in humans, whereas an increase in high-density lipoproteins in the serum (associated with increased physical exercise) may be protective.

The total role of diet and other environmental factors, compared with genetic factors, in the development of cardiovascular disease (atherosclerosis) in humans and animals is incompletely understood, but such knowledge is of great concern to animal and human nutritionists in charting the course of future efforts in altering the chemical composition of animal products for optimum use by humans (National Research Council, 1988).

During the period 1950 through 1993, the proportion of animal fat has decreased and that of plant oils (high in PUFA) has increased in the food consumed by the United States population. The effects of this shift on health and longevity remain to be seen, although deaths from cardiovascular disease have declined steadily since the 1960s. The total lipid and cholesterol content and saturated and unsaturated fatty acid content of a variety of animal and plant products are listed in Table 8.7. One can get an estimate of the relative amounts of saturated (S) and monounsaturated plus polyunsaturated (M + P) fatty acids in the diet by expressing the values as a ratio of one to the other. The average American diet has an (M + P)/S ratio of about 0.5/1.0. Most clinical studies show significant reductions in plasma cholesterol when this ratio is changed to 1.0/1.0 or greater by increasing the amount of plant oils such as soybean and corn oil and decreasing animal fat.

Recent evidence shows an apparent special protective effect of fish oils against the atherogenic process (Parks and Rudel, 1990). The unique fatty acids in fish oils that set them apart from other fats containing high amounts of polyunsaturated fatty acids are the so-called omega-3 fatty acids, consisting of linolenic acid (Cl8:3), eicosapentaenoic acid (EPA) (C20:5), and docosahexaenoic acid (DHA) (C22:6).

At least one role of these omega-3 acids appears to be increasing the clotting time of blood. Marine fish have a higher content of omega-3 fatty acid than freshwater fish. The content varies with species, season of the year, water temperature, and the food eaten by the fish.

❑ SUMMARY

- Lipids are organic compounds containing C, H, and O, but unlike carbohydrates, they are insoluble in water.

- Lipids serve important biochemical and physiological functions in plant and animal tissues.

- Simple lipids are esters of fatty acids with alcohols; compound lipids are esters of fatty acids containing groups in addition to an alcohol and fatty acids.

- Compound lipids include phospholipids, glycolipids, and lipoproteins.

- The most important lipids in nutrition include fatty acids, glycerol, mono-, di-, and triglycerides (triacylglycerol, TAG) and phospholipids.

- Fatty acids consist of chains of C atoms (2 to 24 or more Cs) and are characterized by a carboxyl group on one end of the C chain.

- Most fatty acids in animal tissues are straight chained and have an even number of Cs and are esterified to glycerol to form TAG.

- The fatty acid composition of TAG in tissues of nonruminant animals resembles that of the dietary fat; that in tissues of ruminant animals is less variable and contains mostly saturated fatty acids owing to the action of rumen microbes.

- The degree of softness of TAG depends on the number of double bonds (degree of unsaturation) of the constituent fatty acids and on chain length.

- Lipids (1) supply energy for normal maintenance and production (1 gram of fat yields about 9.45 kcal of heat compared with about 4.1 kcal for carbohydrates); (2) serve as a source of essential fatty acids (linoleic, linolenic); and (3) function as a carrier of the fat-soluble vitamins.

- Absorption of lipids depends on emulsification and formation of micelles containing bile salts, phospholipids, and other lipids and on hydrolysis by lipase enzymes elaborated into the intestinal lumen by the host animal.

- Absorption of long-chain fatty acids occurs mostly via the lymphtaic system, whereas short-chain fatty acids may enter the portal vein.

- Absorbed lipids are carried in the blood as chylomicrons and as lipoproteins varying in proportions of lipid and protein (very low-density

TABLE 8.7 Lipid composition of selected foods (mg/100 g).

| FOOD | CHOLES-TEROL | TOTAL FAT | FATTY ACIDS | | | | | |
			MONO-SATURATED	POLYUN-SATURATED		18:3	20:5	22:6
Beef steak, choice grade	71	26.1	11.2	11.7	1.0	0.3	—	—
Wheat, hard red winter	0	2.5	0.4	0.3	1.2	0.1	—	—
Cheese, cheddar	105	33.1	21.1	9.0	0.9	0.4	—	—
Soybean oil	0	100.0	14.4	23.3	57.9	6.8	—	—
Chicken, skin only	109	32.4	9.1	13.5	6.8	0.3	—	—
Tallow	102	100.0	47.3	40.6	7.8	2.3	—	—
Pork leaf fat	110	94.2	45.2	37.2	7.3	0.9	—	—
Wheat germ oil	0	100.0	18.8	15.1	61.7	6.9	—	—
Avocados	0	17.3	2.6	11.2	2.0	0.1	—	—
Raspberries	0	0.4	trace	trace	0.3	0.1	—	—
Lamb loin	71	27.4	12.8	11.2	1.6	0.5	—	—
Beans	0	1.5	0.2	0.1	0.9	0.6	—	—
Peas	0	2.4	0.4	0.1	0.4	0.2	—	—
Walnuts, black	0	56.6	3.6	12.7	37.5	3.3	—	—
Pork, ham	74	20.8	7.5	9.7	2.2	0.2	—	—
Chicken, dark w/o skin	80	4.3	1.1	1.3	1.0	trace	—	—
Chicken, white w/o skin	58	1.7	0.4	0.4	0.4	trace	—	—
Broccoli	0	0.4	trace	trace	0.2	0.1	—	—
Herring, Pacific	77	13.9	3.3	6.9	2.4	0.1	1.0	0.7
Herring, Atlantic	60	9.0	2.0	3.7	2.1	0.1	0.7	0.9
Perch, white	80	2.5	0.6	0.9	0.7	0.1	0.2	0.1
Pike, walleye	86	1.2	0.2	0.3	0.4	trace	0.1	0.2
Salmon chum	74	6.6	1.5	2.9	1.5	0.1	0.4	0.6
Smelt, rainbow	70	2.6	0.5	0.7	0.9	0.1	0.3	0.4
Smelt, sweet	25	4.6	1.6	1.2	1.0	0.3	0.2	0.1
Snapper, red	—	1.2	0.2	0.2	0.4	trace	trace	0.2
Sunfish	67	0.7	0.1	0.1	0.2	trace	trace	0.1
Swordfish	39	2.1	0.6	0.8	0.2	—	0.1	0.1
Trout, brook	68	2.7	0.7	0.8	0.9	0.2	0.2	0.2
Trout, lake	48	9.7	1.7	3.6	3.4	0.4	0.5	1.1
Trout, rainbow	57	3.4	0.6	1.0	1.2	0.1	0.1	0.4
Tuna, bluefin	38	6.6	1.7	2.2	2.0	—	0.4	1.2
Whitefish, lake	60	6.0	0.9	2.0	2.2	0.2	0.3	1.0
Crab, blue	78	1.3	0.2	2.2	0.5	trace	0.2	0.2
Crayfish	158	1.4	0.3	0.4	0.3	trace	0.1	trace
Lobster, northern	95	0.9	0.2	0.2	0.2	—	0.1	0.1
Shrimp, Atlantic white	182	1.5	0.2	0.2	0.6	trace	0.2	0.2
Shrimp, Atlantic brown	142	1.5	0.3	0.3	0.5	trace	0.2	0.1
Oyster, eastern	47	2.5	0.6	0.2	0.7	trace	0.2	0.2
Scallop, unspecified	45	0.8	0.1	0.1	0.3	trace	0.1	0.1

lipoproteins, VLDL; low-density lipoproteins, LDL; high-density lipoproteins, HDL).

- Lipids mobilized from depot storage and those synthesized in body tissues are transported as lipoproteins.

- Biosynthesis of fatty acids from carbohydrates and amino acids occurs mainly in liver, mammary tissue, and adipose tissue.

- Synthesis of fatty acids progresses from acetyl-CoA (two Cs) to longer chain fatty acids by successive additions of two-C fragments.

- Catabolism of fatty acids proceeds by stepwise removal of two carbon fragments.

- Sterols such as cholesterol are synthesized from acetyl-CoA. High dietary intake of cholesterol depresses cholesterol synthesis.

- Lipid deposition is related to both the number and size of individual adipocytes and is clearly a dynamic phenomenon, emphasizing anatomical (cell size and number) and biochemical (lipogenesis and lipolysis) variables.

❏ REFERENCES

Allen, R. S. 1977. In *Duke's Physiology of Domestic Animals* (9th ed.), Chapter 28, p. 336, M. J. Swenson (ed.). Cornell University Press, Ithaca, NY.

Babatunde, G. M., et al. 1967. *J. Nutr.* **92**: 293.

Baile, C. A., M. A. Dela-Fera, and R. J. Martin. 2000. *Annu. Rev. Nutr.* **20**: 105–127.

Ballard, F. J., R. M. Hanson, and D. S. Kronfeld. 1969. *Fed. Proc.* **28**: 218.

Bauman, D. E., B. A. Corl, and D. G. Peterson. 2003. The biology of conjugated linoleic acids in ruminants. In *Recent Advances in Conjugated Linoleic Acid Research 2*, pp. 146–173, J. L. Sebedio, W. W. Christi, and R. O. Adlof (eds.). AOCS Press, Champaign, IL.

Bauman, D. E. and A. L. Lock. 2005. Conjugated linoleic acid. In *Encyclopedia of Animal Science,* W. G. Pond and A. W. Bell (eds.). Marcel Dekker, New York (In Press).

Baumgardt, L. H., J. K. Sanger, and D. E. Bauman. 2001. *J. Nutr.* **131**: 1764–1769.

Belury, M. A. 2002. *Annu. Rev. Nutr.* **22**: 505–531.

Bensadoun, A. 1991. *Annu. Rev. Nutrition* **11**: 217.

Birchbauer, A., G. Knippling, B. Juritsch, H. Aschhauer, and R. Zechner. 1993. *Genomics* **15**: 643.

Chapman, M. J., and S. Goldstein, 1976. *Atherosclerosis* **24**: 141.

Coleman, R. A., T. M. Lewin, and D. M. Muoio. 2000. *Annu. Rev. Nutr.* **20**: 77–103.

Crandall, D. L., G. J. Hausman, and J. G. KRAL. 1997. *Microcirculation* **4**: 211–232.

Green, D. E. 1960. *Sci. Am.,* February.

Harris, R. B. S. 2000. *Annu. Rev. Nutr.* **20**: 45–75.

Hausman, G. J., and D. B. Hausman. 2005. Ontogeny adipose tissue. In *Encyclopedia of Animal Science,* W. G. Pond and A. W. Bell (eds.). Marcel Dekker, New York (In Press).

Holman, R. T. 1960. *J. Nutr.* **70**: 405.

Hwang, D. 2000. *Annu. Rev. Nutr.* **20**: 431–456.

Jensen, J. K. 1997. *Annu. Rev. Nutr.* **17**: 127–139.

Johnson, P. R., M. C. R. Greenwood, B. A. Horwitz, and J. S. Stern. 1991. *Annu. Rev. Nutrition* **11**: 325.

Kass, M. L., W. G. Pond, and E. F. Walker, Jr. 1975. *J. Anim. Sci.* **41**: 804.

Katan, M. B., and P. L. Zock. *Annu. Rev. Nutr.* **15**: 474–493.

Kim, K.-H. 1997. *Annu. Rev. Nutr.* **17**: 77–99.

Lands, W. E. M. 1991. *Annu. Rev. Nutrition* **11**: 41.

Lee, D. M., T. Mok, J. Hasler-Rapacz, and J. Rapacz. 1990. *J. Lipid Res.* **31**: 839.

Masoro, E. J. 1968. *The Physiological Chemistry of Lipids in Mammals.* W. B. Saunders Co., Philadelphia, PA.

McWhinney, V., H. J. Mersmann, and W. G. Pond. 1993. *FASEB J.* **7**(3): A64.

Mersmann, H. J. 1986. Lipid Metabolism in Swine. In: *Swine in Cardiovascular Research,* pp. 76–97, H. C. Stanton and H. J. Mersmann (eds.). CRC Press, Boca Raton, FL.

Mersmann, H. J. 1991. In *Swine Nutrition.* Chapter 4, pp. 75–89, E. R. Miller, D. E. Ullrey, and A. J. Lewis (eds.). Butterworth-Heinemann, Boston, MA.

National Research Council. 1988. *Designing Foods: Animal Product Options in the Marketplace.* National Academy Press, Washington, DC, 367 pp.

Ogilvie, B. M., G. L. McClymont, and F. B. Shorland. 1961. *Nature* **190**: 725.

Ostrowska, E., et al. 1999. *J. Nutr.* **129**: 2037–2042.

Pariza, M. W. 1999. In *Advances in Conjugated Linoleic Acid Research,* Vol. 1, pp. 12–20, M. P. Yurawecz, et al. (eds.). AOCS Press, Champaign, IL.

Parks, J. S., and L. L. Rudel. 1990. *Atherosclerosis* **84**: 83.

Pond, W. G. 2003. *Comments Theoretical Biology* **8**: 37–68.

Pond, W. G., and H. J. Mersmann. 1996. Genetically diverse pig models in nutrition research related to lipoprotein and cholesterol metabolism. In *Advances in Swine in Biomedical Research,* pp. 843–863. M. E. Tumbleson, and L. B. Schook (eds.). Plenum Press, New York.

Prior, R. L., and R. A. Scott. 1980. *J. Nutr.* **110:** 2011.

Pond, W. G. 1986. In *Swine in Cardiovascular Research,* Vol. II, pp. 1–31, H. Stanton and H. R. Mersmann (eds.). CRC Press, Boca Raton, FL.

Schaefer, E. J., and R. I. Levy. 1985. *New Engl. J. Med* **312:** 1300.

Scott, R. A., S. G. Cornelius, and H. J. Mersmann. 1981. *J. Anim. Sci.* **53:** 977.

Small, D. M. 1991. *Ann. Rev. Nutrition* **11:** 413.

Sul, H. S. and D. Wang. 1998. *Annu. Rev. Nutr.* **18:** 331–351.

Wong, W. D., D. L. Hachey, A. Feste, J. Leggitt, L. Clarke, W. G. Pond, and P. D. Klein. 1991. *J. Lipid Res.* **32:** 1049.

Wong, W. W., D. L. Hachey, W. Insull, A. R. Opekin, and P. D. Klein. 1993. *J. Lipid Res.* **34:** 1403.

Zeisel, S. H. 1993. *J. Nutritional Biochem.* **4:** 550.

9

Proteins and Amino Acids

PROTEINS ARE ESSENTIAL ORGANIC constituents of living organisms and are the class of nutrients in highest concentration in muscle tissues of animals. All cells synthesize proteins for part or all of their life cycle, and without protein synthesis life could not exist. Except in animals whose intestinal microflora can synthesize protein from nonprotein N sources, protein or its constituent amino acids must be provided in the diet to allow normal growth and other productive functions. All cells contain protein, and cell turnover is very rapid in some tissues such as epithelial cells of the intestinal tract. The percentage of protein required in the diet is highest for young growing animals and declines gradually to maturity, when only enough protein to maintain body tissues is required. Productive functions such as pregnancy and lactation increase the protein requirement because of increased output of protein in the products of conception and in milk and because of an increased metabolic rate. In this chapter we describe the structure, functions, digestion, absorption, and metabolism of proteins and amino acids.

❏ STRUCTURE OF PROTEINS AND AMINO ACIDS

Proteins vary widely in chemical composition, physical properties, size, shape, solubility, and biological functions. All proteins are composed of simple units, amino acids. Although there are more than 200 naturally occurring amino acids, only about 20 are commonly found in most proteins, and up to 10 are required in the diet of animals because tissue synthesis is not adequate to meet metabolic needs. The basic structure of an amino acid is illustrated by glycine, the simplest amino acid:

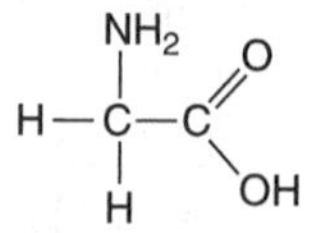

GLYCINE

The essential components are a carboxyl group

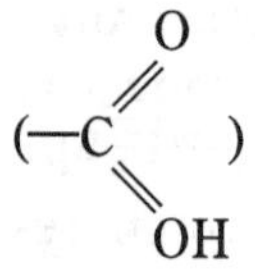

( ) and an amino group (NH_2) on the C atom adjacent to the carboxyl group. This NH_2 group is designated the α-amino group.

The general structure representing all amino acids can be represented as follows:

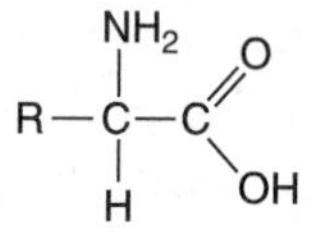

GENERAL STRUCTURE FOR AMINO ACIDS

where R is the remainder of the molecule attached to the C atom associated with the α-amino group of the amino acid. The chemical structure of 20 of the amino acids is given in Fig. 9.1.

All proteins can be classified on the basis of their shape; their solubilities in water, salt, acids, bases and alcohol; and other special characteristics. Such a broad classification includes the following:

A. Globular proteins—soluble in water or in dilute acids or bases or in alcohol
Albumins—soluble in water
Globulins—soluble in dilute neutral solutions of salts of bases and acids
Glutelins—soluble in dilute acids or bases
Prolamines (gliadins)—soluble in 70 to 80% ethanol
Histones—soluble in water
Protamines—soluble in water

B. Fibrous proteins—insoluble in water, resistant to digestive enzymes
Collagens—can be converted to gelatin
Elastins—similar to collagens but cannot be converted to gelatin
Keratins—insoluble in water, resistant to digestive enzymes, contain up to 15% cystine

C. Conjugated proteins—proteins that contain a wide array of compounds of a nonprotein nature. Some examples of protein–lipid and protein–carbohydate complexes should help to clarify their importance in animal tissues.

Protein–Lipid Complexes (Lipoproteins)

An egg yolk phospholipid–protein complex was recognized more than a century ago; now hundreds of lipid–protein complexes are known. Electrostatic, hydrophobic, and hydrogen bonds as well as other forces contribute to the stability of lipoproteins. A few protein–lipid complexes of importance in animal production are discussed here.

A very important type of protein–lipid complex is represented by the membrane proteins of animal cells. Biological membranes act as a permeability barrier, transport substances across the boundary between the interior and exterior of the cell, act as supports for catalytic functions, and probably perform other important, though less well-defined, functions. These membranes are composed of proteins, lipids, and carbohydrates in various proportions. The chemical composition of some important cell membranes is summarized in Table 9.1.

Myelin is a lipoprotein abundant in the nervous system as a sheath around nerve fibers. Peripheral and central nervous tissues contain about 80% and 35% myelin, respectively. Erythrocyte membranes contain mucolipids, phospholipids, and loosely bound proteins. Mitochondria contain structural protein, matrix protein, and about 30% lipids, mainly phospholipids. Fatty acids and other lipids are adsorbed to the surface of blood serum proteins to form lipoproteins. The protein content of plasma lipoproteins ranges from 2% in chylomicrons to about 50% in high-density lipoproteins (HDL). The structures and functions of serum

ALIPHATIC

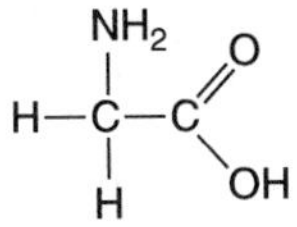

GLYCINE
(amino acetic acid)

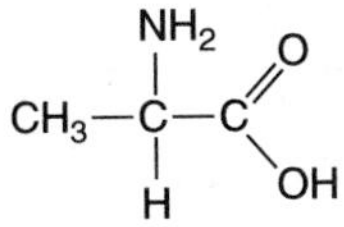

ALANINE
(α-amino propionic acid)

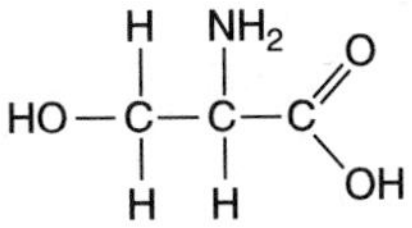

SERINE
(β-hydroxy α-amino propionic acid)

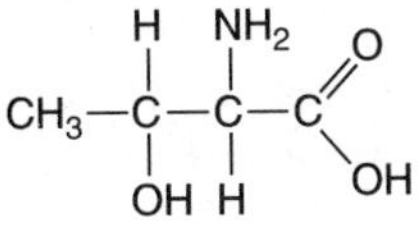

THREONINE
(α-amino β-hydroxy-N-butyric acid)

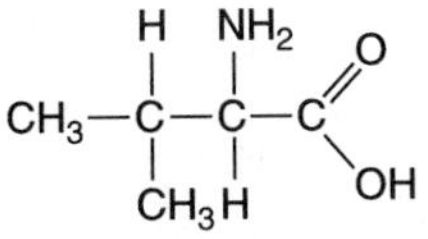

VALINE
(α-amino isovaleric acid)

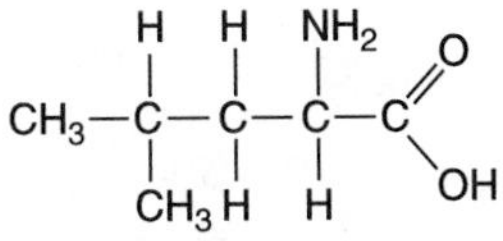

LEUCINE
(α-amino isocaproic acid)

ALIPHATIC

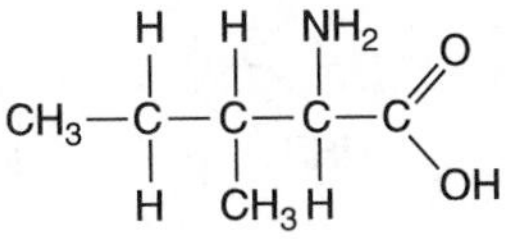

AROMATIC

ISOLEUCINE
(β-methyl α-amino valeric acid)

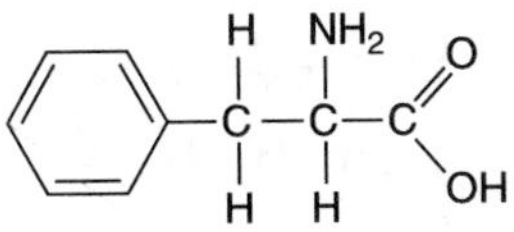

PHENYLALANINE
(β-phenyl α-amino propionic acid)

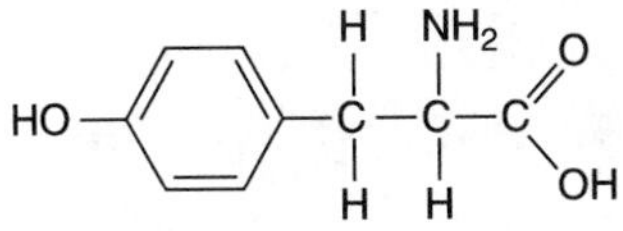

TYROSINE
(β-para hydroxy phenyl α-amino
propionic acid)

SULFUR-CONTAINING

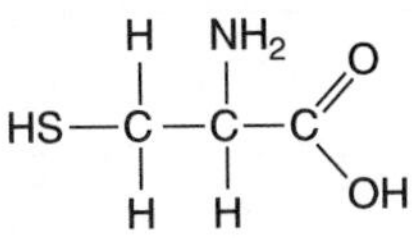

CYSTEINE
(β-thiol α-amino propionic acid)

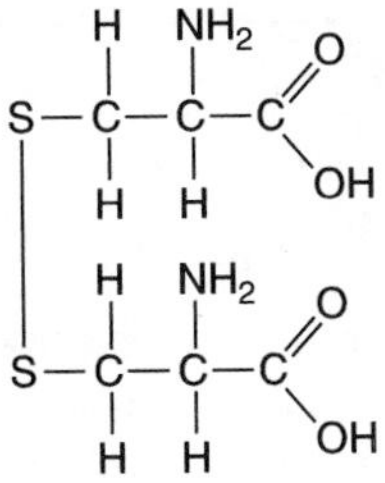

CYSTINE
di-(β-thiol α-amino propionic acid)

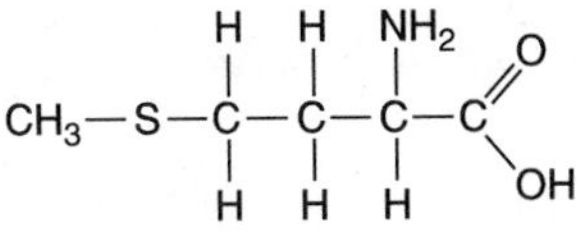

METHIONINE
(gamma methyl thiol α-amino-
N-butyric acid)

Figure 9.1   *Chemical structure of amino acids.*

HETEROCYCLIC

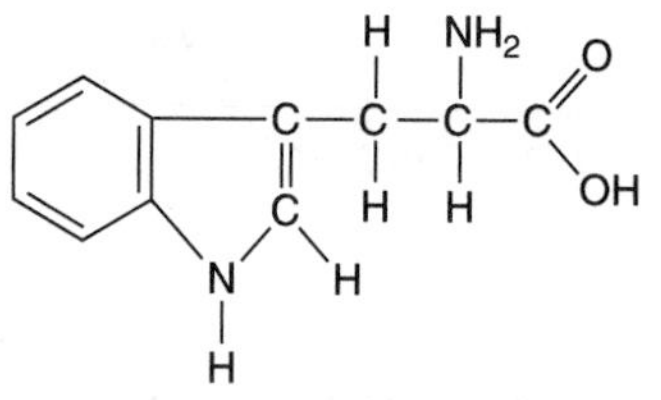

TRYPTOPHAN
(β-3-indole α-amino propionic acid)

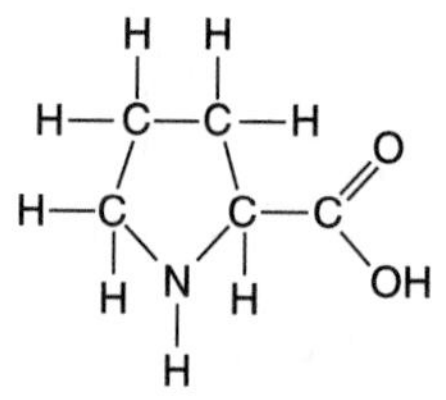

PROLINE
(pyrolidine-2-carboxylic acid)

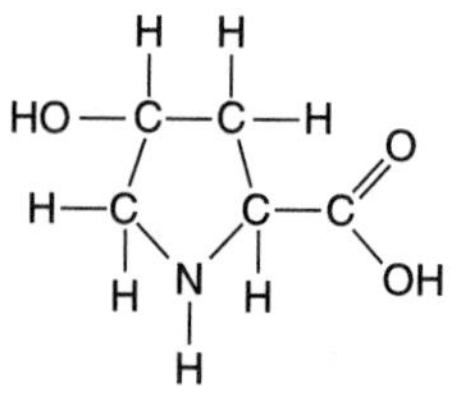

HYDROXY PROLINE
(4-hydroxy pyrolidine-2-carboxylic acid)

ACIDIC

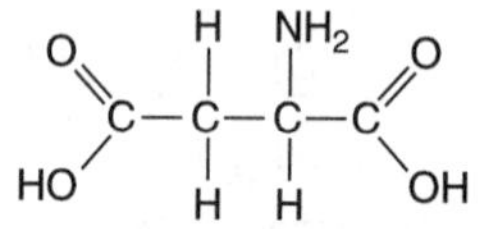

ASPARTIC ACID
(α-amino succinic acid)

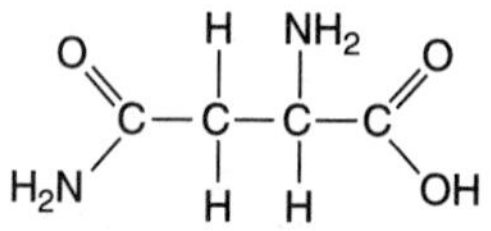

ASPARAGINE
(α-amino succinamic acid)

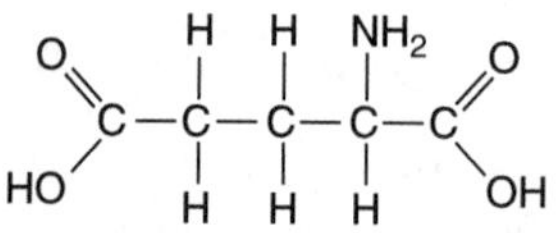

GLUTAMIC ACID
(α-amino glutaric acid)

BASIC

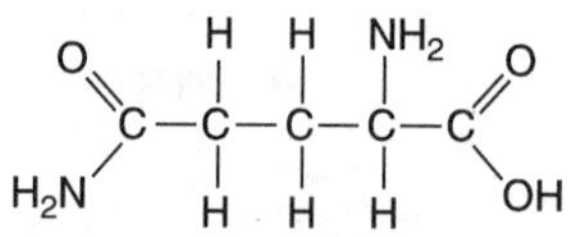

GLUTAMINE
(α-amino glutaramic acid)

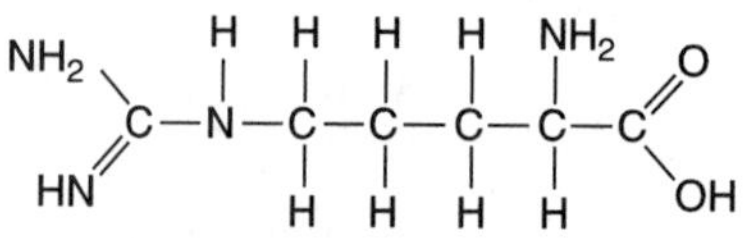

ARGININE
(β-guanidino α-amino valeric acid)

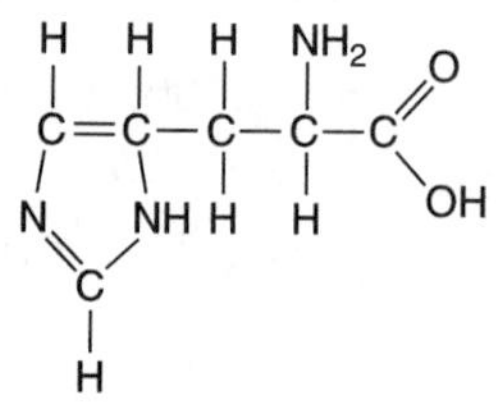

HISTIDINE
(β-imidazol α-amino propionic acid)

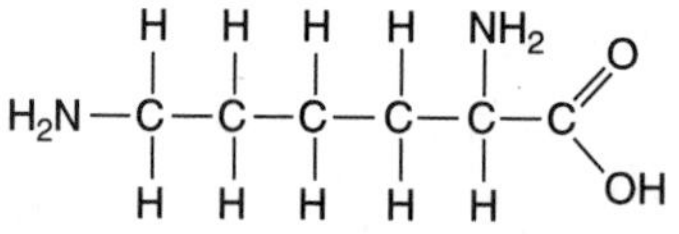

LYSINE
(α, epsilon diamino caproic acid)

Figure 9.1—Continued.

lipoproteins are well known. The large number of apoproteins (protein carriers of lipids) and their genetic variations in relation to susceptibility of animals and humans to atherosclerosis were reviewed by Breslow (1991). More recently, Bruce et al. (1998) described a family of plasma lipid transfer proteins (PLTP) and reverse cholesterol transport proteins (CETP) that shuttle between lipoproteins to redistribute lipids so as to modulate atherogenesis.

Protein–Carbohydrate Complexes (Glycoproteins)

Proteins can complex with carbohydrate to form glycoproteins. These complexes arise from the accept-

TABLE 9.1 Chemical composition of some cell membranes.

MEMBRANE	PROTEIN, %	LIPID, %	CARBOHYDRATE, %	RATIO OF PROTEIN TO LIPID
Myelin	18	79	3	0.23
Blood platelets	33–42	51–58	7.5	0.7
Human red blood cell	49	43	8	1.1
Rat liver cells	58	42	5–10	1.4
Nuclear membrane, rat liver cell	59	35	2.9	1.6
Mitochondrial outer membrane	52	48	2.4	1.1
Mitochondrial inner membrane	76	24	2–3	3.2
Retinal rods, cattle	51	49	4	1.0

Source: Guidotti (1972).

ance of sugars by amino acid residues in the polypeptide chain. Protein complexes of sulfated polysaccharides occur in three types, designated chondroitin sulfate A, B, and C. Cartilage, tendon, and skin are high in chondroitin sulfates complexed with protein. Mucoproteins are complexes of protein with the amino sugars glucosamine and galactosamine; the hexoses mannose and galactose; the pentose, fucose, and sialic acid. The enzyme serum cholinesterase is a mucoprotein, as is the hormone gonadotropin. Mucous secretions contain abundant amounts of mucoproteins. The mucoproteins of submaxillary salivary gland secretions are potent inhibitors of agglutination of red blood cells by influenza viruses. Ovalbumin is a mucoprotein containing glucosamine and mannose. Ovomucoid, the trypsin inhibitor in egg white, is a glycoprotein distinguished by its high-heat stability and high-carbohydrate content. It contains about 14% hexosamine and 7% hexose. Aspartic acid is apparently the amino acid immediately concerned in linkage of the protein to the carbohydrate part of the complex.

Essential and Nonessential Amino Acids

Amino acids not synthesized in animal tissues of most species in sufficient amounts to meet metabolic needs without being added to the diet are termed **essential** or **indispensable,** whereas those generally not needed in the diet because of adequate tissue synthesis are termed **nonessential** or **dispensable.** The array of amino acids that cannot be synthesized in sufficient amounts in the tissues of some animal species or at certain stages of the life cycle varies slightly from Rose's original classification (1948), but this list of dietary essential amino

acids still serves as a good guide. Although proline, generally considered to be a dispensable amino acid, can be synthesized to a limited degree in birds, the evidence clearly establishes it as a dietary essential amino acid for the hen. Berthold et al. (1991) fed hens diets containing algae uniformly labeled with a stable isotope of carbon, ^{13}C. Eggs were collected for 27 days, and the enrichment pattern of ^{13}C in all of the amino acids of egg and tissue proteins was determined. The patterns were divided according to known nutritional amino acid essentiality or nonessentiality, except that proline accretion was derived entirely from the diet, showing that this amino acid is an indispensable amino acid for egg production in the chicken.

Dietary essential amino acids for growth of rats as suggested by Rose (1948) are as follows:

ESSENTIAL (INDISPENSABLE)	NONESSENTIAL
Arginine	Alanine
Histidine	Aspartic acid
Isoleucine	Citrulline
Leucine	Cystine
Lysine	Glutamic acid
Methionine (can be replaced partially by cystine)	Glycine
	Hydroxyproline
Phenylalanine (can be replaced partially by tyrosine)	Proline
	Serine
	Tyrosine
Threonine	
Tryptophan	
Valine	

Arginine is required in the diet of some species for maximum growth. In most species, adults do not require dietary arginine for maintenance. However, mature dogs and cats fed arginine-deficient diets show severe signs of deficiency, including hyperammonemia, tremors, and high levels of urinary orotic acid (Visek, 1984).

Although glutamine is generally considered to be a dispensable amino acid, there is evidence (Souba, 1991) that it may be a conditionally essential amino acid for supporting the metabolic requirements of the intestinal mucosa during illness. Glutamine may be important in normal GI-tract immune function, and in normal metabolism of the pancreas and liver (Souba, 1991).

Citrulline can completely replace arginine in the diet of the cat and other species (McDonald et al., 1984). The cat has a high sulfur amino acid requirement and is the only species known to require taurine. Taurine is a β-amino sulfonic acid that is not present in protein, but it occurs as a free amino acid in the diet. Cats fed taurine-deficient diets develop degeneration of the retina of the eye (National Research Council, 1981), which appears at a higher frequency when dietary sulfur-containing amino acids are also low. Dietary histidine is needed for adult humans (Visek, 1984), contrary to the earlier assumption that synthesis is sufficient in adults.

Asparagine is required for maximum growth during the first few days of consumption of a crystalline amino acid diet. Certain other dispensable amino acids also may be required for maximum growth under some conditions.

In animals whose gastrointestinal microflora synthesize protein from nonprotein N sources (ruminants and other herbivores), the amino acid balance of the diet is of little nutritional consequence except for high-producing animals. Mainly, the quantities of N and readily available carbohydrate are important.

The synthesis of protein from amino acids occurs by the joining of individual amino acids to form long chains. The length of the chain and the order of arrangement of amino acids within the chain are two of the main factors determining the characteristics of the protein.

The linkage between one amino acid and another, as shown in A in the accompanying illustration is called a peptide bond. The dipeptide alanylglycine would be formed as shown in B.

All naturally occurring amino acids are in the L

configuration, which, with few exceptions, is the most biologically active form. Synthetic amino acids are usually found as the racemic mixture of L and D isomers. The schematic representation of L- and D-amino acids is shown in C.

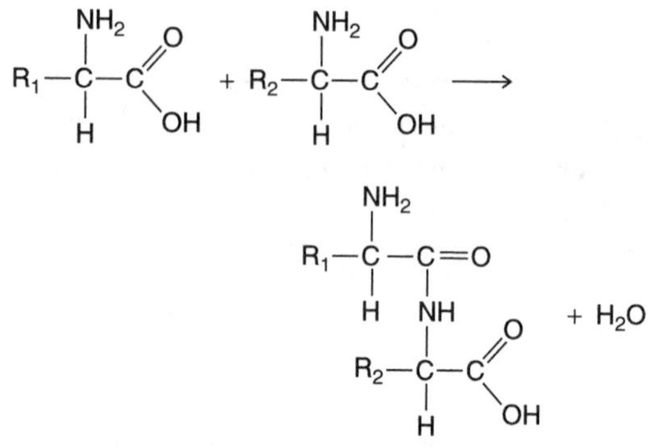

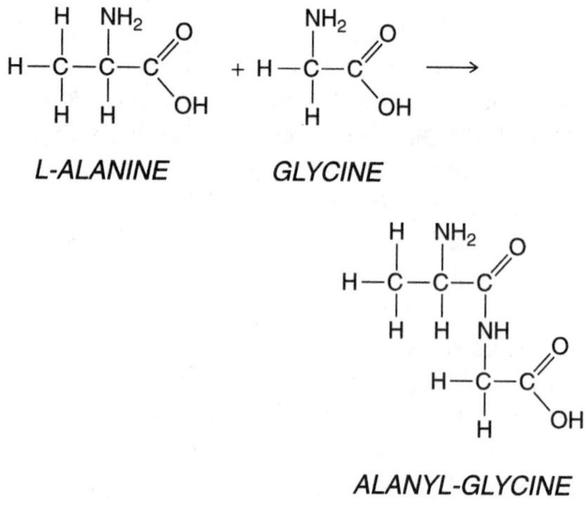

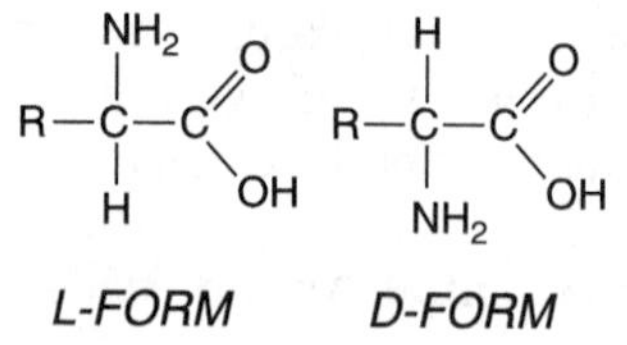

Elongation of the chain by additional amino acids proceeds from tripeptides to polypeptides and eventually to a complete protein molecule of specific amino acid content and sequence. The amino acid composition and sequence of hundreds of proteins have been determined. The amino acid sequences of two enzymes—one of plant origin and one of animal origin (wheat germ cytochrome c and human cytochrome c)—are remarkably similar in their structure, illustrating the commonality associated with evolutionary processes. These two enzymes are very small compared with most proteins. The molecular weight of most proteins is 35,000 to 500,000 daltons, with 350 to 5000 amino acid residues.

The amino acid compositions of some common proteins of plant and animal origin are shown in Table 9.2. Egg albumin is considered the most nearly perfect protein for meeting animal needs because of its nearly ideal amino acid composition and its high digestibility. A single feedstuff may be composed of several distinct proteins, and the adequacy of its protein will depend on the composite amino acid mixture supplied by the individual protein fractions. This is illustrated in Table 9.2 by the wide differences in the amino acid composition of four pro-

tein fractions from corn (maize) endosperm. Zein is notably low in lysine and tryptophan. A genetic mutation discovered by Purdue University scientists involves a gene for high lysine and tryptophan (referred to as the opaque-2 gene). The proportion of zein in opaque-2 corn is lower, and that of glutelin and other protein fractions is higher than in common corn, resulting in a grain with a more favorable balance of amino acids for most species.

Similar genetic mutations in barley and other cereal grains have resulted in seeds with an altered amino acid composition. Some amino acids are present in animal tissues in higher quantities than in plant tissues. For example, hydroxyproline is absent from the plant proteins in Table 9.2 but is abundant in animal connective tissue, especially collagen. A sensitive measure of the nutritive value

TABLE 9.2 Amino acid composition of some plant and animal products.

| | EGG ALBUMIN | CORN ENDOSPERM[a] | | | | | | | | BEEF[d] | PORK[d] | LAMB[d] | BOVINE INTRA-MUSCULAR COLLAGEN[e] |
| | | ALBUMIN | | GLOBULIN | | ZEIN | | GLUTELIN | | | | | |
		N[b]	O[c]	N	O	N	O	N	O				
Alanine	5.7	9.8	7.9	8.4	7.3	10.9	10.5	4.5	4.7	6.4	6.3	6.3	10.4
Arginine	5.9	11.6	11.7	9.6	8.9	2.1	2.3	5.8	5.3	6.6	6.4	6.9	4.5
Aspartic acid	9.2	11.1	10.8	8.2	8.0	5.9	5.7	6.8	6.9	8.8	8.9	8.5	3.8
Cystine/2	3.0	5.3	1.8	1.8	1.0	1.4			1.4	1.3	1.3		
Glutamic acid	15.7	16.5	13.2	19.2	18.2	27.4	24.9	12.9	14.6	14.4	14.5	14.4	7.7
Glycine	3.2	8.8	7.9	5.3	5.5	1.6	2.0	3.3	3.8	7.1	6.1	6.7	32.8
Histidine	2.4	3.2	2.6	3.4	3.8	1.4	1.4	3.8	4.2	2.9	3.2	2.7	0.6
Hydroxyproline	0	0	0	0	0	0	0	0	0				10.5
Isoleucine	7.1	4.6	3.9	4.2	4.3	4.2	4.6	3.4	3.4	5.1	4.9	4.8	1.3
Leucine	9.9	6.3	6.3	13.2	10.7	22.4	21.2	8.1	8.6	8.4	7.5	7.4	2.5
Lysine	6.4	6.4	5.5	4.4	4.6	0.1	0.2	3.7	3.6	8.4	7.8	7.6	2.3
Methionine	5.4		1.6	2.0	1.9	1.7	1.3	1.4	1.1	2.3	2.5	2.3	0.6
Phenylalanine	7.5	2.9	4.6	5.8	6.3	7.6	8.0	3.8	3.8	4.0	4.1	3.9	1.4
Proline	3.8	6.6	5.7	7.2	7.6	10.7	10.5	8.7	10.1	5.4	4.6	4.8	11.8
Serine	8.5	7.0	5.4	5.9	5.4	6.2	5.6	3.7	3.7	3.8	4.0	3.9	4.0
Threonine	4.0	5.4	4.7	4.0	4.1	3.2	2.9	3.3	3.4	4.0	5.1	4.9	1.8
Tryptophan	1.2	1.1	3.0	0.3	1.3		0.2		0.5	1.1	1.4	1.3	
Tyrosine	3.8	3.2	5.0	4.7	4.9	5.6	6.0	2.9	3.1	3.2	3.0	3.2	0.4
Valine	8.8	12.6	6.0	5.8	7.1	4.5	3.4	5.4	5.6	5.7	5.0	5.0	2.3

[a]Expressed as g/100 g protein.
[b]Normal maize.
[c]Opaque-2 maize.
[d]Expressed as % of crude protein (6.25) (assumes proteins contain an average of 16% N).
[e]Expressed as residues/1000 residues.

of protein is the balance among its essential amino acids. The essential amino acid composition of proteins is as variable as the number of proteins present in nature; no two proteins have an identical amino acid composition.

❑ FUNCTIONS

Proteins perform many different functions in the animal body. Most body proteins are present as components of cell membranes, in muscle, and in other supportive capacities such as in skin, hair, and hooves.

In addition to their role as structural components, proteins serve important specialized functions in the body, including a role in gene expression, enzyme catalyzed reactions, muscle contraction, metabolic regulation,and immune function (Table 9.3). All enzymes and many hormones produced by animals are proteins critical to survival, even though they do not contribute greatly to the total protein content of the body (Wu and Self, 2005). For example, erythropoietin, the first growth factor to be discovered, is a hormone produced by specialized kidney cells (now used in treating anemia (Fried, 1995), and lactoferrin, an iron-binding glycoprotein of milk, plays a role in Fe uptake by intestinal cells and acts as a bacteriostatic agent by withholding Fe from Fe-requiring bacteria (Lonnerdal and Iyer, 1995).

Tissue Proteins

Collagen.   The collagen molecule consists of a triple helix (2800 Å long and 15 Å in diameter) arranged in parallel and quarter-staggered to give the characteristic banded appearance of collagen. It has a compact structure and great strength. Collagen content increases with aging of the animal, and, because of its characteristic shrinkage on heating as a result of its high proline and hydroxy-proline content, the toughness of cooked meat from older animals is a well-known phenomenon. It is insoluble in water and resistant to digestive enzymes.

Elastin.   The elastin molecule resembles denatured collagen and consists of long, randomly ordered polypeptide chains. It is rubberlike in its response to stretching, and, when stretched to the elastic limit, it breaks more easily than collagen. It is always found in combination with a large proportion of collagen even in ligaments and artery walls, where it is most effective in restoring the tissue to its original shape or position. It is only a minor component of musculature but, like collagen, is insoluble in water and resistant to digestive enzymes.

Myofibrilar Proteins.   The myofibrilar proteins are the proteins of sarcoplasm, which is the material extracted from finely homogenized muscle with dilute salts. Most of the protein in this extract is in solution and contains more than 20 enzymes involved in muscle metabolism as well as mitochondrial fragments and particles of sarcoplasmic reticulum.

Contractile Proteins.   Three proteins—actin, tropomyosin B, and myosin—take part in muscle contraction. Myosin is the major component of the thick filaments of striated muscle. Its most impor-

TABLE 9.3   Roles of proteins in animals.

ROLES	EXAMPLES OF PROTEINS
Muscle contraction	Actin, myosin, tubulin
Enzyme-catalyzed reactions	Dehydrogenase, kinase, synthase
Gene expression	DNA-binding proteins, histones, repressor proteins
Hormone-mediated effects	Insulin, somatotropin, placental lactogen
Protection	Blood-clotting factors, immunoglobulins, interferon
Regulation	Imodulin, leptin, osteopontin
Storage of nutrients and O_2	Ferritin, metallothionein, myoglobin
Cell structure	Collagen, elastin, proteoglycans
Transport of nutrients and O_2	Albumin, hemoglobin, plasma lipoproteins

Source: Wu and Self (2004).

tant feature is its enzyme activity in an ATPase. Actin, when extracted from myosin, is in the globular form when ATP is present. The globular form polymerizes on the addition of neutral salts to give chainlike molecules of fibrous actin with the simultaneous change of the bound ATP to ADP and liberation of a molecule of inorganic P. The reverse process occurs in the absence of ATP. Tropomyosin B has no enzyme properties and does not combine in solution with either actin or myosin. The complicated interactions among the contractile proteins and their unique chemical structure are vital to normal muscle metabolism.

Keratins. Proteins of hair, wool, feathers, hooves, horns, claws, and beaks are similar in composition in that they are resistant to acid, alkali, and heat treatment and especially resistant to breakdown by digestive enzymes. Therefore, their value as protein supplements for animals is understandably limited unless they are processed adequately.

Blood Proteins. The chief proteins of the blood are albumin and a series of globulins (alpha, beta, gamma), along with thromboplastin, fibrinogen, the conjugated protein hemoglobin, and apoproteins A, B, and E (each existing in various isoforms). Most of these proteins are synthesized in the liver. The rate of synthesis and turnover varies from minutes to days; acute-phase proteins such as apoproteins and fibrinogen have very short half-lives (minutes or hours), whereas proteins such as albumin and hemoglobin turn over more slowly (days or weeks). The composition of the serum proteins of the normal pig at various ages is shown in Table 9.4. Blood contains a large array of conjugated proteins, including lipoproteins, enzymes, and protein–hormones and peptides whose distinct composition and structure and specific physicochemical properties al-

low separation and isolation by electrophoresis, centrifugation, extractions, and other means.

Enzymes. Enzymes, sometimes referred to as organic catalysts, are all protein in nature and are all relatively specific in their reactions. Some are hydrolytic (as the digestive enzymes), others are involved in other degradative metabolic reactions, and still others are involved in synthetic processes. Animal cells contain hundreds of enzymes, each with a specific structure and distinct reactive group. Net protein accretion during growth of the animal is the balance between protein synthesis and protein degradation, both dynamic metabolic processes. Between 15 and 22% of total energy expenditure is devoted to maintaining these processes (Reeds et al., 1985). Protein synthetic enzymes, under autocrine, paracrine, and endocrine control, regulate muscle protein synthesis. Tissues differ in their inherent rates of protein synthesis, mainly as a reflection of the RNA concentration. These differences in protein synthetic rate among tissues in several animal species are shown in Table 9.5 (Reeds, 1989). Examples of enzymes involved in body protein synthesis are shown in Table 9.6 (Goll et al., 1989). Protein degradation is controlled by many enzymes, some of which are listed in Table 9.6. The concepts of protein turnover and net protein accretion that result from these reactions are discussed under Synthesis and Degradation of Protein later in this chapter.

Hormones. Hormones, like enzymes, are produced by cells in minute quantities and have profound effects on metabolism. The action of enzymes is usually restricted within the same cell or in close proximity to the site of elaboration. In contrast, hormones are carried by the blood from the site of release to the target organ, often far re-

TABLE 9.4 Approximate serum protein concentrations of pigs at various ages.

	NEWBORN	6 WEEKS	3 MONTHS	1 YEAR
Total serum protein, g/dL serum	2 to 3	4 to 5	5 to 6	7 to 8
Albumin, g/100 g serum protein	18	47	45	52
α-globulin, g/100 g serum protein	60	25	29	18
β-globulin, g/100 g serum protein	16	19	16	13
γ-globulin, g/100 g serum protein	6[a]	9	10	16
Albumin: globulin ratio	0.2	0.90	0.82	1.1

[a]At 24 h (after sucking) the value is 45–48 g/100 g serum protein because of colostrum ingestion.

TABLE 9.5   Fractional rates of protein synthesis (% of protein pool synthesized/day) in several tissues of different animals.

ANIMAL	TISSUE				
	SKELETAL	HEART	LIVER	KIDNEY	INTESTINE
Rat	15	12	59	32	78
Pig	17.4	7	69	20	67
Sheep	17		95		87
Cattle	2		32		53

Source: Reeds (1989).

moved from the site of release. Not all hormones are proteins. Some important protein hormones are insulin, growth hormone, gonadotrophic hormone, parathyroid hormone, and calcitonin.

Metabolically Active Peptides and Polypeptides. A large number of peptides and polypeptides have been identified as important in modulating growth and other metabolic activities. Such compounds include insulinlike growth factors (IGF-I and IGF-II), transforming growth factor beta (TGF-β), fibroblast growth factor (FGF), nerve growth factor (NGF), and others. The exact functions of these compounds are not entirely elucidated, but it is clear that they play a major role in metabolism. Their concentration in tissues and their turnover rates are influenced by age, diet, and many physiological factors. Some apparently exert their action locally within the cell in which they are produced (autocrine effect), or in close proximity to their site of production (paracrine effect), or have classical hormone action (endocrine effect).

TABLE 9.6   Some proteolytic enzymes in animal tissues.

PROTEINASES (ENDOPEPTIDASES)	EXOPEPTIDASES
Serine proteinases	Amino peptidases
Cysteine proteinases	Tripeptidases
Aspartic proteinases	Dipeptidyl peptidases
Metalloproteinases	Tripeptidyl peptidases
	Carboxypeptidases
	Dipeptidases
	Omega peptidases

Source: Goll et al. (1989).

Immune Antibodies. Like enzymes, hormones, and peptides and polypeptides, antibodies constitute only a minute proportion of total body protein, but they perform a vital role in protecting the animal against specific infections. Antibodies against specific infections can be acquired (passive immunity) by placental transfer to the fetus from the blood of the dam, by ingestion and absorption of antibody-rich colostrum by the newborn, or by parenteral injection into the susceptible animal of purified sources of antibodies from other animals. Exposure of a susceptible animal to an antigen (pathogen) stimulates antibody production against the pathogen, resulting in active immunity.

Prions

Prions are proteinaceous infectious particles that lack nucleic acids (Prusiner, 1997; CDC, accessed March 2004; Larson, 2004, accessed March 2004) and are composed mostly, perhaps entirely, of an abnormal form of a normal cellular protein. They appear to multiply by converting normal protein molecules into abnormal ones through changes in their shape. Prions have a molecular size of about 55 kDa and are composed of an abnormal, pathogenic isoform of the prion protein (PrP) designated PrP-Sc, a term that was initially derived from the sheep disease, scrapie, the prototypic prion disease. (Scrapie has been recognized as a neurological disease of sheep for more than two centuries but is not believed to cause neuropathology in other species.)

Several neurological diseases of mammals have been described, in which prions are known to be the causative agent. As a group, these diseases are referred to as the transmissible spongiform encephalopathies (CAST, 2000). Some of the most notable are Creutzfeldt-Jakob disease (CJD) of humans, bovine spongioform encephalopathy (BSE), also called "mad cow disease," of cattle, scrapie of sheep and goats, and chronic wasting disease (CWD) of mule deer and elk. Other prion diseases are known to occur in mink, cats, and humans. Prions are associated with brain and other nervous tissue. Therefore, it is believed that consumption of food and feed products containing nervous tissue from prion-infected animals is a likely mode of transmission. The use of slaughter house meat by-products in the diets of ruminant animals in the United States has been banned recently because of this possible route of transmission.

The risk of infection of humans by BSE and the risk of spreading a prion from one animal species to another is unclear. However, the potential danger of such transmission is of concern. The recent discovery of BSE in cattle in Canada and the first known case of BSE in a cow imported to the United States from Canada has revived the threat of a repeat of the earlier experience with BSE and human CJD disease in the United Kingdom. Mice and hamsters are the animals of choice for studies of prion disease (CDC, accessed March 2004). The discovery of this group of abnormal proteins in animals and humans illustrates a variant of abnormal proteins in animal tissues.

Modulation by Specific Amino Acids of Immune Function

In addition to antibodies, consisting of intact proteins, in immune protection against pathogens, several amino acids have specific roles in immune function (Johnson and Escobar, 2005): (1) dietary methionine increases humoral and cell-mediated immunity; (2) arginine is a direct precursor of nitric oxide (NO), a cytotoxic agent produced by macrophages and neutrophils to kill bacterial pathogens; and (3) glutamine, though not essential in the diet for animal growth, is essential for normal functioning of macrophages and lymphocytes whose requirement for glutamine is increased during an immune response to pathogens due to this induced increase in metabolic activity.

❏ METABOLISM

Protein metabolism will be considered in two phases here: catabolism (degradation) and an-abolism (synthesis). Both processes proceed simultaneously in animal tissues. Individual amino acids, which are the basic units required in metabolism by the animal, normally are present in the diet as constituents of intact proteins that must be hydrolyzed to allow their component amino acids to be absorbed into the body. Thus, the conversion of dietary protein to tissue protein (or egg or milk protein) involves the following:

Intact dietary protein hydrolysis in GI tract
  (catabolism)
Amino acids in intestinal lumen absorption from GI
  tract
Amino acids in blood synthesis in tissues
  (anabolism)
Intact tissue proteins

Hydrolysis of dietary proteins is accomplished through proteolytic enzymes (exopeptidases; see Table 9.6) elaborated by epithelial cells lining the lumen of the GI tract and by the pancreas. The efficiency with which hydrolysis occurs determines the degree of absorption of individual amino acids and contributes to the nutritional value of the dietary protein. The other important factor contributing to nutritional value is the balance of absorbable essential amino acids. Proteins can be characterized nutritionally on the basis of digestibility and absorbability (protein hydrolysis in the GI tract and subsequent absorption of amino acids) as well as on the degree of utilization of amino acids after absorption. Dietary proteins not containing the proportion of essential amino acids to meet the animal's needs cannot be used efficiently for tissue protein synthesis. Even proteins hydrolyzed easily in the GI tract do not have a high nutritional value if they have a deficiency or an imbalance of one or more amino acids. This is exemplified by the growth failure observed in young animals fed a diet deficient in one of the essential amino acids.

Digestible protein refers to that disappearing from the ingesta as it passes down the GI tract. Apparent protein digestibility of feed represents the difference between what is in the feed and what is in the feces, which includes both unabsorbed and metabolic (endogenous) fecal N. Metabolic fecal N arises from normal metabolism of tissue protein and includes N from cells sloughed from the intestinal lining and residues of digestive enzymes and other substances secreted into the lumen of the GI tract. After absorption, amino acids are subject to further losses in utilization through metabolism in the liver and other tissues. These losses of N occur in the mammal mainly as urinary urea N and in birds as uric acid. Because the degree of utilization of a feed protein depends not only on its digestibility and absorbability but also on utilization of its component amino acids after absorption, protein sources can be ranked in a quantitative way as to their nutritive value. Nutritive value of proteins and its application to feeding practice are described under Measures of Nutritive Value of Proteins later in this chapter.

Absorption of Amino Acids

The intestinal epithelium is an effective barrier to diffusion of a variety of substances. There is very limited transfer of proteins, polypeptides, or even dipeptides across the intestinal epithelium except in early postnatal life (in most species during the first 24 h), when intact protein is absorbed by pinocytosis.

Certain dipeptides and tripeptides are absorbed from the lumen of the intestine of growing nonruminant animals (Leibach and Ganapathy, 1996) and from the rumen and omasum of ruminants (Webb et al., 1992). Specific peptide transporters are present in the intestine and kidney, suggesting a biological function for absorbed peptides.

In prematurely born human infants (but not in full-term infants), intact lactoferrin, a normal constituent of colostrum and milk, is absorbed from the intestine and appears intact in the urine. It is believed to stimulate intestinal mucosal growth and improve immune function (Nichols, 1988) and Fe uptake by intestinal cells (Lonnerdal and Iyer, 1995).

Absorption of amino acids takes place through active transport. The brush border membrane of the small intestine contains at least two active transport systems, one for neutral amino acids and one for basic amino acids. The amino acid moves across the intestinal cell membrane against a concentration gradient, requiring energy supplied by cellular metabolism. The naturally occurring L forms of amino acids are absorbed preferentially to the D forms, probably as a result of specificity of active transport systems. Removal of the carboxyl group by formation of an ester, removal of the charge on the amino group by acetylation, or introduction of a charge into the side chain of a va-

riety of amino acids prevents active transport, emphasizing the highly specific nature of the carrier system. Some amino acids may compete with each other for transport. For example, a high concentration of leucine in the diet increases the requirement for isoleucine. Arginine, cystine, and ornithine inhibit lysine transport, and arginine, lysine, and ornithine inhibit cystine transport. Some neutral amino acids inhibit basic amino acid transport; for example, methionine inhibits lysine transport. Apparently, basic amino acids do not inhibit neutral amino acid transport. Other amino acids— for example the three basic amino acids, ornithine, arginine, lysine—share a common transport system, with cystine. Proline and hydroxyproline apparently share a common transport system along with sarcosine and betaine, and have a high affinity for the neutral amino acid transport system as well as for their own. Whether a common transport system for glycine, alanine, and serine exists in intestinal epithelium as it does in some microorganisms is unknown. Current knowledge of amino acid transport in mammals has been reviewed (Kilberg et al., 1993).

Sites of Amino Acid Synthesis

Amino acids in the digestive tract of animals have three main sources: (1) those deriving from dietary ingestion, (2) those recycled along with other nitrogenous substances returned to the lumen of the intestinal tract from the body (digestive enzymes, bile, sloughed intestinal mucosal cells), and (3) those synthesized by microorganisms residing in the lumen of the digestive tract (the rumen in ruminants and the intestine in ruminants and nonruminants). Amino acids from all three sources are available for absorption, as described in the previous section. Here we focus on amino acid synthesis by microorganisms in the digestive tract and by body tissues in the host animal.

Microorganisms.   Bacteria and protozoans residing in the rumen and lower intestinal tract of ruminants and in the lower intestinal tract of nonruminants can synthesize all amino acids in the presence of ammonia, sulfur, and a carbon source. Thus, a symbiotic relationship exists between the host animal and the digestive tract microbes, wherein the microbes supply amino acids for use by the host animal and the animal provides a friendly environment for the microbes. Ruminants and herbivorous nonruminants such as rabbits and horses, whose large intestine and cecum contain large populations of microbes, therefore can derive most of their amino acid requirements from microbial production of amino acids from inorganic N. In rabbits and horses, microbially produced amino acids (along with vitamins) may be obtained from absorption from the lower intestinal tract and from ingestion of feces (coprophagy) containing unabsorbed amino acids. Nonruminant animals such as pigs, poultry, and humans, whose digestive tract does not contain large microbial populations, must rely on dietary intake of essential amino acids for survival. In ruminants, the microbial synthesis of amino acids from nonprotein N in the rumen is of major importance in meat and milk production and will be discussed more fully later in this chapter.

Body Tissues. Wu and Self (2005) describe the tissue synthesis and degradation of amino acids in animals. The nonessential amino acids (alanine, asparagine, aspartate, cysteine, glutamate, glutamine, glycine, proline, serine, and tyrosine) and their carbon skeletons can be supplied from tissue synthesis. Skeletal muscle, liver, brain, adipose tissue, intestine, kidney, lung, placenta, lactating mammary gland, and other tissues take part to varying degrees in these synthetic conversions.

Sites of Amino Acid Degradation

Microbes in the digestive tract degrade all amino acids. Major products of degradation are ammonia, S, fatty acids (including acetate, propionate, butyrate, and branched-chain fatty acids), and carbon dioxide. In animals, amino acids are degraded by cell- and tissue-specific pathways. Liver is the principal organ degrading all amino acids except branched-chain amino acids (BCAA) and glutamine. Degradation of many essential and non-essential amino acids also occurs extensively in small intestinal cells. Amino acid degradation in animals occurs by a variety of enzymatic reactions, including transamination of BCAA by transaminase, decarboxylation of ornithine to putrecine by ornithine decarboxylase, hydrolysis of arginine to ornithine by arginase, and hydroxylation of arginine to citrulline by NO synthase (Wu and Self, 2005). Further description of the complexity and variety of metabolic conversions in amino acid metabolism is not necessary here.

Nitrogen Cycling in the Intestine

In addition to the absorption of amino acids and di- and tripeptides containing nitrogen (N) from the lumen of the intestine into the blood, there is also recycling of N into the lumen from the body, occurring concurrently (Fuller and Reeds, 1998). This endogenously produced N, including urea, entering the intestinal tract lumen is subjected to the same microbial action as that ingested. Because these digestive tract microbes possess enzymes that the host does not (for example, urease), they can utilize substrate such as urea entering the lumen (by both secretion and diffusion) by releasing ammonia from urea. The ammonia is used for incorporation into amino acids by microbes to produce microbial protein, which can then be digested by the proteolytic enzymes produced by the host. The absorbed amino acids are then used by the host animal to contribute to its amino acid supply. The amounts of amino acids made available to the animal through this recycling scheme are considerable for ruminants and for nonruminant herbivores (e.g., horses and rabbits).

The contribution of endogenous sources of N to the nutrition of the animal can be estimated in several ways, including (1) feeding a protein-free diet (assumes that all N in digesta is of endogenous origin); and (2) labeling endogenous protein by intravenous infusion of N-15 tracer and measuring the fractional contribution of endogenous amino acid N compared to total N in the ileal digesta or feces. Such approaches are subject to error and are difficult to interpret, but still they suggest important contributions of endogenous sources of N to the total amino acid nutrition of the animal. These types of stable isotope labeling studies (reviewed by Fuller and Reeds, 1998) have revealed that a high proportion of intestinal mucosal cell amino acid supply for enterocyte growth and maintenance is derived from endogenously supplied amino acids rather than from amino acids absorbed directly from the lumen of the intestine.

Mass spectrometry is widely used to measure the stable isotopes, C-13 and N-15, in serially collected blood samples to estimate protein turnover and monitor recycling of N in the intestine of animals and humans (Fuller and Reeds, 1998). Incorporation of N-15 into plants grown in an N-15 enriched environment, as done experimentally in laboratories such as the U.S. Department of Agriculture's Children's Nutrition Research Center in

Houston, Texas, offers the opportunity to track protein and amino acid metabolism in animals and humans fed plant constituents enriched with stable isotopes.

New Concepts of Nitric Oxide in Animal Nutrition and Health

Nitric oxide (NO) has a critical role in virtually every cell and organ function in the body (Wu and Meininger, 2002). It is a mediator of immune function, a neurotransmitter, a signaling molecule, and an endothelium-derived relaxing factor, and has other positive roles in metabolism, but it is also a cytotoxic free radical. NO is synthesized from arginine through action of the enzyme nitric oxide synthase (NOS). Several nutrients, including protein and amino acids, though generally beneficial to health, may modulate NO production by affecting NOS. For example, protein deficiency or specific amino acid deficiencies reduce inducible NOS, which lowers NO synthesis, and may result in impaired immune or neurological function in animals. Relatively little is known about the effects of interactions among dietary factors on NO synthesis or underlying molecular mechanisms. The total practical impact of NO in normal metabolism and nutritional manipulation in animals is yet to be determined.

Ruminant Nitrogen Metabolism

Unlike nonruminant animals such as swine and poultry, whose nitrogen needs are met largely by ingestion of essential and nonessential amino acids, ruminants such as cattle and sheep are able to use nonprotein N to derive a major part of their protein requirements due to the large microbial populations residing in the rumen. Rumen anaerobic microbes that produce the enzyme cellulase are able to degrade cellulose present in fibrous plants and to produce short-chain fatty acids (see Chapters 7 and 8) to provide energy for the host animal. Rumen microbes also synthesize amino acids and protein from inorganic N provided in the diet of the animal or from reentry of urea and ammonia into the rumen from salivary secretions or passage into the rumen from the body. Thus, ruminant feeding centers on providing a diet that serves not only the N needs of the microbial population in the rumen but also the amino acid needs of the host animal. An excessively high-protein diet fed to a ruminant results in ammonia release, which is absorbed and wasted as urea lost in the urine, whereas a low-protein diet can be supplemented by microbial protein derived from endogenous recycling of urea.

Manipulation of ruminant diets tends to emphasize the feeding of protein that largely escapes fermentation and ensures passage of dietary protein and amino acids into the small intestine. If protein and amino acids enter the small intestine intact, their digestion and absorption is similar to that of nonruminants.

The protein that reaches the lower digestive tract of the ruminant is the sum of the protein that escaped rumen fermentation and the microbial yield (washout of microbes from rumen). Escaped protein may be beneficial because it will be utilized efficiently in the small intestine if it contains an abundance of essential amino acids. Rumen escape is likely variable and depends on the type of protein and its rate of degradation, which, in turn, depends on whether it is soluble and moves with liquid or is insoluble and moves with particulate matter (Van Soest, 1999). The concentration of rumen undegraded protein (RUP) will depend on the dietary protein concentration and level of intake. Therefore, high-quality feeds with high-protein content will be more associated with the escape of larger quantities of dietary protein than those with low-protein content.

Van Soest (1999) described the metabolism of N compounds in ruminants fed high- and low-protein diets. There is an increase in rumen ammonia production and its transfer to liver in animals fed high protein and an accompanying increase in urinary area.

Active research continues to focus on elucidating the interactions among feed composition, level of intake, physiological state of the animal, rumen microbial populations, and other factors that affect the protein and amino acid nutrition of ruminant animals. Nutrition of ruminants is intimately involved with symbiotic relationships with microbial populations in the rumen. It is an exciting and important field of animal nutrition.

Fate of Amino Acids after Absorption

The fate of amino acids after absorption can be divided broadly into three categories: (a) tissue protein synthesis; (b) synthesis of enzymes, hormones, and other metabolites; and (c) deamination or transamination and use of the carbon skeleton for energy. The first two fates are discussed together because synthesis involves similar metabolic processes. Deamination and transamination are described on page 130.

Synthesis and Degradation of Protein. Protein synthesis in animal tissues requires the presence of nucleic acids. All living cells contain many different nucleic acids that play vital roles. Deoxyribonucleic acid (DNA), a chromosomal component of cells, carries the genetic information in the cell and transmits inherited characteristics from one generation to the next. It is the blueprint of protein synthesis. DNA controls the development of the cell and the organism by controlling the formation of ribonucleic acid (RNA). There are three different kinds of RNA in cells: ribosomal RNA, transfer RNA, and messenger RNA. All three types are involved in the synthesis of proteins. Ribosomal RNA is part of the structure of the ribosome, which is the site of formation of proteins in the cell. Transfer RNA carries specific amino acids to the ribosome, where they interact with messenger RNA. Messenger RNA determines the sequence of amino acids in the protein being formed. Thus, synthesis of each protein is controlled by a different messenger RNA. This basic information in protein synthesis is needed to appreciate the overall role of proteins in normal growth and development of animals.

DNA is composed of phosphate-linked deoxyribose and four nitrogenous bases: adenine, cytosine, guanine, and thymine. Their two- and three-dimensional structures and the structure of DNA are shown in Fig. 9.2. The DNA molecule is in the form of a long double-helix chain of nucleotides composed of phosphate-linked deoxyribose sugar groups, to each of which is attached one of the four bases. The bases always are paired, adenine with thymine and guanine with cytosine. The sequence of pairs can vary infinitely, and this sequence determines the exact protein to be synthesized by the cell (Watson et al., 1983). The paired arrangement is illustrated diagrammatically in Fig. 9.2 (Felsenfeld, 1985). Changes in DNA occur in mutation and subsequent selection.

The transfer of information from DNA in the nucleus to the site of protein synthesis in the cytoplasm is accomplished by RNA. The composition of RNA is similar to that of DNA, except that ribose replaces deoxyribose and uracil replaces thymine. The nucleotides of RNA are linked through their phosphate groups to form long chains as in DNA.

Protein synthesis occurs through transfer of amino acids to ribosomes, particles attached to membrane surfaces and to which amino acids are linked in sequence predetermined by the sequence of nitrogenous bases in DNA and, in turn, in RNA. Ribosomes have molecular weights approximating 4 million daltons, 40% of which is RNA, with the remainder being protein. The ribosomes act as templates for the orderly array of amino acid linkage during protein synthesis. The addition of amino acids to a growing polypeptide chain in this way occurs very rapidly. For example, a molecule of hemoglobin is synthesized in about 1.5 min.

All of the information needed to determine the three-dimensional structure of a protein is associated with the amino acid sequence. Amino acids are classified on the basis of the size of the side chain and the degree to which the amino acid is polarized (separated regions of plus and minus electric charge).

Translation of messenger RNA (mRNA) into protein at a ribosome as described by Darnell (1985) is briefly summarized here. A transfer RNA (tRNA) recognizes a codon by means of a complementary nucleotide sequence called an anticodon. The addition of one amino acid to a protein chain involves the following sequence. An incoming tRNA molecule carrying a particular amino acid binds to the codon exposed at a binding site on a ribosome and then forms a peptide bond with the next amino acid on the protein chain, for example, serine. As the ribosome advances one codon, exposing the binding size to the incoming tRNA, the serine tRNA is released. This transfer procedure continues until the translation of mRNA into protein is complete. Through all of the differences in functions of proteins (antibodies, hormones, peptide growth factors, enzymes, structural proteins) there is a commonality: they work by selectively binding to receptor molecules on cell membranes.

Active research in live animals and in cells, tissues, and organs grown in culture media promises to provide insight into the control mechanisms governing rate of muscle growth, fetal growth and differentiation, and lactation in animals. Isolated organ, tissue, and cell culture techniques allow testing the effects of individual components, such as peptide growth factors, and other serum factors singly and in combination on rate of protein synthesis and degradation.

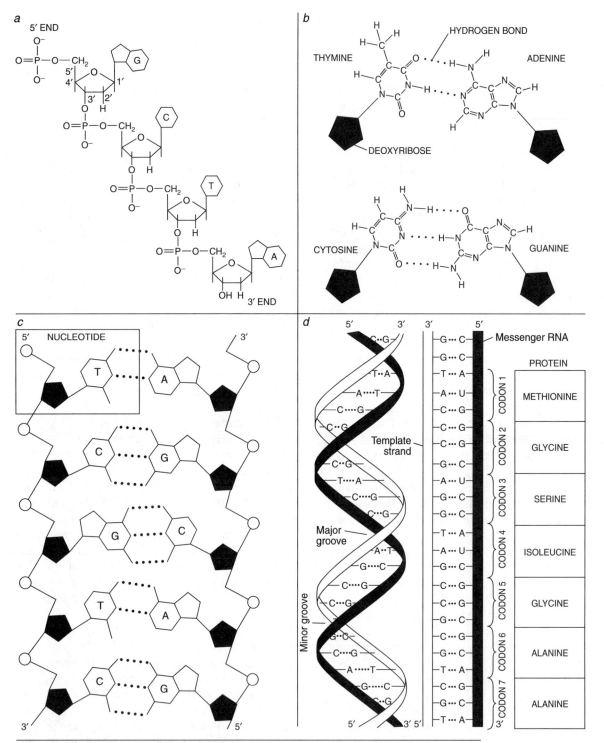

Figure 9.2 *Structure of DNA has a backbone (a) made up of bonded sugar and phosphate groups, to each of which is attached cone of four bases: guanine (G), cytosine (C), thymine (T), or adenine (A). The phosphate group is represented by the structures with the P at the center, the sugar by the pentagon with an oxygen atom (O) at the top. A phosphate group connects the 5′ carbon atom of one sugar to the 3′ carbon atom of the next. The combination of sugar–phosphate group and base constitutes a nucleotide (b). (The distances between atoms are not to scale.) The nature of the hydrogen bonding of the bases is such that the thymine always pairs with adenine and cytosine always pairs with guanine. The structure that results is shown in two dimensions (c) and in three: the double helix (d). In conveying the genetic message of DNA, the sequence of the coding strand is transcribed into a strand of messenger RNA, which serves to make a variety of proteins. The U in the strand of messenger RNA stands for uracil, the RNA counterpart of thymine. Reproduced with permission from* Scientific American **253** *(4): 58–67 (page 60), October 1985.*

Protein degradation in muscle tissue is controlled by cathepsins and calcium-dependent proteinases (calpains) that are important determinants of the net rate of muscle growth in animals. Knowledge of protein turnover and the function of cellular organelles is growing rapidly.

Protein synthesis and degradation in living animals is important in developing a more complete understanding of the growth process and in controlling it for efficient meat production. The relative importance of changes in rate of muscle protein synthesis versus muscle protein degradation in contributing to net protein accretion during growth is an area of active research. The impact of dietary protein quality and the amounts of protein on body protein synthesis, degradation, and deposition are illustrated by data obtained in the pig (Reeds, 1989) in Table 9.7. Knowledge of the mechanisms of muscle protein synthesis and degradation generated from basic research in animals used for meat production also may find application to humans.

Whole-Body Protein Turnover

Protein turnover in the whole body is a dynamic process involving continuous and simultaneous protein synthesis and protein degradation. The relative rates of each process within a given body tissue determine the net change in protein mass of that tissue. Each organ or tissue in the body has its own rate of protein turnover. Muscle is a major constituent of body mass. In growing animals, the rate of whole body muscle growth is therefore affected by the rate of protein synthesis relative to protein degradation. Measurement of protein turnover (reviewed by Waterlow, 1995 and by Fuller and Reeds, 1998) includes as a valuable tool the use of stable isotopes of nitrogen and carbon (N-15 and C-13), to follow enrichment of precursors and end products of amino acids and recycling of labeled amino acids. Such an approach allows one to determine such factors as physiological state, level of dietary protein or energy intake, tissue injury, or disease on protein turnover in the whole body or in individual tissues, for example, liver and muscle. Kinetic studies of protein turnover using stable isotopes and mathematical modeling procedures are expensive and difficult, but advances in this methodology continue to provide new knowledge of importance in animal and human nutrition.

The growth process is fundamental to all living organisms; it is not restricted to young growing animals but is characteristic of maintenance of tissues whose cellular turnover continues throughout life, such as skin and intestinal mucosa. As discussed earlier in this chapter, a series of peptides and polypeptides has been discovered and characterized with respect to their role in cellular metabolism (Clemmons and Underwood, 1991). These substances take part in growth regulatory processes and probably far outnumber the classical polypeptide hormones such as growth hormone and insulin, whose structures and general functions have been known for years. Polypeptide growth factors (PGFs), which have been characterized in detail with regard to amino acid sequence, include epidermal growth factor (EGF), insulinlike growth factors I and II (IGF-I and IGF-II), nerve growth factor (NGF), and platelet-derived growth factor (PDGF), among others. At least 20 other PGFs have been partially or initially characterized. Full appreciation and knowledge of

TABLE 9.7 Differential effects of protein quality and quantity on protein balance and turnover in growing pigs.

DIET	PROTEIN BALANCE (G PROTEIN DAY^{-1})	PROTEIN SYNTHESIS (G PROTEIN DAY^{-1})	PROTEIN DEGRADATION (G PROTEIN DAY^{-1})
Low protein	49	655	606
Low protein + lysine	104	637	533
High protein	128	706	578
High protein + lysine	117	757	640

Source: Reeds et al. (1989).

the role of this broad array of PGFs in animal growth and biology await further research. It does seem clear, however, that protein metabolism and animal development and maintenance at the cellular, tissue, and organ levels are influenced greatly by these PGFs. Actions vary from endocrine (elaborated in a specific cell type and transported to target cells by the blood) to paracrine (elaborated in a specific cell type and transported to target cells in close proximity by diffusion) to autocrine (elaborated in a specific cell type and active in the same cell in which produced).

In food animal production, scientists have devoted considerable attention to the role of IGF-I and IGF-II in fetal development and in muscle growth and lean tissue accretion. The role of nutrition in PGF metabolism and the overall effects of such actions on animal growth and productivity are of ongoing interest.

Deamination and Transamination

Deamination involves removal of the amino group from the carbon skeleton of the amino acid and entrance of the amino group into the urea (ornithine) cycle. The process is illustrated as follows:

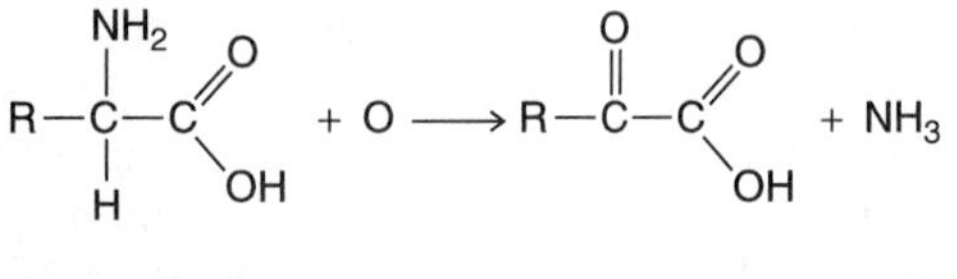

Transamination involves the transfer of an amino group from one amino acid to the carbon skeleton of a keto acid. The process is illustrated in the following:

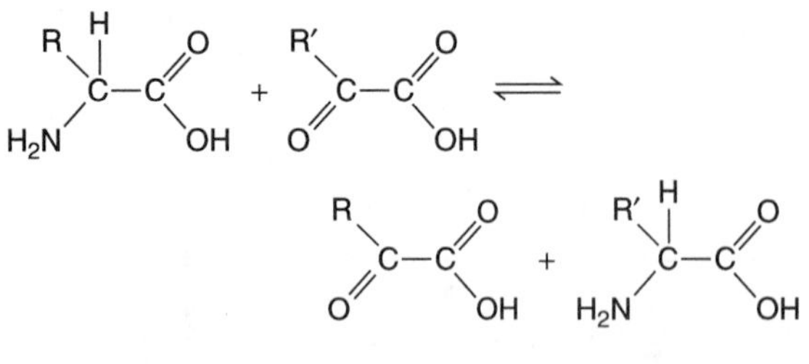

Amino acids available for protein synthesis at the tissue level arise either from absorption from the GI tract or from synthesis within the animal tissues by transamination. The ability of animal tissues to synthesize amino acids from other compounds is the basis for their classification as essential or nonessential. Of the 10 amino acids listed

as essential for most simple-stomached animals, several can be replaced by their corresponding α-hydroxy or α-ketoanalogs, illustrating that it is the carbon skeleton that the animal is unable to synthesize. For example, the α-keto analogs of arginine, histidine, isoleucine, leucine, methionine, phenylalanine, tryptophan, or valine promote normal growth of rats fed diets devoid of the corresponding amino acid but adequate in amino group donors. The α-hydroxy analogs of isoleucine and tryptophan partially can replace the amino acid, whereas those of threonine and lysine are not utilized for growth by rats. Several amino acids, though not themselves dietary essentials, are synthesized from essential amino acids. Cystine is produced from methionine and can replace approximately half the dietary methionine; tyrosine is produced from phenylalanine and can replace approximately one-third the dietary phenylalanine. Many amino acids are precursors or supply part of the structure of other metabolites. For example, methionine supplies methyl groups for creatine and choline and is a precursor of homocysteine, cystine, and cysteine; arginine, when urea is removed, yields ornithine; histidine is decarboxylated to form histamine; tyrosine is iodinated in the thyroid gland to form the hormone thyroxine and is used in the formation of adrenaline, noradrenaline, and melanin pigments; tryptophan is a precursor of serotonin (5-hydroxytryptamine) and the vitamin, niacin; lysine and methionine are the precursors of carnitine [β-hydroxy-(γ-N-trimethylammonia)-butyrate].

Urea Cycle

Ammonia resulting from deamination is joined by carbon dioxide and a phosphate group from ATP to form carbamyl phosphate, which in turn combines with ornithine to form citrulline. Through a series of reactions involving formation successively of citrulline, arginosuccinate, and arginine, the formation of urea takes place (Fig. 9.3). The urea or ornithine cycle is a key metabolic phenomenon in protein metabolism.

The cycle is composed of five enzymes, three of which involve arginine metabolism (argininosuccinate synthetase, ASS; argininosuccinate lyase, AS; and arginase) (Morris, 2002). The urea cycle also requires other enzymes and mitochondrial amino acid transporters to fully function. Liver and to a lesser degree intestinal cells (enterocytes)

One Turn of the Cycle

Carbomyl Phosphate + Aspartate $-\!-$ Urea + Fumarate + 2 ADP + 2 P$_i$
          + 3 ATP + 2 H$_2$O $-\!>$ + AMP + PP$_i$

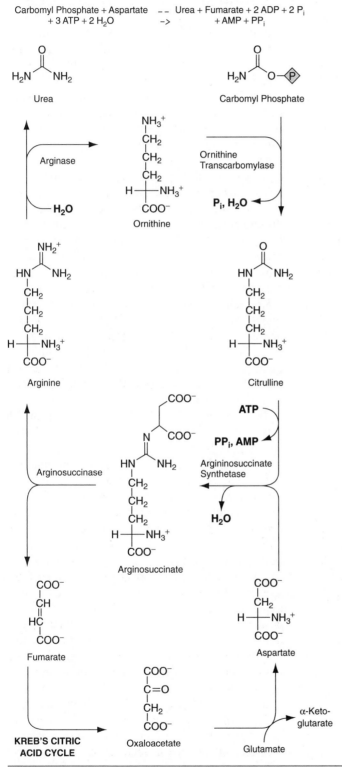

Figure 9.3 *The urea (ornithine) cycle that occurs in mammals. Note the interface with the Krebs citric acid cycle shown at the bottom. (Source: www.gwu.edu/mpb/urea.htm, accessed December 4, 2003).*

operate the urea cycle. The hormones glucagon, insulin, and glucocorticoids regulate the expression of urea cycle enzymes in the liver. The recent recognition that widespread arginase activity exists in nonhepatic cells throughout the body suggests possible roles for arginases, including regulation of arginine availability for nitric oxide (NO) synthesis. Therefore, interest in arginase beyond its role in the urea cycle has been raised in relation to its role in NO function mentioned previously in this chapter. The breakdown of arginine produces urea and ornithine, making ornithine available to repeat the cycle. Urea is excreted in urine and is the chief route of N excretion in mammals. In birds, ornithine synthesis does not occur, and the main route of N excretion is as uric acid. Uric acid is the principal end product of purine metabolism in humans and other primates, but in other mammals the principal end product is the oxidation product of uric acid, allantoin. The structural formulas of urea and uric acid are:

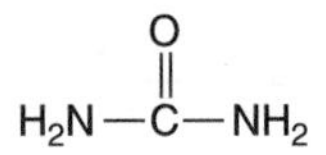

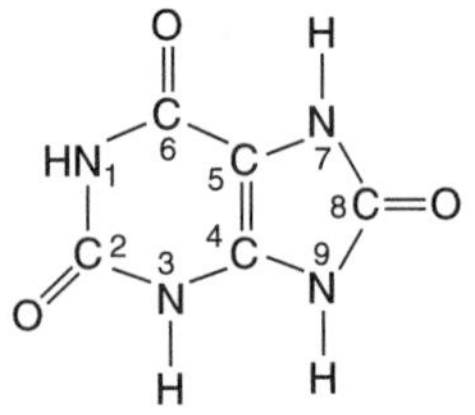

The N in position 1 comes from aspartate, that in positions 3 and 9 from glutamine, and that in position 7 from glycine. This high metabolic requirement for glycine, which is used for synthesis of uric acid, probably accounts for the fact that during periods of rapid growth the chick may require dietary glycine, even though some tissue synthesis occurs.

Hippuric acid is a common constituent of the urine of herbivorous animals, where it acts as a detoxification product of benzoic acid, which often is present in high quantities in plant diets. Its structural formula is:

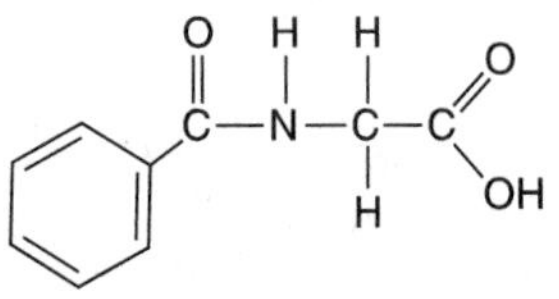

PROTEIN AND AMINO ACID REQUIREMENTS AND DEFICIENCIES

There is no evidence of a metabolic requirement for dietary protein per se, but only for amino acids. Intact proteins such as immune globulins and lactoferrin enter the body without hydrolysis in unsuckled neonates, and peptides are absorbed directly from the rumen and omasum of ruminants, but there is no proof of a metabolic requirement for these proteins and peptides per se. Nonruminant omnivores such as the pig, chicken, and human require specific dietary amino acids (essential amino acids), whereas some nonruminant herbivores such as horses and rabbits are able to utilize nonprotein N for microbial amino acid synthesis in the lower part of the GI tract. Adult ruminants, such as cattle and sheep, can depend entirely on nonprotein N in the diet by virtue of their rumen microbes, which can synthesize amino acids and protein from nonprotein N and which are themselves subsequently utilized as a source of protein by the host animal. Quantitative requirements for N and amino acids for different species performing various productive functions are discussed in more detail in later chapters that address the feeding and nutrition of individual classes of animals.

Inadequate protein (N or amino acids) probably is the most common of all nutrient deficiencies because most energy sources are low in protein and protein supplements are expensive. The quantitative protein requirement is greater for growth than for maintenance and is affected by sex (males tend to have a higher requirement) and by species, and probably by genetic makeup within species.

The ratio of calories to protein in the diet is important. Most animals tend to eat to satisfy energy requirements. Growing pigs and chickens fed a diet containing a marginal level of protein can be made protein deficient if caloric density of the diet is increased by fat. This occurs because the reduced daily intake of a high-calorie diet provides insufficient protein intake if the percentage of protein is marginal. Protein is diverted to energy only when it is provided in excess of the metabolic requirement or when calorie intake is insufficient.

Signs of protein deficiency include anorexia, reduced growth rate, reduced or negative N balance, reduced efficiency of feed utilization, reduced serum protein concentration, anemia, fat accumulation in the liver, edema (in severe cases), reduced birth weight of young, reduced milk pro-

Figure 9.4 *Amino acid deficiency in growing pigs. Littermates fed an adequate diet or the same diet deficient in tryptophan.*

duction, and reduced synthesis of certain enzymes and hormones. A young pig fed a diet deficient in tryptophan is shown in Fig. 9.4 alongside a littermate fed a similar diet adequate in all amino acids.

The small stature, low serum protein concentration, anemia, and edema ("potbelly" appearance) of infants suffering from kwashiorkor are typical manifestations of protein deficiency in humans. In severe protein or amino acid deficiency, growth is arrested completely.

Deficiencies of individual essential amino acids generally produce the same signs listed above because a single amino acid deficiency prevents protein accretion in the same way that a shortage of a particular link in a chain prevents elongation of the chain. Thus, individual amino acid deficiencies result in deamination of the remaining amino acids, loss of the ammonia as urea, and use of the carbon chain for energy. Certain amino acid deficiencies produce specific lesions. For example, tryptophan deficiency produces eye cataracts; threonine or methionine deficiency produces fatty liver; lysine deficiency in birds produces abnormal feathering.

Most individual feedstuffs are inadequate in one or more amino acids for growing animals. For example, corn is especially deficient in lysine and tryptophan (Fig. 9.5) and supports very slow growth. However, by providing soybean meal rich in lysine and tryptophan, an adequate balance of amino acids is provided. Thus, although soybean meal is deficient in methionine–cystine and will not support normal growth by itself, when combined with corn, the two feedstuffs complement each

other by compensating for their individual amino acid deficiencies. The same dramatic complementary effect of two proteins is seen at the breakfast table when milk is mixed with corn flakes or another cereal to provide a balanced protein meal. The percentages of the dietary requirements for essential amino acids provided by a diet containing 95% corn are shown in Fig. 9.5, along with the amino acid requirements for the 10- to 20-kg pig. Mertz et al. (1964), discovered a mutation in corn that resulted in high-lysine content of the affected variety. Pigs fed this high-lysine corn as the sole source of dietary protein grow faster than those fed normal corn. When a protein supplement such as soybean meal is provided to make a mixed diet containing adequate protein, the requirement for all of the essential amino acids is met. Crystalline amino acids may be used to balance amino acid deficiencies in practical diets. During the past few years, the microbial production of large quantities of amino acids by recombinant DNA technology has stimulated the feed use of tryptophan, threonine, and other amino acids that were previously available only at prohibitively high cost for use in animal diets. This technology permits the use of lower protein diets to provide adequate protein and amino acids for growth and other productive functions.

The concept of ideal protein diets for nonruminant animals (Baker, 1992) is based on the formulation of biologically available amino acid mixtures that exactly match the quantitative amino acid requirements of the animal in a given physiological state of growth and with a particular genotype.

The delicate balance among amino acids is promoting growth is shown in Table 9.8, in which the importance of the level of limiting amino acids on growth of pigs is illustrated. The addition of the first limiting amino acid, lysine, does not produce a maximum growth response unless the second and third amino acids, threonine and tryptophan, are supplemented. Such relationships must be taken into account in formulating diets for nonruminant animals.

❏ USE OF D-AMINO ACIDS BY NONRUMINANTS

The unnatural D isomer of some essential amino acids can be used for growth by most animals. D-lysine, D-threonine, D-cystine, D-arginine, and

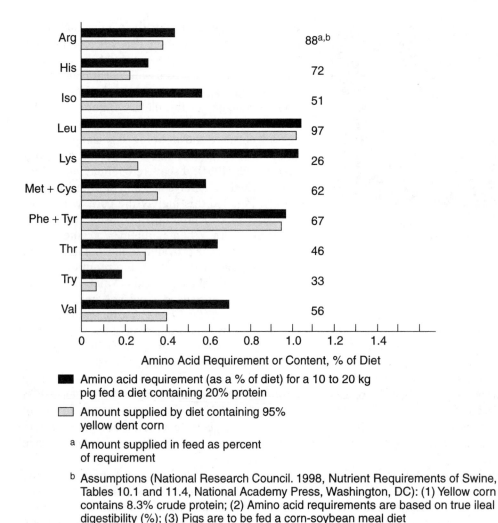

Figure 9.5 *Amino acid requirements of a 10- to 20-kg pig and the amounts supplied by a diet containing 95% yellow corn.*

D-histidine cannot be converted to the L isomer, whereas other D-amino acids can be converted to the L form by D-amino acid oxidase and transaminase (Wu and Self, 2005). The efficiency of conversion varies from 20 to 100%, depending on animal species. Similarly, most alpha-keto acids can form the corresponding L-amino acids by transamination.

❑ USE OF NONPROTEIN NITROGEN (NPN) BY NONRUMINANTS

Urea utilization by nonruminant animals for nonessential amino acid synthesis depends on its hydrolysis to ammonia. Current evidence indicates

that mammalian cells do not produce urease. Thus, any incorporation of N of urea into body tissues depends on microbial urease in the lumen of the GI tract for hydrolysis of urea. Microorganisms can use ammonia for synthesis of amino acids and proteins, but the restriction of ureolytic activity to the area of the GI tract beyond the site of maximal amino acid absorption by the host would limit the availability of amino acids that the microbes synthesize, unless coprophagy occurs. Therefore, in nonruminant animals a response to urea is possible only if the basal diet meets essential amino acid requirements, but not total N requirement, and the ammonia released by bacterial urease is incorporated into nonessential amino acids by body tis-

TABLE 9.8 Importance of amino acid balance in promoting growth of pigs.

DIET	NUMBER OF PIGS[a]	DAILY BODY WEIGHT GAIN, G	FEED PER GAIN, G	SERUM UREA N, MG/DL
Basal (B)[b]	16	490	3.21	12.1
B + tryptophan (tr)	16	530	3.02	10.4
B + threonine (th)	16	440	3.29	10.2
B + tr + th	16	610	2.69	8.9
High-protein diet[c]	16	690	2.66	14.6

[a]Pigs weighed 24 kg initially and were fed the diets for 28 days.
[b]Diet contained 86.7% corn, 10.2% sunflower meal, 0.52% added L-lysine monohydrochloride, plus minerals, vitamins, and an antibiotic supplement; total protein content was 12%.
[c]Diet contained 74.7% corn, 22.7% sunflower meal, 0.31% added L-lysine monohydrochloride, plus minerals, vitamins, and an antibiotic supplement; total protein content was 16%.
Source: Wahlstrom et al. (1985).

sues. Reports of such an effect in humans and other species suggest that some urea N is incorporated into body tissues if the above conditions are met.

❑ USE OF NONPROTEIN NITROGEN (NPN) BY RUMINANTS

In ruminants, where the rumen microbial activity is anatomically "upstream" from the sites of most active absorption of nutrients, NPN can be the main dietary N component. Usually, however, performance of ruminant animals fed NPN is improved by supplementation with intact protein.

Nitrogenous compounds can be classified as protein or NPN. In the rumen, most ingested food proteins are hydrolyzed to peptides and amino acids, most of which will be degraded further to organic acids, ammonia, and carbon dioxide. Most rumen bacteria utilize ammonia, and many require it. When protein well balanced in essential amino acids is fed to the host, it is largely degraded and new microbial protein is synthesized from dietary amino acids, as well as from ammonia and organic acids present in the rumen. The net result is that the mixture of amino acids supplied to the lower part of the GI tract for absorption is remarkably constant. Some protein sources, such as those in certain processed plant byproducts and in some animal protein products, are incompletely digested in the rumen and pass into the small intestine partially intact, where they are subjected to further hydrolysis by the proteolytic enzymes of the host an-

imal. Such proteins are called bypass or escape proteins (see the earlier section, Ruminant Nitrogen Metabolism).

It has been proposed that some bypass protein is necessary for maximum growth and milk production. Proteins can be protected from rumen degradation by lipid encapsulation, chemical derivitization such as by formaldehyde treatment, or use of inhibitors of microbial amino acid deamination. Protective coating can be used as a means of preserving proteins of high biological value by inhibiting their hydrolysis by rumen microorganisms.

The ammonia released from deamination of amino acids or from hydrolysis of dietary urea may be absorbed through the rumen or lower part of the GI tract wall for use at the tissue level in amino acid synthesis or for urea synthesis in the liver. An appreciable amount of urea formed by the liver is excreted into the rumen via saliva or through the rumen wall and thereby made available to the rumen microflora for amino acid synthesis.

Urea is only one of many NPN compounds that may be ingested. Many forages have large amounts of NPN in the form of amino acids, peptides, amines, amides, and nucleic acids. Such NPN compounds (with the possible exception of nucleic acids) are readily available to rumen microorganisms. The animal's efficient utilization of NPN compounds depends on the solubility of the NPN and on the availability to the microflora of readily available carbohydrates. The classical work of Loosli et al. (1949) established that milk production can be sustained in cattle fed diets containing only NPN, which is then used by rumen microorganisms for amino acid synthesis. Beef cattle can also survive and grow on all NPN diets through this process of amino acid synthesis by rumen microflora (Oltjen, 1969).

Microbial protein formed from NPN compounds has a high nutritive value. Lambs and calves fed NPN and little or no protein grow well. Cows fed NPN and no protein or amino acids can lactate and reproduce normally (Loosli, 1949; Virtanen, 1966). However, it is not possible to achieve maximum production with animals fed diets with only NPN as an N source because microbial synthesis of limiting amino acids is insufficient to meet the needs for production of muscle protein and milk in genetically superior animals.

Supplementation of ruminant diets with methionine, lysine, and other amino acids can augment milk production in dairy cattle whose productivity is limited by insufficient microbial protein synthesis.

❏ AMINO ACID ANTAGONISM, TOXICITY, AND IMBALANCE

Amino acid antagonism refers to growth depression that can overcome by supplementation with an amino acid structurally similar to the antagonist. Excess lysine, for example, causes a growth depression that, in chicks, can be reversed by additional arginine. Antagonism differs from imbalance in that the supplemented amino acids need not be limiting (Harper et al., 1970).

The term *amino acid toxicity* is used when the adverse effect of an amino acid in excess cannot be overcome by supplementation with another amino acid. Methionine, if added to the diet in excess, produces growth depression that is not overcome by supplementation with other amino acids.

Amino acid imbalance has been defined as any change in the proportion of dietary amino acids that has an adverse effect preventable by a relatively small amount of the most limiting amino acid or acids. A simple method for detecting amino acid imbalances is to offer a choice of an amino acid imbalanced diet and a protein-free diet. Rats and pigs will reject a diet that is unbalanced but will support life, and consume largely the protein-free diet that will not support life. The rejection of the imbalanced diet is probably a result of some biochemical or physiological disturbance. It may be related to the change in plasma-free amino acid pattern

that, in turn, alters the satiety center in the hypothalamus or the release or metabolism of one or more peptides associated with food intake behavior.

Adverse effects of excess consumption of individual amino acids have been reviewed (Benevenga and Steele, 1984). Interest in this subject has arisen partly because of the existence of inherited disorders such as phenylketonuria, in which affected individuals lack the necessary enzymes to metabolize phenylalanine properly, and tyrosinemia II, in which tyrosine aminotransferase is lacking. In each of these diseases special diets low in the respective amino acid must be provided to allow normal development. Tryptophan, histidine, and methionine toxicity have been studied extensively in animals. Feed intake and growth are severely depressed by excesses of any of these three amino acids and to a lesser degree by excesses of other amino acids. Adverse effects of excess consumption can vary from a slight suppression of food intake followed by a return to normal intake, to marked suppression of intake, tissue damage, and death (Harper et al., 1970). The D-isomers of amino acids are also toxic when consumed in excess and in some cases may be more toxic than the natural L-forms.

Exercise and Protein Metabolism

The effect of strenuous exercise on protein and amino acid metabolism has been of interest in athletes and among those working with horses and other draft and racing animals. Rennie and Tipton (2000), in a review of this subject, reported that sustained exercise stimulates amino acid oxidation (chiefly branched chain) for gluconeogenesis and possibly affects regulation of acid-base balance, resulting in a short-term net loss of tissue protein due to decreased synthesis and increased degradation. Protein synthesis rebounds quickly following rest. The data indicate that exercise does not increase the dietary protein requirement; in fact, efficiency of protein metabolism may be improved as a result of exercise.

Excessive Protein Intake

It is important to be aware of any adverse effects of feeding excesses of any nutrients, and protein is no exception. Sugahara et al. (1969) fed growing pigs 16, 32, or 48% protein and observed a linear depression in weight gain with increasing protein. Feed intake was depressed, and hair became dull and coarse. High-protein diets produced changes in liver water and protein content. Other work has shown that a high-protein diet reduces activity of several adipose tissue enzymes associated with fatty acid synthesis in swine. Short-term (2 weeks) ingestion of a diet containing twice the required level of protein increases liver and kidney weight in 2-week-old pigs (Schoknecht and Pond, 1993). This differential effect on growth of organs with high metabolic rates suggests that efficiency of nutrient utilization for growth is reduced. The extent to which the reduced feed intake and weight gain associated with high-protein intake is due to increased blood and tissue ammonia concentration or to other metabolic changes is unknown.

Ammonia toxicity is a practical problem in ruminants fed urea as a NPN source. Ammonia is absorbed from the rumen as well as from the omasum, small intestine, and cecum of ruminants. Absorption is increased at higher pH, probably as a result of an increase in the proportion of ammonia in relation to ammonium ion, the former passing freely across membranes, the latter not. Bartley et al. (1976) found that high rumen ammonia concentration may exist without producing toxicity if the diet is readily fermentable and lowers rumen pH into the acid range. Toxic symptoms in ruminants include uneasiness, labored breathing, excessive salivation, muscle and skin tremors, incoordination, tetany, and death within 1 or 2 h of onset of symptoms. Emptying the rumen of animals showing toxic signs results in a rapid decrease in blood ammonia concentration and quick recovery of the animal. Similar signs have been reported in ammonia-intoxicated pigs. Peripheral blood contains 1 to 4 mg ammonia/dL at the height of toxic symptoms. Much higher levels would be found in portal blood before reaching the liver for metabolism.

Blood glucose, lactate, pyruvate, pentoses, and ketones all rise during ammonia toxicity, suggesting a drastic effect on energy metabolism, perhaps by inhibition of the citric acid cycle. Visek (1978) has provided evidence that ammonia causes cell death and has suggested that excessive protein intake of humans and animals may increase the incidence of cancer of the GI tract by increasing cell turnover rate. Such an increase would result in greater exposure to ammonia and thereby increase the probability of mutations resulting in neoplastic cell formation. Negative effects of ammonia on tissue metabolism may go unrecognized when peripheral blood ammonia concentrations are in the normal range. Under such conditions, urinary orotic acid excretion has been suggested as a reliable index of ammonia load in relation to urea cycle capacity in some animal species (Visek, 1979).

Some or all of the above effects of ammonia may affect the animal as a result of excessive protein intake. The growth depression associated with excessive levels of protein in the diet is the most visible sign of protein toxicity, whether the effects are produced through its catabolism to ammonia or by other means.

❏ MEASURES OF NUTRITIVE VALUE OF PROTEINS

The degree of utilization of a feed protein depends not only on its absorbability, but also on its utilization after absorption. Protein sources often are evaluated on the basis of their biological value (BV). BV is defined as that percentage of N absorbed from the GI tract that is available for productive body functions. BV is determined experimentally by measurement of total N intake and N losses in urine and feces. A formula for calculating BV is:

$$\frac{N \text{ intake} - [\text{fecal N} + \text{urinary N}]}{N \text{ intake} - \text{fecal N}} \times 100$$

Some N is lost in the feces as a result of endogenous losses (metabolic fecal N), and the urinary loss of N involves both excess dietary N and end products of metabolism, such as urea involving obligatory losses (endogenous N). The Thomas Mitchell method of determining BV takes these metabolic and endogenous N losses into account. It provides an estimate of the efficiency of use of the absorbed protein for combined maintenance and growth. The BV of several plant and animal proteins for growing and adult rats are shown in Table 9.9 (Mitchell and Beadles, 1950).

Among the natural sources of protein, egg protein is considered to have the highest BV. Single protein sources that have very poor BV when fed alone may yield a BV similar to that of a single high-quality protein when combined with other proteins. Thus, a complementary effect is exemplified in the use of a mixture of corn and soybean meal to supply the total amino acid requirement of growing pigs; corn alone or soybean alone would not promote maximum growth.

TABLE 9.9   Biological value of proteins for growing and adult rats.

PROTEIN	GROWING	ADULT
Egg albumin	97	94
Beef muscle	76	69
Meat meal	72–79	—
Casein	69	51
Peanut meal	54	46
Wheat gluten	40	65

Source: Mitchell and Beadles (1950).

Other measures of protein adequacy are the protein efficiency ratio (PER) and net protein utilization (NPU) or net protein value (NPV). PER is by definition the number of grams of body weight gain of an animal per unit of protein consumed. Conventionally, this index is obtained by feeding laboratory rats, but the same calculations could be made for any animal species fed a particular protein mixture (or a specific diet). It is important in any measure of protein utilization to maintain a low or marginal level of dietary protein because even poor-quality protein may allow reasonable growth and productive performance when fed at a higher level of total N than required by the animal. Also, an unrealistically low PER may be obtained with a good protein source if fed at a level above the amount needed for maximum weight gain.

NPU measures efficiency of growth by comparing body N resulting from feeding a test protein with that resulting from feeding a comparable group of animals a protein-free diet for the same length of time (Miller and Bender, 1955). Thus,

$$NPU = \frac{\left(\begin{array}{c}\text{Body N with}\\\text{test protein}\end{array}\right) - \left(\begin{array}{c}\text{Body N with}\\\text{protein-free diet}\end{array}\right)}{\text{Total N intake}}$$

A large number of values can be obtained over a brief test period of 1 to 2 weeks. The NPU of several plant and animal protein sources are shown in Table 9.10. NPU values are correlated highly with values obtained by PER and also agree closely with results using other methods of protein quality assay, including chemical score, relative nutritive value (RNV), dye binding capacity, and pepsin digestible N. Each of these methods has merit, but

TABLE 9.10   Net protein utilization (NPU) of plant and animal protein sources.[a]

	NPU
Animal protein sources	
Whole egg	91.0
Fish (cod)	83.0
Egg albumin	82.5
Whey, dried	82.0
Milk, dried	75.0
Beef muscle	71.5
Beef heart	66.6
Beef liver	65.0
Casein	60.0
Meat meal	35.5–48.3[b]
Fish meal	44.5–54.6[b]
Feather meal	21.2–35.6[b]
Hair meal	11.4–33.4[b]
Blood	3.8
Blood + 0.85 isoleucine	30.5
Gelatin	2.0
Plant protein sources	
Wheat germ	67.0
Cottonseed meal	58.8
Soybean meal	56.0
Linseed meal	55.8
Bran	55.3
Corn (maize)	55.0
Peanut meal	42.8
Dried yeast	42.3
Seaweed	42.0
Wheat gluten	37.0
Rice gluten	36.0

[a]From Miller and Bender (1955) with rats.
[b]From Johnston and Coon (1979) with chicks.

the details of the procedures used for each are beyond the scope of this discussion.

When utilizing digestibility as well as BV data, it is possible to compute a net protein value (NPV) that is simply the product of the two values. Thus, NPV = BV × digestion coefficient.

With the development of automated methods for assay of free amino acid concentrations of blood plasma (column chromatography and gas–liquid chromatography), it is possible to explore the possibility of using changes in amino acid patterns following ingestion of the test protein to assess protein quality. Identification and standardization of such factors as optimum fasting interval and level and duration of feeding eventually may provide a more useful predictive tool. Changes in the ratio of nonessential to essential amino acids in the plasma in protein deficiency also offer encouragement for further refinements in this concept of protein evaluation. Blood plasma concentrations of a specific amino acid in animals fed increments of that amino acid provide a useful aid in establishing the requirement for that amino acid (Fig. 9.6), or in determining its bioavailability as a simple and useful index of protein adequacy and amino acid balance. At levels of dietary protein either above or below the requirement or with diets deficient or imbalanced in one or more amino acids, blood urea N is elevated (Eggum, 1972). Animals fed diets providing optimum levels of protein and essential amino acids have a minimum blood urea N concentration. Figures 9.6 and 9.7 depict the changes in plasma amino acid and blood urea N concentration, respectively, in relation to dietary protein and amino acid adequacy. Note that the serum urea values of pigs fed amino acid supplements (Table 9.8) fit well with this concept.

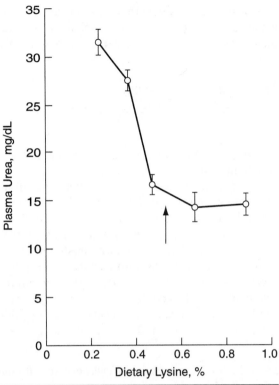

Figure 9.7   *Relationship of plasma urea to dietary protein or lysine adequacy.*

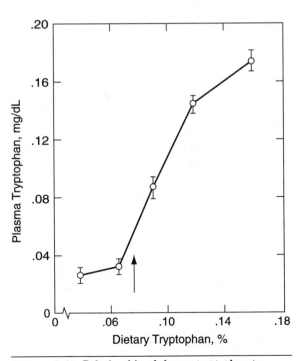

Figure 9.6   *Relationship of plasma tryptophan to dietary protein and tryptophan adequacy.*

All of the estimates of protein utilization described have their limitations, and no one estimate is superior to all others under all conditions. If one is interested in simplicity and a minimum of facilities and analytical work, perhaps PER provides the best estimate of protein value for growth, since only weight gain and protein consumed in a particular time period of 2 to 3 weeks are needed.

Biological Availability of Amino Acids

Several methods exist for estimating amino acid availability to the animal from a variety of feedstuffs. These methods include microbiological assay, fecal analysis, analysis of terminal ileum contents, growth assay, and plasma-free amino acids.

In the microbiological assay, the test material is digested by enzymatic or by acid or alkaline hydrolysis, and the amino acid composition of the hydrolysate is determined by microbiological assay. The rate and degree of release of amino acids from the protein is taken as an index of the availability of the animal.

The fecal analysis method is a balance trial in which percentage amino acid availability is estimated by the following formula (Bragg et al., 1969):

$$\text{Amino acid (AA) availability, \%} = \frac{AA_{total} - (TFAA_{protein} - TFAA_{protein\text{-}free})}{AA_{total}}$$

where TFAA is total fecal amino acids in diets adequate in protein or protein-free.

The extent of absorption of amino acids from the large intestine and the degree of their degradation in the lower intestinal tract are not defined well. Therefore, estimates of amino acid absorption in the small intestine can be made only with animals fitted surgically with a cannula at the terminal end of the ileum (Knabe, 1991). The amino acid content of ileal contents collected in this way can be used to calculate the availability of individual amino acids from the small intestine (ileal

analysis) using the same formula as for fecal analysis. Estimates of quantitative amino acid absorption also can be made with a surgical preparation, allowing the collection of blood from the hepatic portal vein and vena cava coupled with simultaneous determination of rate of flow through the portal vessel.

In the chick, in which urinary and fecal N are excreted together, the method combines urinary and fecal losses. However, the same formula can be applied to mammals by collecting feces if the assumption is made that urinary loss of free amino acids is negligible. This method allows calculation of availability of individual amino acids. Values greater than 90% availability are commonly obtained. Bioavailability of amino acids from several feedstuffs as estimated by ileal or fecal analysis by Tanksley and Knabe (1980) is shown in Table 9.11. The growth assay can be used to study the availability of one or a series of amino acids from a test protein. The procedure is to feed the test diet

TABLE 9.11   Apparent absorbability of essential amino acids in several protein sources estimated by ileal versus fecal analysis.

| | SBM[a] | | MBM[b] | | PM[c] | | CSM[d] GLANDLESS | | CSM[d] SOLVENT | | CSM[d] SCREW PRESS | |
COMPONENT	I[e]	F[f]	I	F	I	F	I	F	I	F	I	F
Nitrogen, %	81	89	59	72	78	85	86	92	72	73	75	78
Amino acid, %												
Arg	92	94	74	76	93	95	96	97	87	87	90	90
His	89	93	65	72	80	88	92	96	79	80	81	83
Iso	83	87	59	66	79	80	85	88	66	65	70	69
Leu	83	88	62	69	79	83	85	89	69	68	73	72
Lys	89	89	61	68	69	78	87	89	62	58	64	61
Met	77	81	72	67	88	77	83	85	65	65	66	66
Phe	88	89	66	70	89	88	91	93	77	78	82	81
Thr	77	85	50	66	61	75	79	87	62	64	65	67
Try	82	91	55	67	70	82	83	91	69	71	68	74
Val	81	86	61	70	76	80	85	90	68	68	71	72
Average, %	85	88	62	69	78	83	86	90	70	70	73	73

[a]Soybean meal, solvent 44% protein.
[b]Meat and bone meal, 50% protein.
[c]Peanut meal, mechanically extracted, 50% protein.
[d]Cottonseed meal; glandless 55% protein, solvent 41% protein, screw press 41% protein.
[e]Ileal.
[f]Fecal.
Source: Tanksley and Knabe (1980).

alongside diets of known amino acid availability such as a crystalline amino acid diet. By comparing the growth curve of animals fed the test protein with that of animals fed the amino acid diet containing the amino acid of concern at several incremental levels, it is possible to estimate the proportion of amino acid in the test protein utilized for growth.

Measurement of free amino acids in the plasma of animals at intervals following a meal of the test protein allows a means of estimating the availability of one or more amino acids. Stockland and Meade (1970) found differences in the availability to the rat of isoleucine, threonine, and phenylalanine from several sources of meat and bone meal. Using the same method, they found isoleucine, methionine, and threonine to vary in availability from different samples of meat and bone meal for the pig. The basis for judging availability is the change in relative concentrations of amino acids in the plasma following a meal of the test protein. The duration of fasting and the selection of appropriate intervals for blood sampling are of importance and must be established for each species used and for the conditions of a particular experiment.

Processing methods including grinding, pelleting, drying, oil extraction, and heating have been developed by the feed industry to improve the value of a large number of feedstuffs. Probably the greatest single factor affecting amino acid availability from feedstuffs is proper heating of feedstuffs during processing. The application of heat must be a balance between beneficial and destructive effects. Reductions in amino acid availability can result from their destruction or their delayed release during digestion owing to change in linkages between amino acids and other diet components.

Biosynthesis, metabolism, and functions of carnitine, derived from lysine and methionine and present in large amounts in animal products, are under extensive study in premature infants and in individuals fed parenterally. It appears that carnitine biosynthesis may be insufficient to meet metabolic needs under some conditions (Rebouche and Paulson, 1986).

❑ SUMMARY

- Proteins and their constituent amino acids are present in all living organisms. About 20 amino acids are found in most proteins, and up to 10 are required in the diet (commonly called essential or indispensable) because tissue synthesis is not adequate to meet metabolic needs.

- Amino acids contain a carboxyl group and an amino group on the carbon atom adjacent to the carboxyl group. The 10 essential amino acids for most nonruminant animal species are arginine, histidine, isoleucine, leucine, lysine, methionine (can be replaced partially by cystine), phenylalanine (can be replaced partially by tyrosine), threonine, tryptophan, and valine.

- Arginine and histidine are required by adults of some species, and proline appears to be required for egg production in hens and for growth in chicks and some mammals.

- All proteins can be classified on the basis of their shapes and their solubilities in water, salt, acids, bases, and alcohol. Many protein–lipid (lipoproteins) and protein–carbohydrate (glycoproteins) complexes exist in biological systems.

- Proteins and their complexes serve many body functions. They are constituents of cell membranes, muscle, skin, hair, hooves, blood plasma, enzymes, hormones, and immune antibodies.

- Ingested proteins are hydrolyzed to their constituent amino acids before absorption into the body from the gastrointestinal tract.

- In ruminant animals, microorganisms inhabiting the rumen can synthesize amino acids and protein from nonprotein N and carbohydrates, and this microbial protein provides the amino acids needed by the host animal.

- Absorbed amino acids are used for synthesis of tissue protein and of enzymes, hormones, and other metabolites, or are deaminated and the carbon skeleton is used as a source of energy by the animal.

- Tissue protein synthesis requires the presence of nucleic acids; deoxyribonucleic acid (DNA) is the blueprint of protein synthesis and controls the development of the cell and the organism by controlling the formation of ribonucleic acids (RNAs).

- Three kinds of RNA, ribosomal, transfer, and messenger, are involved in the sequence of events in tissue protein synthesis.

- Net rate of muscle protein synthesis is governed by a balance of synthetic and degradative processes.

- A series of polypeptides are recognized as factors in growth modulation in animals.

- N released from protein and amino acid degradation in the body is excreted mainly in the form of urea in mammals and uric acid in birds. Blood urea rises in animals fed protein in excess of needs or in animals fed diets deficient in total protein or in one or more essential amino acids.

- Signs of protein deficiency include anorexia, reduced birth weight and reduced growth rate, hypoproteinemia, anemia, liver fat accumulation, edema (in severe deficiency), impaired immune function, and reduced synthesis of certain enzymes and hormones.

- Deficiencies of individual essential amino acids generally produce the same signs because a single amino acid deficiency prevents protein synthesis, just as a shortage of a particular link in a chain prevents elongation of the chain.

❏ REFERENCES

Baker, D. H. 1992. Ideal protein for swine and poultry. *BioKyowa Technical Review* 4. Nutri-Quest, Chesterfield, MO.

Bartley, E. E., et al. 1976. *J. Anim. Sci.* **43**: 835.

Benevenga, N. J., and R. D. Steele. 1984. *Annu. Rev. Nutr.* **4**: 157.

Berthold, H., et al. 1991. *Proc. Natl. Acad. Sci.* **88**: 8091.

Bragg, D. B., C. A. Ivy, and E. L. Stephenson. 1969. *Poult. Sci.* **48**: 2135.

Breslow, J. L. 1991. *Annual Rev. Med.* **42**: 200.

Bruce, C., et al. 1998. *Annu. Rev. Nutr.* **18**: 297–330.

Council for Agricultural Science and Technology (CAST). 2000. Transmissible Spongiform Encephalopathies. Council for Agricultural Science and Technology Task Force Report No. 136, Ames, IA.

Center for Disease Control. 2004. www.cdc.gov/od/ohs/biosfty/bmbl4s7d.htm (Accessed March 2004).

Clemmons, D. R., and L. E. Underwood. 1991. *Annu. Rev. Nutr.* **11**: 393.

Darnell, J. E., Jr. 1985. *Sci. Am.* **253**: 68.

Eggum, B. O. 1972. *Brit. J. Nutr.* **24**: 983.

Felsenfeld, G. 1985. *Sci. Am.* **253**: 58.

Fried, W. 1995. *Annu. Rev. Nutr.* **15**: 353–357.

Fuller, M. F., and P. J. Reeds. 1998. *Annu. Rev. Nutr.* **18**: 385–411.

Goll, D. E., et al. 1989. In *Animal Growth Regulation,* Chapter 9, p. 141, D. R. Campion et al. (eds.). Plenum Press, New York.

Guidotti, G. 1972. *Annu. Rev. Biochem.* **41**: 731.

Harper, A. E., et al. 1970. *Physiol. Rev.* **50**: 428.

Johnson, R. W., and J. Escobar. 2005. Immune system: nutrition effects. In *Encyclopedia of Animal Science,* W. G. Pond and A. W. Bell (eds.). Marcel Dekker, New York (In Press).

Johnston, J., and C. N. Coon. 1979. *Poult. Sci.* **58**: 919.

Kilberg, M. S., et al. 1993. *Annu. Rev. Nutr.* **13**: 137.

Knabe, D. A. 1991. In *Swine Nutrition,* Chapter 19, p. 327, E. R. Miller et al. (eds.). Butterworth-Heinemann, Boston, MA.

Larson, J. 2004. Bovine spongioform encephalopathy (BSE) and other animal related transmissible spongioform encephalopathies. A WIC Series #2001-01 (revised January 2004) www.nal.usda.gov/awic/pubs/bsebib.htm (Accessed March 2004).

Leibach, F. H., and V. Ganapathy. 1996. *Annu. Rev. Nutr.* **16**: 99–119.

Lonnerdal, B., and S. Iyer. 1995. *Annu. Rev. Nutr.* **15**: 93–110.

Loosli, J. K., et al. 1949. *Science* **110**: 144.

McClain, et al. 1965. *Proc. Sci. Exp. Biol. Med.* **119**: 493.

McDonald, M. L., et al. 1984. *Annu. Rev. Nutr.* **4**: 521.

Mertz, E. T., L. S. Bates, and O. E. Nelson. 1964. *Science* **145**: 279.

Miller, D. S., and A. E. Bender. 1955. *Brit. J. Nutr.* **9**: 382.

Mitchell, H. H., and J. R. Beadles. 1950. *J. Nutr.* **40**: 25.

Morris, S. M. 2002. *Ann. Rev. Nutr.* **22**: 87–105.

National Research Council. 1981. *Taurine Requirement of the Cat.* National Academy Press, Washington, DC.

Nichols, B. L., et al. 1988. *Pediatr. Res.* **27**: 112A.

Oltjen, R. R. 1969. *J. Anim. Sci.* **28**: 673.

Prusiner, S. B. 1997. *Science* **278**: 245.

Rebouche, C. J., and D. J. Paulson. 1986. *Annu. Rev. Nutr.* **6**: 41.

Reeds, P. J., et al. 1985. In *Substrate and Energy Metabolism in Man,* p. 46, J. Garrow and D. Holliday (eds.). John Libbey, London.

Reeds, P. J. 1989. In *Animal Growth Regulation,* Chapter 9, p. 183, D. R. Campion et al. (eds.). Plenum Press, New York.

Rennie, M. J., and K. D. Tipton. 2000. *Annu. Rev. Nutr.* **20**: 457–483.

Rose, W. C. 1948. *J. Biol. Chem.* **176**: 753.

Schoknecht, P. A., and W. G. Pond. 1993. *PSEBM.* **203**: 251.

Souba, W. W. 1991. *Annu. Rev. Nutr.* **11**: 285.

Stockland, W. L., and R. J. Meade. 1970. *J. Anim. Sci.* **31**: 1156.

Sugahara, M., et al. 1969. *J. Anim. Sci.* **29**: 598.

Tanksley, T. D., and K. Knabe. 1980. *Proc. Georgia Nutr. Conf.,* p. 157. University of Georgia, Atlanta, GA.

Van Soest, P. J. 1999. *Nutritional Ecology of the Ruminant* (2nd ed.), pp. 1–476. Cornell University Press. Ithaca, NY.

Virtanen, A. I. 1966. *Science* **153:** 1603.

Visek, W. J. 1978. *J. Anim. Sci.* **46:** 1447.

Visek, W. J. 1979. *Nutr. Rev.* **37:** 273.

Visek, W. J. 1984. *Annu. Rev. Nutr.* **4:** 137.

Wahlstrom, R. C., et al. 1985. *Anim. Sci.* **60:** 720.

Waterlow, J. C. 1995. *Annu. Rev. Nutr.* **15:** 57–92.

Watson, J. D., et al. 1983. *Recombinant DNA: A Short Course.* W. H. Freeman and Co., New York, 260 pp.

Webb, K. E., et al. 1992. *J. Anim. Sci.* **70:** 3248.

Wu, G., and C. J. Meininger. 2002. *Annu. Rev. Nutr.* **22:** 61–86.

Wu, G., and J. T. Self. 2005. Amino acids. In *Encyclopedia of Animal Science,* W. G. Pond and A. W. Bell (eds.). Marcel Dekker, New York (In Press).

Wu, G., and J. T. Self. 2005. Proteins. In *Encyclopedia of Animal Science,* W. G. Pond and A. W. Bell (eds.). Marcel Dekker, New York (In Press).

10

Energy Metabolism

THE TOPIC OF ENERGY and its metabolism by animals is known as **bioenergetics.** In the overall subject of animal nutrition, it is very important because energy is, quantitatively, the most important item in an animal's diet and all animal feeding standards (Chapter 18) are based on energy needs. As such, an appreciable amount of effort has been expended to study the metabolism by animals, particularly in the first part of the twentieth century.

While the reader may feel that nonruminant animals are unduly slighted in this chapter, the fact is that considerably more work, particularly on net energy, has been done on ruminant species than on nonruminant species. Very little information is available on wild species. For excellent historical coverage of the topic, the reader is referred to some older books such as Brody (1945), Kleiber (1961), Mitchell (1962), and Blaxter (1967). These books are outdated in some respects but remain comprehensive texts on this field. The most recent NRC publications on nutrient requirements of dairy cattle and beef cattle (2001, 1996) also contain information that may be of interest.

Modern laboratory procedures allow us to analyze feedstuffs, animal tissues, or other items for their content of nutrients or other chemical fractions, and we can isolate proteins, lipids, different minerals, and vitamins, all of which we can weigh, see, smell, or taste. However, study of energy metabolism requires a different approach because energy may be derived from almost all organic compounds ingested by an animal. The animal derives energy by partial or complete oxidation of organic molecules, which are absorbed from the diet or from metabolism of tissues, primarily fat or protein and, to a lesser extent, from carbohydrates.

The mechanisms by which biological organisms cope with energy transfer and oxidation are outside the scope of this chapter. Biochemists have defined the different compounds and enzyme systems that accomplish these reactions (see Chapter 7). Energy transfer from one chemical reaction to another occurs primarily by means of high-energy bonds found in compounds such as ATP (adenosine triphosphate) and other related compounds. It is sufficient for our purposes here to say that all animal functions and biochemical processes require a source of energy to drive the various processes to completion. This applies to all life processes and animal activity such as chewing, digestion, maintenance of body temperature,

liver metabolism of glucose, absorption from the GI tract, storage of glycogen or fat, or protein synthesis.

In normal body metabolism a tremendous transfer of energy occurs from one type to another—for example, from chemical to heat (oxidation of fat, glucose, or amino acids); from chemical to mechanical (any muscular activity); or from chemical to electrical (glucose oxidation to electrical activity of the brain). Based on biochemical laboratory data, the energy cost of many of these reactions can be estimated with reasonable precision, but other animal functions such as excretion and digestion have an energy input from so many different tissues and chemical reactions that it is difficult to evaluate their cost to the animal.

❑ ENERGY TERMINOLOGY

Energy may be defined as the potential to perform work where work is the product of a given force acting through a given distance. However, such a broad definition is not directly applicable to animals because we usually are more concerned with the utilization of chemical energy. Chemical energy may be measured in terms of heat and expressed as calories (or BTUs, British Thermal Units), although, physicists maintain that a joule is a more precise means of expression. In international usage, a calorie (cal) is defined as the amount of heat required to raise the temperature of 1 g of water from 14.5 to 15.5°C and is equivalent to 4.1855 joules. A kilocalorie (Kcal) is equal to 1000 cal, and a megacalorie (Mcal), or therm, is equal to 1000 Kcal. Several European countries have adopted the joule as the standard unit of measurement though the unit of measurement is of no actual consequence (i.e., meters or yards) because one measurement can be converted to any other desired.

The manner in which energy is partitioned into various fractions in terms of animal utilization is shown schematically in Fig. 10.1. Discussion of most of these fractions follows.

Gross Energy

Gross energy (GE) is the quantity of heat resulting from the complete oxidation of food, feed, or other substances such as fuel. The term *heat of combustion,* which is used in chemical terminology, means the same thing. GE is measured in an apparatus called an oxygen bomb calorimeter (see Chapter 3). To study animal energy utilization, we would need to collect GE values on feed and excretory products such as feces. Energy values of different feedstuffs or nutrients may vary, but typical values are (Kcal/g): carbohydrates, 4.10; proteins, 5.65; and fats, 9.45. The differences between these nutrients primarily reflect the state of oxidation of the initial compound. The chemical energy varies inversely with the C:H ratio and the O and N content. For example, a typical monosaccharide such as glucose has an empirical formula of C H O, or one atom of oxgyen/atom of C, whereas in a fat molecule such as tristearin, there are 6 atoms of O and 57 atoms of C for a ratio of 1:9.5. Thus, the fat requires more oxygen for oxidation and gives off more heat in the process.

Examples of GE values of some selected feedstuffs, nutrients, tissues, or compounds are shown in Table 10.1. Note that as the length of the

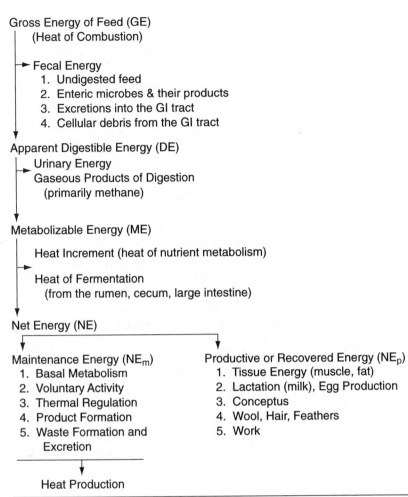

Gross Energy of Feed (GE)
   (Heat of Combustion)

→ Fecal Energy
   1. Undigested feed
   2. Enteric microbes & their products
   3. Excretions into the GI tract
   4. Cellular debris from the GI tract

Apparent Digestible Energy (DE)
   Urinary Energy
   Gaseous Products of Digestion
      (primarily methane)

Metabolizable Energy (ME)

   Heat Increment (heat of nutrient metabolism)

   Heat of Fermentation
      (from the rumen, cecum, large intestine)

Net Energy (NE)

Maintenance Energy (NE$_m$)
   1. Basal Metabolism
   2. Voluntary Activity
   3. Thermal Regulation
   4. Product Formation
   5. Waste Formation and
      Excretion

Productive or Recovered Energy (NE$_p$)
   1. Tissue Energy (muscle, fat)
   2. Lactation (milk), Egg Production
   3. Conceptus
   4. Wool, Hair, Feathers
   5. Work

Heat Production

Figure 10.1   Schematic diagram of energy utilization by animals. The reader should be aware that some publications have chosen to use different abbreviations and some added terminology. Those used in this figure have been in use many years, and, thus, the authors see no reason to change. Losses of energy by way of the feces, urine, and gaseous products cannot be utilized or recovered by the animal (unless it practices coprophagy). Heat produced in the animal body—such as those items listed under ME or NE maintenance—can be useful in maintaining a stable body temperature unless the animal is in a heat-stress situation. Recovered energy may range from almost nil (hair, wool) to more than 30% of GE for high-producing dairy cows.

carbon chain in a molecule increases—that is, compare glycine to alanine or acetic acid to propionic acid—the energy value increases. Also note that a 2-carbon acid such as acetic has a higher energy value than a 2-carbon amino acid such as glycine. For feed ingredients, observe that poor quality material such as oat straw has about the same value as a high-quality feed such as corn grain. This information shows that GE values, by themselves, are of little use for animals eating high-fiber diets.

If we have exact chemical knowledge of a diet's composition, we can compute its GE value. In human nutrition, we express the caloric value of foods as GE. This distinction is more reasonable for humans than for domestic animals because humans do not eat the quantities of indigestible material that most animals consume.

Digestible Energy (DE)

The intake of food energy (IE) minus energy lost in the feces (FE) is called apparent digested en-

TABLE 10.1   Gross energy (GE) values (dry basis) of various tissues, metabolites, or feedstuffs.

ITEM	GE, Kcal/G
Carbohydrates	
Glucose	3.74
Sucrose	3.94
Starch	4.18
Cellulose	4.18
Glycerol	4.31
Fats, fatty acids	
Average fat	9.45
Butterfat	9.10
Beef fat	9.40
Corn oil	9.40
Coconut oil	8.90
Acetic acid	3.49
Propionic acid	4.96
Butyric acid	5.95
Palmitic acid	9.35
Stearic acid	9.53
Oleic acid	9.50
Nitrogen sources	
Average protein source	5.65
Beef muscle	5.30
Casein	5.90
Egg albumin	5.70
Gluten	6.00
Alanine	4.35
Glycine	3.11
Tyrosine	5.91
Urea	2.52
Uric acid	7.74
Ethyl alcohol	7.11
Methane	13.3
Feeds	
Corn grain	4.4
Wheat bran	4.5
Grass hay	4.5
Oat straw	4.5
Soybean meal	5.5
Linseed meal	5.1

the feces accounts for the single largest loss of ingested nutrients (Fig. 10.2). Depending on the species of animal and the diet, fecal losses may range from 10% or less in milk-fed animals to 80% or more in animals consuming very poor quality roughage.

Apparent DE is not a true measure of the digestibility of a given diet or nutrient because the gastrointestinal (GI) tract of an animal is an active site for excretion of various products that end up in the feces. Also, there may be considerable sloughing of cellular debris from cells lining the GI tract. In addition, undigested microbes and their metabolic byproducts may constitute a large portion of the feces of some species. Although some of these microbes may be digestible if passed through the stomach and small intestine, much of their growth occurs in the cecum and large intestine, where there are no enteric enzymes and where much less absorption occurs than in the small intestine.

True DE is determined by measuring, in addition, the energy in fecal excretions (metabolic fecal energy) of an animal that is fasting or being fed a diet presumed to be completely absorbed, such as milk or eggs (refer to Chapter 5), or in some cases, where the animal is fed intravenously. The amount of excretion is then subtracted from total

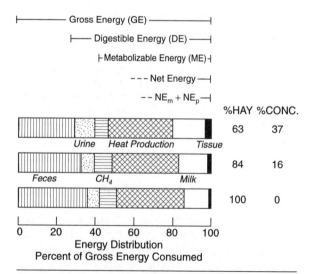

Figure 10.2   *A graphic illustration of energy terminology and the different systems of expressing energy value of feeds. The bar chart shows relative energy losses when a mixed diet is fed to a lactating dairy cow. Reproduced by permission of P. W. Moe, USDA, ARS, Beltsville, MD.*

ergy (DE). In practice, the IE of an animal is carefully measured over a period of time accompanied by collection of fecal excretion for a representative period of time. Both feed and feces are analyzed for energy content, and DE is calculated as DE = IE − FE. In most typical situations, energy lost in fecal excretion of the fed animal. This determination is not feasible with most herbivorous animal species and is done only in practice with poultry.

If we look at data collected from dairy cows (Fig. 10.2) that were fed from 63 to 100% hay, we notice that fecal energy (left side of figure) increases as the percentage of hay fed increases. This is typical of natural diets. In this case fecal energy increased from about 28 to 37% of feed energy as the amount of hay in the diet increased. Somewhat similar predicted values are shown in Fig. 10.3 with dry matter intakes ranging from 5 to 25 pounds/day. Note that all heat losses increased as the level of feed intake increased.

Total Digestible Nutrients (TDN)

TDN is not shown in the scheme in Fig. 10.1, but it is a measure of energy still used directly or indirectly in the United States for ruminants and swine. TDN is roughly comparable to DE but is expressed in units of weight or percent. When conversion of TDN to DE is desired, the values used

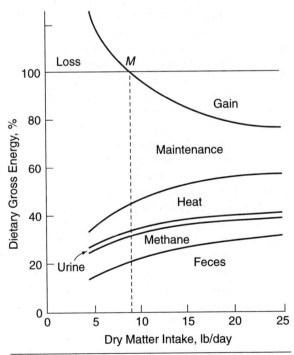

Figure 10.3 *A theoretical example of the proportional changes in energy balance with increasing intake by a ruminant. The proportion of energy lost in methane declines somewhat with increasing intake. Fecal losses increase because of reduction of digestibility with increased consumption. From Reid (1962).*

usually are 2000 Kcal of DE/lb of TDN, or 4.4 Kcal/g. TDN is determined by carrying out a digestion trial and summing the digestible protein (DP) and carbohydrates (NFE and crude fiber) plus 2.25 times digestible ether extract (crude fat). Fat is multiplied by 2.25 in an attempt to account for its higher caloric value, which is about 2.25× that of digestible carbohydrates. The formula for calculating TDN is: TDN = DCP + DNFE + DCF + 2.25(DEE). As compared to DE, TDN undervalues protein because protein is not oxidized completely by the animal body, whereas it is in a bomb calorimeter. Multiplication of DP by 1.25 would put TDN on a more comparable basis with DE.

Using TDN values for developing animal feeding standards or in formulating diets poses several problems. This topic is discussed in some detail in the most recent NRC (2001) publication on nutrition of dairy cattle. One major problem is that most computed TDN values in the literature are from experiments conducted many years ago, and many were done with sheep. In most cases, the animals were fed at lower levels than for producing livestock, and it has been shown that an increased feed intake results in lower digestibility. In addition, many values were computed with the difference method. In such experiments, a digestibility trial is conducted with animals fed one feed or diet, such as alfalfa hay, then another with the hay plus the test feed—perhaps corn grain or another feed that cannot be fed by itself for a long enough period to obtain the data. The digestibility of the alfalfa is computed, ascribing any differences for the alfalfa plus corn to the corn itself. The results may be additive, or a negative or positive associative effect may be found (see Chapter 5). Other problems involve the change in composition of feeds resulting from crop improvement, fertilization practices, and so on. Improved or different analytical data may also be a factor. Many laboratories now do analyses on detergent fiber rather than crude fiber; such values would alter the computation of energy values computed from chemical composition.

Metabolizable Energy

Metabolizable energy (ME) is defined as GE of the feed (or intake energy, IE) minus energy lost in the feces (FE), urine (UE), and combustible gases. Thus, values for ME account for further losses that normally occur with consumed food. ME is commonly used to evaluate feedstuffs and to establish feeding standards for poultry because feces and urine are excreted together. Making complete collections of urine and doing energy values on urine are tedious and time-consuming operations, and so few data are available on species other than poultry. Most ME values found in tabular data for feedstuffs evaluation are calculated values (except for poultry). An appropriate value for calculating ME for swine where DE is known is:

$$ME \text{ (in Kcal/kg)} = \frac{DE \text{ (in Kcal/kg)} \times 0.96 - (0.202 \times \text{protein \%})}{100}$$

In ruminants, gaseous products of digestion include huge amounts of carbon dioxide. Combustible gases are made up primarily of methane. Hydrogen, carbon monoxide, acetone, ethane, and hydrogen sulfide are produced in trace amounts, although under certain dietary conditions they may reach significant quantities. Low-quality diets result in larger proportions of methane, and, generally, the percentage of GE lost as methane declines as feed intake increases. Usual values cited are about 8% of GE at maintenance and 6 to 7% at higher levels of feeding. As shown in Fig. 10.2, less methane is produced on a diet with 63% hay than on one with 100% hay. In nonruminant species, losses of combustible gases are negligible and are normally ignored in computing ME, although some losses occur as a result of fermentation in the cecum and large intestine. In species such as the horse and rabbit with large cecums, gas production may be higher. A number of formulas have been published for calculation of methane; several of these formulas are given in previous editions.

Urinary energy losses usually are relatively stable in a given animal species, although they reflect differences in diet, particularly when excess protein is fed or when forages are consumed that may contain essential oils or detoxification products such as hippuric acid. For ruminants a correction factor of 7.45 Kcal/g of N has been used to account for energy excreted from amino acids metabolized, and a factor of 8.22 has been used for poultry. Actual energy losses run on the order of 3% of GE in ruminants or 12 to 35 Kcal/g of N excreted. As seen in Fig. 10.2, urine losses increased as the diet went from 100% hay to 63% hay. In Fig. 10.3, urinary losses show an increase as the level of feeding increased.

ME for ruminants is often calculated by the formula: ME = DE × 0.82. Many of the NRC tabular values are calculated with this formula. However, this is only an approximation as the ME/DE ratio may vary considerably, being affected by the nature of the diet and the level of feeding. Energy workers generally agree that ME is the most descriptive and reproducible measurement of feeds, especially at the maintenance level. ME values are seldom determined in practice, however, since few laboratories have the facilities and budgets to collect and analyze respiratory gases and urine.

❏ NET ENERGY

As shown in Fig. 10.1, NE is equal to ME minus the heat increment (HI) and the heat of fermentation (HF), or the added heat production in the fed animal as compared to the fasting animal. Obtaining a reliable value for HP in the fasting animal is not easy because ruminants must be fasted for an appreciable time, and collecting such data requires the use of large animal calorimeters.

The NE of food is that portion that is available to the animal for maintenance or various productive purposes. The portion used for maintenance is used for muscular work, maintenance and repair of tissues, maintenance of stable body temperature, and other body functions; most of it will leave the animal body as heat. The portion used for productive purposes may be recovered as energy in the tissues or in some product such as milk, or it may be used to perform work.

The HI, also called the specific dynamic effect when referring to a specific nutrient, may be defined as the heat production associated with nutrient digestion and metabolism over and above that produced prior to food ingestion. The resulting heat is produced by oxidative reactions that are (a) not coupled with energy transfer mechanisms (high-energy bonds) or (b) the result of incomplete transfer of energy; and are (c) partly due to heat production resulting from work of excretion by the kidney and (d) to increased muscular activity of the GI tract, respiratory, and circulatory systems resulting from nutrient metabolism.

Estimates made many years ago indicate that 80 + % of the HI originates in the viscera. Short-term experiments with animals that have had their livers removed show very little additional heat production after food ingestion, indicating that metabolism in the liver accounts for most of the HI. When individual products are fed to rats or dogs, feeding lean meat results in a prolonged period of heat production that amounts to 30–40% of GE. In dogs, the HI from other foodstuffs are: fat, 15%, sucrose, 6%; starch, 20–22%. In cattle, according to Blaxter (1967), the HI of a complete diet is about 3% of GE at half maintenance and increases to 20 + % at 2X maintenance and to higher levels in high-producing animals (Fig. 10.2). Examples of values for different species are shown in Table 10.2. Note that some of the values in the table are expressed as a percentage of ME rather than GE; using ME increases the percentage values relative to those mentioned above. For example, the values on sheep fed chopped hay and fresh grass (Table 10.2) would be about 24 and 27%, respectively, if expressed on the basis of GE.

The HI is not a constant for a given animal and a given foodstuff, but depends on how the nutrient is utilized. For example, if most of the material is absorbed and deposit in the tissues, the HI is very low. Incomplete proteins or amino acid mixtures fed to nonruminant animals result in oxidation of most of the amino acids and a high HI; a deficiency of an essential nutrient required in metabolic reactions, such as Mg or P, will result in a high HI. Frequent feeding results in a lower HI than infrequent feeding, and an increased feed intake results in a larger HI. The reader should note that the HI

is not the same as total body heat production because the body will produce heat regardless of whether the animal is fasting or fed. With lactating dairy cows, Tyrrell and Moe (1972) found that heat production accounted for 35 to 38% of GE intake.

The HF, as mentioned earlier, is poorly quantified. A nongrowing yeast culture comparable in size to that of an adult man has been estimated to produce 100 times the heat production of the man. However, Blaxter (1967) estimated the HF in ruminant animals to be 5 to 10% of GE, and Webster (1979) determined that it is about 65 Kcal/Mcal of fermented energy. In simple stomached species, some HF would originate from fermentation in the lower small intestine, cecum, and large gut, but quantitative information is not available.

Both the HI and HF may serve useful purposes to the animal in a cold environment. The heat resulting from the HI and HF may be used to warm the body just as well as that produced by more controlled metabolism of nutrients (thermogenesis). However, at temperatures that result in heat stress to the animal, the HI is detrimental, requiring additional expenditure of energy to dissipate it by various means. In ruminant species, limited data indicate that feeding of urea in place of protein tends to reduce heat production, and that heat produc-

TABLE 10.2 Heat increment of feeding for some dietary components or diets for different species.

NUTRIENT	RAT	SWINE	SHEEP	CATTLE	RABBIT
			SPECIES		
ME basis, % at maintenance[a]					
Fat	17	9	29	35	
Carbohydrate	23	17	32	37	
Protein	31	26	54	52	
Mixed rations	31	10–40	35–70	35–70	
ME basis, % ~ 2X maintenance[b]					
Chopped hay			45		
Fresh grass			47		
GE basis, % at 2X maintenance[c]					
Mixed diet				16–19	
Mixed diet					26

[a]Data from Armstrong and Blaxter (1957).
[b]Webster (1979).
[c]Data from Brody (1945).

tion is less when minimal amounts of protein are fed. Increasing the fat content of a ration appears to be helpful because fat has a low HI, and reducing the fiber content may also be helpful in hot climates.

Other Methods of Expressing Energy Utilization

As shown in Fig. 10.1, energy may be subdivided into NE of maintenance (NEm) and NE for production (NEp). For dairy cattle NE for lactation (NEl) is now in common usage. These methods will be discussed later. In addition, a method called Starch Equivalents (SE) is still in use in some areas of Europe, although it has largely been replaced by ME. The SE system was devised in Germany by Kellner many years ago. In effect, energy retention of fattening animals is measured by the carbon-nitrogen balance method, and feed values are expressed in relation to the value of starch rather than in Kcal as with the NE method.

The Scandinavian Food Unit System is also used in Europe. With this method, feedstuffs are evaluated in feeding trials by replacing barley with the test feed, and feed value is expressed relative to barley. There are fodder units, fattening feed units, and milk production units. Thus, this method essentially is the same as the SE method.

❏ METHODS OF MEASURING HEAT PRODUCTION

To study the utilization of ME, it is necessary to measure either (a) the animal's heat production or (b) energy retained in the tissues or that used for productive work, or energy deposited in a product such as milk. If one of these quantities (a or b) is known, then the other can be determined by subtracting the known one from ME. Some of these methods are discussed briefly here. For more detail, see the references cited at the beginning of this chapter.

Calorimetry

Techniques of calorimetry have been discussed in detail by Brody (1945) and Blaxter (1967), so only a brief description is included in this chapter. Animals lose heat to the environment as sensible heat or as evaporative heat. Sensible heat is lost through convection, conduction, and radiation.

Evaporative heat is lost through the excreta or via the skin and respiratory tract. Heat loss can be measured directly (direct calorimetry) using either heat sink or gradient layer calorimeters. In the heat sink calorimeter, sensible heat loss is measured as a rise in temperature of an absorbing medium such as the air stream ventilating the chamber or water circulating outside the walls. Evaporative loss can be determined from the increase in humidity of the ventilating air. The gradient layer calorimeter measures sensible heat loss from the temperature difference across a conducting layer between the animal and a constant temperature source. Evaporative heat loss can be measured accurately as the heat balance across the air conditioning system for the chamber. Because of the extremely high costs, both in construction and in operation, few direct calorimeters for farm animals are presently in use. It should be noted that heats of fermentation, and the like, are included in heat production as measured by direct calorimetry. Thus, direct calorimetry does not yield a true measure of metabolism of the body proper.

Indirect calorimetry is based on the principle that metabolic heat production is the result of oxidation of organic compounds. Thus, if all compounds are completely oxidized, heat production can be readily calculated from the amounts of oxygen consumed and the amount of carbon dioxide produced. However, in the animal, incomplete oxidation of protein results in combustible nitrogenous compounds, such as urea, that are excreted. In addition, anaerobic fermentation yields combustible gases, primarily methane. For ruminants, the most commonly used equation to estimate total heat production (HE, heat energy) for indirect calorimetry is:

$$HE = 3.886O_2 + 1.200CO_2 - 0.518CH_4 - 1.431N$$

where HE is in Kcal; O_2, CO_2, and CH_4 refer to gaseous exchange in liters; and N refers to urinary nitrogen in grams (Brouwer, 1965). Contributions of methane and N to the above equation are normally small. Thus, it is often sufficient to estimate HE from O_2 and CO_2 or even from O_2 alone. In fed ruminants with active ruminal fermentations, CO_2 is produced by microbes in the rumen and from bicarbonates. In such situations, the respiratory quotient (RQ) and CO_2 production in particular has limited significance as a measure of animal metabolism. Under such situations, O_2 consumption is the best measure of heat production.

Indirect or respiration calorimeters may be of the closed or open circuit type (Fig. 10.4). In the closed circuit type, the animal is enclosed in a temperature-controlled chamber. Air in the chamber is continuously circulated through an absorbent such as silica gel or KOH, which removes water and carbon dioxide. Constant pressure is maintained within the system by a supply of oxygen. Methane is allowed to accumulate within the chamber, and production is calculated as the concentration difference between the beginning and the end of the test times the volume of the system. Oxygen use is determined as the amount of oxygen supplied to maintain pressure, and carbon dioxide production is determined from the amount collected by the absorbent. When CO_2 production of grazing animals is of interest, equipment involving a removable tracheal cannula can be used (Fig. 10.5).

The most common type of calorimeter is the open circuit, indirect calorimeter. In this type of system, a mask, hood, or animal chamber may be used. Air is drawn past the animal at a precisely determined rate. Oxygen, CO_2, and CH_4 concentrations must be determined accurately in incoming and outgoing air. Rate of consumption or production of these gases is calculated as the concentration difference between incoming and

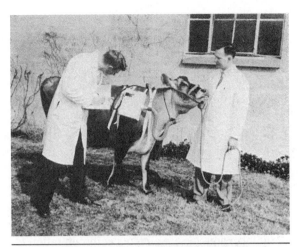

Figure 10.5 *This photograph illustrates the use of portable equipment to measure CO_2 production by a cow with a cannulated trachea. Similar equipment can be used to measure oxygen consumption when a face mask can be used. Courtesy of P. W. Moe, USDA, ARS, Beltsville, MD.*

outgoing air times air flow rate. These types of systems are relatively inexpensive and easy to construct but are subject to error because of the high degree of accuracy required in the measurement of air flow and gas concentration.

Carbon–Nitrogen Balance

Carbon–nitrogen balance is frequently calculated in association with indirect calorimetric measurements. These methodologies are based on the recognition that the main forms in which energy is accumulated in the animal are protein and fat; accumulation of carbohydrate is very low. Carbon–nitrogen balance depends on accurate measurement of carbon and nitrogen intake and losses of carbon and nitrogen from the animal in urine, feces, and respired gases (CO_2 and CH_4). Body protein accretion can be calculated from nitrogen balance. The commonly used assumption is that body protein contains 16% N and 51.2% C. By use of this assumption, N balance, calculated as differences between intake (I) and losses (L) times 6.25, $[(I - L)\ 6.25]$, yields an estimate of body protein accretion. Body protein accretion times 0.512 yields an estimate of C accretion in body protein. The remainder of carbon balance is stored as fat; thus, C balance minus C stored as protein divided by 0.746 (assuming fat contains 74.6% C) yields an estimate of fat accretion. Energy accretion can

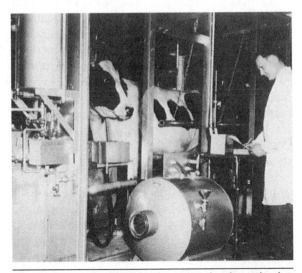

Figure 10.4 *An example of an open circuit respiration chamber used at the USDA Experiment Station at Beltsville, MD. Courtesy of P. W. Moe.*

then be calculated from protein and fat accretion. The most important limitation of this approach is that it is very difficult to measure all losses of C and N from the animal. Therefore, this method generally results in an overestimate of energy, C, or N accretion in the animal.

Comparative Slaughter

In contrast to calorimetry, in which ME intake and HE are determined and RE (retained energy) is estimated by difference, comparative slaughter requires that ME intake and RE be determined and HE can be estimated by difference. Briefly, a group of uniform animals are all fed a common ration for about 2 weeks. At the end of the adaptation period, a sample of animals is slaughtered and the body composition and energy content are determined. The remaining animals undergo predetermined treatments for a period of time and are then slaughtered. Body composition and energy contents are determined. RE is then calculated as the difference in body energy contents between the initial and the final slaughter groups. These techniques have advantages over calorimetric techniques in that they allow experiments to be conducted under situations more similar to those commonly found in practice. However, they require larger numbers of animals and must be conducted over an extended period of time to allow accurate assessment of body energy changes.

Body composition and energy content have often been determined by the accurate but expensive method of whole-body grindings and chemical analysis. These techniques are expensive, laborious, and destructive; that is, each animal can be used only once. As a result, these techniques are often used to calibrate alternative methods to estimate body composition indirectly, such as various water dilution methods, carcass density or specific gravity, ultrasonic scanning, ^{40}K counting, nuclear magnetic resonance, and computer-assisted tomography.

❏ SOME IMPORTANT CONCEPTS IN ENERGY METABOLISM

Most animals and birds are classed as homeotherms. Such animals maintain a stable body temperature, although minor fluctuations of relatively short duration may occur as a result of extensive chilling, heat stress, fevers, or vigorous exercise. With a constant body temperature, heat production over a period of time is equal to heat loss by an animal fed at maintenance. In the nutritional physiology of animals, we are concerned with heat production (or loss) if we wish to relate information obtained with calorimetric studies to more normal environmental conditions. Calorimetric data are obtained under very specific conditions, and it is not feasible to attempt to duplicate all of the situations—diet, temperature, humidity, activity, and disease—that animals encounter in their normal environment or to account for differences in age, size, species, and breed.

Relationship of Heat Production to Body Size or Weight

Early nutrition research showed that heat production was not well correlated to body weight of animals (see Table 10.3), and much research effort was expended to develop means of predicting heat production and to establish some overall "law" that would apply to animals in general. With inanimate objects, it is known that rate of cooling is proportional to surface area. Furthermore, surface area varies with the square of linear size or to the two-thirds power of weight if specific gravity is constant. Thus, surface area varies with the square of linear size or two-thirds power of volume, so heat production can be related to body surface or volume.

The body surface of a living animal is quite difficult to measure. A variety of different methods have been attempted, but repeatability of such measurements is low, even that of those methods that measure the surface of the skin after it is removed. This is because the surface area of a living animal is not constant. It may change with environmental temperature when the animal stretches out, rolls up in a ball, fluffs up its feathers, or otherwise changes its posture. Furthermore, the ability of most animals to constrict or dilate blood vessels in the skin effectively alters the normal skin temperature and heat loss. Insulation of the body by subcutaneous fat, thick skin, hair, wool, or feathers has a marked effect on heat loss as well. For example, at 16–17°C, shearing increased the feed requirement 18% for housed sheep and 24% for exposed sheep, and requirements were increased considerably at colder temperatures (Hafez and Dyer, 1969). In addition, heat loss has

TABLE 10.3 Typical values for heat production of fasting mature animals of different species.

SPECIES	BODY WEIGHT, KG	HEAT PRODUCTION, KCAL/DAY			
		PER ANIMAL	PER KG BODY WEIGHT	PER KG $BW^{0.67}$	PER KG $BW^{0.75}$
Mice	0.0276	5.0	181	55	74
Rat	0.29	28.1	97	65	71
Guinea pig	0.70	63.7	91	81	83
Cat	2.50	196	78	106	99
Rabbit	3.5	189	54	82	74
Dog	5.0	266	53	90	80
Dog	30.7	807	26	81	62
Sheep	70	1,440	21	84	60
Human	70	1,700	24	99	70
Swine	200	2,780	14	80	52
Horse	88	2,028	23	101	71
Horse	650	8,188	13	107	64
Cow	500	6,600	13	103	62
Elephant	3,833	30,924	8	123	63

Source: Brody (1945).

been shown to be related to the animal's profile; thus, a long-legged, long-necked animal, such as a giraffe, would have a markedly different exposed area as compared to an animal with a short compact body such as a pig.

Even though these various factors are involved in heat loss, they can be related reasonably well to surface area estimated by multiplying body weight by a fractional power and thus to body heat production. Body weight (BW) multiplied by a fractional power is referred to as **metabolic weight** or **metabolic size.** Although this subject has been very controversial, all current feeding standards in the United States, Europe, New Zealand, and Australia are now based on BW multiplied to the 0.73 or 0.75 power to estimate heat production. As shown in Table 10.3, heat production of different species ranges from 181 Kcal/kg of BW down to 8 for animals ranging in size from the mouse to the elephant. However, when heat production is expressed on the basis of surface area (estimated from BW), the heat production/kg of BW 0.75 is much more uniform, ranging from 52 to 99 Kcal/day. These observations apply to adult animals in a quiet state. Other factors that may affect HP are discussed in a later section of this chapter.

The point to remember is that energy metabolism may be similar in widely diverse animals, but it is not identical and may be altered by many different factors.

Basal Metabolism

Basal metabolism may be defined as the condition in which a minimal amount of energy is expended to sustain the body. Determinations are carried out under standardized conditions, and such information provides comparative base values where energy requirements are not confounded by other factors. In order to meet the requirements for basal metabolism, the animal should be in (a) a postabsorptive state, (b) a state of muscular repose but not asleep, and (c) in a thermoneutral environment. The postabsorptive state is used so that the HI and HF do not add to the body heat production. Only a few hours (overnight for humans) of fasting are required in most nonruminant species for the GI tract to approach this condition. However, in ruminants food requires many hours to pass out of the GI tract and a true state of postabsorption is rarely obtained, although it may be approached as indicated by a very low methane production. A state of muscular repose is needed so that an unknown amount of activity does not increase heat production. This condition may be difficult to obtain in animals, particularly those that have not had extensive training for such measurements.

Though perhaps not very precise, estimates of the needs for basal metabolism are that about 25% of energy needs are required for circulation, respiration, secretion, and muscle tonus, and the remaining portion represents the cost of maintaining electrochemical gradients across cell membranes and synthetic processes involved in replacing proteins and other macromolecules. These values would not apply in situations where energy expenditure is required for thermoregulation, metabolism of food, and so on. The values are not necessarily minimal because sleeping, prolonged fasting or starvation, torpidity, or hibernation can reduce energy expenditure.

Factors Affecting Basal Metabolism

Age. Age has a pronounced effect on basal metabolism in species that have been studied. For example, heat production in humans is about 31 Kcal per square meter of body surface at birth; this increases to about 50–55 at about 1 year of age and then declines gradually to 35–37 during the early to middle 20's. Further declines occur in older people. The effect of age on heat production in sheep is shown in Table 10.4. Note the marked de-

TABLE 10.4 Fasting metabolism of lambs and sheep as affected by age.

AGE	BODY WT, KG	KCAL/KG$^{0.73}$
1 wk	5.8	132
3 wk	9.1	111
6 wk	13.0	119
9 wk	18.0	116
15 wk	24.3	68
6 mo.	28.2	63
9 mo.	33.9	62
1–2 yr		63[a]
2–4 yr		58[a]
4–6 yr		55[a]
Over 6 yr		52[a]

[a]Wether sheep.
Source: Blaxter (1965).

cline in heat production, particularly between 9 and 15 weeks of age, during which period the lambs were weaned. Other data on dairy cattle (Moe and Tyrrell, 1975) suggest that fasting HP of cattle declines from 140 Kcal at one month of age to 80 Kcal (BW$^{0.73}$) at 48 months of age. Part of the change with age may be related to differential development of tissues that have different oxygen requirements.

Neuroendocrine Factors. It is well known that energy expenditure may be different in the two sexes. In humans, basal metabolism of the male typically is 6 to 7% higher than that of the female, a difference that shows up at 2 to 3 years of age. In domestic animals, castration results in a 5 to 10% depression in basal metabolism. Thyroid activity has a pronounced effect, for hypothyroid individuals may have a very low basal metabolism. Nervous, hyperactive animals have a high heat production. Related to hormonal changes are marked seasonal differences in the basal energy needs of some species of wild animals.

Species and Breed Differences. A basal metabolism value of 70 Kcal BW$^{0.75}$ is considered to be an average value where BW is in kg. Note, however, in Table 10.3, that sheep tend to be about 15% below this value and cattle about 15% above. Considerable variation also exists between breeds and within species. For example, cattle, *Bos indicus* (e.g., Africander, Barzona, Brahman, Sahiwal), appear to have fasting HP and maintenance requirements that are about 10% less than those of European *Bos taurus* breeds, with crossbreds being intermediate. Conversely, dairy or dual-purpose breeds of *Bos tarus* cattle (e.g., Ayrshire, Brown Swiss, Braunvieh, Friesian, Holstein, Simmental) apparently require about 20% more than beef breeds, with crossbreds being intermediate. With respect to the differences between sheep and cattle, Blaxter (1967) argues that cattle originated and developed in cold northern climates where heat production was a critical factor for survival, whereas sheep probably originated in subtropical areas where low-heat production has survival value.

Marked exceptions to the general interspecies value of 0.73 include marsupials and primitive echidna which are much lower; conversely, a number of animal species apparently have elevated metabolic rates. Some of the differences may be artifacts because it is quite difficult to have identical conditions when measuring heat production for all animals.

Miscellaneous Factors. Other factors that have been shown to have some effect on basal metabolism include adaptation to fasting where heat production per unit of surface area decreases with length of fast; muscular training (hypertrophy of muscles), which results in increased heat production; and mental effort, which causes a slight increase.

Maintenance

Maintenance may be defined as a condition in which a nonproductive animal neither gains nor loses body energy reserves. In modern-day agriculture, only rarely are we interested in just maintaining animals; rather, we are usually interested in keeping animals for some productive purposes. In practice, the term *maintenance* has been used frequently to apply to productive animals that are growing, pregnant, and/or lactating animals, as well as to nonproductive adults. To establish an animal's maintenance needs we must consider several factors. In addition to needs for basal metabolism, we must account for energy losses that occur during nutrient metabolism; we must, in some manner, account for the animal's increased physical activity associated with normal functions such as grazing and for environmental factors that may alter energy needs. Studies with beef cattle suggest that 65 to 70% of the ME needed for normal production is utilized to meet maintenance needs.

In biochemical terminology, energy required for maintenance is primarily for the production of ATP. This energy, along with wasted energy (that not converted to ATP), is eventually lost as heat.

Energy for maintenance is used to carry out various functions that tissues and organs perform for the benefit of the entire organism, including circulation and respiration, liver and kidney, and nervous functions. These account for 35 to 50% of the energy needed for maintenance. Cell maintenance is the other major need. Such functions as ion transport (particularly Na and Ca), and protein and lipid turnover account for 30 to 50% of maintenance energy requirements. Other metabolic processes such as glycogen turnover, gluconeogenesis, ketogenesis, urea synthesis, and RNA and DNA synthesis, among many others, require the expenditure of energy and thus contribute to the animal's maintenance energy expenditures. The energy costs of some physical activities have been estimated and are shown in Table 10.5. Standing is one of the more costly activities.

TABLE 10.5 Energy costs associated with various kinds of physical activity in ruminants.

ACTIVITY	ENERGY COST PER KG WEIGHT
Standing (compared to lying)	2.39 Kcal
Changing position (lying down and standing)	0.06 Kcal
Walking (horizontal component)	0.62 Kcal/km
Walking (vertical component)	6.69 Kcal/km
Eating (prehension and chewing)	0.60 Kcal/h
Ruminating	0.48 Kcal/h

Source: CSIRO (1990).

Maintenance energy expenditures may vary with age, body weight, breed or species, sex, physiological state, season, temperature, and previous nutrition. A complete explanation is not available at this time for all variations known, but part of this variation may be explained by differences in rates of substrate cycles. For example, energy expenditure for ion transport varies among tissues and is apparently greater both in lactating than in nonlactating animals and in cold-adapted animals than in those not cold-adapted (Milligan and McBride, 1985). Similarly, protein turnover rates vary tremendously among tissues, are higher in young than in mature animals, and decrease in response to lower planes of nutrition.

Some variation in ME needs may be explained by variations in proportions of various body tissues or organs. A change in the proportions of these tissues may have a large impact as illustrated in Table 10.6. In this example, nutritional treatment influenced body weights, the weights of organs such as liver, kidney, and heart, and fasting heat production.

In calorimetric experiments, the maintenance requirements of nonproductive animals may be measured with precision. However, in such situations the animal is much less active than under more normal conditions, and it has proved to be quite difficult to put precise estimates on maintenance requirements that are reliable under different environmental conditions. Information on sheep indicates that the maintenance requirements of wethers are about 60 to 70% greater for grazing animals than for those housed in inside pens, primarily because of the additional activity associated with grazing (Young and Corbett, 1972). The exact amount depends on the condition

TABLE 10.6 Effect of nutritional treatment on organ weight and fasting heat production of lambs.

NUTRITIONAL TREATMENT	BODY WEIGHT, KG	DIGESTIVE TRACT, G	LIVER, G	KIDNEY, G	HEART, G	FASTING HEAT PRODUCTION, KCAL/DAY
High	44.0	1,889	668	121	155	1,674
Medium	47.2	1,653	625	114	143	1,549
Low	39.9	1,304	428	93	126	1,143
Very low	34.4	1,162	350	83	130	966

Source: Adapted from Ferrell and Jenkins (1988).

of the sheep, the environment, and the availability of herbage. If one assumes that heat production of maintenance for sheep in pens is about 70 Kcal/kg $BW^{0.73}$, then this would indicate heat production when grazing of about 115 Kcal/kg $BW^{0.73}$. Work with Holstein dairy cows in calorimeters indicates a maintenance requirement on the order of 114 to 122 Kcal ME/kg $BW^{0.75}$, a value that is about equal to heat production of 100 Kcal/kg $BW^{0.75}$.

Energy requirements are frequently estimated with the use of feeding trials with animals under normal farm conditions. Although data of this type are less precise in that less information is available on actual tissue gain or loss of energy, such information may have more practical value than calorimetric data. Most feeding standards are based on the assumption that maintenance requirements under normal conditions are appreciably higher than basal or fasting metabolism rates, and the basal rate × 2 is used frequently, or 1.25 to 1.35 × fasting metabolism values are used when calculating maintenance requirements. Current NRC recommendations for cattle are based on maintenance estimates that were estimated by the comparative slaughter technique (NRC, 1996). In this work, the maintenance requirements were estimated to be 77 Kcal/kg $BW^{0.75}$, a value that would seem to be too low as compared to calorimetric experiments on cattle. If this is the case, then NE of maintenance values given for feedstuffs would overestimate the relative value as compared to NE of gain.

❏ ENERGY EXPENDITURE AND THE ENVIRONMENT

Environmental influences on energy expenditures have been discussed in detail (CSIRO, 1990; NRC, 1981). Heat production in animals arises from tis-

sue metabolism and from fermentation in the digestive tract, as noted previously. Animals dissipate heat by evaporation, radiation, convection, and conduction. Both heat production and heat dissipation are regulated in homeotherms to maintain a nearly constant body temperature. Within the thermoneutral zone (Fig. 10.6), HE (heat production) is essentially independent of temperature and is determined by normal animal metabolism, feed intake, and efficiency of energy use.

Within this temperature range, body temperature is controlled primarily by regulation of heat dissipation. When effective ambient temperature increases above the thermoneutral zone—that is,

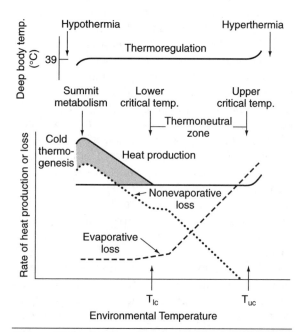

Figure 10.6 *A schematic representation of the effect of environmental temperature or thermoregulation by the animal.*

is greater than the upper critical temperature (T_{uc})—the animal promotes heat loss by evaporation from the skin (i.e., sweating) and from the lungs by increased respiration rate (panting). Productivity decreases primarily as a result of reduced feed intake. In addition, elevated body temperature results in increased tissue metabolic rate and increased work of dissipating heat (e.g., increased respiration and heart rate). Consequently, energy requirements increase. Conversely, when effective ambient temperature decreases below the thermoneutral zone, that is, below the lower critical temperature (T_{lc}), HE produced from normal tissue metabolism and fermentation is inadequate to maintain body temperature. As a result, animal metabolism must increase to provide adequate heat to maintain body temperature. This increase is referred to as cold thermogenesis (Fig. 10.6). The maximal attainable heat production (summit metabolism) is about eight times H_eE but can be maintained for only a few hours, whereas half-maximal rates of heat production may be maintained for several days. Both T_{uc} and T_{lc} vary with the rate of heat production and the animal's ability to dissipate or conserve heat. An animal's heat production may differ substantially as functions of acclimatization, feed intake, body condition, physiological state, genotype, sex, and activity.

The term *acclimatization* is used to describe adaptive changes in response to changing climatic conditions. Adaptive changes may be behavioral as well as physiological. Modifications in behavior include use of variation in terrain or other topographical features such as windbreaks, huddling in groups, or changes in posture to minimize heat loss in cold, and decreased activity, seeking shade to decrease exposure to radiant heat, seeking a hill to increase exposure to wind, or wading in water to increase heat dissipation in high temperatures. Physiological adaptations include changes in basal metabolism, respiration rate, distribution of blood flow to skin and lungs, feed and water consumption, rate of passage of feed through the digestive tract, hair coat, and body composition or distribution of fat. Physiological changes usually associated with acute temperature changes include shivering and sweating as well as acute changes in feed and water consumption, respiration rate, heart rate, and activity. Animals differ greatly in their behavioral responses and in their ability to adapt physiologically to the thermal environment. Genotype differences in cattle, for example, are particularly evident in this regard.

In general, animals have greater potential to protect themselves from cold than from hot temperatures. In a cold environment, the temperature difference between the animal and its environment may exceed 100°F. Obviously, animals cannot survive if environmental temperature reaches 100°F above normal body temperature. In fact, heat stress to the point of exceeding an animal's ability to dissipate heat may occur at relatively mild temperatures if the animal is adapted to cold and is consuming large amounts of feed. High humidity reduces rates of evaporative cooling and thereby accentuates the effects of high temperatures.

❏ EFFICIENCY OF ENERGY UTILIZATION

Efficiency of energy utilization is of practical as well as academic concern to people involved in animal production because efficiency is a vital factor in profitable production. The most common means of expressing efficiency is in terms of units of production/unit of feed required (weight gain, eggs or milk/unit of feed consumed). When diets and type of production are similar, a satisfactory comparison of gross efficiency is possible. However, if we compare gain produced by a high-concentrate ration to that produced on a high-roughage ration, the comparison is poor in terms of utilization of available energy (DE, ME, or NE) because the roughage is apt to be much lower in available energy than a concentrate.

Another factor must be considered when efficiency is measured in terms of units of production, particularly when dealing with body gain or loss in weight. Although the energy value of milk can be calculated easily if the fat percentage is known or that of eggs can be estimated with reasonable precision, the energy content of body gain may vary widely. In studies with dairy cows, Reid (1970) points out that the caloric value of body gain ranged from 4.8 to 9.4 Mcal/kg and that for body loss from 6.3 to 7.9 Mcal/kg; he suggested that body tissue gain or loss could range from as much as 100% water to about 90% fat. In growing animals, most of the tissue gain in early postnatal growth is proteinaceous (lean tissues contain about 75% water), but during fattening, added tissue may contain 90% fat. For these extremes, the combustion values of 1 g of added tissue would range from 1.4 to 8.5 Kcal, or a ratio of about 1:6, indicating that weight gain may be of little value in estimating caloric efficiency. If gain resulted only from water retention, the spread would be considerably greater.

Information from a variety of sources indicates that caloric efficiency is greatest for maintenance, followed by milk production, and then by growth and fattening. The relative efficiency of maintenance may be due largely to the more efficient utilization of the HI and HF in maintaining body temperature. The less efficient use of energy for production as compared to maintenance is due in part to a decline in the digestibility of a given diet as feed intake increases, particularly in ruminants.

Early postnatal growth is quite efficient because it approaches that of maintenance, but the efficiency declines with age, partly because of gradually increased fat deposition. Nevertheless, research on swine, sheep, and cattle suggests that the energy cost of protein deposition is greater than that for fat deposition. The lower efficiency of protein may be due to its dynamic state and more rapid turnover in the body than is the case for lipid tissues.

Gross efficiency (caloric value of product/caloric intake) is affected greatly by age, as indicated, and by level of production. Young animals generally eat considerably more per unit of metabolic size than older animals, so that maintenance requirements represent a smaller percentage of dietary intake. The same comment applies to thin animals as opposed to fat animals, or to any situation where animals are fed at appreciably less than maximum intake. As a result, in many production situations it is economically feasible to feed for maximum intake and rate of production.

Net efficiency (caloric value of product/caloric intake *above* maintenance) is less affected by level of intake and perhaps more by genetic capability of the animal. In high-producing animals, size is not a factor.

Examples of the range in efficiency that might be expected in swine and ruminants as suggested by Reid (1962) are shown in Table 10.7. Relative efficiencies for utilization of NE suggested by Kleiber (1961) are: fat production in adult steers, 100; maintenance in cattle, 120; milk production by cow, 119; maintenance in swine, 145; growth and fattening in swine, 125; and fattening in swine, 130. Reid (1962) has made some interesting calculations on the efficiency of production of food energy by different domestic species of animals. In these calculations, he has considered the cost of maintaining the dam and has included credit for the carcass of the dam toward total food production. His

TABLE 10.7   Efficiency of energy utilization by pigs and ruminants ingesting diets of the usual range of quality fed in practice.[a]

ITEM	PIGS	RUMINANTS
Digestible energy, %	75–90	50–87
Utilization of ME above maintenance, %		
Body gain	75–80	30–62
Milk production	75–85	40–75
Fattening during lactation[b]		73

[a]From Reid (1962). Data from a variety of sources. That for ruminants include diets ranging from fair quality forages to high-concentrate rations.
[b]Data from Moe et al. (1971).

calculated values in terms of percentage of DE recovered in food are: pork, 18; eggs, 12; broilers, 12; milk (3600 kg/year), 22; milk (5400 kg/year) 27; and beef, 6. One reason for the low overall efficiency of beef production is that a beef cow produces at a very low level compared to a dairy cow; thus, maintenance is a much greater proportion of total feed intake, and gross efficiency drops considerably.

A graphic illustration of changes in energy utilization with increasing feed consumption is shown in Fig. 10.3. The changes in energy losses from a consumption of 8 lb (maintenance) to 20 lb/day were: feces, 20 to 27; methane, 10 to 7, urine, 2–3 to 3; and heat production, 13 to 18 (as a % of GE consumption). Thus, although losses increased from 45 to 55% of GE intake, maintenance becomes a smaller proportion of the total (from about 55 to 21%), with the net result that the gross and net efficiency improve at the higher rate of feeding.

The data shown in Fig. 10.3 are more applicable to ruminants than to nonruminants because generally less depression in digestibility takes place as feed consumption increases in nonruminant animals. Calculations by the Agricultural Research Center (1980) indicate that the change in digestibility of a diet, expressed as a multiple of maintenance, can be described by the equation

Change in digestibility
$$= (L - 1)(0.107 - 0.113 \times DE)$$

This equation (where L = feeding level) implies that a diet with a digestibility of 70% DE at maintenance would have a digestibility of 67.2% at twice maintenance [0.70 − (0.107 − 0.113 × 0.70)] and 58.8% at five times maintenance. Although DE values may decrease, the increased energy lost in feces tends to be compensated for by decreased methane and urinary energy losses. The net effect is that ME is less affected than DE by intake. For example, the ME of a diet having a DE of 70% is expected to be 57% at maintenance. At two and five times maintenance, the ME is expected to be about 56% and 53%, respectively. Thus, an increase in feed consumption of a diet of this type from maintenance to five times maintenance is expected to result in a decrease in ME of less than 7% but a decrease in digestibility of 20% in DE.

In the latest NRC publication on dairy cattle (2001), the committee has adopted a procedure to discount the change in digestibility with increased intake. For example, if we have a diet that has a computed value of 74% TDN when fed at maintenance, when the diet is fed at three times maintenance, it is calculated to have a TDN value that is reduced to 91.8% of the maintenance value, or 67.9%. Numerous other formulas are given in this publication for computing the value of DE, ME, and NE for feeds based on updated chemical composition of various feeds used in modern dairies.

Efficiency may also be altered by various feed additives. For example, two antibiotic feed additives, monensin and lasalocid, generally result in reduced feed consumption with little, if any, effect on daily gain. This occurs, in part, because of an increased production in the rumen of propionic acid (used more efficiently than acetic or butyric acids) and a reduction in methane production without having any appreciable effect on digestibility.

❏ ENERGY TERMINOLOGY USED IN RATION FORMULATION AND FEEDING STANDARDS

Many different factors are clearly involved in energy utilization by animals. One major factor is variation in feed ingredients. Roughages, grains, or other energy sources may vary from truckload to truckload or according to source from field to field, region to region, and so on. By the time the user has had some of these sources analyzed for nutrients of interest, the feed in question may be long gone. In addition to these factors affecting animal feed, all of the various factors discussed in previous sections may alter feed utilization or animal requirements. The point of this discussion is to remind the reader that we cannot be too precise in calculating feed values or animal requirements.

As noted before, GE is of little direct value for evaluating feedstuffs, so other energy values are used today. DE and TDN are sometimes used, particularly with swine and horses. ME values are used for poultry. DE, TDN, or ME values are usually given for sheep or goats. For cattle, some form of energy is usually given in tabular data.

With beef cattle, a system developed at the California Experiment Station has been in use since 1984 and is continued in the latest NRC for beef (1996). Animal requirements or feed values are expressed as NEm or NEg. For NEm, requirements are assumed to be 77 Kcal × $BW^{0.75}$ which may be adjusted for breed or species. This value is based on data from a variety of calorimetric studies. Since other needs such as for milk production or pregnancy needs (conceptus energy) are used with similar efficiency as maintenance, these needs are all expressed as NEm. Thus, to formulate a diet for an animal, one simply sums the requirements for maintenance, milk, and pregnancy (all expressed as NEm to obtain an estimate of total NEm required. Where live weight gain is involved, it is necessary to calculate an additional amount of energy required for the gain. Use of two values, NEm and NEg, complicates ration formulation somewhat, but most computer programs will handle it. For dairy cattle, essentially all of the requirements are calculated as NEl because of the similar efficiencies. Additional information on formulas used to calculate the energy value of feeds or the energy needs of different animals can be found in the various NRC bulletins.

❏ SUMMARY

- Information collected on many different species shows that heat production (or loss) can be equated with reasonable accuracy to body surface area, which can, in turn, be estimated by multiplying body weight by the 0.73 or 0.75 power.
- Body heat production usually is estimated under standard conditions such as fasting, at maintenance, or some higher level of activity. Heat production measurements for active free-ranging animals are fraught with many uncertainties, especially in animals that are difficult to train in

the use of equipment or in species that fly or swim.

- Various methods have been devised to describe and quantify animal utilization of feed energy. All have some faults that reduce the precision used to estimate feed values or animal requirements. Nevertheless, even with some of the problems that have been mentioned, energy requirements for the domestic species are reasonably well quantified and are satisfactory guides for practical livestock production in most situations.

❏ REFERENCES

Agricultural Research Council (ARC). 1980. *The Nutrient Requirements of Ruminant Livestock.* Agric. Res. Council, Commonwealth Agric. Bureaux, London, England.

Armstrong, D. G., and K. L. Blaxter. 1957. *Brit. J. Nutr.* **11**: 247.

Blaxter, K. L. 1967. *The Energy Metabolism of Ruminants* (2nd ed.). Hutchinson & Co., London.

Blaxter, K. L., and J. L. Clapperton. 1965. *Br. J. Nutr.* **19**: 511.

Brody, S. 1945. *Bioenergetics and Growth.* Hafner Publishing Co., New York.

Brouwer, E. 1965. In *Proc. Energy Metabolism of Farm Animals,* EAAP Publ. No. 11: 441. Academic Press, London.

Commonwealth Scientific and Industrial Research Organization (CSIRO). 1990. *Feeding Standards for Australian Livestock.* Ruminants. CSIRO Publications, East Melbourne, Victoria, Australia.

European Association of Animal Production (EAAP). 1958 through 1994. *Triennial Symposia, Energy Metabolism of Farm Animals.* European Association of Anim. Production.

Ferrell, C. L. 1988. *The Ruminant Animal.* Prentice-Hall, Englewood Cliffs, NJ.

Hafez, E. S. E. and I. A. Dyer (eds.). 1969. *Animal Nutrition and Growth.* Lea & Febier Publishing Co., Philadelphia, PA.

Kleiber, M. 1961. *The Fire of Life.* John Wiley & Sons, New York.

Milligan, L. P., and B. W. McBride. 1985. *J. Nutr.* **115**: 1374.

Mitchell, H. H. 1962. *Comparative Nutrition of Man and Domestic Animals,* Vols. 1 & 2. Academic Press, New York.

Moe, P. W., and H. F. Tyrrell. 1975. *J. Dairy Sci.* **58**: 602.

Moe, P. W., H. F. Tyrell, and W. P. Flatt. 1971. *J. Dairy Sci.* **54**: 548.

National Research Council (NRC). 1981. *Nutritional Energetics of Domestic Animals and Glossary of Energy Terms.* National Academy Press, Washington, DC.

National Research Council. 1984. *Nutrient Requirements of Beef Cattle* (6th ed.). National Academy Press, Washington, DC.

National Research Council. 1989. *Nutrient Requirements of Dairy Cattle* (5th ed.). Academy Press, Washington, DC.

National Research Council. 1996. *Nutrient Requirements of Beef Cattle* (7th ed.). National Academy Press, Washington, DC.

National Research Council. 2001. *Nutrient Requirements of Dairy Cattle* (6th ed.). National Academy Press, Washington, DC.

Reid, J. T. 1962. In *Animal Nutrition's Contributions to Modern Animal Agriculture.* Cornell Univ. Spec. Publ.

Reid, J. T. 1970. In *Physiology of Digestion and Metabolism in the Ruminant.* Oriel Press, Newcastle-Upon-Tyne, England.

Tyrrell, H. F., and P. W. Moe. 1972. *J. Dairy Sci.* **55**: 1106.

Webster, A. J. F. 1979. In *Digestive Physiology and Nutrition of Ruminants,* Vol. 2 (2nd ed.), D. C. Church (ed.). O & B Books, Corvallis, OR (out of print).

Young, B. A., and J. L. Corbett. 1972. *Austral. J. Agr. Res.* **23**: 77.

11

Macromineral Elements

IT HAS BEEN SHOWN that at least 22 inorganic elements (hereafter referred to as minerals or mineral elements) are required by some animals species, and an additional five or more elements may be essential metabolically based on indirect or limited data. The required mineral elements can be divided into two groups based on the relative amounts needed in the diet, namely, macrominerals and micro- or trace minerals. The macrominerals are: calcium (Ca) phosphorus (P); sodium (Na); chlorine (Cl); potassium (K); magnesium (Mg); and sulfur (S).

Some macrominerals—such as Ca, P, and Mg—are required as structural components of the skeleton, whereas others—such as Na, K, and Cl—function in acid-base balance. Mg, in addition to being a constituent of bone, is required for activation of several enzymes. All macrominerals have more than one function.

Animal tissues contain many minerals in addition to those recognized as dietary essentials. Some of these minerals may have a metabolic function not yet recognized, but others may be present as innocuous contaminants. Still others are toxic to the animal at relatively low concentrations. Even the required minerals may be toxic if they are present in the diet in excess.

The National Research Council (NRC) Subcommittee on Mineral Toxicity in Animals (NRC, 1980) reviewed the literature on maximum tolerable levels of minerals in the diets of domestic animals and emphasized that "all mineral elements, whether essential or non-essential, can affect an animal adversely if included in the diet at excessively high levels." The NRC has recently convened a new subcommittee to review and update the existing knowledge of mineral tolerances in domestic animals (National Research Council, 2005).

In this chapter, we discuss the nutrition and metabolism of the macrominerals and address some of the important metabolic interactions between and among them. Each required macromineral element is discussed with respect to distribution in body tissues, functions, metabolism, deficiency signs, and toxicity. The Ca, P, Mg, Na, K, and Cl content of a range of newborn and adult mammals are shown in Table 11.1 (Widdowson and Dickerson, 1964).

TABLE 11.1   Ca, P, Mg, Na, K, and Cl content of newborn and adult animals (amounts/kg of fat free body tissue).[a]

	MAN	PIG	DOG	CAT	RABBIT	GUINEA PIG	RAT	MOUSE
Newborn								
Body wt, g	3560	1250	328	118	54	80	5.9	1.6
Water, g	823	820	845	822	865	775	862	850
Ca, g	9.6	10.0	4.9	6.6	4.6	12.3	3.1	3.4
P, g	5.6	5.8	3.9	4.4	3.6	7.5	3.6	3.4
Mg, g	0.56	0.32	0.17	0.26	0.23	0.46	0.25	0.34
Na, meq	82	93	81	92	78	71	84	—
K, meq	53	50	58	60	53	69	65	70
Cl, meq	55	52	60	66	56	—	67	—
Adult								
Body wt, kg	65	125	6.0	4.0	2.6	—	0.35	0.027
Water, g	720	750	740	740	730	—	720	780
Ca, g	22.4	12.0	—	13.0	13.0	—	12.4	11.4
P, g	12.0	7.0	—	8.0	7.0	—	7.5	7.4
Mg, g	0.47	0.45	—	0.45	0.50	—	0.40	0.43
Na, meq	80	65	69	65	58	—	59	63
K, meq	69	74	65	77	72	—	81	80
Cl, meq	50	—	43	—	32	—	40	—

[a]To convert meq (milliequivalents) to mg, multiply by 23 for Na, 39 for K, and 35.5 for Cl; to convert g to meq, divide by 20,000 for Ca, 31,000 for P, and 12,000 for Mg.
Source: Widdowson and Dickerson (1964).

❏ CALCIUM, PHOSPHORUS, AND MAGNESIUM CONTENTS OF ANIMALS

The animal skeleton contains most of the Ca, P, and Mg present in the whole body. The proportions of each macromineral, expressed as amount/kg of fat-free body tissue, are similar among species in adults. The Ca, P, and Mg content of various organs and tissues (expressed as a percent of fresh tissue) of adult humans are shown in Table 11.2. Note the preponderance of these three elements in the skeleton and teeth relative to other tissues. Newborn animals also have similar composition, but differences among species in the degree of physiological maturity at birth are reflected in the amounts of Ca, P, and Mg present. Humans and pigs are more mature physiologically at birth than rabbits and mice and have higher concentrations of bone minerals. The Ca: P: Mg ratios of both newborn and adults are remarkably similar among species.

TABLE 11.2 Calcium, phosphorus, and magnesium content of organs and tissues of adult humans.

TISSUE OR ORGAN	Ca	P	Mg
	(% OF FRESH TISSUE)		
Skin	0.015	0.083	0.0102
Skeleton	11.51	5.19	0.191
Teeth	25.46	13.24	0.618
Striated muscle	0.014	0.116	0.0198
Nerve tissue	0.015	0.224	0.0107
Liver	0.012	0.127	0.0081
Heart	0.018	0.123	0.0168
Lungs	0.017	0.228	0.0069
Spleen	0.010	0.169	0.0124
Kidney	0.019	0.124	0.0169
Alimentary tract	0.009	0.111	0.131
Adipose tissue	0.009	0.055	0.0060
Remaining tissue (solid)	0.047	0.163	0.0170
Remaining tissue (liquid)	0.008	0.088	0.0084
Total composition			
Whole body	1.98	1.06	0.045
Fat-free	2.07	1.11	0.047
Bone ash	39.44	17.80	0.654

Source: Forbes et al. (1956).

❏ CALCIUM (Ca)

Tissue Distribution

About 99% of the Ca stored in the animal body is in the skeleton as a constituent of bones and teeth. It occurs in about a 2:1 ratio with P in bone, primarily as hydroxyapatite crystals:

$$Ca_{10}^{++}x(PO_4)_6(OH^-)_2(H_3O^+)_{2x}$$

where x may vary from 0 to 2. When $x = 0$, the compound is called octacalcium phosphate; when $x = 2$, it is called hydroxyapatite.

Ca is present in blood mostly in the plasma (extracellular) at a concentration of about 10 mg/dl and exists in three states (D'Sousa and Flock, 1973): as the free ion (60%), bound to protein (35%), or complexed with organic acids such as citrate or with inorganic acids such as phosphate (57%).

Functions

Skeletal Structure. The most obvious function of Ca is as a structural component of the skeleton.

A brief description of normal bone metabolism is needed to develop a full appreciation of the function of Ca as a major skeletal component. Bone is a metabolically active tissue with continuous turnover and remodeling in both growing and mature animals. The physiological control of bone metabolism is related to both endocrine and nutritional factors. Blood is the transport medium by which Ca is moved from the GI tract to other tissues for utilization. Elaborate controls exist to maintain a relatively constant Ca concentration in the plasma. A decline in plasma Ca concentration triggers the parathyroid gland to increase secretion of parathyroid hormone (PTH), which stimulates the biosynthesis of the metabolically active form of vitamin D (1,25-dihydroxy cholecalciferol or 1,25-(OH)2D3) by the kidney (see discussion of vitamin D in Chapter 14), which, in turn, causes increased Ca absorption from the GI tract and increased resorption of Ca from bone. An increase in plasma Ca concentration triggers "C cells" in the thyroid gland to release calcitonin, a hormone that depresses plasma Ca by inhibiting bone resorption. Thus, dietary factors that affect Ca absorption have an effect on the endocrine system in direct response to the amount of Ca reaching the blood from the GI tract (Tanaka et al., 1981). The amount of Ca absorbed from the GI tract depends on the amount ingested and the proportion absorbed. As the percentage of Ca in the diet increases, the proportion absorbed tends to decline. This is related to the fact that Ca absorption is an active process under the control of a Ca-binding protein (CaBP; see vitamin D, Chapter 14), which, at least in most species, is vitamin-D dependent. The absolute amount absorbed generally is greater in animals fed high-Ca diets, despite the lower percentage absorption. In vitamin D deficiency, Ca absorption is reduced because of impaired CaBP formation, so that skeletal abnormalities can occur even in the presence of adequate dietary Ca.

Bone apposition occurs through the activity of osteoblasts (bone-forming cells) at growth plates and in other areas of rapid bone growth. Bone resorption occurs concomitantly by two processes—one by resorption of bone surfaces by activity of osteoclasts (bone resorbing cells) and the other by resorption deep within formed bone (osteocytic resorption). The latter process is now generally believed to be the major process of bone resorption. The continuous apposition and resorption of bone (bone remodeling) is the basis for bone growth and for changes in its shape and density.

The continuous process of bone remodeling resulting from simultaneous bone apposition and bone resorption regulates bone growth in the young animal and bone homeostasis in the adult animal under physiological conditions of pregnancy, lactation, or maintenance (Prentice, 2000). Recent evidence (Takeda et al., 2003) indicates that the homeostatic control of bone mass may also involve the action of the protein hormone, leptin, produced by fat cells. Leptin is known to signal the central nervous system about the energy status of the peripheral tissues (Ahima et al., 2000) and also appears to operate through the central nervous system to inhibit bone formation. The overall impact of leptin as a modulator of bone metabolism in obese animals is unknown. However, leptin appears to be an important factor linking obesity and bone density.

Cellular Metabolism.   Ca controls the excitability of nerve and muscle. Reduced Ca++ concentration produces increased excitability of pre- and postganglionic nerve fibers. Higher than normal Ca++ concentrations have the opposite effect on nerves and muscles, causing them to be hypoexcitable. Ca is required for normal blood coagulation. The great importance of Ca in cellular metabolism has recently become appreciated through knowledge of its role in sarcoplasmic reticular function in skeletal and cardiac muscle and in the physiological functions of the nervous system. In cardiac and skeletal muscle, Ca release from the sarcoplasmic reticulum is controlled by a series of relationships beyond the scope of this discussion. Suffice it to say that Ca sequestration by a Ca-stimulated ATPase is functionally important for muscle relaxation, and Ca release provides a source of Ca for myofilament activation, the importance of which is still uncertain (Sutko, 1985). Furthermore, skeletal muscle protein synthesis appears to be closely associated with a calculated factor whose nature and physiological importance continue to be under study.

Ca++ regulates the level of phosphorylation of endogenous proteins in the nervous system, especially through activation of calmodulin-dependent protein kinases (Fujisawa et al., 1984). Ca is a key ion in calmodulin metabolism with respect to neurotransmitter synthesis in the overall function of the nervous system. This field of investigation has revealed that the Ca++ ion serves vital functions not previously appreciated. The implications of

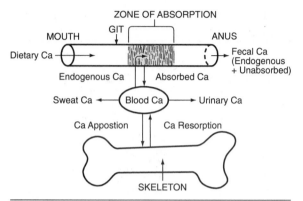

Figure 11.1   *Schematic diagram of overall metabolism of Ca.*

these functions in applied animal nutrition are still unclear but provide for an exciting future for research in Ca metabolism. The overall movement of Ca in metabolism is diagrammed schematically in Fig. 11.1. In the growing animal, net retention of Ca occurs in the body; that is, the amount stored in bone and other tissue exceeds that lost in feces, urine, and sweat. In nonpregnant, nonlactating adults, the amount of Ca ingested equals the amounts lost if the metabolic requirement is met. Ca participates as a major constituent of the skeleton, which is composed of bone, cartilage, and tooth. The details of skeletal development are outside the realm of our discussion, but it is important to be aware of the dynamic nature of Ca accretion in the skeleton. Early skeletal development begins in embryonic life and involves morphogenesis (formation and migration of new cells) and differentiation (the appearance of chondroblasts and osteoblasts, the precursors of chondrocytes of cartilage and osteocytes of bone). The general scheme of skeletal growth and development of animals is comparable to that described in humans (Oklund and Prolo, 1990).

Absorption.   Dietary Ca is absorbed largely in the duodenum and jejunum of most animals. An exception appears to be the hamster in which most absorption is from the ileum. Absorption occurs both by active (energy-dependent) and passive (diffusion) transport. In rats and man, and presumably in other species, about half of the dietary Ca is absorbed by active transport. The importance of a vitamin D-dependent protein carrier of Ca in active transport has been elucidated as described in Chapter 14.

Other factors in addition to vitamin D affect the efficiency of Ca absorption. Increased dietary Ca concentration decreases the percentage of Ca absorbed, although the absolute amount absorbed tends to be relatively constant within the normal range of Ca concentration of the diet. Lactose as well as lysine and, to a lesser extent, several other amino acids have been shown to enhance Ca absorption in rats, but phytic and oxalic acids decrease its absorption by forming insoluble complexes of Ca oxalate and Ca phytate. Such factors as high pH of the intestinal contents, high levels of dietary fat, which may form insoluble fatty acid-Ca soaps, and high-fiber levels in the diet, probably are not of significance.

Bone Formation. Ossification of the skeleton to form hydroxyapatite crystals requires that the product of Ca ions and P ions in the fluid surrounding the bone matrix exceed a critical minimum level. Thus, if the product $(Ca++) (PO_4)$ falls below the concentration required to precipitate Ca phosphate in the crystal lattice structure, ossification fails to occur and rickets or osteomalacia results, whether the deficiency is one of Ca or P, or both. Calcification is an active process, specifically requiring ATP.

Excretion. The three major routes of Ca excretion are feces, urine, and sweat. Fecal output includes both an unabsorbed fraction and an endogenous fraction, arising largely from secretions of the intestinal mucosa. The endogenous fraction probably is reabsorbed partially, as illustrated in Fig. 11.1. Therefore, that appearing in the feces is termed fecal endogenous Ca and represents about 20 to 30% of total fecal Ca. The apparent Ca absorbability (feed Ca minus fecal Ca) generally approximates 50%, although the percentage tends to decline as intake increases.

Urinary output of Ca generally is considerably less than that of fecal output in most species (Table 11.3). About half of the plasma Ca, mainly ionized Ca, is filtered in the kidney; more than 99% of this amount is reabsorbed under normal conditions. Diuretics generally do not affect Ca excretion, but Ca chelators, such as large intravenous doses of Na-citrate, NA-EDTA or CA-EDTA, greatly increase Ca excretion. Data obtained from human studies show that increasing dietary protein intake above the metabolic requirement causes a striking increase in urinary Ca excretion.

Loss of Ca in sweat is of only minor significance in most species, but in humans, horses, and other species in which sweating is prominent, large amounts of Ca can be lost by this route. Sweat loss of more than 1 g of Ca/day occurs in adult men doing heavy physical work at high environmental temperatures. Sweat losses equal or exceed fecal and urinary losses in these men.

Signs of Deficiency

The main effect of Ca deficiency is on the skeleton. In young, growing animals, a simple Ca deficiency or vitamin D deficiency results in rickets, and, in adults, the disease is called osteomalacia. In each case the bones become soft and often deformed due to failure in calcification of the cartilage matrix. The familiar skeletal changes associated with rickets are illustrated in Fig. 11.2. Simple Ca deficiency or vitamin D deficiency, which results in poor utilization of dietary Ca, even when

TABLE 11.3 Distribution of fecal and urinary Ca excretion in humans, cattle, and rats.

SPECIES	AGE	Ca INTAKE	FECAL Ca	URINE Ca
		(MG/DAY)		
Human				
Male	11–16 yr	1,866	1,018	127
	23	1,461	1,229	72
Female	14–16 yr	874	655	194
	55–63	713	586	500
Cattle	Young adult	29,000	27,000	500
Rats	12 wk	44.1	20.8	0.9

Source: Bronner (1964).

called nutritional secondary hyperparathyroidism, features a generalized effect on the entire skeleton. "Big head" disease of horses (Fig. 11.3), simian bone disease of monkeys, and twisted snouts of pigs (Fig. 11.4) may result from feeding excess P. High-P diets depress intestinal absorption of Ca in horses. The histological picture of the nasal turbinate is similar to that often observed in growing pigs with atrophic rhinitis (AR) (Brown et al., 1966).

Lameness and spontaneous bone fractures often, but not necessarily, accompany both osteomalacia and nutritional secondary hyperparathyroidism in adult animals. Reproduction and lactation are adversely affected in rats fed Ca-deficient diets. The demands of the fetus for Ca are tremendous during late gestation. In rats, fetal uptake of Ca/h in

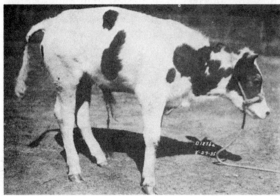

Figure 11.2   Top. Bowed legs in a pig with Ca-deficiency rickets. Courtesy of L. Krook, Cornell University. Bottom. An advanced stage of rickets in a calf. Courtesy of J. W. Thomas, Michigan State University.

the Ca level of the diet is adequate, may produce such abnormal bone development. The histological picture in rickets is one of reduced calcification. The degree of change is related to the growth rate of the animal. In species such as the pig, dog, and chick, in which body weight may be doubled in a few days and the skeletal mineral turnover rate is therefore very rapid, a Ca deficiency may produce profound changes in bone after only a few days. In other species, such as sheep and cattle, a longer period is required to show signs of deficiency.

A deficiency of Ca (or even normal amounts) in the presence of excess P also causes abnormal bone, but, in this instance, excess bone resorption by osteocytic osteolysis (resorption deep within bone) results in fibrous osteodystrophy. This condition is characterized histologically by replacement of osseous tissue with fibrous connective tissue. The parathyroid gland is hyperactive in an attempt to maintain normal blood serum Ca. The disease,

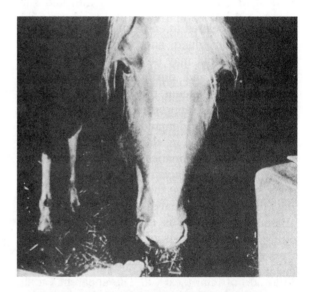

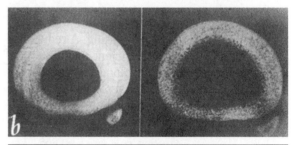

Figure 11.3   Top. Nutritional secondary hyperparathyroidism (NSHP) in a horse, 'big head,' a spontaneous case. The facial bones are enlarged and the facial crest is no longer visible. Bottom. Radiographs of a section of the metacarpus of a horse (left) fed a normal Ca:P diet; right, horse fed excess P. In NSHP the cortex is thinner and the radiographic density is markedly decreased. Courtesy of L. Krook, Cornell University.

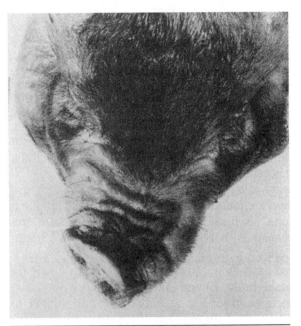

Figure 11.4 *A pig with a distorted snout caused by nutritional seondary hyperparathyroidism (NSHP) that was induced by a dietary Ca–P imbalance.*

late gestation is equal to the total maternal Ca blood content. Thus, inadequate dietary intake necessitates resorption of Ca from the maternal skeleton to meet fetal needs.

Although blood serum Ca concentration may decline slightly during the early weeks of dietary Ca deficiency, control by parathyroid hormone (increases bone resorption) and calcitonin (inhibits bone resorption) renders this a relatively useless index of Ca nutrition. Only by frequent serial blood sampling over an extended period of weeks or months is monitoring of serum Ca meaningful.

A reduction in the content of bone ash occurs in all forms of dietary Ca deficiency or Ca–P imbalance. All bones of the skeleton are affected, although the magnitude of change may vary. The proportion of Ca and P remains constant (about a 2:1 Ca–P ratio). The bone ash may be determined directly by ashing the fat-free bone or indirectly by determining specific gravity or by expressing density as g ash/unit volume. This latter method eliminates the need to ether-extract the bone before ashing. In the live animal, changes in bone density can be detected through a bone mineral density scan (Dual Energy X-Ray Absorptiometry), as reviewed by Mitchell (2004). Large-magnitude changes can be identified by radiography, but this method is too insensitive to perceive differences not exceeding about 30%.

In humans increasing concern has arisen over the high incidence of osteoporosis in the elderly. It is estimated that at least 10% of women over 50 years of age in the United States have bone loss severe enough to cause hip, vertebrae, and long-bone fractures. At least part of this high frequency is believed to be due to suboptimum intake of dietary Ca during prepubertal adolescence and extending into adulthood. Surveys of daily Ca intake in the United States indicate that after age 35 the value declines gradually in both men and women to levels well below recommended levels. Osteoporosis is characterized by reduction in bone mass, with resulting tendency to fracture. Osteoporosis and osteomalacia are common in sows following lactation, resulting in "downer" (not ambulatory) sows. Osteomalacia is the adult counterpart of rickets in growing animals, reulting from dietary Ca or vitamin D deficiency. The nutritional factors in osteoporosis, in addition to Ca intake, include P, vitamin D, vitamin K, and the minerals Cu, F, and B. Also, excess intakes of protein and Na, and interactions among nutrients, have been implicated in osteoporosis (Heaney, 1993). It appears that Ca is the overriding nutrient in this debilitating bone disease in both humans and animals. Physical inactivity increases Ca excretion. Striking increases in Ca losses in urine and feces were noted in astronauts confined to small space capsules in the early space flights. A twofold increase in Ca excretion has been reported in healthy human subjects after three weeks of bed rest. These results from studies in humans have important implications for animal production in close confinement systems.

Calcium Tetany (Milk Fever)

Severe Ca deficiency may produce hypocalcemia, which results in tetany and convulsions. The pathogenesis of Ca tetany in animals is as follows: Ca deficiency: deficient intake; disturbance in absorption, with vitamin D deficiency or diarrhea; excessive kidney excretion; formation of complexes; parathyroid hypofunction. Ca tetany is related to the requirement of Ca in normal transmission of nerve impulses and in muscle contraction. Ca tetany may be transient or may culminate in sudden death. Presumably, death results from failure in normal heart muscle contractions, which require Ca in association with Na and K.

The classical example of Ca tetany is "milk fever," or parturient paresis of dairy cattle and sheep. The condition usually occurs early in lactation during the period of large drains on body Ca reserves for milk production. The etiology probably is related to endocrine function. Both the parathyroid gland and calcitonin-secreting cells of the thyroid may be involved. Dietary Ca intake prior to and during early lactation is important in influencing the capability of these glands to respond appropriately to the sudden change in metabolic demands for Ca imposed by lactation. Ca tetany in dairy cattle, sheep, and other species responds dramatically to intravenous injection of Calcium chloride, Ca lactate, or other Ca salts to elevate serum Ca above the threshhold concentration of 5 or 6 mg/dL that is required to prevent tetany.

Blood-clotting time is influenced by Ca. The clotting process consists of a complicated series of reactions as outlined in Chapter 14. Such substances as oxalate, citrate, and ethylene diamine tetraacetic acid (EDTA) are commonly used to prevent blood coagulation in vitro; these same compounds, if administered in large quantities, may also form salts with Ca in vivo and interfere with blood clotting. A reduction in ionized Ca in blood of sufficient magnitude to become a limiting factor in blood clotting is not always seen in simple dietary Ca deficiency. Widespread hemorrhage has been observed in tissues of rats and dogs fed diets severely deficient in Ca.

Toxicity

Acute Ca toxicity has not been reported, but chronic ingestion of Ca in excess of metabolic requirements results in abnormalities in bone which can be regarded as manifestations of toxicity. The tendency toward hypercalcemia resulting from continued absorption of excess Ca stimulates calcitonin production by the thyroid. The inhibitory effect of calcitonin on bone resorption is a homeostatic mechanism aimed at minimizing the sources of serum Ca of other than dietary origin. Calcitonin has a stimulatory effect on $1,25(OH)_2D_3$ accumulation in kidney tubules. Sustained calcitonin secretion leads to excess bone mass in response to inadequate bone resorption relative to bone apposition. This abnormal thickening of bone cortex is termed osteopetrosis. It has been produced in growing dogs (Fig. 11.5) by feeding 2.0% Ca–1.4% P diets from weaning to young adulthood, and in

mature dairy bulls by prolonged feeding of high Ca diets designed for lactating dairy cows. Tumors of the ultimobranchial (calcitonin-producing cells) tissue of the thyroid of these bulls were observed, presumably in response to the sustained hyperactivity resulting from continued hypercalcemia.

Calcification of soft tissues may occur in high-Ca feeding, but such calcification only occurs in sites of cellular damage such as in atherosclerosis or inflammation. Tissue damage from Mg deficiency is associated with soft tissue calcification, but high Ca in itself usually does not induce it. A diet containing 2.5% Ca and 0.4% available P has been shown to produce nephrosis, visceral gout, and Ca urate deposits in the ureters of growing pullets. Increasing the P level to achieve a more

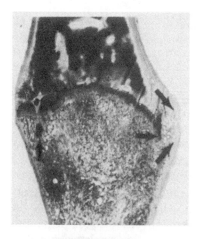

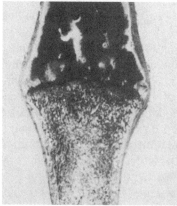

Figure 11.5 *Excess bone mass (osteopetrosis) in dog fed excess Ca (top) or normal (bottom) level, showing costochondral junction with irregular growth of cartilage at horizontal arrow, cartilage cells island at vertical arrow, and thick perichondral ring at oblique arrows. Courtesy of L. Krook and Cornell Veterinarian.*

nearly optimum Ca-P ratio in such high Ca diets prevented the kidney lesions. The Ca requirement of laying hens is much higher than in most animals and birds because egg shell is composed mostly of CaCO3. The hen requires about 3.3 g Ca/day for egg production. The amount in relation to body size would be considered toxic to mammals, but is required by the caged hen to prevent cage fatigue. This condition results in excess removal of Ca phosphate from medullary bone and proneness to fracture. Thus, the level of a mineral required to produce toxicity is dependent on species as well as on physiological (productive) state of the animal. This principle generally applies to most minerals.

Urinary calculi (kidney stones) which block the kidney tubules or ureters may occur in animals. The calculi are not believed to be formed by high Ca alone, but rather to require imbalances of other minerals or formation of abnormal complexes with cholesterol or other steroids. One form of calculi is composed of struvite (an insoluble magnesium ammonium phosphate crystal structure). In ruminants, excess P in relation to Ca is more likely to cause calculi. A Ca to P ratio of 2:1 or greater is normally recommended to prevent urinary calculi in rams.

Excess Ca in the diet reduces the absorption and utilization of other minerals. The classical example of a nutritional disease precipitated by high dietary Ca in pigs is the Zn deficiency disease, parakeratosis. The Zn deficiency may not appear in animals fed a normal level of Ca but can become a serious problem when dietary Ca is increased without changing Zn intake. Similar reductions in mineral utilization induced by excess Ca have been reported for Mg, Fe, I, Mn, and Cu.

❑ PHOSPHORUS (P)

Tissue Distribution

The P content of adult humans approximates 1.1% of the fat-free body, of which about 80% is in the skeleton. Bone ash contains about 18% P. The percentage of P in the body and the proportion of total P in the skeleton increase throughout prenatal and postnatal life as ossification of the skeleton progresses to maturity. The P in the skeleton is present as part of the hydroxyapatite crystal (see the Ca section), while that in the soft tissues is present mostly in organic forms. In blood serum, P exists in both inorganic and organic form, the organic form being a constituent of lipids. Of the inorganic P, about 10% is bound to serum proteins and 50 to 60% is ionized. The P in red blood cells is present as inorganic P, organic acid-soluble P, lipid P, and RNA P, the proportions varying with age and species. Total serum P concentration under normal conditions in most species is 6 to 9 mg/dL. The transfer rates of P among spatially distinct compartments of the body have been calculated (Lax et al., 1956), assuming that each compartment exchanges P with others through a central compartment (blood). The exchange rate of P among several of the important organs was highest for the skeleton, followed in descending order by muscle and heart. The average time spent by a single P atom in each compartment also was calculated (2.8 h for blood cells vs. 393 h for brain). The length of time in a compartment was found to be related to the metabolic activity of the tissue. Thus, P movement within the body clearly is in a dynamic state.

Functions

As with Ca, the most obvious function of P is as a component of the skeleton. In this role, it provides structural support for the body. Because Ca and P occur together in bone, the discussion of the function of Ca (see the Ca section) in bone also applies to P. P is a component of phospholipids, which are important in lipid transport and metabolism and cell-membrane structure. Therefore, P is present in virtually all cells (see Chapter 8). P functions in energy metabolism as a component of AMP, ADP, and ATP and of creatine phosphate. The importance of high-energy phosphate bonds in normal life processes was discussed in Chapter 8. P is a component as phosphate of RNA and DNA, the vital cellular constituents required for protein synthesis (see Chapter 9). It is a constituent of several enzyme systems (e.g., cocarboxylase, flavoproteins, NAD).

Metabolism

The metabolism of P can be discussed in terms of bone metabolism, the metabolism of phospholipids, and high-energy phosphate compounds such as ATP, ADP, and creatine phosphate. Bone metabolism was discussed earlier in this chapter, and phospholipid metabolism and metabolism of high-energy phosphate compounds are discussed in Chapter 8 and in other sections relating to energy utilization.

Absorption

Absorption of P from the GI tract is rapid, as demonstrated by radioisotope studies with P-32. Much of the labeled P is incorporated into phospholipids in the intestinal mucosal cells. P absorption occurs by active transport and passive diffusion. Vitamin D apparently has an effect on P absorption. In vitro work has shown that P may traverse the intestinal cell membrane against a concentration gradient in the presence of Ca and requires Na.

Phytate P

P absorption is related directly to dietary P concentration and the proportion of soluble inorganic forms of P relative to organically bound P presented in the form of phytin P (phytate P). Phytate P is biologically unavailable to nonruminant animals because they do not produce phytases (enzymes that hydrolyze phytate P to release inorganic P). Ruminants can utilize phytate P because their rumen microbes produce phytases. The low bioavailability for nonruminants of the P bound as phytate P requires inorganic P supplemetation of plant-based diets, resulting in a level of total dietary P that is considerably higher than the metabolic requirement of the animal so as to compensate for the low bioavailability of P in plant seeds. The unabsorbed phytate P is excreted in the feces and contributes to contamination of the environment. This environmental concern has stimulated interest in improvement of dietary P bioavailability for nonruminant animals.

Microbial Phytases to Increase P Bioavailability

Major efforts are underway to develop commercial sources of microbial phytases for use in high-phytate P diets of nonruminant animals, including swine, poultry, and humans. The use of phytases to improve organic P utilization by nonruminants (reviewed by Lei and Porres, 2004) is briefly described here. The chemical structure and characterization of phytases were reviewed by Wyss et al. (1999). The release of P and other minerals from phytate is depicted diagrammatically in Fig. 11.6.

Phytate P is the hexaphosphorylated ester of inositol, a constituent of plant seeds and other plant parts. Phytases produced by plants and by microbes (e.g., bacteria, yeasts) are capable of hydrolyzing the bond between inositol and P, thereby releasing the inorganic P for absorption by the animal. Phytases have emerged as effective tools to improve P bioavailability and reduce environmental pollution (Ullah, 2000) in animal agriculture and to enhance bioavailability of trace minerals such as Zn and Fe bound to phytate in feedstuffs of plant origin. Lei and Stahl (2001) described the biotechnological

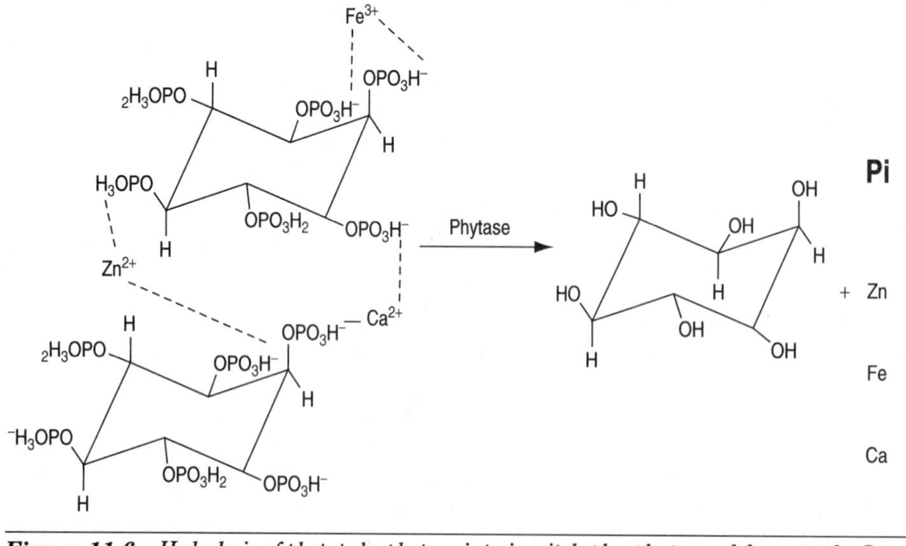

Figure 11.6   *Hydrolysis of phytate by phytase into inositol, phosphate, and free metals. Source: Lei and Porres (2004) with permission by Marcel Dekker, N.Y.*

development of microbial phytases for improving mineral nutrition and environmental protection.

The first phytase to be characterized was obtained from the fungus *Aspergillis niger*. This and other fungal phytases have a molecular weight range of 80 to 120 kDaltons, whereas bacterial phytases have a range of 40 to 55 kDaltons and plant phytases isolated from corn, wheat, and other plant seeds have a range of 47 to 55 kDaltons. Phytases differ in their mode of splitting the phosphate group on the inositol ring. Some initiate splitting at the C-1 or C-3 carbon, whereas others act at the C-6 carbon. Phytases differ in their optimum pH, temperature, and other properties. Therefore, the choice of a phytase depends on the purpose for which its use is intended. For example, if pelleted feed is intended as the feed to which a phytase is to be added, a heat-stable phytase will be needed. The future of commercial use of phytases in animal nutrition seems certain, and this active area of research will continue because of the important role of P and other minerals in animal nutrition and environmental stability.

Ca absorption is decreased in the presence of high-dietary phytate P. This may be because of formation of insoluble Ca phosphate salts and/or to Ca being bound by phytic acid (the hexaphosphoric acid ester of inositol).

Many plant seeds are high in P, much of which is present as phytic acid, decreasing the bioavailability of P for nonruminants. Bioavailability of phytate-P for swine and poultry is influenced by phytase that is present in plant materials, by the pH of the GI tract, and the ratio of Ca to P in the diet. Estimates of bioavailability of total plant P range from 20 to 60%.

Excretion of Dietary P

Secretion of P into the intestinal lumen (endogenous fecal P) occurs, but this loss does not represent as high a proportion of the daily loss as for Ca. In ruminants, large amounts of P are secreted in the saliva and enter the rumen during rumination. In general, most of the P excretion occurs through the kidneys, and renal excretion appears to be the main regulator of blood P concentration. When intestinal absorption of P is low, urinary P falls to a low level, with reabsorption by the kidney tubules approaching 99%. Kidney excretion of P is under control of the parathyroid hormone and 1,25-dihydroxy vitamin D as part of the overall blood homeostatic mechanicsm for Ca and P.

Signs of Deficiency

The most common sign of P deficiency in growing animals is rickets (see Fig. 11.2). The gross and histological changes in rickets are described earlier in the Ca section and also in Chapter 14). Fecal output of P tends to remain relatively unchanged while urinary excretion is reduced, but the total excretion may still exceed intake when Ca intake is relatively high. Ca excretion in both urine and feces is increased in P deficiency as a manifestation of reduced calcification of bone. As the deficiency progresses, appetite fails and growth is retarded. Deficient animals (Fig. 11.7) often have a depraved appetite and may chew on wood and other inappropriate objects. This abnormal behavior of eating or chewing is termed *pica* (Fig. 11.8). Adults fed low-P diets may exhibit pica, and bone density is decreased as in rickets (Fig. 11.9). Impaired fertility has been reported in P-deficient cattle. Blood-serum Ca is increased, and serum P is decreased by P deficiency.

Blood P homeostasis is more complicated than that for blood Ca because blood P is in equilibrium not only with bone P but also with several organic P compounds. Nevertheless, kidney excretion of P is sufficiently controlled by parathyroid hormone

Figure 11.7 *P-deficient cow. Note stiffness and thin condition. Bones were depleted to such an extent that one rib was broken in a casual examination. Courtesy R. B. Becker, Florida Agricultural Experiment Station.*

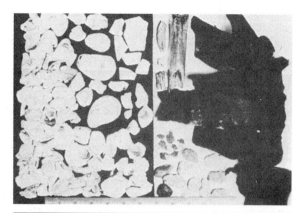

Figure 11.8 *Pica. A P deficiency, as well as deficiencies of other nutrients, may result in a depraved appetite (pica). This example shows some of the material recovered from the stomach of a deficient cow. It includes oyster shells, porcelain, teeth, a section of cannon bone inner tube, tire casing, pieces of metal, and pebbles. Courtesy of R. B. Becker, Florida Agricultural Experiment Station.*

secretion and 1,25-dihydroxy vitamin D to result in a relatively stable serum P concentration even with severe dietary P deficiency.

Toxicity

An excess of dietary P results in nutritional secondary hyperparathyroidism manifested in excessive bone resorption (fibrous osteodystrophy). This condition may result in lameness and spon-

Figure 11.9 *A P deficiency. Note enlarged joints and crooked hind legs. This pig was fed a diet with 0.3% P. Courtesy of W. M. Beeson, Purdue University.*

taneous fractures of long bones. Growing pigs with severe bone resorption produced by low-Ca–high-P diets may suffocate because of softening of the ribs to the extent that normal respiratory movements are inhibited. High-phosphate diets depress intestinal absorption of Ca, plasma Ca, and Ca retention in horses. Lean meat and many cereal grain byproducts (notably wheat bran) contain several times as much P as Ca (see section on Ca deficiency. Failure to correct this dietary Ca–P imbalance leads to fibrous osteodystophy caused by nutritional secondary hyperparathyroidism in growing and adult animals (see Figs. 11.3 and 11.4).

High P has a laxative effect so that dietary excesses result in diarrhea and high fecal loss of P as well as other nutrients.

❏ MAGNESIUM (Mg)

Mg is distributed widely in the body and, except for Ca and P, is present in larger amounts in the body (see Table 11.1) than any other mineral. About half of body Mg is in bone at a concentration of 0.5–0.7% of the bone ash (Wacker and Vallee, 1964). Mg in soft tissues is concentrated within cells; the highest concentration is in liver and skeletal muscle. Blood Mg is distributed about 75% in red blood cells (6 meq/L) and 25% in serum (1.1 to 2.0 meq/L). The concentration in serum seems to vary among species as shown in Table 11.4. Of the serum Mg, about 35% is protein-bound in mammals and birds, even though total Mg is variable among species.

Functions

Mg is required for normal skeletal development as a constituent of bone; it is required for oxidative phosphorylation by mitochondria of heart muscle and probably by mitochondria of other tissues. Mg is required for activation of enzymes that split and transfer phosphatases and the many enzymes concerned in the reactions involving ATP. As discussed elsewhere (see Chapters 7, 8, 9, 10), ATP is required in such diverse functions as muscle contraction, protein, nucleic acid, fat, and coenzyme synthesis; glucose utilization; methyl-group transfer; sulfate, acetate, and formate activation; and oxidative phosphorylation, to name but a few. By inference, the activation action of Mg extends to all of these functions. Mg is a cofactor in de-

TABLE 11.4 serum Mg concentration of man and animals.

SPECIES	Mg CONCENTRATION (MEQ/L)
Cow	1.6
Dog	1.6
Goat	1.9
Horse	1.5
Hen	1.6
Human	2.0
Mouse	1.1
Pig	1.3
Rabbit	2.1
Rat	1.6
Sheep	1.7

Source: Wacker and Vallee (1964).

carboxylation and is required to activate certain peptidases. Specific examples of enzyme systems requiring Mg for activity were described by Wacker and Vallee (1964).

Metabolism

Mg metabolism is complex and varied. Absorption from the GI tract occurs mostly from the ileum. No carrier is known for Mg absorption, nor has vitamin D been shown to enhance its absorption. Although some have suggested a common pathway for Ca and Mg absorption, the fact that Ca is absorbed mainly from the upper small intestine and is associated with a vitamin D-dependent binding protein suggests no common pathway.

Homeostatic control of blood and tissue Mg is not understood clearly. Hyperparathyroidism is associated with increased urinary excretion and reduced serum Mg, but a specific effect on Mg aside from the concomitant release of Mg from bone when Ca is released in response to parathyroid hormone has not been shown. Some evidence suggests an effect of plasma Mg on parathyroid activity.

Mg excretion occurs via the feces and urine. About 55 to 60% of ingested Mg is absorbed, and the absolute amount absorbed is proportional to dietary intake. Urinary excretion accounts for about 95% of losses of absorbed Mg, and fecal excretion accounts for most of the remainder. Endogenous fecal excretion is largely into the proximal small in-

testine, so that, as with Ca, probably some reabsorption occurs as it traverses the GI tract.

Signs of Deficiency

Mg deficiency in growing rats results in anorexia, reduced weight gain, reduced serum Mg, hypomagnesemic tetany, and, within 3 to 5 days, characteristic hyperemia of the ears and extremities. Kidney Ca elevation in Mg deficiency is accompanied by a decrease in total serum Mg but an increase in the ratio of free to total Mg (Bunce and Bloomer, 1972). Concentration of Mg in liver is not depressed. Severe leukocytosis develops concurrently with hyperemia of extremities. Mg deficiency in pigs results in weak and crooked legs, hyperirritability, muscular twitching, reluctance to stand, tetany, and death.

Recent evidence (Rude et al., 2004) suggests that Mg deficiency may be a causative factor in osteoporosis in animals and humans. Plasma and urinary histamine are elevated in Mg deficiency, and serum and liver glutamic oxalacetic transaminase activity are increased. Calcification and necrosis of the kidney occur in Mg deficiency. High dietary Ca and P appear to aggravate Mg deficiency, probably owing to depressed Mg absorption, and accentuate the calcification of soft tissues associated with inadequate Mg.

Red blood cell Mg concentration is reduced as well as plasma Mg concentration, but the decline is more gradual and reaches 50% of control values later than is the case for plasma Mg concentration. Although liver Mg concentration is unaffected, skeletal muscle concentration is reduced by 25% or more. Bone Mg content is reduced in Mg deficiency when expressed either as a percentage of Mg in whole bone or in the bone ash and as skeletal deformities are apparent.

A decline in tissue K and a rise in tissue Ca and Na occur in Mg-deficient animals. Activities of several Mg-dependent enzymes, including plasma alkaline phosphatase, muscle enolase, and pyruvate phosphokinase, are depressed in Mg deficiency. Mitochondria of kidney tubule cells are swollen in Mg deficiency as shown by electron microscopy. Mitochondrial swelling and loss of intramitochondrial Mg and PO_4 ions also occur in liver from ammonia-intoxicated rats. It has been proposed (Head and Rook, 1955) that high ammonia interferes with Mg absorption by forming the insoluble Mg ammonium phosphate (struvite) at alkaline pH.

Grass Tetany. A common problem of grazing cattle is a syndrome called grass tetany (also called grass staggers, Mg tetany, or wheat-pasture poisoning). It occurs most frequently in cattle grazing cereal forages or native pastures in periods of lush growth (usually in spring months), but at times it may also be a problem in cattle fed conventional wintering rations. The symptoms of tetany, which have been ascribed to hypomagnesemia (low blood Mg), were first described in the 1930s. The etiology of Mg tetany is not understood completely, although it is generally agreed that hypomagnesemia, whatever its underlying cause, is the triggering factor. The high level of K and protein usually present in lush pastures has suggested the possibility of an antagonism with Mg. Newton et al. (1972) found that sheep fed high K tended to excrete more Mg in urine and feces and that high K interfered with Mg absorption, but not its reexcretion into the GI tract. Elevated urinary Mg excretion has been reported in sheep fed high-N (urea) diets, and an increased severity of Mg deficiency occurs in rats fed high-Ca, -K, or -protein diets.

High concentrations of trans-aconitate, an intermediate in the citric acid cycle, have been observed in early-season forages and have been suggested as having a role in Mg tetany (Burau and Stout, 1965). Intravenous administration of either trans-aconitic or citric acid into cattle produces tetany resembling clinical Mg tetany. Rumen microflorea convert trans-aconitate to tricaballylate, which binds Mg and disrupts the TCA cycle. It seems clear that grass tetany is more complicated than a simple dietary Mg deficiency. Rather, it is a complex syndrome involving changes in content of K and other minerals in forage as well as rumen microorganism-induced conversions of organic acids that result in reduced Mg bioavailability. Some evidence implicates high-Al levels (2000–6000 ppm) in some samples of rapidly growing forage which may alter solubility and affect the incidence of Mg tetany.

In practice, cows grazing fall-seeded small grain (oats, wheat, rye, ryegrass) pasture are at the highest risk of developing grass tetany because the plants accumulate K faster than Mg. A high concentration of K in the forage is highly correlated with the incidence of grass tetany (Apley and Hilton, 2003). Heavy fertilization of pastures in the spring with N and K increases the risk. Supplemental Mg from inorganic sources such as magnesium oxide (MgO) included in a standard mix (equal parts of MgO, salt, dicalcium phosphate and ground corn) is effective in reducing but not eliminating the incidence of Mg tetany. The problem is one of overcoming the practical difficulties in administering Mg to free-ranging cattle and sheep on pasture. Cattle consuming mineral supplement free-choice vary considerably in voluntary intake, resulting in some consuming much more than required, whereas others consume none or much less than is needed to prevent grass tetany (Apley and Hilton, 2003). Chambliss and Kunkle (2003) recommend management practices such as placement of mineral supplements close to watering and resting areas, removal of animals from pasture during rapid plant growth, and fertilization of soils with Mg. When liming is needed for pH adjustment, they recommend using dolomite limestone (limestone containing Mg) to reduce the risk of grass tetany in lactating cattle. It has been suggested (Greene et al., 1989) that variation in efficiency of Mg absorption and susceptibility to Mg deficiency observed between different breeds of cattle may be due to genetic variation in the Mg absorptive mechanisms and other factors associated with gastrointestinal tract physiology.

Toxicity

Mg toxicosis in animals includes depressed feed intake, diarrhea, loss of reflexes, and cardiorespiratory depression. Severe diarrhea and reduced feed intake and growth have been observed in calves fed diets containing more than 2.3% Mg, and accidental feeding of high-Mg diets to sheep results in diarrhea and anorexia.

Intravenous injection of $MgSO_4$ was shown more than a century ago to result in peripheral muscle paralysis of dogs. Subsequent work confirmed this effect and led to the use of Mg as an anesthetic for surgery. Mg decreases the release of acetylcholine at the neuromuscular junction and sympathetic ganglia.

Mg induces a drop in blood pressure, and high-serum concentrations (greater than 5 meq/L) affect the electrocardiogram, sometimes causing the heart to stop in diastole.

❑ POTASSIUM, SODIUM, AND CHLORINE (K, Na, Cl)

Potassium, sodium, and chlorine are considered together because they are all electrolytes that play a vital role in maintaining osmotic pressure in the extracellular and intracellular fluids, and in maintaining acid-base balance. In addition, each has its own special functions, which will be discussed separately.

Distribution in Body Tissues

The total body contents of K, Na, and Cl in newborn and adults of several mammals including man are shown in Table 11.1. Normal ratios among electrolytes are remarkably constant among species. Houpt (1977) has described the tissue distribution and interrelationship of K, Na, and Cl in maintaining acid-base balance. K is present mainly within cells (about 90% of body K is intracellular) and is readily exchangeable with the extracellular fluid. On the other hand, Na is present largely in extracellular fluid with less than 10% within cells. Of the other 90%, about half is adsorbed to the hydroxyapatite crystal of bone and half is present in plasma and interstitial fluids. Cl acts with bicarbonate (HCO3) to balance electrically the Na of the extracellular fluid. Excess excretion of Na by the kidney is accompanied by Cl excretion. Cl is present almost exclusively in the extracellular fluid. The concentrations of electrolytes and other constituents in blood plasma, intercellular fluid, and intracellular fluid are depicted in Fig. 11.10 (Tortora and Anagnostakos, 1990). The meq of cations/L within each compartment exactly equals the meq of anions/L to ensure electrical neutrality. The composition depicted for intracellular fluids is only an approximation because some individual tissues have cell electrolyte composition that is considerably different from the mean.

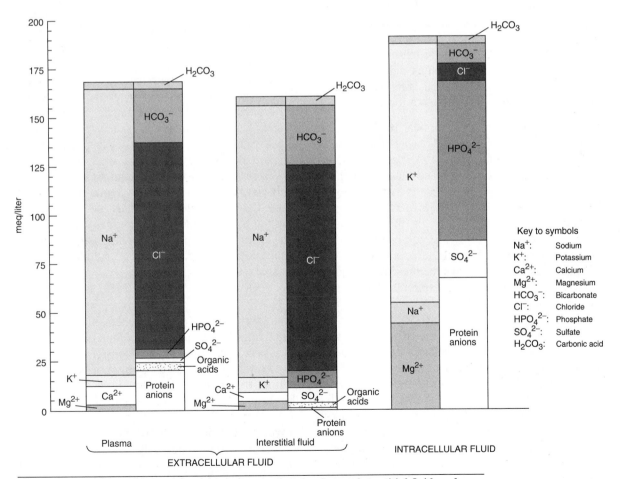

Figure 11.10 *Comparison of electrolyte concentrations in plasma, interstitial fluid, and intracellular fluid. The height of each column represents the total electrolyte concentration. Source: Tortora and Anagnostakos (1990).*

Functions

K is located mostly within cells, and, by means of an energy-requiring system related to Na movement, it influences osmotic equilibrium. K functions in maintenance of acid-base balance in the body. It is required in enzyme reactions involving phosphorylation of creatine, and it is required for activity of pyruvate kinase. K facilitates uptake of neutral amino acids by cells, and it influences carbohydrate metabolism by affecting the uptake of glucose into cells. It is required for normal tissue protein synthesis in protein-depleted animals, as well as for normal integrity of the heart and kidney muscle and for a normal electrocardiogram.

Na functions as the extracellular component through an energy-dependent Na "pump," along with K and Mg and intracellular components, in maintaining osmotic pressure. It functions in maintaining acid-base balance in the body and in nerve impulse transmission by virtue of the potential energy associated with its separation from K by the cell membrane.

C1 functions in regulation of extracellular osmotic pressure and in maintaining acid-base balance in the body. It is the chief anion of gastric juice where it unites with H ions to form HC1.

Acid-Base Balance. The critical roles of Na, K, and Cl in maintaining acid-base homeostasis (Tortora and Anagnostopos, 1990) are summarized here. The overall acid-base balance in the body is maintained by controlling the hydrogen ion (H+) concentration of body fluids, particularly extracellular fluid. In healthy animals, the pH of the extracellular fluid is 7.3 to 7.5; survival requires maintenance of pH within this narrow range.

Homeostasis of pH depends on three major mechanisms: buffer systems, respirations, and kidney excretion.

Buffer systems include (1) the carbonic acid–bicarbonate system, (2) the phosphate system, and (3) the hemoglobin–oxyhemoglobin system. The carbonic acid–bicarbonate buffer system is the most important regulator of blood pH and the most abundant buffer in extracellular fluid. The system is based on the bicarbonate ion (HCO_3^-), which acts as a weak base, and carbonic acid ($H_2CO_3^-$), which acts as a weak acid. Therefore, the system can compensate for either an excess or a shortage of H+ ions. If there is an excess of H+ ions (acid condition), HCO_3^- functions as a weak base and removes excess H+:

$$H+ + HCO_3^- \rightleftarrows H_2CO_3 \rightleftarrows H_2O + CO_2$$

If there is a shortage of H+ ions (alkaline condition), H_2CO_3 functions as a weak acid and provides

$$H+ :H_2CO_3 \rightleftarrows H+ + HCO_3^-$$

The phosphate buffer system operates similarly as a regulator of pH in erythrocytes as well as in kidney. The components of the system are depicted as follows:

DIHYDROGEN PHOSPHATE ION BUFFERS STRONG BASES

$$NaOH + NaH_2PO_4 \rightleftarrows H_2O + Na_2HPO_4$$

MONOHYDROGEN PHOSPHATE ION BUFFERS STRONG ACIDS

$$HCl + Na_2HPO_4 \rightleftarrows NaCl + NaH_2PO_4$$

The hemoglobin–oxyhemoglobin system buffers H_2CO_3 in erythrocytes. CO_2 produced by body cells enters eythrocytes and combines with H_2O to form H_2CO_3. Some of the oxygen released from oxyhemoglobin becomes reduced hemoglobin, which carries a negative charge and attracts H+ ion from H_2CO_3. The bicarbonate ion (HCO_3^-) combines with K+ inside the cell or with Na+ outside the cell, and in this way CO_2 is carried to the lungs as $KHCO_3$ or $NaHCO_3$.

Respirations play a role in maintaining pH in the body by changing the rate and depth of breathing. More CO_2 is exhaled with increased respiration rate and blood pH increases; decreased respiration rate lowers blood pH. The relationship between respiration rate and blood pH is as follows:

$$CO_2 + H_2O \rightleftarrows H_2CO_3 \rightleftarrows H+ + HCO_3$$

The *kidney* excretes excesses of Na, K, and thereby contributes to maintenance of blood pH homeostasis. These three mechanisms for pH homeostasis could not function, and life could not be sustained without the presence of adequate dietary levels of the three electrolytes, Na, K, and Cl.

In dairy cattle, health and productivity are improved by balancing the dietary cation–anion difference (DCAD). Block (2004) reviewed the factors to be considered in DCAD balancing to contol blood-buffering capacity and blood pH in dairy cattle. Producers can control blood pH by balancing the ration to achieve specific DCAD levels, leading to improved animal performance. The DCAD level should be adjusted by adding appropriate commercial rumen buffers according to the physiological status of the cow. For lactating cows, a highly positive DCAD level of between +35 and +40 milliequivalents (Na + K) − (Cl + S) per 100 g of feed dry matter should be achieved. In dry cows, a negative DCAD of −10 should be used to eliminate hypocalcemia (milk fever) and reduce udder edema. Cows fed high-forage diets with high K should be supplemented with Na to prevent K overload, which may impede blood pH control and increase incidence of grass tetany. Appropriate adjustment of DCAD in dairy cattle and other livestock and poultry promotes animal health and productivity.

Metabolism

K, Na, and Cl ions are not absorbed in appreciable amounts from the stomach, but considerable absorption takes place from the upper small intestine and lesser amounts from the lower small intestine and large intestine. The daily secretion of fluids into the GI tract from saliva, gastric juice, bile, and pancreatic juice contributes four to five times the daily oral intake of these ions. Thus, large variation in intake has a relatively small effect on the total load of fluids and electrolytes entering the GI tract. About 80% of the NaCl load and 50% of the K load on the GI tract are from secretions. Cl is secreted in large quantities in the stomach, Na is secreted into the upper small intestine, and K mainly into the ileum and large intestine. All three ions are absorbed by both passive and active processes; Na and K cross the mucosa by active transport in the intestine but largely by diffusion in the stomach, whereas Cl is transferred by active processes in the stomach and upper intestine, but by passive diffusion in the large intestine.

Concentration of K, Na, and Cl ions is regulated within rather narrow limits in extra- and intracellular compartments of the body. Ingestion of more of each mineral than is needed results in ready excretion by the kidneys. Thus, toxicity is not likely unless water intake is restricted, which impedes urine output. Na is excreted by the kidney, and partial, but not complete, reabsorption occurs from the kidney tubules. The portion not reabsorbed is lost in the urine. The plasma level of the Na tends to be controlled by the action of the hormone aldosterone, secreted by the adrenal cortex in response to endocrine signals from the kidney, including secretion of rennin and other hormones. Its action is to increase Na reabsorption from the kidney tubule. A fall in plasma Na, as a result of reduced intake, results in increased aldosterone release. Aldosterone production is, in turn, under the control of the adrenocorticotrophic hormone of the anterior pituitary gland; its secretion is impaired in hypophysectomized animals.

The antidiuretic hormone of the posterior pituitary also plays a part in Na excretion through its response to changes in osmotic pressure of the extracellular fluid induced by water deprivation. Osmoconcentration results in increased antidiuretic-hormone release as a means of conserving water. Thus, the complex relationships between body Na content, aldosterone secretion, and antidiuretic hormone secretion are of utmost importance in maintaining fluid electrolyte homeostasis. In animals that perspire extensively, such as the human and the horse, large amounts of Na are lost from the body by this route. K intake usually exceeds by several times the metabolic requirements for it, yet K toxicity does not ordinarily occur because of the kidney's ability to regulate its excretion. Aldosterone also affects K excretion; high K in extracellular fluid stimulates aldosterone secretion in the same way that low Na does. A rather constant ratio of Na to K in the extracellular fluid is maintained. In K deficiency, some Na is transferred inside the cell to replace K, and in that way osmotic and acid-base equilibrium are preserved. When intake of either Na or K is inadequate, the deficiency is aggravated if intake of the other is increased. Cl concentration in the extracellular fluid tends to be controlled in relation to Na. Excess kidney excretion is also affected by bicarbonate ion (HCO_3^-) concentration. If plasma bicarbonate rises, an equal amount of Cl is excreted in the urine to maintain an equal concentration of cations and anions in the plasma.

K and Cl homeostasis are closely related. A deficiency of one leads to a metabolic deficiency of the other. K reabsorption by the kidney tubule requires the presence of Cl. Thus, KCl is more effective than other K salts in repletion from K deficiency.

Deficiency Signs

K.   Deficiency of K results in abnormal electrocardiograms in most or all species. Postmortem examination may not always reveal pathological changes in heart muscle, but in rats tiny gray opacities on the ventricles of the heart and loss in striations have been observed. Kidney lesions also have been observed. Growth retardation, unsteady gait, general overall muscle weakness, pica (e.g., wool biting), and emaciation, followed by death, are the symptoms of K deficiency in animals and birds. Mg deficiency results in failure to retain K and, in this way, may lead to K deficiency in animals and man. Diarrhea is associated with loss of electrolytes, notably K, in abnormally high quantities in the feces, thereby upsetting osmotic pressure relationships and acid-base balance.

Na.   The main signs of Na deficiency are reduced growth rate and efficiency of feed utilization in growing animals and reduced milk production and weight loss in adults. Na deficiency in growing pigs is associated with an immediate decline in Na urinary excretion, hemoconcentration, and decreased plasma volume.

Smith and Aines (1959) studied salt (NAC1) deficiency in lactating dairy cows and found that Na, K, and C1 concentration of milk remained unchanged, as did the plasma Na concentration. Urinary excretion of Na declined to near zero within a month, and appetite and milk production dropped sharply. Animals deprived of Na display a great craving for it and have been observed to drink urine in an effort to satisfy their craving. The Na ion appears to be the critical mineral in NaC1 deficiency.

C1.   Depressed growth rate seems to be the major sign of C1 deficiency. Low dietary C1 results in a decline in urinary C1 concentration to near zero, and C1 concentration of skin, muscle, liver, kidney, brain, viscera, and total carcass is reduced. Kidney lesions develop within one month of feeding on a C1-deficient diet. Leach and Nesheim (1963) described a syndrome in chicks fed C1-deficient diets. The chicks show a characteristic nervous reaction induced by sudden noise; they fall forward with their legs extended backward. In practice, Na and C1 are supplied in the diet together as common salt; it appears that the metabolic requirement for each is proportional to the amounts contributed by NaC1.

Toxicity

The kidney normally regulates its excretion of K, Na, and C1 in accord with variations in dietary intake. Therefore, a toxicity of any of the three electrolytes is unlikely except when water intake is restricted, drinking water is saline (high in these and other ions), or animals have kidney malfunction. Chronic K excess induced by one of the above means results in hypertrophy of the adrenal cortex, with accompanying changes in aldosterone output. A high level of dietary K reduces the apparent absorption of Mg in ruminants. Studies with goats indicated that supplementary K may increase the cellular uptake and retention of Mg because of increased cellular K levels. Lentz et al. (1976) suggested that prolonged elevation of K in blood plasma of ruminants may lead to a series of metabolic disturbances, including elevated insulin secretion, which play an important role in the etiology of grass tetany. This metabolic disorder was discussed briefly earlier in this chapter. The complex metabolic relationships between dietary K and Mg in ruminant nutrition remain incompletely understood.

Chronic, excess Na ingestion results in hypertension associated with degenerative vascular disease. The glomeruli of the kidney are affected. Excess salt ingestion has been implicated in the etiology of hypertension in humans with a genetic propensity to the problem (Papper, 1982). There is evidence that average salt intake among humans in the United States is higher than desirable. Continued high intakes of salt increase total extracellular fluid volume; this is the basis for the use of low-salt diets in hypertension and congestive heart failure.

Acute salt toxicity has been produced experimentally in growing pigs by severely restricting water intake while simultaneously providing feed containing 2% NaC1 for 1 or 2 days. Symptoms include staggering, marked weakness, paralysis of hind limbs or general paralysis, violent convulsions, and death. Postmortem examination may reveal no lesions except a slight increase in volume of fluid in the pericardium and small hemorrhages in the liver.

Excess C1 is not likely, but the use of purified diets containing amino acids and other salts added in the hydrochloride form may increase the total acidity of the diet. In such cases, extra K and Na should be added to the diet to assure maintenance of acid-base balance.

❑ SULFUR (S)

Tissue Distribution

Animals require sulfur mainly as a component of organic compounds, including the amino acids methionine, cystine, cysteine; the vitamins biotin and thiamin; certain mucopolysaccharides, including the chondroitin sulfate and mucoitin sulfates; heparin; glutathione; and coenzyme A. Because proteins are present in every cell of the body, and S-containing amino acids are components of virtually all proteins (usually 0.6 to 0.8% of the protein), S is distributed widely throughout the body and in every cell and makes up about 0.15% of body weight.

Functions

Sulfur functions mainly through its presence in organic metabolites. Inorganic sulfate from exogenous dietary sources and from endogenous release from S containing amino acids are used in synthesizing the chondroitin matrix of cartilage, in biosynthesis of taurine, heparin, cystine, and other organic constituents in the animal body (Baker, 1977). In birds, sulfate is incorporated into feathers, gizzard lining, and muscle. Its functions in organic compounds will be listed in terms of the metabolic function of the compounds containing it.

Inorganic SO_4 functions in acid-base balance as a constituent of intracellular, and to a lesser extent, extracellular fluid; as a component of S-containing amino acids, it is required for protein synthesis (see Chapter 9); as a component of biotin, it is important in lipid metabolism (see Chapter 15; and as a component of thiamin, it is important in carbohydrate metabolism (see Chapter 15); as a component of coenzyme A, it is important in energy metabolism (see Chapters 7, 8, 10). As a component of mucopolysaccharides, S is important in collagen and connective tissue metabolism; as a component of heparin, it is required for blood clotting. S is a component of ergothioneine of red blood cells and of glutathione, a universal cell constituent; and of hormones containing S-amino acids.

Metabolism

Absorption of inorganic sulfate from the GI tract is inefficient. Active transport of sulfate takes place from the upper small intestine. Both inorganic and organic forms of S are used for sulfation of cartilage mucopolysaccharides.

Organic forms of S are absorbed readily, as discussed elsewhere (Chapters 7, 8, 9, 15) in relation to the compounds that contain S. Inorganic S is excreted via the feces and urine. Unabsorbed S is probably reduced in the lower GI tract and excreted as sulfate.

Endogenous fecal S enters the GI tract largely through the bile as a component of taurocholic acid. Urinary S is present mainly as inorganic sulfate, but also as a component of thiosulfate, taurine, cystine, and other organic compounds. Because the bulk of body S is present in amino acids, it is not surprising that urinary S excretion tends to parallel urinary N excretion. High-protein diets are associated with large amounts of urinary S and N.

In ruminants fed nonprotein N, a growth response may be obtained by inorganic S supplementation. Rumen microflora incorporate N into cellular protein, but synthesis is limited if S is not available in sufficient quantities for formation of S-amino acids. The N to S ratio in rumen microbial protein is about 15:1. The commonly recommended N to S ratio in diets containing high levels of urea is 10:1. Supplementation of S to a semipurified diet for sheep improves rumen microbial activity and rumen thiamin status. It appears that inorganic S may be important in ruminant nutrition, but only in terms of its requirement by microflora of the rumen for amino acid and thiamin synthesis. Sheep, which produce wool high in S-containing amino acids, need a higher S:N ratio than non–wool-producing ruminants. Birds, whose feathers are high in S-amino acids, also have a higher S requirement than most mammals.

Deficiency Signs

Inorganic S has not been shown to be essential for animals for normal maintenance or productive functions. Deficiency of any of the organic metabolites containing S of course produces functional and morphological lesions, as discussed elsewhere (Chapters 7, 8, 9, 15), for S-containing amino acids, vitamins, glycoproteins, and lipoproteins. In addition, the absence of inorganic S from the diet may increase the requirement for S-containing amino acids, implying that in the absence of inorganic S the sulfur from these sources is used for synthesis of other organic forms of S.

Sheep fed nonprotein N to replace protein without concomitant supplementation with S show reduced wool growth, and weight gain of sheep and cattle is depressed in S deficiency. These effects on animal performance are manifestations of inadequate microbial nutrition on which the host depends for synthesis of organic metabolites and therefore cannot be considered as direct S deficiency.

Assuming a dietary S requirement for dairy cattle to be 0.20% (NRC, 2001), a sulfur deficiency may occur in cattle fed diets high in corn or sorghum grain or silage, whose content is 0.10 to 0.14%. Type of forage in the diet may also influence S requirement. For example, cattle grazing sorghum-sudan grass pasture have an increased S requirement because S is required in the detoxification of cyanogenic glucosides present in most sorghum forsges, and cattle fed fescue hay, in which S bioavailability is low, may need S supplementatation (Berger, 2004).

Toxicity

Because the intestinal absorption of inorganic S compounds is low, S toxicity is not a major problem. However, a serious neurological disease, polioencephalomalacia (PEM), can occur in ruminants fed high levels of S, which leads to excessive production of sulfides in the rumen (Lowe et al., 1996). Affected animals have necrosis in the brain cortex and develop characteristic patterns of "stargazing" (head turned upward and backward) and of head-pressing against a post or wall. Injection of thiamin relieves the syndrome, but if treatment is not prompt, death occurs within 48 hours. PEM may occur in animals consuming bracken fern, which is high in thiaminase activity, and in animals changed rapidly from an all-forage diet to a high-concentrate diet. In this latter case, PEM is caused by the sudden dietary change to concentrates resulting in shifts to rumen microbial populations that produce thiaminase (Berger, 2004).

Excesses of S-amino acids cause anorexia and growth depression, but such effects are observed with excesses of amino acids in general, not specifically the S-amino acids. The toxicity of S is determined, to a large extent, by the enzyme systems of the exposed animal and by whether the animal has the capacity to form H_2S from the inorganic sulfate sources presented (National Research Council,

1980). Ultimately, the toxicity of S depends on its form. While elemental S is considered one of the least toxic elements, H_2S rivals cyanide in toxicity. In ruminant animals, the formation of a thiomolybdate complex composed of S and Mo produces an effect on Cu absorption (Suttle, 1991). Cu toxicity or Cu deficiency may result, depending on the ratio of the three ions in the diet. The significance of this interaction in ruminant nutrition and the current knowledge concerning it have been reviewed (Suttle, 1991). This three-way interaction among inorganic ions illustrates the complex nature of inorganic mineral nutrition in animals. Berger (2004) noted that S reduces Cu absorption by formation of insoluble copper sulfide in the rumen independent of the formation of thiomolybdate. High intakes of S (in feed or water) reduce copper absorption and may induce copper defiency. High intakes of S that result from feeding high-S byproduct feeds or from drinking water containing high concentrations of S (2200 to 2800 ppm) may precipitate Cu deficiency. In such cases, Cu supplementation of the diet may be required (copper carbonate rather than copper sulfate) (Berger, 2004).

❏ SUMMARY

- The required macrominerals are calcium (Ca), phosphorus (P), magnesium (Mg), sodium (Na), chlorine (Cl), potassium (K), and sulfur (S).

- Ca, P, and Mg are required as structural components of the skeleton, while Na, K, and Cl function in acid-base balance and S functions mainly through its presence in organic metabolites; all macrominerals have other functions as well.

- Ca metabolism is under delicate endocrine control; parathyroid hormone, calcitonin, and the metabolically active form of vitamin D work in concert to maintain Ca homeostasis.

- Ca deficiency results in reduced bone calcification (rickets in growing animals, osteomalacia in adults) and in hypocalcemia in lactating animals.

- P is a component of phospholipids, which are important in lipid transport and cell membrane structure and is a constituent of several enzyme systems.

- Mg functions as a skeletal constituent (about one-half) and is required for oxidative phosphorylation by mitochondria and for activation of many enzyme systems.

- Na (present mainly in extracellular fluids) and Cl are electrolytes vital in maintaining osmotic pressure in the intracellular and extracellular fluids and in maintaining acid-base balance.

- Na functions in transfer of nerve impulses by virtue of the potential energy associated with its separation from K by the cell membrane.

- Cl acts with bicarbonate (HCO_3^-) to balance electrically the Na of the extracellular fluid.

- K (present mainly within cells) is vital in maintaining osmotic pressure and acid-base balance.

- S is required mainly as a constituent of organic metabolites, including the amino acids methionine and cystine; the vitamins biotin and thiamin; certain mucopolysaccharides; and Coenzyme A.

- Interactions between and among macrominerals and between macrominerals and micro-(trace) minerals are increasingly recognized as important in nutrition.

❏ REFERENCES

Ahima, R. S., et al. 2000. *Front. Neuroendocrinol.* **21:** 263.

Apley, M., and W. M. Hilton. 2003. Watch for grass tetany. *Beef,* April.

Baker, D. H. 1977. *Sulfur in Nonruminant Nutrition.* National Feed Ingredients Association, Des Moines, IA, 122 pp.

Berger, L. L. 2004. Sulfur Nutrition Affects Copper Requirement. www.saltinstitute.org/STM-8.html (accessed January 2004)

Block, E. 2004. DCAD Balancing. www.ahdairy.com/research/techup_dcad.html (accessed January 2004)

Bronner, F. 1964. In *Mineral Metabolism,* Vol. 2, Part A. C. L. Comar and F. Bronner (eds.). Academic Press, New York.

Brown, W. R., et al. 1966. *Cornell Vet.* **56** (Suppl. 1), 108 pp.

Bunce, B. E., and J. E. Bloomer. 1972. *J. Nutr.* **102:** 863.

Buraw, R., and P. R. Stout. 1965. *Science* **150:** 766.

Chambliss, C. G., and W. E. Kunkle. 2003. *Grass Tetany in Cattle.* SS-AGR-64, Florida Coop. Ext. Service, Inst. Food and Agric. Sci., University of Florida.

D'Sousa, A., and M. N. Flock. 1973. *Am. J. Clin. Nutr.* **26:** 352.

Forbes, R. M., et al. 1956. *J. Biol. Chem.* **223:** 969.

Fujisawa, H., et al. 1984. *Fed. Proc.* **43:** 301.

Greene, L. W., et al. 1989. *J. Anim. Sci.* **67:** 3463.

Head, M. J., and J. F. Rork. 1955. *Nature* **176:** 262.

Heaney, R. P. 1993. *Annu. Rev. Nutr.* **13:** 287.

Houpt, T. R. 1977. In *Duke's Physiology of Domestic Animals,* 9th ed., p. 443. Cornell University Press, Ithaca, NY.

Lax, L. C., et al. 1956. *J. Physiol.* **132:** 1.

Leach, R. M., and M. C. Nesheim. *J. Nutr.* **81:** 193.

Lei, X. G., and J. Porres. 2004. Phytases. In *Encyclopedia of Animal Science,* W. G. Pond and A. W. Bell (eds.). Marcel Dekker, New York (In Press).

Lei, X. G., and C. H. Stahl. 2001. *Applied Microbiol. Biotechnol.* **57:** 474.

Lentz, D. E., et al. 1976. *J. Anim. Sci.* **43:** 1082.

Lowe, J. C., et al. 1996. *Vet. Rec.* **138:** 327.

Mitchell, A. D. 2004. Body Composition: Indirect Measurement. In *Encyclopedia of Animal Science,* W. G. Pond and A. W. Bell (eds.). Marcel Dekker, New York (In Press).

National Research Council. 1980. *Mineral Tolerances of Domestic Animals.* National Academy of Sciences, Washington, D.C.

National Research Council. 2001. Nutrient Requirements of Dairy Cattle. Seventh Edition. National Academy Press, Washington, D.C.

National Research Council. 2005. *Tolerances of Minerals and Toxic Substances in Feed and Water of Domestic Animals.* National Academy Press, Washington, DC (In Press).

Newton, G. L., et al. 1972. *J. Anim. Sci.* **35:** 440.

Oklund, S., and D. J. Prolo. 1990. In *Handbook of Human Growth and Developmental Biology,* Vol. II, Part B, E. Meisami and P. Timeras (eds.). CRC Press, Boca Raton, FL.

Papper, S. 1982. *Sodium: Its Biological Significance.* CRC Press, Boca Raton, FL.

Prentice, A. 2000. *Annu. Rev. Nutr.* **20:** 249.

Rude, R. K., et al. 2004. *J. Nutr.* **134:** 79.

Smith, S. E., and P. D. Aines. 1959. Salt deficiency in dairy cattle, *Cornell Agric. Exper. Sta. Bul.* No. 939.

Sutko, J. L. 1985. *Fed. Proc.* **44:** 2959.

Suttle, J. L. 1991. *Annu. Rev. Nutr.* **11:** 121.

Takeda, S., et al. 2003. *Annu. Rev. Nutr.* **23:** 403.

Tanaka, Y., et al. 1973. *J. Nutr.* **102:** 1569.

Tortora, G. J. and N. P. Anagnostakos. 1990. Principles of Anatomy and Physiology (Sixth Ed.). Harper and Row Publishers, New York. Fig. 27.2, page 864.

Ullah, A. H. J. 2000. *Biochem. Biophys. Res. Comm.* **275:** 279.

Wacker, W. E. C., and B. L. Vallee. 1964. In *Mineral Metabolism.* Academic Press, New York.

Widdowson, E. M., and J. W. T. Dickerson. 1964. In *Mineral Metabolism,* Vol. 2, Part A. Academic Press, New York.

Wyss, M. L., et al. 1999. *Applied and Environ. Microbiol.* **65:** 359.

12

Micro- (Trace)
Mineral Elements

THE DISTINCTION BETWEEN MACRO-MINERALS and micro or trace elements is based on the relative amounts required in the diet for normal life processes. Modern analytical techniques permit detection of minute traces of elements in feeds and animal tissues. For example, more than 25 different trace elements have been reported in human blood serum and kidney tissue, some of them in concentrations lower than those detectable by previous methods. Of these elements, only about half are required in the diets of animals. Fox and Tao (1981) tabulated responses in mineral content of animal tissues in relation to low or high dietary intake. Elements shown to be required in the diet as well as those for which no metabolic or dietary requirement has been established tend to concentrate in certain tissues in direct proportion to the amount supplied in the diet. Excessive accumulation even of required elements can produce toxic signs. This principle will be illustrated often in this chapter.

❏ TRACE MINERAL ELEMENTS REQUIRED BY ANIMALS

The list of trace mineral elements known to be required by animals continued to grow throughout most of the twentieth century. The latest addition to the list is boron (B), which is required for normal bone metabolism and for Ca transport across the cell membranes (Nielsen, 1986; Nielsen and Poellot, 1993). The trace minerals known to be or possibly required by one or more species for normal life processes are shown in Table 12.1.

Each of the trace minerals whose essentiality is established is discussed in terms of tissue distribution, functions, metabolism, deficiency signs, and toxicity. Numerous reviews of trace element nutrition are available (National Research Council, 1980; Mertz, 1986; Pearson and Dutson, 1990; Miller, 1991; Hill and Spear, 2001; Lei and Yang, 2004). Recent progress in molecular biology provides the tools needed to determine the effects of specific trace elements on metabolic functions at the molecular, cellular, and genomic levels and on gene and protein expression. The availability of transgenic and gene-knockout animal models permits the identification of specific functions of trace element-dependent or related proteins, for example, the use of Se-dependent glutathione peroxidase (GPx1; see discussion of Se later in this chapter) knockout mice to determine the role of this enzyme in the total function of Se (Zhu et al., 2004).

Kubota (1980) described the geographic distribution of trace element deficiency and toxicity in North America and developed Co, Se, and Mo maps of the United States (see Figs. 12.1a, 12.1b, and 12.1c, respectively) that are useful in identifying problem areas for trace mineral deficiencies and toxicities in crop and animal production and nutrition. Many mineral elements are ubiquitous in the environment and may be detected in animal tissues, but are of doubtful or unknown biological importance to the animal. Some are presently not recognized as essential but eventually may be identified as such. Others may prove to be toxic. Some trace minerals that are known to be essential are of practical concern because of their wide distribution in the environment or their toxic properties when ingested in relatively small amounts. Such minerals include Pb, Cd, Hg (mercury), F, Se, and Mo. The National Research Council (1980; 2005, in press) and Mertz (1986) published detailed

Table 12.1   Trace mineral elements reported to be required by one or more animal species.

KNOWN TO BE REQUIRED	POSSIBLY REQUIRED FOR ONE OR MORE SPECIES[a]	
Boron (B)	Aluminum (Al)	Carlisle and Curran (1993) Greger (1993)
Cobalt (Co)	Arsenic (As)	Nielsen (1977)
Chromium (Cr)	Bromine (Br)	Nielsen (1984)
Copper (Cu)	Lithium (Li)	Nielsen (1984)
	Lead (Pb)	Nielsen (1984)
	Nickel (Ni)	Nielsen and Sandstead (1974)
Iodine (I)	Tin (Sn)	Schwarz et al. (1970)
Fluorine (F)	Vanadium (V)	Schechter and Kerlich (1980)
Iron (Fe)		
Manganese (Mn)		
Molybdenum (Mo)		
Selenium (Se)		
Silicon (Si)		
Zinc (Zn)		

[a]Authors and dates of reporting a possible dietary requirement for each element are shown in this column.

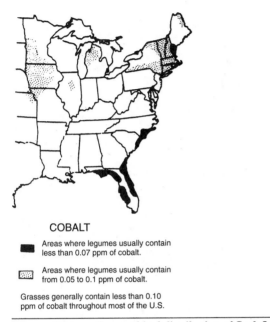

COBALT

█ Areas where legumes usually contain less than 0.07 ppm of cobalt.

▨ Areas where legumes usually contain from 0.05 to 0.1 ppm of cobalt.

Grasses generally contain less than 0.10 ppm of cobalt throughout most of the U.S.

Figure 12.1a *Geographical distribution of Co-deficient areas in the eastern United States.*

information on the effects of ingesting excessive levels of required mineral elements.

Analytical Detection of Trace Minerals

Modern analytical techniques permit detection of minute amounts of elements in feeds and animal tissues. Techniques such as neutron activation, spark-source spectrometry, and inductively coupled plasma emission atomic absorption spectroscopy allow previously impossible approaches to trace mineral element nutrition, metabolism, toxicity, and nutritional interactions among minerals. Some of these instruments were illustrated in Chapter 3.

General Functions of Trace Minerals

The trace minerals function as activators of enzyme systems or as components of organic compounds, and, as such, they are required in small amounts. A vast array of enzyme systems require trace elements for activation.

Examples of enzymes requiring specific trace minerals are given for each mineral. Note that Mg,

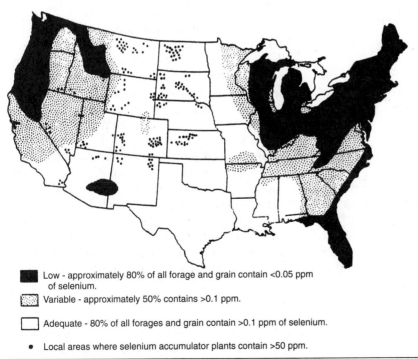

█ Low - approximately 80% of all forage and grain contain <0.05 ppm of selenium.

▨ Variable - approximately 50% contains >0.1 ppm.

☐ Adequate - 80% of all forages and grain contain >0.1 ppm of selenium.

• Local areas where selenium accumulator plants contain >50 ppm.

Figure 12.1b *Geographical distribution of low-, variable-, and adequate-Se areas in the United States.*

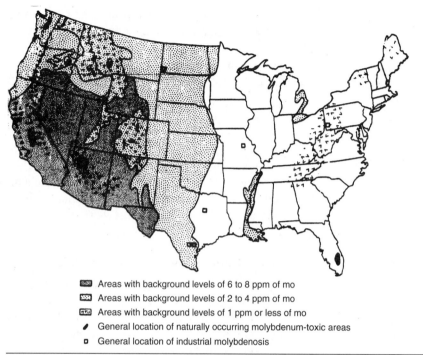

Areas with background levels of 6 to 8 ppm of mo
Areas with background levels of 2 to 4 ppm of mo
Areas with background levels of 1 ppm or less of mo
General location of naturally occurring molybdenum-toxic areas
General location of industrial molybdenosis

Figure 12.1c   *Generalized regional pattern of Mo concentration in legumes and its relation to MO-toxic areas in the United States. Figure 12.1a, b, and c courtesy of J. Kubota, USDA.*

a macromineral discussed in Chapter 11, functions in many respects like the trace elements.

Specific Trace Elements

Each of the trace elements whose essentiality is established firmly is discussed in terms of tissue distribution, functions, metabolism, deficiency signs, and toxicity.

❏ COBALT (Co)

The only known animal requirement for cobalt (Co) is as a constituent of vitamin B_{12} (see Chapter 15). Co was first recognized as an important diet constituent as early as 1929 when large amounts were shown to be capable of stimulating red blood cell synthesis in rats. Later, Underwood and Filmer (1935) showed that "wasting disease" of sheep and cattle in Australia could be cured or prevented by Co, and Smith et al. (1951) showed that vitamin B_{12} injections brought about dramatic improvement in sheep showing Co deficiency.

Tissue Distribution

Liver, kidney, adrenal, and bone tissue contain Co in the highest concentration in all species studied. The forms in which it exists in tissues other than as a constituent of vitamin B_{12} are not known completely, although other bound forms do exist. Dry liver from normal sheep contains a concentration (150 ppb) of Co about seven times greater than that of Co-deficient sheep; dried spleen, kidney, and heart of normal sheep also contain more Co.

Co supplementation of gestation diets of cows results in a higher concentration of Co in the tissues of calves. Similarly, Co content of milk is increased by supplementation of the lactation diet. This extra Co apparently is not present as a constituent of vitamin B_{12}, for it is not utilized until the young animal has developed a functioning rumen.

Functions

All evidence indicates that Co functions only as a constituent of vitamin B_{12}. Co deficiency, in the presence of adequate vitamin B_{12} intake, has never been produced in nonruminants. Similarly, ruminants with normal rumen function apparently do not require dietary Co greater than that needed for microbial vitamin B_{12} synthesis. Although Co can activate certain enzymes, there is no evidence that it is required for this purpose because the enzymes will function in the presence of other divalent metals.

Metabolism

Inorganic Co is absorbed poorly from the GI tract in animals, although adult humans may be an exception. About 80% of an orally administered dose of radioactive Co appears in the feces of rats and cattle. Most of the parenterally administered Co is excreted through the kidney. Although some loss occurs via the GI tract, the amount is small and probably via the bile and intestinal wall, but not from the pancreas. Tissue Co appears to be eliminated slowly. No synthesis of vitamin B_{12} in animal tissues from inorganic Co is known, even though tissue levels are high.

Large oral doses of Co induce polycythemia (increased RBC concentration) in several species. The mechanism apparently is through production of anoxia, possibly by binding SH compounds, with increased RBC synthesis as a compensatory response.

Deficiency Signs

Because Co functions as a constituent of vitamin B_{12}, the deficiency signs described for vitamin B_{12} (see Chapter 13) apply for Co. Ruminants grazing in Co-deficient areas show loss of appetite, reduced growth or loss in body weight followed by emaciation, normocytic normochromic anemia, and eventually death. Fatty degeneration of the liver and hemosiderosis of the spleen occur. Microbial synthesis of B_{12} in the rumen of deficient ruminants is depressed greatly. The first notable response to Co feeding is an improved appetite, followed by increased blood hemoglobin concentration. Critical levels of Co in diets for ruminants below which Co deficiency signs appear are 0.07 to 0.11 ppm of dry matter.

Soil maps (Mertz, 1986; Kubota, 1980) show large areas in the United States, Australia, and other parts of the world where soil is deficient in Co, resulting in low Co content of plant tissues and deficiencies in animals consuming the plants.

Co deficiency in grazing ruminants is treated by orally administering small dense pellets of Co ox-

ide and Fe. The pellets lodge in the rumen and are dissolved gradually over a period of months, yielding a constant supply of Co for vitamin B_{12} synthesis. Such a technique eliminates the need for a mineral supplement, which may not be voluntarily consumed in appropriate amounts by free-ranging animals. Work done in Australia indicates that two pellets offer enough abrasion to provide a steady supply of Co for vitamin B_{12} synthesis for more than five years in sheep.

Toxicity

Because of its low absorption rate, Co toxicity is not likely. A daily intake of 3 mg Co/kg body weight for 8 wk is tolerated by sheep without harmful effects. Higher doses result in appetite depression and anemia. Co given as a soluble salt to provide 300 mg Co/kg body weight is lethal to sheep. Cattle may be less tolerant than sheep. A concentration of about 2 mg of Co/kg of diet is apparently nontoxic for growing pigs.

❏ IODINE (I)

It was suggested nearly 150 years ago that iodine (I) deficiency caused goiter in humans, but the theory was rejected until about 1900 when it was observed that I was concentrated in the thyroid gland and the concentration was reduced in persons with goiter. Tragically, endemic goiter remains a widespread problem in the developing countries today.

Thyroxin was isolated in 1919 and synthesized in 1927. The structure of thyroxin (3,3', 5,5'-tetraiodothyronine or T_4) is as follows.

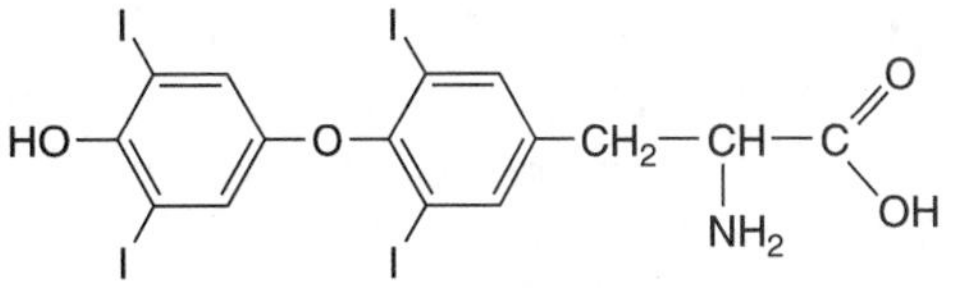

THYROXIN

Of the several compounds showing thyroid activity, thyroxin is present in blood in the highest concentration. The thyronine nucleus (structure shown above, minus I's) is essential for activity. Other compounds with thyroid activity include T_3 or 3,5-3'-triodothyronine; 3,5-diiodo-3'-5'-dibromothyronine (nearly as active); and 3,5,3'-triiodohyropropionic acid (300 times as active as thyroxin). The conversion of T_4 to T_3 and other thyroid active forms is catalyzed by the selenocysteine-containing enzymes, Type 1 and 3 iodothyronine deiodinase (Larsen and Berry, 1995). The importance of Se in many other metabolic functions is discussed later in this chapter.

Tissue Distribution

The thyroid gland contains the highest concentration of I (0.2 to 5% of dry weight) and in the largest amount (70 to 80% of total body I). It also is preferentially concentrated in the stomach (or abomasum), small intestine, salivary glands, skin, mammary gland, ovary, and placenta. Species differences exist in this capacity. The total I content of the adult human is 1020 mg. The concentration of inorganic I in most tissues is 1 to 2 mcg/ 100 g, and that of organically bound I is about 5 mcg/100 g for muscle, with higher concentrations in tissues that concentrate I. The only known function of I is as a constituent of thyroxin and other thyroid-active compounds. Thus, it is intimately associated with basal metabolic rate.

Metabolism

Inorganic I is absorbed from the GI tract by two processes, one in common with other halides (Cl, Br) and one specific for I. The specific I transport system is present in the stomach as well as the small intestine. The stomach and the midsection of the small intestine secrete I from the serosal to the mucosal surface. The gastric juice reaches an I concentration up to 40 times that of the plasma in humans. The I-specific transport mechanism is saturated by high-I concentrations. This is in common with the similar system in the thyroid gland but differs in that GI tract absorption is not affected by thyroid-stimulating hormone (TSH) from the anterior pituitary gland, although thyroid tissue is affected profoundly.

Salivary secretion of I is an active process in most species, as evidenced by the 40-fold concentration of radioactive I in the saliva as compared to plasma of animals given radioiodine. The rat apparently is an exception.

The I supply to the developing embryo is enhanced by at least two mechanisms. The ovary and placenta both concentrate I by an active process. The role of the ovary is best illustrated in birds.

Radioiodine injected into the hen is taken up by the yolk of eggs laid afterward, indicating that the yolk is an I store for the developing chick embryo. Similarly, placental tissue of mammals concentrates I in late pregnancy. A second mechanism favoring fetal I uptake is the presence in fetal serum of a specific thyroxin-binding protein that increases in concentration in late fetal development and has a higher affinity for thyroxin than the thyroid-binding protein of the maternal plasma. This favors the fetus receiving an adequate I supply if the combined thyroxin content of the maternal and fetal plasma is low. The mammary gland also concentrates I, and the transfer of inorganic I is through an active process, resulting in a 40-fold concentration of I in milk as compared to plasma. Although some T4 and T3 are found in milk, the amounts are small under normal conditions.

The key organ for I metabolism is the thyroid gland. It concentrates I by an active process that is enhanced by thyroid-stimulating hormone (TSH) secreted by the anterior pituitary gland. The I-concentrating ability of the thyroid is expressed as a ratio of the I concentration of the thyroid to the I concentration of the serum (T/S value). Normal animals have a T/S value of 20. The T/S value is decreased by hypophysectomy and increased by stimulating TSH release.

Iodine reaching the thyroid from the plasma is concentrated in the lumen of follicles of the thyroid, each of which is composed of a single layer of cells arranged as a sphere. The I so stored is contained in a colloid protein, thyroglobulin. Inorganic radioiodine given intravenously appears largely as protein-bound I (thyroproteins) in the thyroid gland shortly after injection. The iodinated proteins of the thyroid include mainly thyroglobulins (the thyroactive fraction), but also small amounts of others.

Iodine is oxidized by peroxidase to a reactive form for thyroglobulin formation. The steps in biosynthesis of the thyroid active compounds—thyroxin (T_4), 3,5,3'-triiodothyronine (T_3), 3-mono-iodotyrosine (MIT), and 3,5-diiodotyrosine (DIT)—are described briefly as follows: tyrosine present in the thyroglobulin molecule can be iodinated to form MIT, which is in turn used to form 3,3'-diiodothyronine, or MIT can be iodinated further to form DIT, which is in turn used to form T_3 or T_4. The amount of I in thyroglobulin, therefore, is dependent on the proportions of these tyrosine derivatives present.

Thyroglobulin does not appear in the plasma but is hydrolyzed by thyroid proteases in the thyroid follicle. On hydrolysis, only the iodothyronines are detected in the plasma. Free iodothyronines are deiodinated by enzymes in the thyroid, and the free I is available for recycling through thyroglobulin.

Iodothyronines are transported in the plasma bound to a globulin (thyroxin-binding globulin or TBG) or to a thyroxin-binding prealbumin. Some binding of thyroxin by plasma albumin also occurs. TBG, however, binds most of the iodothyronines at normal plasma concentrations. No evidence has been shown that thyroxin and its derivatives leave the plasma, except for entrance into lymph. Thus, the concentration of free iodothyronines in the blood plasma and extracellular fluids probably controls the rate of transfer to sites of action. Only about 0.05% of total plasma thyroxin is present in the free state.

The protein-bound I (PBI) level of the serum varies with level of thyroid activity as well as with species and age. Plasma PBI levels are increased with hyperthyroid activity and generally are higher in young animals and in pregnancy. Although plasma PBI levels provide a general assessment of thyroid activity, consistent results have not been obtained in attempts to relate them to growth rate or milk production.

About 80% of thyroid hormones entering the tissue are broken down by deiodinization by liver, kidney, and other tissues with the liberated I recycled for further use and the tyrosine residues catabolized or used for tissue protein synthesis. The remaining 20% is lost to the body through excretion via the bile, by conjugation to form glucuronides or sulfate esters, or by oxidative deamination. Inorganic I is excreted mainly via the kidneys. Smaller amounts are lost to sweat and in feces. The salivary gland secretes large amounts, most of which is reabsorbed from the GI tract.

Deficiency Signs

Because I functions as a constituent of thyroid-active compounds, which, in turn, play a major role in controlling the rate of cellular oxidation, it is not surprising that a dietary deficiency of I has profound effects on the animal. Dietary I deficiency reduces basal metabolic rate (BMR). I deficiency in young animals is called cretinism and in adults, myxedema. Tissues of I-deficient animals consume less oxygen and the reduced BMR is associated with reduced growth rate and reduced gonadal activity. Skin becomes dry and hair brittle. Reproductive problems associated with I deficiency include resorbed fetuses, abortions, stillbirths, irregular or suppressed estrus in females, and a decreased libido and deterioration of semen quality in males. Lambs, pigs, and calves produced by I-deficient dams may be hairless with dry, thick skin (Fig. 12.2), or, if hair is present, the coat is harsh and sparse, or in sheep the wool is scanty. The fleece of adult sheep recovered from I deficiency in early life may be of poor quality because of interference with normal development of the wool-producing cells.

Thyroid enlargement (goiter) is induced by an attempt of the thyroid gland to secrete more thyroxin in response to TSH stimulation. TSH is released in response to a reduced thyroxin production. Thyroid hormones, in a negative feedback arrangement, inhibit release of thyrotropic hormone-releasing factor (TRF) by the hypothalamus and thyroid-stimulating hormone (TSH) by the anterior pituitary. In the absence of adequate thyroxin for inhibiting TSH release, the thyroid gland becomes hyperactive and enlarged. Goiter is a common problem in human and animal populations living in inland areas in many parts of the world. The use of iodized salt has reduced the problem, but endemic goiter still remains a major nutritional disease in many areas.

In addition to simple I deficiency, several goitrogeric substances are present in common

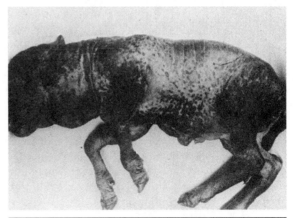

Figure 12.2 *An iodine-deficient lamb showing hairlessness, thick, wrinkled skin, and a marked enlargement of the thyroid gland. Photo courtesy of W. M. Hawkins, Montana State University.*

foods. These antithyroid compounds act by interfering with the iodinization of tyrosine, thus blocking iodothyronine synthesis.

Natural goitrogens present in plants consumed by animals and humans include thiocyanate in cassava; glucosinolates, which on hydrolysis release thiocyanate, in cabbage, rape, and mustard; a glucopeptide in soybeans; and metabolites of anthocyanin pigments in peanuts (Matovinovic, 1983). Synthetic compounds with goitrogenic properties include polychlorinated biphenyls (PCBS) used as plasticizers, polybrominated biphenyls (PBBs) used as fire retardants, organochlorine compounds (DDT, DDD, and Dieldren) used as insecticides, and various animal and antibacterial agents such as ethylene carbamates (EBDC), sulfanamides, and tetracyclines. The mechanisms of action of natural and synthetic goitrogenic agents have as a common basis the inhibition of uptake of inorganic *I* by the tyrosine skeleton to form iodothyronines.

Toxicity

Long-term chronic intake of large amounts of I reduces thyroid uptake of I, but there are marked species differences in tolerance of high intakes. Fertility of male rats fed 2400 ppm I for 200 days is unaffected; reproduction of female swine is unaffected by the same intake during gestation, but rabbits show increased prenatal mortality. Egg production of hens was reduced markedly by feeding 312 ppm and terminated with 5000 ppm 1, and hatchability was reduced (Perdomo et al., 1966). An effect on thyroxin production seems unlikely, as egg production was resumed within 1 week after withdrawal of the I from the diet.

Instances of goiter resulting from consumption of large amounts of high-iodine plants such as kelp have been reported in humans and in horses. Excess iodine disturbs all thyroid functions, including transport of I, synthesis of thyroxin, and release of the hormone (Wolff, 1969).

Apparently, the levels of I normally encountered in nutrition are far less than levels necessary to cause toxic symptoms. Single oral massive doses of I are toxic and may be lethal, but such toxicity must be categorized as poisoning in the general sense.

❏ ZINC (Zn)

Zinc (Zn) was shown to be an essential nutrient for animals in 1934 when Todd et al. (1934) produced a deficiency in the rat. In farm animal nutrition, the disease of swine termed *parakeratosis* was shown later to result from Zn deficiency (Tucker and Salmon, 1955). Subsequently, Zn deficiency has been produced experimentally in other farm animals, and Zn deficiency in humans has been reported as a practical problem. The nutrition, physiology, and metabolism of Zn has been reviewed in detail (Prasad, 1979; Cousins, 1979a, b and c; O'Dell, 1998; Hambidge and Krebs, 2001).

Tissue Distribution

Zn is distributed widely in body tissues but is in highest concentration in liver, bone, kidney, muscle, pancreas, eye, prostate gland, skin, hair, and wool. Radioactive Zn given orally or intravenously to normal or Zn-deficient cattle reaches a peak concentration in the liver within a few days, but concentrations in red blood cells, muscle, bone, and hair do not peak until several weeks later. Deficient animals retain more Zn-65 in skin, testes, scrotum, kidney, muscle, heart, lung, and spleen than normal animals, suggesting tissue specificity in meeting needs when Zn supply is short. Zn is a constituent of a large number of enzymes, and the tissue distribution of Zn is associated roughly with the tissue distribution of enzyme systems to which it is related. For example, when bone Zn is high, alkaline phosphatase of bone is high. The high concentration of Zn in the pancreas probably is related both to its presence in digestive enzymes and to its association with the hormone insulin, which is secreted by the pancreas.

The Zn concentration in blood is divided between the cells and plasma in a 9:1 ratio. Plasma Zn is bound loosely to albumins (1/3) and more firmly to globulins (2/3) and is responsive to dietary levels. Most of the Zn in red blood cells (RBCs) is present as a component of carbonic anhydrase.

Amounts of Zn in heart, kidney, liver, and muscle of normal cattle, chickens, sheep, and swine have been summarized (Doyle and Spaulding, 1978). Tables summarizing the Zn content of foods of animal and plant origin have been prepared (Murphy et al., 1975).

Functions

Zn is a constituent of numerous metalloenzymes, including carbonic anhydrase, carboxypeptidases A and B, several dehydrogenases, alkaline phosphatase, ribonuclease, and DNA polymerase. Zn activates some enzymes and plays a role in the configuration of DNA and RNA. Its biochemical functions have been reviewed (Chester, 1978). Zn is also involved in immune function (Keen and Gershwin, 1990).

The biochemical functions of Zn are related to the functions of the enzymes of which it is a constituent. Zn also is an activator of several metalloenzyme systems and probably shares with other metal ions, which it can replace, the function of binding reactants to the active site of the enzyme. Zn is required for nominal protein synthesis and metabolism; as a component of insulin it functions in carbohydrate metabolism.

Metabolism

Absorption of Zn from the GI tract occurs throughout the small intestine and amounts to 5 to 40% of the intake. Regulation of Zn absorption occurs in the intestinal cell. Isolated intestines from rats deficient or adequate in Zn nutrition absorb greater or lesser amounts of Zn, thereby contributing to Zn homeostasis (DiSilvestro and Cousins, 1983). Transfer of Zn out of the intestinal mucosal cells to the plasma is controlled closely by metallothionein, a low-molecular-weight binding protein

that is synthesized in response to a rise in plasma Zn concentration. Thus, the overall process of Zn absorption appears to be regulated by intracellular compartmentalization as well as by endogenous secretion of Zn in excess of immediate metabolic needs from the intestinal cells into the intestinal tract lumen. The regulatory pathway of Zn movement across the intestinal cell is depicted in Fig. 12.3. Details of the biochemical and physiological aspects of zinc absorption have been reviewed (O'Dell, 1998; Hambidge and Krebs, 2001).

The absorption of Zn is affected adversely by high dietary Ca concentration, and the presence of phytate further aggravates it. The complexing of Zn by phytate forms an insoluble and unabsorbable compound and is one mechanism whereby Zn availability to animals is reduced. A chelating agent, ethylene diamine tetra acetate (EDTA), improves Zn availability by competing with phytate to form an EDTA-Zn complex that allows absorption of Zn either as free Zn ion or as the EDTA-ZN complex. The presence of low-molecular-weight binding ligands such as histidine and cystine in feedstuffs such as soybean meal and corn decrease it. Some research suggests a role for picolinic acid in enhancing Zn absorption. Diets containing similar amounts of Zn may produce a variable incidence

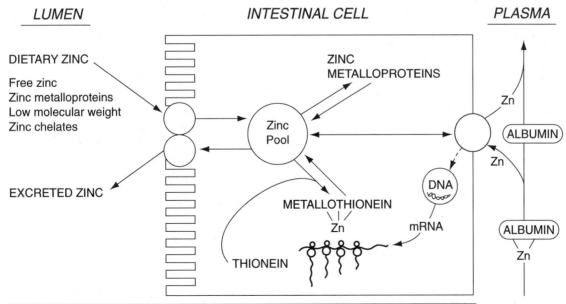

Figure 12.3 *Regulatory pathway of dietary zinc processing by intestinal cells. (Dashed line denotes induction of metallothionein mRNA; thionein is metal-free metallothionein). Courtesy R. J. Cousins (1979a).*

of parakeratosis in swine and variable weight gain in chicks and rats. Phosphate also binds Zn and may be a factor in the observed difference among protein sources in Zn availability.

Zn metabolism and movement in tissues have been studied with the radioactive isotope of Zn-65, and more recently, stable Zn has been used to study Zn homeostasis (Hambidge and Krebs, 2001). Zn is removed from the blood rapidly by the tissues, and tissues saturated with Zn (muscle) translocate Zn to unsaturated tissues (liver, pancreas, and kidney). Saturated tissues were characterized by no net change in Zn content, and unsaturated tissues showed a marked increase in Zn content. Injected Zn is excreted largely in the feces. Thus, fecal Zn in animals on an adequate Zn intake includes both unabsorbed and endogenously secreted Zn. The pancreatic juice appears to be the main route of excretion of endogenous Zn. Endogenous losses are reduced during dietary Zn deficiency, serving to conserve body stores. Except in abnormal conditions such as nephrosis or hypertension, urinary losses of Zn are very low. Excretion can be increased 10-fold by administration of EDTA.

Placental transfer of Zn is directly related to maternal dietary intake (Mertz, 1986; Apgar, 1985).

In animals that sweat freely, Zn loss by this route can be extensive in hot environments. Humans may lose 5 mg Zn/day on a diet adequate in Zn; Zn-deficient individuals lose less than half that amount under the same environmental conditions, again illustrating the homeostatic control of body Zn stores.

Marked fluctuations in liver Zn content may occur in response to varied Zn intake. Zn in excess

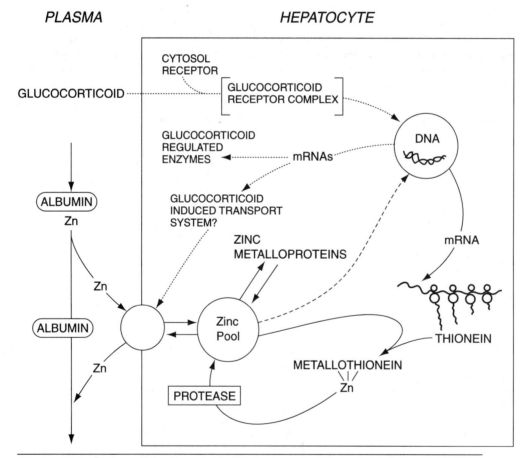

Figure 12.4   *Pathway of zinc processing by hepatocytes: role of plama zinc and glucocorticoids. (Dashed line denotes induction of metallothionein mRNA by zinc; dotted lines denote glucocorticoid-related events; thionein is metal-free metallothionein). Courtesy R. J. Cousins (1979a).*

of current needs is bound in the liver to metallothonein, which, as in the intestinal cell, is synthesized in response to a rise in plasma Zn. Glucocorticoids cause liver to accumulate Zn with a concomitant decrease in plasma Zn. The role of the liver cell in processing Zn is depicted in Fig. 12.4. Although the biological significance of this hormone-induced shift in tissue Zn is not fully understood, it has been suggested that any stress involving an increase in glucocorticoid activity may lead to an increase in liver metallothionein synthesis.

Deficiency Signs

The most conspicuous signs of Zn deficiency are growth retardation (Fig. 12.5) and anorexia in all species studied and a reduction in plasma alkaline phosphatase activity and in plasma Zn concentration. Thickening or hyperkeratinization of the epithelial cells is common, as exemplified by parakeratosis in swine. Rats show scarring and cracking of the paws, rough hair coat, and alopecia (loss of hair); sheep show abnormal changes in wool and horns; and poultry exhibit poor feathering and dermatitis.

Zn deficiency retards bone formation and is associated with reduced division and proliferation of cartilage cells in the epiphyseal growth plate. Bone alkaline phosphatase is reduced, bone density is depressed, and Zn content of bone and of liver are reduced. Apgar (1985) reviewed the effects of Zn deficiency on reproduction in animals. Clinical manifestations of Zn deficiency in humans were first described more than 30 years ago (Prasad et al., 1963).

Figure 12.5 *Effect of inadequate Zn on growth of pigs. From left to right, pigs received 46, 36, and 24 ppm of Zn. Note the severe parakeratosis in pig on the far right. Courtesy of W. M. Beeson, Purdue University.*

Hens fed Zn-deficient diets for several months show no abnormalities, but the chicks produced show poor livability and a high incidence of malformations. Chicks from hens fed adequate levels of Zn develop a perosis-like leg abnormality when fed Zn-deficient diets. This defect has been prevented by feeding histidine or histamine. The mechanism of this protective effect is unknown. Feeding female rats Zn-deficient diets during gestation results in high early mortality as well as difficulty in parturition and abnormal changes in maternal behavior of the dams. The effects of Zn deficiency are aggravated by intraperitoneal injection of EDTA (a chelating agent) in late pregnancy. EDTA increases Zn excretion and is itself excreted rapidly. Sheep fed a Zn-deficient diet during pregnancy and lactation show reduced plasma and wool Zn, and their lambs are affected likewise at 6 weeks of age. Pups from Zn-deficient dams show lower liver and total body Zn concentrations than controls and have a smaller birth weight.

Zn deficiency has drastic effects on the male reproductive organs. Hypogonadism is observed in Zn-deficient males of all species studied. In humans, hypogonadism, suppressed development of secondary sex characteristics, and dwarfism have been observed in young men (Prasad, 1963) ingesting suboptimum Zn. Recovery of testes size and sperm production are achieved in young animals repleted with Zn.

Wound healing of Zn-deficient animals is delayed severely. The exact mode of action of Zn in tissue repair is unknown, but the role of Zn in normal protein synthesis probably provides a clue to the relationship. Decreased activity of hepatic enzymes, leucine aminopeptidase, and ornithine transcarbamylase has been noted in Zn-deficient swine.

Zn deficiency causes impairment of glucose tolerance in rats, supporting the known relationship between Zn and insulin. The glucose intolerance is not related to blood insulin concentration but to peripheral resistance to insulin action (Park et al., 1986).

A striking feature of Zn deficiency is the dramatic remission of clinical signs when Zn is administered. This is illustrated in parakeratosis in Zn-deficient pigs. The skin of animals with severe lesions shows marked improvement after a few days of Zn refeeding and complete disappearance of the lesion within 2 to 3 weeks; feed intake improves immediately after Zn is added to the diet.

Toxicity

A wide margin of safety exists between the required intake of Zn and the amount that will produce toxic effects. Although the requirement for most species is less than 50 mg/kg of diet, levels of 1 to 2.5 g/kg have been fed to rats with no deleterious effects. Anemia, growth failure, and death have been observed in rats fed 1% Zn. Zn fed at 1 g/kg (0.1% of the diet) show no ill effects in pigs, but levels of 4 and 8 g/kg may produce depressed growth, stiffness, hemorrhages around bone joints, and excessive bone resorption. Birds are similar to swine in their tolerance to Zn, but sheep and cattle are less tolerant. Levels of 0.9 to 1.7 g/kg of Zn depress appetite and induce depraved appetite manifested by wood-chewing and excessive consumption of mineral supplements by sheep. The lower tolerance of ruminants to high dietary Zn may be related to changes in rumen metabolism brought about by a toxic effect of Zn on the rumen microflora.

Because the anemia produced by excess Zn can be prevented by extra Cu and Fe, it has been suggested that the anemia is an induced Cu and Fe deficiency as a result of interference with their absorption from the GI tract in the presence of high Zn.

In most studies reviewed (National Research Council, 1980), no adverse effects occurred when dietary Zn concentration was below 600 ppm. Many factors influence Zn toxicity; dietary Pb, Cu deficiency, marginal Se intake, and low Ca intake exacerbate it, while soybean protein appears to protect against excess Zn compared with casein, perhaps due to its phytate content. Details of the toxic effects of Zn have been reviewed (National Research Council, 1980, 2005).

❏ IRON AND COPPER: COMPARATIVE METABOLISM

Before discussing iron (Fe) and copper (Cu) separately, it will be helpful to address briefly the comparative metabolism of these two trace minerals and their complex interactions. Cu exerts important control over the movement and metabolism of Fe. This subject was reviewed by Winzerling and Law (1997). Both elements exist in living organisms in both the oxidized (Fe+++ and Cu++) and reduced (Fe++ and Cu+) ionic states. Fe++ ions are more soluble at physiological pH than are FeIII ions, but both ionic forms occur in cells. Both forms are complexed with various small molecules and proteins. Without such organic complexing, the free ion would readily react with molecular oxygen to produce deadly free radicals. Some of the Fe-containing and Cu-containing enzymes and proteins are discussed later in this chapter. The interplay of Fe and Cu in metabolism will be evident in the separate descriptions of Fe and Cu nutrition that follow.

❏ IRON (Fe)

Iron has been recognized as a required nutrient for animals for more than 100 years. Nonetheless, Fe deficiency remains a common disease affecting nearly half of the human population in some areas of the world and persisting as a major problem in livestock production. Reviews of Fe metabolism and nutrition are numerous (Mertz, 1986; Dallman, 1986; Leibold and Guo, 1992; Mascotti et al., 1995; Eisenstein, 2000).

Tissue Distribution and Function

Sixty to 80% or more of body Fe is present in hemoglobin in red blood cells and myoglobin in muscle; 20% is stored in labile forms in liver, spleen, and other tissues and is available for hemoglobin formation; the remainder is fixed firmly in unavailable forms in tissues as a component of muscle myosin and actomyosin and as a constituent of enzymes and associated with metalloenzymes. The absolute amounts of Fe present in various forms in several species are summarized in Table 12.2 (Moore, 1951).

Fe is present in hemoglobin, myoglobin, and heme enzymes, cytochromes, catalases, and peroxidases as heme, an organic compound consisting of Fe in the center of a porphyrin ring. In hemoglobin, which contains 0.34% Fe, an atom of ferrous Fe in the center of the porphyrin ring connects heme, the prosthetic group with globin, the protein. Hemoglobin contains four porphyrin rings and combines reversibly with atmospheric oxygen brought into the blood via the lungs. Hemoglobin is found within the red blood cell. Myoglobin, which has one-fourth the molecular weight of hemoglobin (16,500 daltons), is present in muscle, and has an affinity for oxygen greater than that of hemoglobin. The heme enzymes—catalases and peroxidases—contain Fe in the ferric state (Fe+++).

Table 12.2 Amounts and distribution of Fe in the rat, dog, calf, horse, and human.

SPECIES	BODY WT.	IRON CONTENT, G				
		TOTAL BODY	HEMOGLOBIN	MYOGLOBIN	CYTOCHROMIC	STORAGE
Calf	182.0	11.13	7.55	1.060	0.0053	2.517
Dog	6.35	0.44	0.300	0.040	0.00059	0.100
Horse	500.0	33.0	19.85	6.45	0.0715	6.617
Human	70.0	4.26	3.10	0.120	0.00336	1.03
Rat	0.25	0.015	0.011	0.0003	0.00006	0.0036

Source: Adapted from Moore (1951).

Each of these classes of enzyme liberates oxygen from peroxides. Cytochrome enzymes are heme enzymes of the cell mitochondria and act in electron transfer by virtue of the reversible oxidation of Fe ($Fe^{++} = Fe^{+++}$). Cytochromes a, b, and c, cytochrome oxidase, and others have been identified, but only cytochrome c is readily extractable from tissues. Other enzymes containing Fe include xanthine oxidase, succinic dehydrogenase, and NADH-cytochrome reductase.

Fe in blood plasma is bound in the ferric state (Fe^{+++}) to a specific protein, transferrin, a B1-globulin. Transferrin is the carrier of Fe in the blood and is saturated normally only to the extent of 30 to 60% of its total Fe-binding capacity.

Fe is stored in liver, spleen, and bone marrow as an Fe-protein complex, ferritin, and as a component of hemosiderin. Fe makes up 20% of the ferritin-Fe complex and is present in the ferric state. Hemosiderin contains 35% Fe as ferric hydroxide. It is present in tissues as a brown, granular pigment. Ferritin can be considered the soluble and hemosiderin the insoluble form of storage Fe. Under normal conditions and in Fe deficiency, Fe is stored in about equal amounts in each form, but in Fe excess, hemosiderin Fe predominates.

Iron overload (excess hemosiderin deposition) also has a genetic basis, as reviewed by Wessling-Resnick (2000).

Metabolism

Current evidence (Mascotti et al., 1995; Eisenstein, 2000; Miret et al., 2003) indicates several physiologic mechanisms that maintain Fe homeostasis. Iron regulatory proteins (IRP) are central to Fe metabolism because they regulate the synthesis of proteins required for the cellular uptake, storage, and use of Fe. Further description of these IRPs is beyond the scope of our discussion, but recognition of their importance is needed as a basis for understanding the complexities of Fe metabolism to be addressed here.

Absorption. Fe is absorbed only from the duodenum in the ferrous state (Fe^{++}), and usually only to the extent of 5 to 10%. The body holds absorbed Fe tenaciously for reuse. Thus, the Fe released from hemoglobin breakdown associated with RBC destruction (RBC life is 60 to 120 days in most species) is recycled for resynthesis of hemoglobin. This conservation of Fe is illustrated in the diagrammatic outline of Fe metabolism in Fig. 12.6.

Absorption is more efficient under acid conditions; thus, the amount of Fe absorbed from the stomach and duodenum, where HCl from stomach secretion results in a low pH, is greater than from the ileum. In rats there is a gradient from the upper to lower end of the small intestine in absorption of Fe, with Fe uptake more than 10 times greater in the proximal duodenum than in the distal ileum. Iron absorption is increased by certain amino acids (valine, histidine); ascorbic acid; organic acids (lactic, pyruvic, citric); and certain sugars (fructose, sorbitol), probably by forming soluble chelates with Fe.

Tophan et al. (1981) showed that xanthine oxidase in intestinal mucosa promotes incorporation of iron into transferrin for transport in blood. Thus, the oxidation of Fe^{++} to Fe^{+++} for transport in blood (which is associated with ceruloplasmin for Fe mobilization from liver) is accomplished by xanthine oxidase in the intestinal mucosa. Maximum absorption occurs at pH 2 to 3.5. Fe absorption is greater in Fe-depleted animals than in animals receiving adequate Fe, and the absolute amount of Fe absorbed is increased as the size of an oral dose

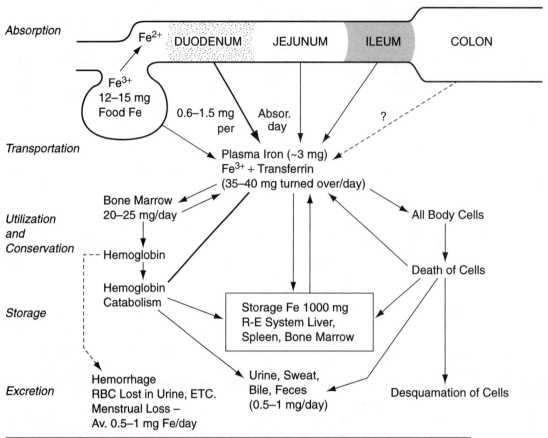

Figure 12.6   *Schematic representation of iron metabolism in humans. RBC, red blood cells;
R-E system, reticuloendothelial system. From Moore, C. V. and R. Dubach. 1962. In* Mineral
Metabolism, *Vol. II, Part B, C. L. Komar and F. Bronner (eds.). Academic Press, New York.*

increases, but the percentage absorption decreases. Fe present in hemoglobin and myoglobin
is absorbed readily as heme Fe. High levels of inorganic phosphates reduce Fe absorption by forming insoluble salts; phytate also has been reported
to reduce Fe absorption, but the practical importance in normal diets is unknown. High levels of
other trace elements, including Zn, Mn, Cu, and
Cd, also reduce Fe absorption, presumably by
competing at protein-binding sites in the intestinal
mucosa. Neonatal pig intestinal cell membranes
and cytoplasm contain Fe-binding substances. The
serosal as well as the mucosal epithelium regulate
Fe transfer from the intestinal lumen of the pig.

The ultimate regulation of Fe absorption is apparently dependent on the Fe concentration of the
intestinal mucosal cells. That is, of the Fe taken up
by the mucosal cells, only a small part is transferred from the cell to the blood. The bulk is retained in the cell and lost in the intestinal lumen
when the cell is sloughed off in the normal process
of regeneration of intestinal epithelium. In deficient animals, Fe is transferred more readily from
the mucosal cell to blood until saturation of the tissues results in a return to normal retention of Fe
in the mucosal cell. Parenterally administered Fe
decreases Fe absorption in some species by transfer into the intestinal epithelial cell from the serosal
side (from the blood). Thus, the "mucosal block"
theory of Fe absorption remains an important general concept, even though the mechanisms may
not be exactly as originally envisioned, based on
recent knowledge (Eisenstein, 2000).

The foregoing discussion pertains primarily to
nonheme Fe absorption. The mechanism of heme
Fe absorption is less understood, although it is well
known that heme Fe (derived from meat) is highly
bioavailable (Miret et al., 2003). It is thought that

the iron-porphyrin ring is split from the globin moiety in the intestinal lumen and that intact porphyrin crosses the enterocyte brush border instead of being further split into porphyrin and free Fe. The heme that has entered the cell is either split into ferrous Fe and bilirubin by the enzyme, heme oxygenase, or transported as intact porphyrin ring. The ferrous iron then likely exits the enterocyte along with Fe absorbed as nonheme Fe (Miret et al., 2003).

Excretion. Body Fe is retained tenaciously (see Fig. 12.6). Fecal Fe is mainly unabsorbed dietary Fe, but a small amount (0.3 to 0.5 mg/kg in humans) is lost through bile and sloughed intestinal mucosal cells. Even when Fe is injected, very little of it is excreted in either feces or urine, although urinary Fe loss does occur when parenteral Fe is given in excess of the plasma Fe-binding capacity or when chelating agents are given. Small amounts of Fe are lost in sweat in humans. In human females and other primates, considerable amounts of Fe are lost during menstruation; this loss partially accounts for the high incidence of anemia in adult women.

Placental and Mammary Transfer. Considerable species variability exists in the efficiency of transfer of Fe to the fetus across the placenta. Fe is transported to the fetus by an active process, and concentration in the fetal circulation exceeds that in maternal plasma. Transferrin does not cross the placenta. Rather, Fe is dissociated from transferrin on the maternal side of the placenta and is reassociated with transferrin on the fetal side. As gestation progresses, more and more Fe is transferred to the fetus. Although the newborn of some species have a relatively high liver Fe concentration, the newborn pig is not supplied well with Fe and is especially prone to Fe deficiency. The low placental transfer of Fe is not increased appreciably even by high levels of supplementation of the maternal diet or by parenteral administration to the sow. Milk of all species is low in Fe. Attempts to increase the Fe concentration of sow milk by parenteral Fe or by feeding extra Fe to the sow during lactation have not resulted in appreciable increases in milk Fe content.

Storage and Mobilization. Fe is stored within the cells of the liver, spleen, bone marrow, and other tissues as ferritin and hemosiderin in roughly equal proportions. Incorporation of plasma Fe (transferrin) into ferritin in liver cells is energy-dependent (ATP) and is related to the reduction of $Fe+++$ of transferrin to $Fe++$, making it available for ferritin formation. Release of $Fe++$ in liver ferritin to plasma is catalyzed by xanthine oxidase. Similar reactions are presumed to occur in other Fe-storage tissues. Ferritin and transferrin receptor, a protein involved along with ferritin in the regulation of Fe homeostasis (Leibold and Guo, 1992), shuttles Fe into the erythrocyte or other cell by endocytosis of transferrin. Following delivery to the cell, Fe is released and transferred to the Fe-storage protein, ferritin, or to other cellular proteins requiring Fe. Although Fe is an essential element and plays a role in numerous biological processes, its toxicity as free Fe requires that it be sequestered in the ionic form; ferritin serves this function.

The turnover rate of Fe in the plasma is very rapid; about 10 times the amount of Fe in the plasma at any one time is transported each day. Most of this is used for hemoglobin synthesis. Inorganic radioiron appears almost entirely as a component of hemoglobin in 7 to 14 days in humans. The continuous redistribution of body Fe can be summarized as follows:

Quantitatively most important:
    plasma → erythroid marrow → red cell →
    senescent red cell → plasma
Less important:
    plasma → ferritin and hemosiderin → plasma →
    myoglobin and Fe-containing enzymes → plasma

Enzymes contain Fe bound to the protein, to porphyrin, or to flavins (metalloenzymes) or as a loosely bound activator (metal-activated enzymes).* A partial list of Fe-containing enzymes in mammals and birds, along with some important Fe-containing proteins, is given in Table 12.3.

Deficiency Signs

The most common sign of Fe deficiency is a microcytic, hypochromic anemia, which is charac-

---

*Metalloenzymes have the active metal firmly bound in a constant stoichiometric ratio to the protein of the enzyme. In metal-activated enzymes, the activating metal is loosely bound and readily lost on processing. Often, metals loosely bound in this way can be replaced by other metal ions without loss of the activity of the enzyme.

Table 12.3   Fe-containing enzymes and proteins in animals.

Metalloporphyrin enzymes

  Cytochrome oxidase
  Cytochrome *c*
  Other cytochromes (P-450, b_5)
  Peroxidase
  Catalase
  Aldehyde oxidase

Metalloflavin enzymes

  NADH cytochrome *c* reductase
  Succinic dehydrogenase
  Lactic dehydrogenase
  α-Glycerophosphate dehydrogenase
  Choline dehydrogenase
  Aldehyde dehydrogenase
  Xanthine oxidase

Metalloproteins other than enzymes

  Hemoglobin (>10% of body iron)
  Myoglobin (10% of body iron)
  Transferrin
  Ovotransferrin
  Lactotransferrin
  Ferritin

Source: Adapted from Bowen (1966) and Dallman (1986).

terized by smaller than normal red cells and less than normal hemoglobin. Fe-deficiency anemia is a common problem in newborn animals because of inefficient placental and mammary transfer. In pigs unsupplemented with Fe, blood hemoglobin declines from about 10 g/dl at birth to 3 or 4 g/dl at 3 weeks. The extremely rapid growth rate of young pigs (five times birth weight at 3 weeks) results in a dilution effect of total body Fe stores unless Fe is fed or injected. Intramuscular injection of 150 to 200 mg of Fe-dextrin at two or three days of age maintains normal hemoglobin level at 3 weeks of age, at which time consumption of dry feed provides ample Fe. Anemic pigs show pallor, labored breathing, rough hair coats, poor appetite, reduced growth rate, and increased susceptibility to stresses and infectious agents. Heavy infestation with internal parasites of the blood-sucking type leads to an induced Fe-deficiency anemia in many animal species, including sheep, pigs, cattle, and humans.

Suckling lambs and calves also become anemic if fed exclusively on milk. Veal calves have light-colored muscle because of low myoglobin content and low blood hemoglobin. This is a desired characteristic of veal in most markets and has led to the practice of feeding low-Fe milk replacers.

Fe deficiency anemia in humans is most common in children and among women of child-bearing age because of large losses of Fe during menstruation. Menstrual loss accounts for 16 to 32 mg of Fe during the cycle or 0.5 to 1.0 mg/day, in addition to the 0.5 to 1.0 mg excretion by other routes. Thus, the Fe loss in adult women may be twice that of adult men. Most surveys indicate that Fe deficiency anemia in humans is a major disease problem worldwide; in some countries 30 to 50% of the population may be affected. Fe deficiency in humans is associated with pallor, chronic fatigue, and general lack of the sense of well-being. The fact that the response to Fe supplementation includes a rapid return to a sense of well-being, which occurs before increased hemoglobin synthesis can take place, suggests that Fe-containing enzymes are affected. Although changes in activities of Fe-containing enzymes in Fe deficiency in humans are not well documented, reduced liver catalase in Fe-deficient swine has been recorded.

Fe is required for adequate myelination of brain and spinal cord and other aspects of neural development involved in behavior. Delayed behavioral development in preschool and older children may be reversed following Fe repletion but appears to persist in Fe-deficient infants even after correction of Fe status (Beard and Connor, 2003). The relationship between iron deficiency anemia and behavioral changes that may accompany it is of interest in humans (Dallman, 1986; Pollitt, 1993; Beard and Connor, 2003) and could have implications in animals.

Toxicity

Fe overload has been produced in animals by injection or by long periods of excessive intake and in humans by repeated blood transfusions or after prolonged oral administration of Fe. Chronic Fe toxicity causes diarrhea and reduced growth and efficiency of feed utilization, and it may also produce signs of P deficiency. Acute toxicity, including vascular congestion of tissues and organs, metabolic acidosis, and death, has been produced in pigs and rabbits given oral doses of ferric ammonium citrate or ferrous sulfate. Excess Fe is found in the tissues as hemosiderin. Transferrin concentration is normal, and plasma Fe is increased only until transferrin is saturated. Liver fibrosis is common in some cases of human Fe toxicity because of a genetic defect in control of Fe absorption and excretion (idiopathic hemochromatosis), but it is not normally associated with Fe toxicity in animals. In the genetic defect, Fe accumulates in the parenchyma, but in excess Fe intake in humans and animals, it accumulates in reticuloendothelial cells.

Baby pigs have been administered 10 times the amount of Fe normally injected for prevention of anemia with no ill effects. Hemoglobin tends to be elevated, but growth rate is not affected. However, some forms of organic Fe, such as ferric ammonium citrate, may cause death of newborn pigs when administered orally to supply 200 mg of Fe. Generally, when metabolic Fe demand is high as in rapidly growing suckling animals, a wide margin of safety exists between adequate and toxic dosages.

Fe toxicity may be reduced by dietary Cu, P, and vitamin E, while certain amino acids (valine and histidine), ascorbic acid, simple carbohydrates, and several organic acids (lactic, pyruvic, citric) increase Fe absorption. The enhancement of Fe absorption is believed to be due to formation of complexes with Fe that render it soluble during its transit down the GI tract. It follows that precipitation as the insoluble hydroxide would decrease Fe toxicity. It is on this basis that milk of magnesia is used in treatment of Fe toxicosis.

❑ COPPER (Cu)

Copper (Cu) was first shown to be a dietary essential in 1928. Numerous reviews are available on the role of Cu in nutrition (Mertz, 1986; DiSilvestro and Cousins, 1983; O'Dell, 1976; Vulpe and Packman, 1995; Harris, 2000; Hellman and Gitlin, 2002).

Tissue Distribution

The liver, brain, kidneys, heart, pigmented part of the eye, and hair or wool are highest in Cu concentration in most species; pancreas, spleen, muscles, skin and bone are intermediate; and thyroid, pituitary, prostate, and thymus are lowest. The concentration of Cu in tissues is highly variable within and among species. Young animals have higher concentrations of Cu in their tissues than adults, and dietary intake has an important effect on the Cu content of liver and blood. Cu in blood is 90%

associated with the alpha 2-globulin, ceruloplasmin, and 10% in the red blood cells as erythrocuprein. Intracellular Cu and Cu-proteins are found in all cellular organelles (Vulpe and Packman, 1995). Pregnancy is associated with increased plasma Cu in the form of ceruloplasmin, apparently in response to elevated blood estrogens. Maternal ceruloplasmin is not transferred across the placenta to the fetus.

Functions

Cu is required for the activity of enzymes associated with Fe metabolism, elastin and collagen formation, melanin production, and integrity of the central nervous system. It is required for normal red blood cell formation (hematopoiesis), apparently by allowing normal Fe absorption from the GI tract and release of Fe from the reticuloendothelial system and the liver parenchymal cells to the blood plasma (Fig. 12.6). This function appears to be related to the required oxidation of Fe from the ferrous to ferric state for transfer from tissue to plasma. Ceruloplasmin is the Cu-containing enzyme required for this oxidation. The metabolism and functions of ceruloplasmin have been reviewed (Hellman and Gitlin, 2003). A critical physiological role for ceruloplasmin involves its control of the rate of Fe efflux from cells with mobilizable Fe stores.

Cu is required for normal bone formation by promoting structural integrity of bone collagen and for normal elastin formation in the aorta and the remainder of the cardiovascular system. This requirement appears to be related to the presence of Cu in lysyl oxidase, the enzyme required for removal of the epsilon amino group of lysine in the normal formation of desmosine and isodesmosine, key cross-linkage groups in elastin. Cu is required for normal myelination of the brain cells and spinal cord as a component of the enzyme cytochrome oxidase, which is essential for myelin formation. Numerous enzymes, including lysyl oxidase, cytochrome C oxidase, ferroxidase, and tyrosinase, are Cu-dependent (Table 12.4). Some of the metabolic pathways of copper in humans are represented schematically in Figure 12.7.

Cu is required for normal hair and wool pigmentation, presumably as a component of polyphenyl oxidases, which catalyze the conversion of tyrosine to melanin, and for incorporation of disulfide groups into keratin in wool and hair (see Fig. 12.8).

Table 12.4   Metalloprotein enzymes, metalloporphyrin enzymes, and nonenzyme metalloproteins containing Cu in animals.

Metalloprotein enzymes

  Tyrosinase
  Monoamine oxidase
  Ascorbic acid oxidase
  Ceruloplasmin
  Galactose oxidase

Metalloporphyrin enzymes

  Cytochrome oxidase

Metalloproteins other than enzymes

  Erythrocuprein
  Hepatocuprein
  Cerebrocuprein
  Milk copper protein

Source: From Bowen (1966).

Metabolism

Absorption.   Site of Cu absorption from the GI tract varies among species. In the dog, it is absorbed mainly from the jejunum, in humans from the duodenum, and in the rat from the small intestine and colon. Degree of absorption is also variable, exceeding 30% in humans but less in other species studied. A Cu-binding protein has been found in the duodenal epithelial cells of the chick, which presumably plays a part in Cu absorption. The pH of the intestinal contents affects absorption; Ca salts reduce Cu absorption by raising pH. Absorption is also affected by other elements. Ferrous sulfide reduces Cu absorption by forming insoluble CuS; Hg, Mo, Cd, and Zn all reduce Cu absorption. Cd and Zn have been shown to displace Cu from a Cu-binding protein in the intestinal mucosa of chicks. The mode of action of Hg and Mo is not clear, although the formation of insoluble $CuMoO_4$ has been suggested as the explanation for the effect of Mo.

Some forms of Cu are absorbed more readily than others. For example, cupric sulfate is more readily absorbed than cupric sulfide; and cupric nitrate, chloride, and carbonate are more readily absorbed than cuprous oxide. Metallic copper is very poorly absorbed.

Transport and Tissue Utilization.   Cu transport in animal cells involves three steps: uptake, intracellular distribution, and export (Vulpe and Packman, 1995). Cu uptake, whether bound or un-

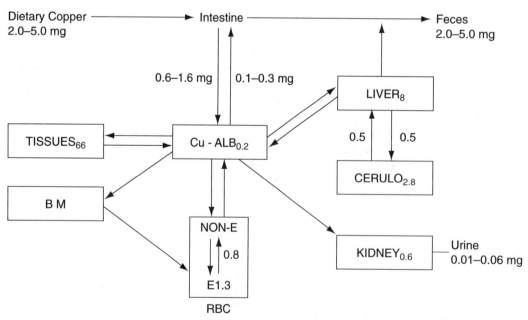

Figure 12.7   *Schematic representation of some metabolic pathways of copper in humans. Numbers in the boxes refer to mg of Cu in the pool. The numbers next to the arrows refer to mg of Cu transversing the pathway each day. Cu-ALB, direct-reading fraction; CERULO, ceruloplasmin; NON-E, nonerythrocuprein; BM, bone marrow; RBC, red blood cell. From Cartwright and Wintrobe (1964).*

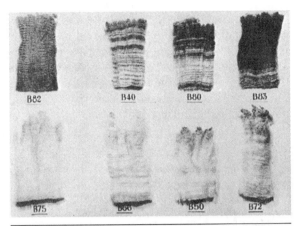

Figure 12.8 *Wool from Cu-deficient (or excess Mo)*
sheep. Normal wool on the left. The other samples show
banding of black wool, loss of definition of crimp, and,
in some, secondary waves. By permission of
G. L. McClymont.

bound to ceruloplasmin, is controlled by various membrane proteins and peptides organized in an integrated system to maintain Cu homeostasis. The membrane-bound Cu-transporting ATPase enzymes, ATP_7A and ATP_7B, work in combination with a group of peptides, the copper chaperones, to exchange Cu or incorporate it into the structure of Cu-dependent enzymes, including cytochrome c oxidase and Cu, Zn superoxide dismutase. It has been suggested (Harris, 2000) that ceruloplasmin, Cu-albumin, and Cu-histidine may be equal in their ability to mediate delivery of Cu to the cell and that transcuprein and perhaps other transporters may behave as Cu transporters. It appears that the carrier of Cu to cells is not decisive in determining intracellular Cu metabolism. Known internal Cu transport determinants of Cu metabolism include glutathione (GSH) and the copper chaperone noted above. The details of their actions are beyond the scope of this discussion.

Absorbed Cu is distributed to the tissues and taken up by the bone marrow in red blood cell formation where it is present partly as erythrocuprein. Cu reaching the liver is taken up by parenchymal cells and stored or released to the plasma as Cu-albumin and in larger quantities as a component of ceruloplasmin, or is used for synthesis of a large array of Cu-containing enzymes (see Table 12.4) and other Cu-containing proteins.

Excretion. Bile is the major pathway of Cu excretion. The Cu excreted via the bile appears in the feces (see Fig. 12.7). Smaller amounts are lost in the feces as intestinal cell and pancreatic secretions; in urine; and negligible amounts in sweat. Radiocopper studies indicate that the main source of urinary Cu is that loosely bound to albumin in the plasma.

Deficiency Signs

Dietary Cu deficiency is associated with a gradual decline in tissue and blood Cu concentration. Blood Cu levels below 0.2 micrograms/mL result in interference with normal hematopoiesis and anemia. The anemia is hypochromic microcytic in some species (rat, rabbit, pig), but in other species it is hypochromic macrocytic (cattle, sheep) or monochromic (chicks, dogs). Cu deficiency shortens RBC lifespan and reduces Fe absorption and utilization. Thus, the anemia appears to be related in part to a direct effect on RBC formation arising from the need for Cu as a component of RBCs and in part to an indirect effect related to the reduced ceruloplasmin concentration of plasma, which in this way reduces Fe absorption and utilization. Apparently, no interference with heme biosynthesis occurs in Cu deficiency.

A widespread problem in lambs that has been traced to a Cu deficiency results in incoordination and ataxia. Low levels of Cu in pastures used for grazing, coupled with high intakes of Mo and S, precipitate the condition that is known as swayback or enzootic neonatal ataxia. Newborn lambs most often are affected, but a similar condition can also be produced in the young of goats, guinea pigs, pigs, and rats. Degeneration and failure in myelination of the nerve cells of the brain and spinal cord are the basis for the observed nervous disorder. The Cu content of the brain is reduced, leading to a reduction in cytochrome oxidase activity, which is necessary for phospholipid synthesis. Intramuscular injection of Cu-glycine, Cu-EDTA, or Cu-methaionine complexes into pregnant ewes has been used successfully to prevent swayback in lambs. Genetic diseases of mice and human infants associated with abnormal Cu metabolism and resembling dietary Cu deficiency have been described.

Cu deficiency results in bone abnormalities in many species—pigs, chicks, dogs, horses, and rabbits. A marked failure of mineralization of the cartilage matrix occurs. The cortex of long bones is thin, although the Cu, P, and Mg concentrations of the ash remain normal. The defect appears to be related to a change in the cross-link structure of collagen, rendering it more soluble than collagen from normal bones. The lysyl oxidase activity of bones from Cu-deficient chicks is reduced 30 to 40%.

The hair and wool of Cu-deficient animals fail to develop normally, resulting in alopecia and slow growth of fibers (Fig. 12.8). Wool growth in sheep is sparse, and normal crimping is impeded, leading to straight, hairlike fibers termed *steely wool*. The change in wool texture is related to a decrease in disulfide groups and increase in sulfhydryl groups and an interference with arrangement of the polypeptide chains. The pigmentation process is extremely sensitive to Cu. Levels of dietary Cu that fail to produce anemia, nerve damage, or bone changes can produce pigmentation failure in the wool of black sheep or in the hair of pigmented cattle. Achromotrichia (loss of hair pigmentation) can be produced in wool in alternating bands by feeding a Cu-deficient or Cu-adequate diet to sheep in alternating intervals. Blockage of the pigmentation process is achieved within two days by feeding Mo and inorganic sulfate to sheep in the presence of marginal Cu. Presumably, the effect of Cu deprivation on pigmentation is related to its role in conversion of tyrosine to melanin.

Cu-deficient chicks, pigs, and cattle show cardiovascular lesions and hemorrhages. Cu deficiency causes aortic rupture in turkey poults. More than 30 years ago, falling disease in cattle in Australia was found to be related to a progressive degeneration of the myocardium of animals grazing on Cu-deficient plants. Later, histological changes in blood vessels of chicks and pigs were found to be related to a derangement of the elastic tissue of the aorta and other blood vessels. There is some evidence for a role of Cu deficiency in coronary heart disease in humans; there is no evidence that cardiovascular disease in animals is associated with suboptimum intakes of Cu.

Fetal death and resorption in rats and reduced egg production and hemorrhage and death of the embryos in poultry have been shown to be produced by Cu deficiency. Reproduction failure of cattle in Cu-deficient locations has also been reported, but a specific effect of Cu in contributing to the syndrome has not been established. In poultry, the primary lesion appears to be a defect in RBC and connective tissue formation in the embryo, possibly induced by a reduction in monoamine oxidase activity.

In ruminants, there appears to be a larger problem of Cu deficiency than of Cu toxicity worldwide (Cooke, 1983). There is wide variation in Cu content in plants (legumes have higher concentrations than grasses) and in the bioavailability of their Cu. Therefore, Cu concentration alone may be of little nutritional significance (Cooke, 1983).

Toxicity

Sheep and calves appear to be more susceptible to Cu toxicity than other species. Hemoglobinuria, jaundice, and tissue necrosis have been observed in calves fed a milk substitute containing 115 ppm Cu. Death of sheep, with accompanying hemoglobinuria from excess Cu, has been reported on pasture because of continued free-choice consumption with a trace mineral salt mixture containing the recommended Cu level. Cu toxicity occurs in sheep when the Cu content of soil and pasture is high, when Mo of plant is low, or when liver damage from consumption of certain poisonous plants predisposes to Cu poisoning by decreasing the liver's ability to dispose of Cu. Plants such as lupin and *Heliotropium europium,* which contain toxic alkaloids, produce Cu toxicity in ruminants by impairing hepatic capacity to metabolize ingested Cu (National Research Council, 1980).

Much research has been done on supplementation of growing swine diets with high levels of Cu since many reports have been issued showing improved growth rate with 250 ppm Cu added to practical diets. The mode of action of Cu in producing this growth response still is not known, and the occasional toxicity associated with its use at this level has resulted in concern about the use of levels above 150 ppm. Signs of toxicity vary from slight growth depression and mild anemia to sudden death accompanied by liver damage and hemorrhage. Levels greater than 425 ppm Cu produce marked anemia, jaundice, and liver damage. When a level of 250 ppm of Cu has resulted in depressed weight gain, there is a microcytic hypochromic anemia associated with it but usually no liver damage. Gipp et al. (1973) reported reduced liver Fe and mean corpuscular volume and hemoglobin concentration in growing pigs fed a semipurified diet containing 250 ppm of Cu. These changes were prevented by feeding extra Fe, indicating an induced Fe deficiency. Source of protein is an important factor in the response to added Cu; milk protein is associated with more severe anemia and growth depression than soybean protein in swine and poultry. Presumably, both the interaction between Cu and Fe and the effect of protein source on Cu toxicity are related to effects on absorption of Cu and Fe from the GI tract. Variations among animals in Cu toxicity probably are related in part to differences in dietary levels of S, Mo, Fe, Se, and Zn (National Research Council, 1980).

Cu toxicity in humans seems highly unlikely, except by accidental contamination of foodstuffs. No cases of chronic disorders traceable to excessive Cu intakes have been reported. However, genetically induced metabolic Cu intoxication in humans is a well-known pathological condition termed Wilson's disease.

❏ MANGANESE (Mn)

Manganese (Mn) was recognized as a dietary essential for animals in 1931. Although the widespread occurrence of Mn in foodstuffs makes a deficiency less likely than for many other mineral elements, sufficient examples exist of practical problems with Mn deficiency in animals and birds to justify detailed consideration. The metabolism, functions and deficiency, and toxicity signs of Mn have been reviewed in detail (Leach and Harris, 1997; National Research Council, 1980).

Tissue Distribution

The total body supply of Mn is less than that of most other required minerals. For example, the total Mn content of adult humans is only 1% of the amount of Zn and 20% of Cu. It is widespread throughout the body and, in contrast to most other trace elements, tends not to accumulate in liver and other tissues in high concentrations when ingested in large amounts. Bone, kidney, liver, pancreas, and pituitary gland have the highest concentration. Bone Mn content is more responsive to dietary intake than that of other tissues. It is less mobilizable from the skeleton than from liver, skin, feathers, and muscle. A large proportion of soft tissue Mn is present in labile intracellular forms, but in bone it is associated primarily with the inorganic fraction.

Functions

Mn is essential in the formation of chondroitin sulfate, a component of mucopolysaccharides of bone organic matrix. Thus, it is essential for normal bone formation. Mn is necessary for prevention of ataxia and poor equilibrium in newborn animals and birds. Biochemical and histological evidence relating to the role of Mn in this area is lacking except for the presence of a structural defect in the inner ear of affected animals. This abnormality appears to be related to defective cartilage mucopolysaccharide formation. Many glycosyl transferases, enzymes of importance in synthesis of polysaccharides and glycoproteins, require Mn for activity. The effects of Mn deficiency on bone and cartilage formation appear to be explained on this basis.

Mn is a component of the metalloenzyme, pyruvate carboxylase, mitochondrial SOD, and arginase and activates phosphoenolpyruvate carboxykinase and other enzymes. As such, Mn plays a role in carbohydrate metabolism. Mn is necessary as a cofactor in the enzyme that catalyzes the conversion of mevalonic acid to squalene and stimulates synthesis of cholesterol and fatty acids in rat liver. Thus, it plays a vital role in lipid metabolism.

Metabolism

Absorption. The mode and control of Mn absorption from the GI tract are poorly understood. The amount absorbed appears to be proportional to the amount ingested, and absorption usually is less than 10%. Excessive dietary Ca or P reduces Mn absorption. However, both urinary and fecal excretion of parenterally administered radioactive Mn is greater in rats previously fed a low-Ca diet than those fed a high-Ca diet. Although this mainly may reflect the greater bone resorption of rats fed low Ca, thereby resulting in less Mn retention by bone, the results illustrate that dietary Ca and P content affects not only Mn absorption but also Mn tissue losses. Divalent Mn is absorbed about equally well in the form of its oxide, carbonate, chloride, and sulfate salts. Mn absorption is increased in Fe deficiency.

Transport and Storage. Mn is absorbed from the GI tract as Mn++, oxidized to Mn+++, transported rapidly to all tissues, and finally concentrated in mitochondria-rich tissues. Liver, pancreas, and brain mitochondria have been shown to accumulate radioactive Mn in many species. Mn is carried in blood as Mn+++, loosely bound to a plasma beta-1 globulin (other than transferrin), and is removed from blood quickly, first appearing in mitochondria and later in the nuclei of cells. The exact role of Mn in the metabolism and function of mitochondria is unknown, but Mn deficiency is associated with alterations in their structure and metabolism. Mn content of milk and fetuses and of eggs closely reflects dietary intake, illustrating the dissimilarity between Mn and such minerals as Fe and Cu in transport across mammary and reproductive tissues.

Excretion.   The main route of Mn excretion is via the bile, with smaller amounts lost through pancreatic secretions and sloughed intestinal mucosal cells and still smaller amounts in urine and sweat. Much of the bile Mn is reabsorbed. Bile concentration parallels that of blood; this may explain the failure of body tissues to accumulate large concentrations of Mn. Thus, excretion rather than absorption is the primary means by which body Mn homeostasis normally is maintained.

Deficiency Signs

Many different skeletal abnormalities are associated with Mn deficiency in several animal species. All these abnormalities undoubtedly are related to the role of Mn in mucopolysaccharide synthesis in bone organic matrix. Lameness, shortening and bowing of the legs, and enlarged joints are common in Mn-deficient rodents, pigs, cattle, goats, and sheep, and perosis or slipped tendon is common in poultry. Choline deficiency also produces perosis (see Chapter 15). This condition includes malformation of the tibiometatarsal joint, bending of the long bones of the leg, and slipping of the gastrocnemius tendon from its condyle. Parrot beak, shortened and thickened legs and wings (manifestations of the same defect in bone formation and often termed chondrodystrophy) are common signs of Mn deficiency in birds. Impairment in calcification of bones in Mn-deficient animals is not a factor in the bone malformations. Although plasma alkaline phosphatase sometimes is reduced in Mn-deficient animals, no change occurs in the composition of bone ash, except an occasionally observed decline in Mn concentration. Eggshell thickness and strength also are reduced in Mn deficiency.

The ataxia and poor coordination and balance commonly observed in newborn mammals and newly hatched birds whose dams have been fed Mn-deficient diets are irreversible and are related to abnormal bone formation of the inner ear bones during prenatal life. In chicks, the characteristic head retraction sometimes is referred to as the "star-gazing" posture. A genetically related congenital ataxia of mice caused by defective development of inner ear bones has been shown to be preventable by Mn supplementation of the maternal diet at a specific time in gestation. This represents a unique example of an interaction between a nutrient and a mutant gene affecting bone development.

Rats fed Mn-deficient diets during growth and pregnancy produce pups whose liver activities of two gluconeogenic enzymes, pyruvate carboxylase and phosphoenolpyruvate carboxykinase, are less than in pups of control dams. This finding suggests that glucose homeostasis may be compromised in neonates by Mn deficiency (Baly et al., 1985). Newborn guinea pigs from Mn-deprived dams show hypoplasia of the pancreas, and adult Mn-deficient guinea pigs show similar changes in pancreas histology and impaired glucose tolerance. Whether these effects of Mn deficiency on rats and guinea pigs are related to a specific effect on insulin production or to an indirect action through the presence of Mn as a constituent of gluconeogenic liver enzymes is unclear.

Delayed estrus, poor conception, decrease in litter size, and livability have been reported in farm and laboratory animals, and decreased egg production and hatchability occur in birds deprived of Mn. In males, Mn deficiency produces absence of libido and failure in spermatogenesis (National Research Council, 1980).

The blood of diabetics is low in Mn, and administration of Mn to these subjects has a hypoglycemic effect. This information, together with the data cited above in Mn-deficient rats and guinea pigs, provides strong evidence that Mn plays a role in glucose metabolism, but the practical importance of this role in animal nutrition is not known.

Mn deficiency has been reported to be associated with reduced liver and bone lipids in rats, but the exact role of Mn in lipid metabolism remains unknown. Supplemental Mn increased milk Mn, pig birth weight, and serum cholesterol concentration of sows (Christenson et al., 1990), suggesting that Mn plays a role in lipid metabolism in pregnant swine.

Although Mn deficiency is not considered probable in humans, Schroeder et al. (1966) suggested the possibility of a relationship between several human disorders and Mn nutrition. The relatively high levels of Mn in many foods commonly available to humans in relation to metabolic requirements would suggest that such relationships are unlikely but worthy of examination.

Toxicity

High dietary levels of Mn are tolerated well by most species; the toxic effects of Mn appear to be related more to interference with utilization of other minerals than to a specific effect of Mn itself. Rats, poultry, and calves show no ill effects of diets containing 820 to 1000 ppm Mn, but pigs show reduced growth at 500 ppm. The reduced growth rate associated with excess Mn is a reflection mainly of reduced appetite.

Ca and P utilization are affected adversely by excess Mn. Rickets has been produced in rats fed 1.73% Mn in the diet, and hypoplasia of tooth enamel has been observed in rats, cows, and pigs fed excess Mn. Fe deficiency anemia is produced in cattle, pigs, rabbits, and sheep fed 1000 to 5000 ppm Mn. In the presence of adequate dietary intake of all nutrients, 1000 ppm Mn appears to be the maximum tolerable level, while for poultry and swine 2000 and 400 ppm are the upper limits (Mertz, 1986). Neurological disorders and behavioral changes have been observed in workers mining Mn ores.

❏ SELENIUM (Se)

Selenium was recognized as a toxic mineral many years before its essentiality for animals and birds was discovered. Schwarz and Foltz (1957) first reported that Se prevents liver necrosis in rats and in the same year it was shown also that it prevents exudative diathesis in chicks. The following year, muscular dystrophy was cured in lambs with supplemental Se. All of these lesions previously had been known to be produced by vitamin E deficiency (see Ch. 14). Thus, a new era of nutrient interrelationships unfolded. Knowledge of Se and vitamin E in nutrition and metabolism has been reviewed extensively (Mertz, 1986; Ganther et al., 1976; National Research Council, 1980). Ullrey (1992) reviewed the regulatory developments associated with the use of Se as a supplement to animal diets. The relatively low quantitative dietary requirement for Se by animals (i.e., less than 4 ppm

in most cases), the incomplete knowledge of species differences in Se requirements, and the rather narrow margin of safety between adequacy and toxicity are factors in the controversy surrounding the use of inorganic Se as a dietary supplement in animal nutrition. However, it seems clear that supplemental Se is needed for adequate dietary Se intake under many conditions among animals in the U.S. and many other parts of the world.

Tissue Distribution

Se is present in all cells of the body, although the concentration generally is less than 1 ppm; thus, the total body supply is relatively low. Liver, kidney, and muscle generally contain the highest concentrations of Se, and the values for these and other tissues are affected by dietary intake (Tables 12.5, 12.6). Concentration of Se in liver and kidney of animals fed toxic levels (5 to 10 ppm) may be as high as 5 to 7 ppm.

Functions

The first known function of Se was as a component of the enzyme glutathione peroxidase (GSH-Px). In this role it is involved in catabolism of peroxides arising from tissue lipid oxidation. Thus, it

Table 12.5 Selenium content of the longissimus muscle of pigs fed diets from different sources in the USA.

SOURCE[a]	DIET SE, PPM	MUSCLE SE, PPM, DRY BASIS
Arkansas	0.152	0.817
Idaho	0.086	0.392
Illinois	0.036	0.223
Indiana	0.052	0.232
Iowa	0.235	0.977
Michigan	0.040	0.206
Nebraska	0.330	1.178
New York	0.036	0.163
North Dakota	0.412	1.430
South Dakota	0.493	1.893
Virginia	0.027	0.118
Wisconsin	0.178	0.501
Wyoming	0.158	1.098

[a]Corn–soybean meal diets of similar composition.
Source: Ku et al. (1972).

Table 12.6   Selenium content of tissues of pigs fed Se at different levels from sodium selenite.

TISSUE	DIETARY SE, PPM				
	0.05[a]	0.10	0.15	0.25	1.05
	... tissue Se ppmp fresh basis ...				
Longissimus muscle	0.05	0.08	0.09	0.11	0.13
Myocardium	0.11	0.21	0.21	0.23	0.33
Liver	0.14	0.51	0.57	0.59	0.80
Kidney	1.37	2.07	2.06	1.95	2.10

[a]Corn–soybean meal basal diet.
Source: Groce et al. (1973).

plays a central role in maintaining the integrity of cellular membranes. GSH-Px is present in all tissues, with activity highest in liver and red blood cells, intermediate in heart, kidney, lung, stomach, adrenal glands, pancreas, and adipose tissue and lowest in brain, skeletal muscle, eye lens, and testis (Ganther et al., 1976). Since then, many selenoproteins with known functions and some with unknown functions have been identified (Allan et al., 1999; Behne and Kyriakopoulos, 2001; Driscoll and Copeland, 2003). Se is a constituent of three groups of selenium-containing proteins: 1) those containing nonspecifically incorporated Se, 2) specific Se-binding proteins, and 3) specific selenocysteine selenoproteins. The selenoproteins with known functions in mammals are listed in Table 12.7.

Since the original discovery of the presence of Se as a component of glutathione peroxidase (now known as cytosolic or classical GPx1), four other seleno-cysteine-containing glutathione peroxidases (GPx's) have been identified, each present in one or more tissues. GPx1 is found in nearly all tissues, whereas others are associated with specific tissues, designated in their name. The iodothionine deiodinases are of physiological importance as catalysts in thyroid hormone conversions (see Iodine earlier in this chapter). The thioredoxin reductases and selenophosphate synthetase2 selenoproteins have unique metabolic functions not discussed here.

In addition to the selenoproteins with known functions (Table 12.7), several selenoproteins with unknown functions have been discovered, including selenoproteins P and W, and 15-kD and 18-kD. Thus, it is clear that Se as a required trace mineral for animals has multiple functions, some of which await identification.

Se is required for normal pancreatic morphology and through this effect on pancreatic lipase production is responsible for normal absorption of lipids and tocopherols from the GI tract. The mechanism of action of Se in this protective role on pancreatic morphology and function is still not entirely clear.

Table 12.7   Mammalian selenoproteins with known functions.

SELENOPROTEIN	ABBREVIATIONS USED
Glutathione peroxidases	GPxs
Cytosolic or classical GPx	cGPx, GPx1
Gastrointestinal GPX	GI-GPx, GPx-GI, GPx2
Plasma GPx	pGPx, GPx3
Phosphlipid	PHGPx, GPx4
hydroperoxidase GPx	
Sperm nuclei GPx	snGPx
Iodothyronine deiodinases	
Type 1 deiodinase	D1, 5′DI
Type 2 deiodinase	D2, 5′DII
Type 3 deiodinase	D3, 5′DIII
Thioredoxin reductases	TrxRs
Thioredoxin reductase 1	Trx1
Thioredoxin reductase 2	Trx2
Thioredoxin reductase 3	Trx3
Selenophosphate synthetase 2	SPS2

Source: Adapted from Behne and Kyriakopoulos (2001).

Metabolism

Absorption.   The main site of Se absorption is from the duodenum. It is not absorbed from the rumen or abomasum of sheep or the stomach of pigs. It is absorbed relatively efficiently either from natural Se-containing feedstuffs or as inorganic selenite (35 to 85%). Absorption of Se in ruminants probably is largely as selenomethionine and selenocysteine, as a result of the incorporation of inorganic dietary Se into amino acids by rumen microflora. Because Se-depleted animals retain more Se than adequately fed animals, it has been suggested that increased absorption occurs in response to tissue needs. However, excretion also is responsive to tissue needs. Pigs fed Se as a component of sweet clover grown on coal ash high in Se retain more tissue Se than pigs fed equal amounts of inorganic Se, suggesting that plant Se is more available (Mandizodsa et al., 1979).

Transport and Storage. After absorption, Se is transported in the plasma in association with a plasma protein and enters all tissues where it is stored mainly as selenomethionine and selenocysteine. Se is incorporated into red blood cells, leukocytes, myoglobin, nucleoproteins, myosin, and several enzymes, including cytochrome c and aldolase.

Excretion. Tissue Se is relatively labile, as illustrated by its rapid loss from tissues of animals fed low-Se diets following consumption of high-Se diets. Loss occurs via the lungs, feces, and urine. The proportion excreted by each route depends on the route of administration, tissue levels, and animal species. In sheep, injected Se is excreted mainly in the urine in proportion to the amount administered; fecal loss is small and constant with dosage level; expired Se is similar to fecal loss at low levels of administration but rises with dosage level and exceeds fecal loss by several fold at high dosage levels. Orally administered Se is excreted in feces in largest quantities at low-intake levels. As intake rises, fecal loss remains relatively stable and expired Se increases steadily. Urinary loss rises at moderate levels of supplementation, then declines.

Fecal losses represent mostly unabsorbed Se, but also that excreted via the bile, pancreatic duct, and intestinal mucosal cells. Arsenic (As), which is known to reduce absorption of Se from the GI tract, also increases bile and urinary excretion of Se when As is injected.

The form in which Se is provided in the diet has an important effect on its absorption. The generally superior utilization of plant Se appears to be related to the form in which it is present, compared with its form in animal tissues. Animal tissues contain Se as selenomethionine and selenocysteine.

The sulfate ion has important effects on Se metabolism under some conditions. Urinary excretion of Na selenate is increased by parenterally or only administered sulfate. The effect of sulfate is less marked when Se is provided as selenite than as selenate; sulfate probably has no effect on utilization of Se provided in the organic form. The similar chemical and physical characteristics of Se and S in relation to metabolism have been reviewed (Levander, 1976).

Placental transfer of Se is extensive in all species studied, as evidenced by prevention of musuclar dystrophy in lambs and calves by Se administration to the pregnant dam and by radioactive tracer studies. Selenomethionine is transferred more readily to the fetus than inorganic Se.

Deficiency Signs

Several deficiency diseases of animals and birds that were earlier considered to be from vitamin E deficiency more recently have been shown to be preventable by Se as well. These include nutritional muscular dystrophy (NMD) in lambs, poultry, pigs, and calves; exudative diathesis (ED) in poultry; and liver necrosis in rats and pigs. Encephalomalacia in chicks is cured or prevented by vitamin E or antioxidants but not by Se.

Se-responsive NMD is common in calves and lambs in many parts of the world. Allaway et al. (1966) mapped geographic areas with soil Se deficiency and Se excess in the United States. Undoubtedly, widely differing levels of soil Se are present in other parts of the world as well. Affected lambs may show a stiff gait and arched back, giving rise to the name *stiff lamb disease*. Severe cases result in death. Injection or oral administration of vitamin E or Se to the pregnant or lactating dam or the newborn protects against NMD. Postmortem examination reveals whitish streaks in the striated muscle, which are caused by degeneration of the muscle fiber (Zenker's degeneration) and account for the common name *white muscle disease*. If the heart muscle is affected, sudden death may occur. This is an increasingly common problem in growing pigs and often is referred to as *mulberry heart disease*. Animals with NMD show drastic rises in blood plasma concentrations of several enzymes that normally are intracellular but are released into the plasma when tissue damage occurs. Such enzymes include glutamic oxaloacetic transaminase (GOT) and lactic dehydrogenase isoenzymes (LDH).

In pigs, the chief target organ, in addition to the heart muscle, is the liver. Massive liver necrosis and sudden death is a common sign of vitamin E–Se deficiency in growing pigs. Serum ornithine carbamyltransferase is elevated because of liver cell damage, which releases the intracellular enzyme into the blood. Vitamin E supplementation of the diet of pigs fed marginal Se prevents this liver damage. The syndrome in pigs is called *hepatosis dietetica* and has become a practical field problem in Scandinavia, New Zealand, and the United States. This problem is due, presumably, to soil depletion of Se or changes in methods of fertilization or to harvesting and processing of cereal grains which may affect their tocopherol or Se content or biological availability. Corn grown on low-Se soil fertilized with Se contains a higher level of Se than that grown on adjacent unfertilized plots, and the Se is available biologically for chicks.

Chicks fed diets deficient in vitamin E and Se develop exudative diathesis (ED), which is prevented by supplementation with either nutrient. ED is characterized by accumulation of subcutaneous fluids on the breast, resulting from escape of blood constituents to the extracellular spaces after capillary damage. Glutathione peroxidase activity in plasma of Se-deficient chicks declines to less than 10% of normal within about 5 days. The activity of this enzyme correlates well with the protection by Se of chicks from ED. Vitamin E, though it protects against ED in the absence of Se, has no effect on glutathione peroxidase activity. Thus, the metabolic role of Se in protecting against ED is distinct from that of vitamin E. Plasma glutathione peroxidase acts as the first line of defense against peroxidation of lipids in the capillary membrane by destroying peroxides formed in the absence of vitamin E. Peroxidation of membrane lipids is prevented with adequate vitamin E, thus blocking the destruction of the capillary membranes. In its absence, Se functions to prevent peroxidation by its presence in glutathione peroxidase. Thus, the disruption of capillary membrane integrity, which results in ED, is prevented by either vitamin E or Se, but through slightly different mechanisms.

Se deficiency in chicks fed adequate vitamin E results in pancreatic degeneration and fibrosis, which impedes vitamin E absorption from the GI tract. Gries and Scott (1972) showed that the pancreas is the target organ for Se deficiency in chicks, as skeletal muscle, heart muscle, liver, and other tissues are unaffected if vitamin E is adequate.

Infertility in cattle, sheep, and birds has been reported to respond favorably to Se feeding or injection. Knowledge is incomplete as to whether such effects are related distinctly to Se or whether a sparing effect on the vitamin E requirement is involved, although experiments with rats over a successive generation have demonstrated reduced fertility in females and males even in the presence of adequate vitamin E. Thus, inconsistent results in farm animals may reflect differences in tissue stores of Se and vitamin E as well as species differences.

The addition of inorganic Se to the diets of swine, chickens, beef cattle, and sheep in the United States has been approved by the Food and Drug Administration as a means of controlling Se-deficiency diseases. The approved levels vary with animal species and stage of the life cycle. Only limited information is available on the incidence of Se deficiency in humans. Modern food distribution ensures that foods produced and distributed in a wide geographgic area contribute to the diet of most people. This minimizes the likelihood of Se deficiency even in areas whose soils are known to be Se deficient and from which crops are produced that may be low in Se content. However, surveys in some countries suggest the possibility of suboptimum Se intake among subgroups of the population in those countries. Additional research is needed to confirm and quantify these observations.

Toxicity

Animals grazing on seleniferous soils or fed crops grown on these soils develop a fatal disease known as blind staggers or alkali disease. This syndrome was described in horses as early as 1856 and since then has been studied in other species as well. Soils containing more than 0.5 ppm Se potentially are dangerous to livestock, for the plants harvested from them may contain 4 ppm or more. Blind staggers is characterized by emaciation, loss of hair, soreness and sloughing of hooves, and eroding of the joints of long bones leading to lameness; heart atrophy; liver cirrhosis; anemia; excess salivation; grating of the teeth; blindness; paralysis; and death.

In poultry, egg production and hatchability are reduced, and deformities are common, including lack of eyes and deformed wings and feet. Low hatchability of eggs from wild birds (in at least one case, an endangered species) has been reported in locations in which excessive levels of Se have accumulated in the food supply of these birds as a result of industrial or agricultural Se pollution. Abnormal embryonic development also is common in rats, pigs, sheep, and cattle fed excess Se.

Diets containing 5 ppm of Se produce toxic signs in most species. High-protein diets tend to protect against Se toxicity, and inorganic sulfate has been reported to relieve the growth depression in rats fed 10 ppm Se. Arsenic (As) added to the feed or drinking water alleviates Se toxicity in dogs, pigs, cattle and chicks. Liver succinic dehydrogenase activity in rats is reduced by high Se and restored by As. This is the only enzyme known to be affected by excess Se. The mode of action by which toxicity symptoms are produced is unknown.

Se toxicity in humans, even those living in seleniferous areas, is not considered likely because of food distribution patterns and the wide variety of foodstuffs consumed by most human populations.

❏ OTHER TRACE ELEMENTS

In addition to those elements already discussed in detail, other elements may be considered essential, based on the broad definition that their absence from the diet results in metabolic or functional abnormalities in one or more tissues of one or more animal species. These are chromium (Cr), fluorine (F), molybdenum (Mo), boron (B), and silicon (Si), and possibly aluminum (Al), arsenic (As), bromine (Br), lithium (Li), cadmium (Cd), nickel (Ni), lead (Ph), tin (Sn), and vanadium (V). The experimental basis for some of these findings has been reviewed (Schechter and Kerlich, 1980; Schwarz et al., 1970; Mertz, 1986; Greger, 1993; Nielsen and Mertz, 1984).

A detailed description of the tissue distribution, functions, metabolism, and deficiency signs of these minerals in animals is beyond the scope of this chapter. Excellent detailed coverage has been provided by Mertz (1986) and in the books edited by Comar and Bronner (1962–1972). The complex subject of interactions between and among mineral elements has been addressed by Mills (1985). Only selected references are included in the present discussion to explain the reasoning on which each of the mineral elements has been interpreted

as being essential. Much more information is available on toxic effects, especially of As, Cd, F, and Mo. Toxic aspects of these and other minerals are considered in the last major section of this chapter.

Chromium (Cr)

Evidence that Cr might be required by animals was reported in 1954 when synthesis of cholesterol and fatty acids by rat liver was shown to be increased by Cr. Schwarz and Mertz (1959) demonstrated the importance of trivalent Cr in glucose utilization. Since then, this function has been shown to be related to the presence of Cr as a cofactor in insulin. Cr utilization depends not only on its valence (trivalent is utilized; hexavalent is not) but on its chemical form. Mertz (1969) showed that radioactive Cr transfer to the rat fetus was accomplished only when Cr was incorporated into yeast and not when it was administered to the mother as inorganic salts. Mertz (1967) and Schroeder (1968) both concluded that trivalent Cr is required for normal carbohydrate and lipid metabolism. The requirement for Cr is increased in humans with unpaired glucose tolerance. No clear evidence has been found for a practical need for Cr supplementation of diets for animals fed typical natural ingredients (National Research Council, 1974). The addition of 200 ppb of Cr as chromium tripicolinate has been reported to increase lean and decrease fat in growing pigs (Page et al., 1993; Amoikon et al., 1995). The mode of action of Cr appears to be its influence on the cellular action of insulin (Lukaski, 1999).

Fluorine (F)

Fluorine (F) can qualify as an essential nutrient for animals and humans mainly on the basis of its preventative effect against dental caries. Navia et al. (1976) concluded that supplemental F in the drinking water should be provided during or shortly after tooth eruption for maximum cariostatic effects. No reports of a positive growth response to F addition to F-low diets appeared until Schwarz and Milne (1972) obtained improved growth in rats by applying strict control of dietary and environmental contamination. Although low levels in the diet or in drinking water (1 ppm) protect against dental caries (tooth decay), higher levels produce mottled tooth enamel and enlarged bones. Low dietary F has been reported to protect against osteoporosis in humans, but this point remains controversial inasmuch as animal experiments have failed to support the observation. Because growth and remodeling of bones are affected by the level of dietary F, the conditions under which F is beneficial or harmful to bone and teeth may be related not only to level of F intake but to other unexplained factors. High dietary Ca depresses F uptake of bone; probably other dietary variables also are important.

Kleerekoper and Balena (1991) concluded that F has both beneficial and detrimental effects on the skeleton and that the therapeutic value of Fl in osteoporosis in humans is still uncertain. The adverse effects of industrial F pollution on animals have been reviewed and documented (Krook and Maylin, 1979). Excellent reviews of F in nutrition are available (Mertz, 1986; National Research Council, 1974, 1980; Spencer et al., 1980; Kleerekoper and Balena, 1991).

Molybdenum (Mo)

Evidence that Mo is essential for animals was reported in 1953 when it was shown to be a component of the metalloenzyme, xanthine oxidase (Rickert and Westereld, 1953). Later, chicks, turkey poults, and lambs were shown to perform more favorably when Mo was added to semipurified diets. The growth response in lambs was obtained using a basal diet containing 0.36 ppm Mo, which resulted in a 2.5-fold improvement in daily gain. Because most feedstuffs contain considerably more Mo, it is not surprising that Mo generally has not been recognized as an essential nutrient. Mo in animal nutrition has been reviewed (Underwood, 1976). Its presence in feeds at toxic levels is of greater practical concern. Dietary levels of Cu, Zn, S, Ag, Cd, and S-containing amino acids have major effects on the susceptibility of animals to Mo toxicosis.

Silicon (Si)

Silicon (Si) has been recognized as an essential nutrient for chicks since 1972 (Carlisle, 1972a, 1972b). It is the most abundant element in the earth's crust, next to oxygen, and is present in large amounts in plants. As a result, it is ingested in large quantities by animals. Absorption occurs as monosilicic acid, which represents only a small fraction of the total Si ingested. Carlisle (1973) first reported a growth response in chicks fed Si by using

purified diets prepared carefully to minimize Si contamination and by raising the animals in a rigidly controlled environment low in Si. Subsequently, Schwarz and Milne (1972), using similarly controlled conditions, obtained a 25 to 34% increase in growth of rats by Si supplementation to a low-Si purified diet. Si appears to be involved in some way in initiating the mineralization process in bones. From a practical viewpoint, adverse effects of high-Si intake appear to be of greater importance than Si deficiency. Urinary calculi appear more frequently in grazing cattle as Si content of the forage increases; prairie hay, which contains more Si than alfalfa hay, increases the incidence of calculi. Increasing water intake and urine volume by feeding a high Na-containing diet reduces or eliminates siliceous calculi (National Research Council, 1980). The essentiality and toxic roles of Si in animal and human nutrition have been reviewed (National Research Council, 1980; Nielsen, 1976).

❏ ADDITIONAL POSSIBLY REQUIRED MINERALS

Many mineral elements are ubiquitous in nature, and for many of those commonly present in animal and human tissues no metabolic role has been discovered. Some appear to be harmless, others toxic. Among these elements, Al, As, Br, Cd, Li, Ni, Pb, Sn and V have been reported to be required by one or more animal species.

Aluminum. The concentration of Al in tissues changes in a circadian rhythm and with other changes in biological activity. It accumulates in regenerating bone, stimulates enzyme systems involved with succinate metabolism, and has been reported essential for fertility in female rats (National Research Council, 1980). These observations provide indirect evidence for an essential role of Al in nutrition, but if a metabolic requirement does exist, it has not been quantified. The role of Al in nutrition has been reviewed thoroughly (Life Science Research Office, 1975; Greger, 1993).

Arsenic. Supplementation of As to purified diets has been reported to increase growth of chicks, decrease neonatal mortality in rats, and improve birth weight and decrease neonatal mortality in goats.

Cadmium. Schwarz and Spallholz (1985) found that rats fed a highly purified diet containing less than 0.4 ppb Cd showed a growth depression when maintained in a metal-free environment. This is the first evidence of a metabolic requirement for Cd; the overwhelming research emphasis has been on toxicology of Cd ingestion.

Nickel. A dietary requirement for Ni first was reported in chicks; deficiency has also been produced in pigs, goats, rats (Nielsen and Sandstead, 1974), and sheep (Spears et al., 1978). Deficiency signs have been varied but include decreased hematocrit, abnormalities in the liver, reproductive problems, high neonatal mortality and depressed growth, serum proteins, liver and serum lipids, and dietary N utilization. The metabolic bases for these diverse effects of Ni deficiency have not been established. All reports taken together suggest the possibility of inadequate Ni when the diet contains less than 40 ppb Ni and when environmental contamination is controlled with filtered air and carefully prepared Ni-free diets are fed throughout a lifetime or through more than one generation (National Research Council, 1980).

Tin. A single report (Schwarz et al., 1970) of a growth response to dietary Sn in rats kept in plastic isolators to prevent environmental contamination is the basis on which Sn is considered a possibly required element for animals. The wide uses of Sn in food containers and in industry has stimulated far more research on Sn toxicosis than on Sn requirements.

Vanadium. It has been reported that a deficiency of V impairs reproductive efficiency. Several reports have shown beneficial effects of V on rats, chicks, and others have described tissue uptake and movement of V (National Research Council, 1980). V stimulates the rate of glucose transport into rat adipocytes, possibly by activating the insulin-sensitive transport system for glucose. Shechter and Kerlich (1980) reported that V ions mimic the effect of insulin on glucose oxidation in rat adipocytes. They also suggested that the effect is due mainly to the presence of vanadyl tetravalent ions produced within the cells and not primarily to inhibition of the Na pump. An effect of V on glucose metabolism in intact animals remains to be demonstrated, but such a possibility seems reasonable. As with many other trace elements, V toxicity has received attention; its toxic action appears to be exerted through inhibition of an array of enzymes.

Barium. Barium (Ba) may be required for growth of some species.

Bromine. Bromine (Br) may be required for growth of mice and chicks.

Rubidium (Rb) and Cesium (Cs) may replace part of the K requirement. None of these effects qualifies Ba, Br, Rb, or Ce, on the basis of present knowledge, as dietary essentials according to the criteria outlined by Cotzias (1967).

Future research with the aid of more sensitive analytical methods and more carefully controlled nutritional and environmental variables may enable the reclassification of one or more of the ubiquitous and, in some instances, toxic minerals as essential nutrients.

Bioavailability of Mineral Elements

Associated with the complexities of interactions between and among mineral elements and between mineral elements and other dietary constituents are the problems of bioavailability. Further advances in knowledge of essential elements and their identification will require careful methodology and interpretation of data obtained under conditions in which efficiency of entrance into and excretion from the body may be affected by bioavailability (Forbes and Erdman, 1983). The advancing technology of stable isotope use (Matthews and Bier, 1983) in animals and humans is proving to be highly useful in addressing these nutritional voids. Current knowledge and advances in mineral bioavailability for animals have been reviewed (Lei, 2004).

More sensitive analytical methods, closer control of environmental contamination, and a greater understanding of mineral interrelationships may well lead to discovering additional essential minerals for animals and humans.

❏ SUMMARY

- The following trace minerals are known to be required by one or more animal species for normal life processes: boron (B), cobalt (Co), chromium (Cr), copper (Cu), fluorine (F), iodine (I), iron (Fe), manganese (Mn), molybdenum (Mo), selenium (Se), silicon (Si), and zinc (Zn).
- Trace minerals function as activators of enzyme systems or as constituents of organic com-

pounds. Many complex interactions have been identified among and between trace minerals, between and among trace and macrominerals, and between trace minerals and vitamins.

- Geographic distribution in the United States of Co-deficient areas, low; variable-, or high-Se areas, and toxic Mo areas have been mapped. Such information is valuable in identifying areas of mineral deficiencies and excesses in animals.

- Trace minerals most likely to be deficient in animals fed commonly used feedstuffs are Fe, Mn, Co (required as a constituent of vitamin B_{12}), I (supplied easily as iodized salt), Zn (requirement increased by high dietary Ca and plant phytate P content), Cu (only in regions where soil Mo is high), and Se (only in regions where soil Se is low). Feed manufacturers often add trace minerals to macromineral supplements to ensure an adequate intake of each.

- Tissue distribution, functions, metabolism, and deficiency signs of each required trace mineral have been described.

❏ REFERENCES

Adler, P. 1970. In *Fluorine and Human Health*. World Health Organization, Geneva, Switzerland.

Alfrey, A. C. 1986. In *Trace Elements in Human and Animal Nutrition* (5th ed.), p. 399, W. Mertz (ed.). Academic Press, Orlando, FL.

Allan, C. B., et al. 1999. *Annu. Rev. Nutr.* **19**: 1.

Allaway, W. H., et al. 1966. *J. Nutr.* **88**: 411.

Ammerman, C. B., et al. *J. Anim. Sci.* **44**: 485.

Amoikon, E. K., et al. 1995. *J. Anim. Sci.* **73**: 1123.

Apgar, J. 1985. *Annu. Rev. Nutr.* **5**: 43.

Baly, D. L., et al. 1985. *J. Nutr.* **115**: 872.

Beard, J. L., and J. R. Connor. 2003. *Annu. Rev. Nutr.* **23**: 41.

Behne, D., and A. Kyriakopoulos. 2001. *Annu. Rev. Nutr.* **21**: 453.

Bhussry, B. R., et al. 1970. In *Fluorine and Human Health*. World Health Organization, Geneva, Switzerland.

Bowen, H. J. M. 1966. *Trace Elements in Biochemistry*. Academc Press, New York.

Browning, E. 1961. *Toxicity of Industrial Metals*. Butterworth and Co., London.

Burau, R., and P. R. Stout. 1965. *Science* **150**: 766.

Carlisle, E. M. 1972a. *Fed. Proc.* **31**: 700.

Carlisle, E. M. 1972b. *Science* **178**: 619.

Carlisle, E. M. 1986. In *Trace Minerals in Human and Animal Nutrition* (5th ed.), Vol. 2, Chapter 7, p. 375, W. Mertz (ed.). Academic Press, New York.

Carlisle, E. M., and M. S. Curran. 1993. *FASEB J.* **7**: A205.

Cartwright, G. E., and M. M. Wintrobe. 1964. *Am. J. Clin. Nutr.* **14**: 224; **15**: 94.

Chester, J. K. 1978. *World Rev. Nutr. Dietetics* **32**: 135.

Christenson, S. L., et al. 1990. *J. Anim. Sci.* **68**(Suppl. 1): 368.

Clarkson, T. W. 1976. In *Trace Elements in Human Health and Disease*, Vol. II, *Essential and Toxic Elements*, pp. 453–475, A. S. Prasad and D. Oberleas (eds.). Acdemic Press, New York.

Comar, C. L., and F. Bronner. 1962–1972. *Mineral Metabolism*, Vols. I, II, III. Academic Press, New York.

Cooke, B. C. 1983. In *Recent Advances in Animal Nutrition*, W. Haresign (ed.). Butterworths, London, England.

Cotzias, G. C. 1967. *Proc. First Annual Trace Substances in Environmental Health*, p. 5, D. C. Hemphill (ed.). University of Missouri, Columbia.

Cousins, R. J. 1979a. *Nutr. Rev.* **37**: 97.

Cousins, R. J. 1979b. *Am. J. Clin. Nutr.* **32**: 339.

Cousins, R. J. 1979c. *Environ. Health Perspect.* **28**: 131.

Dallman, P. R. 1986. *Annu. Rev. Nutr.* **6**: 13.

DiSilvestro, R. A., and R. J. Cousins. 1983. *Annu Rev. Nutr.* **3**: 261.

Doyle, J. J., and J. E. Spaulding. 1978. *J. Anim. Sci.* **47**: 398.

Driscoll, D. M., and P. R. Copeland. 2003. *Annu. Rev. Nutr.* **23**: 17.

Eisenstein, R. S. 2000. *Annu. Rev. Nutr.* **20**: 627.

Flegal, K., and W. G. Pond. 1979. *J. Nutr.* **110**: 1255.

Forbes, R. M., and J. W. Erdman, Jr. 1983. *Annu. Rev. Nutr.* **3**: 213.

Forbes, R. M., and G. C. Sanderson. 1978. In *The Biogeochemistry of Lead in the Environment*, Chapter 16. Elsevier/North-Holland Biomedical Press, New York.

Forsyth, D. M., et al. 1972. *J. Nutr.* **102**: 1639.

Fox, M. R. S., and S. H. Tao. 1981. *Fed. Proc.* **40**: 2130.

Friberg, L. F. 1985. *Cadmium and Health: A Toxicological and Epidemiological Appraisal*, Vol. 1. CRC Press, Boca Raton, FL.

Ganther, H. E., et al. 1976. In *Trace Elements in Human Health and Disease*, Vol. II, pp. 165–234, A. S. Prasad and D. Oberleas (eds.). Academic Press, New York.

Ganther, H. E. 1980. *Ann. NY Acad. Sci.* **35**: 212.

Gipp, W. F., et al. 1973. *J. Nutr.* **103**: 713.

Goyer, R. A. 1997. *Annu. Rev. Nutr.* **17**: 37.

Greene, L. W., et al. 1989. *J. Anim. Sci.* **67**: 3463.

Greger, J. L. 1993. *Annu. Rev. Nutr.* **13:** 43.

Gries, C. L., and M. L. Scott. 1972. *J. Nutr.* **102:** 1287.

Groce, A. W., et al. 1973. *J. Anim. Sci.* **37:** 948.

Hambridge, M., and N. F. Krebs. 2001. *Annu. Rev. Nutr.* **21:** 429.

Harris, E. D. 2000. *Annu. Rev. Nutr.* **20:** 291.

Hellman, N. E., and J. T. Gitlin. 2002. *Annu. Rev. Nutr.* **22:** 439.

Hill, G. M., and J. W. Spears. 2001. In *Swine Nutrition* (2nd ed.), A. J. Lewis and L. L. Southern (eds.). CRC Press, Boca Raton, FL.

Houpt, T. R. 1977. In *Duke's Physiology of Domestic Animals* (9th ed.), p. 443. Cornell University Press, Ithaca, NY.

Jowsey, J. R., et al. 1968. *J. Clin. Endocrine Metab.* **28:** 869.

Keen, C. L., and M. E. Gershwin. 1990. *Annu. Rev. Nutr.* **10:** 415.

Kleerekoper, M., and R. Balena. 1991. *Annu. Rev. Nutr.* **11:** 309.

Kostial, K. 1986. In *Trace Elements in Human and Animal Nutrition* (5th ed.), W. Mertz (ed.). Academic Press, New York.

Krook, L., and Maylin, G. A. 1979. *Cornell Vet.* **69** (Suppl. 8): 1.

Ku, P. K., et al. 1972. *J. Anim. Sci.* **34:** 208.

Kubota, J. 1968. *Arch. Environ. Health* **16:** 7.

Kubota, J. 1980. In *Applied Soil Trace Elements,* Chapter 12, B. E. Davies (ed.). John Wiley & Sons, New York.

Larsen, P. R., and M. J. Berry. 1995. *Annu. Rev. Nutr.* **15:** 323.

Leach, R. M., Jr., and E. D. Harris. 1997. In *Handbook of Nutritionally Essential Mineral Elements,* B. L. O'Dell and R. A. Sunde (eds.). Marcel Dekker, New York.

Leach, R. M., and M. C. Nesheim. 1963. *J. Nutr.* **81:** 193.

Lei, X. G., and C. Yang. 2004. Mineral elements: micro (trace). *Encycloped. Anim. Sci.,* W. G. Pond and A. W. Bell (eds.). Marcel Dekker, New York.

Leibold, E. A., and B. Guo. 1992. *Annu. Rev. Nutr.* **12:** 345.

Levander, O. A. 1976. In *Trace Elements in Human Health and Disease,* Vol. II, pp. 135–163, A. S. Prasad and D. Oberleas (eds.). Academic Press, New York.

Levander, O. A., and L. Cheng (eds.). 1980. Micronutrients interactions: vitamins, minerals, and hazardous elements. *Ann. NY Acad. Sci.* **355:** 1372.

Life Sciences Research Office. 1975. Evaluation of the Health Aspects of Aluminum Compounds and Food Ingredients. Life Sciences Research Office, Federation of American Societies for Experimental Biology, Bethesda, MD SCOG-43 Contract No. FOA 223-75-2004.

Lukaski, H. C. 1999. *Annu. Rev. Nutr.* **19:** 279.

Mandizodsa, K. T., et al. 1979. *J. Anim. Sci.* **49:** 535.

Mascotti, D. P., et al. 1995. *Annu. Rev. Nutr.* **15:** 239.

Matovinovic, J. 1983. *Annu. Rev. Nutr.* **3:** 341.

Matthews, D. E., and D. M. Bier. 1983. *Annu. Rev. Nutr.* **3:** 309.

Mertz, W. 1967. *Fed. Proc.* **26:** 186.

Mertz, W. 1969. *Physiol Rev.* **49:** 163.

Mertz, W. (ed.). 1986. *Trace Elements in Human and Animal Nutrition* (5th ed.), Vols. I and II. Academic Press, New York.

Miller, E. R., et al. 1991. In *Micro Nutrients in Agriculture* (2nd ed.), Chapter 16, p. 593, Soil Service Society of America, Book Series No. 4, Ames, IA.

Mills, C. F. 1985. *Annu. Rev. Nutr.* **5:** 173.

Miret, S., et al. 2003. *Annu. Rev. Nutr.* **23:** 283.

Montague, P., and K. Montague. 1971. *Mercury.* Gruen Co., New York.

Moore, C. V. 1951. *Harvey Lect.* **55:** 67.

Moore, C. V., and R. Dubach. 1962. In: *Mineral Metabolism,* Vol. II, Part B, C. L. Comar and F. Bronner (eds.). Academic Press, New York.

Murphy, W., et al. 1975. *J. Am. Dietet. Assoc.* **66:** 345.

National Research Council. 1971. *Fluorides, Biological Effects of Atmospheric Pollutants.* National Academy Press, Washington, DC.

National Research Council. 1974. *Chromium.* NAS-NRC, Washington, DC.

National Research Council. 1980. *Mineral Tolerances of Domestic Animals.* National Academy of Sciences, Washington, DC.

National Research Council. 2005. *Mineral Allowances and Tolerances of Domestic Animals.* National Academy Press, Washington, DC (In Press).

Navia, T. M., et al. 1976. In *Trace Elements in Human Health and Disease,* Vol. II, p. 249, A. S. Prasad and D. Oberleas (eds.). Academic Press, New York.

Nielsen, F. H. 1976. In *Trace Elements in Human Health and Disease,* Vol. II, pp. 379–399, A. S. Prasad and D. Oberleas (eds.). Academic Press, New York.

Nielsen, F. H. 1984. *Annu. Rev. Nutr.* **4:** 21.

Neilsen, F. H., and W. Mertz. 1984. In: *Present Knowledge of Nutrition* (5th ed.), Chapter 2, R. E. Olson et al. (eds.). Nutrition Foundation, Washington, DC.

Nielsen, F. H., and R. A. Poellot. 1993. *FASEB J.* **7:** A204.

Nielsen, F. H., and H. H. Sandstead. 1974. *Am. J. Clin. Nutr.* **27:** 515.

Neilsen, F. H., et al. 1977. In *Trace Element Metabolism in Man and Animals,* Vol. III, pp. 244–247, M. Kirchgessner (ed.). Technical University, Munich, Germany.

Neilsen, F. M., et al. 1992. *FASEB J.* **6:** A1946.

O'Dell, B. L. 1976. In *Trace Elements in Human Health and Disease,* Vol. I, A. S. Prasad (ed.). Academic Press, New York.

O'Dell, B. L. 1998. *Annu. Rev. Nutr.* **18:** 1.

Page, T. G., et al. 1993. *J. Anim. Sci.* **71:** 656.

Parizek, J., and J. Kalouskova. 1980. *Ann. NY Acad. Sci.* **355:** 347.

Park, J. H., et al. 1986. *Am. J. Physiol.* **251:** E273.

Pearson, A. M., and T. R. Dutson. 1990. *Meat and Health: Advances in Meat Research,* Vol. 6, Elsevier Applied Science, New York, 554 pp.

Perdomo, J. T., et al. 1966. *Proc. Soc. Exp. Biol. Med.* **122:** 757.

Pollitt, E. 1993. *Annu. Rev. Nutr.* **13:** 521.

Prasad, A. S. 1979. *Zinc in Human Nutrition.* CRC Press, Boca Raton, FL.

Prasad, A. S., et al. 1963. *J. Lab. Clin. Med.* **62:** 84.

Quarterman, J. 1986. In *Trace Elements in Human and Animal Nutrition* (5th ed.), p. 281, W. Mertz (ed.). Academic Press, New York.

Rickert, D. A., and W. W. Westerfeld. 1953. *J. Biol. Chem.* **203:** 915.

Schecter, Y., and S. J. D. Kerlich. 1980. *Nature* **284:** 556.

Schroeder, H. A. 1968. *Am. J. Clin. Nutr.* **21:** 230.

Schroeder, H. A., and J. J. Balassa. 1967. *J. Nutr.* **92:** 245.

Schroeder, H. A., and M. Mitchener. 1971. *J. Nutr.* **101:** 1431.

Schroeder, H. A., and M. Mitchener. 1975. *J. Nutr.* **105:** 421.

Schroeder, H. A., et al. 1963. *J. Nutr.* **80:** 39; 1964—**83:** 239; 1965—**86:** 31, 51; 1968—**95:** 95.

Schroeder, H. A., et al. 1966. *J. Chronic Dis.* **19:** 545.

Schuette, S. A., and H. M. Linkswder. 1984. In *Present Knowledge in Nutrition* (5th ed.). Nutrition Foundation, Washington, DC.

Schwarz, K., and C. M. Foltz. 1957. *J. Am. Chem. Soc.* **79:** 3293.

Schwarz, K., and W. Mertz. 1959. *Arch. Biochem.* **85:** 292.

Schwarz, K., and D. B. Milne. 1972a. *Nature* **239:** 333.

Schwarz, K., and D. B. Milne. 1972b. *Bioinorg. Chem.* **1:** 331.

Schwarz, K., and J. E. Spallholz. 1985. In *Cadmium and Health: A Toxicological and Epidemiological Appraisal,* Vol. 1, L. F. Friberg (ed.). CRC Press, Boca Raton, FL.

Schwarz, K., et al. 1970. *Biochem. Biophys. Res. Commun.* **40:** 22.

Smith, S. E., and P. D. Aines. 1959. *Cornell Agr. Expt. Sta. Bul.* **938.**

Smith, S. E., et al. 1951. *J. Nutr.* **44:** 455.

Spears, J. W., et al. 1978. *J. Nutr.* **108:** 313.

Spencer, H., et al. 1980. *Ann. NY Acad. Sci.* **355:** 181.

Suttle, N. F. 1991. *Annu. Rev. Nutr.* **11:** 121.

Todd, W. R., et al. 1934. *J. Physiol.* **107:** 146.

Tophan, R. W., et al. 1981. *Biochem.* **20:** 319.

Tucker, H. F., and W. D. Salmon. 1955. *Proc. Soc. Exp. Biol. Med.* **88:** 613.

Ullrey, D. E. 1992. *J. Anim. Sci.* **70:** 3922.

Underwood, E. J. 1976. In *The Biology of Molybdenum,* W. Chapple and K. Peterson (eds.). Marcel Dekker, New York.

Underwood, E. J., and J. F. Filmer. 1935. *Austral. Vet. J.* **11:** 84.

Valdivia, R., et al. 1978. *J. Anim. Sci.* **47:** 1351.

Vulpe, C. D., and S. Packman. 1995. *Annu. Rev. Nutr.* **15:** 293.

Wacker, W. E. C., and B. L. Vallee. 1964. In *Mineral Metabolism.* Academic Press, New York.

Wasserman, R. H. 1977. In *Duke's Physiology of Domestic Animals,* Chapter 34, pp. 413–432, M. J. Swendson (ed.). Cornell University Press. Ithaca, NY.

Wessling-Resnich, M. 2000. *Annu. Rev. Nutr.* **20:** 129.

Whanger, P. D., et al. 1980. *Ann. NY Acad. Sci.* **355:** 333.

Widdowson, E. M., and J. W. T. Dickerson. 1964. In *Mineral Metabolism,* Vol. 2, Part A. Academic Press, New York.

Winzerling, J. J., and J. H. Law. 1997. *Annu. Rev. Nutr.* **17:** 501.

Wolff, J. 1969. *Am. J. Med.* **47:** 101.

Zhu, J., et al. 2004. *FASEB Journal* **18**(5): A850 (abstract 571.7).

13

Mineral Toxicities and Organic Toxins in the Food Chain

ALL STABLE AND SOME radioactive elements in the Periodic Table are present in the earth's crust. Unlike organic compounds whose presence in nature depends on synthesis by living organisms, mineral elements are finite and are not distributed uniformly on the planet. Translocation of minerals through human intervention (e.g., mining, agriculture) and physical disturbances (e.g., glaciers, earthquakes, hurricanes, vulcanic eruptions) has contributed to differences in geographic distribution and concentrations of individual minerals. In addition to mineral toxicities, many organic toxins are encountered by animals and humans in feeds and foods.

❏ THE ENIGMA OF MINERAL TOXICITY

All inorganic elements, indeed all nutrients, may be toxic to animals when ingested in excess amounts. The margin of safety between the minimum amount required in the diet and the amount that produces adverse effects varies among minerals and according to conditions. For example, NaCl may produce convulsions and death in pigs if fed at only four to five times the required concentration in the diet when access to water is restricted, but the tolerance is much greater with adequate water intake. On the other hand, although the amount of Zn required by pigs is about 25 to 50 ppm in the diet, 20 to 40 times this concentration is required to produce toxic signs.

Although addressing toxic minerals in a separate chapter may seem inappropriate, their potential impact on the environment and on animal and human health demands our attention.

This chapter deals mainly with those inorganic elements that are innately toxic to animals and for which there is little or no evidence of a metabolic requirement. Some are ubiquitous in the environment, and their presence in animal tissues is of unknown biological importance.

The advances in capabilities for multielement analysis of animal and plant tissues (see Chapter 3) have allowed approaches to trace element toxicities and interactions that were never before possible. Schroeder et al. (1964, 1968) and Schroeder and Mitchener (1971, 1975) as well as Schroeder and Balassa (1967) discussed the concepts of mineral toxicity and suggested the possibility of toxicity to animals from a wide array of trace minerals. Here we discuss some of the minerals that represent more pressing practical problems for animals and humans and about which considerable knowledge is available. These minerals, including Pb, Cd, Hg, F, and Mo, are inexorably linked to animal nutrition, even though they may serve no nutritional role. Other minerals that may be hazardous to health under some conditions also are discussed. A review of mineral tolerances of domestic animals (National Research Council, 1980; 2005) provides detailed narrative and tabular information on the effects of ingesting toxic and excessive levels of required minerals. These reviews and the book edited by Mertz (!986) should be consulted for specific information on all of the required and highly toxic minerals as well as on those widespread in the environment but of limited concern as toxicants. Maximum tolerable levels of required and toxic minerals in the diet of domestic animals are shown in Table 13.1 (National Research Council, 1980; 2005)).

Goyer (1997) reviewed the interactions between toxic and required mineral elements in animals. Some of these interactions of importance in animal nutrition are described in the discussions of individual toxic minerals that follow.

TABLE 13.1   Maximum tolerable levels of dietary minerals for domestic animals.[a]

	SPECIES					
ELEMENT	CATTLE	SHEEP	SWINE	POULTRY	HORSE	RABBIT
Aluminum,[b]	1,000	1,000	200*	200	200*	200*
Antimony, ppm	—	—	—	—	—	70–150
Arsenic, ppm						
Inorganic	50	50	50	50	50*	50
Organic	100	100	100	100	100*	100*
Barium,[b] ppm	20*	20*	20*	20*	20*	20*
Bismuth, ppm	400*	400*	400*	400*	400*	2,000
Boron, ppm	150	150*	150*	150*	150*	150*
Bromine, ppm	200	200*	200	2,500	200*	200*
Cadmium,[c] ppm	0.5	0.5	0.5	0.5	0.5*	0.5*
Calcium,[d] %	2	2	1	0.4 (laying hen) 1.2 (other)	2	2

TABLE 13.1   Maximum tolerable levels of dietary minerals for domestic animals—*Continued.*[a]

| | SPECIES | | | | | |
ELEMENT	CATTLE	SHEEP	SWINE	POULTRY	HORSE	RABBIT
Chromium, ppm						
Chloride	1,000*	1,000*	1,000*	1,000	1,000*	1,000*
Oxide	3,000*	3,000*	3,000*	3,000	3,000*	3,000*
Cobalt, ppm	10	10	10	10	10*	10*
Copper, ppm	100	25	250	300	800	200
Fluorine,[e] ppm	40 (young)	60 (breeding)	150	150 (turkey)	40*	40*
	40 (mature dairy)	150 (finishing)		200 (chicken)		
	50 (mature beef)					
	100 (finishing)					
Iodine, ppm	500[f]	50	400	300	5	—
Iron, ppm	1,000	500	3,000	1,000	500*	500*
Lead,[c] ppm	30	30	30	30	30	30*
Magnesium, %	0.5	0.5	0.3*	0.3*	0.3*	0.3*
Manganese, ppm	1,000	1,000	400	2,000	400*	400*
Mercury,[c] ppm	2	2	2	2	2*	2*
Molybdenum, ppm	10	10	20	100	5*	500
Nickel, ppm	50	50*	100*	300*	50*	50*
Phosphorus,[d] %	1	0.6	1.5	0.8 (laying hen)	1	1
				1.0 (other)		
Potassium, %	3	3	2*	2*	3*	3*
Selenium, ppm	2*	0.2	2	2	2*	2*
Silicon,[b] %	0.2*	0.2	—	—	—	—
Silver, ppm	—	—	100*	100	—	—
Sodium chloride, %	4 (lactating)	9	8	2	3*	3*
	9 (nonlactating)	—	—	—	—	—
Strontium, ppm	2,000	2,000*	3,000	30,000 (laying hen)	2,000*	2,000*
				3,000 (other)		
Sulfur, %	0.4*	0.4*	—	—	—	—
Tin, ppm	—	—	—	—	—	—
Titanium,[g] ppm	—	—	—	—	—	—
Tungsten, ppm	20*	20*	20*	20	20*	20*
Uranium, ppm	—	—	—	—	—	—
Vanadium, ppm	50	50	10*	10	10*	10*
Zinc, ppm	500	300	1,000	1,000	500*	500*

[a]The text (NCR, 1980) should be consulted prior to applying the maximum tolerable levels to practical situations. Continuous long-term feeding of minerals at the maximum tolerable levels may cause adverse effects. The listed levels were derived from toxicity data on the designated species. The levels identified by an * were derived by interspecific extrapolation. Dashes indicate that data were insufficient to set a maximum tolerable level (NRC, 1980; Table 1).

[b]As soluble salts of high bioavailability. Higher levels of less soluble forms found in natural substances can be tolerated.

[c]Levels based on human food residue considerations.

[d]Ratio of calcium to phosphorus is important.

[e]As sodium fluoride or fluorides of similar toxicity. Fluoride in certain phosphate sources may be less toxic. Morphological lesions in cattle teeth may be seen when dietary fluoride for the young exceeds 20 ppm, but a relationship between the lesions caused by fluoride levels below the maximum tolerable levels and animal performance has not been established.

[f]May result in undesirably high iodine levels in milk.

[g]No evidence of oral toxicity has been found.

❏ LEAD (Pb)

Lead (Pb) poisoning in humans was recognized several hundred years ago and became a clinical problem in Western Europe in the seventeenth and eighteenth centuries where cooking utensils, other household articles, and water pipes contained high concentrations of Pb. In more recent times, Pb poisoning in children became more common as a result of ingesting residues of high-Pb paint or eating and drinking from utensils made of high-Pb materials. Chronic Pb poisoning in people of all ages is a threat whenever the water supply passes through tanks and pipes soldered with Pb (Quarterman, 1986). Pb toxicity presently is considered the most common cause of accidental death by poisoning in humans and animals worldwide (Quarterman, 1986). Pb toxicity may also occur in areas where contamination results from agricultural spray residues or near smelting plants, mines, and roadways. Contamination on roadways is a result of tetraethyl lead from motor fuels. (The use of low-Pb gasoline in late-model automobiles in the United States is based on this hazard.)

Signs of Pb toxicity include pallor, lassitude, anorexia, and irritability in both children and adults, but the most definitive symptoms are anemia, elevated blood Pb levels, and urinary excretion of α-amino-levulinic acid in humans and animals. Pb toxicity affects several organs and tissues. Microcytic, hypochromic anemia is produced in Pb toxicity as a result of a decrease in survival time of red blood cells (excessive hemolysis) and a decrease in red cell formation from a block in heme synthesis. In the kidney, pathological changes occur, resulting in amino aciduria, glycosuria, and hyperphosphaturia. Necrosis, hemorrhage, and ulceration of the stomach and small intestine occur in animals, and, in the nervous system, petechial hemorrhages and loss of myelin from nerve sheaths in the brain, accompanied by cerebrocortical softening after prolonged exposure, may occur. Pb toxicity also affects the skeleton, causing osteoporosis and reduced bone matrix formation and excess resorption of mineralized bone. Enlarged joints of long bones are common in swine and horses. High dietary Ca rduces Pb absorption from the intestinal tract; the mechanism is not completely understood. The National Research Council (1980) has listed the main clinical signs of toxicity and the main pathological effects for various animal species.

Pb may be absorbed into the body via the GI tract, lungs, and skin. Absorption from the GI tract is about 20% of that ingested. Retention of inhaled Pb in the body is 37 to 47% (Levander and Cheng, 1980). Dietary factors influencing Pb toxicity have been enumerated and discussed (National Research Council, 1980; Quarterman, 1986; Forbes and Sanderson, 1978; Ammerman et al., 1977; and Smith, 1976).

Pb is transported in blood as an aggregate of Pb phosphate, which is adsorbed to the surface of red blood cells. Major routes of Pb excretion are the GI tract and kidneys. Whatever is lost through the GI tract is excreted largely via the bile. Urinary Pb output is proportional to blood Pb concentration.

Various reports indicate that increased dietary Ca, P, Fe, Mg, Zn, and vitamin E reduce Pb toxicity, while Hg, Cd, Mo, Se, F, and vitamin D aggravate it. Although there are conflicting reports on the effect of dietary protein on Pb toxicity, the weight of evidence suggests that signs of Pb toxicity are reduced in animals fed a marginal level of protein. Much still remains unknown concerning the importance of nutrient interrelationships in the response to high-Pb intake in humans and animals.

❏ CADMIUM (Cd)

Cadmium (Cd) is toxic to a wide range of animal life and has specific adverse effects on the testes and kidney. Cd metabolism has been reviewed (National Research Council, 1980; Kostial, 1986). Cd occurs geologically in Zn ores and is used widely in industry in the production of batteries, pigments, and stabilizers and in electroplating. The atmosphere around some cities of the United States contains a significant amount of Cd (as high as 6.2 mg Cd/m3) and is a significant source of water pollution. Food contamination with Cd is variable among food sources.

Total Cd in tissues of adults in the United States has been estimated at 30 mg, with 10 mg in kidney and 4 mg in liver. Kubota et al. (1968) detected Cd in less than half of the human blood samples tested; the concentration of Cd in samples with detectable amounts was less than 1 mcg/dL in 83% of the cases. Cd concentration of testes, liver, spleen, kidneys, teeth, and hair of calves is increased by high levels of dietary Cd, but mammary transfer is low.

Specific adverse effects of Cd, when fed to experimental animals, include kidney damage and hypertension, as well as microcytic, hypochromic anemia. The anemia and growth depression can be alleviated partially by concomitant oral or parenteral administration of additional Fe.

Cd toxicity is alleviated partially by high dietary Zn, Fe, and Ca; a complex interaction is apparent between levels of dietary Cu, Fe, and Zn in protecting against Cd toxicity. Low dietary Ca exacerbates the effects of Cd in reducing bone Ca content. Metallothionein in liver, kidneys, and intestinal mucosa is involved with the complex interactions between Cd, Cu, and Zn (Cousins, 1979; Whanger et al., 1980).

A series of unrelated compounds protect the testes against injected Cd. These include Zn, Se, cysteine, glutathione, and estrogen. The mode of protection is unknown, although it has been suggested that Se may protect by forming relatively unstable Se salts of Cd that are stored harmlessly in the body. In this regard, there is evidence that Cd, in turn, may protect against Se toxicity (Flegal and Pond, 1979). Diet supplementation with Zn, Cu, and Mn produces a reduced uptake of Cd by liver, kidney, and whole body in Japanese quail fed a diet containing Cd at a level similar to that ingested normally by humans and domestic animals. More information is needed to ascertain the importance of environmental Cd contamination in human and animal health.

❑ MERCURY (Hg)

The toxicity of mercury (Hg) has been recognized for many years; it is a hazardous environmental contaminant because of its uses in industry and agriculture. Its occurrence in nature and its toxic properties have been reviewed in depth (National Research Council, 1980; Montague and Montague, 1971; Clarkson, 1986).

Hg combines preferentially with -SH groups and in this way inhibits enzyme systems containing such groups. Hg accumulates in the lysosomes within cells and has been associated with their rupture, resulting in hydrolytic enzyme release. Hg poisoning produces kidney necrosis and death. Concomitant administration of Se protects against the necrosis and improves survival (Parizek and Kalouskova, 1980). The protection may be related to formation of relatively insoluble compounds of low toxicity, such as $HgSeO_3$, which are stored in the body, or the possibility that Se catalyzes Hg to change to a less damaging form.

Organic Hg compounds such as methyl Hg and phenyl Hg are more toxic than inorganic Hg compounds such as $HgCl_2$. All of these compounds cause accumulation of Hg in liver and kidney, severe growth depression in rats, and decreased egg production and hatchability in chickens. Vitamin E protects against the toxic effects of methyl Hg in mammals ansd birds. Whether the protective effects of vitamin E and of Se against Hg toxicity have a common metabolic basis is unknown. The known functions of vitamin E and Se are related to the prevention of cellular oxidative damage (Ganther, 1980).

Hg content of hair, fingernails, and teeth gives a useful index of the degree of exposure of humans and animals. Kidneys consistently have a higher Hg concentration than all other tissues in humans. Urinary excretion is less than fecal excretion, although genetic differences have been observed among chickens. Some of the fecal Hg is a result of excretion through the bile.

Knowledge of the forms in which Hg is present in body tissues is incomplete. Living organisms can methylate Hg compounds, and methyl Hg is retained in tissues longer than inorganic forms. The conversion of Hg compounds to methyl Hg by microorganisms has been suggested as a factor in the high concentration of Hg present in fish. The high concentration of Hg in fish apparently is not always a result of recent environmental contamination because samples of fish stored for many years also have been shown to be high in Hg. The specific antidote for acute Hg poisoning is dimercaprol, which, in conjunction with proteins such as those in milk and eggs, binds Hg still in the GI tract (National Research Council, 1980).

❑ FLUORINE (F)

Aside from its role in preventing dental caries, fluorine (F) can be considered a toxic mineral for humans and animals. The tissue distribution, metabolism, and toxic effects of F have been described in detail (National Research Council, 1971, 1974; Mertz, 1986; Kleerekoper and Balena, 1991). Chronic fluorosis has been reported from several parts of the world mostly as a result of F-containing dusts from steel mills and other industries processing F-rich substances (such as reduction of aluminum ore), and secondarily in association with the use of high-F rock phosphate as mineral supplements for animals. Krook and Maylin (1979) described chronic F poisoning in cattle raised adjacent to an industrial aluminum smelting plant in Canada. Stained teeth of fluorotic cattle are shown in Fig. 13.1. Toxic effects in humans exposed to F in air, water, or food are as follows (Bhussry et al., 1970): 1 ppm F in water reduces dental caries, but 2 ppm or more F induces mottled enamel, 8 ppm in water induces osteosclerosis in some subjects, and 110 ppm in food or water produces growth retardation and kidney changes; 20–80 mg/day or more ingested from water or air produces crippling fluorosis; and 2.5 g or more in an acute oral dose is fatal. Detailed discussion of the effects of F on teeth is inappropriate here but may be found in several reviews (Bhussry et al., 1970; Adler, 1970).

The effects of high F on bone are somewhat contradictory; reported effects of fluorosis in animals range from osteosclerosis to osteoporosis. Jowsey et al. (1968) reported increased bone formation, and Romberg et al. (1970) reported increased bone resorption in response to F. Forsyth et al. (1972) who fed pigs diets containing up to 450 ppm of F in the presence of 1.2% or 0.5% Ca (Ca:P ratio kept constant at 1.2:1) to pigs through two generations, found that high F did not increase bone density but did cause mottled bone and interfered with collagen metabolism. The lower level of Ca–P was associated with greater F retention in bone, but there was no evidence of soft tissue lesions or of hyperostosis. Growth rate was depressed by high F but only in the presence of low dietary Ca. Newborn

pigs from sows fed high F had decreased length, volume, and weight and increased F concentrations of humerus, demonstrating the effective placental transfer of F.

Fluorine is a potent enzyme inhibitor. Fatty acid oxidase activity in kidney is decreased in fluorosis, suggesting impaired lipid metabolism. Carbohydrate metabolism is also disturbed. In general, all animals can tolerate slightly higher intakes of F from rock phosphate, or phosphatic limestones than from more water-soluble sources. High dietary Ca, Mg, or Al or a high concentration of Ca in the water, when F is administered in the water, inhibit the toxic effects of F (Spencer et al., 1980).

Almost complete absorption of soluble F from the GI tract occurs. F tends to follow the distribution of Cl in the body; it crosses cell membranes freely and thus leaves the blood quickly after absorption. Some F enters red blood cells so that about 25% of total blood F is within the red cells. The skeleton takes up F readily. Absorbed F that escapes retention in the skeleton is excreted mainly in the urine, but also to a smaller extent in sweat and by reexcretion into the GI tract. Urinary excretion in cattle and sheep normally represents 50 to 90% of the dietary intake, depending on the solubility and the proportion of F absorbed. Milk F concentrations are affected only slightly by dietary F (National Research Council, 1980), although pigs nursing sows fed high F do show increased bone F concentration (Forsyth et al., 1972). Although the beneficial effect of small amounts of F on teeth is recognized, ingestion of excessive F induces characteristic bone and teeth lesions and should be avoided in animals and humans.

❏ MOLYBDENUM (Mo)

Although molybdenum (Mo) is recognized as a required nutrient, it is of more concern as a toxic mineral. Mo toxicity was first recognized in England when it was reported that cattle grazing certain pastures high in Mo developed severe diarrhea. The syndrome became known as *peat scours* or *teart* and later was found to be prevented by Cu supplementation of the diet. Subsequently, it was discovered that high Mo and sulfate intake produced Cu deficiency in ruminants. Thus, the primary toxic effect of Mo intake is the precipitation of Cu deficiency.

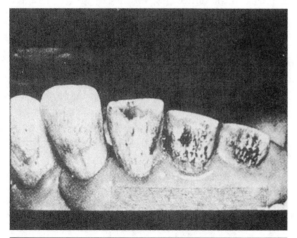

Figure 13.1 *Stained teeth of fluorotic cattle. Vegetative staining is typical of excessive fluorine intake at moderate levels. University of Tennessee photograph.*

Striking species differences exist in susceptibility to high-Mo intakes. Cattle appear to be the most susceptible, sheep less so, and horses and pigs still less. Mo has been fed at 1000 ppm in the diet of pigs for 3 mo with no ill effects; levels of 50 to 100 ppm produce severe diarrhea in cattle. Poultry, guinea pigs, rabbits, and rats are intermediate between pigs and cattle in their tolerance to high Mo. Signs of toxicity, in addition to diarrhea in cattle and rats, include anemia, dermatosis, anorexia, deformed front legs in rabbits, and reduced lactation and testicular degeneration in several species (Underwood and Filmer, 1935). Manifestations of Mo toxicity in sheep include decreased liver and plasma Cu, reduced crimp and pigmentation of the wool, anemia, and alopecia. Newborn lambs from affected ewes show enzootic ataxiz (swayback), resulting from demyelination of nerves (National Research Council, 1980).

Liver alkaline phosphatase activity is decreased in kidney and intestine of Mo-intoxicated rats, and liver sulfide oxidase activity is decreased. This interference with S metabolism has been suggested as a factor in the mechanism of Mo toxicity. The observed relationship of dietary sulfate level to Mo tolerance fits this suggestion. Mo-Cu-sulfate interrelationships have been reviewed in detail by Suttle (1991). The reduction in liver sulfide oxidase activity in Mo toxicity may lead to precipitation of insoluble cupric sulfide, rendering Cu physiologically unavailable. Species differences in both absorption and excretion of Mo probably account for differences in tolerance to high dietary Mo. Some of this apparent species variability in susceptibility to excess Mo may be due to variations in concurrent dietary intakes of Cu, Zn, S, Al, Cd, and S-containing amino acids (National Research Council, 1980). Hexavalent forms of Mo present in forages are water soluble and well absorbed by cattle. Mo is absorbed more rapidly by pigs than by cattle but is lost quickly in the urine of pigs. Mo content of the milk of ruminants is increased by high-Mo forage. Milk Mo of ewes grazing forage containing 1 ppm Mo was 10 mcg/L compared to 980 mcg/L in milk of ewes grazing forage containing 3 ppm Mo (Hogan and Hutchinson, 1965). The Mo of milk apparently is bound entirely to xanthine oxidase, so that Mo concentration is proportional to enzyme activity.

The chemical form in which Mo exists in blood is unknown, although it is present both in red cells and in plasma and is a constituent of the enzyme xanthine oxidase. The proportion of Mo in red cells is about 70% at low-Mo intake but decreases progressively as dietary Mo increases. The absorption of Mo from the GI tract is reduced greatly by the presence of high dietary sulfate; this in turn affects the concentration of Mo in blood and tissues.

The practical importance of Mo as a toxic mineral appears to relate mainly to cattle and sheep grazing pastures in areas of the world with high soil Mo contents. Such areas are widespread and of significant worldwide concern; molybdenosis has been reported in cattle grazing in such widely separated geographic locations as England, New Zealand, Ireland, and the western United States. Ironically, sheep grazing pasture low in Mo and S and moderate Cu content (12 to 20 mg Cu/kg dry matter) can succomb to Cu toxicity, while others grazing pasture with similar Cu content but high in Mo and S can bear lambs suffering from the Cu deficiency disease swayback (Suttle, 1991).

❏ OTHER MINERALS

Schroeder et al. (1963; Schroeder and Balassa, 1967); National Research Council (1980); and Browning (1961) are characterized several trace mineral elements as to their innate toxic effects as judged by growth, lifespan, and carcinogenic activity in mice. Based on records of weight gain, longevity, and incidence of malignant tumors in rats and mice, the following minerals appear to have innate toxicity: Cd, Pb, F, Se, Te, Ti, As, Ge, Sn, Nb, Sb, Ni, Sc, Cr ++++++, Ga, Y, Rb, Pd, and In. Schroeder and Mitchener (1971) suggested that suppression of growth rate induced in mice by some of the toxic elements is in the following order: Ga > Y > Sc > In > Cr > Pd > Rh. The following have been shown to be tumerogenic: Se, Ni, Pd, and Rb.

The following elements have had additional toxic properties ascribed to them (Mertz, 1986; Nielsen, 1986; Valdivia et al., 1978). Aluminum (Al) interferes with P availability in nonruminants, although it has only minor effects in ruminants. Excess Al has several potential toxic effects: decreased absorption of a number of elements, including Ca, P, F, Mg, and Fe, and of vitamin D and of lipids from decreased motility of the GI tract; pulmonary damage if inhaled; systemic toxicity if absorbed, including brain damage; skeletal decalcification; and anemia. Evidence linking Al toxicity to Alzheimer's disease in humans is suggestive but inconclusive (Alfrey, 1986; Greger, 1993). Antimony (Sb) reduces mean lifespan of female rats with no evidence of carcinogenesis or tumorigenesis. Arsenic (As) attaches to sulfhydryl groups of protein, and trivalent As specifically blocks lipoate-dependent enzymes (Ammerman, 1977) and results in vomiting, diarrhea, cellular necrosis, and death in a few days after a single toxic dose. Inorganic As ingested by ruminants and nonruminants is methylated and excreted in the urine. Chronic As toxicity is unlikely and seldom reported, probably because As is bound loosely to tissue protein and excreted quickly by the kidney. Barium (Ba) is highly toxic when absorbed, but it never occurs freely in nature, so the danger of ingesting soluble forms of Ba is small. Barium sulfate is used in taking X-ray photographs of the GI tract, but it is so nearly insoluble that it is nontoxic. Barium chloride, carbonate, and sulfide are toxic, and their ingestion results in paralysis of the central nervous system, intense myocardial stimulation, skeletal muscle tumors, and death.

Beryllium (Be), bismuth (Bi), boron (B), bromine (Br), rubidium (Rb), silicon (Si), silver (Ag), strontium (Sr), thallium (Tl), tungsten (W), uranium (U), and vanadium (V) have also been shown to be toxic when consumed in excess amounts (Nielsen, 1986; Schroeder et al., 1968). Rubium (Rb) and cesium (Cs) can replace K as a nutrient for growth of yeast and sea urchin eggs; thus, as in the case of other trace elements, the distinction of toxicity and essentiality is often difficult. Tungsten (W) depresses growth and decreases lifespan in chicks but not rats (Schroeder and Mitchener, 1975) and causes death by respiratory paralysis.

❏ RADIOISOTOPES

Considerable public and medical concern has been voiced about the effects of exposure to radioactive isotopes of minerals on human and animal health. The public response to the accidental escape of radioactive substances from nuclear power plants at Three Mile Island in Pennsylvania in 1980 and at Chernobyl in the former Soviet Union in 1986, and to the threat of the use of nuclear weapons by rogue nations and terrorist groups, typifies the high degree of anxiety and concern about the dangers of radioactivity in the environment. This general subject is outside the realm of our discussion, but there is a need to address the widespread concern about contamination of the food chain with dangerous radioisotopes. For example, the presence of I-131 and Sr-90 in food products, especially milk, has led to an appreciable body of literature on the physiological effects.

❏ POTENTIAL POLLUTION OF THE ENVIRONMENT BY INORGANIC MINERAL ELEMENTS

Based on amounts of inorganic elements present in the earth's crust and their use patterns in industry, Bowen (1966) divided them into four categories with respect to relative pollution potential: very high, high, moderate, and low. Minerals considered to have very high potential for pollution were Ag, Au, Cd, Cr, Cu, Mg, Pb, Sb, Sn, Ti, and Zn. Nutritionists and biologists concerned with controlling pollution, while providing an adequate feed and food supply, must recognize the concept of patterns of movement in the environment of elements required in the diet (e.g., Cu, Se, Zn), toxic elements, and macrominerals such as N and P required in the diet in large quantities (CAST, 2002), in response to changing technology.

❏ TOXIC MINERSALS AND MOLDS

Many different poisonous plants are recognized, and relatively large numbers of animals are affected adversely by consumption of these plants. Utilization of calories or of specific nutrients may be compromised by the presence of some of these toxins. Poisoning in humans is less apt to occur, but a significant number of cases occur every year. Animals, in general, have the ability to detect and reject many plants that may have adverse effects on them. In many instances, poisoning occurs when food availability is reduced. However, animals cannot always detect potentially toxic plants. Two examples of potent toxins that animals do not appear to detect are (1) nitrates (often found in toxic or lethal amounts in grass species fertilized heavily with N or in drought-stunted plants) and (2) prussic acid (toxicity that frequently occurs following regrowth of sorghum species after they have been stunted by drought or the plants have been frosted).

Contamination of cereal grains and seed legumes (e.g., soybeans, peanuts) with fungi (molds) that produce mycotoxins that are highly toxic to animals and humans creates major problems for the livestock industry and for humans. Aflatoxins, a group of toxic metabolites from the fungi, *Aspergillus flavus* and *A. parasiticus*, are of great concern for animals and humans because of their extreme toxicity and carcinogenicity. Consumption of peanuts contaminated with aflatoxins has been associated with liver cancer in experimental animals and humans. The mycotoxins zearalenone and vomitoxin are produced by several *Fusarium* species, for example, *F. roseum, F. graminearum (Gibberella zeae);* they interfere with reproduction in animals as a result of their estrogenic activity. Effects of mycotoxins on health and productivity of animals have been reviewed (CAST, 1989, 2003; Morasas, 1984). Mycotoxins can be classified according to the organ systems they affect. Several mycotoxins affect more than one system (e.g., the immune system), and a particular mycotoxin may affect several systems simultaneously. Trichothecenes (such as T-2 toxin) are produced by *Fusarium* species and by species of other genera. They affect hematopoiesis and cause kidney and nervous system damage. Ochratoxin A (produced by *Aspergillus* and *Penicillium* species) chelates Ca and other metal cations, causing abnormal egg shells in chickens. Fescue alkaloids, present on tall fescue (a perennial grass grown widely in the United States) cause several symptoms in cattle grazing the infected grass. The syndromes cause serious economic losses owing to impaired growth, increased body temperature, gangrenous necrosis of tissue of feet, tail, and ears, and reduced conception rate. The underlying physiological disturbance involves vasoconstriction and reduced peripheral blood flow. The fungus *Neotyphodium coenophialum* and other ergot-like alkaloids (e.g., ergovaline) are involved in the syndrome.

Molds may be difficult to detect, and toxins frequently are not inactivated by normal cooking methods. Details of fungal species, mycotoxins produced, and their effects on animals and humans have been reviewed (Morasas, 1984; CAST, 1989, 2003). Many food plants contain organic toxins (Hall, 1977). Legume seeds contain inhibitors, including antitrypsins, hemaglutinins, and goitrogens; *Brassica* species contain goitrogens; toxic metabolites of vitamin D are present in the shrub *Cestrum diurnum* and in similar species not nor-

mally consumed by humans but occasionally eaten by animals in amounts sufficient to cause toxicity. Other common food plants contain potent toxins. For example, carrots contain carotatoxin, myristicin, and estrogens; onions a mixture of toxic allyl di- and trisulfides; olives, tannins, and benzo(a)pyrene; and potatoes, especially if exposed to sunlight, solanine. If a careful analysis is done with sensitive analytical instrument techniques, toxins can be identified in almost any food or feed. Fortunately, in most cases, the toxins are present in such low concentrations, or the plants make up such a small proportion of the diet, that usually no harm results from their ingestion.

Although not directly related to this discussion of mycotoxins, it is important to recognize that a large number of antimicrobial compounds occur naturally in a wide range of plants used for food (CAST, 2000). Many of these antimicrobials have antifungal activity and are believed to have developed as defense mechanisms against invasive agents.

❏ SUMMARY

- Many mineral elements are ubiquitous in the environment but have no known physiological requirement for animals. Yet, they may enter the food chain.

- Trace minerals commonly present in animal tissues, but for which no metabolic requirement has been discovered, include aluminum (Al), arsenic (As), barium (Ba), bromine (Br), cadmium (Cd), cesium (Ce), nickel (Ni), rubinium (Rb), tin Sn), and vanadium (V).

- The trace minerals Cd, F, Hg, and Mo represent practical problems of toxicity. Although F is known to be required for its role in preventing dental caries, it can be considered toxic for animals and humans. Mo, though a required nutrient, is also of concern as a toxic mineral because of its interference with Cu bioavailability. Many other trace minerals in nature have known toxic properties but no known dietary requirement.

- Pollution of the environment with trace elements, both required (e.g., Cu, Se) and not required (e.g., Hg, Pb, Cd), and macrominerals (e.g., N, P) resulting from changes in agricultural and industrial technology is of increasing concern to society.

- Many different poisonous plants are recognized, and large numbers of animals are affected adversely by consumption of these plants.

- Contamination of cereal grains, forages, and seed legumes (e.g., soybeans, peanuts) that produce mycotoxins are toxic to livestock and humans.

❏ REFERENCES

Adler, P. 1970. In *Fluoride and Human Health.* World Health Organization, Geneva, Switzerland.

Alfrey, A. C. 1986. In *Trace Elements in Human and Animal Nutrition* (5th ed.), p. 399, W. Mertz (ed.). Academic Press, Orlando, FL.

Ammerman, C. B., et al. 1977. *J. Anim. Sci.* **44:** 485.

Bhussry, B. R., et al. 1970. In *Fluoride and Human Health.* World Health Organization, Geneva, Switzerland.

Bowen, H. J. M. 1966. *Trace Elements in Biochemistry.* Academic Press, New York.

Browning, E. 1961. Toxicity of Industrial Metals. Butterworth and Company, Ltd., London.

CAST. 1989. Task Force Report 116, Council for Agricultural Science and Technology, Ames, IA, 99 pp.

CAST. 2002. *Animal Diet Modification to Decrease the Potential for Nitrogen and Phosphorus Pollution.* Issue Paper No. 21, Council for Agricultural Science and Technology, Ames, IA, 16 pp.

CAST, 2003. *Mycotoxins: Risks in Plant, Animal and Human Systems.* Task Force Report No. 139, Council for Agricultural Science and Technology, Ames, IA, 199 pp.

Clarkson, T. W. 1976. In *Trace Elements in Human Health and Disease,* Vol. II, *Essential and Toxic Elements,* pp. 453–475, A. S. Prasad and D. Ogerleas (eds.). Academic Press, New York.

Cousins, R. J. 1979. *Environ. Health Perspect.* **28:** 131.

Flegal, K., and W. G. Pond. 1979. *J. Nutr.* **110:** 1255.

Forbes, R. M., and G. C. Sanderson. 1978. In *The Biogeochemistry of Lead in the Environment,* Chapter 16. Elsevier/North-Holland Biomedical Press, New York.

Forsyth, D. M., W. G. Pond, and L. Krook. 1972. *J. Nutr.* **102:** 1639.

Ganther, H. E., et al. 1976. In *Trace Elements in Human Health and Disease,* Vol. II, pp. 165–234, A. S. Prasad and D. Oberleas (eds.). Academic Press, New York.

Goyer, R. A. 1997. *Annu. Rev. Nutr.* **17:** 37.

Greger, J. L. 1993. *Annu. Rev. Nutr.* **13:** 43.

Hall, R. L. 1977. *Nutr. Today* **12**(6): 6.

Hogan, K. G., and A. J. Hutchinson. 1965. *New Zealand J. Agric. Res.* **8:** 625.

Jowsey, J. R., et al. 1968. *J. Clin. Endocrin Metab.* **28:** 869.

Kleerekoper, M., and R. Balena. 1991. *Annu. Rev. Nutr.* **11:** 309.

Kostial, K. 1986. In *Trace Elements in Human and Animal Nutrition.*

Krook, L., and G. A. Maylin. 1979. *Cornell Vet.* **69**(Suppl. 8): 1.

Kubota, J. 1980. In *Applied Soil Trace Elements,* Chapter 12, B. E. Davies (ed.). John Wiley & Sons, New York.

Kubota, J., et al. 1968. *Arch. Environ. Health* **16:** 7.

Levander, O. A. 1976. In *Trace Elements in Human Health and Disease,* Vol. II, pp. 135–163, A. Prasad and D. Oberleas (eds.). Academic Press, New York.

Levander, O. A., and L. Cheng (eds.). 1980. Micronutrients interactions: vitamins, minerals and hazardous elements. *Ann. NY Acad. Sci.* **355:** 1372.

Mertz, W. (ed.). 1986. *Trace Elements in Human and Animal Nutrition* (5th ed.), Vols. 1 and 2. Academic Press, Orlando, FL.

Montague, P., and K. Montague. 1971. *Mercury.* Gruen Co., New York.

Morases, W. F. O., et al. 1984. *Toxicogenic Fusarium Species, Identity, and Mycotoxicology.* Pennsylvania State University, University Park, PA.

National Research Council. 1974a. *Nutrients and Toxic Substances in Water for Livestock and Poultry.* National Academy of Sciences, Washington, DC.

National Research Council. 1974b. *Effects of Fluoride in Animals National Academy of Sciences.* Washington, DC.

National Research Council. 1977. *Medical and Biological Effects of Environmental Pollutants: Copper.* National Academy of Sciences, Washington, DC.

National Research Council. 1980. *Mineral Tolerances of Domestic Animals.* National Academy of Sciences, Washington, DC.

National Research Council. 2005. *Mineral Tolerances in Feed and Water for Domestic Animals.* National Academy Press, Washington, DC (In Press).

Nielsen, F. H. 1986. In *Trace Elements in Human and Animal Nutrition* (5th ed.), p. 415, W. Mertz (ed.). Academic Press, Orlando, FL.

Parizek, J., and J. Kaloskova. 1980. *Ann. NY Acad. Sci.* **355:** 347.

Quarterman, J. 1986. In *Trace Elements in Human and Animal Nutrition* (5th ed.), p. 281, W. Mertz (ed.). Academic Press, Orlando, FL.

Romberg, C. F., Jr., et al. 1970. *J. Nutr.* **100:** 981.

Schroeder, H. A. 1968. *Am. J. Clin. Nutr.* **21:** 230.

Schroeder, H. A., and J. J. Balassa. 1967. *J. Nutr.* **92:** 245.

Schroeder, H. A., and M. Mitchener. 1971. *J. Nutr.* **101:** 1431.

Schroeder, H. A., and M. Mitchener. 1975. *J. Nutr.* **105:** 421.

Schroeder, H. A., et al. 1963. *J. Nutr.* **80:** 39; 1965—**86:** 51; **86:** 51; **92:** 334.

Schroeder, H. A., et al. 1964. *J. Nutr.* **83:** 239.

Schroeder, H. A., et al. 1966. *J. Chronic Dis.* **19:** 545.

Schroeder, H. A., et al. 1968. *J. Nutr.* **95:** 95.

Smith, J. C. 1976. In *Trace Elements in Human Health and Disease,* Vol. II, *Essential and Toxic Elements,* p. 443, A. S. Prasad and D. Oberleas (eds.). Academic Press, New York.

Spencer, H., et al. 1980. *Ann. NY Acad. Sci.* **355:** 181.

Suttle, N. F. 1991. *Annu. Rev. Nutr.* **11:** 121.

Underwood, E. J., and J. F. Filmer. 1935. *Austral. Vet. J.* **11:** 84.

Valdivia, R., et al. 1978. *J. Anim. Sci.* **47:** 1351.

Whanger, P. D., et al. 1980. *Ann. NY Acad. Sci.* **255:** 333.

Fat-Soluble Vitamins

THE TERM ***VITAMIN*** WAS coined nearly a century ago by Funk (1912), who suggested that food contained special organic constituents that prevented certain of the classical human diseases of that time—beriberi, pellagra, rickets, and scurvy. Since then, a long list of vitamins has been identified and characterized.

Vitamins are required in minute quantities for normal body metabolism, yet each has its own specific function, and the omission of a single vitamin from the diet of a species that requires it produces specific deficiency symptoms and may result ultimately in death of the animal. Although many of the vitamins function as coenzymes (metabolic catalysts), others have no such role but perform other essential functions. The known vitamins can be divided on the basis of solubility properties into fat soluble and water soluble. The fat-soluble vitamins—A, D, E, and K—are considered in this chapter, and the water-soluble vitamins are considered in Chapter 15. The various names of the vitamins sometimes cause confusion. Therefore, the nomenclature has been standardized in accordance with the policy of the *Journal of Nutrition* as developed from recommendations of the International Committee on Nomenclature and the Committee on Nomenclature of the American Institute of Nutrition (AIN). The details of nomenclature for each vitamin are published in *Journal of Nutrition* **120:** 12–19, 1990; this nomenclature is used in this chapter.

❏ VITAMIN A

Structure

Vitamin A is required in the diet of all animals so far studied. It can be provided as the vitamin or as its precursor, carotene. The nomenclature and formulas for vitamin A and many different carotenes have been described in detail (Goodwin, 1986; Harris, 1967; American Institute of Nutrition Committee on Nomenclature, 1990). The structures of vitamin A alcohol and β-carotene are shown (Fig. 14.1) [International Union of Nutritional Sciences (IUNS) Committee, 1979]. Vitamin A is composed of a β-ionone ring and an unsaturated side chain. β-Carotene is composed of two vitamin A molecules joined as shown.

Vitamin A can occur as the alcohol (retinol), as shown in the figure, as the aldehyde (retinal), or as the acid (retinoic acid) in the free form or esterified with a fatty acid (for example, as vitamin A palmitate). It can occur as the all-trans form (as shown), the all-cis form, or as a mixture of cis and trans forms. The trans form of retinol is considered to have 100% biological activity. Retinoids are a class of compounds consisting of four isoprenoid units joined in a head-to-tail configuration; retinol, retinal, and retinoic acid are all retinoids. Biological activities of two isomers of retinal, expressed as a percentage of potency of all-trans retinol are: all-trans, 91%; 2-mono-cis (neo), 93%.

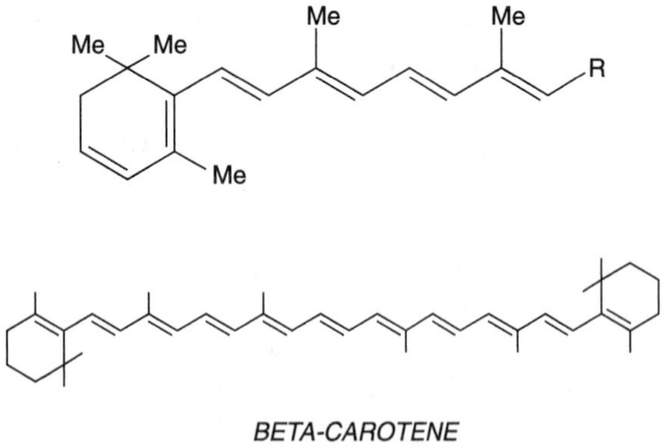

More than 500 carotenoid pigments exist in nature in addition to β-carotene, but plants contain no vitamin A. Carotenoids differ from each other in the configuration of the ring portion of the molecule. They include α- and γ-carotene, cryptoxanthin, zeaxanthin, and xanthophyll. The vitamin A precursors must be modified to release biologically available vitamin A. Zeaxanthin and xanthophyll possess no vitamin A activity, but others have some activity, the amount depending on the animal species. For the rat the relative biopotency of retinol is 100; β-carotene, 50; α-carotene, 25; γ-carotene, 14; and cryptoxanthin, 29.

Vitamin A activity originally was expressed as international units (IU), determined in a bioassay with β-carotene as a standard. Later, an additional standard was established based on retinyl acetate; thus, there were two standards, one for preformed vitamin A and one for provitamin A. One IU of preformed vitamin A was 0.344 mcg of retinyl acetate (0.3 mcg of retinol) and one IU of provitamin A was 0.6 mcg of β-carotene. Bieri and McKenna (1981) reviewed the history of expressing values for fat-soluble vitamins and described the currently accepted expression of dietary vitamin A, D, E, and K. Vitamin A is expressed in retinol equivalents (mcg). One retinol equivalent = $0.167 \times$ β-carotene or $0.084 \times$ other provitamin A. Thus, the total vitamin A value of a food or mixed diet is:

$$\begin{aligned} \text{Retinol equivalents (mcg)} \\ = \text{mcg retinol} + \frac{\text{mcg } \beta\text{-carotene}}{6} \\ + \frac{\text{mcg other provitamin A}}{12} \end{aligned}$$

The conjugated double bonds in carotenoids cause a characteristic yellow color. Exposure to ultraviolet light destroys the double bonds and the biological activity of vitamin A and its precursors. Some enzymes present in natural feedstuffs destroy carotenoids, but esterified vitamin A is more stable than retinol or retinal. The stability of vitamin A added to feeds can be increased by covering minute droplets of vitamin A with gelatin or wax or by adding an antioxidant such as ethoxyquin to the feed; the antioxidant is oxidized in preference to vitamin A. In current nutrition practice in the United States, most vitamin A is supplied by synthetic sources, which can be produced very economically.

BETA-CAROTENE

Figure 14.1   *Upper: Structure of vitamin A, the term used as the generic descriptor of all β-ionone derivatives. Me refers to methyl group. The compound (R = —CH₂OH), also known as vitamin A, vitamin A alcohol, vitamin A₁, vitamin A1 alcohol, axerophthol, or axerol, should be designated all-trans retinol. Lower: The structure of β-carotene (provitamin A). (IUNS Committee, 1990).*

Absorption, Transport, and Storage of Vitamin A and Carotenoids

Retinoids are aggregated into lipid globules in the intestinal lumen and dispersed by bile acids as esters of vitamin A in lipid emulsions and incorporated into mixed micelles containing other types of lipids. These mixed micelles containing vitamin A are passively absorbed (80 to 90%) into the intestinal mucosal cells. Most of the vitamin A is reesterified with fatty acids and phospholipids within the enterocyte and incorporated into chylomicrons, then carried into the lymph system and eventually to the blood. Vitamin A is transported in the blood largely bound to retinol-binding protein (RBP). About 90% of the vitamin A is stored in the liver in perisinusoidal stellate cells (lipocytes) as a lipoglycoprotein complex, 96% as retinyl esters and 4% as free retinol. Liver vitamin A content is directly related to dietary intake of the vitamin. Release of stored vitamin A for use in the body is transported by RBP.

Carotenoid metabolism in animals has been reviewed (van den Berg, 1999). Carotenoids are transported in the blood largely by lipoproteins (very-low-density lipoproteins, VLDL, in the fasted state, and low-density lipoproteins, LDL, in the fed state (see Table 14.2). Liver is the main storage site.

Functions

Several distinct functions of vitamin A have been identified. Vitamin A is required for normal night vision (formation of rhodopsin or visual purple in the eye). Vitamin A combines as a prosthetic group of rhodopsin, which breaks down on exposure to light. This reaction is part of the physiological process of sight. Vitamin A is required for normal epithelial cells, which line or cover body surfaces or cavities—respiratory, urogenital, and digestive tracts, and skin; and it is required for normal bone growth and remodeling (normal osteoblastic activity). Vitamin A and carotenoids appear to protect against cancer.

Darroch (1998) summarized the role of retinoids in affecting cell proliferation, differentiation, and gene expression. These cellular functions involve binding of retinoids to cellular-binding proteins (cytosolic retinol-binding protein, CRBP, and cytosolic retinoic acid-binding protein, CRABP), and their transport from cytoplasm to nucleus, stimu-lating messenger RNA (mRNA) and protein synthesis. Suffice it to say that the multiple functions of vitamin A continue to be revealed through studies of the regulation of cellular metabolism and gene expression.

The carotenoids have several important actions in biological systems (Krinsky, 1993). They act as antioxidants and have antimutagenic and anticarcinogenic properties. The importance of these protective actions in humans and animals is not completely understood, but the available evidence suggests that the presence of carotenoids such as lycopene and canthoxanthin in foods protects against the actions of ingested mutagenic and carcinogenic agents. This active area of research has been reviewed (Bloomhoff et al., 1992; Byers and Perry, 1992; Hill and Grubbs, 1992; Krinsky, 1993; Darroch, 1998). In addition to its other functions, vitamin A may play a primary role in the synthesis of glycoproteins.

Deficiency Signs

The practical significance of hypovitaminosis A (vitamin A deficiency) is greater perhaps than for any other vitamin. Because one fucntion of vitamin A is to allow formation of rhodopsin in the eye, night blindness is a symptom of vitamin A deficiency in all animals studied. The degree of failure in dark adaptation has been used as a measure of the quantitative needs in humans and some animals. Underwood (1990) reviewed the methods of assessment of vitamin A status. Common indices in humans are serum concentration of the vitamin and dark adaptation ("night blindness"). New and developing indicators of vitamin A status include (1) the RDR (relative dose response) test—the principle of the test relates to the depletion of liver reserves of vitamin A in vitamin A deficiency and the degree of mobilization of the carrier protein, retinol-binding protein (RBP) in response to a meal containing a known amount of the vitamin; (2) the CIT (conjunctival impression cytology) test—the principle underlying the test is that mucus-secreting epithelial tissues, when receiving insufficient vitamin A, develop thickening, which may involve skin keratinization, and changes in intestinal mucosa, lungs, and conjunctiva; (3) liver vitamin A concentration; (4) serum RBP concentration; and (5) isotope dilution procedures in which isotopically labeled vitamin A is administered and the dilution of the label in blood, relative to the amount administered, is used to calculate the total body content of vitamin A (a constant liver content is assumed).

The essentiality of vitamin A for normal epithelium creates a wide variety of deficiency symptoms in animals deprived of the vitamin. Some of the common deficiency signs in various animals are xerophthalmia in children and in growing animals (this condition is characterized by dryness and irritation of the cornea and conjunctiva of the eye and results in cloudiness and infection), keratinization of respiratory epithelium, resulting in greater severity of respiratory infections; reproduction difficulties, including abortions and birth of weak offspring, and associated thickening of vaginal epithelium; reproductive failure in males because of effects on spermatogenic epithelium; embryonic death in chicks and embryos; poor growth in surviving young; uric acid deposits in kidneys, heart, liver and spleen and keratinization of epithelium of respiratory tract of chicks; and xerophthalmia in chicks and mammals. Xerophthalmia remains a major public health problem in developing countries. Oomen (1976) classified the stages of xerophthalmia in human infants and presented photographs depicting the appearance of the eye in each case, ranging from night blindness to the more permanent scarring effects involving the cornea.

The importance of vitamin A in normal bone formation relates to a variety of deficiency signs that, although seemingly unrelated, have a common basis involving abnormal skeletal development. Gallina et al. (1972) provided evidence in calves that vitamin A deficiency produced increased osteoblastic activity. This agrees with earlier reports in other species. Nervous disorders such as unsteady gait, ataxia, and convulsions occur as a result of partial occlusion of the spinal cord by the surrounding vertebral column in growing animals. Exophthalmia (bulging eyes) and elevated cerebrospinal fluid pressure exist in vitamin A deficiency, apparently as a result of excess pressure on the brain stem associated with a constricted spinal column and optic foramen. Blindness and skeletal abnormalities occur in deficient newborn pigs. A wide range of additional manifestations (Table 14.1) (Roels, 1967) of vitamin A deficiency have been reported, but there is no definitive knowledge as to the exact mode of action on which the metabolic changes are based. Bone changes occurring in vitamin A deficiency are associated with changes in chondroitin sulfate, mucopolysaccharide synthesis, and increased urinary excretion of inorganic sulfate in rats. Vitamin A deficiency increases bone thickness but results in less Ca deposition and more glycosaminoglycan (GAG). This increase in GAG is apparently due to an increase in the amount of sulfated GAG in bone, resulting from a defect in degradation of GAG (Navia and Harris, 1980). Vitamin A metabolism has been linked with vitamin E (as an antioxidant in the stability of biological membranes); vitamin D (in bone metabolism); sterols (deficiency reduces cholesterol synthesis); squalene (deficiency increases squalene synthesis); and coenzyme Q or ubiquinone (deficiency increases ubiquinone synthesis in liver). Vitamin A deficiency in rats causes adrenal gland atrophy and reduced gluconeogenesis. Vitamin A is involved in some manner with biosynthesis of adrenal steroids and of glycogen. Vitamin A deficiency may be related to kidney stone formation in rats, based on the observed reduction in urinary Ca excretion, and is associated with altered Fe metabolism, including reduced plasma Fe and sometimes anemia (Hodges et al., 1980). The effect does not appear to be caused by increased red blood cell destruction.

Metabolism

Vitamin A from synthetic sources or animal tissues in feedstuffs is present primarily as the palmitate ester that is hydrolyzed by pancreatic enzymes activated by bile salts. Free vitamin A is incorporated into the lipid micelles (see Chapter 8) and reaches the microvilli of the upper jejunum, where it is transferred into the mucosal cell by active transport as retinol. Within the mucosal cell it is reesterified to palmitate and other esters, incorporated into chylomicrons and transported to the lymph (Ullrey, 1972) for storage in the liver parenchymal cells as retinyl esters. Release of vitamin A from the liver is preceded by hydrolysis to free retinol by retinyl palmitate hydrolase. This enzyme is increased in activity 100-fold during vitamin A deficiency, suggesting that it is important in maintaining serum vitamin A concentrations. Vitamin A is transported from the liver to peripheral tissues as free retinol bound to retinol-binding protein (RBP). RBP is synthesized primarily in the liver, but also in the kidney and other tissues.

Carotenoids are split within the rat intestinal mucosal cell by a specific enzyme to form retinal,

TABLE 14.1 Vitamin A deficiency signs in animals.

ABNORMALITY	ANIMALS STUDIED	ABNORMALITY	ANIMALS STUDIED
General		*Liver*	
Anorexia	Rat, fowl, farm animals	Metaplasia of bile duct	Rat
Growth failure and weight loss	Rat, fowl, farm animals	Degeneration of Kupffer cells	Rat
Xerosis of membranes	Rat, fowl	*Nervous system*	
Roughened hair or feathers	Rat, birds, farm animals	Incoordination	Rat, bovine, pig
Infections	Rat, birds, farm animals	Paresis	Rat, pig
Death	Rat, birds, farm animals	Nerve degeneration or twisting	Rat, dog, rabbit Bovine, bird, pig
Eyes		Constriction of optic foramina	Bovine, dog
Night blindness	Rat, farm animals	Bone formation	
Xerophthalmia	Rat, bovine	Defective modeling	Dog, bovine
Keratomalacia	Rat	Restriction of brain cavity	Dog
Opacity of cornea	Rat, bovine	*Reproduction*	
Loss of lens	Rat, bovine	Degeneration of testes	Rat
Papilledema	Bovine	Abnormal estrus cycle	Rat, bovine
Constriction of optic nerve	Bovine, dog	Resorption of fetuses	Rat
Respiratory system		*Congenital abnormalities*	
Metaplasia of nasal passages	Fowl	Anophthalmia	Pig, rat
Pneumonia	Rat, bovine	Microophthalmia	Pig, rat
Lung abscesses	Rat	Cleft palate	Pig, rat
GIT		Aortic arch deformation	Rat
Metaplasia of forestomach	Rat	Kidney deformiites	Rat
Enteritis	Rat, farm animals	Hydrocephalus	Rabbit, bovine
Urinary system		*Miscellaneous*	
Thickened bladder wall	Rat	Increased cerebrospinal fluid pressure	Bovine, pig
Cystitis	Rat	Cystic pituitary	Bovine
Urohthiasis	Rat		
Nephrosis	Rat		

Source: Adapted from Roels (1967).

which is reduced to retinol. Some retinol is converted to retinal and retinoic acid and absorbed into the portal blood as a glucuronide. Tissues other than intestinal mucosal cells are capable of splitting carotenoids to vitamin A. Liver contains an enzyme with the same properties as the β-carotene-splitting enzyme of the intestine and lung and kidney also may be involved. The details of absorption in other species are less well known. Species differences seem certain, as some animals deposit very little carotene in depot fat even though their diets are high in carotenoids, but others have

appreciable amounts of carotene in their depot fat and milk. Rats, cats, dogs, sheep, goats, pigs, and guinea pigs apparently convert most or all carotene to vitamin A, but cattle (especially Guernseys and Jerseys), horses, some rabbits, chickens, and humans have blood and depot fats high in carotenoids if dietary carotenoids are high. In fact, chickens absorb only hydroxycarotenoids unchanged and deposit them in tissues. A wide variation in efficiency exists even among species considered to be efficient converters of carotenoids to vitamin A. For example, the pig is only one-third as efficient as the rat in converting β-carotene to vitamin A. Some recycling of vitamin A by enterohepatic circulation occurs, but this probably is not a major conservation mechanism.

The degree of absorption of vitamin A and its precursors varies, depending on the animal species and type of diet. Apparent absorbability of vitamin A in dairy cattle fed a variety of forages averaged 78% in one report (Wing, 1969). There is considerable degradation in the rumen. In humans, vitamin A absorption and storage may be improved by adding fat to a low-fat diet.

The distribution of vitamin A, its esters, and its carotenoids in human blood serum during absorption from the GI tract is summarized in Table 14.2 (Krinsky et al., 1958). The liver hydrolyzes the retinyl ester and free retinol is carried by the blood complexed to RBP to tissues that require it for normal function. In vitamin A deprivation, liver stores are depleted even though the release of RBP by the liver is inhibited. Total liver vitamin A can be estimated, within limits, from plasma vitamin A concentration in several species of animals fed diets containing vitamin A levels ranging from deficient to toxic.

Many nutrients affect vitamin A concentrations in plasma and liver. Protein deficiency and Zn deficiency (Smith, 1980; Tielsch and Sommer, 1984) decrease vitamin A concentrations in plasma and liver; excess vitamin E interferes with the conversion of β-carotene to vitamin A. Liver concentrations, in general, are of more diagnostic value in assessing vitamin A status than are blood concentrations.

Liver biopsy techniques have been developed for use in cattle as a means of studying vitamin A

requirements. The turnover rate of vitamin A in liver of beef cattle has been estimated. Protein level of the diet does not affect vitamin A turnover rate significantly. Kohlmeier and Burroughs (1970) suggested that cattle entering finishing lots with 20 to 40 mcg/g of vitamin A in the liver have sufficient reserves for 90 to 120 days under normal feeding conditions and that no dietary vitamin A is required for good feedlot performance if plasma vitamin A is maintained above 25 mcg/mL and liver vitamin A exceeds 2 mcg/g. This claim is still controversial. Such factors as vitamin A destruction in feedstuffs and initial stores of liver vitamin A affect the level of dietary vitamin A needed to maintain these minimum tissue levels. The data apply here to cattle, but the same type of consideration must be applied to other species when determining appropriate levels of supplementation.

Protein ingestion affects vitamin A utilization. Protein deficiency causes reduced vitamin A concentrations of plasma and reduced liver storage. Signs of vitamin A deficiency may appear in protein-deficient animals even in the presence of adequate liver vitamin A storage. This has been suggested to result from reduced transport of vitamin A from liver because of reduced serum albumin, the carrier protein for vitamin A in blood. Impaired conversion of carotene to vitamin A in protein deficiency also may occur, but the dominant factor appears to be a defect in transport.

There is evidence that fat-soluble vitamins play a role in the release of hormones and in their cellular functions and that growth hormone may exert a positive effect on tissue uptake and storage of vitamin A (Ahluwalia et al., 1980).

Thyroid function is affected by vitamin A intake. Vitamin A deficiency reduces thyroxin secretion and causes thyroid hyperplasia. Conversely, thyroxin stimulates conversion of carotenoids to vitamin A and increases storage of vitamin A, but also increases depletion of vitamin A reserves when a vitamin A-deficient diet is fed.

There is increasing evidence that β-carotene and the carotenoid canthaxanthin inhibit the growth of skin tumors. The protective effect has been reported with β-carotene addition to the diet or perhaps by topical application. The effect may be related to the fact that carotenoids quench free radical formation. The use of carotenoids in treating photosensitization is also related to the oxygen quenching property. The anticancer effect of carotenes and retinoids has been studied in long-term

TABLE 14.2 Distribution of retinol, vitamin A esters, and carotenoids in human blood serum during absorption.

| SERUM FRACTION | VITAMIN A | | |
	RETINOL, %	ESTER, %	CAROTENOIDS, %
Chylomicrons	5.3	7.5	0
Lipoprotein Sf 10–100	3.9	79.4	0
Lipoprotein Sf 3–9	20.2	8.6	78.3
Other proteins	70.6	4.4	21.7

Source: Krinsky et al. (1958).

human investigations (Byers and Perry, 1992; Hill and Grubbs, 1992; Dorgan et al., 1998).

Oral contraceptives in humans elevate blood vitamin A concentrations, possibly by an effect of the estrogen component of the formulation on increasing the levels of retinol-binding protein (RBP). The rapidly advancing capabilities of hormone assays of blood and tissues offer exciting possibilities for identifying more completely the role of the endocrine system in affecting the metabolism of vitamins and other nutrients in humans and animals.

Vitamin A is transported readily by the mammary gland to the milk of swine and goats. Cattle transfer both vitamin A and carotene to the milk in response to dietary intake, the proportions depending partly on breed. Vitamin A concentration of human milk tends to be related to maternal intake of the vitamin.

Toxicity

Vitamin A is not excreted readily, so long-term ingestion of amounts larger than needed or acute dosage with a large excess may result in toxic symptoms. The toxic range is reached when daily intake reaches 50 to 500-fold the metabolic requirement. Death has been reported in humans after a single dose of 500,000 to 1,000,000 IU of vitamin A. Chronic toxicity manifests itself as anorexia, weight loss, skin thickening, scaly dermatitis, swelling and crusting of the eyelids, patchy hair loss, hemorrhaging, decreased bone strength, spontaneous bone fractures, thinning of the bone cortex, and death. Excess mucus forms and normal keratinization is inhibited in hypervitaminosis A. The bone changes described in young pigs fed excess vitamin A have been attributed to destruction of epiphyseal cartilage and decreased matrix formation in the presence of normal remodeling. Excess vitamin A may cause disruption of lipoprotein membranes. Plasma and liver levels of vitamin E are reduced in animals fed excess vitamin A, but the effect seems to be related to reduced absorption of vitamin E rather than to interference with tissue metabolism (Smith, 1980).

The dietary levels causing toxic symptoms will vary in accordance with species, age, body store, degree of absorbability, and degree of conversion of carotene to vitamin A where the free vitamin is not fed. In pigs, toxic symptoms include rough hair coat, hyperirritability and sensitivity to touch, pe-

techial hemorrhages over the legs and abdomen, cracked, bleeding skin above the hooves, blood in urine and feces, loss of control of legs, periodic tremors, and death.

Excess vitamin A during pregnancy results in malformed young in rats, mice, and pigs, but less so for guinea pigs and rabbits. During early gestation an excess induces embryonic death but, if begun later in gestation, abnormalities may occur, the severity and type differing according to species. In pigs, a single excess dose of vitamin A injected on day 18 or 19 of pregnancy causes cleft palate, abnormal skulls and skeleton, and, sometimes, eyelessness in the newborn. Injection of the same dose before or after this stage of development has no such effect. Clearly, the effect is related to the relative rate of growth and differentiation of a particular organ at the time the excess vitamin A is given. Retinoic acid and other metabolites of vitamin A are involved in cell differentiation in the embryo as well as in adult epithelial, connective, and hematopoietic tissues (Petkovich, 1992). The identification of retinoic acid nuclear receptors, which belong to the broad family of steroid receptors, suggests that retinoic acid produces cellular responses through the regulation of gene expression via these receptors (Petkovich, 1992). This field of research into vitamin A metabolism and action promises to bring a much more complete understanding of the mechanisms by which vitamin A and its metabolites exert their actions.

❏ VITAMIN D

Structure

Several sterols have biological vitamin D activity, but only two, vitamin D_2 (irradiated ergosterol or calciferol) and vitamin D_3 (irradiated 7-dehydrocholesterol) are of major importance. Most mammals can use either vitamin D_2 or D_3 efficiently, but birds utilize D_2 only one seventh as efficiently as they do D_3. The structures of vitamins D_2 and D_3 and their sterol precursors are shown in Fig. 14.2 (DeLuca, 1979).

Ergosterol is the chief plant source and 7-dehydrocholesterol is found in animal tissues. Ultraviolet light converts each provitamin to its respective biologically active form. Exposure of harvested green forage to sunlight for several hours converts plant sterols to vitamin D_2. Similarly, exposure of

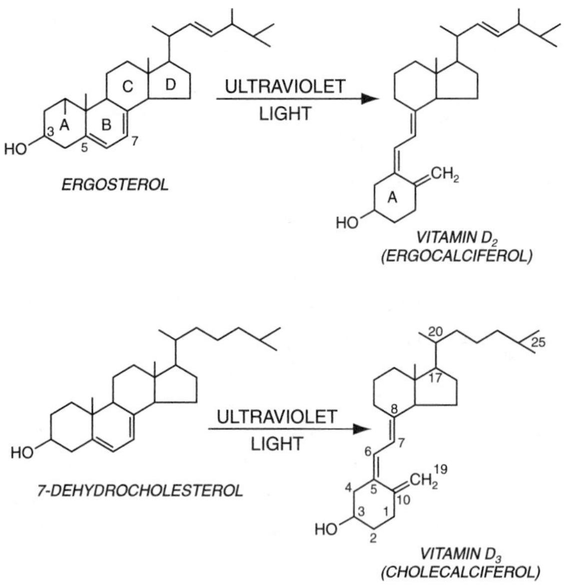

Figure 14.2 *The structures of vitamins D_2 and D_3 and sterol precursors. Courtesy H. F. DeLuca, University of Wisconsin (1979).*

animals to sunlight for a few minutes a day is sufficient to convert skin sterols to vitamin D_3, thus eliminating the need for a dietary source under most conditions.

Functions

Vitamin D was named by E. V. McCollum, who showed that it differed from vitamin A, which he also named (McCollum and Davis, 1913). Vitamin A was named on the basis of the presence of a substance present in butter fat and cod liver oil that was required for growth of animals fed what was otherwise a chemically defined diet. McCollum realized that the antirachitic activity of cod liver oil was distinct from its antixerophthalmic activity and subsequently determined that vitamin A and D were two distinct fat-soluble compounds with different curative functions. Shortly thereafter, Steenbock and Black (1924) discovered that ultraviolet light caused an alteration of some substances in animals and that ultraviolet irradiation not only of animals but of their feed prevented or cured rickets. This fascinating account of early classical research discoveries of vitamin D and its functions is detailed by DeLuca (1979). Historical perspective, physiology of vitamin D action, discovery of the vitamin D endocrine system, current knowl-

edge of vitamin D metabolism, and functions of its metabolites have been summarized (DeLuca, 1979; Henry and Norman, 1984; Holick, 1990; Norman, 1980).

The general functions of vitamin D are to elevate plasma Ca and P to levels that allow normal bone mineralization and that prevent the tetany that results if plasma Ca falls appreciably below normal. Vitamin D, in conjunction with parathyroid hormone (PTH), prevents tetany by elevating plasma Ca concentration. Normal plasma Ca levels are achieved by adjusting intestinal transport of Ca from ingested sources and by release of Ca from bone. Stimulation of the active transport of Ca and P across the intestinal epithelium involves the active form of vitamin D. PTH stimulates Ca absorption indirectly by stimulating the production of an active form of vitamin D, 1,25-dihydroxycholecalciferol [1,25-$(OH)_2D_3$] under conditions of hypocalcemia. Reabsorption of Ca from kidney tubules is stimulated by 1,25-$(OH)_2D_3$. Thus, intestinal absorption, bone resorption, and renal tubule reabsorption of Ca and P represent the three reservoirs available for maintenance of plasma Ca and P within limits compatible with normal bone mineralization and neuromuscular tone.

In addition to the role of 1,25-$(OH)_2D_3$ in regulation of Ca/P metabolism and bone mineralization, it affects the immune system. Immunity is enhanced by 1,25-$(OH)_2D_3$ in several ways, including effects on T-cells (Griffin et al., 2003). The apparent ability of some or all cell types to produce 1,25-$(OH)_2D_3$ suggests a paracrine effect of 1,25-$(OH)_2D_3$ on immunomodulation. Thus, locally stimulated 1,25-$(OH)_2D_3$ production serves a survival function in immune responses.

One or more Ca-binding proteins are involved in Ca transport. Wasserman and colleagues (Wasserman and Faher, 1977) isolated a Ca-binding protein from intestinal mucosa of several species (birds, rats, dogs, cattle, pigs, monkey, guinea pigs) that requires vitamin D. However, there appears to be at least one other factor functioning in Ca transport. DeLuca (1978) suggested that Ca-dependent ATPase, alkaline phosphatase, and actin likely may be factors induced by 1,25-$(OH)_2D_3$ to function in Ca transport.

Metabolism

Although the importance of vitamin D in normal Ca and P metabolism has been recognized for many decades, great strides have been made during the past few years in understanding the metabolic reactions of vitamin D and the importance of vitamin D metabolites in various tissues and organs of the body. Current knowledge of vitamin D metabolism has been detailed (DeLuca, 1979; Haussler, 1986; Henry and Norman, 1984; Norman, 1980; Holick, 1990). The sequential alterations in vitamin D in the body are illustrated in Fig. 14.3 (Henry and Norman, 1984). Vitamin D absorbed from the intestines or made in the skin by ultraviolet radiation is carried to the liver, where it is hydroxylated to produce 25-hydroxyvitamin D_3 [25-(OH)D_3], the main circulating form of vitamin D. Significant hydroxylation also occurs in other tissues, including lung, intestine, and kidney (Henry and Norman, 1984). There does not appear to be a direct action of 25-(OH)D_3 on any target tissue. Instead, further transformation is needed;

metabolism to 1,25-(OH)$_2$D$_3$ and 24,25-(OH)$_2$D$_3$ occurs exclusively in the kidney. These final products are delivered by the blood to the target tissues of intestine, bone, and elsewhere in the kidney where they carry out their functions. Thus, the metabolically active forms of vitamin D are considered to be hormones. Both steps in the conversion of vitamin D to 1,25-(OH)$_2$D$_3$ and 24,25-(OH)$_2$D$_3$ are under the control of mixed function or monoxygenase enzymes in the microsomes of the liver and in the mitochondria of the kidney and other organs.

Vitamin D is converted to its metabolically active forms by three hydroxylase enzymes, 25-hydroxylase (expressed mostly in liver), 1-hydroxylase (expressed in kidney, skin, intestine, macrophages, and bone), and 24-hydroxylase (expressed in nearly all cells) (Omdahl, 2002). The formation of hydroxylated metabolites of vitamin D_3 from skin and diet

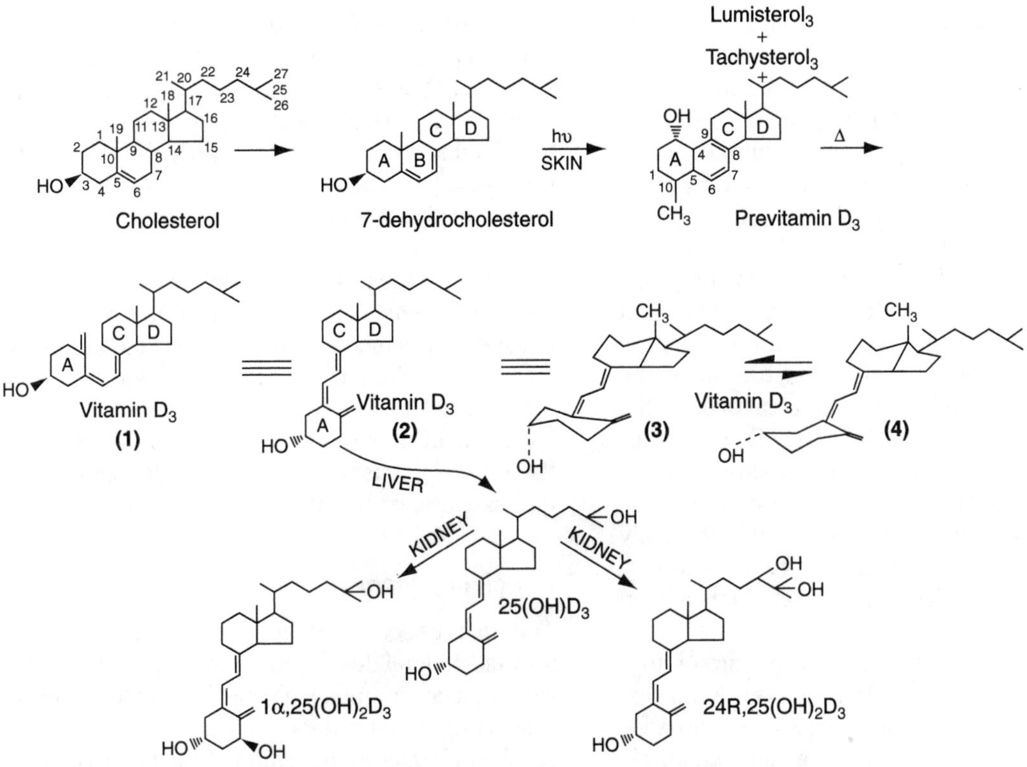

Figure 14.3 *Key vitamin D sterols. Top line: Vitamin D is produced from 7-dehydrocholesterol via previtamin D as a consequence of the UV-mediated opening of the B ring. Middle line: Summary of the evolution of the conformational representations of vitamin D. Structure 1 or 2 resulted from original chemical analysis; structures 3 and 4 represent results of recent studies indicating the conformational mobility of the A ring. (Bottom line: Structure of the three principal metabolites; hν = ultraviolet light irradiation. By permission from Henry and Norman (1984).*

has been summarized in Fig. 14.2 (DeLuca, 1979) and Fig. 14.3 (Henry and Norman, 1984). In addition to these metabolites of vitamin D, more than 20 other compounds are produced under certain conditions; their physiological roles are unknown. They include $1,25,25-(OH)_3D_3$; $25-(OH)D-26,23-$ lactone; calcitroic acid; and cholecalcitroic acid.

High dietary Ca decreases the production of $1,25(OH)_2D_3$ by the kidney, and low dietary Ca stimulates it. Regulation of $1,25-(OH)_2D_3$ synthesis is related also to serum Ca concentration through the action of parathyroid hormone (PTH), which catalyzes the conversion of $25-(OH)-D_3$ to $1,25-(OH)_2D_3$. Thus, $1,25-(OH)_2D_3$ appears to be the ideal compound for treatment of diseases related to parathyroid insufficiency. Various analogs of this very expensively isolated and purified compound are receiving attention for use in clinical problems of Ca homeostasis in humans and animals.

Nephrectomized animals cannot make significant amounts of $1,25-(OH)_2D_3$ and, therefore, do not respond to physiological doses of vitamin D in terms of increased Ca absorption or serum Ca concentration.

Although it is generally agreed that the kidney is the principal site of production of $1,25-(OH)_2D_3$, a number of other cell types, including bone, placenta, intestine, and yolk sac, have been reported to make the conversion of $25-(OH)_2D_3$ to $1,25-(OH)_2D_3$ or $24,25-(OH)_2D_3$.

Administration of $1,25-(OH)_2D_3$ to nephrectomized animals induces the same response as in normal animals, suggesting that this compound does not have to be metabolized further for metabolic activity. The specific biochemical events that occur after localization of $1,25-(OH)_2D_3$ in the target tissue to produce the physiological response are currently under study. The overall homeostatic mechanism of vitamin D movement in the body and functions of its metabolites in relation to hormonal control are diagrammed schematically in Fig. 14.4 (Holick, 1990).

The concept of vitamin D as a precursor to a steroid hormone, $1,25-(OH)_2D_3$, has been extended to the increased knowledge of receptors for $1,25-(OH)_2D_3$ in target organs, analogous to receptors for insulin, estrogen, and other hormones in target tissues (Haussler, 1986). Bone cells apparently respond to $1,25-(OH)_2D_3$ by modulating an array of proteins, including collagen and alkaline phosphatase, required for bone mineralization and remodeling. There is also new evidence that

$1,25-(OH)_2D_3$ may be involved in immunology by modulating production of antibody-producing B cells or helper T cells.

In addition to dietary vitamin D status as a regulator of $25-(OH)D_3$ metabolism, dietary Ca and P, estrogen, parathyroid hormone, calcitonin, and pituitary hormones, including growth hormone, have been shown to play a role. Obviously, the complex regulation of vitamin D metabolism is still incompletely understood and complete knowledge of all of the interacting forces will require further study, the details of which are beyond the scope of this discussion.

Vitamin D is stored mainly in the liver, but also in kidney and lung and perhaps in other tissues. Placental and mammary transfer are limited, when compared with vitamin A, but sufficiently high levels are present in the newborn and milk of most species to prevent early rickets. Only limited data are available on quantitative transfer of vitamin D across placental and mammary tissue. DeLuca (1980) summarized evidence suggesting that Ca can be mobilized from bone and intestine of some animal species during pregnancy and lactation by a vitamin D-independent mechanism and that the young vitamin D-deficient rat pup is not responsive to the administration of $1,25-(OH)_2D_3$. Therefore, it appears that factors other than vitamin D may be involved in Ca mobilization, during pregnancy and lactation.

Although the vast majority of research has been directed at determining the effect of vitamin D on Ca and P metabolism, available evidence also indicates that vitamin D promotes absorption from the GI tract of Be, Co, Fe, Mg, Sr, Zn, and, perhaps, still other elements. It is not known whether the effect is caused by vitamin D-dependent protein carriers, or by other mechanisms.

Deficiency Signs

The main effect of a deficiency of vitamin D is abnormal skeletal development. Normal calcification cannot occur in the absence of adequate Ca and P. Therefore, either a deficiency of vitamin D, which results in impaired utilization of Ca, or a deficiency of Ca or of P, will produce similar abnormalities in the skeleton (the roles of Ca and P are discussed in Chapter 11).

The term applied to vitamin D deficiency in young, growing animals is rickets; the comparable condition in adults is osteomalacia. In each in-

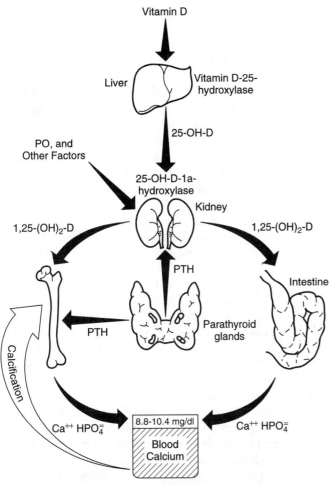

Figure 14.4 *Schematic representation of the hormonal control loop for vitamin D metabolism and function. A reduction in the serum calcium below ~8.8 mg/mL prompts a proportional increase in the secretion of parathyroid hormone, which enhances the mobilization of calcium stores from bone. Parathyroid hormone also promotes the synthesis of 1,25(OH)₂D in the kidney, which, in turn, stimulates the mobilization of calcium from the bone and its absorption from the intestine. Reproduced with permission (Holick, 1990).*

stance inadequate calcification of the organic matrix occurs, which results in lameness, bowed and crooked legs (Fig. 14.5), spontaneous fractures of long bones and ribs, and beading of the ribs. In growing animals the bone ash is reduced (Ca:P tends to remain consstant), and weight gain may be depressed. In adults, negative mineral balance occurs, and bone ash concentration declines.

Species differences exist in the response to deprivation of vitamin D by dietary means or protection from ultraviolet light (sunlight). Calves and pigs can grow normally on a level of vitamin D in the diet that quickly produces rickets in chicks. Pheasants and turkeys have a higher vitamin D re-

quirement than chicks. The difference in species requirements may be related to growth rate, as animals and birds with very rapid growth tend to be more susceptible to rickets. Protein source affects the vitamin D requirement of pigs. Soybean protein, high in phytate, is associated with a higher vitamin D requirement than milk protein.

Serum Ca concentration tends to be reduced in vitamin D deficiency, although hormonal mechanisms (parathyroid hormone and calcitonin) are quite efficient in maintaining a relatively constant range. Although parathyroid hormone is under the influence of vitamin D, calcitonin is not. Calcitonin produces hypocalcemia and hypophosphatemia by

Figure 14.5 *Lameness and sore joints in pig fed a vitamin D-deficiency diet and kept indoors without access to sunlight.*

inhibiting bone resorption. During growth, lactation, and pregnancy, calcitonin and $1,25\text{-}(OH)_2D_3$ concentrations in plasma are high. It has been suggested that by opposing the resorptive action of $1,25\text{-}(OH)_2D_3$ on bone, calcitonin preserves the integrity of the skeleton and directs the action of $1,25\text{-}(OH)_2D_3$ to the GI tract to meet the need for Ca. Serum alkaline phosphatase may be increased in vitamin D-deficiency rickets. This enzyme is present in bone and is associated with bone resorption. There appears to be a mechanism of preferential translocation by a binding protein of vitamin D_3 formed in the skin of animals by photosynthesis whereby the synthesis, storage, and slow, steady release of vitamin D_3 from skin to the circulation is accomplished. Thus, the skin serves as the site of synthesis, as a reservoir for storage of pro-D_3 and is the organ where the slow dermal conversion of pro-D_3 to D_3 occurs.

Vitamin D deficiency can be prevented by only a few minutes of exposure to direct sunlight, although skin pigmentation affects the amounts of sunlight required to prevent rickets; white-skinned animals require less sunlight than dark-skinned ones. Loomis (1967) suggested that skin color in humans is an adaptation that provides for protection against rickets in inhabitants of northern regions as well as protection against excess vitamin D synthesis in skin of inhabitants of equatorial regions.

Toxicity

Excess vitamin D causes abnormal deposition of Ca in soft tissues. This Ca is resorbed from bone, resulting in brittle bones subject to deformation and fractures. Ca deposits are frequent in kidneys, aorta, and lungs. Such lesions can be produced in rats with dosages of 300,000 to 600,000 IU, in chicks with 4,000,000 IU/kg of diet, and in pigs at 250,000 IU per animal per day for 30 days. Human infants show toxic signs with levels as low as 3000 to 4000 IU/day (only 10 times the requirement). Hypervitaminosis D can lead to death, usually from uremic poisoning resulting from severe calcification of kidney tubules. Excess vitamin D during pregnancy apparently does not cause severe abnormalities in the fetus, but it is not harmless, because premature closing and shortening of the skull bone and abnormal teeth have been observed in newborn rabbits whose dams were given excess vitamin D.

During the past few years several plants indigenous to widely different geographical areas have been found to produce signs resembling those of vitamin D toxicity in animals. Wasserman and Nobel (1980) summarized the isolation and identification of the toxic substances in these carcinogenic plants and their physiological effects on animals. The problem was first recognized more than 20 years ago in Argentina in cattle and eventually was found to be due to consumption of a plant, *Solarium malacoxylon,* which contains glycosides of $1,25\text{-}(OH)_2D_3$. Subsequently, cattle and horses grazing a shrub in Florida, *Cestrum diurnum,* showed the same clinical signs, and the plant was found to contain the same or similar metabolite of $1,25\text{-}(OH)_2D_3$. Cattle and sheep in Germany and Austria are affected similarly by consuming a plant called *Trisetum flavescens,* and cattle in Jamaica, Hawaii, India, and New Guinea are similarly afflicted after consuming plants of unknown identity. The signs in all cases include weight loss, stiffness of forelimbs, arching of back, emaciation, hypercalcemia and hyperphosphatemia, and death. The ingestion of a source of $1,25\text{-}(OH)_2D_3$ bypasses the normal regulatory step (the kidney), thereby stimulating excess synthesis of intestinal Ca-binding protein and causing excessive amounts of Ca and P to be retained. Haschek et al. (1978) described the pathogenesis of vitamin D toxicity and concluded that the osteocytic death and osteonecrosis is a direct toxic action of vitamin D, since hypoparathyroidism and hypercalcitoninism, which occur secondarily to hypercalcemia, could not account for the appearance of bone cell death, which occurred within 1 day of vitamin D overdose.

❏ VITAMIN E

Vitamin E first was recognized as a fat-soluble factor necessary for reproduction in rats (Evans and Bishop, 1922). α-Tocopherol, the most active biological form of vitamin E, was isolated by Evans et al. (1936). Other forms of tocopherol have been designated with prefixes such as beta and gamma. The structure of α-tocopherol is shown in Fig. 14.6 (American Institute of Nutrition Committee on Nomenclature, 1979).

Other compounds with chemical structures similar to tocopherols are found in animal tissues but have limited biological activity. The chemistry and biochemistry of vitamin E have been detailed (Machlin, 1980). Vitamin E is very unstable; its oxidation is increased by the presence of minerals and of polyunsaturated fatty acids (PUFA) and decreased by esterification to form tocopheryl acetate. The D isomer is more active than the L-form. Most commercially available vitamin E is available as DL-α-tocopheryl acetate.

Knowledge of vitamin E nutrition and metabolism has been summarized in several other reviews (Bieri, 1984; Boguth, 1969; Horwitt et al., 1968; McCay, 1985).

Functions

In 1969 the activity of α-tocopherol was classified under two main headings, effects attributable to the hydroxy group of the molecule (antioxidant effect) and effects brought about by metabolites of α-tocopherol. Current concepts of the biochemical functions of vitamin E include its role as a biological free radical scavenger (McCay, 1985; McCay and King, 1980), in nucleic acid and protein metabolism (Catignani, 1980), and in mitochondrial metabolism (Corwin, 1980).

The antioxidant effect of vitamin E and its action as a free radical scavenger can explain most of the effects of vitamin E deficiency in animals. It is important in maintenance of the integrity of cellular membranes. The recognition that the mineral, Se, can protect against most vitamin E-deficiency signs has led to an enormous body of literature aimed at elucidating the relationship between the two nutrients. These interrelationships are described in greater detail in Chapter 12, and have been reviewed by Mahan (2005).

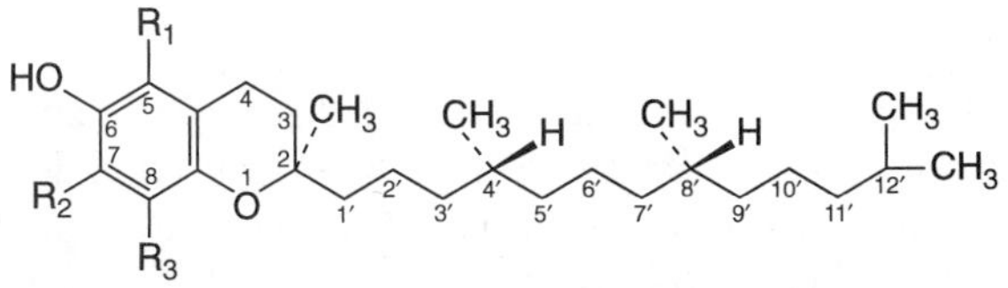

There is growing evidence that vitamin E influences the synthesis of specific proteins. However, the exact role of vitamin E in protein synthesis still is unknown. The activity of several enzymes is affected by vitamin E deficiency. In many cases, activity is increased, suggesting that the vitamin may serve as a repressor for the synthesis of certain enzymes. Some 30 enzymes have been reported to increase in activity during vitamin E deficiency, whereas a decrease in activity has been observed for at least 10. It has been suggested the vitamin E may function in regulation of gene transcription, but such a possibility requires further testing.

Vitamin E has been shown to affect several mitochondrial and microsomal functions, some of them associated with the ability to oxidize. Corwin (1980) suggested that part of the vitamin E molecule can serve a reduction–oxidation function to protect S and Se proteins and, perhaps, to transfer single electrons at the mitochondrial membrane and at other cell organelles. By its membrane localization, vitamin E helps to maintain compartmentalization and permeability of membranes. Vitamin E modulates synthesis of prostaglandin synthesis; in some tissues synthesis is increased, and in others it is decreased by vitamin E deficiency. Low vitamin E decreases the production of prostaglandins by microsomes from muscle, testes, and spleen while it increases their production by blood platelets. The mechanism by which vitamin E shows these effects is unknown.

Massive vitamin E supplementation of well-balanced diets increases antibody production [especially immunoglobulin G (IgG)] against a variety of antigens in several species of animals. It also increases protection of chickens against *Escherichia coli* infection and mice against some types of pneumonia infection (Tengerdy, 1980). The mode of action of these observed effects of vitamin E on the immune response in the laboratory and the practical importance of these effects under field conditions are not completely understood. Detailed biochemical studies of vitamin E metabolism presently underway in many laboratories can be

Figure 14.6 *Structure of vitamin E [International Union of Nutritional Sciences (1990)].*

expected eventually to clarify the uncertainties still remaining in the total role of vitamin E in nutrition.

Deficiency Signs

The manifestations of vitamin E deficiency are varied, and some are species specific, but they can be divided into three broad categories as follows: reproductive failure, derangement of cell permeability, and muscular lesions (myopathies).

Reproductive failure associated with vitamin E deficiency can be related to embryonic degeneration, as in the rat and bird, or to sterility from testicular atrophy, as in the chicken, dog, guinea pig, hamster, and rat, or to ovarian failure in female rats.

Derangement of cell permeability may affect liver, brain, kidney, or blood capillaries. Liver necrosis occurs in vitamin E deficiency of rats and pigs (Fig. 14.7), but Se can prevent it, even in the absence of supplemental vitamin E. Red blood cell hemolysis occurs in vitamin E-deficient rats, chicks, and human infants; kidney degeneration in deficient mink, monkeys, and rats; and steatitis (inflammation of adipose tissue) in chicks, mink, and pigs. Kidney degeneration and steatitis were prevented by Se in the species studied.

Two common manifestations of vitamin E deficiency in chicks are encephalomalacia (abnormal Purkinje cells in the cerebellum) and exudative diathesis (ED). Se prevents ED but not encephalomalacia. In encephalomalacia, incoordination and ataxia result from hemorrhages and edema in the cerebellum. Synthetic antioxidants, such as ethoxyquin, prevent encephalomalacia.

Polyunsaturated FA enhance the incidence of encephalomalacia in chickens fed diets marginal in vitamin E. In ED, severe edema results from increased capillary permeability. A bluish-green subcutaneous exudate resulting from blood protein loss is evident on the breast of affected birds preceding death.

Nutritional muscular dystrophy (NMD, stiff-lamb disease, white muscle disease, muscular dystrophy) is common in growing lambs, calves, pigs, chicks, turkeys, and rabbits fed vitamin E-deficient diets. The lesions involve degeneration of skeletal muscle fiber (Zenker's degeneration), which is seen grossly as whitish streaks on the surface of muscles. Degeneration of heart muscle (mulberry heart disease) and of liver (liver necrosis) are common in vitamin E-deficient pigs. Sudden death is a common occurrence in calves and in affected pigs, which show the heart and liver lesions on necropsy. In NMD, peroxidation of muscle tissues and activities of lysosomal enzymes are increased. One or more transaminases [including alanine aminotransferase (ALT), aspartate aminotransferase (AST), and γ-glutamyltransferase (GGT)] in blood serum are increased drastically in animals with NMD. This is an index of damage to muscle or heart tissue resulting in release of these enzymes into the blood, although they are not specific for NMD. A common sign of vitamin E deficiency in the turkey is myopathy of the gizzard as well as of heart and skeletal muscles. Se can prevent these lesions as well as those of liver, heart, and skeletal muscle in most types of NMD in mammals. NMD of birds apparently is only partly responsive to Se. The amino acid cysteine has a primary role in prevention of NMD in chicks as NMD is increased by factors that deplete body cysteine and decreased by factors that spare it. It has been suggested that cysteine and vitamin E act two different ways to prevent NMD in chicks. The complete story on vitamin E, cysteine, Se, and other factors as agents in the prevention of NMD in birds and animals is being unraveled gradually.

Metabolism

Absorption, Transport, and Storage. The absorption of vitamin E has been reviewed by Gallo-Torres (1980a, b). The site of greatest absorption is in the jejunum in mammals. Tocopherols are absorbed mainly by micelle formation in the presence of bile salts. Both the D and L iso-

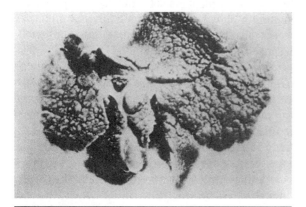

Figure 14.7 *Necrotic liver from pig fed a vitamin E–Se-deficient diet.*

mers are absorbed, but the D isomer has a relative potency 1.2 times that of the racemic mixture. Esterified tocopherols are absorbed less efficiently than the unesterified vitamin; hydrolysis of esters is nearly complete in the GI tract lumen. Absorption of tocopherol is enhanced by solubilization in medium-chain triglycerides compared with long-chain triglycerides, and unsaturated fatty acids reduce tocopherol absorbability. Degree of absorption of ingested tocopherols has been estimated at 10 to 36% in humans and probably approaches this range of values in animals. Both bile and pancreatic secretions are needed for optimum tocopherol absorption. Tocopherols are absorbed into the lymphatics and are transported as a part of lipoproteins.

Plasma concentration of tocopherols is affected by rate of removal from plasma and retention in individual tissues; total plasma lipid and β-lipoprotein contents are positively correlated with plasma tocopherol level. Vitamin E also is transported in the red blood cell, localized in the cell membrane. Rapid exchange of tocopherol occurs between plasma and red blood cells. The dynamics of vitamin E transport between tissues have been studied thoroughly. Storage occurs in liver, skeletal muscle, heart, lung, kidney, spleen, and pancreas in similar amounts and in pituitary, testes, and adrenals in even higher concentrations. The wide tissue distribution of tocopherol would be expected in view of its role in preventing peroxidation and in maintaining cell membrane integrity. Because tocopherols are fat soluble, concentrations are expressed best as units per gram of fat rather than as units per gram of tissue.

Interrelationships between Vitamin E and Other Nutrients. Several authors (Amrich and Arthur, 1980; Draper, 1980; Machlin and Gabriel, 1980; McCay, 1985) have summarized the knowledge of interrelationships of vitamin E with other nutrients. The following nutrients have been shown to affect the vitamin requirement of animals: Se, polyunsaturated fatty acids (PUFA), S-amino acids, Fe, vitamin A, vitamin C, choline, and Zn.

Se—Several vitamin E-deficiency diseases, including muscular dystrophy, liver necrosis, myocardial lesions, and chick exudative diathesis (ED) can be prevented by Se. The metabolic basis for the protective action of Se appears to be related to the prevention by Se of lipid peroxidation in vitamin E-deficient animals. Se is a component of the enzyme glutathione peroxidase, which catalyzes the destruction of peroxides, which are a product of lipid peroxidation in the absence of the antioxidant action of vitamin E (see Chapter 12 for details).

PUFA—Vitamin E requirement is increased as the concentration of PUFA in body tissues increases. This relationship is based on the antioxidant properties of vitamin E and the greater susceptibility of PUFA to peroxidation because of the higher proportion of double bonds in the unsaturated carbon chain of the FA. Attempts to define vitamin E requirements in terms of an optimum ratio to PUFA have not yielded a satisfactory value, presumably because of differences in tissue stores of vitamin E and PUFA and, perhaps, because of a lack of linearity in the optimum ratio with changes in concentration of vitamin E. Young dogs fed vitamin E-deficient diets with safflower oil (high in PUFA) have been reported to develop browning of the fat around the intestine, hemolysis of RBCs, and depressed plasma tocopherol in direct relation to dietary fat level; however, other work has failed to demonstrate that high levels of vitamin E can prevent this pigment formation. Testes, heart, adipose tissue, adrenal, brain, skeletal muscle, and bone marrow accumulate large amounts of fluorescent pigment in vitamin E-deficient animals fed PUFA. There is strong evidence that vitamin E is a preventative. For example, Reddy et al. (1973) found that rats fed 10% lard–1% cod liver oil diets without vitamin E for 4 months accumulate twice the fluorescent pigments in adipose tissue as those fed the same diet but supplemented with vitamin E. Furthermore, adipose tissue of rats fed cod liver oil, which is higher in PUFA than corn oil, had three times the fluorescence of adipose tissue from those fed corn oil. Fluorescent pigments in adipose tissue extracts may be the products of in vivo lipid peroxidation. Dietary supplementation of vitamin E has been shown (Tsai et al., 1978) to improve the oxidative stability of pork from pigs fed the fortified diet.

The apparent increase in consumption of fats high in PUFA in the United States increases the possibility that vitamin E may tend to be a limiting nutrient in human nutrition. Harman (1969) suggested that peroxidation (free radical reactions) of adipose tissues play a significant role in aging. He hypothesized that human productive lifespan could be extended by dietary manipulation to reduce peroxidation in body tissues.

Lifespan of mice is shortened as much as 10% by increasing the amount and/or degree of unsaturation of dietary fat; several antioxidants reduce the incidence of congenital malformation (mostly skeletal deformities) in vitamin E-deficient rats.

S-Amino acids—S-amino acids prevent liver necrosis in animals fed vitamin E-deficient diets. The protective effect is due apparently to the reducing (antioxidant) properties of the sulfhydryl group of S-amino acids.

Fe—Neonatal animals and humans given supplemented Fe in the presence of a suboptimal intake of vitamin E or of a low-plasma tocopherol level induced by a high-PUFA diet may develop vitamin E-deficiency anemia. Hemolytic anemia, characterized by low hemoglobin, high reticulocyte count, and changes in red blood cell morphology has been reported in premature infants with low plasma vitamin E. Newborn pigs marginally deficient in vitamin E may show signs of Fe toxicity and death when given a normal intramuscular dose of Fe for prevention of Fe-deficiency anemia. The adverse effects of Fe on vitamin E status may be due to increased peroxidation of PUFA in red blood cells.

Vitamin A—Vitamin E reduces depletion of liver vitamin A and appears to spare vitamin A by its antioxidant properties, although this suggested mechanism is not accepted universally; high levels of vitamin A increase the vitamin E requirement. Signs of vitamin A toxicity may be allevaited by increased vitamin E ingestion. The mechanisms involved in vitamin A–E interactions have been reviewed recently (Bieri, 1984). The relationships may be of practical importance only outside the range of normal vitamin A intakes.

Vitamin C—Some inhibition of lipid peroxidation in tissues of vitamin E-deficient animals has been shown by vitamin C administration. It reduces the incidence of ED in chickens. However, a sparing effect of vitamin C on vitamin E is not a consistent finding, and its apparent protective role may be one of sparing Se by improving its absorption from the GI tract which, in turn, protects tissues from lipid peroxidation. The nutritional roles of vitamins E and C and their possible interactions are not completely understood.

Choline—The liver lesions associated with choline deficiency have been reported to be prevented by supplemental vitamin E. It has been suggested that lipid peroxidation may be a lesion of choline deficiency, which would then be expected to respond favorably to vitamin E.

Vitamin B_{12}—Recent evidence suggests that vitamin E is required for the conversion of vitamin B_{12} to its coenzyme, S′-deoxyadenosyl-cobalamin, which is necessary for the metabolism of methylmalonate. This could explain the inverse relationship between serum tocopherol and vitamin B_{12} concentration reported in human patients.

Zn—Deficiency of either Zn or vitamin E is associated with testes degeneration and increased arachidonic acid in testes lipids (Machlin and Gabriel, 1980). Data obtained in rats fed diets of differing Zn and vitamin E content showed that peroxidation precedes hemolysis in vitamin E-deficient red blood cells and that Zn stabilizes the red cell membrane against cellular changes following peroxidation. Other data from chicks fed a low-Zn diet (Bettger et al., 1980) support the hypothesis that Zn plays a role analogous to that of vitamin E in stabilizing cell membrane structure and thereby reducing peroxidative damage. Further research is needed to determine the biochemical interactions between these two nutrients and to quantify the effect of dietary level of one on the dietary requirements of the other.

Tengerdy (1980) has suggested the possibility that vitamin E, through its stimulatory effect on the immune system, may have a beneficial effect on aging, which is known to be accompanied by a decreased functioning of the immune system. Such a relationship deserves further study as a possible mechanism in addition to that of reduced peroxidation of body tissues, to explain a potential benefit of vitamin E in longevity.

Placental and Mammary Transfer of Vitamin E

The placental transfer of vitamin E is inefficient, whereas fatty acids cross the placenta more efficiently. Therefore, a reduction in vitamin E:PUFA ratio in fetal tissues may predispose the newborn to vitamin E deficiency, but no information is available on the question. Some placental transfer of vitamin E does occur, but malformations have been observed in young rats fed low vitamin E even though vitamin E was found in liver, blood, and carcasses of the young.

The tocopherol content of the milk of both ruminant and nonruminant animals is affected by concentration in the diet. Thus, susceptibility of newly weaned animals to vitamin E deficiency is dependent on the body stores they have accumulated in response to the diet of the dam during gestation and lactation.

Toxicity

Few reports hav been made of vitamin E toxicity in animals or humans. Scattered descriptions of hemorrhagic syndromes, nervous disorders, edema, changes in endocrine glands, and antagonism of vitamin K have appeared in experimental animal research, but detailed pathology of hypervitaminosis E is not available. Human adults have tolerated oral doses of 1 g/day for months without undesired effects. Apparently the range of safe level of intake is wider than for other fat-soluble vitamins, but the wholesale consumption of vitamin E as well as of all nutrients should be avoided.

❑ VITAMIN K

Structure

A hemorrhagic syndrome was described in 1929 in chicks fed a diet low in sterols, and it was shown (Dam, 1935) that the missing factor was a fat-soluble vitamin which Dam named vitamin K. Vitamin K really is a group of compounds. The structure of the two most important natural sources of vitamin K, phylloquinone (vitamin K_1) and menaquinone (vitamin K_2), are shown in Fig. 14.8

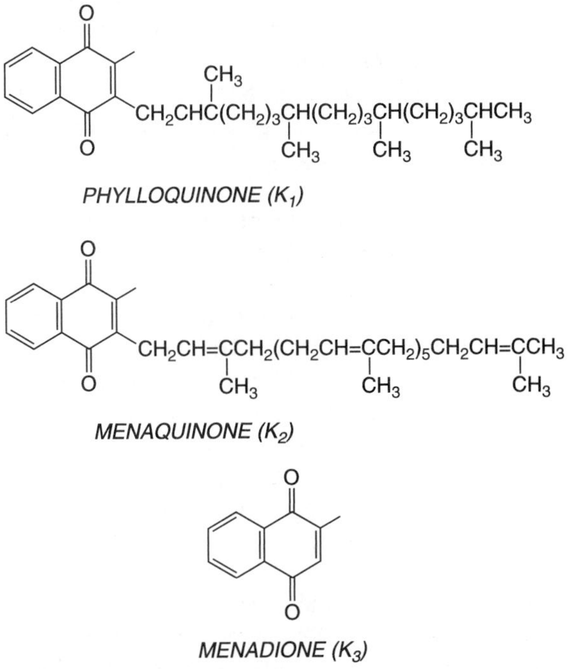

PHYLLOQUINONE (K_1)

MENAQUINONE (K_2)

MENADIONE (K_3)

Figure 14.8 *Structure of vitamin K.*

(American Institute of Nutrition Committee on Nomenclature, 1979). Vitamin K_1 is common in green vegetables and vitamin K_2 is a product of bacterial flora in the GI tract of animals and humans. A synthetic, menadione (vitamin K_3), is used widely commercially; its structure also is shown.

Menaquinone also can occur as menaquinone-6, -7, -8, or -9; that is, the side chain may contain more than the four isoprene units shown. Apparently, the liver converts these forms of vitamin K to vitamin K_2, suggesting that this is the metabolically active form.

Functions

Vitamin K is required for normal blood clotting. Specifically, it is required for synthesis of prothrombin in the liver. It is not a component of prothrombin, but acts on enzyme systems involved in prothrombin synthesis and in the synthesis of other factors involved in the total blood clotting mechanism. It has been suggested that vitamin K acts by influencing messenger RNA formation needed for prothrombin synthesis. Such a role could help to explain the rapid synthesis of prothrombin that occurs in vitamin K-deficient animals treated with vitamin K (2–6 h). Vitamin K is involved in at least four steps in clot formation: a plasma thromboplastin component (Factor IX), a tissue thromboplastin component (Factor VII), "Stuart" factor (Factor X), and prothrombin (Factor II).

The clotting mechanism (Table 14.3) is stopped by oxalate and citrate, which precipitate Ca^{2+}, and by heparin, which apparently blocks formation of the Stuart factor.

Current knowledge of vitamin K metabolism and the details of its function in blood coagulation has been summarized by Olson (1984). He described the biosynthesis of prothrombin; the distribution, requirements, purification, and mechanism of action of vitamin K-dependent γ-glutamylcarboxylase; the absorption and metabolism of vitamin K; and the action of vitamin K antagonists. The unique role of the amino acid, γ-carboxyglutamic acid as a component of prothrombin in vitamin K metabolism was described. New knowledge on the importance of the utilization and reutilization of vitamin K for the carboxylation of peptide-bound glutamate residues in the vitamin K-dependent proteins was emphasized. Eight of these proteins are involved in blood coagulation. Suffice it to say

TABLE 14.3   Vitamin K involvement in blood clotting.

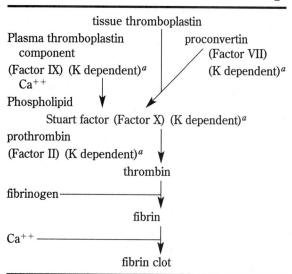

[a]Synthesis of each is inhibited by vitamin K antagonists.

that, although the end result of vitamin K deficiency in animals and humans is a failure in blood coagulation, the biochemical complexities are being clarified by the tools of molecular biology.

Deficiency Signs

A deficiency of vitamin K results in prolonged blood-clotting time, generalized hemorrhages, and death in severe cases. Often, subcutaneous hemorrhages appear over the body surface, giving a blotchy, bluish, mottled appearance to the skin. The clotting time of the blood is a good index of vitamin K status. A normal clotting time of a few seconds may be extended to several minutes in vitamin K-deficient animals.

The microflora of the intestinal tract normally produce adequate vitamin K to meet the metabolic needs of the host. The vitamin is obtained either by absorption from the lower GI tract after synthesis or by coprophagy (feces eating), which serves as an important means of supplying nutrients to many animal species. Prevention of coprophagy in rats produces vitamin K deficiency. Nutritionists have not generally concerned themselves greatly with supplying supplemental vitamin K in the diet, because the GI tract microflora play such an important role in synthesizing vitamin K. However, newer technology has created problems with deficiency of vitamin K under prac-

tical conditions. Chicks fed sulfaquinoxaline to control coccidiosis develop vitamin K deficiency. This sulfa drug is an antagonist of vitamin K. Pigs fed in wire floor cages, in which access to feces is reduced, develop increased prothrombin clotting times and growing–finishing pigs fed 60 to 65% cane sugar developed deficiency symptoms. Field reports of a hemorrhagic syndrome preventable by supplemental vitamin K in growing–finishing swine suggest that vitamin K deficiency may be a serious practical problem under some conditions of husbandry. Several factors may be contributing to this increase. Mycotoxins (from moldy grain) may antagonize vitamin K; increased confinement feeding of swine has been associated with less use of pasture and forages that are high in vitamin K; and the use of slotted floors lessens the opportunity for coprophagy. Clearly, much remains to be discovered concerning factors that affect vitamin K requirement and that produce deficiencies.

Metabolism

Absorption and Utilization.   The naturally occurring and synthetic vitamin Ks are absorbed most efficiently in the presence of dietary fat and bile salts, and menadione requires bile salts for absorption. The efficiency of absorption of vitamin K depends on the form in which the vitamin is combined as well as on the specific isomer involved. The mechanisms by which vitamin K is transported in blood are not defined clearly, probably because of the variable solubility in water vs. fats of the several compounds having vitamin K activity.

The biological activity of several compounds showing vitamin K potency is listed in Table 14.4. The activity of menadione and its derivatives depends on the realtive stabilities of the preparations used and on whether or not sulfaquinoxaline is present in the test diet. Menadione and its derivatives are less effective than phylloquinone in counteracting the antagonistic effects of sulfaquinoxaline. Menadione is very reactive and may be toxic; for this reason, and because of its lower biological activity, it is used far less commonly in feeds than are its derivatives.

Vitamin K compounds are very unstable under alkaline conditions. Menadione dimethylpryimidinol bisulfate is more stable as well as higher in biological activity than menadione.

TABLE 14.4 Relative biological activity of several forms of vitamin K.

	ACTIVITY RELATIVE TO THAT OF NATURAL VITAMIN K_1 (PHYLLOQUINONE-4), %
Phylloquinone-1	5
Phylloquinone-2	10
Phylloquinone-3	30
Phylloquinone-4	100
Phylloquinone-5	80
Phylloquinone-6	50
Menaquinone-2	15
Menaquinone-3	40
Menaquinone-4	100
Menaquinone-5	120
Menaquinone-6	100
Menaquinone-7	70
Menadione	40–150[a]
Menadione sodium bisulfate	50–150[a]
Menadione dimethylpyrimidinol bisulfate	100–160[a]

[a]Activity depends on relative stabilities of preparations used and on presence or absence of sulfaquinoxaline in the test diet.
Source: P. Griminger. 1966. *Vitamins and Hormones* **24:** 605.

Vitamin K Antagonists and Inhibitors

A well-known natural antagonist of vitamin K is dicoumarol, often present in weather-damaged sweet clover hay. The presence of dicoumarol in hay or silage causes massive internal hemorrhages and death in calves. Another important antagonist of vitamin K is Warfarin, which is a competitive inhibitor of menaquinone (K_2). The inhibition of vitamin K function by Warfarin has resulted in the large-scale marketing of Warfarin as a rat poison. The effect of Warfarin in increasing clotting time can be reversed by simultaneously increasing the amount of K_2 given to the animal.

The structures of dicoumarol and of Warfarin are shown here.

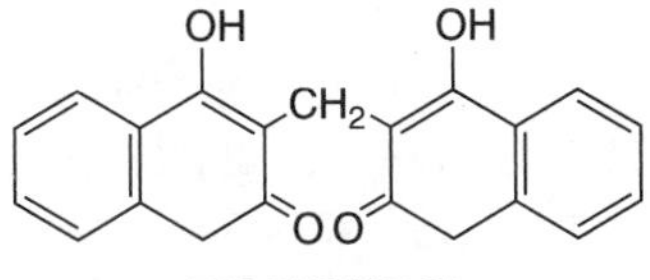

DICOUMAROL

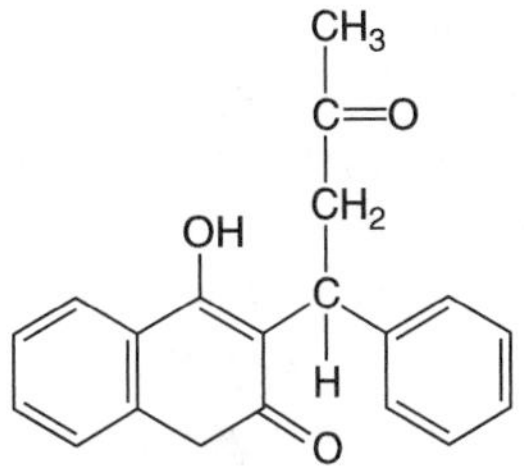

WARFARIN

Other vitamin K antagonists, including α-tocopherylquinone, sulfaquinoxaline (previously discussed), and some napthoquinone derivatives, as well as other substances, including butylated hydroxtoluene, salicylate, and high doses of vitamin A and E, may be of importance in affecting the dietary requirement for vitamin K.

Toxicity

Phylloquinone and menaquinone derivatives are nontoxic even in high dosage levels, but menadione is toxic to the skin and respiratory tract of several animal species; its bisulfate derivatives are not. Menadione given in prolonged high doses produces anemia, porphyrinuria, and other abnormalities in animals and chest pains and shortness of breath in humans.

❑ SUMMARY

- Vitamins are organic compounds required in minute quantities for normal body functions. The known vitamins can be divided on the basis of solubility properties into fat soluble and water soluble.

- The fat-soluble vitamins are A, D, E, and K. Storage in body tissues allows consumption of deficient diets over a longer period of time than for water-soluble vitamins before deficiency signs appear.

- Vitamin A can be provided as the vitamin or its precursor, carotene. Many carotenoid pigments exist in nature, but β-carotene is biologically the most active. Vitamin A activity is expressed in retinol equivalents (mcg). One retinol equivalent $= 0.167 \times$ β-carotene or $0.084 \times$ other provitamin A.

- Vitamin A is required for normal night vision, normal epithelial cells, and normal bone growth and remodeling, and it may play a primary role in the synthesis of glycoproteins. Vitamin A deficiency results in xerophthalmia (irritation of the cornea and conjunctiva of the eye), keratinization of respiratory epithelium, reproductive failure in males, embryonic death, and poor growth in surviving young.

- Vitamin D_2 (irradiated ergosterol or calciferol) and vitamin D_3 (irradiated 7-dehydrocholesterol) are the two major sterols with vitamin D activity. Most animals use either vitamin D_2 or vitamin D_3 efficiently, but birds require vitamin D_3. Ultraviolet light converts each provitamin to its biologically active form.

- Vitamin D functions to elevate plasma Ca and P levels for normal bone mineralization and prevention of tetany produced by hypocalcemia. The active metabolite of vitamin D (1,25-dihydroxycholecalciferol) is produced by the kidney under conditions of hypocalcemia and stimulates intestinal absorption, bone resorption, and possibly kidney tubular reabsorption of Ca and P to maintain plasma Ca within normal limits.

- Deficiency of vitamin D results in inadequate bone calcification (rickets in growing animals and osteoporosis in adults); thus, dietary Ca deficiency and vitamin D deficiency produce the same signs. Vitamin D deficiency can be prevented by only a few minutes of daily exposure to direct sunlight.

- Vitamin E was first recognized as a fat-soluble factor needed for reproduction in rats. α-tocopherol is the most biologically active form of vitamin E. Functions of vitamin E include its role as a biological free radical scavenger and as an antioxidant. There is evidence that vitamin E influences the synthesis of specific proteins and of prostaglandins.

- Vitamin E deficiency results in reproductive failure, derangement of cell permeability, and muscular lesions. Deficiency causes encephalomalacia and exudative diathesis in chicks, muscular dystrophy in several animal species, and liver necrosis and cardia cmuscle degeneration in pigs and calves. Muscular dystrophy and liver necrosis caused by vitamin E deficiency can be prevented by dietary Se.

- Other nutrients shown to affect the vitamin E requirement include polyunsaturated fatty acids, S-amino acids, Fe, vitamin A, vitamin C, choline, and Zn.

- Vitamin K is really a group of compounds; the two most common forms are phylloquinone (vitamin K_1) and menaquinone (vitamin K_2). Vitamin K_1 is common in green vegetables; vitamin K_2 is a product of bacterial flora in the gastrointestinal tract of animals and humans.

- Vitamin K is required for normal blood clotting (for synthesis of prothrombin by the liver). Vitamin K deficiency results in prolonged blood-clotting time and generalized hemorrhages.

- The microflora of the gastrointestinal tract normally produce adequate vitamin K to meet metabolic needs of the host animal, but newer technology, such as feed additives that alter intestinal microflora and the use of wire floor cages to prevent access to feces, may create the need for greater attention to meeting the vitamin K needs of animals.

❏ REFERENCES

Ahluwalia, G. S., L. Kaid, and B. S. Ahluwalia. 1980. *J. Nutr.* **110:** 1185.

American Institute of Nutrition Committee on Nomenclature. 1990. *J. Nutr.* **120:** 12.

Amrich, L., and V. A. Arthur. 1980. *Ann. NY Acad. Sci.* **355:** 109.

Bettger, W. J., et al. 1980. *J. Nutr.* **110:** 50.

Bieri, J. G. 1984. In *Present Knowledge in Nutrition,* 5th ed., Chapter 16, pp. 226–240, R. E. Olson (ed.). Nutrition Foundation, Washington, DC.

Bieri, J. G., and M. C. McKenna. 1981. *Am. J. Clin. Nutr.* **34:** 289.

Bloomhoff, R., et al. 1992. *Annu. Rev. Nutr.* **12:** 37.

Boguth, W. 1969. *Vitamins and Hormones.* **27:** 1.

Byers, T., and G. Perry. 1992. *Annu. Rev. Nutr.* **12:** 139.

Catignani, G. L. 1980. In *Vitamin E. A Comprehensive Treatise,* Section 2, p. 318, L. J. Machlin (ed.). Marcel Dekker, New York.

Corwin, L. M. 1980. In *Vitamin E. A Comprehensive Treatise,* Section 3, p. 32. L. J. Machlin (ed.). Marcel Dekker, New York.

Dam, H. 1935. *BIochem. J.* **29:** 1273.

Darroch, C. S. 1998. In *Swine Nutrition,* 2nd ed., A. J. Lewis and L. L. Southern (eds.). CRC Press, Boca Raton, FL.

DeLuca, H. F. 1978. *Ann. NY Acad. Sci.* **307:** 356.

DeLuca, H. F. 1979. *Nutr. Rev.* **37**(6): 161.

DeLuca, H. F. 1980. *Nutr. Rev.* **38**(5): 169.

Dorgan, J. B., et al. 1998. *Cancer Causes Control* **9:** 89.

Draper, H. H. 1980. In *Vitamin E. A Comprehensive Treatise,* Part 5D, p. 272. L. J. Machlin (ed.). Marcel Dekker, New York.

Evans, H. M., and K. S. Bishop. 1922. *J. Metabolic Res.* **1:** 319, 335.

Evans, W. M., et al. 1936. *J. Biol. Chem.* **113:** 319.

Fraser, D. R. 1984. In *Present Knowledge of Nutrition,* 5th ed., Chapter 15, p. 209, R. E. Olson (ed.). Nutrition Foundation, Washington, DC.

Funk, C. 1912. *J. State Med. London,* **20:** 341.

Gallina, A. M., et al. 1972. *J. Nutr.* **100:** 129.

Gallo-Torres, H. E. 1980a. In *Vitamin E. A Comprehensive Treatise,* Part 5A, p. 170, L. J. Machlin (ed.). Marcel Dekker, New York.

Gallo-Torres, H. E. 1980b. In *Vitamin E. A Comprehensive Treatise,* Part 5B, p. 193, L. J. Machlin (ed.). Marcel Dekker, New York.

Goodwin, T. W. 1986. *Annu. Rev. Nutr.* **6:** 273.

Griffin, M. D., et al. 2003. *Annu. Rev. Nutr.* **23:** 117.

Griminger, P. 1966. *Vitamins and Hormones* **24:** 605.

Harman, P. 1969. *J. Am. Geriatrics Soc.* **17:** 721.

Harris, R. S. 1967. In *The Vitamins* (2nd ed.), Vol. 1. Academic Press, New York.

Haschek, W. M., L. Krook, F. A. Kallfelz, and W. G. Pond. 1978. *Cornell Vet.* **68**(3): 324.

Haussler, M. R. 1986. *Annu. Rev. Nutr.* **6:** 527.

Henry, H. L., and A. W. Norman. 1984. *Annu. Rev. Nutr.* **4:** 493.

Hodges, R. E., R. B. Rucker, and R. H. Gardner. 1980. *Ann. NY Acad. Sci.* **355:** 58.

Hill, D. L., and C. L. Grubbs. 1992. *Annu. Rev. Nutr.* **12:** 161.

Holick, M. F. 1990. *J. Nutr.* **120**(Suppl.): 1464.

Horwitt, M. K., et al. 1968. *Vitamins and Hormones* **26:** 487.

International Union of Nutritional Sciences Committee. 1990. *J. Nutr.* **120:** 12.

Kayden, H. J. and L. Bjomson. 1972. *Ann. NY Acad. Sci.* **203:** 127.

Kohlmeier, R. H., and W. Burroughs. 1970. *Ann. NY Acad. Sci.* **30:** 1012.

Krinsky, N. I. 1993. *Annu. Rev. Nutr.* **13:** 561.

Krinsky, N. I., et al. 1958. *Arch. Biochem. Biophys.* **73:** 233.

Loomis. W. F. 1967. *Science* **157:** 501.,

Machlin, L. J. 1980. *Vitamin E. A Comprehensive Treatise.* Marcel Dekker, New York. 660 pp.

Machlin, L. J., and E. Gabriel. 1980. *Ann. NY Acad. Sci.* **355:** 98.

Mahan, D. C. 2005. In *Encyclopedia of Animal Science,* W. G. Pond and A. W. Bell (eds.). Marcel Dekker, New York.

McCay, P. B. 1985. *Annu. Rev. Nutr.* **5:** 323.

McCay, P. B. and M. M. King. 1980. In *Vitamin E. A Comprehensive Treatise,* Part 5E, p. 289, L. J. Machlin (ed.). Marcel Dekker, New York.

McCollum, E. V., and M. Davis. 1913. *J. Biol. Chem.* **15:** 167.

McCollum, E. V., et al. 1982. *J. Biol. Chem.* **53:** 293.

Molennaar, I., C. E. Hulstaert, and M. J. Hardonk. 1980. In *Vitamin E. A Comprehensive Treatise,* p. 372, L. J. Machlin (ed.). Marcel Dekker, New York.

Morii, H. and H. F. DeLuca. 1967. *Am. J. Physiol.* **213:** 358.

Navia, J. M. and S. S. Harris. 1980. *Ann. NY Acad. Sci.* **355:** 45.

Nitowski, H. M., et al. 1962. *Vitamins and Hormones* **20:** 559.

Norman, A. W. 1980. *Vitamin D. Molecular Biology and Clinical Nutrition.* Marcel Dekker, New York.

Olson, J. A. and O. Hayaishi. 1965. *Proc. Natl. Acad. Sci. U.S.* **54:** 1364.

Olson, R. E. 1964. *Science* **145:** 926.

Olson, R. E. 1984. *Annu. Rev. Nutr.* **4:** 281.

Omdahl, J. L., et al. 2002. *Annu. Rev. Nutr.* **22:** 139.

Oomen, H. A. P. C. 1976. In *Present Knowledge in Nutrition,* 4th ed., Chapter 9, R. E. Olson (ed.). Nutrition Foundation, Washington, DC.

Petkovich, M. 1992. *Annu. Rev. Nutr.* **12:** 443.

Reddy, K., B. Fletcher, and A. Tappel. 1973. *J. Nutr.* **103:** 908.

Roels, O. A. 1967. In *The Vitamins,* 2nd ed., Vol. 1. Academic Press, New York.

Smith, J. C., Jr. 1980. *Ann. NY Acad. Sci.* **355:** 62.

Steenbock, H., and A. Black. 1924. *J. Biol. Chem.* **61:** 405.

Tengerdy, R. P. 1980. In *Vitamin E. A Comprehensive Treatise,* Chapter 8, p. 429, L. J. Machlin (ed.). Marcel Dekker, New York.

Tielsch, J. M. and A. Sommer. 1984. *Annu. Rev. Nutr.* **4:** 183.

Tsai, T. C., G. H. Wellington, and W. G. Pond, 1978. *J. Food Sci.* **43:** 193.

Ullrey, D. E. 1972. *J. Anim. Sci.* **35:** 648.

Underwood, B. A. 1990. *J. Nutr.* **120**(Suppl.): 1459.

van den Berg, H. 1999. *Nutr. Rev.* **57:** 1.

Wasserman, R. H., and J. J. Faher. 1977. In *Calcium Binding Proteins and Calcium Function,* pp. 292–302, R. H. Wasserman et al. (eds.). Elsevier North-Holland, New York.

Wasserman, R. H., and T. A. Nobel. 1980. In *Vitamin D. Molecular Biology and Clinical Nutrition,* p. 455, A. W. Norman (ed.). Marcel Dekker, New York.

Wing, J. M. 1969. *J. Dairy Sci.* **52:** 479.

Wright, K. E. and R. C. Hall, Jr. 1979. *J. Nutr.* **109:** 1063.

15

Water-Soluble Vitamins

UNLIKE THE FAT-SOLUBLE vitamins, water-soluble vitamins (except B_{12}) are not stored in appreciable quantities in body tissues. They must be supplied in the diet on a day-to-day basis for those animals having a GI tract in which microbial synthesis is not a prominent feature. In ruminants, under normal conditions the water-soluble vitamin requirement is met almost entirely from microbial synthesis in the rumen and lower GI tract; in herbivores such as the horse and rabbit, microbial synthesis occurs in the colon or cecum. Thus, the nutritionist need not be concerned about providing routinely a dietary source of water-soluble vitamins for ruminants and nonruminant herbivores. However, recent evidence suggests that high-producing dairy cows and feedlot cattle may benefit from dietary niacin supplementation and hoof health in dairy cattle may be improved by supplemental biotin. Pigs, poultry, and many other nonruminant animals, including humans, require a dietary source of water-soluble vitamins (Table 15.1).

Most of the water-soluble vitamins are required in minute amounts. They are all organic compounds but are unrelated to each other in structure. Most function as metabolic catalysts, usually as coenzymes. There are many examples of free and bound forms of coenzymes. Coenzymes, which are of organic nature, generally occur in freely dissociable form, but a few are covalently bound factors only. Several cofactors, such as the flavoproteins, occur in the covalently attached form. Decker (1993) reviewed the biosynthesis and function of enzymes with covalently bound flavin.

All water-soluble vitamins produce profound aberration in metabolism if unavailable to the tissues in

251

TABLE 15.1 Water-soluble vitamins.

Ascorbic acid
Biotin
Choline
Folacin
Myoinositol[a]
Niacin
Pantothenic acid
Para-aminobenzoic acid (PABA)[a]
Riboflavin
Thiamin
Vitamin B6
Vitamin B12

[a]Myoinositol and para-aminobenzoic acid (PABA) are both synthesized in abundance by microorganisms residing in the gastrointestinal tract of animals, including humans. No clear evidence exists for a dietary requirement under normal conditions.

sufficient amounts. Because of their ready excretion by the kidney, acute toxicity of the water-soluble vitamins is unlikely. The history, structure, metabolism, signs of deficiency and toxicity, and sources of water-soluble vitamins have been reviewed (Dove and Cook, 2001; Dove, 2005).

Each water-soluble vitamin will be discussed as to structure, function, and deficiency signs. The nomenclature used here is that of the International Union of Nutritional Science Committee (1980) as described in the *Journal of Nutrition* **120**: 12–19, 1990. These rules are needed to minimize confusion and misinterpretation in discussing vitamin nutrition.

❏ THIAMIN

Structure

Thiamin consists of one molecule of pyrimidine joined with one of thiazole. Williams and Klein (1936) established its structure, as shown:

THIAMIN

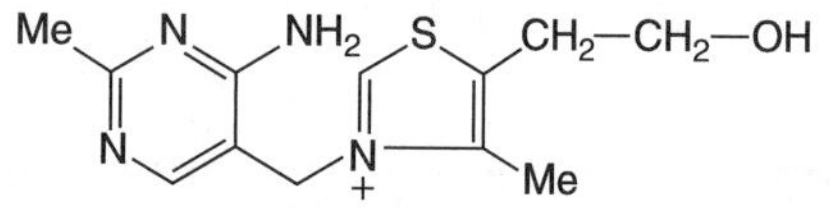

The compound, 3-(4-amino-2-methylpyrimidinyl-5-methyl)-5-(2-hydroxyethyl)-4-methylthiazolium, formerly known as vitamin B_1, vitamin F, aneurin (e), or thiamine, should be designated *thiamin*. Its use in phrases such as thiamin activity and thiamin deficiency represents preferred usage.

Absorption, Transport, and Excretion

Thiamin is absorbed from the small intestine by passive diffusion at high concentrations and by active transport at low concentrations. It is phosphorylated after entering the mucosal cell. Folic acid deficiency decreases thiamin absorption. Free thiamin is carried in the portal blood to the liver for coenzyme

synthesis and to other tissues for small amounts of storage. The total amount of thiamin in the human body is about 30 mg, half of which is in muscle. It is excreted via the urine as free thiamin and its metabolites (Tanphaichitr and Wood, 1984).

Functions

Thiamin is phosphorylated in the liver to form coenzymes, cocarboxylase or thiamin pyrophosphate (TPP), and lipothiamide (LTPP). TPP is involved in the decarboxylation of alpha-keto acids. Among the reactions in which TPP participates, the decarboxylation of pyruvic acid to acetaldehyde is important and can be illustrated as:

$$\text{Pyruvic acid} \xrightarrow{\text{TPP}} \text{acetaldehyde} \quad CO_2$$

LTPP is involved in oxidative decarboxylation of alpha-ketoglutarate and other metabolites of the citric acid cycle. The pig is somewhat unique in that it stores appreciable thiamin in its tissues. This accounts for the fact that pork is an excellent source of dietary thiamin.

A thiamin-binding protein has been identified in chicken egg and liver and in pregnant rat blood plasma (Adiga and Muniyappa, 1978; Muniyappa et al., 1978). Its presence in plasma may play a role in the transfer of thiamin across the placenta. Estrogen induces its synthesis in liver and modulates the concentration in blood plasma.

Deficiency Signs

In thiamin deficiency, blood pyruvic acid and lactic acid concentrations increase; this increase has been used in the assessment of thiamin status. Because pyruvic acid is a key metabolite in energy utilization in the citric acid cycle (see Fig. 7.2), it is evident that deficiency seriously disturbs carbohydrate and lipid metabolism.

The immediate effect of a thiamin deficiency is a reduction in appetite (anorexia). In humans, the thiamin deficiency syndrome, beriberi, has been recognized for centuries. The syndrome includes enlargement of the heart, numbness of extremities, weakness and stiffness in the thighs, edema of the feet and legs, unsteady gait, rigidity and paralytic symptoms, and pain along the spine. Wernicke's encephalopathy is generally associated with thiamin deficiency in humans, usually in alcoholics. It is characterized by ataxia and mental confusion. Brain glial cells have impaired ability to synthesize fatty acids and cholesterol in thiamin deficiency; this impairment has been suggested as the basis for degenerative changes seen in glial cells in deficient animals. In animals, the classical disease is polyneuritis in chicks (retraction of the head; Fig. 15.1) and rats (walking in circles). Bradycardia (slow heart rate) is

a common manifestation in all species studied, and myocardial damage results in dilation of the heart and heart failure in swine. Body temperature is reduced and the adrenal gland is enlarged. These symptoms of deficiency, including the heart lesion, may result from pyruvic or lactic acid accumulation. It has been suggested that changes in liver aminotransferase activities in thiamin-deficient animals may be due to accumulation of pyruvic acid or other metabolites. The most common assessment of thiamin nutritional status is the measurement of erythrocyte transketolase activity.

Antagonists

Several compounds resemble thiamin in chemical structure but have no thiamin activity. Addition to the diet or injection of these antimetabolites* results in thiamin deficiency, which can be overcome by increasing the thiamin intake. One such thiamin antagonist is pyrithiamin.

Many foods, especially certain fish and sea foods, contain significant amounts of a group of enzymes, thiaminases, which split the thiamin molecule and render it ineffective biologically. Chastek paralysis, a disease in silver foxes first reported in 1940, later was shown to be caused by thiaminases in raw fish being fed. These enzymes are heat labile, so that cooking of foods containing them destroys the antithiamin activity.

Thiamin is heat labile, especially in alkaline conditions. Toxicity has not been reported, presumably because of the rapid loss of excess thiamin in the urine.

❏ RIBOFLAVIN

Structure

Riboflavin is a yellow fluorescent pigment consisting of ribose and isoalloxazine. The compound 7,8-dimethyl-10-(l'-D-ribityl)isoalloxazine, formerly known as vitamin B_2, vitamin G, lactoflavin(e), or riboflavine, should be designated riboflavin. Its use in phrases such as riboflavin activity and riboflavin deficiency represents preferred usage. The structure of riboflavin is shown on the following page:

***Figure 15.1** Polyneuritis in the chick showing the typical retraction of the head and rigidity of the legs, as well as the marked effect on growth. Courtesy of Poultry Science Department Cornell University.*

---

*The concept of antimetabolites in nutrition was reviewed by Woolley (1959). Antimetabolites are used in clinical medicine (for example, in control of certain types of cancer) and in nutritional research.

RIBOFLAVIN

$$CH_2-(CHOH)_3-CH_2OH$$

Riboflavin was synthesized in 1935. It is heat stable but is readily destroyed by light. Riboflavin, riboflavin 5-phosphate (also called flavin mononucleotide or FMN), and flavin adenine dinucleotide (FAD) are the only naturally occurring substances with riboflavin activity, although several synthetic derivatives show some activity (Horwitt, 1972).

Absorption, Transport, and Excretion

Riboflavin and FMN are absorbed in the upper GI tract by a specialized active transport mechanism, which appears to be saturable. The degree of riboflavin absorption is increased by the presence of food in the lumen of the GI tract. The upper limit of riboflavin absorption in humans is about 25 mg. Several binding proteins are involved with transport of riboflavin in serum. A specific genetically controlled binding protein is present in chicken serum and eggs; in birds lacking this protein, riboflavin is lost in the urine, and the eggs become riboflavin deficient (Rivlin, 1984). Other specific riboflavin-binding proteins have been identified in different tissues of several species. Riboflavin passes across the renal tubule in both directions by active transport. Free riboflavin as well as small amounts of its coenzyme derivitives appear in urine. Small amounts of riboflavin are also found in bile and sweat.

Functions

Riboflavin functions in the coenzymes FAD and FMN, which occur in a large number of enzyme systems. The interconversions of these biologically active forms can be illustrated as follows:

flavokinase

Riboflavin + ATP $\longrightarrow$ FMN + ADP

FAD pyrophosphorylase

FMN + ATP $\longrightarrow$ FAD + pyrophosphate

FMN and FAD are present in most, if not all, animal cells and occur as coenzymes in the flavoprotein enzyme systems such as oxidases (aerobic dehydrogenases) and anaerobic dehydrogenases. Riboflavin metabolism is affected by the endocrine system. Both thyroid hormones and adrenocorticotropic hormone enhance the conversion of riboflavin to its coenzyme derivitives (Rivlin, 1984). Both FAD and FMN are related closely in several reactions with niacin coenzymes, coenzyme I (NAD), and coenzyme II (NADP).

Deficiency Signs

The primary general effect, as with most of the water-soluble vitamin deficiencies, is reduced growth rate in young animals. Various pathological lesions accompany a deficiency. In rats, conjunctivitis, corneal opacity, and vascularization of the cornea are common signs of deficiency; skin lesions and hair loss related to a failure in regeneration of hair follicles may occur. Corneal opacity (cataracts) associated with riboflavin deficiency in rats appears to occur only on high-galactose diets. Humans with cataracts have a higher incidence of biochemical riboflavin deficiency and abnormal galactose tolerance (Bhat and Gopalan, 1974). The exact biochemical basis for the formation of cataracts in animals fed riboflavin-deficient high-galactose diets remains uncertain. Galactose metabolism does not seem to be affected by riboflavin deficiency. Mature female rats fed riboflavin-deficient diets have abnormal estrous cycles and give birth to a high proportion of pups with congenital malformations of the skeleton. In deficient dogs, fatty infiltration of the liver and corneal opacity have been reported. Changes in the cornea also have been reported as typical of riboflavin deficiency in pigs. In addition, hemorrhages of the adrenals, kidney damage, anorexia, vomiting, and birth of weak and stillborn piglets have been reported in deficient swine. Erythrocyte glutathione reductase (EGR), a flavin-dependent enzyme, is widely accepted as an indicator of riboflavin status (Sauberlich, 1984). Riboflavin-deficient swine have a high EGR activity coefficient and produce stillborn and weak piglets (Frank et al., 1984). A specific sign of deficiency in poultry is curled-toe paralysis (Fig. 15.2), which results from degenerative changes in the myelin sheaths of the nerve fibers.

Fatty acid oxidation by liver mitochondria of riboflavin-deficient rats is decreased rapidly, apparently as a result of a reduction in acyl-CoA dehydrogenase activity. Riboflavin has not been shown to be required for activity of other enzymes

Figure 15.2 *Curled-toe paralysis in a riboflavin-deficient chick, a typical sign of a severe deficiency. Courtesy of Chas. Pfizer Co.*

involved in fatty acid beta-oxidation. Riboflavin deficiency in baby pigs results in depressed liver glutathione peroxidase activity (Brady et al., 1979). The role of riboflavin in vitamin E–Se nutrition (Se is a constituent of glutathione peroxidase) in animals and humans is not well defined.

In mature ruminants, the microflora of the rumen synthesize sufficient riboflavin and other water-soluble vitamins to eliminate a dietary need under most conditions. Young preruminant animals (lambs and calves) have a dietary requirement early in life, however, and deficiency signs can be induced with synthetic milk diets. Diets low in riboflavin produce lesions in the corners of the mouth and on the lips, anorexia, loss of hair, and diarrhea in calves and lambs. The type of diet fed to immature ruminants appears to be a controlling factor in the development of rumen microbial populations that synthesize riboflavin.

Riboflavin deficiency has also been reported in monkeys, mice, cats, horses, and other species; the symptoms resemble in one or more ways those described for other animals. Riboflavin deficiency has been reported to cause moderate suppression of antibody production in rats and pigs (Harmon et al., 1963).

In humans, riboflavin deficiency causes lesions of the lips and mouth (angular stomatitis), insomnia, irritability, scrotal dermatitis, conjunctivitis, and burning of the eyes. There is evidence for a need for riboflavin-dependent oxidoreductase systems in the activation of vitamin B_{12}. A reduction in pyridoxine enzyme systems is associated with riboflavin deficiency. These are just two among several interactions that are known to exist among pairs of B-vitamins.

Metabolism

Riboflavin is absorbed rapidly by phosphorylation in the intestinal mucosal cell and is used directly by all cells in the body. The maintenance requirement is dependent mainly on excretion rather than decomposition. A half-life of 16 days (the time required for half of a given dose of riboflavin to be excreted) has been reported for the rat under normal conditions; turnover is accelerated with high-riboflavin diets (Yang et al., 1967).

Diet composition and environmental temperature may affect the dietary requirement. High-fat diets or low-protein diets tend to increase the requirement, and the requirement of growing pigs has been reported to be increased by high environmental temperature.

Toxicity

Riboflavin toxicity is extremely unlikely because of its rapid loss in urine. Rats given 560 mg/kg of body weight in an intraperitoneal injection show some mortality from kidney damage, but such a massive dose is not physiological. High-protein diets may be toxic in riboflavin deficiency. This may be related to the reduction in D-amino acid oxidase activity of liver observed in riboflavin deficiency.

Analogs and Antagonists

Some analogs show riboflavin activity. At least four derivatives of isoalloxazine have about half the biological activity of riboflavin for rats. A long list of other isoalloxazine derivatives have no activity.

Several analogs of riboflavin have been shown to be inhibitors of riboflavin. These include isoriboflavin (5,6-dimethyl-9-D-l'-ribitylisoalloxazine), dinitrophenazine, galactoflavin, and D-araboflavin. Many riboflavin antagonists inhibit growth of tumors and/or pathogenic bacteria and, in this role, can be extremely valuable medically. Chlorpromazine, phenothiazine derivative with structural similarities to riboflavin, inhibits the conversion of riboflavin to FAD. Boric acid increases urinary excretion of riboflavin and thereby increases its requirement (Rivlin, 1984).

❑ NIACIN (NICOTINIC ACID)

The human disease pellagra was traced to a deficiency of a factor in yeast. Since then, the importance of niacin as the specific nutrient involved has been clarified further in animals as well as in humans (Roe, 1973; Darby et al., 1975).

Structure

Niacin (nicotinic acid) has the following structure:

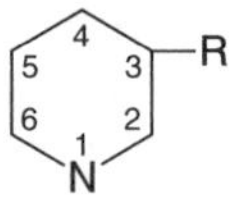

The term *niacin* should be used as the generic descriptor for pyridine 3-carboxylic acid and derivatives exhibiting qualitatively the biological activity of nicotinamide. Thus, phrases such as niacin activity and niacin deficiency represent preferred usage.

The compound pyridine 3-carboxylic with (R = −COOH), also known as niacin or vitamin PP, should be designated *nicotinic acid*.

The compound with (R = −CONH$_2$), also known as niacinamide or nicotinic acid amide, should be designated *nicotinamide*.

Niacin and niacinamide are equivalent to each other in biological activity for animals and humans. Both are very stable in heat, light, and alkali, and, therefore, are stable in feeds.

Niacin is the accepted generic description for pyridine 3-carboxylic acid and derivatives showing the qualitative nutritional activity of nicotinic acid. Nicotinimide is the amide of nicotinic acid. All forms of niacin are expressed as nicotinamide equivalents in terms of milligram or microgram equivalents of nicotinamide.

Absorption, Transport, and Excretion

Niacin and niacinamide are readily absorbed from the small intestine by simple diffusion. In humans, niacin is absorbed with equal efficiency from the stomach and upper small intestine (Rao and Gopalan, 1984). The efficiency of riboflavin absorption from the stomach in other species is less well known. The mechanism of absorption of niacinamide bound to nucleotides in natural feedstuffs is unknown; hydrolysis may occur in the lumen of the GI tract or within the mucosal cell, but the exact site of hydrolysis is unclear. Free niacin entering the intestinal mucosa cell from the GI-tract lumen is converted to niacinamide in the cell and transported in that form via the blood to tissues where it is incorporated into its coenzymes. Excretion of free niacin and niaciamide in the urine is minimal; most of the vitamin is excreted as N'-methylnicotinamide (N'MN),N'-methyl-4-pyridone-5-carboxamide, and N'-methyl-2-pyridone-5-carboxamide. Only about 3% of absorbed niacin is oxidized to carbon dioxide (Rao and Gopalan, 1984).

Functions

Niacin functions as a constituent of two important coenzymes that act as codehydrogenases. These enzymes are NAD (nicotinamide adenine dinucleotide) and NADP (nicotinamide adenine dinucleotide phosphate). These niacin-containing enzymes are important links in the transfer of hydrogen from substrates to molecular oxygen, resulting in water formation. As such, they are vital in several chemical reactions that occur in animal tissues.

NAD and NADP probably combine with a variety of protein carriers (apoenzymes) with specificity for a particular reaction. Thus, a large number of substrates can be dehydrogenated. All animal cells contain NAD and NADP; the ratio of the two varies among tissues. NAD is convertible to NADP and vice versa. Similarly, many of the reactions above that involve NAD and NADP are reversible. The enzymes (NAD and NADP) are reduced to the dihydro forms, which are, in turn, dehydrogenated by flavin (riboflavin-containing) enzymes.

Deficiency Signs

The general sign of niacin deficiency is reduced growth and appetite. Specific deficiency signs in animals include diarrhea, vomiting, dermatitis, unthriftiness, and an ulcerated intestine (necrotic enteritis) in swine; poor feathering, a scaly dermatitis and sometimes "spectacled eye" in chicks; and a peculiar darkening of the tongue (black tongue) and mouth lesions in dogs. In humans, a bright red tongue, mouth lesions, anorexia, and nausea are the common syndrome of pellagra.

Niacin present in most cereal grains is not available biologically for the pig and other nonruminant animals (Chadhuri and Kodicek, 1960; Luce et al., 1966; Yen et al., 1977). There are several bound forms of niacin containing peptides and complex carbohydrates. Some of these complexes are biologically available; others are not. A bound form of niacin, niacinogen, composed of niacin and a peptide, found in cereal grains, is biologically available (Rao and Gopalan, 1984). One of the bound forms in wheat is termed *niacytin,* which contains peptides, hexoses, and pentoses. The presence of biologically unavailable bound forms of niacin in cereal grains and maize helps to account for the high incidence of pellagra among human populations subsisting on diets composed mainly of these foods.

Adult ruminants are not fed supplemental niacin under normal conditions because of the synthesis by rumen microflora. Niacin deficiency has been produced in calves, and some evidence suggests that milk production of high-producing dairy cows, together with daily weight gain and feed utilization of feedlot cattle, may be improved, at least during the first few weeks of the finishing period, by 50 to 500 ppm of supplemental niacin. The environmental and feeding conditions conducive to a beneficial response to niacin and other B-vitamin supplementation need to be identified. Robinson (1966) summarized the evidence supporting the view that intestinal synthesis of niacin in humans may provide some of the metabolic requirement for the vitamin under some conditions.

Niacin–Tryptophan Interrelationships

If animals are deprived of dietary niacin, the metabolic requirement for the vitamin can be met by conversion of tryptophan, an amino acid, to niacin (see Chapter 9). If tryptophan is provided in the diet in slight excess of the amount needed for tissue protein synthesis, the excess can be used to satisfy the niacin requirement. The synthesis of 1 mg of nicotinamide in this way requires about 60 mg of tryptophan for the pig and human. The efficiency of conversion probably varies with species

and with dietary and metabolic states. For example, Harmon et al. (1969) showed that the dietary niacin requirement of growing pigs was higher when the diet contained corn protein instead of milk protein, presumably because of the lower tryptophan availability from corn in addition to the known lack of availability of niacin from corn.

N1-methylnicotinamide and N1-methyl-6-pyridine-nicotinamide are the main excretory products of niacin in the pig, dog, rat, and humans, but in the chick it is dinicotinyl-ornithine. Ruminants appear to excrete the vitamin largely unchanged. Levels of niacin metabolites in urine often are used as indices of niacin adequacy in studies of niacin requirements and metabolism.

Analogs and Antagonists

Several compounds related structurally to niacin possess some vitamin activity. These include several fatty acid esters of niacin and 3-hydroxyxanthranilic acid, an intermediate in the conversion of tryptophan to niacin. The structures of 13 known metabolites of niacin were illustrated by Darby et al. (1975).

Mice fed 3-acetyl-pyridine show signs of niacin deficiency, although it appears to have some vitamin activity for normal dogs. Considerable variation exists among species in utilization of or antagonism by various analogs of niacin.

Effects of Massive Doses of Niacin

Large doses of niacin sometimes are prescribed for schizophrenia and in efforts to induce cerebrovascular dilation in patients with senile ataxia. Similar doses (3 to 9 g daily) of nicotinic acid reduce serum cholesterol in patients with hypercholesterolemia. These high intakes of niacin are effective in decreasing mobilization of fatty acids from adipose tissue and increasing utilization of muscle glycogen stores. Similar effects have been reported on cardiac muscle metabolism. Long-term results of niacin doses of 3 g per day given for therapy of coronary heart disease in men with previous heart attacks failed to show a clear reduction in mortality. Furthermore, long-term high-dose niacin therapy may be associated with liver damage, multiple enzyme changes, gastrointestinal disturbance (peptic ulcer), and elevated serum uric acid and glucose. Therefore, the wisdom of dietary supplementation of niacin at these levels appears in doubt.

❑ PANTOTHENIC ACID

A growth factor for yeast was first identified as pantothenic acid (pantoyl-beta-alanine). Later (Elvehjem and Koehn, 1934; Lepkovski and Jukes, 1936), it was prepared from liver, and since then it has been shown to be required by a large number of animal species.

Structure

The structure of pantothenic acid, the peptide of a butyric acid derivative and beta-alanine (determined by Williams and Major, 1940), is as follows:

PANTOTHENIC ACID

$$
\begin{array}{c}
\text{H}_3\text{C} \quad \text{OH} \\
| \quad\quad | \\
\text{HOCH}_2-\text{C}-\text{CH}-\text{CO} \\
| \quad\quad\quad\quad | \\
\text{H}_3\text{C} \quad\quad \text{HN}-\text{CH}_2-\text{CH}_2-\text{COOH}
\end{array}
$$

The compound, N-(2,4-dihydroxy-3,3-dimethyl-1-oxobutyl)-beta-alanine, formerly known as pantoyl-beta-alanine, should be designated *pantothenic acid*. Its use in phrases such as pantothenic acid activity and pantothenic acid deficiency represents preferred usage.

Usually, it is present as the Ca or Na salt of the d-isomer (active form) of pantothenic acid. The Ca salt is the most common form in which the vitamin is added to diets because it is less hygroscopic than the Na salt.

Absorption, Transport, and Excretion

Pantothenic acid is present in feeds in various forms, including acyl carrier protein, CoA, and CoA esters (Olson, 1984). It is readily absorbed and is transported in the blood serum as the free vitamin and as CoA; it is present in eythrocytes and other tissues as CoA. The vitamin is excreted in the urine; the amount excreted is related directly to intake. Therefore, low concentrations in the urine may indicate pantothenic acid deficiency.

Functions

Pantothenic acid functions as a component of coenzyme A (CoA), the coenzyme required for acetylation of numerous compounds in energy metabolism. The role of CoA in metabolism of fatty acids was discussed in Chapter 8. CoA is required in the formation of two-C fragments from fats, amino acids, and carbohydrates for entry into the citric acid cycle and for synthesis of steroids. Selected biochemical reactions catalyzed by CoA are summarized in Table 15.2 (Olson, 1984).

The synthesis and metabolism of CoA has been reviewed (Wright, 1976; Olson, 1984). The pathway of CoA synthesis involves pantothenate →4'-

TABLE 15.2   Selected biochemical reactions catalyzed by coenzyme A.

ENZYME	PANTOTHENATE DERIVATIVE	REACTANT(S)	PRODUCT(S)	SITE
Pyruvic Dehydrogenase	CoA	Pyruvate	Acetyl CoA	Mitichondria
α-Ketoglutarate Dehydrogenase	CoA	α-Ketoglutarate	Succinyl CoA	Mitochondria
Fatty Acid Oxidase	CoA	Palmitate	Acetyl CoA	Mitochondria
Fatty Acid Synthetase	Acyl Carrier Protein	Acetyl CoA Malinyl CoA	Palmitate	Microsomes
HMG-Co-A Synthetase	CoA	Acetyl CoA Acetoacetyl CoA	HMG-CoA	Microsomes
Propionyl CoA Carboxylase	CoA	Propionyl CoA Carbon Dioxide	Methyl-malonyl CoA	Microsomes
Acyl CoA Synthetase	Phosphopantetheine	Succinyl CoA GDP + P_i	Succinate GTP + CoA	Mitochondria

From Olson (1984).

phosphopantothenic acid → 4′phosphopantothenyl-cysteine → 4′-phosphopantetheine → diphospho CoA → CoA. The degradation of CoA is not well known.

A deficiency of pantothenic acid precludes the synthesis of CoA because the CoA molecule contains pantothenic acid at 10% of its weight. The structure of CoA is as shown:

ADENOSINE PHOSPHATE

PANTOTHENIC ACID

β-MERCAPTOETHYLAMINE

COENZYME A

The transformations of lipids and carbohydrates that involve S-acylated CoA include hydrolysis of the thiol ester bond; racemization; dehydrogenation; reduction; hydration; carboxylation; condensation reactions; and transferases.

The concentration of CoA in liver is increased by fasting, and the proportion that is acylated is increased. This increase in the proportion of acylated CoA is brought about by the increase in tissue-free fatty acids resulting from fasting. Endocrine changes such as the decrease in ratio of insulin to glucagon which occurs in diabetes also increases acylated CoA in liver through the effect on mobilization of free fatty acids from tissue lipid stores. The data illustrate that nutritional and endocrine changes have a striking effect on CoA metabolism, which in turn provides a mechanism by which the liver can adjust quickly to changing needs for tissue fatty acid synthesis and oxidation.

Deficiency Signs

In addition to causing reduced growth rate as a generalized effect in all animals studied, pantothenic acid deficiency results in dermatitis in chicks, graying of the hair (achromotrichia) in rats and foxes, hemorrhaging and degeneration of the adrenal cortex of rats and mice, fetal death and resorption in rats, nervous derangement, skin lesions, lymphoid cell necrosis in the thymus, degeneration of duodenal mucosal cells, and vacuolation of pancreatic acinar cells in chicks (Gries and Scott, 1972). In guinea pigs, diarrhea, rough hair coat, anorexia, and enlarged and hemorrhagic adrenals have been reported. A striking feature of pantothenic acid deficiency is fatty infiltration of the liver. Other signs in dogs are vomiting, gastritis, enteritis, and hemorrhage of the thymus and kidney. A prominent feature of deficiency in pigs is the effect on the nervous system. In addition to the impaired growth, loss of hair, and enteritis noted in other species, the deficient pig develops a peculiar gait called goose stepping (Fig. 15.3). The abnormal gait is a result of nerve degeneration (Follis, 1958). Slow growth, anorexia, and impaired sow productivity result when pantothenic acid is deficient in sow diets.

Ruminants normally do not require dietary pantothenic acid because of rumen synthesis, but deficiency has been produced in calves by feeding a synthetic milk diet low in the vitamin. Signs of de-

Figure 15.3 *"Goose-stepping" in a pig fed a pantothenic acid-deficient diet. Courtesy of W. M. Beeson, Purdue University.*

ficiency are rough hair coat, dermatitis, anorexia, loss of hair around the eyes (spectacle eye), and demyelinization of the sciatic nerve and spinal cord.

Pantothenic acid appears to be related closely to antibody production. Reduced formation of antibodies to several types of antigens in rats and pigs and to tetanus toxoid in humans has been observed in pantothenic acid deficiency.

Although clearcut deficiency of pantothenic acid has not been described in humans, deficiency signs have been induced by feeding a pantothenic acid antagonist (omega-methyl-pantothenic acid). Signs include fatigue, apathy, gastrointestinal disturbances, cardiovascular instability (tachycardia and variable blood pressure), and increased susceptibility to infections. Caution must be used in interpreting results of experiments in which vitamin antagonists are used to induce deficiency of the vitamin. It is important to confirm that biochemical or morphological lesions are due to the vitamin deficiency and not to the effects of the antagonist itself.

❑ VITAMIN B$_6$

Birch and Gyorgy (1936) first established the properties of a component of a crude concentrate that was found to cure dermatitis in rats fed diets supplemented with thiamin and riboflavin. This compound was named pyridoxine. The International Committee on Nomenclature (1990) now recommends (*J. Nutrition* **120:** 12, 1990) that the term *Vitamin B$_6$* be used for all 3-hydroxy-2-methyl pyridine derivatives showing qualitatively the biological activity of pyridoxine. The structures of the major compounds with vitamin B$_6$ activity (pyridoxine, pyridoxal, and pyridoxamine) are as shown:

Phrases such as vitamin B$_6$ activity and vitamin B$_6$ deficiency represent preferred usage.

The compound with (R = Ch2OH), 3-hydroxy-4,5-bis(hydroxy-methyl)-2-methyl-pyridine, formerly known as vitamin B$_6$, adermin, or pyridoxol,

should be designated *pyridoxine*. The term *pyridoxine* is not synonymous with the generic term *vitamin B$_6$*.

The compound with formula (R = HCO), also known as pyridoxaldehyde, should be designated *pyridoxal*.

The compound with (R = CH2NH2), 3-hydroxy-4-methylamino-5-hydroxymethyl-2-methylpyridine, should be designated *pyridoxamine*.

The chemistry of each of these forms of vitamin B$_6$ and the biosynthesis and metabolic interconversions of them have been described (Harris, 1968; Sauberlich, 1968; Ink, 1984). A specific enzyme is involved in each conversion, and both niacin and riboflavin are associated with one or more reactions as components of coenzymes (NADP and FMN, respectively). Vitamin B$_6$ is stable to heat, light, and all solutions and maintains its activity well in mixed diets.

Absorption, Transport, and Excretion

Dietary vitamin B$_6$ occurs as pyridoxine (PN), pyridoxal (PL), and pyridoxamine (PM). Pyridoxine is absorbed by a nonsaturable process (Henderson, 1984). It is absorbed at a higher efficiency than either of the other forms; PM is absorbed more efficiently than PL. Transport from the absorption site in the small intestinal mucosal cells to the liver is in all three forms. PL, the major form of the vitamin in blood plasma, is tightly bound to albumin. The details of vitamin B$_6$ transport throughout the body and processed by tissues and cells were reviewed (Henderson, 1984). Suffice it to say that transport of any form of vitamin B$_6$ into animal cells is by simple diffusion. The high content of PL in muscle tissue as a component of the enzyme glycogen phosphorylase indicates that up to 70 or 80% of total vitamin B$_6$ storage in the body is as a constituent of muscle phosphorylase (Henderson, 1984). Most of the excretion of vitamin B$_6$ occurs in the urine as pyridoxic acid. Measurement of this metabolite in urine reflects current vitamin B$_6$ intake but is not a good indicator of vitamin B$_6$ storage.

Functions

Vitamin B$_6$ functions as a coenzyme of a vast array of enzyme systems associated with protein and nitrogen metabolism. Metabolic roles of vitamin B$_6$ and human requirements for it have been reviewed

(Ink, 1984). Methods used in vitamin B$_6$ analysis and in assessment of vitamin B$_6$ status of humans and animals have been summarized (Sauberlich, 1984; Lehlem and Reynolds, 1981).

As early as 1945, the B$_6$ requirement was known to increase with high dietary protein. This is explained by the fact that pyridoxal phosphate is the coenzyme of transaminases, which must be increased in activity to metabolize increased quantities of dietary protein. Sauberlich (1968) tabulated a list of more than 50 pyridoxal-51-phosphate-dependent enzymes and the reactions catalyzed and has listed the major types of enzymatic reactions involving pyridoxal phosphate-dependent enzymes as follows: transamination; decarboxylation, racemization (not in animal tissues); amine oxidation; aldol reaction; cleavage; dehydration (deamination); and desulfhydration.

Aspartic transaminase (glutamic oxaloacetic transaminase) is the most abundant transaminase in mammalian tissues. Amino acid decarboxylases are widespread in microorganisms, but the following are important in animal tissues: glutamic acid decarboxylase, cysteine-sulfonic add decarboxylase, and aromatic L-amino add decarboxylases. Tryptophan metabolism is dependent to a great extent on pyridoxal phosphate dependent enzymes, and a vitamin B$_6$ deficiency results in urinary excretion of abnormal metabolites of tryptophan: xanthurenic acid, kynurenine, and 3-hydroxykynurenine. Tyrosine and phenylalanine metabolism also involve a number of pyridoxal phosphate-dependent enzymes, and deaminases (dehydrases) requiring pyridoxal phosphate are important in catabolism, especially of serine, threonine, and homoserine. Many phosphorylases in animal tissues contain pyridoxal phosphate, and total phosphorylase activity in skeletal muscle in rats is decreased in vitamin B$_6$ deficiency.

Vitamin B$_6$ is absorbed chiefly as pyridoxal after hydrolysis of pyridoxal phosphate by alkaline phosphatase in the lumen of the GI tract (Mehansho et al., 1979). Pyridoxal crosses the intestinal wall and accumulates in tissue more rapidly than pyridoxine and pyridoxamine. Vitamin B$_6$ circulates in the plasma as pyridoxal phosphate complexed to albumin; albumin-bound pyridoxal phosphate is not transported into cells.

The bioavailability (absorption from the GI tract) of vitamin B$_6$ varies widely among feed sources and may be as low as 50% in dry-heated cereals. It has been postulated that the reduced bioavailability of vitamin B$_6$ may be the result of the formation of complexes with proteins, amino acids, or reducing sugars, and of degradation products during food processing. Yen et al. (1976) have suggested that a chick growth assay is more sensitive than a plasma enzyme assay as an indicator of biologically available B$_6$ in heated corn and soybean meal.

Vitamin B$_6$ is involved somehow in red blood cell formation, probably because pyridoxal phosphate is required for porphyrin synthesis. B$_6$ is also important in the endocrine system; deficiency affects activities of growth hormone, insulin, and gonadotrophic, adrenal, thyroid, and sex hormones, but the mechanisms are not understood fully. The defective cellular transport of amino acids in vitamin B$_6$-deficient animals is believed to be due to the reduced pituitary growth hormone levels associated with the vitamin B$_6$ depletion (Heidel and Riggs, 1978). Growth hormone secretion is stimulated by dopamine, which is produced from dopa decarboxylase of which pyridoxal phosphate is a coenzyme.

Deficiency Signs

The most common sign of B$_6$ deficiency involves the nervous system. Convulsions have been observed in all species studied. Demyelinization of the peripheral nerves, swelling and fragmentation of the myelin sheath, and, later, other degenerative changes occur. The convulsive seizures observed in B$_6$-deficient animals have resulted in many studies of brain metabolites and enzymes. The administration of gamma-amino-butyric acid temporarily controls the seizures in B$_6$-deficient animals.

Vitamin B$_6$ deficiency results in reduced antibody response to various antigens in rats, guinea pigs, and pigs, and reduces immunocompetence in progeny of depleted female rats. The mechanisms of immunosuppression induced by B$_6$ deficiency are unclear.

In some animal species, vitamin B$_6$ deficiency results in skin lesions on the feet, around the face and ears (acrodynia), atrophy of the hair follicles and abscesses and ulcers around the sebaceous glands from secondary infection in rats, mice, and monkeys. Skin lesions have not been reported in B$_6$-deficient swine, calves, or poultry.

Vitamin B_6 is required for normal reproduction in the rat and for normal egg laying and hatchability in hens. It is synthesized by rumen bacteria, and evidence has been shown for some intestinal synthesis in simple-stomached animals. Although deficiency of B_6 is unlikely in farm animals because of its wide distribution in feedstuffs, the pig fed a B_6-deficient experimental diet exhibits the classical symptoms of deficiency, including convulsive seizures, poor growth, a brown exudate around the eyes, and low urinary excretion of the vitamin. The deficient chick shows depressed growth and abnormal feathering. The B_6 requirement of humans is increased during pregnancy and in women receiving oral contraceptives. Oral lesions similar to those of riboflavin deficiency were alleviated by vitamin B_6 therapy, and low plasma pyridoxal phosphate levels were increased by vitamin B_6 (Cleary et al., 1975). It has been proposed that pyridoxal phosphate may function in steroid hormone action by serving as a ligand for binding the transporting steroid-receptor complexes between the cytoplasm and nucleus where the hormones influence gene expression. A need for increased vitamin B_6 intake during pregnancy of domestic animals has not been established.

Vitamin B_6 deficiency produces a wide variety of changes related to lipid metabolism, including reduced carcass fatness, fatty liver, and elevated plasma lipids and cholesterol. The metabolic mechanism by which these effects occur in B_6 deficiency is unclear.

Oxalic acid is excreted in large quantities in the urine of B_6-deficient animals and humans; it is derived from metabolism of glycerine, serine, alanine, and other compounds that, under normal conditions, are metabolized to compounds other than oxalic acid.

Pyridine-2-carboxylic acid (picolinic acid) content of the rat pancreas is increased by increasing dietary vitamin B_6 content. Picolinic acid, a metabolite of tryptophan, appears to be an important Zn-binding ligand in the intestinal tract and pancreas, although this function is controversial.

Toxicity and Analogs

Vitamin B_6 toxicity is very unlikely; rabbits, rats, and dogs tolerate doses up to 1 g/kg body weight (Coates, 1968). High doses given orally or parenterally produce convulsions, impaired coordination, paralysis, and death. All three forms of the vitamin are absorbed readily from the GI tract and are distributed widely in tissues. Most of an administered dose appears in the urine as 4-pyridoxic acid (derived from pyridoxal), with negligible amounts found in feces or sweat.

Structural analogs, including 4-deoxypyridoxine and 2-methyl-pyridoxine, function as B_6 antagonists in animals, and agents such as hydrazones, hydroxylamines, and semicarbazides form inactive complexes with pyridoxal or pyridoxal phosphate and produce vitamin B_6 deficiency signs (Gregory and Kirk, 1981), including convulsions in humans and animals. Gregory (1980) showed that epsilon-pyridoxyl lysine, a complex formed between vitamin B_6 aldehydes and lysine during food storage, produces vitamin B_6 deficiency signs when fed at low levels to rats. Further work is needed to determine the practical significance of the formation of complexes of B_6 with other feed constituents in processed feeds and foods.

❏ VITAMIN B_{12}

Vitamin B_{12} is the most recently discovered vitamin. It was first known as the animal protein factor (APF) because animals fed diets not containing animal protein were prone to the hematological and neurological signs that were later shown to be prevented or cured by vitamin B_{12}. It has been shown to be present in animal tissues and excreta, although it is probably not formed by animal or plant tissues. The only known primary source of vitamin B_{12} is from microorganisms; it is synthesized by a wide range of bacteria but not in appreciable amounts by yeast and fungi. The chemistry and properties of B_{12} and the early history of the association of vitamin B_{12} with Co and with folacin have been reviewed (Underwood, 1975; Shabe and Stokstad, 1985). Current knowledge of the neuropathological effects of vitamin B_{12} deficiency was reviewed by Metz (1992).

Structure

The compound, Co alpha-[alpha-(5,6-dimethyl-benzimidazolyl)[-Co beta-cyanocobamide], formerly known as vitamin B_{12} or cyanocobalamine should be designated cyanocobalamin. Several derivatives, including hydroxycobalamin, aquacobalamin, and nitritocobalamin, also have B_{12}

activity. The structure for cyanocobalamin is as shown:

CYANOCOBALAMIN

The term *vitamin B$_{12}$* should be used as the generic descriptor for all corrinoids exhibiting qualitatively the biological activity of cyanocobalamin. Thus, phrases such as vitamin B$_{12}$ activity and vitamin B$_{12}$ deficiency represent preferred usage (International Union of Nutritional Science, 1990).

Vitamin B$_{12}$ is a dark red crystalline compound, unstable at temperatures above 115°C for 15 min, unstable to sunlight, and stable within the pH range of 4 to 7 at normal temperatures. It contains Co and P in a 1:1 molar ratio.

The vitamin was first isolated in 1948 (Rickes, 1948; Smith and Parker, 1948). Since then, a variety of isolation sources have been used with high yields coming from fermentation liquors of streptomycin, aureomycin, or special fermentations carried out with selected microorganisms.

Functions

Vitamin B$_{12}$ functions as a coenzyme in several important enzyme systems. These include isomerases (mutases), dehydrases, and enzymes involved in metathionine biosynthesis. The oxidation of propionate in animal tissues involves a series of reactions requiring B$_{12}$ as well as pantothenic acid as components of coenzymes. Vitamin B$_{12}$ is a component of a coenzyme for the enzyme, methythamalonyl-CoA isomerase, which catalyzes the conversion of methylmalonyl-CoA to succinyl-CoA, which in turn is converted to succinate for entrance into the TCA cycle.

Vitamin B$_{12}$ is linked closely with another water-soluble vitamin, folacin, in the synthesis of methionine in both bacterial and animal cells. Vitamin B$_{12}$ deficiency leads to secondary folacin deficiency, and methionine alleviates it.

The effect of vitamin B$_{12}$ deficiency on folate metabolism is explained by the reduced reaction of a folic acid derivative (CH3-H4Pte-Glu) with homocysteine by the vitamin B$_{12}$-dependent enzyme, methyltransferase, and by its effect on liver folate levels. Methionine also increases liver folacin uptake and concentration. The interrelationships between folacin, vitamin B$_{12}$, and methionine in metabolism have been described (Shabe and Stockstad, 1985; Stockstad, 1976). Niacin (as NAD) is involved in the series of reactions as well.

Either vitamin B$_{12}$ or folacin deficiency interferes with intestinal absorption of nutrients. Changes in the epithelial cells of the intestine, along with shortened villi, are observed. In the vitamin B$_{12}$ deficiency of pernicious anemia, increased bacterial populations occur in the upper small intestine due to achlorhydria.

A more complete understanding of the role of vitamin B$_{12}$ in animal metabolism and its relationship to folacin awaits further study, but the general statement can be made that it is a component of coenzymes required for methyl-group synthesis and metabolism and that it is required for nucleoprotein synthesis in tandem with folic acid. Methionine appears to be the key factor regulating cellular uptake of folate; in the presence of low methionine, folate becomes trapped as 5-methyltetrahydrofolate, as in vitamin B$_{12}$ deficiency.

Several amino acid interconversions involve enzyme systems associated with vitamin B$_{12}$ and folacin. Folate coenzymes serve as acceptors or donors of one-carbon units in an array of reactions involved in amino acid and nucleotide metabolism, which are diagrammed in the review by Shane and Stokstad (1985). Methionine synthetase is one of the three mammalian enzymes known to require vitamin B$_{12}$ as a cofactor, along with methylmalonyl-CoA mutase and leucine 2,3-aminomutase.

Absorption, Transport, Storage, and Excretion of Vitamin B$_{12}$

Vitamin B$_{12}$ absorption requires the presence of an enzyme secreted by the mucosal cells of the stomach and upper small intestine (cardiac region of the stomach in humans; pyloric region of stomach and upper duodenum of pigs and probably other animals) (Reisner, 1968). This enzyme has been designated the intrinsic factor (IF), and in its absence B_{12} deficiency occurs. IF from one species may inhibit absorption of B_{12} in another species. For example, IF from pig stomach renders orally administered B_{12} less available to chicks and rats.

The B_{12}–IF complex passes down the GI tract to the ileum where it is complexed further with Ca and Mg ions and is adsorbed to the surface of the mucosa. It then is disassociated from the Ca or Mg by an apparently specific releasing enzyme contained in the intestinal secretions, and the B_{12} is absorbed. The exact mechanism by which it passes through the mucosal cell membrane is not known, but it has been suggested that it passes through the cell membrane by pinocytosis before the complex is broken by the releasing enzyme. Receptor-mediated endocytosis (uptake by intestinal mucosal cells) of vitamin B_{12} has been reviewed and described (Seetharam, 1999). After reaching the blood, B_{12} is bound to an alpha-globulin designated as transcobalamin I, or if injected, to a beta-globulin designated transcobalamin II, from which it is subsequently shifted to transcobalamin I. The exact role of each form in the transport and storage of B_{12} is not known.

The amount of B_{12} not needed for immediate use is stored in liver and other tissues. In humans, 30 to 60% is stored in liver, 30% in muscle, skin, and bone, and smaller amounts in other tissues (Reisner, 1968). When the protein-binding capacity of the blood serum for B_{12} is surpassed, excretion of the free vitamin occurs through the kidney and bile. More B_{12} is excreted daily in the bile in humans than is contained in the blood. The daily fecal excretion is less than the daily bile excretion, showing that reabsorption of a high proportion of B_{12} from the GI tract occurs.

Deficiency Signs

Although B_{12} is distributed widely in animal tissues and products, a deficiency is likely when simple stomached animals are maintained for long periods of time on diets entirely of plant origin. Sufficient microbial synthesis of vitamin B_{12} occurs in the upper and lower intestine of vegetarians to protect partly against a vitamin B_{12} deficiency (Albert, 1980). Hygenic practices in food handling also may have an effect on the vitamin B_{12} status of humans.

In ruminants, whose rumen microflora are capable of synthesizing the vitamin for use by the host, a B_{12} deficiency can be induced by feeding a Co-deficient diet. Because Co is a constituent of the B_{12} molecule, the vitamin cannot be synthesized when Co is not available. The relationships between B_{12} and Co in ruminants have been reviewed (Underwood, 1977; Church, 1979) and are discussed in Chapter 12.

Growth failure is a general symptom of deficiency in all animal species studied. In B_{12}-deficient chicks, poor feathering, kidney damage, impaired thyroid function, lower level of sulfhydryl groups in the thyroid, perosis, depressed plasma proteins, and elevated blood nonprotein N, and glucose occur (Reisner, 1968). Eggs from B_{12}-deprived hens fail to hatch; embryos die at about day 17 and show multiple hemorrhages, fatty livers, enlarged hearts and thyroids, and lack of myelination of the sciatic nerves and spinal cord. Humans with pernicious anemia associated with vitamin B_{12} deficiency signs may have an abnormal electroencephalogram.

Baby pigs deprived of B_{12} show rough hair coats, uncoordinated hind leg movements, normocytic anemia, and enlarged liver and thyroids. Sows fed B_{12}-deficient diets during gestation have a high incidence of abortion, small litters, abnormal fetuses, and inability to rear the young. In deficient rats, retarded heart and kidney development, fatty liver, reduced blood sulfhydryl compounds, a decrease in activities of liver cytochrome oxidase and dehydrogenases, and an increase in liver CoA activity occur. The increased pantothenic acid content of liver in B_{12} deficiency is a general observation among the animal species studied. Vitamin B_{12} deficiency in humans is manifested by a megaloblastic anemia (pernicious anemia), but is observed mainly when absorption is impaired by the absence of IF.

Breast-fed human infants nursing vegetarian mothers may show signs of vitamin B_{12} deficiency, including anemia, skin pigmentation, apathy, retardation, involuntary movements, and methylmalonic aciduria (Higginbottom et al., 1978). Mothers of these infants have mild anemia, megaloblastic bone marrow, and low plasma and milk vitamin B_{12} concentrations.

In humans consuming vitamin B_{12}-deficient diets for extended periods, severe neurological disease, including degeneration of the spinal cord and breakdown of myelin in spinal cord, brain, and peripheral nerves occur. The lesions are associated with defective methylation in the nervous system. Metz (1992) reviewed the use of animal models for studies of vitamin B_{12} deficiency; nitrous oxide accelerates vitamin B_{12} depletion in tissues by oxidizing the bioactive form of vitamin B_{12}, resulting in its excretion. The use of nitrous oxide to hasten the onset of vitamin B_{12} deficiency provides a useful tool to advance the knowledge of vitamin B_{12} nutrition.

Co-deficient cattle and sheep develop deficiency signs reversible by B_{12}. Signs include reduced appetite, emaciation, anemia, fatty liver, birth of weak young, and reduced milk production. Ketosis, a relatively common disease of lactating dairy cattle, may be related to a metabolic deficiency of B_{12}, which interferes with the metabolism of propionate. Only a small proportion of B_{12} synthesized in the rumen is absorbed and of the total vitamin B_{12} analogs produced in the rumen, only a small fraction have biological activity. Vitamin B_{12} analogs also may be produced in significant amounts by small intestinal bacteria and result in vitamin B_{12} insufficiency for the host.

Blood plasma contains several vitamin B_{12} binding proteins (human plasma contains three, trancobalamin I, II, and III) whose role is not entirely clear. One suggested role is that of providing a selective mechanism for removal of analogs of vitamin B_{12} from the body by excretion in the bile. Another role is that of leaching out vitamin B_{12} from intra- and extracellular bacteria, thereby imparting antibacterial action. Much remains unknown about the importance of vitamin B_{12} carrier proteins in nutrition.

The literature reports no evidence that vitamin B_{12} given in large doses causes acute or chronic toxicity.

❏ FOLACIN

Structure

The compound monopteroylglutamic acid (folic acid) has the structure shown below:

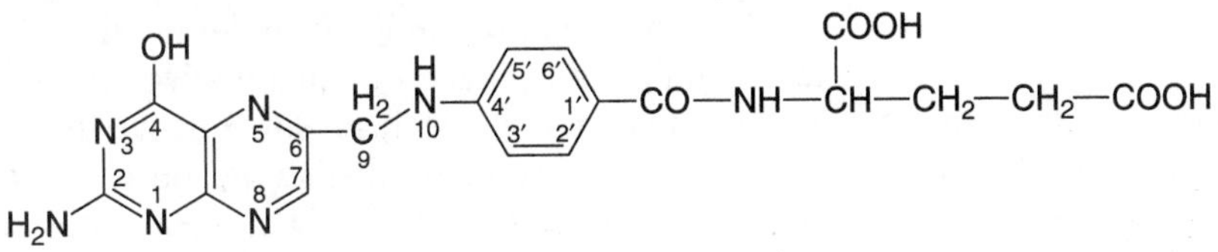

The term *folacin* should be used as the generic description for folic acid and related compounds with biological activity of folic acid (International Union of Nutritional Sciences Committee, 1980). Folic acid derivatives with the glutamic acid residue combined through a peptide bond with one or more additional residues are designated folic acid derivatives with biological activity: tetrahydrofolic acid; 5-formyltetrahydrofolic acid; 10-formyltetrahydrofolic acid; or 5-methyltetrahydrofolic acid.

Functions

The relationship of folacin to B_{12} metabolism was emphasized in the previous section. Folic acid and its derivatives take part in a variety of metabolic reactions involving incorporation of single-C units into large molecules. The metabolically active form of folacin is tetrahydrofolic acid, which is a constituent of a coenzyme associated with metabolism of single-C fragments. The in vivo kinetics of folacin metabolism have been described (Gregory and Quinlivan, 2002). Folate receptors affect its intracellular movement (Anthony, 1996). A dietary source of folacin is converted by the body tissues to tetrahydrofolic acid, which is known to be required for the biosynthesis of purines, pyrimidines, glycine, serine, and creatine. Folacin also may be concerned with synthesis of the enzymes, choline oxidase, and xanthine oxidase, and is involved in choline and methionine metabolism. Other suggested roles of folacin in metabolism, which are less well documented, include ascorbic acid biosynthesis in rats; porphyrin portions of metal porphyrin enzyme synthesis; and regulation of metal ions in the body.

Absorption, Storage, and Excretion

Considerable intestinal synthesis of folic acid and its derivatives occurs, and humans and animals may absorb significant quantities. Folacin is absorbed freely from the GI tract and is carried to all tissues of the body. Absorption is an active process. The liver contains high concentrations of folacin and apparently is the main site of conversion of folic acid into 5-formyltetrahydrofolic acid, along with the bone marrow. Vitamin B_{12} enhances the conversion.

Folic acid excretion occurs in urine, feces, and sweat. Sweat constitutes a major route of loss of folic acid in humans, but in animals whose sweat glands are less well developed (pig, cow, sheep), most of the excretion occurs in feces and urine. Methyltetrahydrofolate (MTHF), the main form of folacin in plasma, is transported to nonhepatic tissues, where it is demethylated and then returned to the liver as nonmethyl folate. Some of the nonmethyl folate is excreted in the bile and thus is subject to reabsorption. Thus, the maintenance of normal plasma folate levels depends on enterohepatic circulation. Of course, much of the fecal excretion is of microbial origin in the GI tract.

Deficiency Signs

The most prominent sign of folacin deficiency in humans and animals, aside from reduced growth rate, is a macrocytic, hyperchromic anemia, leucopenia, and thrombocytopenia. The anemia produced by folacin deficiency is indistinguishable from that of vitamin B_{12} deficiency (Shabe and Stockstad, 1985). Resistance of animals to infections is affected by folacin. For example, monkeys show increased resistance to experimental poliomyelitis when suffering from chronic deficiency, and deficient rats show impairment of antibody response to murine typhus.

Folacin deficiency causes abnormal fetuses in pregnant rats, and folacin supplementation during pregnancy may prevent birth defects (Butterworth and Bendich, 1996). The central nervous system and the skeleton appear to be the most affected. Immature ruminants have been shown to require folacin in the diet, although a deficiency is unlikely because of the high content of most feedstuffs.

Folic acid antagonists such as aminopterin and amethopterin have been shown to inhibit cell division in normal and abnormal tissues, presumably because of their effect on purine and pyrimidine synthesis. Because folacin is distributed widely in nature, it is unlikely to be deficient in common diets for birds and animals. However, antagonists create the possibility for deficiency when they are present in the diet by accident or as antimicrobials.

❏ BIOTIN

Biotin was first recognized as a growth factor for yeast. Later, rats fed raw egg white developed skin lesions and loss of hair, which were found to be cured by a protective factor in liver, now known to be biotin. Biotin was first isolated from egg yolks (29) and later from liver. The chemistry, isolation, and biosynthesis of biotin have been reviewed (Harris, 1968; Gyorgy and Langer, 1968).

Structure

The structure of biotin, formerly known as vitamin H or coenzyme R, is as follows:

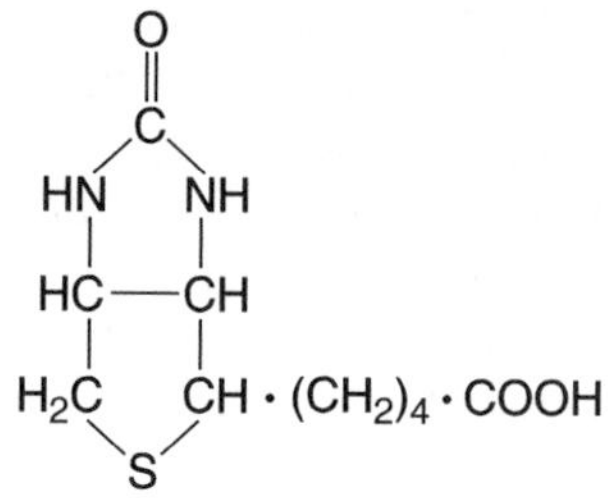

The compound hexahydro-2-oxo-1H-thieno[3,4d] imidazole-4-pentonoic acid, formerly known as vitamin H or coenzyme R, should be designated *biotin*. Its use in phrases such as biotin activity and biotin deficiency represent preferred usage (International Union of Nutritional Sciences Committee, 1980).

Functions

Biotin is a component of several enzyme systems and, as such, participates in the following reactions: conversion of propionyl CoA to methylmalonyl-CoA; degradation of leucine; fat metabolism; and transcarboxylation reactions. Its metabolic role has been reviewed (Sweetman and Nyhan, 1986; Dove and Cook, 2001).

Biotin has been suggested to be involved in aspartic acid synthesis, in deamination of amino acids, and somehow in the activities of malic enzyme and omithine transcarboxylase. The effects of biotin deficiency on protein, carbohydrate, and lipid metabolism probably can be explained on the basis of its involvement in the reactions listed above rather than on some general basis.

Deficiency Signs

A deficiency of biotin is unlikely under nominal dietary conditions because the quantitative requirement of most animal species is low relative to the biotin content of common feedstuffs and because microbial synthesis in the GI tract provides a source of biotin for ruminants and for simple-stomached animals that practice coprophagy.

Deficiency in rats is characterized by a progressive scaly dermatitis and alopecia (Gyorgy and Langer, 1968). Biotin is needed for normal reproduction in female and male rats. Deficiency in females results in birth of young with abnormalities of the heart and circulatory system, and in deficient males development of the genital system is retarded. Dermatitis and perosis are the chief signs of biotin deficiency in chicks and turkey poults. Biotin deficiency is aggravated when pantothenic acid also is deficient in the diet of rats and chicks. Biotin deficiency has been produced experimentally in young pigs, resulting in alopecia, seborrheic skin lesions, hind leg spasticity, and cracked hooves. Dietary biotin supplementation in cattle may improve hoof health. Experimental biotin deficiency signs among healthy humans eating normal diets are unconvincing. However, biotin deficiency signs have been described in human patients given total parenteral nutrition of a mixture not containing biotin (Mock et al., 1981). Signs included skin rash, hair loss, pallor, lethargy, and irritability. Within two weeks following the beginning of daily ingestion of 10 mg of biotin, the rash was gone, hair growth had started, and irritability was relieved. Inherited deficiencies of each of the three mitochondrial biotin-containing carboxylases have been reported in humans (Higginbottom et al., 1978).

Deficiency signs have been produced in chicks, rats, pigs, fish, hamsters, mink, guinea pigs, and other species fed raw egg white. A proteinlike constituent of egg white, avidin, forms a stable complex with biotin in the GI tract and renders the biotin unavailable for absorption. Avidin is destroyed by moist heat, and the avidin-biotin complex also is destroyed by steaming for 30 to 60 min, so that biotin deficiency is not likely in animals fed cooked egg protein.

Biotin deficiency symptoms have been reported recently in swine under field conditions. It has been suggested that the increased use of slotted floors may be a factor, because coprophagy may be reduced. If so, we have an example of the effects of changing technology on the need for continued reappraisal of the nutrient requirements of animals under practical conditions of production.

There is no evidence for significant toxicity signs from oral or parenteral administration of massive doses of biotin in animals.

❏ CHOLINE

Choline was first isolated from hog bile in 1849, and its structure was reported in 1867. The chemistry and metabolism of choline have been reviewed (Kuksis and Mookerjea, 1978). Choline is distributed widely in animal tissues as free choline, acetylcholine, and as a component of phospholipids, including lecithin and sphingomyelin. The structure of free choline is as follows:

$$HO-CH_2CH_2-\overset{+}{N}\begin{smallmatrix}CH_3\\[2pt]CH_3\\[2pt]CH_3\end{smallmatrix}$$

The compound with formula $(CH_3)_3$ N + CH_2CH_2OH should be designated *choline*. Its use in phrases such as choline activity and choline deficiency represents acceptable usage (International Union of Nutritional Sciences Committee, 1980).

Choline added to feeds usually is supplied as choline chloride or choline dihydrogen citrate. Choline chloride is extremely hygroscopic and therefore is often added to the diet as a 70% aqueous solution.

Functions

Choline has the following broad functions: it is a structural component of tissues (in lecithin, sphingomyelin); it is involved with transmission of nerve impulses as a component of acetylcholine; it supplies biologically labile methyl groups; it has lipotropic effects (prevents accumulation of liver fat); it facilitates the formation and secretion of chylomicrons in the intestine. Unlike most other water-soluble vitamins, there is no good evidence that choline or its derivatives are required as cofactors in enzymatic reactions, although lysophosphatidyl-choline has been found to be required for activation of glycosyltransferases in membranes.

The importance of choline as a component of phospholipids is clear from the discussion of lipids in Chapter 8. Because phospholipids are essential components of cell membranes and lipoproteins, it is not surprising that choline deficiency causes abnormalities in cell structure and function. Therefore, the effect of choline in modifying phospholipid level and composition seems to be the best available explanation for much of its broad impact on cell physiology and biochemistry.

Choline's role as a component of acetylcholine is vital, as acetylcholine is the compound responsible for transmitting nerve impulses. Brain acetylcholine levels are directly related to blood levels. It has been suggested that dietary choline may affect brain acetylcholine concentration not only by increasing acetylcholine synthesis, but also by affecting the acetylcholine receptor membranes that are rich in choline and ethanolamine phosphatides. The main physiological effects of acetylcholine are peripheral vasodilation, contraction of skeletal muscle, and slowing of the heart rate. Acetylcholine is hydrolyzed to choline and acetic acid by cholinesterases. The acetylation of choline to acetylcholine is driven by an enzyme system requiring CoA.

Choline serves as a donor of methyl groups in transmethylation reactions. Labile methyl groups may come from choline or from its oxidation products, betaine aldehyde, betaine, and dimethylglycine.

Labile methyl groups, though for a time considered essential in the diet, are synthesized in body tissues from other one-C fragments. Folic acid is involved in the metabolism of these one-C fragments from which methyl groups are derived, and vitamin B_{12} is required in the transfer of methyl groups from one metabolite to another once they are formed.

Deficiency Signs

The manifestations of choline deficiency probably are related to interference with all of its distinct functions. As choline is a component of phospholipids, a deficiency is associated with fatty liver. Other signs of choline deficiency are hemorrhagic kidneys and other tissues of rats and other species and perosis in chicks. The accumulation of lipids in the liver (fatty liver syndrome) in choline deficiency is common to all species studied. Liver fat accumulation is associated with a depressed level of serum lipids and with a change in membrane and in conformational structure of lipoproteins needed for lipid transport from liver to other tissues. Such an action of choline would explain the reduced serum lipid levels observed in choline deficiency.

Dietary methionine can replace choline completely for prevention of fatty liver in rats and pigs. The choline requirement consists of two parts—one that is indispensable and one for which methionine can be substituted. In young animals, choline synthesis from methionine can occur only if methionine is provided in excess of that needed for tissue protein synthesis; that available above the amount needed for tissue growth can supply labile methyl groups for choline synthesis from ethanolamine.

The choline requirement is reduced when the diet limits growth by a deficiency of another nutrient. Fat increases the requirement, and caloric restriction reduces it. The fatty liver and hemorrhagic kidneys produced in rats fed a choline-deficient diet are prevented by restricting the fat and carbohydrate of the diet to reduce growth.

In perosis of choline-deficient chicks, the symptoms are the same as those seen in Mn and biotin deficiencies. In swine, choline deficiency causes an abnormal gait in growing pigs and reproductive failure in adult females, in addition to the fatty liver and hemorrhages noted for other species.

No direct evidence exists for choline deficiency in humans, but a considerable effort has been devoted to studying the possible relationship between fatty infiltration of the liver in alcoholics and choline nutrition. Pharmacological doses of choline have been given successfully to patients with brain disorders associated with deficiencies of cholinergic tone. The relationship between alcoholic liver cirrhosis and nutrition is a complex one in which choline appears to be only one of a number of factors.

Reports of choline toxicity have not been found in the literature.

❏ ASCORBIC ACID (VITAMIN C)

A factor in citrus fruits capable of preventing scurvy in humans was recognized many years ago. The isolation of L-ascorbic acid and its characterization as a water-soluble vitamin have been reviewed (Englard and Seifter, 1986).

Structure

The structure of L-ascorbic acid is as shown:

The term *vitamin C* should be used as the generic descriptor for all compounds exhibiting qualitatively the biological activity of ascorbic acid. Thus, phrases such as vitamin C activity and vitamin C deficiency represent preferred usage (International Union of Nutritional Science, 1990).

The compound, 2,4-didihydro-L-threo-hexano-1,4-lactone, formerly known as vitamin C, cevitamic acid, or hexuronic acid, should be designated L-as*corbic acid* or *ascorbic acid*. The generic name *ascorbic acid* represents preferred usage.

The compound, L-threo-hexano-1,4-lactone, should be designated L-*dehydroascorbic acid* or *dehydroascorbic acid*. The generic name *dehydroascorbic acid* represents preferred usage. Both the reduced and oxidized forms of L-ascorbic acid are physiologically active.

Functions

L-ascorbic acid is formed from D-glucose by the following reactions: D-glucose → D-glucuronic acid → L-gulonic acid → L-gulono-gamma-lactone → L-ascorbic acid. Animals that cannot synthesize vitamin C lack the enzyme L-gulonolactone oxidase, which is responsible for the conversion of L-gulono-gamma-lactone to L-ascorbic acid.

Ascorbic acid is involved in several important metabolic converions, which may be summarized as follows: it is involved directly with a number of enzymes catalyzing oxidation and reduction (electron transport) reactions. Ascorbic acid, itself easily oxidized and reversibly reduced, serves as a reducing agent. It is required to maintain normal tyrosine oxidation and for normal collagen metabolism. Specifically, it is required for the formation of hydroxyproline from proline (Mapson, 1967) and hydroxylysine from lysine (Barnes and Kodicek, 1972). The hydroxylation reaction occurs only when the amino acids are in peptide linkage in collagen. Ascorbate does not participate in the hydroxylation reaction but is required to keep enzyme-bound Fe^{++} in the reduced form.

It is required as a co-substrate in certain mixed-function oxidations, as in the conversion of dopamine to norepinephrine (Barnes and Kodicek, 1972). L-ascorbic acid can be replaced by D-ascorbic acid, isoascorbic acid, or glucoascorbic acid in this reaction. It is required along with ATP for the incorporation of plasma Fe into ferritin. ATP, ascorbate, and Fe appear to form an activated complex, which allows release of ferric Fe from a tight complex with plasma transferrin for incorporation into tissue ferritin.

Because ascorbic acid is not stored in appreciable quantities in body tissues, it must be supplied almost on a day-to-day basis for those species unable to synthesize it. Most mammalian and avian species are able to synthesize it from glucose in adequate amounts, but a growing list of species is known to develop a deficiency when dietary ascorbate is withheld. Humans, other primates, and the guinea pig are the classical examples of animals unable to synthesize vitamin C. Other species now known to require a dietary source are the Indian fruit bat, red-vented bulbul, the flying fox, and some nonmammals, such as rainbow trout, Coho salmon, two species of locust, and the silkworm. Although the pig synthesizes ascorbic acid, some evidence (Cromwell et al., 1970; Yen and Pond, 1981) suggests that a growth response may be obtained by supplemental vitamin C in growing-pig diets under some conditions.

Deficiency Symptoms

The first sign of vitamin C deficiency is a depletion of tissue concentrations of the vitamin. Early signs, known as scurvy in humans, include edema, weight loss, emaciation, and diarrhea. Specific structural defects occur in bone, teeth, cartilage, connective tissues, and muscles. These defects can be explained largely by a failure in collagen formation in these tissues, including defective matrix formation in bones. Hemorrhages are commonly seen in muscles and in the gums as a result of increased capillary fragility. Hemorrhage, fatty ko-tration, and necrosis may occur in the liver, and the spleen and kidney are damaged by hemorrhage. The adrenal gland is enlarged, congested, and infiltrated with fat. Adrenal hormone secretions are affected markedly. Biochemical changes in vitamin C deficiency include increased formation of mucopolysaccharides in connective tissue, increased hyaluronic acid in repair tissue of animals with scurvy, and decreased incorporation of sulfate into mucopolysaccharides. Vitamin C deficiency is associated with decreased serum protein concentration and anemia, prolonged blood-clotting tune, and delayed wound healing.

The role of ascorbic acid in Ca and P metabolism apparently is related to its association with bone matrix formation. Ca and P cannot normally be deposited in bone when matrix formation is impaired by vitamin C deficiency. Ascorbic acid enhances Fe absorption and has less predictable, though generally beneficial, effects on utilization of S, F, and I.

Deficiency signs in humans resemble those reported in animals. Because of the generally recognized reduction in plasma ascorbic acid concentration and reduced urinary excretion of ascorbic acid in the presence of infectious disease, some researchers have suggested that higher than recommended levels should be ingested in times of illness or stress. Pauling (1971) suggested that levels several times the NRC recommended levels are effective in preventing the common cold. This claim has not been supported by the research of others and must be rejected. Although the toxicity level of ascorbic acid apparently is many-fold greater than the requirement, evidence suggests that chronic ingestion of 5 to 50 times those levels recommended may increase susceptibility to scurvy when vitamin C is withdrawn (Sorensen et al., 1974). Moreover, there is evidence of impaired phagocyte bactericidal activity associated with the administration of large doses of vitamin C. Claims of benefits in humans consuming megavitamin doses of vitamin C (several times the recommended level of 45 mg/day) have been made; it appears that the evidence does not justify the practice. The vitamin C content of plasma of farm animals able to synthesize the vitamin averages 0.3 to 0.4 mg/dl; this level is similar to that of plasma of humans consuming 45 mg of vitamin C daily. Adverse effects of high levels in the diet of animals not considered to have a dietary ascorbic acid requirement may reduce utilization of Cu and perhaps other trace elements to a marked degree. Therefore, indiscriminate ingestion of large doses of ascorbic acid in animals and humans should be avoided, even though tissue storage is limited and tolerance is high.

❑ MYOINOSITOL (Inositol)

Inositol is a common and abundant constituent of plant materials. Phytic add, a major source of P in many cereal grains and other seeds, is the hexaphosphoric acid ester of inositol, but phytase enzymes needed to hydrolyze phytic acid to release free inositol are absent or limiting in normal digestive secretions. Nevertheless, inositol is present in abundant amounts in available form. Intestinal synthesis by microflora of the GI tract also provides considerable inositol. The structure of myoinositol is as shown:

MYOINOSITOL

The compound cyclohexitol, also known as inositol or mesoinositol, should be designated *myoinositol*. Its use in phrases such as myoinositol activity and myoinositol deficiency represents acceptable usage (International Union of Nutritional Science Committee, 1990).

Myoinositol (inositol) is essential for the growth of cells in tissue cultures and has been reported to be required for the growth of gerbils, rats, and other animals under certain conditions (Holub, 1986). Tissue synthesis of inositol has been demonstrated in mammals. The structure of inositol is shown. Early work on dietary inositol was directed toward its role as a lipotropic factor for animals. Recently, the emphasis has been on the influence of dietary inositol on the levels of free inositol and its phospholipid derivatives in mammalian cells and the metabolic basis for the accumulation of fat in the liver of deficient animals. In mammalian cells, inositol exists mainly in its free form and is bound to phospholipid as phosphatidylinositol. Its concentration in animal tissues is increased by dietary inositol. Impaired release of plasma lipoprotein, increased fatty acid mobilization from adipose tissue, and enhanced liver fatty acid synthesis are all suggested as factors in dietary inositol deficiency. Cellular phosphoinositides tend to be high in arachidonic acid, which suggests special importance for inositol in biological membranes. Available data (Holub, 1986) lead to the conclusion that inositol has important functions in lipid metabolism not heretofore appreciated in animal nutrition.

The chemistry, biogenesis, and occurrence of deficiency effects in animals, and the pharmacology and toxicology of inositol have been reviewed (Alam, 1971; Cunha, 1971; Milhorat, 1971; Levander, 1980).

❏ PABA

Para-aminobenzoic acid (PABA) was first discovered as an essential growth factor for microorganisms. Although isolated reports have been made of a beneficial effect on animals by addition of PABA to the diet, no clearcut evidence for a dietary requirement exists. Thus, PABA shares with inositol the status of being a questionable candidate for classification as a vitamin. It is synthesized abundantly in the intestine, and the fact that it is a growth factor for some microorganisms suggests that a shortage in the diet of some animals may affect performance indirectly by limiting the synthesis of other vitamins. The structure of PABA is as shown:

p-AMINOBENZOIC ACID (PABA)

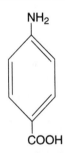

The compound should be designated *p-aminobenzoic acid*. Its use in phrases such as P-aminobenzoic acid activity and P-aminobenzoic acid deficiency represents acceptable usage (International Union of Nutritional Sciences Committee, 1980).

Interactions of Vitamins with Other Nutrients

Although nearly 50 years have passed since the discovery of the newest member of the water-soluble vitamin group (B_{12}), sporadic reports persist in the literature of unidentified factors that improve performance of animals and birds fed diets adequate in all known nutrients. Possibly one or more undiscovered organic compounds in nature are dietary essentials for animals and, therefore, will be classified as vitamins. The nature of such compounds, their distribution in nature, and the amounts and conditions in which they are required have not been determined.

Nutritional interrelationships of vitamins with trace elements were discussed in Chapter 12 and with other vitamins in previous sections of this chapter. Detailed accounts of these interactions are available for in-depth knowledge. Suffice it to say that the field of nutrition has reached heightened complexity, and an increased awareness of the importance of delicate interrelationships between and among specific nutrients must continue to be a high-priority goal in animal and human nutrition. Progress in identifying and quantifying these interactions can be expected to be enhanced by the availability of newer laboratory methods and greater analytical capabilities.

❏ SUMMARY

- Water-soluble vitamins are all organic compounds but are unrelated to each other in structure. Unlike the fat-soluble vitamins, they are not (except B_{12}) stored in appreciable quantities in body tissues and therefore must be supplied in the diet daily for those animals whose gastrointestinal tract does not provide appreciable microbial synthesis.

- Water-soluble vitamins include the following compounds: thiamin, riboflavin, niacin (nicotonic acid), pantothenic acid, vitamin B_6, vitamin B_{12}, folacin (folic acid or monopteroylglutamic acid), biotin, choline, vitamin C (ascorbic acid), myoinositol, and p-aminobenzoic acid.

- Most of the water-soluble vitamins function as metabolic catalysts, usually as coenzymes. Profound aberrations in metabolism occur if they are unavailable to the tissues in sufficient amounts. For example, thiamin deficiency produces neurological disturbances, including beriberi in humans (numbness of extremities, unsteady gait, edema of feet and legs), polyneuritis in chicks (retraction of the head), and bradycardia (slow heart rate) in all species studied.

- Rapid kidney excretion of water-soluble vitamins ingested in excess of requirements makes day-to-day administration necessary and toxicity signs unlikely.

- Some water-soluble vitamins such as biotin may be synthesized by microflora in the gastrointestinal tract of nonruminant animals in amounts sufficient to meet requirements; ruminant animals generally do not need supplemental dietary water-soluble vitamins owing to synthesis by rumen microflora.

- Vitamin C is synthesized in the tissues of most animals, but some species, including humans, monkeys, and guinea pigs, require a dietary source.
- Nutritional interrelationships among water-soluble vitamins and between vitamins and macro- and trace mineral elements are numerous. More complete knowledge of these interrelationships must be a high-priority goal.

❏ REFERENCES

Adiga, P. R., and K. Muniyappa. 1978. *J. Steroid Biochem.* **9:** 829.

Alam, S. Q. 1971. In *The Vitamins,* Vol. III (2nd ed.). Academic Press, New York.

Albert, M. J., et al. 1980. *Nature* **283:** 781.

Anthony, A. C. 1996. *Annu. Rev. Nutr.* **16:** 501.

Barnes, M. J., and E. Kodicek. 1972. *Vitamins and Hormones* **30:** 1.

Bhat, K. S., and C. Gopalan. 1974. *Nutrition Metabol.* **17:** 1–8.

Birch, T. W., and P. Gyorgy. 1936. *Biochem. J.* **30:** 304.

Brady, P. S., et al. 1979. *J. Nutr.* **109:** 1615.

Butterworth, C. E., and A. Bendich. 1996. *Annu. Rev. Nutr.* **16:** 73.

Chadhuri, D. K., and E. Kodicek. 1960. *Brit. J. Nutr.* **41:** 35.

Church, D. C. (ed.). 1979. *Digestive Physiology and Nutrition of Ruminants,* Vol. 2—*Nutrition* (2nd ed.). O & B Books, Corvallis, OR.

Cleary, R. E., et al. 1975. *Amer. J. Obstet. Gynecol.* **121:** 25.

Coates, M. E. 1968. In *The Vitamins.* Academic Press, New York.

Cromwell, G. L., et al. 1970. *J. Animal Sci.* **31:** 63.

Cunha, T. J. 1971. In *The Vitamins* (2nd ed.), Vol. III, W. H. Sebrell Jr. and R. S. Harris (eds.). Academic Press, New York.

Decker, K. F. 1993. *Annu. Rev. Nutr.* **13:** 17.

Dove, C. R. 2005. In *Encyclopedia of Animal Science,* W. G. Pond and A. W. Bell (eds.). Marcel Dekker, New York (In Press).

Dove, C. R., and D. A. Cook. 2001. In *Swine Nutrition* (2nd ed.), A. J. Lewis and L. L. Southern (eds.). CRC Press, Boca Raton, FL.

Elvehjem, C. A., and C. J. Koehn. 1934. *Nature* **134:** 1007.

Follis, R. H. 1958. *Deficiency Disease.* C. C. Thomas, Springfield, IL.

Frank, G. R., et al. 1984. *J. Animal Sci.* **59:** 1567.

Gregory, J. F., III. 1980. *J. Nutr.* **110:** 995.

Gregory, J. F., III, and J. R. Kirk. 1981. *Nutr. Rev.* **39:** 1.

Gregory, J. F., III, and E. P. Quinlivan. 2002. *Annu. Rev. Nutr.* **22:** 199.

Gries, C. L., and M. C. Scott. 1972. *J. Nutr.* **102:** 12.

Gyorgy, P., and B. W. Langer. 1968. In *The Vitamins,* Vol. II (2nd ed.). Academic Press, New York.

Harmon, B. G., et al. 1969. *J. Animal Sci.* **28:** 848.

Harris, S. A. 1968. In *The Vitamins,* Vol. II. Academic Press, New York.

Heindel, J. J., and T. R. Riggs. 1978. *Amer. J. Physiol.* **235:** E316.

Henderson, L. M. 1984. In *Present Knowledge of Nutrition* (5th ed.). Chapter 21, p. 303, Nutrition Foundation, Washington, DC.

Higginbottom, M. C., et al. 1978. *New Eng. J. Med.* **299:** 317.

Holub, B. J. 1986. *Ann. Rev. Nutr.* **6:** 563.

Horwitt, M. K. 1972. In *The Vitamins* (2nd ed.). Academic Press, New York.

Ink, J. L. 1984. *Annu. Rev. Nutr.* **3:** 289.

International Union of Nutritional Sciences. Committee on Nomenclature. 1990. *J. Nutr.* **120:** 12–19.

Kuksis, A., and S. Mookerjea. 1978. *Nutr. Rev.* **36:** 201.

Lehlem, J. E., and R. D. Reynolds. 1981. *Methods in Vitamin B-6 Nutrition,* pp. 1–401. Plenum Press, New York.

Levander, O. A., and L. Cheng. 1980. Micronutrient interactions: vitamins, minerals and hazardous elements. *Ann. NY Acad. Sci.* **355:** 1–372.

Luce, W. G., et al.. 1966. *J. Nutr.* **88:** 39; 1967—*J. Animal Sci.* **26:** 76.

Mapson, L. W. 1967. In *The Vitamins,* Vol. 1. Academic Press, New York.

Mehansho, H., M. W. Hamm, and L. M. Henderson. 1979. *J. Nutr.* **109:** 1542.

Metz, J. 1992. *Annu. Rev. Nutr.* **12:** 59.

Milhorat, A. T. 1971. In *The Vitamins,* Vol. III (2nd ed.). Academic Press, New York.

Mock, D. M., et al. 1981. *New Eng. J. Med.* **304:** 820.

Muniyappa, K., U. S. Murthy, and P. R. Adija. 1978. *J. Steroid Biochem.* **9:** 888.

Olson, R. A. 1984. In *Present Knowledge of Nutrition* (5th ed.). Chapter 26, p. 377. Nutrition Foundation, Washington, DC.

Pauling, L. 1971. *Vitamin C and the Common Cold.* W. H. Freeman, San Francisco.

Rao, B. S. N., and C. Gopalan. 1984. In *Present Knowledge of Nutrition,* 5th ed., Chapter 22, p. 318. Nutrition Foundation, Washington, DC.

Reisner, E. H. 1968. In *The Vitamins.* Academic Press, New York.

Rickes, E. L., et al. 1948. *Science* **107:** 396.

Rivlin, R. S. 1984. In *Present Knowledge of Nutrition* (5th ed.), Chapter 20, p. 285. Nutrition Foundation, Washington, DC.

Robinson, F. A. 1966. *The Vitamin Co-Factors of Enzyme System.* Pergamon Press, London.

Roe, D. A. 1973. *A Plague of Corn: The Social History of Pellagra.* Cornell University Press, Ithaca, NY.

Sauberlich, H. E. 1968. In *The Vitamins,* Vol. II. Academic Press, New York.

Sauberlich, H. E. 1984. *Annu. Rev. Nutr.* **4:** 377.

Seetharam, B. 1999. *Annu. Rev. Nutr.* **19:** 173.

Shabe, B., and E. L. R. Stokstad. 1985. *Annu. Rev. Nutr.* **5:** 115.

Smith, E. L., and L. E. J. Parker. 1948. *Biochem. J.* **43:** viii.

Sorensen, D. I., et al. 1974. *J. Nutr.* **104:** 1041.

Stokstad, E. L. R. 1976. In *Present Knowledge in Nutrition* (4th ed.), Chapter 20, p. 204. Nutrition Foundation, New York.

Sweetman, L., and W. L. Nyhan. 1986. *Annu. Rev. Nutr.* **6:** 317.

Tanpaichitr, V. and B. Wood. 1984. In *Present Knowledge of Nutrition* (5th ed.), Chapter 19, p. 273. Nutrition Foundation, Washington, DC.

Underwood, E. J. 1975. *Nutr. Rev.* **33:** 65.

Underwood, E. J. 1977. *Trace Elements in Human and Animal Nutrition.* Academic Press, New York.

Williams, R. R., and J. K. Cline. 1936. *J. Amer. Chem. Soc.* **58:** 1504.

Williams, R. J., and R. T. Major. 1940. *Science* **91:** 246.

Woolley, D. W. 1959. *Science* **129:** 615.

Wright, L. D. 1976. In *Present Knowledge of Nutrition* (4th ed.), Chapter 22, p. 226. Nutrition Foundation, Washington, DC.

Yang, Chung-Shu, and D. B. McCormick. 1967. *J. Nutr.* **93:** 445.

Yen, J. T., and W. G. Pond. 1981. *J. Animal Sci.* **53:** 1292.

Yen, J. T., et al. 1976. *J. Animal Sci.* **42:** 866.

Yen, J. T., et al. 1977. *J. Animal Sci.* **45:** 269.

16

Regulation of Nutrient Partitioning: An Overview of the Hormones That Are Affected by Nutrition or That Directly Alter Metabolism

ANIMALS EAT TO OBTAIN nutrients for a variety of reasons. Of greatest priority is their need to meet their maintenance requirement—the basic bodily functions necessary to sustain life. Young animals eat greater than maintenance intake so that they can grow. Lactating animals eat more than their maintenance requirements so that they can produce milk. Working animals consume in excess of maintenance needs to fuel muscle movement. And most animals will eat above their maintenance needs of the moment so that they can store nutrients for later use—after all, you never know when your next meal might be. We store nutrients for the long term because there may be a famine next week, but we also store nutrient for the short term so that we don't have to eat continuously. Although digestion rates vary for different foods and different animals, absorption of nutrients generally peaks within an hour or two after a meal. As time after the meal increases, nutrient absorption slows down until a new meal is consumed. This pattern of nutrient uptake results in changes in metabolism called the *feast and fast cycle* (Fig. 16.1).

After a meal (the feast), the body rapidly stores absorbed nutrients so that they will be available for later use (the fast). Liver glycogen is an efficient short-term store for absorbed carbohydrate. However, much more energy can be stored per kg body

275

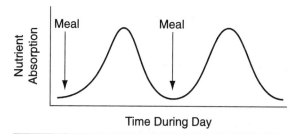

Figure 16.1   *The feast and fast cycle: changes in nutrient absorption relative to a meal.*

weight with triglycerides than with glycogen; thus, storage of energy as triglyceride in adipose depots is also a metabolic priority after a meal. In addition, although muscle is much more than just a storage depot for protein, muscle protein synthesis increases after a meal to store the excess amino acids. During periods of fasting (even if the fast is only a few hours between the absorptive phases of two meals), glycogen serves as a glucose source, stored triglycerides are used as an energy source, and muscle proteins serve as a source of amino acids for building other proteins and as a source of carbon to make glucose. All of these processes for the storage and release of nutrients in many different tissues are closely regulated by the animal's endocrine system. Hormones and growth factors regulate many of the key steps in metabolism, including transport into cells, activation to phosphorylated intermediates, transcription of metabolic enzymes, transport into mitochondria, and rate-limiting catabolic or anabolic reactions. Regulation is tissue specific, and some metabolites are more closely regulated than others. Two key concepts underlying metabolic regulation of the balance between anabolic and catabolic processes are homeostasis and homeorhesis.

Homeostasis is the maintenance of the normal body state. Most metabolites have a narrow range of concentrations in blood that is considered normal. If the concentration in blood of a certain metabolite increases as it is being absorbed after a meal, the regulatory system adjusts its use or storage so that the concentration is relatively stable. For example, blood glucose concentrations are regulated so that they are ~100 mg/dL. Glucose is absorbed after a high-starch meal, and its concentration in blood increases. Shortly thereafter, glucose uptake by body tissues is stimulated so that the blood glucose concentration drops and within a few hours is back to ~100 mg/dL. Blood calcium is also closely regulated. When mammals begin to produce milk at parturition, calcium rapidly leaves blood for the mammary gland. A drop in blood calcium stimulates release of calcium from bone and active absorption of calcium from the gut so that blood calcium returns to its normal concentration of ~10 mg/dL.

Homeorhesis is the coordination of metabolism in support of a dominant physiological process, such as growth, pregnancy, or milk production (Table 16.1). As nutrients are absorbed, they are partitioned to various body tissues. If a young growing animal is fed at 80% of ad libitum intake, it will grow at a slower rate, but the most affected area will be adipose tissue storage because muscle growth is a higher metabolic priority for the young animal. Metabolism is thus coordinated so that the limited nutrient supply is partitioned toward muscle and away from adipose. If a mature animal is in early lactation, the metabolic priority is milk synthesis. Thus, if a lactating sow does not eat enough grain to meet the

Table 16.1   Metabolic priorities of animals in various physiological states.

PHYSIOLOGICAL STATE	METABOLIC PRIORITIES
Early growth	Muscle and bone growth
Late growth	Adipose and muscle accretion
Late pregnancy	Gravid uterus and mammary development
Early lactation	Milk synthesis
Late lactation	Milk synthesis and adipose tissue repletion

needs of all body tissues, she may supply glucose and other nutrients to the mammary gland at the expense of other body tissues. Because she must meet her requirements for maintenance (circulation, respiration, etc), she will lose body fat and perhaps even muscle to ensure nutrient availability for maintenance and milk production. In this case, some stored nutrients were repartitioned to milk synthesis. Homeorhesis also exists for minerals; for example, at calving, calcium is sent to the mammary gland at the expense of bone. In a sense, an animal's tissues are competing with each other for nutrients. By adjusting the milieu of metabolic hormones and growth factors, the regulatory pathways governing nutrient flow are coordinated so that specific tissues win in this competition.

Both homeostasis and homeorhesis are affected by nutrient intake and integrated by the endocrine system, which, in turn, is governed by long-term nutritional status and physiological state. To com-

plete the cycle, the endocrine system and the dominant physiological processes will affect feed intake, which is a major determinant of nutrient intake and hence nutritional status. An overview of the control of nutrient partitioning is shown in Fig. 16.2 and Table 16.2.

❏ METABOLIC REGULATION

There are several levels of regulation in metabolism. Metabolites themselves can regulate metabolic pathways through end-product inhibition or allosteric regulation of key rate-limiting enzymes. This level of regulation may be important for homeostasis, but endocrine regulation is most important for nutrient partitioning. Endocrine factors are bioactive compounds secreted by an endocrine gland into the bloodstream to alter the function of a target cell in a different tissue. Some bioactive compounds are active within the tissues producing them through

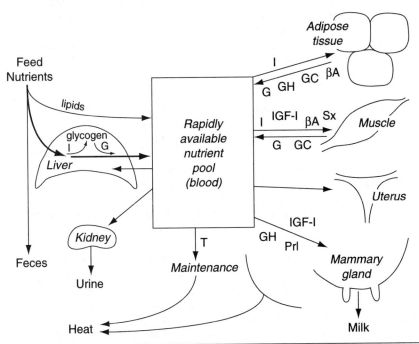

Figure 16.2 *The regulation of nutrient partitioning. Maintenance is the most important prioirty; maintenance requirement is increased by the thyroid hormones (T). Depending on the physiological state, the mammary gland, gravid uterus, or muscle will be the next priority. Nutrients are partitioned toward muscle by insulin (I), IGF-I, beta-agonists (βA), and sex steroids (Sx) and away from muscle by glucagon (G) and glucocorticoids (GC). Partitioning toward mammary tissue is enhanced by prolactin (Prl), growth hormone (GH), and probably IGF-I. Adipose tissue is usually lowest priority; insulin promotes net lipogenesis and glucagon, growth hormone, glucocorticoids, and beta-agonists promote net lipolysis.*

Table 16.2   The principal hormones that regulate metabolism.

HORMONE	STRUCTURE	SOURCE	REGULATION	ACTIONS
Insulin "Storage hormone"	51-AA peptide	β cells of pancreas	↑ rapidly by glucose, AA, ketoacids, propionate, butyrate	↑ glucose use by most tissues ↑ glycogen synthesis ↑ lipogenesis, ↓ lipolysis ↑ net protein synthesis
Glucagon "Mobilizing hormone"	29-AA peptide	α cells of pancreas	↑ by AA, butyrate, propionate ↓ by glucose, FFA, ketones	↓ glucose use by most tissues ↑ glycogenolysis ↑ gluconeogenesis ↑ net lipolysis ↑ muscle protein mobilization
Cortisol "Mobilizing hormone"	Steroid	Adrenal cortex	↑ by ACTH ↑ by stress	↓ glucose use by tissues ↑ gluconeogenesis ↑ glycogen synthesis ↑ net lipolysis ↑ muscle protein mobilization
Somatotropin "Anti-obesity production hormone"	191-AA Peptide	Anterior pituitary	↑ by G ↓ by somatostatin inversely correlated with nutritional status	↑ IGF-I and thus promotes skeletal and muscle growth ↓ lipogenesis (↑ net lipolysis)
Insulin-like growth factor-I "Anabolic hormone"	70-AA peptide	Most tissues, especially liver	↑ by GH ↓ by low insulin long term ↓ by malnutrition positively correlated with nutritional status	↑ net protein synthesis ↑ DNA synthesis (so–cell number in most body tissues) ↑ bone elongation, muscle growth (might ↑ milk synthesis and ovarian function)
Estrogens and androgens "Sex steroids"	Steroid	Gonads, also adrenal gland	↑ by pituitary gonadotropins	↑ bone elongation and maturation ↑ GH cascade ↑ muscle protein synthesis
Thyroid hormones "Metabolic pacesetter"	Small tyrosine-derived compound	Thyroid gland	↑ by cold temperature, ↑ by TSH	↑ metabolic rate within tissues (oxygen consumption) regulates metabolic rate, needed for normal growth and lactation
Catecholamines "Fight or flight" hormones	Small tyrosine-derived compound	Adrenal medulla	Diet is normally not an important regulator Rapidly secreted in response to stimulation by the nervous system	α-receptors: ↑ metabolic rate β-receptors: ↓ AA release from skeletal muscle → ↑ protein synthesis ↑ net lipolysis → ↑ serum non-esterified fatty acids ↑ glycogenolysis
Leptin "Fuel gauge"	146-AA peptide	Several tissues, esp. adipose	↑ by greater body fat mass ↑ by greater food intake	↓ food intake in adults ↑ metabolic rate and thermogenesis in neonates likely important for onset of puberty, fertility, and mammary development

paracrine or autocrine actions. Paracrine factors affect neighboring cells, and autocrine factors affect the cell that secretes them. Generally, hormones act by endocrine pathways, and growth factors act by paracrine or autocrine pathways. However, some bioactive compounds may have both endocrine and paracrine/autocrine actions. Moreover, we now know that many body tissues, not just the classical endocrine glands, produce bioactive compounds to send signals to the rest of the body. A good example is leptin, a "hormone" produced by adipose tissue that informs the brain and other tissues about the status of body energy reserves.

The Role of Insulin and Glucagon in Glucose Metabolism

Several hormones regulate carbohydrate metabolism, but the two major short-term regulators are insulin and glucagon. The glucocorticoids and catecholamines also play important roles in regulating glucose metabolism, but these will be discussed later in the context of their roles in growth. Oxytocin and the endogenous opioids (endorphins and enkephalins) also play small roles in regulating glucose metabolism.

Insulin was discovered early in the twentieth century and is well known for its role in glucose metabolism and the disease diabetes. The action of glucagon is nearly opposite that of insulin, and the ratio of insulin to glucagon is probably more important than the concentrations of either hormone alone. After a high carbohydrate meal, glucose is absorbed into blood and increases the blood concentration of insulin. Insulin then stimulates glucose use by liver, muscle, and adipose tissues, and thus removes glucose from blood. Consequently, blood glucose concentration remains relatively stable throughout the day, even though its entry into blood is much faster after a meal. In a person with insulin insufficiency, glucose concentrations after an oral glucose load will reach levels threefold that of normal and some glucose will be lost in the kidney to urine (see Fig. 7.6).

The mechanisms regulating glucose use and insulin secretion provide a good example of the layers of metabolic regulation (Barthel and Schmoll, 2003). There are several types of facilitated glucose transporters with varying affinities for glucose and capacities for glucose transport (Olson and Pessin, 1996). These transporters all move glucose across membranes with the concentration gradient, but a high-affinity transporter will transport glucose into cells even when blood glucose concentration is low, whereas a low-affinity transporter is not very active unless blood glucose concentrations are high. The capacity of a transporter indicates how quickly it can move glucose. GLUT2 is the glucose transporter of the pancreatic β cells that produce insulin; GLUT2 is a low-affinity high-capacity transporter. Thus, glucose can be transported into the β cells rapidly if blood glucose concentrations are elevated, which of course happens after an animal eats a meal with lots of starch or sugar. As blood glucose concentration rises, the concentration of glucose in the β cells increases rapidly as well. The glucose is rapidly phosphoryated and oxidized in the glycolytic and TCA cycles, and consequently the ratio of NADH to NAD in the β cells is increased. This changes the membrane voltage potential within the cells so that the secretory vesicles containing insulin migrate to the cell membrane and release insulin into blood. An increased concentration of insulin relative to glucagon directly alters the activity of several key enzymes in various tissues so that it promotes storage of glucose as glycogen in muscle and liver, conversion of glucose to fatty acids in liver or adipose tissue, and storage of fatty acids as triglycerides in adipose tissue. Muscle and adipose tissues contain the insulin-responsive, high-capacity transporter GLUT4. Greater concentrations of insulin in blood stimulate translocation of GLUT4 from microvesicles in the cell (where it is inactive) to the plasma membrane to support the greater use of glucose by these tissues. Glucose uptake is absolutely critical all the time for other tissues in the body. Red blood cells, for example, must have glucose to produce ATP; the glucose transporters in red blood cells are active even at low blood glucose concentrations and are not insulin-responsive. In ruminants, the milk-secretory cells of the mammary gland are not insulin-responsive. Thus, when glucose is in short supply, the body partitions available glucose to the tissues that most need it, but when glucose is abundant, the body is able to rapidly remove excess glucose to muscle, adipose, and liver tissues for storage.

Insulin secretion also is stimulated by amino acids, ketoacids, propionate, butyrate, and ketones. Because insulin is stored upon its synthesis, it can be rapidly secreted when needed. Thus, it is a fast-acting, short-term regulator of metabolism. It increases glucose use by most tissues, increases lipogenesis, decreases lipolysis, and increases amino acid uptake and protein synthesis in muscle. Thus, insulin promotes energy storage as glycogen, adipose tissue triglycerides, and muscle protein.

Glucagon, on the other hand, increases glucose concentrations and generally has actions that oppose insulin (Hippen et al., 1999). Glucagon is produced by the α cells of the pancreas when blood glucose concentrations are low. As time after a meal increases, absorption of glucose from the digestive tract declines, and although glucose use by tissues also declines, eventually blood glucose concentrations would drop too low for support of red blood cells and brain metabolism. Glucagon is released into blood as a signal to liver cells to begin glycogenolysis—cleaving individual glucose molecules from stored glycogen so that the glucose can be exported into blood and glucose concentrations will be maintained. Glucagon also activates the enzymes that promote gluconeogenesis so that as glycogen reserves become depleted, the body can use amino acids and glycerol to make glucose. In addition to glycogenolysis and gluconeogenesis, glucagon also stimulates mobilization of lipids from adipose tissue and mobilization of protein from muscle. Many of the mobilized amino acids serve as the primary glucogenic precursors, and the mobilized fatty acids serve as an alternate fuel when glucose is in short supply.

Insulin and glucagon are counterregulated with varying levels of carbohydrate intake, but in some situations both hormones are increased simultaneously. For example, eating a meal high in protein and low in carbohydrate stimulates both glucagon and insulin release into blood. The glucagon stimulates conversion of some of the absorbed amino acids to glucose, and the insulin promotes both storage of the glucose as glycogen and use of the glucose by tissues. Some species, such as cats, are especially adapted to this type of diet and obtain most of their glucose needs from gluconeogenesis using dietary protein as the substrate. Similarly, when ruminants eat a meal, even a meal high in starch, little glucose is absorbed because of ruminal fermentation. Instead, propionate is absorbed; propionate stimulates increases in blood concentrations of both hormones. Consequently, the propionate is converted to glucose, and the glucose is stored or used by tissues.

❑ REGULATION OF GROWTH AND FATTENING

Skeletal growth of an animal is primarily the increase in mass and length of bone and skeletal muscle. Along with this increase in skeletal mass, there is a concurrent increase in the mass of many other tissues and organs, including adipose depots. As an animal grows closer to its mature body size, nutrients are partitioned less toward muscle tissue and more toward adipose tissue. Hence, the ratio of protein to energy in the diet can be decreased as the amount of energy accretion as muscle protein becomes proportionately reduced.

The regulation of growth is complex. Physiological maturity, nutrition, environmental signals such as photoperiod and temperature, and genotypic differences in the propensity for lean and fat accretion all impact growth rates and partitioning of nutrients. Most of these effects are mediated by the endocrine system; in turn, the endocrine system can impact feed and nutrient intake. Thus, unless food or environmental constraints limit growth, the endocrine system, under the control of higher brain centers, integrates metabolism so that an animal grows to its genetic potential.

Postnatal muscle growth is regulated by several hormones that impact muscle mass through the regulation of protein synthesis, protein degradation, amino acid transport, and the proliferation and differentiation of muscle satellite cells. Fat accretion also is regulated by several hormones, which impact fat mass through the regulation of lipogenesis and fatty acid esterification, lipolysis and fatty acid oxidation, fat and glucose transport, and cell proliferation and differentiation. The most important hormones controlling muscle mass are insulin, somatotropin, and insulin-like growth factor-I. Muscle mass is also regulated by other hormones, including glucagon, the catecholamines, glucocorticoids, and sex steroids. The most important hormones controlling fat accretion are insulin and somatotropin, with additional controls from glucagon, the catecholamines, and the glucocorticoids. It is especially important in metabolic regulation, however, that two of these hormones, somatotropin and the catecholamines, have differential effects on muscle and adipose tissues—they favor deposition of muscle and mobilization of adipose. Somatotropin in particular is perhaps the major player in diverting absorbed nutrients toward muscle or mammary gland and away from adipose tissue.

The Somatotropic Axis—Somatotropin and IGF-I

The somatotropic "axis" or "cascade" was originally described as being comprised of the hypothalamus, pituitary, and liver. It included somatotropin (or growth hormone, GH), along with two upstream hypothalamic hormones, growth hormone releasing factor (GRF), which stimulates GH release, and somatostatin, which inhibits GH release, and one downstream hepatic hormone, insulin-like growth factor-I (IGF-I, also called somatomedin-C). The original somatomedin hypothesis was that hepatic IGF-I acted on target tissues in an endocrine fashion to mediate the effects of GH. Based on research in the last 20 years, however, the role of hepatic IGF-I is unclear. Most tissues synthesize IGF-I locally, and this IGF-I, acting in an autocrine/paracrine fashion, may in fact be more important in the somatotropic axis than is hepatic IGF-I (Le Roith et al., 2001). The current version of the somatotropic axis is shown in Fig. 16.3.

Somatotropin stimulates the growth of bone and muscle, although most of its effects on growth are likely indirect and mediated by IGF-I. A deficiency of GH or IGF-I results in dwarfism, whereas excess GH (and thus excess IGF-I) in young growing animals or children can cause gigantism. In 1983 scientists made headlines when they demonstrated that inserting GH transgenes into mice caused them to grow over twice the size of normal mice (Palmiter et al., 1983). In adults of most species, the growth plates of long bones are closed so

that bone elongation is no longer possible; thus, extra GH given to adults does not increase skeletal size. In adult humans, excess GH causes acromegaly, a condition in which the facial bones, hands, and feet are enlarged.

Whereas GH promotes the growth of muscle tissue, it demotes the growth of adipose tissue, and its effects in adipose tissue are not mediated by IGF-I. Somatotropin may promote lipolysis, but at least in growing pigs, its major effect is to decrease lipogenesis (Figs. 16.4 and 16.5). By altering the insulin sensitivity of adipose tissue, somatotropin can decrease the rate of fatty acid synthesis within adipose tissue by > 90%. Specifically, GH blunts insulin's stimulation of glucose transporters, acetyl-CoA carboxylase, and fatty acid synthase and decreases transcription of the fatty acid synthase gene (Etherton, 2000). In muscle, however, GH causes no change in insulin action. Therefore, GH diverts glucose from adipose tissue and toward muscle, where it is used to fuel the process of protein synthesis.

Because of its dramatic effects on nutrient partitioning, considerable interest has been shown in using GH as a metabolic modifier for livestock. As a protein hormone, feeding it is not effective, so exogenous GH must be injected directly, or endogenous GH production must be increased by injections of a GH-secretagogue (such as GRF) or genetic modification. To date, the commercial use of GH in food production has been limited to its use in dairy cows for promoting milk production. Although GH has the potential to be an effective tool for increasing the efficiency of meat production, to date the FDA has not approved GH for use

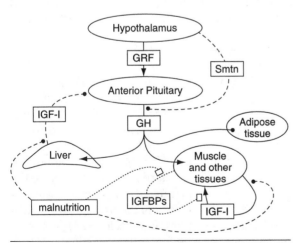

Figure 16.3 *Arrows are stimulatory pathways, closed circles indicate inhibitory actions, and open boxes indicate actions that are sometimes stimulatory and other times inhibitory.*

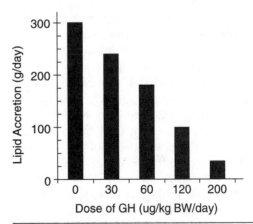

Figure 16.4 *Effect of porcine growth hormone (GH) on deposition of lipid in pigs. Adapted from Boyd and Bauman (1989).*

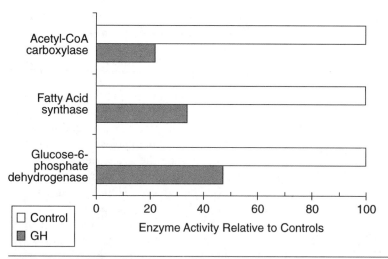

Figure 16.5   *Effect of porcine growth hormone (GH) on the activity of various lipogenic
enzymes in pigs. Acetyl-CoA carboxylase is the step needed to activate acetyl-CoA for elongation
of a fatty acid chain. Glucose-6-phosphate dehydrogenase is the first step for oxidation of glucose
in the pentose phosphate pathway, which is used to generate the reducing equivalents needed to
power fatty acid synthesis. Adapted from Harris et al. (1993).*

in commercial food production. The most success
in GH transgenic food-producing animals has been
with fish, but the FDA has not yet approved any
GH-transgenic animals.

Interestingly, in most animals, average GH concentrations in blood are inversely related to nutritional status. In other words, GH concentrations
are low when an animal is well fed and high when
it is malnourished. Because GH causes mobilization of adipose tissue but not muscle, the high-GH
concentration of an animal fed too little energy, in
conjunction with low insulin, promotes the use of
adipose tissue for body sustenance while minimizing muscle wasting. The ability of GH to increase synthesis and blood concentrations of its
mediator IGF-I is severely diminished in chronically malnourished animals. Moreover, nutrition
has a direct impact on IGF-I. Chronic dietary restriction of energy, protein, or even a single essential amino acid will decrease plasma IGF-I
concentrations. Changes in IGF-I synthesis and
secretion are mediated at the level of transcription,
and consequently several hours must elapse before dietary effects on plasma IGF-I can be observed. Therefore, feeding an animal a diet that
promotes high concentrations of GH will not promote growth; rather, it will result in lower concentrations of IGF-I and lower growth rates. Although GH is secreted episodically and its

concentrations in blood are highly variable, IGF-I
concentrations in blood are relatively stable and
can serve as an indicator of an animal's nutritional
status.

The IGF system is perhaps one of the most complex of the endocrine systems. There are two
IGFs—IGF-I and IGF-II—with two distinct receptors. Of these, IGF-I is the more important for postnatal growth and metabolism. The IGFs can interact with each other's receptors and with the insulin
receptor. Insulin, at high levels, can interact with
the IGF-I receptor. In addition, at least six IGF
binding proteins (IGFBP) have been identified that
alter the bioactivity of the IGFs, and these binding
proteins are themselves regulated by nutrition,
GH, IGF-I, as well as other factors. Furthermore,
some IGFBP may have IGF-I independent actions.

The IGFs are key regulators of the cell cycle;
IGF-I stimulates the progression of cells through
the later stages of the cell cycle and thus stimulates cell proliferation. This is especially important
in bone elongation, where locally produced IGF-I,
which is upregulated by GH, stimulates proliferation of the chondrocytes of the bone growth plates
and causes elongation. In muscle, IGF-I promotes
proliferation of satellite cells and stimulates net
protein synthesis in muscle fibers; both of these
effects contribute to its role in promoting muscle
growth. Systemic administration of recombinant

IGF-I reduces the negative nitrogen balance of postsurgical or burn trauma and corrects the hyperglycemia of insulin-resistant diabetes. Tumors often produce their own IGF-I, which contributes to their unruly cell proliferation. Topical administration of IGF-I promotes wound healing, and local synthesis of IGF-I, which is increased after a tissue wound, is likely an important part of the natural healing process. The IGFs also are involved in regulating apoptosis, or programmed cell death, and they play a role in differentiating cells into their mature function.

The liver is a major site for IGF-I synthesis and a major source for the IGF-I in circulation, but the importance of hepatic IGF-I in growth is not clear. Genetic selection of animals for increased plasma IGF-I concentrations results in increased body weight. However, whereas injections of GH or insertion of a GH transgene with a hepatic promoter enhance growth above normal, injections of IGF-I or insertion of hepatic IGF-I transgenes have only marginal effects on growth and development in vivo, even though IGF-I is downstream in the GH signaling cascade. Furthermore, transgenic animals with targeted reductions in hepatic but not peripheral IGF-I synthesis have nearly normal growth. Perhaps the major function of circulating IGF-I is to keep GH secretion in check and restrain the somatotropic axis. This feedback loop would help to keep growth tightly controlled and in concert with nutritional and environmental signals.

Many other systemic hormones, including the thyroid hormones, the glucocorticoids, the catecholamines, insulin, and leptin, influence the somatotropic axis. The hypothalamus produces other factors that regulate GH secretion—thyrotropin releasing factor increases GH secretion, whereas neuropeptide Y inhibits it. GH secretagogues from peripheral tissues include ghrelin, a 28-AA peptide produced in the stomach, and hexarelin, a 6-AA peptide that may eventually be useful in animal production or medicine. All in all, the somatropic axis is perhaps the most important hormone system for growth regulation. It is tightly controlled by the higher brain centers to partition available nutrients to various functions, and both its regulation and its effects are intertwined with the other major endocrine systems.

The Catecholamines and Glucocorticoids

The hormones of the adrenal gland also play an important role in growth and fattening. These hormones include the glucocorticoids and the catecholamines. Diet is not a major regulator of either of these hormones; instead, they are controlled by the higher nervous centers and are upregulated in response to stress.

The catecholamines include epinephrine and norepinephrine (adrenalin and noradrenalin). They are rapidly secreted from the medulla of the adrenal gland in response to stress and thus are known as the "fight or flight" hormones. There are several types of receptors for the catecholamines, and they can be broadly classified as either α- or β-adrenergic receptors. In general, stimulation of the α-receptors increases metabolic rate and peripheral blood flow, whereas stimulation of the β-receptors increases blood glucose and nonesterified fatty acid concentrations. Together, these responses enable humans or animals to do amazing things like lift cars off of trapped relatives or run faster than lions. Activation of the β-receptors also stimulates net protein synthesis in muscle and net lipolysis in adipose tissue, thus promoting partitioning of nutrients toward muscle and away from fat. Compounds that activate the β-receptors are called β-adrenergic agonists. For many years, scientists tried to produce synthetic compounds that had β-agonistic but not α-agonistic activity and could be fed to animals without leaving residues in tissues. One β-agonist, ractopamine, has been approved by the FDA for commercial use in pigs and beef cattle. Ractopamine is an effective nutrient partitioning agent that, when fed for the last four weeks before slaughter, increases muscle mass and decreases fat mass of pigs and cattle (Fig. 16.6).

The adrenal gland also produces glucocorticoids, such as cortisol. The hypothalamus of the brain responds to stress by increasing release of corticotopin-releasing hormone, which stimulates release from the pituitary gland of adrenocorticotrophic hormone (ACTH), which in turn increases the synthesis and release of cortisol, a glucocorticoid, from the cortex of the adrenal gland. Cortisol functions much like glucagon and favors gluconeogenesis and mobilization of lipid and protein from body reserves, while decreasing glucose use of peripheral tissues. Chronically high concentrations of cortisol in blood lead to muscle wasting; in humans, this is known as Cushing's Syndrome. The glucocorticoids can also inhibit somatotropin release from the pituitary, inhibit longitudinal bone growth, and suppress the immune

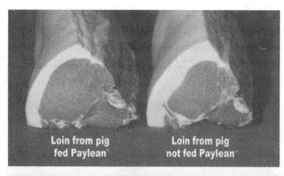

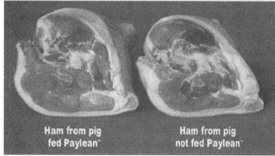

Figure 16.6 *Effect of the β-adrenergic agonist ractopamine (Paylean™) on carcass composition of pigs. Figure provided by Elanco Animal Health.*

system. Animal behaviorists often use blood cortisol concentrations as an indicator of stress. Clearly, high levels of stress and concurrent increases in cortisol are obstacles to the well-being and productivity of animals.

The Role of Puberty and the Sex Steroids in Growth

The development of the gonads and the resulting increase in circulating concentrations of estrogens and androgens around the time of puberty have a major impact on growth. The timing of puberty varies across and within species. Leptin is permissive for the onset of puberty, and extremely thin or underfed animals may experience delayed onset of puberty because they have less circulating leptin (Mann and Plant, 2002). In contrast, when prepubertal animals are fed high-energy diets ad libitum, they generally reach puberty at an earlier age. This is at least part of the reason that children in developed countries of the twenty-first century attain puberty at a younger age than did their ancestors. As puberty progresses, the secondary sex characteristics develop, bones grow and mature, and gender-specific muscling and fattening occur.

Both androgens and estrogens are found in males and females, albeit in different amounts. Although testosterone is considered the male hormone, it is converted to estradiol, considered the female hormone, by aromatase in one step. In both sexes, estrogens play an important role in the accumulation of bone mass during puberty and the maturation of the skeleton after puberty. Bones elongate at the growth plate as the chondrocytes (cartilage cells) proliferate. As these cells proliferate, the growth plate widens. Eventually, those cells that are at each end of the growth plate accumulate mineral and become hardened bone in a process called *ossification*. Eventually, the rate of ossification exceeds that of proliferation, and the growth plate becomes thinner and eventually "closes"; once this occurs, bone elongation ceases and mature skeletal size has been attained. The relatively low-estrogen and elevated GH levels of early pubertal stages stimulate longitudinal bone growth, whereas the higher concentrations of estrogen in later pubertal stages may inhibit elongation through closure of the growth plate. In contrast to estrogens, androgens can stimulate elongation without increasing bone maturation, which is why males typically have a larger mature size than do females (van der Eerden et al., 2003). The three principal hormones that determine bone mineral content are parathyroid hormone, 1,25-dihydroxycholecalciferol Vitamin D (calcitriol), and calcitonin; these are discussed further in Chapters 11 and 14.

The mechanism of action for the sex steroids is multifaceted and involves both direct and indirect effects on bone and muscle. The sex steroids stimulate the somatotropic axis during puberty, and the elevated concentrations of somatotropin and IGF-I are partly responsible for the faster growth rates. In addition, estrogens and androgens stimulate synthesis in the kidney of 1,25-$(OH)_2$ vitamin D, which enhances bone mineralization. The sex steroids also provoke the release of autocrine and paracrine factors at the growth plate. Whereas estrogen may play a more important role in stimulating the somatotropic axis, androgens do have unique effects in stimulating protein anabolism and muscle growth.

Several commercial growth promotants have been developed for beef cattle using combinations of estrogens, androgens, and progestins (Mader, 1998). These are usually administered as slow-release implants (small pellets injected under the skin, usually of the ear), with one implant lasting 100 days. Examples of these products include Compudose, Finaplex, Implus, Ralgro, Revalor, and Synovex. In general, these products induce growth and feed efficiency in steers so that they emulate bulls, but without the negative masculine behavior and composition of intact male cattle. Implants are also available for market heifers, but the relative amounts of estrogens and androgens in the implant usually differ from those for steers. Melengesterol acetate (MGA) is a progesterone analog that is commercially available as a feed additive to suppress estrous activity and enhance growth and feed efficiency in feedlot heifers.

❏ REGULATION OF MILK PRODUCTION

The endocrine system regulates mammogenesis or development of the mammary gland in young animals, lactogenesis or the transformation of the gland from the nonlactating to lactating state, and galactopoiesis or the maintenance and enhancement of an already lactating gland. Because milk synthesis can contribute considerably to an animal's nutrient requirements, the hormones that regulate mammary function will have a major impact on nutrient partitioning. For a more detailed review of the hormones that regulate mammary function, see Tucker (2000).

Before puberty, mammogenesis involves proliferation of mammary epithelial cells and some differentiation into ducts and alveolar-like structures. During pregnancy, proliferation continues, and differentiation is even more important. Because the number of alveolar cells present during lactation is a major determinant of the amount of milk produced, the hormones that govern mammogenesis early in life can impact nutrient flow later in life. Several hormones are involved in mammogenesis. Before puberty, IGF-I (and thus somatotropin) and estrogen are likely most important; the effect of estrogen may be dose dependent, with higher doses inhibiting mammogenesis but lower concentrations stimulating it. During pregnancy, progesterone is also mammogenic, and, in some species, placental lactogen is too. Other mammogenesic hormones include fibroblast growth factor, epidermal growth factor (or betacellulin in cattle), transforming growth factor-α, hepatocyte growth factor, and macrophage colony stimulating factor.

Transforming growth factor-β has been shown to inhibit or stimulate mammary development, depending on its dose. The principal lactogenic hormones are prolactin and estrogen, but the glucocorticoids and insulin are also required. In contrast, progesterone inhibits lactogenesis, and the high concentrations of progesterone that exist for most of gestation prevent the onset of lactation.

The processes of mammogenesis and lactogenesis require only a small part of an animal's nutrient requirement, so nutrient partitioning is not particularly important in a discussion of their control. The only important effect of nutrition on these two processes is that overfeeding energy and underfeeding protein will impair mammogenesis, at least in ruminants. The mechanism for this effect is not understood but may involve leptin or the somatotropic axis and the timing of the onset of puberty.

Once lactation has been established, however, the mammary gland can account for 80% or even more of an animal's net energy requirement. In early lactation, the mammary gland is a very high metabolic priority, and nutrients are portioned to the gland even if feed intake is insufficient to meet an animal's total energy needs. If energy is in short supply, the rate of fatty acid esterification falls below the rate of lipolysis in adipose tissue and the mobilized fatty acids serve as substrate for milk lipid synthesis or fuel for muscle and other tissues. If protein is in short supply, the rate of protein synthesis falls below the rate of proteolysis in muscle, and muscle amino acids become a net contributor to milk protein. For many lactating animals, especially ruminants, glucose is spared from oxidation or lipogenesis so that it can be used to make lactose. In later lactation, especially if the animal is pregnant, the priority of the mammary gland diminishes, and replacement of muscle and adipose tissues begin to take greater priority. Muscle generally has greater priority than adipose, and net accretion of adipose tissue occurs only when the energy needs of all other body functions have been met.

The endocrine system regulates galactopoiesis, at least in part, and thus regulates the partitioning of nutrients in lactating animals. The major galactopoietic hormone in cattle is GH; because GH is so galactopoietic, many scientists prefer the name "somatotropin" to "growth hormone." In nonruminants, GH is also important, but prolactin is the major hormone directing nutrients toward the mammary gland. Long-term effects of GH may be partly mediated by IGF-I. Insulin-like growth factor-I helps to establish and maintain the number of mammary epithelial cells that secrete milk; indeed, mammary-specific IGF-I-transgenes interfere with the normal mammary involution process of the gland in late lactation. However, other evidence suggests that IGF-I has little effect on the metabolic activity of mammary cells and does not mediate the galactopoietic action of GH. Growth hormone may elicit some of its effects on milk production through its lipolytic/anti-lipogenic actions in adipose tissue. Despite the fact that its mechanism of action is not understood, GH clearly has a powerful ability to partition nutrients toward milk synthesis and away from body fat deposition (Fig. 16.7).

Insulin partitions glucose to the mammary gland in some species (such as rats). However, in ruminants, insulin is not needed for glucose uptake of mammary tissues but is needed for glucose uptake of muscle and adipose tissues. Therefore, the low insulin concentrations of early lactation favor use of glucose for milk synthesis, whereas the higher insulin concentrations of later lactation increase the priority of other tissues. Interestingly, if glucose concentrations are maintained in blood by infusions of glucose along with insulin, the infused insulin promotes milk protein synthesis. Thus, the role of insulin in ruminant mammary gland is not clear at this point in time.

The glucocorticoids seem galactopoietic in some species, but the effect of cortisol on milk production in cattle is inconsistent. Actually, cortisol is more likely to suppress milk yield in cows than enhance it. The thyroid hormones may also be galactopoietic; during lactation conversion of thyroxin to triiodothyronine is increased in the mammary gland but decreased in other tissues. This differential regulation at the tissue level would enhance the metabolic activity of mammary glands relative to other body tissues, and thus partition nutrients toward milk. Although the ovarian steroids are important regulators of lactogenesis, they have little impact on an established lactation, except that high doses of estrogen may suppress milk production.

❏ REGULATION OF METABOLIC RATE AND FEED INTAKE

Thyroid Hormones

Thyroxine (T4) and triiodothryronine (T3) are synthesized in the thyroid gland from tyrosine. Nutrition has little impact on the thyroid except when the diet is deficient in iodine, which is a critical component of the thyroid hormones. Synthesis is ultimately under neural control. The hypothalamus produces thyrotropin releasing hormone (TRH), which stimulates release of thyroid stimulating hormone (TSH) from the anterior pituitary gland. Synthesis is increased by cold temperatures. The thyroid hor-

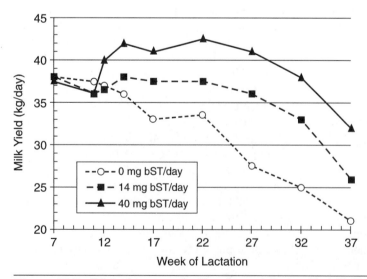

Figure 16.7   *Effect of exogenous bovine somatotropin on milk yield. Treatments began at 12 weeks of lactation and continued for 27 weeks. Adapted from Bauman et al. (1985).*

mones regulate metabolic rate and are thus important for thermoregulation. Unless accompanied by an increase in energy intake, an increase in metabolic rate would result in mobilization of body stores. The thyroid hormones are considered "permissive" for the action of many other hormones; in other words, without T3 or T4, many other hormones would not work well. The thyroid hormones are also crucial for normal bone maturation, and a deficiency of T3 is associated with severe growth retardation. In contrast, hyperthyroidism increases growth rate but can lead to premature fusion of the growth plate and thus decreased mature skeletal size.

Leptin

Leptin is a protein hormone produced by adipose tissue, so its concentrations in blood are correlated with body fatness. High-energy diets also increase circulating leptin levels, independent of their effects on adiposity. Leptin is considered a satiety hormone and plays a central role in the regulation of appetite. This likely explains why an animal can sense its energy status and eat to a specific body fat setpoint (Barb et al., 2001). For example, chronic underfeeding decreases an animal's body fat, which in turn, results in a lower circulating concentration of leptin, which in turn alters the sensitivity of the hypothalamic appetite centers and increases the animal's appetite. As a result, a thin animal tends to eat more and gain faster (compensatory gain phenomenon) than does a fat animal of the same genotype, weight, and age. Leptin is also produced by other tissues, including the placenta and mammary gland, and leptin acts as a body fat sensor for other body functions besides appetite. Of particular importance in the young animal, leptin increases metabolic rate and thus contributes to thermogenesis. Leptin also is important for the onset of puberty and normal fertility.

❏ OTHER HORMONES

Several other hormones deserve mention in a discussion of nutritional endocrinology. The gastrointestinal tract produces several hormones that are influenced by nutrient intake and/or regulate digestion, metabolism, and appetite (see Table 4.6). *Gastrin* is a peptide hormone produced by the stomach in response to distention of the stomach or the presence of amino acids and peptides

in the stomach contents. It is also stimulated by high epinephrine concentrations in blood and is inhibited by high acid concentrations in the stomach. Gastrin increases the secretion of gastric acid and pepsin into the stomach and promotes the growth of the gastric mucosal lining. *Cholecystokinin (CCK)* is a peptide hormone produced by the duodenum in response to the presence of amino acids, peptides, and long-chain fatty acids in the duodenum. It increases the secretion of bile, pancreatic enzymes, and glucagon and decreases the rate of gastric emptying. It plays a role in appetite regulation, and CCK infusion while an animal is eating causes early cessation of the meal. *Secretin* is another peptide hormone produced by the duodenum. Its synthesis is stimulated by the presence of amino acids, peptides, and acid in the duodenum. Secretin increases the secretion of bicarbonate by pancreatic and bile duct cells. *Gastric inhibitory peptide (GIP)* is secreted by the duodenum in response to the presence of glucose and fat in duodenum. It decreases gastric secretion and motility and increases insulin secretion. *Somatostatin* is produced not just by the hypothalamus to regulate GH secretion but also by gut tissues. Its secretion is stimulated by the presence of acid in the intestinal lumen. Gut somatostatin decreases the secretion of gastrin, GIP, secretin, gastric acid, pancreatic enzymes, and bile. It also decreases gastric motility and glucose absorption. These gut hormones also are discussed in Chapter 4. Adipose tissue produces not only leptin but also adiponectin and resistin, which likely play roles in insulin sensitivity, fat metabolism, and obesity.

❏ SUMMARY: IMPACT OF DIFFERENT DIETS ON HORMONES

- When a nonruminant animal is fed a well-balanced diet, high in easily digested starch and with adequate protein and other essential nutrients, absorption of glucose and amino acids is elevated soon after the meal. This results in very high-insulin and low-glucagon concentrations, which, in turn, inhibits gluconeogenesis, promotes glycogen synthesis, and promotes glucose uptake by muscle and adipose tissues, and thus accretion of body fat and protein. In addition, feeding a well-balanced diet for several days promotes maximal synthesis of IGF-I and controlled release of GH, which together support maximal rates of muscle and fat accretion and milk production.

- In contrast, if the same animal is fed a meal low in carbohydrate but high in protein, both glucagon and insulin concentrations will be elevated. Together these hormones will promote hepatic gluconeogenesis using the excess dietary amino acids as the glucogenic substrate. Because insulin will be elevated more than glucagon, and because the newly synthesized glucose will be abundant, glucose use by muscle and adipose tissues and glucose storage as liver glycogen will increase as well. The response of the somatotropic axis to this diet over time will be much the same as if the diet contained more starch and less protein. This type of diet also can result in close to maximal rates of muscle and adipose tissue accretion and milk production.

- If the same animal were fed a meal high in carbohydrate and low in protein so that the protein-to-calorie ratio was considerably below that required, insulin would be elevated and glucagon depressed much the same as with a well-balanced meal. Again, glucose uptake in muscle and adipose tissue would be upregulated, but because the meal had inadequate protein, amino acids would be lacking for net protein synthesis to be positive. Thus, the only significant storage of nutrients would occur as lipid in adipose tissue. Repeated meals of this type would eventually lead to an impairment of GH's ability to stimulate IGF-I synthesis, which would further depress growth and protein accretion or milk production.

- When a ruminant is fed a well-balanced meal, composed mostly of fiber, starch, and protein, very little glucose is absorbed. Instead, rumen microbes produce fermentation acids (especially acetate and propionate) that are absorbed across the rumen wall. Insulin and glucagon concentrations will both be elevated, and together these hormones will promote hepatic gluconeogenesis using propionate and some dietary amino acids as the glucogenic substrate. Because insulin will be elevated more than glucagon, and because the newly synthesized glucose will be abundant, glucose use by muscle and adipose tissues and glucose storage as

liver glycogen also will increase. The response of the somatotropic axis to this diet over time will be much the same as in the well-fed monogastric, and maximal rates of muscle and adipose tissue accretion or milk production will be promoted.

- During a long-term fast (several days to weeks), several metabolic adaptations take place in an effort to conserve glucose for those functions that absolutely require it. Insulin and IGF-I concentrations will be depressed, whereas glucagon, glucocorticoids, and GH concentrations will be elevated. These hormonal changes will promote the mobilization of body fat and protein, and the resulting amino acids and glycerol will serve as substrate for the increased rate of gluconeogenesis. Some of the mobilized nonesterified fatty acids will be converted to ketones in the liver, and tissues will oxidize more ketones and fatty acids and less glucose. Gluconeogenesis peaks within the first couple of days after the initiation of a fast and then declines slightly as the use of glucose declines. Together, these changes will promote the use of body lipid as the major fuel for meeting the maintenance requirement, the use of muscle protein as the glucogenic precursor for the tissues that absolutely require glucose, and the preservation of muscle protein as much as possible. With a prolonged fast, growth and milk production cease, and fertility is impaired as well.

- In cases where animals are eating but at an energy intake substantially below their energy requirements, many of the same changes will occur as with fasting, though to a lesser extent. Chronic underfeeding of calories results in decreased insulin and IGF-I and elevated GH. These changes will result in partitioning of the available nutrients to muscle and bone rather than to adipose tissue in growing animals and to mammary gland instead of to adipose tissue in lactating animals. Consequently, growth of muscle and bone in young animals may continue to occur, albeit at slower rates, while accretion of adipose tissue is negative or nearly zero. Moreover, a lactating animal could be in negative energy balance losing body fat, while still producing a large volume of milk. In adults at maintenance, chronic underfeeding of calories leads to a reduction in body calories, and unless the diet is low in protein or starch, most of the calories lost will be fat.

❏ REFERENCES

Barb, C. R., G. J. Hausman, and K. L. Houseknecht. 2001. Biology of leptin in the pig. *Domest. Anim. Endocrinol.* **21:** 297–317.

Barthell, A., and D. Schmoll. 2003. Novel concepts in insulin regulation of hepatic gluconeogenesis. *Am. J. Physiol. Endocrinol. Metab.* **285:** E685–E692.

Bauman, D. E., P. J. Eppard, M. J. DeGeeter, and G. M. Lanza. 1985. Responses of high-producing dairy cows to long-term treatment with pituitary somatotropin and recombinant somatotropin. *J. Dairy Sci.* **68:** 1352–1362.

Boyd, R. D., and D. E. Bauman. 1989. Mechanisms of action for somatotropin in growth. In *Current Concepts of Animal Growth Regulation,* pp. 257–293, D. R. Campion, G. J. Hausman, and R. J. Martin (eds.). Plenum, New York.

Etherton, T. D. 2000. The biology of somatotropin in adipose tissue growth and nutrient partitioning. *J. Nutr.* **130:** 2623–2625.

Harris, D. M., F. R. Dunshea, D. E. Bauman, R. D. Boyd, S. Y. Wang, P. A. Johnson, and S. D. Clarke. 1993. Effect of in vivo somatotropin treatment of growing pigs on adipose tissue lipogenesis. *J. Anim. Sci.* **71:** 3293–3300.

Hippen, A. R., P. She, J. W. Young, D. C. Beitz, G. L Lindberg, L. F. Richardson, and R. W. Tucker. 1999. Metabolic responses of dairy cows and heifers to various intravenous dosages of glucagon. *J. Dairy Sci.* **82:** 1128–1138.

Le Roith, D., C. Bondy, S. Yakar, J. Liu, and A. Butler. 2001. The somatomedin hypothesis: 2001. *Endocrine Reviews* **22:** 53–74.

Mader, T. L. 1998. Feedlot medicine and management. *Implants. Vet. Clin. N. Am. Food Anim. Pract.* **14:** 279–290.

Mann, D. R., and T. M. Plant. 2002. Leptin and pubertal development. *Semin. Reprod. Med.* **20:** 93–102.

Olson, A. L., and J. E. Pessin. 1996. Structure, function, and regulation of the mammalian glucose transporter gene family. *Annu. Rev. Nutr.* **16:** 235–256.

Palmiter, R. D., G. Norstedt, R. E. Gelinas, R. E. Hammer, and R. L. Brinster. 1983. Metallothionein-human GH fusion genes stimulate growth of mice. *Science* **222:** 809–814.

Tucker, H. A. 2000. Hormones, mammary growth, and lactation: a 41-year perspective. *J. Dairy Sci.* **83:** 874–884.

van der Eerden, B.C.J., M. Karperien, and J. M. Wit. 2003. Systemic and local regulation of the growth plate. *Endoc. Rev.* **24:** 782–801.

17

Factors Affecting
Feed Consumption

THE MANY FACTORS THAT affect feed consumption by animals are of great interest because of the economic factors related to feed intake and cost of production. It is well known that the total efficiency of production can be increased greatly in growing, finishing, or lactating animals if their feed intake can be maintained at a high level without creating health problems (see Table 17.1). The maintenance requirement of animals gaining weight at a slow rate represents a much greater percentage of the total feed consumed than that of animals gaining at more rapid rates. In the example shown in the table, maintenance accounted for 75%, 59%, and 48% of total energy for cattle gaining at rates of 0.45, 0.91, and 1.36 kg per day, respectively.

Additional costs are incurred at low levels of productivity, including greater labor required to maintain animals over a longer period of time, longer tie-up of invested capital, and less efficient use of facilities and equipment. High-level production is not without its costs, however, for some animals, particularly dairy cows and sheep, may be more prone to metabolic diseases such as ketosis, hypocalcemia, and acute indigestion when they are fed at high levels of intake to maximize productivity than when they are fed at lower levels of intake. Although higher quality feed ingredients and more expensive diets usually are required to attain high levels of productivity, the net cost is usually less for high rates of production in intensive animal production systems. The need to attain high levels of productivity provides the impetus for studying and understanding the factors that influence feed intake.

TABLE 17.1   Energy required to produce 340 kg (750 lb) of gain on 500 steers at three rates of gain.

DAILY GAIN, KG (LB)	DAYS ON FEED	AV. WT., KG (LB)	NET ENERGY REQUIRED		CORN EQUIVALENT, KG × 10³
			FOR MAINTENANCE, MCAL	FOR GAIN, MCAL	
0.45 (1.0)	700	340 (750)	2,156,000	708,500	1,707.7
0.91 (2.0)	350	340 (750)	1,078,000	740,250	1,176.6
1.36 (3.0)	234	340 (750)	720,720	782,730	1,018.8

Source: From McCullough (1973).

It is sometimes desirable to limit the intake of part or all of the daily ration. Good examples are limiting the intake of supplemental protein by wintering beef cows to prevent overconsumption of this expensive diet component, or restricting feed consumption by pregnant sows to prevent obesity. The incentive to limit feed intake may apply whenever the objective is to maintain the animal at a constant weight, as in mature, nonpregnant, nonlactating animals, or to cause the animal to lose weight, as in pets or other animals that have overindulged. Consequently, more information is constantly sought on factors that tend to inhibit feed consumption as well as to stimulate it.

❏ PALATABILITY, HUNGER, AND APPETITE

Palatability may be defined as the degree of acceptability of a feed or feedstuff to the taste or the degree of its acceptability to be eaten by an animal. Palatability is, in essence, the result of a summation of many different factors sensed by the animal in the process of locating and consuming food. Palatability is determined by appearance, odor, taste, texture, temperature, and other sensory properties of the food. These properties are, in turn, affected by the physical and chemical nature of the feed. In research, palatability usually is measured by giving animals a choice of two or more feeds so that they can express a preference. Hunger is aroused by a physiological need, whereas appetite is aroused by a desire to repeat a pleasant experience. **Appetite** generally refers to internal factors (physiological or psychological) that stimulate or inhibit hunger in the animal. Hunger can be satisfied by calories; appetite is satisfied by palatability (Houpt and Houpt, 1991).

Taste

The basic tastes are described as sweet, sour, salty, and bitter, in addition to the so-called common chemical sense, that is, the detection of certain chemicals that do not fall in one of these four classes. Most taste research has been done with pure chemicals in water solutions or as constituents of dry diets. However, in the usual environment, the animal is only occasionally exposed to a pure chemical, and most taste perceptions are a result of complex mixtures of organic and inorganic compounds that frequently defy description. In addition, odor frequently has a pronounced effect on taste perception. A large amount of research on taste in humans and laboratory animals has been done, but relatively little published research is available on taste in the domestic species. Research with domestic animals does show clearly that their responses do not always mimic human responses; therefore, sensations perceived by humans for a given chemical may be quite different from those perceived by animals.

Animals are able to taste chemicals that dissolve, partly because of taste buds located primarily on the tongue, but also on the palate, pharynx, and other parts of the oral cavity. Lower animals may have external taste buds on antennae, feet, and other appendages. On the tongue, different areas are sensitive to different tastes. In humans, the tip and front edges of the tongue are particularly sensitive to sweet and salty flavors. On the sides, the tongue is more sensitive to sour, and at the back, bitter. The number of taste buds varies greatly among species. It is said that averages are as follows: chicken, 24; dog, 1700; humans, 9000; pig and goat, 15,000; and cattle, 25,000 (see bibliography covering the years 1556–1966 in Kare and Maller, 1967). The number of taste buds does not necessarily reflect taste sensitivity; the chicken, for example, will reject certain flavored solutions that apparently are imperceptible to cattle.

With the four basic tastants in water solutions, sheep show a positive preference only for a few sugars in water solutions; cattle have a strong preference for sweets and a moderate preference for sour flavors; deer show a strong preference for sweets and a moderate to weak preference for sour and bitter flavors; and goats tend to show a preference for all four taste classes. The concentration of the solution has a marked effect on preference or rejection. Other information indicates that sheep and cattle prefer grasses with high organic acid content. Arnold (1970) reported that a variety of organic compounds common to some plants resulted in reduced intake by sheep. Other data with cattle indicate that some complex volatile biochemicals may have a marked effect on the taste preferences of similar forages (Aderibigbe et al., 1982). Salt-deficient sheep, however, have a marked preference for various Na salts, and it has

been shown that ruminant animals will choose to graze on grasses that have been fertilized with P and N in preference to those not so fertilized. Pigs have a pronounced liking for sweets. Chickens show a positive response to a wide variety of plant or animal tissue components but have little or no liking for some sugars, although they will consume xylose, which is not utilized and may be toxic.

Human taste is affected by a variety of factors such as age, sex, physiological condition (pregnancy, for example), and disease. Information of this type is almost nonexistent in domestic animals, although some is available on rats and dogs. It has been demonstrated that the buck deer has a stronger preference for sour compounds than the doe, and the buck shows a preference for bitter compounds such as quinine, although the doe does not. It is not known whether domestic species show the same type of response.

Great differences are known to exist in the human's ability to detect low concentrations of different tastants and, when detected, in the response. Likes and dislikes for foods and fluids vary widely. The same principle undoubtedly applies to animals. An example of the variability to be expected in similar sheep given different test solutions is presented in Table 17.2. Note the wide range in acceptability of these different solutions, a situation that is typical of animal response to such tests.

A variety of different flavoring agents, which usually have moderate to strong odors also are sold for

TABLE 17.2 An example of variation in response of sheep to different chemical solutions when given a choice between the chemical solution and tap water.[a]

CHEMICAL IN TEST SOLUTION	CONSUMPTION OF TEST SOLUTION, % OF TOTAL					
	SHEEP NO.					
	1	2	3	4	5	AV.
Sucrose, 15%	95	24	30	82	8	48
Lactose, 1.3%	73	54	42	46	68	56
Maltose, 2%	8	61	33	44	21	33
Glucose, 5%	69	88	27	71	90	69
Saccharin, 0.037%[b]	36	34	7	28	57	32
Sodium chloride, 2.5%[b]	8	23	14	22	29	19
Quinine HCl, 0.0063%[b]	51	80	23	27	43	45
Urea, 0.16%[b]	20	46	60	64	48	48

Source: From Goatcher (1969).

use in commercial feeds. A flavoring agent is often added to animal feeds on the assumption that if it smells good to humans it ought to taste and smell good to animals. A substantial amount of money is likely wasted as a result, although some information indicates that certain commercial favoring agents may increase feed consumption in some situations. Flavors (and odors) are likely to be of less consequence when animals do not have a choice of feedstuffs than when a variety of feedstuffs is available at one time. Flavor association may be useful in increasing feed consumption. This has been shown by feeding a sow a flavoring agent that is secreted into milk. Providing the same flavor in dry feed for the young pig may temporarily increase consumption of the dry feed.

Odor

Odors are produced by volatile compounds. Investigators have been unable to agree on a single broad classification of odors such as we find in taste.

It is conceded that most animals have a keener sense of smell than humans do, but quantitative information at this time is not adequate to predict an animal's response. Arnold (1966, 1970) has reported some of the classical research with domestic species (sheep). When various senses—notably, smell, taste, or touch—were impaired by surgery, Arnold found that loss of the sense of smell did not increase consumption of any of five plant species and that intake of two was decreased. Sheep were less apt to consume flowering heads, a fact also noted by others. When various odoriferous compounds were added to feed, Arnold found that the response depended on whether or not the animals had a choice, and the response (increased or decreased total consumption) was not predictable. Of six different compounds, only butyric acid increased consumption with and without a choice of feeds. A wide variety of odoriferous compounds may be objectionable to sheep at first, with the sheep reacting strongly, but eventually they may overcme their objections and eat the feed even when nonodiferous feed is available. This information suggests that odor may serve as an attractant but may not greatly influence total subsequent consumption.

Sight

Many, though not all, animal species have better vision than humans. However, the importance of sight as it affects food consumption in animals is not understood fully. Experimental evidence in humans (Pangborn, 1967) indicates that sight has a pronounced effect on taste because individuals tend to associate different colors, shapes, and other visual cues with known flavors and odors. Sight in animals appears to be used more for orientation and location of food. Research with cattle and sheep shows no effect of coloring feeds red, green, or blue on feed preference.

Texture-Physical Factors

The texture and particle size of feedstuffs may influence acceptability. Most domestic and some wild ruminant species will readily accept pelleted feeds and may prefer them to the same feeds presented in unpelleted form, even though they may be completely unfamiliar with them. Many animals will adapt more readily to rolled or cracked grains than to whole grains. Feed preparatory methods that reduce dustiness usually result in an increased feed intake, as almost all animals discriminate against dust if given a chance to do so. This is probably one reason succulent feeds are more readily consumed than dry feeds.

❑ APPETITE AND REGULATION OF FEED INTAKE

Appetite may be defined as the desire of an animal to eat and **satiety** as the lack of desire to eat. **Hunger** may be defined as the physiological state that results from the deprivation of feed of a general or specific type; hunger is abolished by the ingestion of these feeds. Appetite is frequently measured by recording the intake of feed in a short time span (Fig. 17.1)

❑ CONTROL MECHANISMS OF FEED INTAKE

All current theories agree that feed intake is affected by factors operating to control intake over a long-term period and that other short-term controls operate to control daily feed consumption. The National Research Council (1987) reviewed the factors affecting the feed intake of domestic animals and described the physiological control

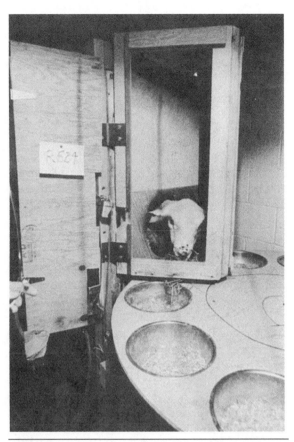

Figure 17.1 Apparatus designed to measure feed consumption of sheep in timed intervals to study periodic intake of food. Courtesy of P. J. Wangsness, Pennsylvania State University.

mechanisms. Regulation of body energy content is interfaced with the external (e.g., environmental conditions, sensory cues, diet composition) and internal (e.g., metabolic, hormonal) factors that control short-term and long-term feed intake. The central nervous system (CNS) is the primary site for the integration of feed intake and energy balance. Peptides found in the CNS have a direct effect on feeding behavior and metabolism. For example, opioid peptides influence onset of feeding, and cholesystokinin influences cessation of feeding. The CNS and peripheral receptor systems that signal information about the metabolic state of the animal coordinate feeding behavior and short-term feed intake. Although meal size and frequency may vary greatly over a 24-hr period, the overall control of feed intake must be under long-term regulation to maintain energy balance in mature animals and energy and nutrient accretion in the body

at an appropriate rate for the age and species of a growing animal.

Variations in Feeding Behavior of Domestic Animals

Houpt and Houpt (2005) have summarized the mechanisms by which animals control short- and long-term feed intake. Feeding behavior and digestive mechanisms of domestic animals vary widely from carnivores such as the cat to omnivores such as the pig and, further, to herbivores such as the cow (a ruminant) and the horse (a nonruminant herbivore). The widely varying eating habits and accompanying differences in the digestive tract are associated with correspondingly diverse physiological mechanisms to match body needs for food with eating behavior. The pig has eating habits similar to those of humans. The pattern of eating consists of periodic short meals separated by several hours of intermeal intervals, during which nutrient deficits develop and are corrected by another meal. These physiological mechanisms use hormonal and neural pathways. The signals the body uses to initiate satiation (stopping a meal) and to determine the meal size depend on the presence of food in the digestive tract. As ingested feed traverses the mouth, pharynx, and esophagus, this tactile stimulation signals a message to reduce meal size; but either an attractive or unpleasant taste may modulate the meal size. In the stomach, distention receptors in its lining inhibit intake. As the ingesta pass into the small intestine, other sensory receptors as well as endocrine cells release the hormone cholecystokinin (CCK) in response to the increasing intestinal content and to glucose absorption. These satiation effects intensify and bring the meal to an end.

In addition to the short-term control systems (hours), long-term controls (days and weeks) also operate. For example, the hormone leptin is produced by and released from fat cells and acts centrally over time to modulate eating. As fat stores slowly increase, levels of leptin produced also increase and tend to depress feed intake. Thus, the controls of feed intake are primarily inhibitory.

The cow and other ruminants as well as nonruminant herbivores have a different eating pattern. The high-cellulose forages consumed by these animals are not readily digested by their digestive tract enzymes. Instead, the cellulosic feed is digested in the rumen (in nonruminant herbivores, the cecum and colon) by the symbiotic microorganisms (mainly bacteria and protozoa) inhabiting the rumen. This is a slow process, the rate of which is decreased by a large volume of ingesta in the rumen(or cecum/colon). The end products of cellulose digestion are volatile fatty acids (VFAs) produced by anaerobic fermentation by the resident microbes. Through the process of rumination, the ingesta are regurgitated and remasticated and are returned to the rumen for further fermentation and release of VFAs. The microbial breakdown of the feed consumed following hours of grazing takes hours (not minutes). It is believed (Houpt and Houpt, 2005) that FFAs act at receptor sites either in the rumen wall or in the vascular bed of the liver and send satiety signals to the central nervous system (CNS). The CNS is informed continuously by chemical signals and nerve impulses that increase in intensity as the meal progresses and cause the meal to end at the appropriate meal size. Thus, mechanisms of control of feed intake in nonruminant carnivores and omnivores compared with ruminant and nonruminant herbivores have some similarities, but the rate of digestive release of energy from the feed for use by the animal is much slower in animals that depend on anaerobic fermentation by symbiotic microbes than it is in other animals.

Long-Term Control

Animals unquestionably have some means of controlling long-term feed intake. This is clearly demonstrated by the fact that, when adequate feed of an acceptable nature is available, wild animals do not starve or overeat; humans and some domestic animals may be exceptions. This stability of feed intake seems to be particularly true for wild species that (except for some arctic or hibernating species) seldom accumulate the amount of body fat seen in domestic species. Intrinsic and extrinsic factors involved in long-term appetite control include the physiological state of the animal such as lactation or estrus, the nitrogen and energy status of the animal, environmental factors such as temperature and humidity, photoperiod, and/or season. The effects of season (low intakes in winter, high intakes in summer) may reflect photoperiod and, hence, may be mediated at least in part by the pineal gland (Suttie et al., 1984).

Short-Term Control

Hunger and satiety centers are located in the hypothalamus (midbrain). Studies with experimental animals have shown that lesions in the appropriate location (lateral hypothalamus) will cause temporary loss of appetite and lack of thirst; this area therefore appears to be responsible for initiation of feeding. Conversely, lesions in the ventromedial hypothalamus may cause an animal to overeat, indicating that this area inhibits appetite. Electrodes strategically placed in these areas can be used to show that electrical stimuli cause the same effect; the animal will eat during stimulation and stop eating as soon as the electrical stimulus is stopped.

Daily food intake is normally consumed in a number of discrete meals. The short-term controls that initiate and end each meal are accomplished, as presently known, either by neural receptors and afferent neurons that relay impulses from the gastrointestinal tract, liver, and perhaps other organs into centers in the midbrain or by humoral (bloodborne) factors whose mechanisms are not fully understood. Humoral correlates of satiety include hormones (e.g., cortisol, insulin, estrogens) and metabolites (e.g., glucose, free fatty acids). Houpt and Houpt (1991) provided a succinct review of appetite and feeding behavior of pigs; much of the information also applies to other species (see Houpt and Houpt, 2005).

Chemostatic Regulation of Appetite

In most nonruminants, blood glucose concentration is related negatively to feed intake over the short term, and hunger contractions of the stomach are more pronounced when blood glucose is low. Administration of glucose solution into the stomach a few minutes prior to a meal may result in a marked delay in solid food intake of pigs and horses, but administration of glucose intravenously has little effect when given at about the same time as the meal starts. Either may affect subsequent meals (Houpt, 1984; Grovum, 1987), so it appears that the effect of glucose on feed intake is more of a short-term type of control. In ruminants, such relationships do not hold because blood glucose concentration has little, if any, relationship to feed intake. Data on ruminants indicate that blood insulin levels increase after feeding and that the level of insulin is influenced by the energy level of the diet and/or the amount of food consumed (Lofgren and Warner, 1972). Some evidence shows that blood propionate levels are related negatively to feed intake, as are rumen volatile fatty acids, where high concentrations in the dorsal sac may inhibit feeding. High osmotic pressure in the duodenum inhibits appetite. This effect can be produced by concentrated solutions of sugar or salt, so it is not an energy-related response (Houpt, 1984).

Growth hormone increases milk production and feed intake, but its normal role in hunger and satiety is unclear. Some of the gastrointestinal peptide hormones (gastrin, cholecystokinin, and secretin) depress intake in a dose-related manner when administered intravenously into the intestinal area (Fig. 17.2). In pigs, cholecystokinin levels in the blood are at least double after eating as before eating (Anika et al., 1981). Evidence is also available that cholecystokinin, which is produced and released within the brain, acts as a signal of satiety. Pentagastrin, a synthetic peptide with the activity of gastrin, appears to act on the brain to depress intake by sheep. Other opioid peptides administered in the central nervous system will stimulate feeding in sheep, but further information is required to understand how they fit into the overall scheme of appetite stimulation and control

(Grovum, 1987). Other unidentified chemical factors probably affect appetite in animals; this is an active area of investigation.

Caloric Density and Physical Limitations of the GI Tract on Appetite

Most adult animals can maintain a relatively stable body weight over long periods of time. Similarly, young animals of a given species tend to grow at uniform rates. Both adults and growing animals do this in spite of marked variation in physical activity and energy expenditure. This ability indicates that the animal can adjust energy intake to energy expenditure by some unclearly defined means of appetite control. If no other problems interfere, such as nutrient deficiencies or disease, animals eat to meet their caloric needs. If the diet is diluted by water, a much greater volume will be consumed. Similarly, if the diet is diluted with undigestible ingredients, the animal will eat more up to the point at which its GI tract can no longer accommodate the bulk in the diet. This principle is illustrated in Table 17.3 with chickens and graphically in Fig. 17.3.

In Table 17.3 note that feed consumption declines gradually as the energy content of the diet is increased over a rather narrow range, although total energy intake increased in this example. Feed conversion also was improved as energy concentration in the diet increased. These principles are recognized in the nutrition of nonruminant animals, and diets are adjusted to provide optimum caloric density for a given production function. For example, if one wishes to control the energy intake of pigs, it can be done by diluting the diet with alfalfa hay or some other high-fiber feedstuff.

The effect of feeding growing–finishing pigs increasing amounts of alfalfa hay in pelleted diets is shown in Table 17.4. In contrast to most instances in which pelleted diets have been fed, especially in high-roughage diets for ruminants, feed consumption did not increase (and even decreased at the highest level of alfalfa). DE consumption decreased, accompanied by less efficient feed use. However, backfat thickness was reduced with added alfalfa.

In ruminant species, in which a large proportion of the diet is composed of roughages (i.e., pasture, hay, silages) the relationship is less clearly understood, probably because so many different factors affect the optimum caloric density for a

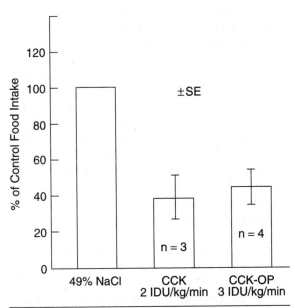

Figure 17.2 *Effect on food intake of CCK (natural cholecystokinin of porcine origin) and CCK-8 (synthetic cholecystokinin) infused into the jugular vein continuously of pigs over a 10-min. test meal time. Food consumption was depressed to about 40% of the controls. From Anika et al. (1981). Courtesy of T. K. Houpt, Cornell University.*

TABLE 17.3   Effect of energy level on performance of finishing broilers from 4.5 to 8 weeks of age.

RATION ENERGY, ME, MCAL/KG	CONSUMPTION OF FEED, KG	ENERGY, MCAL	FEED CONVERSION	DAILY GAIN, G
2.81	2.30	6.46	2.80	34.2
2.98	2.31	6.87	2.65	36.3
3.10	2.31	7.17	2.45	39.3
3.14	2.22	6.98	2.39	38.8
3.18	2.20	7.02	2.35	39.1
3.31	2.18	7.20	2.32	39.1
3.79	2.09	7.90	2.14	40.6

Source: From Combs and Nicholson (1964).

given class and species of animals. Factors such as physical density of the feed, particle size, amount of indigestible residue, solubility of the dry matter of the feed, rapidity of rumen fermentation, and level and frequency of feeding influence rate of passage through the GI tract. This, in turn, influences the amount of space in the stomach and intestinal tract for the next meal. For example, consumption of low-quality roughages such as straw and poor hay can be increased markedly by the addition of protein supplements and, sometimes, *P* or molasses. Such forage usually is deficient in both *N* and *P* for adequate rumen digestion, and small amounts of readily available carbohydrate tend to stimulate cellulose digestion in the rumen. Other nutrients required by rumen microorganisms would be expected to have the same effect (Mg and S, for example). Pelleting of low-quality roughages will almost always increase consumption greatly due to more rapid digestion and passage out of the rumen. With the very low-quality roughages (i.e., straws, stovers), only moderate amounts of supplemental feed will result in depressed roughage consumption. This phenomenon is illustrated in Table 17.5. Note that 1.2 kg of a barley–alfalfa supplement depressed ad libitum consumption of barley straw and that increasing levels continued to depress straw consumption, although total organic matter increased gradually.

As a result of these factors, determining a precise value at which increasing caloric density of a diet would result in a reduced intake of dry matter or energy would be most difficult. Age is a factor also, as young lambs or calves are less able to handle high roughage diets than older animals.

Two sets of data illustrating the effect of caloric density on energy intake are shown in Table 17.6. The data on steers illustrate the curvilinear effect on energy intake as the caloric density goes from low to high. At the low end, physical limitations

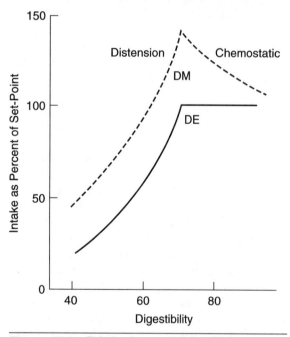

Figure 17.3   *Relation between intake of dry matter (DM) or digestible energy (DE). Data fitting this model are obtained when a concentrate diet is diluted with a bulky filler or coarse forage. Similar results have been obtained for cattle, sheep, and rats. The constancy of energy intake in high-density diets is determined by the set point of metabolism; below the inflection point gut fill, or the time spent eating and/or ruminating, becomes limiting so that the animal is not able to eat up to the level of its appetite. From Van Soest (1982).*

TABLE 17.4   Effect of percentage of roughage in the diet on feed consumption of pigs.

ITEM	DIETARY ALFALFA, %			
	0	20	40	60
Daily feed consumed, kg	3.0	3.0	3.2	2.7
Daily gain, kg	0.86	0.73	0.63	0.41
Feed efficiency	3.6	4.1	5.0	6.7
Daily DE intake, Mcal	8.91	7.80	7.40	5.50
Av. backfat, cm	3.9	3.5	3.2	2.9

Source: From Powley et al. (1981).

of the GI tract are most likely limiting intake to less than would be consumed otherwise. At the two highest energy concentrations, feed and energy intake were reduced considerably, probably because of abnormal rumen fermentation or appetite inhibition by some chemical factor produced in or present in the GI tract. In the example with lambs, dry matter was almost the same with each of these diets which were composed of varying amounts of barley, oats, and oat hulls. However, estimated ME intake and carcass gain were highest with the most concentrated diet (79.5% DE). Note also that dry matter found in the forestomach and intestinal tract after slaughter increased progressively as the caloric density of the diets decreased. These data illustrate the limiting effect of physical properties of a roughage (oat hulls) that is of low digestibility. In the lambs, the high-energy diet apparently had not reached the point of diminishing returns as it had in the steers.

The bulk density (weight per unit of volume) of a diet may have a marked effect on feed consumption. This is illustrated in Fig. 17.4. In this instance, the diets were formulated to be isocaloric but with different densities. Note that volume of feed consumed increased as the density decreased and that energy consumption (TDN) decreased as density decreased. Feed processing such as pelleting, cubing, chopping, or grinding results in increased density of most feeds.

The point (in terms of caloric density) at which an animal will have satisfied its energy demand varies considerably among species and classes of animals and according to productive function (e.g., growth, maintenance, lactation). Thus, a lactating cow will have a greater demand than a pregnant, nonlactating cow, which, in turn, will have a higher requirement than an animal that is neither pregnant nor lactating. Rapid growth is associated with a greater need for energy.

TABLE 17.5   Effect of supplemental feeds on consumption by cattle of barley straw.

ITEM	SUPPLEMENTAL FEED/DAY, KG[a]					
	1	1.5	3.0	4.5	6.0	7.5
Organic matter intake, kg/day						
Straw[b]	4.18	3.66	3.24	3.37	2.91	2.04
Supplement	—	1.20	2.40	3.60	4.80	6.00
Total	4.18	4.86	5.64	6.97	7.71	8.06
Straw OM intake, g/kg BW	12.0	10.6	9.4	9.7	8.4	5.9
Total OM intake, g/kg BW	12.0	14.0	16.3	20.1	22.2	23.3
Crude protein, % of whole diet	4.1	7.0	8.7	9.7	10.7	12.3

[a]Supplement was a 1:2 mixture of barley and dried alfalfa.
[b]Straw was fed ad libitum.
Source: From Horton and Holmes (1976).

TABLE 17.6   Effect of caloric density on feed and energy intake of steers[a] and lambs.[b]

ITEM	EXPERIMENTAL DATA										
Steers											
Roughage in ration, %	100	90	80	70	60	50	40	30	20	10	0
DE, Kcal/g	2.50	2.62	2.81	2.92	3.02	3.08	3.21	3.25	3.36	3.46	3.60
Dry matter intake, g/kg $BW^{0.75}$	94	91	98	97	91	97	98	87	85	71	46
DE intake, Kcal/kg $BW^{0.75}$	235	242	276	282	276	299	316	282	288	245	166

Lambs					
DE, %	65.2	71.6	74.9	76.6	79.5
Calculated ME, Mcal/kg	2.36	2.59	2.71	2.73	2.82
Dry matter intake, kg/day	1.06	1.03	1.00	1.00	1.02
ME intake, Mcal/day	2.50	2.66	2.71	2.73	2.88
Carcass gain, g/day	105	118	121	125	143
Dry matter in forestomach, g	1287	1302	818	816	503
Dry matter in hind gut, g	265	236	226	220	188

[a]Data from Parrott et al. (1968).
[b]Data from Andrews et al. (1969).

❏ INHIBITION OF FEED INTAKE

Control of feed intake is a complex mechanism. In addition to the various established and theorized mechanisms, appetite is affected generally by factors that interfere with the normal functions of the GI tract or that affect other tissues and organs. A high-protein meal (in humans, at least) tends to reduce feed intake, due partly, at least, to the prolonged and relatively high-heat increment resulting from the metabolism of amino acids. Fat also is inhibitory, presumably because a high-fat meal does not pass out of the stomach rapidly and fat entering the duodenum triggers hormonal mechanisms that may cause restriction of the pylorus, resulting in a slow emptying of the stomach (see also the previous section).

In human nutrition great interest has been shown in appetite inhibitors (substances that produce anorexia) that are often used in weight reduction programs. Many different compounds have been used, but most of those currently in use are amines similar to or derived from amphetamines; these compounds are stimulatory to the central nervous system but, at the same time, result in some inhibition of appetite. Undesirable side effects occur frequently, and, in addition, many people develop tolerance to the effect on appetite after a period of a few weeks. Generally, these chemical inhibitors have not been proven to be very successful, and in many conditions their use is contraindicated.

Feed intake for animals is more commonly limited by caloric dilution of diets with feedstuffs of low digestibility or by restriction of the daily feed allowance. Feed intake can be restricted partially in other ways. For example, feed provided in a physical form that is not preferred—such as a dusty meal—nearly always reduces feed intake. Unpalatable ingredients can be used such as di-ammonium phosphate or quinine. With ruminant species, salt (NaCl) has been used successfully to restrict intake, but this would be hazardous with birds and pigs because they are relatively susceptible to salt toxicity. Salt, when mixed with protein supplements such as cottonseed meal, can be fed for several months to cattle or sheep without detrimental results if water is not restricted. Salt

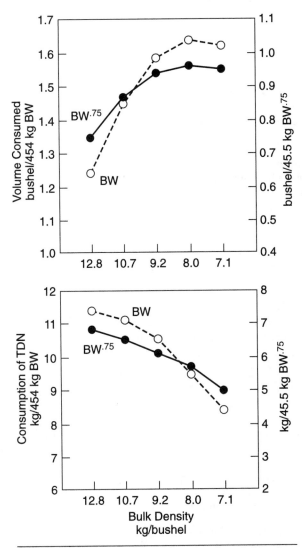

Figure 17.4 *Graphs illustrating the effect of changing bulk density on volume (upper) and energy consumption (lower) by cattle. The diets were presumed to be isocaloric. From Kellems and Church (1981).*

may be needed in a concentration of 20 to 30% of the mixture to restrict protein supplement consumption to about 1 kg per day by cattle.

Predicting Feed Intake

Some knowledge of expected feed intake is required in animal production enterprises. Profitability often depends on attaining a high level of feed intake or on knowledge of the factors associated with the feed and with the animal that affect feed intake. The National Research Council (1987) summarized the knowledge of some of these fac-

tors and provided valuable information and prediction equations that can be used for dry matter intake of beef cattle, sheep, dairy cattle, swine, poultry, and fish. This publication should be consulted for a detailed background on the factors that allow the prediction of feed intake by each of these species. Table 17.7 (National Research Council, 1987) summarizes the factors that affect feed intake by animals and the mechanisms by which they act. General determinants of feed intake include body weight of the animal, individuality, type and level of production, climatic environment, infectious and metabolic diseases, and stressors such as crowding, excessive noise, and excessive handling. Some of these determinants are discussed briefly here.

Body Weight

Energy requirements of adult animals are related to body weight raised to some power (e.g., weight to the 0.75 power). Thus, an increase in body size from, let us say, 100 kg to 110 kg, does not increase energy requirements by 10%, but by a lesser amount, 7.4%.* However, body weight is not itself always highly related to feed intake, particularly when we are dealing with finishing animals. For example, beef animals just starting on feed are apt to have relatively little body fat (e.g., 10%) and may be expected to consume on the order of 2.5 to 2.75% of body weight per day. When they are approaching market condition with, perhaps, 25 to 30% of body tissues being fat, feed consumption is likely to be more nearly 2.2% of body weight per day. This reflects the fact that energy need (and feed consumption) is related more nearly to lean body mass than to total body weight.

Individuality of Animals

Anyone who has had experience with individually fed animals realizes that they do not all eat alike and may readily show pronounced likes and dislikes when they have the opportunity to express themselves. Behavioral or hormonal differences may result in hyperexcitable or phlegmatic animals, with resultant effects on activity and feed consumption. Such differences make it difficult to predict how much some of these animals may consume.

*$100.0^{0.75} = 17.78$; $110.0^{0.75} = 19.10$; the difference, $1.32 = 7.4\%$ increase over 17.78.

TABLE 17.7   Summary of factors influencing food intake.

FACTOR	EFFECT
Sensory	
Olfaction	Controls food intake
Taste	Controls food intake
Temperature	Controls food intake
Brain	
Hypothalamus	Controls energy balance
Pituitary	Controls energy balance
Metabolites and hormones	
Somatomedin	Affects muscle and cartilage
Glucose	Little control of food intake in ruminants; greater control in monogastrics
GH[a]	Decreases insulin; initiates feeding
Insulin	Decreases insulin; initiates feeding
Glucagon	Undefined
FFA	Undefined
Amino acids	None in ruminants
Leptins	Undefined
Digestion	
Meal size	Tension receptors detect rumen distension
Diet digestibility	Duodenal receptors detect absorbed nutrients in sheep
Feeding frequency	Affects rate of ingesta passage
H_2O intake	Controls food intake
Fermentation pH	Affects chemoreceptors in rumen wall
Urea	
Urea, ammonium chloride, and ammonium lactate (injections)	Shortens meal length in goats
Urea (as feed additive)	Decreases length of first meal and meal size, but total intake remains unchanged because spontaneous meal number increases in cows
Acetate	Reduces meal size in cattle, sheep, and goats
Lactate (sodium lactate injections)	Reduces meal size in goats
Propionate	Reduces intake; shows evidence of propionate receptors in ruminal vein walls
Sex hormones	
Estrogenic compounds	Increase food intake in ruminants
Progesterone	Affects other ovarian hormones
DHEA[b]	Decreases weight gain without affecting food intake in mice
PRL[c]	Affects lactation and other physiological responses
FSH[d]	Affects lactation and other physiological responses
LH[e]	Affects lactation and other physiological responses

[a]GH, growth hormone.
[b]DHEA, dehydroepiandrosterone.
[c]PRL, prolactin.
[d]FSH, follicle-stimulating hormone.
[e]LH, luteinizing hormone.
From Houpt and Houpt (2005).
Source: From National Research Council (1987).

Type and Level of Production

All young animals have a so-called biological urge to grow, and growth cannot be impeded greatly except by severe feed restriction or nutritional deficiency. Almost invariably, those animals showing the most rapid growth rates have the best appetites. Pregnancy and lactation result in an increased appetite. With respect to lactating animals, feed intake is not correlated completely with production. Some cows, for example, have the ability to lactate at very high levels, yet do not increase feed intake sufficiently to accommodate the greater milk secretion, resulting in a loss of a considerable amount of body weight during peak lactation. On the other hand, some animals will not deplete body reserves greatly in order to produce milk.

Other Factors

High ambient temperatures, especially along with high humidity, reduce feed intake. On the other hand, cold temperatures usually stimulate feed intake. Most infectious diseases result in reduced feed intake, the degree of which is related to the severity of the infection. Similarly, intestinal parasites such as ascarids usually result in reduced feed consumption, often for prolonged periods of time. Metabolic diseases such as ketosis, bloat, and diarrhea result in restricted feed consumption.

Stressors usually reduce consumption. Crowding, noise and disturbances, and excessive handling all contribute to reduced feed consumption.

Proper design for feed bunks, mangers, and water supplies can encourage increased intake. Cleanliness of feed and water containers may also have a pronounced effect, as most animals object to dirt, sour feed, molds, and manure in or near their feed or water. Inadequate water supplies or water that is contaminated with animal excreta or spoiled feed or has a high dissolved solids content result in reduced feed consumption.

Expected Feed Intake of Cattle, Sheep, Swine, Poultry, and Fishes

The overall factors affecting feed consumption by most food animals and methods of predicting feed intake were thoroughly reviewed by the National Research Council (1987). Selected tabular and graphic data from that publication are reproduced in this section to provide an appreciation of the feed intake characteristics of various animal species. Each species of animal has unique feed consumption patterns based on differences in gastrointestinal tract architecture and many other variables related to physiological and metabolic differences. For example, the total feed consumption of cattle, as with other ruminants, is highly dependent on the quality of the roughage being consumed. These effects are illustrated in Table 17.8.

Lactating cows may consume forage at the upper limits or even exceed the values shown in the table, but those with low requirements will not consume nearly as much. Feed preparation, such as pelleting, or proper supplementation of low-quality roughages may increase consumption greatly. Silages are only rarely consumed at a level approaching that of high-quality hay, even though the

TABLE 17.8 Expected roughage consumption by cattle.

FEEDSTUFF	USUAL RANGE IN ENERGY		DRY MATTER INTAKE % BODY WT/DAY
	TDN, %	DE, MCAL/KG	
Lush, young legume, grass pasture; barn-dried grass	70+	3.10+	2.75–3.5
Well-eared corn silage; high quality sorghum silage	70	3.10	2.0–2.5[a]
Moderate quality, actively growing pasture	60–65	2.65–2.87	2.5–3.2
Grass and grass–legume stage of good quality	55–60	2.43–2.65	2.0–2.5[a]
Good quality legume hays	50–55	2.20–2.43	2.5–3.0
Grass hay from mature plants; regrowth pasture	45–50	1.98–2.20	1.5–2.0
Poor quality grass hays; dormant pasture	40–45	1.76–1.98	1.0–1.5
Cereal and grass straws	35–40	1.54–1.76	1.0 or less

[a]Cattle will only rarely eat as much dry matter from silage as when the same crop is preserved in other forms.

digestibility of the silage may be quite good; the high water content may be partially responsible as well as the acid pH of the silage.

Consumption by dairy cows of rations of low digestibility may be in direct proportion to body weight rather than in proportion to metabolic size (BW to the 0.75 power).

Figure 17.5 illustrates the range in forage consumption that may be expected by dairy cows given various levels of concentrates. Small amounts of concentrates may not depress forage consumption; they may even increase consumption of low-quality roughages. As a general rule, however, increased concentrate consumption will result in a gradual reduction in roughage intake and in total dry matter consumption, although energy intake will nearly always increase as concentrates are added. For dairy cows, most nutritionists recommend 35 to 55% concentrate in the total ration for optimum utilization of roughage and near-maximum energy intake. For finishing beef cattle, the amount of concentrate fed in the United States in current practice is about 60 to 90%, depending on the quality of the roughage and on feed preparation methods used on both roughage and concentrates. Expected daily dry matter intake of dairy cows during and after lactation have been plotted (Fig. 17.6) (National Research Council, 1987). The shape of the time curve is similar at different levels of production, although the absolute daily intake is greater for high-producing cows than for low producers.

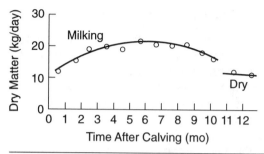

Figure 17.6   *Feed intake of Holstein cows during and after lactation. From National Research Council (1987).*

With respect to total feed consumption, young animals consume more feed per unit body weight than older animals. Young Holstein calves may be expected to consume 3.2 to 3.4% of body weight at 2 months of age of a complete feed and to consume about 3% at 6 months of age. After this their feed intake will decline gradually. The effect of lactation on daily fed intake of dairy cattle is shown in Fig. 17.6 (National Research Council, 1987). Note the gradual increase to peak of lactation, followed by a gradual decline as lactation progresses.

Suggested maximum dry matter intakes for finishing beef cattle (National Research Council, 1994) range from 2.96% for lightweight calves to 2.2% for calves weighing 450 kg. For yearling steers, values given range from 3.2% of body weight for animals weighing 135 kg decreasing to 2.36% for those weighing 454 kg. Of course, the caloric and bulk densities will have a marked effect on maximum consumption. Mature cattle will not normally be fed diets with such high caloric density as finishing cattle.

Experiments carried out under calorimetric conditions indicate that sheep will consume roughages at about the same level/unit of metabolic size ($BW^{0.75}$) as cattle. However, because their actual size is considerably smaller, the amount, when expressed as a percentage of body weight, will be greater. Daily intake of total diet by finishing lambs is about 4.3% of body weight at 30 kg (66 lb) ranging down to 3.5% at 55 kg (121 lb). This level of consumption may be exceeded when lambs are fed high-roughage, pelleted rations.

Feed consumption of piglets weaned to dry diets at 2 or 3 weeks of age may approach or exceed 10% of body weight daily. That of growing pigs from weaning to market weight (18–100 kg; 40–220 lb) may be expected to be about 2.0 to 3.2 kg/day

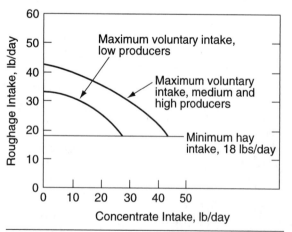

Figure 17.5   *Estimated maximum voluntary intake of roughages (90% dry matter) by 1200-lb cows at varying concentrate intakes. From Dean et al. (1969). By permission of Lea & Febiger.*

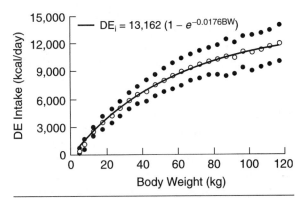

Figure 17.7 *Digestible energy intake as an asymptotic function of BW (body weight). Mean values (○) and standard deviations (●) of 8072 observations of 1490 pens of pigs fed nutritionally adequate corn-soybean meal diets are shown (Ewan, 1983). Summaries were made at 5 kg increments. From National Research Council (1987).*

(4.5–7 lb). Consumption level will be affected markedly by physical and caloric density as well as other factors discussed previously. A graphic presentation of typical digestible energy intake of pigs from 5 to 120 kg body weight is shown in Fig. 17.7 (National Research Council, 1987). Averaged over the entire growing period, daily feed consumption is about 5% of body weight; it is higher for the young pig and decreases gradually as the pig gets older and fatter. Mature animals, except for lactating sows, usually are limit-fed to about 1.8–2.3 kg (4–5 lb) daily to avoid overfattening.

Lactating sows may be expected to consume 3 to 4% of body weight per day of a moderate energy diet.

Expected feed consumption of mixed sexes of broiler chicks fed typical diets and with favorable conditions is shown in Table 17.9.

As in other species, feed consumed per unit body weight declines as the bird grows. Daily feed consumption of laying hens is affected markedly by size of the hen and her level of egg production. Birds weighing 2.3 kg and producing at a level of 60 to 70% consume feed at about 10.9 to 11.3 kg per 100 birds.

Feed intake by fishes depends not only on size, genus, and species, but also on other factors including water temperature. There is an increase in feed consumed per unit body weight as the water temperature rises and a decrease as the body weight increases (The feed intake of beef and dairy cattle, sheep and goats, swine, poultry, fishes, and other animals is addressed in the respective chapters of this book written by experts in the feeding and nutrition of the class of animals they have covered.)

❑ SUMMARY

- Many different factors affect feed consumption by animals. Ruminants and nonruminant herbivores (e.g., horses) depend on symbiotic microorganisms in their digestive tract to release energy from cellulose, whereas nonruminant

TABLE 17.9 Estimated dry matter intake of broilers at different ages.

AGE (DAYS)	BW (G)	DAILY GAIN (G, APPROX.)	ESTIMATED ENERGY NEEDS[a] (ME, KCAL/DAY)			DAILY CONSUMPTION OF AIR-DRY FEED (3.2 ME, KCAL/G)	DAILY DRY MATTER INTAKE (G) (12% MOISTURE)
			MAINTENANCE	GAIN	TOTAL		
7	130	27	47.4	55.3	102.7	32.2	28.3
14	320	34	85.9	69.7	155.6	48.6	42.8
21	560	43	124.4	88.1	212.5	66.4	58.4
28	860	56	165.1	114.8	279.9	87.5	77.0
35	1250	63	211.3	129.2	340.5	106.4	93.6
42	1690	59	257.8	120.9	378.7	118.4	104.2
49	2100	60	297.6	123.0	420.6	131.4	115.6

[a]Based on the following equation of Hurwitz et al. (1978): ME (Kcal/day) = 1.91 BW 0.66 + 2.05ΔW, where BW is body weight (in g) and ΔW is daily gain (in g).
Source: From National Research Council (1987).

omnivores and carnivores depend more on intestinal enzymes for digestion of starch and other carbohydrates. These differences between species are related to differences in feed intake controls, although similar long-term controls of feed intake operate in ruminants and nonruminants. Factors such as taste, odor, physical texture, and chemical composition of the diet may alter consumption. In general, animals tend to regulate short-term and long-term daily feed intake by complex physiological responses to diet and to the environment and by their need for energy.

- Animals that require more feed—such as in lactation, heavy work, and rapid growth—have greater appetites, but the specific mechanisms that control feed intake are not fully understood. Expected feed intake is important when diets are designed for a specific purpose. If consumption is less than anticipated, then digestible energy concentration must be higher in order to meet nutritional needs.

❏ REFERENCES

Aderibigbe, A. O., D. C. Church, R. B. Frakes, and R. G. Petersen. 1982. *J. Animal Sci.* **54**: 164.

Andrews, R. P., M. Kay, and E. R. Orskov. 1969. *Animal Prod.* **11**: 173.

Anika, S. M., T. R. Houpt, and K. A. Houpt. 1981. *Amer. J. Physiol.* **240**: R310.

Anonymous. 1972. Poultry management and business analysis manual. *Maine Ext. Serv. Bul.* 566.

Arnold, G. W. 1966. *Austral. J. Agr. Res.* **17**: 531.

Arnold, G. W. 1970. In *Physiology of Digestion and Metabolism in the Ruminant Oriel Press.* Newcastle-Upon-Tyne, England.

Combs, G. F., and J. L. Nicholson. 1964. *Feedstuffs* **36**(34): 17.

Conrad, H. R., A. D. Pratt, and J. W. Hibbs. 1964. *J. Dairy Sci.* **47**: 54.

Dean, G. W., D. L. Bath, and S. Olayide. 1969. *J. Dairy Sci.* **52**: 1008.

Della-Ferra, Mary Anne, and C. A. Bade. 1984. *J. Animal Sci.* **59**: 1362.

Goatcher, W. D. 1969. M.S. Thesis, Oregon State University, Corvallis, OR.

Grovum, W. L. 1987. In *The Ruminant Animal,* p. 202, D. C. Church (ed.). Prentice-Hall, Englewood Cliffs, NJ.

Horton, G.M.J., and W. Holmes. 1976. *Animal Prod.* **22**: 419.

Houpt, T. R. 1984. *J. Animal Sci.* **59**: 1345.

Houpt, T. R. and K. A. Houpt. 2005. Behavior: feeding. In: *Encyclopedia of Animal Science,* W. G. Pond and A. W. Bell, (eds.). Marcel Dekker, New York (In Press).

Houpt, T. R. and K. A. Houpt. 1991. In *Swine Nutrition,* Chapter 21, E. R. Miller et al. (eds.). Butterworth-Heinemann, Boston, MA.

Hurwitz, S., et al. 1978. *Poult. Sci.* **57**: 197.

Kare, M. R., and O. Maller (eds.). 1967. *The Chemical Senses and Nutrition.* Johns Hopkins University Press, Baltimore, MD.

Kellems, R. O., and D. C. Church. 1981. *Proc. West. Sec. Amer. Soc. Animal Sci.* **32**: 26.

Lofgren, P. A., and R. G. Warner. 1972. *J. Animal Sci.* **35**: 1239.

McCullough, M. E. 1973. *Feedstuffs* **45**(7): 34.

Montgomery, M. J., and B. R. Baumgardt. 1965. *J. Dairy Sci.* **48**: 569.

National Research Council (NRC). 1973. *Nutrient Requirements of Swine.* National Academy of Sciences, Washington, DC.

National Research Council (NRC). 1994. *Nutrient Requirements of Beef Cattle.* National Academy of Sciences, Washington, DC.

National Research Council (NRC). 1987. *Predicting Feed Intake of Food-Producing Animals.* National Academy Press, Washington, DC.

National Research Council (NRC). 1994. *Nutrient Requirements of Beef Cattle.* National Academy Press, Washington, DC (In Press).

Pangborn, R. M. 1967. *The Chemical Senses and Nutrition.* Johns Hopkins University Press, Baltimore, MD.

Parrott, C., H. Loughhead, W. H. Hale, and C. B. Theurer. 1968. *Arizona Cattle Feeders Day Report,* University of Arizona, Tucson.

Powley, J. E., et al. 1981. *J. Animal Sci.* **53**: 308.

Suttie, J. M., R. N. B. Kay, and E. D. Goodall. 1984. *Livestock Prod. Sci.* **11**: 529.

Van Soest, P. J. 1982. *The Nutritional Ecology of the Ruminant.* O & B Books, Corvallis, OR.

18

Feeding Standards and Productivity

❑ WHAT ARE FEEDING STANDARDS?

FEEDING STANDARDS ARE STATEMENTS or quantitative descriptions of the amounts of one or more nutrients needed by animals. Use of such standards dates back to the early 1800s. Currently, after many years of research and experience, nutrient requirements for farm animals may be specified with a relatively high degree of accuracy. Although there are still gaps in our knowledge of nutrient requirements for animals, the nutrient needs of most animals are far better understood than those for humans. This is mainly a result of limitations in the type of experimentation that can be done with humans to collect good quantitative data. The wide variability in cultures and living conditions and the largely uncharted genetic differences in nutrient requirements and metabolism in humans also contribute to the incomplete information on appropriate diet composition and intakes through the life cycle.

In the United States the most widely used standards are those published by the various committees of the National Research Council (NRC), National Academy of Sciences. These standards (see Chapters 22 through 30 for information on each animal species covered in this book) are periodically revised and reissued.

In the United Kingdom, standards in use are developed by the Agricultural Research Council (ARC). Other countries have similar bodies that update information and make recommendations on animal nutrient requirements.

307

In the last 20 to 25 years, personal computers have come into common use in most areas and have found wide application to animal nutrition. If you were to look closely at the 1996 NRC for beef cattle, you would find that computer modeling has been used extensively to predict animal utilization of carbohydrates, proteins, and fats and to suggest how these values may be used in ration formulation. It is hoped that these procedures will improve our ability to predict how animals use feed and what their response will be in many different environmental situations.

Terminology Used in Feeding Standards

Feeding standards usually are expressed in quantities of nutrients required per day or as a percentage of a diet, the former being used for animals given exact quantities of a diet and the latter more commonly when the diet is fed ad libitum. With respect to the various nutrients, most are expressed in weight units, as percentages, or as parts per million (ppm). Some vitamins—A, D, E—may also be given in international units (IU), although carotene, if used, may be given in milligrams and vitamin E, in ppm.

Protein requirements are most frequently given in terms of digestible protein (DP), although amino acids could be substituted for protein in nonruminant species when adequate information is at hand. Crude protein is often used with ruminants, although in the NRC (1996), they have introduced the use of metabolizable protein (MP). MP or absorbed protein is a calculated value and is defined as the true protein absorbed by the intestine; it is supplied by microbial protein coming from the rumen and undegraded intake (dietary) protein (UIP). Values are available on most feedstuffs to calculate how much dietary protein is degraded by rumen microorganisms, represented by the acronym DIP (dietary intake protein). Crude protein (see Chapter 3) is the sum of the UIP and DIP. Sometimes feedstuff values are also given for soluble protein and nonprotein nitrogen (NPN).

When dealing with energy, we find a number of different terms used. The Europeans and Australians generally use metabolizable energy (ME) values for most species. In the United States, ME values are used for poultry, and a considerable amount of data are available from research studies, both public and private. ME values are used for dogs and cats, whereas digestible energy (DE) is used for rabbits and horses. The swine bulletin uses total digestible nutrients (TDN), DE, and ME. For sheep, TDN, DE, and ME values are presented, but the goat bulletin uses these plus an estimated net energy (NE) value. For growing-fattening beef cattle, TDN, NEm, NEg values are given. The same values plus ME are given for replacement heifers and cows. For dairy cows, TDN, and NE1 values are given in the latest bulletin.

As pointed out in Chapter 10, energy (for adult animals) is required in relation to body weight$^{0.75}$ rather than in direct proportion to body weight. All current feeding standards recognize this approach, although not all emphasize it in their descriptions. For the other nutrients, Crampton (1965) suggested that other nutrients should be specified in relation to the amount of available energy consumed. Crampton suggested that all animals have a requirement for maintenance of 19 g of digestible protein/100 Kcal of DE. Although this approach probably is applicable, it has not been used widely in calculating protein requirements, except for poultry where calorie to protein ratios have been shown to have practical significance. Because most animals tend to eat to meet energy needs, if a diet changes from a moderate-energy concentration to a high-energy concentration, feed consumption will decline; therefore, more protein is needed per unit of diet, but the calorie to protein ratio may not change. However, with herbivores, DP is influenced markedly by dry matter (DM) intake. Thus, if DM intake is increased, DP declines, but digestible energy (DE) intake may remain essentially unchanged. As a result, relationships between caloric intake and protein requirements are more obscure when a wide range of diet types is considered.

In the case of mineral requirements, P need is sometimes computed in relation to protein intake. Ca need calculation is based on P requirements and body size, with additional amounts for lactation or egg production of laying birds. Other mineral requirements generally are specified on the basis of metabolic size. Some vitamins, particularly thiamin and niacin, are known to be required in relation to energy needs, whereas riboflavin and B_6 needs are more directly related to protein intake.

Computation of nutrient needs in relation to energy seems to be a very sound approach for nonruminant species. Certainly, if protein is required in direct relation to energy, then it should be feasible to relate the needs of most other nutrients, including vitamins, amino acids, and mineral elements, to energy consumption.

❏ METHODS OF ESTIMATING NUTRIENT NEEDS

Nutrient requirements of animals are evaluated and computed on the basis of many different types of information. With respect to energy, animal needs may be based on calorimetric studies modified as necessary with data from feeding experiments under practical environmental conditions. Although calorimetric data provide useful information on fasting or basal metabolism or energy expenditure of closely confined animals, such data do not appear to give good estimates of the energy expenditure of grazing animals because field studies indicate a considerably higher energy requirement than do calorimetric studies; the increase, in many cases, is caused by increased activity associated with grazing. In addition, the effects of environmental changes in temperature, solar radiation, wind, humidity, and other stressing factors are difficult to quantify. We are therefore faced with the need to conduct feeding experiments under natural environmental and management conditions.

One limitation of feeding trials is that estimating energy gain or loss by the animal is difficult (see Chapter 10). We can easily calculate the energy value of milk or eggs produced, and, in a long experiment, slight changes in body energy may not cause much of an error in estimating the energy needs of the animal, but such errors may be relatively large in short-term studies. This problem can be overcome partially by using data obtained from slaughter experiments (see Chapter 10).

With protein and the major minerals, one method used at times to calculate requirements is designated as the factorial method. This method is based on back calculations of utilization using amounts excreted in urine and feces and, for lactating animals, amounts secreted in milk. The computations used in calculating protein requirements are too complex to justify a complete explanation here. The reader is referred to ARC (1980) for more details, but a simple example will be given for Ca. In this method the net requirement of Ca for maintenance and growth is calculated as the sum of the endogenous losses (fecal + urinary) and the quantity retained in the body. To determine the dietary requirement, the net requirement is divided by an average value of availability (expressed as a decimal). For example, a heifer of 318 kg (700 lb) gaining 0.45 kg/day (1 lb) might have endogenous losses of 5 g of Ca/day and retain 3 g/day; its net requirement would be 8 g Ca/day. For an animal of this weight, average availability of Ca would be about 40%. Thus the animal's dietary requirement for Ca would be $8 \div 0.4 = 20$ g/day.

The factorial approach suffers from the fact that endogenous losses are difficult to estimate because endogenous losses are those originating from body tissues and must be estimated by feeding an animal a diet free of soluble Ca (or whatever nutrient is of interest) or by use of radioisotope procedures. In addition, availability of dietary nutrients is not a fixed biological quantity and may vary with age, source of the nutrient, and other dietary or environmental factors. Factorial estimates of requirements generally are lower than those derived from feeding trials or balance studies.

With respect to vitamins, balance studies are of little value because excretory metabolites of some vitamins are difficult to measure or have not been identified; thus, data must be obtained from feeding trials or some measure of animal productivity. Data on blood levels, tissue storage, freedom from deficiency signs and symptoms, and ability of the animal to produce at maximal levels are used to specify vitamin requirements.

❏ INACCURACIES IN FEEDING STANDARDS

For the nutritionist, feeding standards provide a useful base from which to formulate diets or estimate feed requirements of animals. However, they should not be considered as the final answer on nutrient needs. Rather, they should be used primarily as a guide. Current NRC recommendations are specified in terms believed (by the committees) to be minimum requirements for a population of animals of a given species, age, weight, and productive status. Some of the earlier versions were called allowances and, as such, included a safety factor on top of what was believed to be required.

It is well known that animal requirements vary considerably, even within a relatively uniform herd. The expected distribution of nutrient requirements for animals in a broad population is represented by the typical bell-shaped curve (Fig. 18.1). In the case of an individual producer, he or she is not likely to be dealing with a population of animals with such a symmetrical distribution, and the curve may be skewed in one direction or the other (Fig. 18.2). Even if the population were symmetrical, it is not practical to attempt to provide a nutrient intake that would be sufficient for 100% of the population, even for humans. The end result will be overfeeding a few animals that have low requirements and underfeeding some that have high requirements. With most domestic animals, we would normally cull out those that do poorly for some reason or other, possibly because some of them have extraordinarily high requirements for a specific nutrient. If we can identify individual ani-

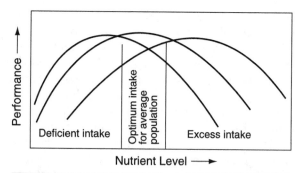

Figure 18.2   *A theoretical example of how nutritional requirements may be shifted for individuals or groups of animals by any of a number of causative factors. The usual objective for animal production would be to supply nutrients within the optimum zone, but it may not be easy to achieve this objective in all cases because the optimum zone will vary with different population and environmental conditions.*

mals (such as dairy cows) that have high requirements and are high producers, then it may be feasible to provide supplementary nutrients without grossly overfeeding the whole herd.

As suggested in Fig. 18.2, excess nutrients may result in depressed animal production and, in a few cases, outright toxicity or death. In any event, from the point of view of cost it is not desirable to feed more than is required.

It is quite obvious from the literature that management and feeding methods may alter an animal's needs or efficiency of feed utilization apart from known breed differences in nutrient metabolism and requirements. In addition, most current recommendations provide no basis for increasing intake in severe weather, and no allowance is made for the effect of other stresses such as disease, parasitism, injury, or surgery. Frequently, beneficial effects of additives or feed preparatory methods are not considered (if known). Thus, many variables may alter nutrient needs and nutrient utilization, and these variables are often difficult to include quantitatively in feeding standards, even when feed quality is well known.

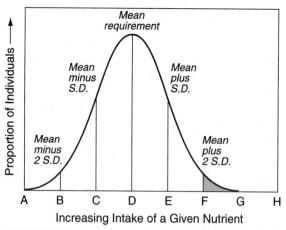

Figure 18.1   *Distribution of the nutrient requirement of a hypothetical population. The statistical unit of variation is known as the standard deviation (SD). An intake of F, obtained by adding 2 SD to the mean value, would cover the requirements for nearly all (97.5%) of the individuals in this theoretical population. The colored region shows the small proportion of healthy individuals (2.5%) not covered by the allowance.*

❏ NUTRITIONAL NEEDS AND THE PRODUCTIVE FUNCTIONS

The remainder of this chapter is devoted to some general relationships between various productive functions and nutrient requirements.

Maintenance

Maintenance may be defined as the condition in which an animal is neither gaining nor losing body energy (or other nutrients). With productive animals, only occasionally is maintenance desired; it is approximated closely or attained in adult male breeding animals other than during the breeding season and, perhaps, for a few days or weeks in adult females following the cessation of lactation and before pregnancy increases requirements substantially. However, as a reference point for evaluating nutritional needs, maintenance is a standard benchmark. Other things being equal, nutrient needs are lowest during maintenance. In field conditions during dry periods during the year, we may find that range animals may need to expend considerable energy just to obtain enough plant material for consumption, as opposed to the amount of energy expended when forage growth is more lush. This does not, however, alter the fact that nutrient needs are less during maintenance than when an animal is performing some productive function.

Growth and Fattening

Growth, as measured by increase in body weight, is at its most rapid rate early in life. When expressed as a percentage increase in body weight, the growth rate declines gradually until puberty, followed by an even slower rate until maturity. As animals grow, different tissues and organs develop at differential rates, and it is quite obvious that the conformation of most newborn animals is different from that of adults. This differential development has, no doubt, some effect on changing nutrient requirements. Growth rate decreases because the biological stimulus to grow is lessened because young animals cannot continue to eat as much per unit of metabolic size and, as measured by increase in body weight, because relatively more of the tissue of older animals is fat, which has a much higher caloric value than muscle tissue.

Nutrient requirements per unit of body weight or metabolic size are greatest for very young animals; these needs taper off gradually as the growth rate declines and as the animal approaches maturity. In young mammals, nutrient needs are so great that, because the capacity of the GI tract is relatively limited in space and function, they must have milk (or a milk replacer) and additional highly digestible food to approach maximal growth

rates. As the young mammal grows, quality of the diet generally decreases as more and more of its food is from nonmilk sources. As a result, digestibility is lower and the dry matter of food is used less efficiently.

Dry matter consumption for all young animals is usually far greater per unit of body weight during their early life than in later periods. This high level of feed consumption provides a large margin above maintenance needs, thus allowing a high proportion to be used for growth and development of the young animal. Because of differences in capability of the GI tract for food utilization and because the rate, duration, and character of body growth vary with age and animal species, nutrient requirements vary greatly among animal species. Nevertheless, it is characteristic for all species that nutrient requirements (in terms of nutrient concentration per unit of diet) are higher for the very young and then decline gradually as the animal matures. Naturally, total food and nutrient consumption are less for young animals because of their smaller size.

Nutrient deficiencies show up quite rapidly in young animals, particularly when the young are dependent in the early stages of life on tissue reserves obtained while in utero. With few exceptions, tissue reserves in newborn animals are low. Milk or other food may be an inadequate nutrient source, so that deficiencies may occur frequently until the food supply changes or the young animal develops a capability to eat the existing food supply. The young pig is an example. Fe reserves are low, and rapid growth soon depletes body reserves. Because milk is a poor source of Fe, young pigs often become anemic unless supplemented with Fe.

Young mammals are dependent on consumption of colostrum early in life. In the first few hours after birth, the intestinal tract is permeable to large molecules such as proteins. Colostrum has a large supply of globulins and other proteins that provide nutrition, as well as a temporary supply of antibodies, which greatly increase resistance to many diseases. In addition, colostrum is a rich source of most of the vitamins and trace minerals, and the young animal's tissues can be supplied rapidly with needed nutrients that may not have been provided adequately in utero.

From a production point of view, nutrient requirements per unit of gain are least and gross efficiency is greatest when animals grow at maximal rates. However, net efficiency (nutrient needs above maintenance) may not be altered greatly. In a number of instances, it may not be desirable or economical to attempt to achieve maximal gain. For example, if we want to market a milk-fed veal calf at an early age, maximum gain and fattening are desired. On the other hand, if the calf is being grown out for a herd replacement, then less than maximal gain will be just as satisfactory and considerably cheaper.

The biological stimulus to grow cannot be suppressed greatly in young animals without resultant permanent stunting. It is possible to maintain young animals for a period during which they do not increase body energy reserves, yet they will—if other nutrients are adequate—continue to increase in stature. Following a period of subnormal growth caused by energy restriction, most young animals will gain weight at faster than normal rates when given adequate rations. This response is termed *compensatory growth*. This phenomenon has practical application when young calves are wintered at low to moderate levels (submaximal). When new grass is available in the spring or if the calves are put in the feedlot, weight gain occurs at a very rapid rate initially. Efficiency for the total period and especially for a given amount of gain is greater, however, if the animal is fed at near maximal rates, but this practice (deferred feeding) may allow the use of much cheaper feedstuffs and, as a result, may be a profitable management procedure.

Work

Experimental studies with humans and animals indicate that work results in an increased energy demand in proportion to the work done and the efficiency with which it is accomplished (Brody, 1945). Carbohydrates are said to be more efficient sources of energy for work than fats (Mitchell, 1962). With respect to protein, balance studies show little, if any, increase in N excretion in horses or humans as a result of muscular exercise (Brody, 1945; Mitchell, 1962). Although this evidence has been obtained in a number of studies, data on adult male humans indicate reduced quality and quantity of work when protein intake is marginal, but still above maintenance levels. It is a common belief that hard workers need more protein, and it is a typical practice in competitive sports for the athlete to eat high levels of meat and other high-protein foods. Perhaps some of the effect on performance is psychological.

If sweating occurs, work may be expected to increase the need of Na and Cl in particular. P intake should be increased during work because it is a vital nutrient in many energy-yielding reactions. Similarly, the B vitamins involved in energy metabolism, particularly thiamin, niacin, and riboflavin, probably should be increased in nonruminants as work output increases, although data on this subject are not clear.

Reproduction

Although nutrient needs of animals for reproduction generally are considerably less critical than during rapid growth, they are more critical than for maintenance. If nutrient deficiencies occur prior to breeding, they may render animals sterile or result in low fertility, or failure to establish or maintain pregnancy.

It has been demonstrated many times that underfeeding (energy, protein) during growth will result in delayed sexual maturity and that both underfeeding and overfeeding (energy) usually will result in reduced fertility as compared to that of animals fed on a moderate intake. Of the two, overfeeding is usually more detrimental to fertility.

Energy needs for most species during pregnancy are more critical during the last third of the term. This fact is illustrated by data on cattle shown in Table 18.1; daily energy deposits increase quite rapidly between day 200 and 280. Although the data in Table 18.1 show that only a relatively small amount of the cow's energy requirements are needed for reproduction, other information indicates a somewhat greater need. Pregnant animals have a greater appetite and will spend more time grazing and searching for food than nonpregnant animals. Furthermore, the basal metabolic rate of pregnant animals is higher; the increase over maintenance is called the heat increment of gestation. By the end of pregnancy, the basal rate is about 1.5-fold that of similar nonpregnant cows.

Protein deposition in the products of conception follows the same trend as energy, but protein is relatively more critical for development of the fetus in the late stages than early in pregnancy as is true for Ca, P, and other minerals and vitamins.

Inadequate nutrition of the mother during pregnancy may have variable results, depending on the species of animal, the degree of malnutrition, the nutrient involved, and the stage of pregnancy. With a moderate deficiency, fetal tissues tend to have a

TABLE 18.1 Deposition of various nutrients in the fetus, uterine tissues, and mammary gland of the cow at different stages of pregnancy.

DAYS AFTER CONCEPTION	UTERINE AND FETAL DEPOSITS (PER DAY)				MAMMARY GLAND, G PROTEIN/DAY
	ENERGY, KCAL	PROTEIN, G	CA, G	P, G	
100	40	5			
150	100	14	0.1		
200	235	34	0.7	0.6	7
250	560	83	3.2	2.7	22
280	940	144	8.0	7.4	44
Approximate net daily maintenance requirement of a 1000-lb cow	7000	300	8	12	

Source: From data of Moustgaard (1959).

priority over the mother's tissues; thus, body reserves of the mother may be withdrawn to nourish the fetus. A very severe deficiency, however, usually will result in partial depletion of the mother's tissues and such detrimental effects as resorption of the fetus, abortion, malformed young, or birth of dead, weak, or undersized young with sometimes long-term effects on the mother. When the mother's tissues are depleted of critical nutrients, tissue storage in the young animal almost always is low, nutrients secreted in colostrum also are low, milk production may be nil, and survival of the young animal may be jeopardized.

Lactation

Heavy lactation probably results in more nutritional stress in mature animals than any other productive function, except heavy work. During a year, high-producing cows or goats typically produce milk with a dry matter content equivalent to four- to fivefold that in the animal's body, and some animals reach production levels as high as seven times the body dry matter. High-producing cows give so much milk that normally it is impossible for them to consume enough feed to prevent weight loss during peak periods of lactation.

Milk of most domestic species contains 80 to 80% water; thus, water is a critical nutrient needed to sustain lactation. All nutrient needs are increased during lactation because milk components are either supplied directly via the blood or synthesized in the gland and, thus, are derived from the animal's tissues or more directly from food consumed. All recognized nutrients are secreted to some extent in milk. The major components of milk are fat, protein, and lactose, with substantial amounts of ash, primarily Ca and P. Milk yield varies widely between and within species. In cows, peak yields usually occur between 60 and 90 days after parturition and then gradually taper off. Thus, the peak demand for nutrients follows the typical milk flow characteristic for the species concerned. Milk composition and quantity in ruminants, particularly the fat content and, to a lesser extent, the protein and lactose, may be altered by the type of diet. In nonruminants, diet may affect fat, mineral, and vitamin composition of milk.

Limiting water or energy intake of the lactating cow (and probably of any other species) results in a marked drop in milk production, whereas protein restriction has a less noticeable effect, particularly during a short period of time. Although mineral deficiencies do not affect milk composition markedly, they will result in rapid depletion of the lactating animal's reserves. The needs of elements such as Cu, Fe, and Se will be increased during lactation, even though they are found in very low concentrations in milk. The effects of marked nutrient deficiencies during lactation often will carry over into pregnancy and the next lactation.

❏ FACTORS AFFECTING NUTRIENT REQUIREMENTS AND NUTRIENT UTILIZATION

Many factors affect the nutrient requirements of animals and/or the extent of nutrient utilization. Several of the most important ones are described here.

Genetics

Genetic background plays an important role in nutrient utilization. More than 270 "inborn errors of metabolism" involving defects in the utilization of specific nutrients or metabolites are recognized in humans, many of which are controlled genetically. About 30 of them can be treated by dietary modifications (Wong and Hsia, 1973). Molecular biology techniques have provided the knowledge required to identify, isolate, and clone specific genes controlling many of these diseases in humans. For example, victims of cystic fibrosis, and several other diseases under the control of single genes, soon may be helped by the insertion of the missing gene into their body to provide the missing gene or replace the defective gene that has caused the disease.

In numerous animals, genetic differences result in higher or lower nutrient requirements: a strain of cattle that has a high Zn requirement, mice with high requirements for some B vitamins, sheep that require larger amounts of Fe and those that tolerate high dietary Cu. Such traits are controlled by several genes.

In recent decades, marked improvement has occurred in the growth and efficiency of nutrient utilization of several domestic animal species. Such genetic improvement is not necessarily only a result of a change in nutrient utilization, but may involve concomitant overall improvement in animal health, in resistance to disease, or in better adaptation to a particular type of environment. Nevertheless, the net result is an improvement in animal productivity. An example with swine is the 12.7% improvement in rate of gain, 16.6% improvement in efficiency of feed utilization, and 45.9% decrease in backfat in pigs achieved over a 20-year period of selection at Iowa State University. Similar and probably more marked changes have occurred in poultry owing to their shorter generation interval and larger number of offspring produced per year. Such changes can be produced in cattle and sheep, but genetic progress is much slower because of their older age at puberty and their lower reproductive rates.

Nutritional Individuality

Not all animals of a particular age, sex, or breed respond in an identical manner to a given diet. Examples of such differences are cited in several chapters throughout this book; we address the issue here to emphasize the importance of the concept. Such variations have been documented in laboratory animals and humans by Williams (1956, 1971, 1979). These examples show clearly that a wide range exists in requirements for various nutrients, even among animals from closely inbred strains of laboratory rats. Obviously, this variability in nutrient utilization and quantitative daily requirements presents challenges in animal nutrition and in human nutrition.

Breed and Species Differences

In addition to the genetic differences already noted, there may be substantial differences between different breeds. For cattle, some of these are recognized, and adjustment factors are found in the appropriate NRC bulletins for calculating animal or ration requirements. Of course, there may be substantial differences between *Bos tarus* and *Bos indicus* species; these can be adjusted for also.

Other Environmental Factors

The NRC requirements always have adjustments for age and body type or size. In more recent years, condition scores have been used extensively to allow for greater feed consumption by animals that have low levels of body fat for whatever reason. Such scores are generally used for adult animals such as beef cows, dairy cows, and ewes. Pictures or drawings are available showing cross sections of the back of animals ranging from emaciated animals to those that are severely overconditioned (very fat). Refer to NRC-2001) for such diagrams.

Environmental factors such as very cold or hot temperatures and such factors as muddy lots or steep terrain can be accounted for to some degree. The effect of processing feed may have marked effects on utilization; so adjustment factors are available for all commonly used feed ingredients.

One of the biggest variables that must be accounted for is feed consumption by an animal. Many different factors influence feed consumption, such as quality of feed, how it is processed, and how it is provided to the animal. The presence of appetite inhibitors such as molds or other toxins, unpalatable feed ingredients, and feed bunk management may all have substantial effects on daily feed consumption and, thus, on the nutrient concentration needed in a diet.

❏ NUTRITION AND DISEASE

Relatively little information is available regarding the nutritional state of domestic animals to the immune system. Immune function is known to be affected by individual nutrients, including the amino acids, arginine, glutamine, and lysine; arachidonic acid, an omega-3 fatty acid; the trace minerals Zn, Cu, Fe, and Se; and vitamins A and E. The relationship between nutrition and immunity has been reviewed (Johnson et al., 2001; Johnson, 2005). The following relationships between nutrition and immune function have been summarized (Johnson et al., 2001): (1) immunological stress (exposure to pathogens) changes the physiological state of the animal and alters nutrient requirements; (2) if the changed requirement for a particular nutrient can be defined, diet composition might be altered to allow acceptable growth performance and/or to promote optimum responses of the immune system. As more knowledge is accrued about the observed relationships between immunity and specific nutrients, it may be possible to adjust nutrient intakes to accommodate changes in immunological stress and improve animal productivity on the farm.

Well-fed animals may be more susceptible than poorly fed animals to certain viral infections. For example, incidence of hoof-and-mouth disease is greatest in well-fed cattle, while the disease occurs in a milder form during periods of starvation. On the other hand, well-fed animals are generally more resistant to bacterial and parasitic infections. Most of the data on this topic are of an epidemiological nature rather than based on research studies designed to bear on the question.

In many research studies, protein deficiencies have resulted in an impaired immune response to bacteria, viruses, and fungi (Wannemacher, 1980). Similarly, vitamin deficiencies, particularly vitamin A, have been linked with increased susceptibility to infectious diseases. In dogs, experimental obesity has been reported to result in reduced resistance to both bacterial and viral infections (Newberne et al., 1969). The relationship between nutrition and chronic disease has been reviewed (Olson, 1978). In a disease such as hemophilia, the genetic determinant accounts for 100% of the illness, and no nutritional control is possible. With phenylketonuria, the affected person has a genetic deletion of the enzyme phenylalanine hydroxylase. Because this enzyme is involved in the synthesis of tyrosine, through dietary intervention it is possible to readjust phenylalanine-tyrosine ratios and effect some control of the disease. At the other end of the spectrum, scurvy is 100% controllable by nutrition, but diseases such as pellagra and rickets, though primarily nutritional, do have genetic components.

❏ EVALUATING THE NUTRITIONAL STATUS OF AN ANIMAL

The evaluation of the nutritional status of animals and humans requires a variety of approaches, including epidemiological surveys of a given population of people or of animals, and an array of physical and biochemical measurements. Here we address the nutritional assessment of animals in broad terms. Human nutrition monitoring and nutrition status assessment has been the subject of a symposium (Olson, 1990) and is not discussed here. Table 18.2 lists briefly the primary deficiencies encountered for most of the required nutrients.

Reproducing animals often are affected by deficiencies that may result in such symptoms as thin-shelled eggs, low hatchability, and embryonic death in poultry or in delayed estrus, low conception rate, fetal resorption, spontaneous abortion, stillbirths, or birth of small, weak, unthrifty young. Nutrient demands are higher during reproduction and/or lactation than for maintenance of the body. Depending on the nutrient involved, the pregnant dam may direct the nutrient toward the fetus and sacrifice her body tissues in the process. However, at some point before the deficiency becomes life threatening to the dam, she may interrupt a pregnancy by resorption or abortion or by production of small, unthrifty young. Even if the fetus survives to

TABLE 18.2   Major nutritional deficiency symptoms.

NUTRIENT	CLINICAL SYMPTOMS[a]
Protein	Kwashiorkor or, with low energy, marasmus, black hair turns red and is brittle (H); poor productivity, low fertility, birth of underweight young with poor livability, poor milk production, low blood albumin and/or blood urea, unkept appearance.
Lysine	White barring of primary wing feathers (T).
Methionine	Fatty liver (YS).
Other amino acids	No specific symptoms have been observed in farm animals.
Essential fatty adds	Poor growth (YC, YS, YCh), skin lesions (YC, YS).
Carbohydrates	No specific symptoms unless associated with low-energy intake.
Energy	Low productivity, loss of body weight, stillborn young, poor or nil milk production, poor fertility. Usually associated with deficiencies of other nutrients such as protein and minerals such as P.
Water-Soluble Vitamins	
Thiamin	Beriberi, edema, heart failure (H); polynuritis (YCh); opistothonus, anorexia, high blood pyruvic acid; blindness (C).
Riboflavin	Facial dermatitis, insomnia, irritability (H); lesions of the eye, anorexia, vomiting, birth of weak or stillborn young (S); curled-toe paralysis (YCh); diarrhea, loss of hair (YC, YSh).
Niacin	Pellegra (H); diarrhea, vomiting, dermatitis around the eye (YCh), poor growth.
Pyridoxine	Convulsions, anorexia, poor growth, brown exudate around eyes (YS); abnormal feathering (YCh).
Pantothenic acid	Poor growth, graying of hair in some species; dermatitis, embryonic death (YCh); loss of hair, enteritis, goose-stepping, incoordination in walking (YS).
Biotin	Dermatitis, perosis (YCh, YT).
Choline	Perosis, fatty liver (YCh, YT); abnormal gait, reproductive failure in females, hemorrhagic kidneys, fatty liver (S).
Folacin	Anemia, gastrointestinal disturbances, impaired coordination (H); anemia in other species.
Cobalamin	Anemia, poor feathering, perosis (YCh); low hatchability, fatty livers, enlarged hearts (Ch, T); rough hair coats, uncoordinated hind leg movements, anemia, abortion, other reproductive problems (S).
Vitamin C	No problems in domestic animals; scurvy (H).
Fat-Soluble Vitamins	
Vitamin A	Xeropthalmia, night blindness, permanent blindness (all species); diarrhea, convulsions, high cerebrospinal fluid pressures (Y); incoordination, reproductive failure in males, abortion or birth of weak or dead young (C).
Vitamin D	Rickets in young, osteomalacia in adults, lameness and sore joints, crooked legs, spontaneous fractures of long bones (Y); negative mineral balance, low bone ash (A).
Vitamin E	Nutritional muscular dystrophy (Y), fetal reabsorption, sterility, testicular atrophy (C, CH, T), liver necrosis (YS), fragile red blood cells (YS, YSh); encephalomalacia, exudative diathesis (YCh).
Vitamin K	Slow blood clotting; subcutaneous hemorrhages.
Macrominerals	
Calcium	Rickets, osteoporosis, poor growth; muscle cramps, convulsions (H).
Phosphorus	Rickets (Y), osteoporosis (A), anorexia, pica, low fertility.
Magnesium	Anorexia, poor productivity, tetany; weak crooked legs (S).
Potassium	Muscular weakness, paralysis (H); abnormal electrocardiograms, unsteady gait, weakness, pica.
Sodium	Anorexia, muscle cramps, mental apathy (H); dehydrated appearance, craving for salt, weight loss.

TABLE 18.2 *Continued.*

NUTRIENT	CLINICAL SYMPTOMS[a]
Macrominerals	
Chlorine	Depressed growth.
Sulfur	Reduced gain or loss of weight (C, Sh), loss of wool (Sh).
Trace Minerals	
Chromium	Impaired ability to metabolize glucose.
Cobalt	Primarily ruminants; symptoms similar to cobalamin emaciation, anemia, fatty degeneration of the liver.
Copper	Anemia; when coupled with high Mo and/or sulfate, swayback or enzootic neonatal ataxia, loss of pigment in hair or wool (C, Sh); bone abnormalities, cardiovascular lesions, reduced egg production, reproductive failure.
Fluorine	Excessive tooth decay (H).
Iron	Anemia and associated poor productivity; very common in young pigs.
Manganese	Lameness and shortening and bowing of the legs, enlarged joints (Y); perosis (YCh, T); reduced egg shell thickness (Ach, T); weakness, poor sense of balance; crooked calf disease (C); poor fertility (C).
Molybdenum	Reduced growth rates (not common).
Nickel	Prenatal mortality, unthriftiness, decreased growth rate (YCh, T, YS), poor N retention (YSh).
Selenium	Nutritional muscular dystrophy (Y), exudative diathesis (YCh, T); liver necrosis (YS); heartfailure (YC); retained placenta (AC).
Zinc	Poor growth, anorexia, parakeratotic lesions on head, neck, belly and legs (C, Sh, S); perosis, abnormal feathering (YCh, T); poor testicular development, slow wound healing (H, other species).

[a]Y, young ; A, adults; H, humans; C, cattle; Ch, chickens; S, swine; Sh, sheep; T, turkeys.

birth, milk production, especially vital colostrums, may not be adequate to provide the necessary immunity, or the milk supply may be insufficient to produce a normal rate of growth. The dam may cease lactation altogether not long after parturition if the deficiency is severe enough.

Nutritional History

If information is available on the type and amount of diet consumed by an animal, often such data can be used to establish a possible cause or to rule out other possibilities. Such information may be obtained on animals fed in confinement, but it is often not available on free-ranging animals. Nevertheless, some attempt should be made to provide background of this kind whenever possible. Feed samples also should be obtained for analysis. Information should be obtained on eating habits such as amount consumed, frequency of eating, consumption of concentrates or roughages, and evidence of pica (abnormal eating habits such as wood chewing).

Nutritional deficiencies result in chronic problems, with a few exceptions, and they may take days, weeks, or months to develop. This is in contrast to the more acute onset of some infectious and metabolic diseases, which may develop in hours or days. What is the best method to evaluate nutritional status? There is no simple answer to that question; the answer will depend on the species of animal involved; the level, stage, and type of production; complications with other deficiencies or excesses; the presence of stressors; and other unknown or unrecognized factors.

Animal Productivity

Human health workers normally base evaluation of growth on standard charts for weight and height at a given age by sex, with some consideration for body type. With domestic animals, height at the withers, birth weights, weaning weights, or weights at other times are useful measures. Body weight gain also may be expressed as weight per day of age or rate of gain during some particular period of rapid growth.

Nutrient deficiencies of any type will, sooner or later, result in some reduction in weight gain, although the specific cause may vary considerably from nutrient to nutrient. Young, rapidly growing animals will normally show deficiency signs in a much shorter period of time than older animals, partly because the tissue reserves are usually low and partly because the need for nutrients during rapid growth is higher than for more mature animals. In fact, mature animals may be quite resistant to developing clinical signs of nutrient deficiencies.

As with growth, other measures of productivity such as work (exercise), milk production, and egg production usually are affected by deficiencies. The deficiency may not be severe enough to produce any specific clinical signs, although a thorough biochemical evaluation may provide information needed to establish a biochemical lesion.

Physical Examinations

Animals that do not feel well will not groom their hair or feathers as often (if at all) as do normal animals. As a result, the hair looks dry (no mucus from the saliva, or for feathers, oil from the oil gland in birds) and unkempt. Neither do they participate in the normal practice of mutual grooming. Also, animals that get up after lying down normally will stretch to relieve muscle strains, whereas sick animals usually do not. Observation of ruminant animals will indicate whether normal rumination is taking place during noneating periods. Deficiencies of some nutrients result in lameness or some neurological interference with muscular movements. Nutrient deficiencies that cause dermatitis can be detected by visual examination of the skin, feathers, hair, or wool. Manifestations such as goose-stepping in swine from pantothenic acid deficiency, retracted head from thiamin deficiency, encephalomalacia in chicks from vitamin E deficiency, or polioencephalomalacia in ruminants from a metabolic deficiency of thiamin are examples of deficiencies, that, because of their specificity, can be detected readily by gross physical examination. Other problems such as vitamin A deficiency may be detected by evidence of exophthalmia, excessive lacrimation, or examination of the interior of the eye with an ophthalmoscope. Goiter is a clearcut indication of thyroid malfunction, although not necessarily of a simple iodine deficiency.

Tissue Analyses

Blood, urine, feces, hair, bone, liver, and other tissues may be sampled from individual animals or from selected individuals within a group for a variety of biochemical analyses as a diagnostic aid. Data on blood may be useful in some instances, but not in all deficiencies. Low blood glucose and high plasma free fatty acids are correlated (although not highly) with inadequate energy consumption in cattle. Plasma albumin and urea are used to identify protein deficiency. Blood mineral levels are not correlated highly with current consumption unless the body is grossly deficient because of the homeostatic mechanisms that favor maintenance of normal concentrations of Ca, P, Mg, Na, K, and most of the trace elements. Ceruloplasmin (a Cu-containing enzyme) is a useful assay for Cu, hemoglobin for Fe, and circulating thyroid hormones for iodine.

Hair. Hair tissue is easily obtained; concentrations of inorganic elements in hair are a reflection of past consumption of some of the trace elements, but it does not appear to be precise enough for common usage.

Bone. Bone analyses—either individual minerals or total bone ash—can serve as a good means of evaluating the Ca–P status of an animal or a group of animals. Of course, bone samples are not so easily taken from a living animal as are blood or hair. X-ray densitometry procedures can be used to evaluate density of bone, but the cost and availability of the equipment make it an expensive method, though one that might be used on highly valued animals.

Other Tissues. Liver and kidney tissues are storage sites for most of the trace minerals and a number of the vitamins. For experimental animals, data from these organs can be very useful in determining the body storage for such nutrients. Liver biopsy techniques are available for animals. Tissue samples can be subjected to histological or chemical evaluations.

Urine. Urine analyses are not used routinely with farm animals, but useful nutritional and metabolic information can be obtained by urinalysis. The presence of ketones in urine in large amounts may be indicative of primary or secondary ketosis, especially in lactating cows or does. The concentrations of mineral elements, vitamins (or their excretory products), nitrogenous constituents, and glucose (indicative of diabetes) in urine may provide important diagnostic information about nutritional status and metabolic diseases in animals.

❏ SUMMARY

- Feeding standards are the outgrowth of vast amounts of experimentation with animals. They are intended to specify quantitative nutritive requirements of different animal classes and species at a given age, rate of growth, and/or specific productive function. In general, current standards are satisfactory guidelines, but they must be revised continually as health or management is improved or genetic changes are made that allow more rapid growth or higher levels of production, which, in turn, may alter nutrient needs. Computer modeling is being used to further improve information for feeding standards and animal utilization.

❏ REFERENCES

Agricultural Research Council. 1980. *The Nutrient Requirements of Ruminant Livestock.* Commonwealth Agricultural Bureaux, London, England.

Brody, S. 1945. *Bioenergetics and Growth.* Hafner Publishing Co., New York (out of print).

Crampton, E. W. 1965. *J. Nutr.* **82:** 353.

CAST. 1998. *Naturally Occurring Antimicrobials in Food.* Task Force Report No. 132, Council for Agricultural Science and Technology, Ames, IA, 103 pp.

Friendship, R. M., et al. 1984. *Can. J. Comp. Med.* **48:** 390.

Johnson, R. W. 2005. Immune system: nutrition effects. In *Encyclopedia of Animal Science,* W. G. Pond and A. W. Bell (eds.). Marcel Dekker, New York (In Press).

Johnson, R. W., et al. 2001. In *Swine Nutrition* (2nd ed., Chapter 24, A. J. Mitchell. H. H. 1962. *Comparative Nutrition of Man and Domestic Animals.* Academic Press, New York.

Moustgaard, J. 1959. In *Reproduction in Domestic Animals,* P. T. Cupps and H. H. Cole (eds.). Academic Press, New York.

National Research Council (NRC). 1996. *Nutrient Requirements of Beef Cattle,* 7th rev. ed. National Academy of Sciences, Washington, DC.

National Research Council (NRC). 2001. *Nutrient Requirements of Dairy Cattle,* 7th rev. ed. National Academy of Science, Washington, DC.

Olson, J. A. 1990. *J. Nutr.* **120**(Suppl): 1431.

Olson, R. E. 1978. *Nutr. Today* **13**(4): 18.

Wannemacher, C. W., Jr. 1980. *Nutr. Today* **13**(4): 18.

Williams, R. J. 1956. *Biochemical Individuality: The Basis for Genetotrophic Concept.* John Wiley, New York.

Williams, R. J. 1971. *Nutritional Against Disease.* Pitman Publishing Corp., New York.

Williams, R. J. 1979. *Nutritional Individuality. J. Nutr. Acad.* **11**(11): 8.

Wong, P. W., and D. Y. Hsia. 1973. In *Modern Nutrition in Health and Disease* (5th ed.), R. S. Goodheart and M. E. Shils (eds.). Lea & Febiger, Philadelphia, PA.

19

Feedstuffs

FEEDSTUFFS AND THE DIETS formulated from them comprise the raw materials for animal production. In this chapter we are concerned with the efficient conversion of feedstuffs to useful food, fiber products, and other products or to maintain companion animals for pleasure. Consequently, an understanding of the chemical and nutritional composition of important classes of feedstuffs will provide the basis for applying that knowledge to feed preparation and processing (see Chapter 20) and diet formulation (see Chapter 21). More than 2000 different products have been characterized to some extent for animal feeds, not counting varietal differences in various forages and grains.

Several important factors determine the acceptability of a given feedstuff for inclusion in the diet of a particular animal species. Cost is relevant in all cases. Generally, products fed to animals are either those that are not edible for humans or those that are produced in excess of human needs in a given location or country. With our imperfect systems of distribution, a grain that is in surplus in the United States may be in high demand for human use in some other area of the world.

Other factors affecting feedstuff acceptability include acceptance by the animal (palatability), digestibility or bioavailability of the energy and nutrients contained in the feedstuff, nutrient content and balance, presence of toxins or nutrient inhibitors, and handling and milling properties.

The number of feedstuffs is too great to cover all of them in this chapter. Instead, we describe (1) the conventional classes of feedstuffs and explain their characteristics and properties and (2) the nutrient composition, nutritive value, and special characteristics of selected common feedstuffs within each class. More detailed information on feedstuffs and their composition is available from numerous sources, for example, Morrison (1956), which enjoyed wide use throughout the first half of the twentieth century; and more recent publications, such as the National Research Council (1982), the International Feed Institute (1984), McDowell et al. (1974), Gohl (1975), Bath et al. (1980), Pond and Maner (1984), Pond et al. (1991), Patience and Thacker (1989), Seerley (1991), Thacker and Kirkwood (1992), and Lewis and Southern (2001). Genetic modification of feedstuffs produced through biotechnology has been reviewed by Owens et al. (2005).

The composition of foods and feed ingredients is tabulated in several publications of the National Research Council and the United States Department of Agriculture. See Chapter 12 of *Nutrient Requirements of Nonhuman Primates*, Second Revised Edition (National Research Council, 2003) for detailed information on energy values, proximate analysis, plant cell wall constituents, and mineral and vitamin concentrations of more than 325 feedstuffs.

Tables 2 through 6 of the Appendix list the amino acid composition of selected feedstuffs (Table 2); vitamin content of selected feedstuffs (Table 3); mineral element composition of commonly used mineral supplements (Table 4); energy values and energy sources of selected feeds (Table 5); and mineral element concentrations of selected feeds used for ruminants (Table 6).

❑ CLASSIFICATION OF FEEDSTUFFS

A **feedstuff** may be defined as any component of a diet that serves some useful function. Most feedstuffs provide an array of nutrients, but ingredients also may be included to provide bulk, to reduce oxidation of readily oxidized nutrients, to emulsify fats, or to provide flavor, color, or other factors related to palatability.

The classification scheme used in this chapter includes fresh forages and silages (high moisture), dry forages, roughages, energy concentrates, protein concentrates, mineral supplements, vitamin supplements, and nonnutritive feed additives. These classes of feedstuffs are outlined as follows.

HIGH-MOISTURE FORAGES

Permanent pasture and range plants
Temporary pasture plants
Soil or green chop
Cannery and food crop residues

SILAGES

Corn
Sorghum

Grass
Grass–legume
Legume
Miscellaneous

DRY FORAGES

Hay
    Legume
    Nonlegume (primarily grasses)
    Cereal crop hays

ROUGHAGES

Straw and chaff
Fodder, stover
Other products with >18% crude fiber
    Corncobs
    Shells
    Sugar cane bagasse
    Cottonseed hulls
    Cotton gin trash
    Animal wastes
Sorghum
Grass

Grass–legume
Legume
Miscellaneous

ENERGY CONCENTRATES

Cereal grains
Milling byproducts (primarily from cereal grains)
Molasses of various types
Seed and mill screenings
Beet and citrus pulps
Animal and vegetable fats
Whey
Brewery byproducts
Miscellaneous
  Waste from food processing plants
  Cull fruits, vegetables, and nuts
  Garbage
  Roots and tubers
  Bakery waste

PROTEIN CONCENTRATES

Oilseed meals
  Cottonseed, soybean, linseed, safflower, sunflower, canola
Animal meat or meat and bone meals
Marine meals
Avian byproduct meals
Seeds (whole) from plants
Milling byproducts
Distillers and brewers dried grains
Dehydrated legumes
Single-cell sources (bacteria, yeast, algae)
Nonprotein nitrogen (urea, biuret, ammonia)
Dried manures

MINERAL SUPPLEMENTS

VITAMIN SUPPLEMENTS

NONNUTRITIVE ADDITIVES

Antibiotics
Antioxidants
Buffers
Colors and flavors

Emulsifying agents
Enzymes
Hormones
Medicines
Miscellaneous

❏ FORAGES AND ROUGHAGES

The terms *forages* and *roughages* are often used interchangeably. They are the primary feeds for all herbivorous animals and provide the major portion of the diet for most or all of the year. Harvested and stored forages (hays, silages, and other forms) provide valuable energy and nutrients for animals from plant sources of limited value for human consumption. Roughages usually refer to high-fiber byproduct plant sources such as corncobs, soybean hulls, and cottonseed hulls; they generally contain 18% or more crude fiber. Forages, if harvested early in the vegetative growth stage, are relatively high in protein and low in fiber unless harvested at late stages of maturity. Forages and roughages are of primary interest for ruminants, horses, rabbits, and other animals that depend on fermentation of insoluble carbohydrates in the digestive tract for a major part of their energy supply. Although other species such as swine can utilize some forage or roughage, productivity in the absence of another source of feed is too low to be economical under most circumstances.

Nature of Roughages and Forages

A **roughage** is a bulky feedstuff that has a low weight per unit of volume. This criterion is rather subjective, but any means of classifying roughages has its limitation because there is great variability in physical and chemical composition. Most feedstuffs classed as roughages have a high-fiber content and a lower digestibility of energy than most concentrates. In general, roughages contain more than 18% fiber; however, there are many exceptions. On the other hand, forages, for example, corn silage, nearly always contains more than 18% fiber, but the digestible energy often exceeds 70% on a dry basis. Lush young grass is another example. Although its weight per unit volume may be relatively low and fiber content relatively high, its digestibility is high. Soybean hulls, considered a roughage, are high in fiber but have a high-energy value for ruminants. These examples illustrate the wide variation in energy value of forages and roughages high in fiber content.

Most roughages and forages have a high content of cell-wall material (see Chapter 3). The cell-wall fraction has a highly variable composition; it contains appreciable amounts of lignin, cellulose, hemicellulose, pectin, polyuronides, silica, and other components.

The amount of lignin present in a roughage or forage is generally inversely related to the digestibility of the energy it contains. Lignin is composed of a series of phenolic compounds that interfere with the utilization by animals of carbohydrates in the cell wall of plants (Jung and Fahey, 1983). Lignin is covalently bound to the hemicellulose fraction of the cell wall of plant tissue. It limits fiber digestibility, probably because of the physical barrier between digestive enzymes and the carbohydrate in question. Removal of lignin with chemical methods increases utilization of the carbohydrate by rumen microorganisms and probably by cecal organisms. Lignin content of plant tissue increases gradually with maturity of the plant, and a high negative correlation exists between lignin content and digestibility, particularly for grasses and somewhat less for legumes. There is evidence that the silica content of plant tissue also is related negatively to fiber digestibility.

The protein, mineral, and vitamin contents of forages and roughages are highly variable. Legumes may contain 20% or more crude protein, one-third or more of which may be in the form of nonprotein N. Roughages, such as wheat straw, may have only 3 to 4% crude protein. Most others fall between these two extremes.

Mineral content is exceedingly variable; most forages are relatively good sources of Ca and Mg, particularly legumes. P content is generally moderate to low, and K content high; the trace minerals vary greatly depending on plant species, soil, and fertilization practices.

In overall nutritional terms, forages and roughages range from very good nutrient sources (lush young grass, legumes, high-quality silage) to very poor sources (straws, hulls, some browse). The

TABLE 19.1 Effect of stage of maturity on composition and digestibility of orchard grass.

	STAGE OF MATURITY			
ITEM	6–7″ HIGH CUT 5/19 PASTURE	8–10″ HIGH CUT 5/31 LATE PASTURE	10–12″ HIGH CUT 6/14 EARLY HAY	12–14″ HIGH CUT 6/27 MATURE HAY
Composition, %				
Crude protein	24.8	15.8	13.0	12.4
Ash	9.3	6.8	7.1	7.2
Ether extract	4.0	3.5	3.9	4.2
Organic acids	6.3	6.0	5.4	5.0
Total carbohydrates[a]	49.9	63.0	64.4	63.1
Sugars	2.1	9.5	5.4	2.4
Starch	1.2	9.5	0.8	0.9
Alpha cellulose	19.5	19.8	19.1	27.7
Beta and gamma cellulose	3.4	5.4	3.8	2.5
Pentosans	15.1	15.8	16.8	18.1
Nitrogen-free extract	35.0	45.7	44.2	41.2
Lignin	5.7	5.0	6.2	8.1
Crude fiber	26.9	28.2	31.8	35.0
Digestibility, %				
Dry matter	73	74	69	66
Crude protein	67	63	59	59
Crude fiber	81	77	71	68

[a]Total carbohydrates = (crude fiber + NFE) − (lignin + organic acids); NFE = nitrogen-free extract.
Source: From Ely et al. (1953).

nutritional value of the very poor sources often can be improved considerably by proper supplementation or by appropriate feed processing methods.

Factors Affecting Forage Composition

Stage of Maturity. A major factor affecting forage composition and nutritive value is the stage of maturity of the plant at harvest (or grazing). An example of the effect of stage of maturity on composition and digestibility is shown for orchard grass (*Dactylis glomerata*) in Table 19.1. In Table 19.1, note the rapid decline in crude protein; in Fig. 19.1, the same type of decline in digestibility is shown for alfalfa and four grass species. A gradual decline also occurs in ash and soluble carbohydrates, and an increase takes place in lignin, cellulose, and crude fiber (CF), all during a six-week period. The magnitude of the changes that occur depends on the plant species and on the environment in which the plant is grown. For example, if the growing season progresses rather rapidly from cool spring weather to hot summer weather, changes in plant composition will be more rapid than when the weather remains cool during plant maturation, especially in the case of a cool season grass. In plants such as alfalfa (*Medicago sativa*), which has different growing patterns from those of grasses, rapid changes take place as the plant matures and

blooms. Crude protein contents given by the National Research Council indicate the following values (% on dry basis) for second cuttings of alfalfa: immature, 21.5; prebloom, 19.4; early bloom, 18.4; midbloom, 17.1; full bloom, 15.9; and mature, 13.6. Corresponding changes in total digestible nutrients (TDN) range from 63 to 55%. A portion of these differences occurs because the plant loses leaves as it matures, and leaves have a higher nutrient value than the remainder of the plant. As plants mature, there is a decline in concentration of Ca, K, and P and of most of the trace minerals.

Fertilization. Soil fertility and fertilization practices also may affect the nutritional value of the roughage produced, but the magnitude of the difference in plant composition related to soil fertility is far less than that related to stage of maturity at harvest. In pastures with mixed plant species, fertilization may alter the vegetative composition because some plant species respond more than others to fertilization. If a grass–legume mixture is fertilized with high levels of N, for instance, the grass may be stimulated much more than the legume. Fertilization of grasses with N tends to increase total, nonprotein, and nitrate N of the plants. Potassium content and, in some cases, other mineral elements may increase in response to N; however, the marked increase in plant growth that may be obtained by high levels of N may be expected to result in some dilution of most mineral elements, particularly during the first few days or weeks of rapid growth after fertilization.

Digestible protein and, in some instances, palatability and dry matter intake may increase in response to fertilization with N. Fertilization studies have shown, in general, that the concentrations of most of the mineral elements may be increased in the plant by fertilization with the element in question.

Harvest and Storage Methods. Harvesting and storage methods markedly affect the nutritive value of roughages. For example, forage cut for hay may be unduly bleached by the sun or damaged by dew and/or rain. Sun bleaching results in a rapid loss of carotenes. Leaching by rain results in losses in soluble carbohydrates and N, and the added handling often required may result in additional losses of leaves. Any harvesting method used for legumes that reduces leaf loss will help to maintain nutritive value. With respect to storage, hays may lose a considerable amount of their

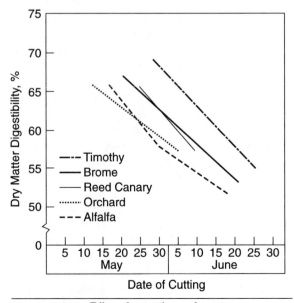

Figure 19.1 *Effect of maturity on dry matter digestibility of first cutting forages.*

original nutritive value if stored before sufficient drying; this may result in heating and mold formation. If hay is properly dried before storage, relatively little nutritive loss takes place if the hay is stored in dry conditions over a period of several years. The same principle applies to silages; well-preserved silages will keep for a number of years without an appreciable decline in nutritive value.

Latitude at which the plant is grown, along with accompanying effects of temperature and light intensity, has an appreciable effect on the composition and nutritive value of the roughage produced from it. Limited data indicate that the nutritive value of forage harvested on a given date is increased progressively as the distance from the equator increases (north or south).

❏ PASTURE AND GRAZED FORAGES

A wide variety of vegetation is utilized directly as pasture by herbivorous animals. With respect to agricultural production, herbage usually is divided into native and cultivated species, the latter being utilized to improve productivity, to extend the period of high-quality grazing, or to improve the versatility of crop production.

Herbage may also be divided into the following classes:

Grasses—members of the family Gramineae
(more than 6000 species):
 Cool season grasses—grasses that make their
 best growth in the spring and fall
 Warm season grasses—grasses that grow
 slowly in the early spring and grow most
 actively in early summer, setting seed in
 summer or fall
Legumes—members of the family Leguminosae
(more than 14,000 species)
Forbs—primarily broadleaf, nonwoody, plants
Browse—woody plants consumed in some degree
by ruminants, particularly selective eaters such as
deer and antelope.

Grasses

Grasses are by far the most important plant family that humans are concerned with agriculturally, because the grass family not only includes all of the wild and cultivated species used for grazing and forage harvesting, but also the cultivated cereal grains and sorghum species. However, in this section we are concerned only with grass as a forage and do not consider it from an agronomic point of view. The discussion is very general because of the tremendous numbers of grasses of importance.

As a food for grazing animals, grass has many advantages. Most grass species are palatable when immature and only a few, such as some sorghums during regrowth following frost (HCN toxicity), are highly toxic for any appreciable part of the grazing season. Grasses representing many species have the ability to grow in most environments in which herbivores can survive (arctic regions are the notable exception). Furthermore, nutrients supplied in grasses provide amounts of most nutrients roughly in parallel with animal needs during a yearly life cycle of reproduction and production in semitropical and tropical environments. Obviously, during midwinter in cold climates harvested forages are required to sustain year-round animal production.

Problems in Grazing Immature Grasses. Early in the growing season, grasses, especially cool season species, have a very high water content and an excess of protein and total N for ruminant animals. This may result in diarrhea and, because of the low dry matter content, in difficulty obtaining an adequate intake of energy. Grass proteins usually are high in the amino acid arginine and also contain appreciable amounts of glutamic acid and lysine. If N application to the pasture is liberal, and particularly if S is deficient, there is likely to be a high level of NPN in the form of nitrate, amino acids, and amides and a relative deficiency of the S-containing amino acids. Toxic symptoms may occur at levels of about 0.07% nitrate N (dry matter basis), and amounts on the order of 0.22% may be fatal to ruminants. However, if ruminants are adapted to and fed on high-nitrate grasses continuously, toxicity is less likely because rumen microorganisms are capable of reducing nitrate to ammonia, which is well utilized.

In comparison with legumes (see next section), such as alfalfa (also known as lucerne in some countries), the protein content of grasses is nearly always lower, particularly in mature plants. Digestible energy is high (more than 70%) in young grass but declines rapidly with maturity. Mature plants, especially those that are weather leached, are low in digestible energy and protein, as well as in soluble carbohydrates, carotene, and some of the minerals. Therefore, they may not meet the animal's needs, even when productive requirements of the animal are low.

The mineral element content of grasses varies considerably, depending mainly on the species and soil fertility. Grasses usually are adequate in Ca, Mg, and K but are likely to be borderline or deficient in P. Trace element concentrations also vary considerably. Canadian research indicates that 80% or more of forage species measured contain levels of one or more elements less than those generally considered adequate for ruminate. Ranges and typical values to be expected are shown in Table 19.2. The bioavailability of mineral elements present in plants is highly variable and not clearly known for most plant species. Also, factors such as stage of plant maturity, and the soil concentration of a specific element and its availability to the plant, may have important effects on bioavailability to the animal. Relatively, little information is available concerning the bioavailability of mineral elements from plant feedstuffs. The important interface between plant and animal scientists needs to be encouraged as a means of expanding the knowledge base related to mineral element nutrition in animals.

Cool-Season and Warm-Season Grasses. The desirability of different grasses depends on their growing habits and on animal needs. Large differences exist in the composition of grasses clas-

sified on the basis of their growth patterns as cool- or warm-season grasses. Generally, cool-season grasses mature at slower rates than warm-season grasses, and their quality deteriorates less rapidly. Lush, young grass usually is highly palatable, but palatability declines for most species as the plants mature, and most animals object to the seed heads of many grass species.

Quality differences between species of grasses become more evident with maturity. Furthermore, regrowth of grass in the fall usually is not as nutritious as spring grass, partly because of the lower concentration of soluble carbohydrates.

Cultivated cool-season and warm-season grasses held in high esteem include perennial ryegrass (*Lolium perenne*), Italian ryegrass (*Lolium multiforum*), orchard grass (*Dactylis glomerata*), blue grass (*Poa* spp.), and smooth brome grass (*Bromus inermus*). Others considered less desirable include Bermuda grass (*Cynodon dactylon*), foxtail (*Alopecurus pratensis*), bent (*Agrostis* spp.), and tall fescue (*Festuca arundinacea*). In addition to varying in growth habits, these species may vary in nutrient composition, palatability, and digestibility. Tropical and subtropical species that have found appreciable use in various countries include Buffelgrass (*Cenchrus ciliarus* L.,), Rhodegrass (*Chloris gayana*), Kikuyugrass (*Pennisetum clandestinum*), Pangola digitgrass (*Digitaria decumbens*), panicgrass (*Panicum maximum*; several

TABLE 19.2 Range and typical mineral concentrations for pasture grasses and alfalfa plants.

MINERAL ELEMENT	GRASSES			ALFALFA		
	LOW	TYPICAL	HIGH	LOW	TYPICAL	HIGH
Major elements, % of dry matter						
Ca	<0.3	0.4–0.8	>1.0	<0.60	1.2–2.3	>2.5
Mg	<0.1	0.12–0.26	>0.3	<0.1	0.3–0.4	>0.6
K	<1.0	1.2–2.8	>3.0	<0.4	1.5–2.2	>3.0
P	<0.2	0.2–0.3	>0.4	<0.15	0.2–0.3	>0.7
S	<0.1	0.15–0.25	>0.3	<0.2	0.3–0.4	>0.7
Trace elements, ppm of dry matter						
Fe	<45	50–100	>200	<30	50–200	>300
Co	<0.08	0.08–0.25	>0.30	<0.08	0.08–0.25	>0.3
Cu	<3	4–8	<10	<4	6–12	>15
Mn	<30	40–200	>250	<20	25–45	>100
Mo	<0.4	0.5–3.0	>5	<0.2	0.5–3.0	>5
Se	<0.04	0.08–0.1	>5	<0.04	0.08–0.1	>5
Zn	<15	20–80	>100	<10	12–35	>50

strains are grown), paragrass (*Brachiaria multica*), Napiergrass (*Pennisetum purpureum*), and molassesgrass (*Melinis minutiflora*). Kikuyugrass is from Hawaii, and the others are from Africa. In the appropriate environment, these grasses may be very productive in a tropical or subtropical climate.

In many regions in North America, some of the cereals are used for pasture during the vegetative stage of growth. Wheat is most commonly used in this way, but so are barley, oats, and rye. These plants can be pastured during the winter and early spring with little or no effect on grain yield, provided soil conditions permit. The forage of these plants is high in readily available carbohydrates (more than 50%), and crude protein is high. Extensive use of such pasture is made in the southwestern states of the United States, particularly for growing calves and lambs.

Several sorghum species are used for pasture or harvested forage. Sudan grass (*Sorghum vulgare sudanense*) is one of the more common ones used in the United States, but others, such as Johnson grass (*Sorghum halepense*), find some use. Sudan–sorghum hybrids have been developed. Some producers utilize these species because they can be sown during early summer in temperate areas and will produce late summer and fall pastures. They are prone to have high levels of glycosides that can be converted to hydrocyanic acid (a highly toxic product), particularly following drought or frost damage, so care must be used if these conditions occur while pasture is in use.

Legumes

Many different legumes are utilized by grazing animals, although the cultivated legumes comprise a much smaller group than do cultivated grasses. In overall usage, alfalfa (*Medicago sativa*) is by far the most common legume used for pasture, hay-crop silage, and hay in North America. Alfalfa finds much favor because of its high yield, its persistence as a perennial crop, and its high palatability and excellent nutrient content.

Other legumes used extensively for pasture or hay include clovers such as ladino or red (*Trifolium pratense*), white (*Trifolium repens*), and subterranean (*Trifolium subterraneum*), as well as common lespedeza (*Lespedeza striata*), lupines (*Lupinus* spp.) and vetches (*Vicia* spp.), and birdsfoot trefoil.

Legumes are higher in protein than grasses, particularly in more mature plants. The leaves are rich sources of protein and other nutrients, but stems, which are high in cellulose and lignin, are of much less value, especially in mature plants. Changes in plant composition with maturity are largely a result of increased lignification and increased fiber in the stems and a reduced leaf:stem ratio. Compared to grasses, legumes have characteristically high concentrations of Ca, Mg, and S (Table 19.2) and, frequently, Cu. They tend to be lower than grasses in Mn and Zn. On the whole, legumes are palatable, although most are bitter and may require some adaptation before they are readily consumed by cattle.

Some legumes, particularly alfalfa and white, ladino, and red clover, are prone to cause bloat in grazing ruminants, especially cattle. Bloat is caused primarily by foam-producing compounds from the plant, of which cytoplasmic proteins and, perhaps, pectins are the most important. Foam in the rumen (see Church, 1988) causes entrapment of normal rumen gases, which cannot be eliminated and so cause a gradual increase in rumen pressures and, if not relieved, eventual suffocation of the animal as a result of encroachment of the distended rumen on the lung cavity. Plant breeding research indicates that alfalfa plants can be selected to have a lower content of the proteins that are involved in bloat production.

Native Pastures and Range

Pastures and rangeland comprised of uncultivated native forage plants cover vast areas of land in regions of the world where the environment, soil, or topography rule out intensive agricultural methods. These areas usually contain a wide range of grasses, sedges, forbs, and browse. The nutrient properties of these various plants vary widely, and, in addition, there are distinct seasonal patterns of use by different grazing animals. The subject is too complex to discuss here; for more information, the reader is referred to Heath et al. (1973), Morley (1981), and Rayburn (1986).

Miscellaneous Forage Plants

In some areas and for some specific seasonal usage, plants such as the cabbage family (*Brassica* spp.) are used extensively. Kale, cabbage, and rape are included in this group. Rape, for example, is planted sometimes for use as fall pasture by sheep. The tops of root crops such as beets and turnips also are used as forage. As a part of the total forage resources, however, these crops account for a negligible amount.

❏ HARVESTED DRY FORAGES

In temperate and subtropical regions, roughages stored in the dried form are the most common type used for feeding during the time of year when grazing is not available or in confinement housing facilities. Forages harvested and stored as long hay or bales require a relatively high labor input for handling. Modern harvesting, processing, and storage equipment allows rather complete mechanization in most large cropping and livestock enterprises (see Figs. 19.2 and 19.3). Continual improvement is being made in such equipment.

Hays

Hay—from grasses or legumes—is grown and harvested almost exclusively for animal use. Although haymaking is the most common method of conserving green crops, its relative importance has de-

Figure 19.2 *Hay baler capable of producing bales weighing 750 to 2000 pounds that also can be surface-wrapped to preserve the hay in storage. Courtesy of John Deere & Co.*

clined somewhat in recent years because of increased use of green chop (feeding of freshly cut plants) and forage crops in silage.

The usual intent in haymaking is to harvest the crop at the optimum stage of maturity in order to

Figure 19.3 *A square hay baler. Such balers can dispense bales on the ground or into a wagon manually or automatically. Courtesy of J. I. Case Implement Co.*

capture a maximal yield of nutrients per unit of land without damage to the next crop. To make good hay, the water content of the plant material must be reduced to a point low enough to allow storage without spoilage or marked nutritional changes. Moisture content of green herbage may range from 65 to 85% or more, depending on maturity and the plant species. For hay to keep satisfactorily in storage, the water content must be reduced to about 15% or less.

Losses in Haymaking.   It is impossible to cut, dry, and move hay into storage without losses occurring in the process. However, it may be possible to harvest more units of nutrients per unit of land by haymaking than could be obtained by grazing, because of selective grazing of preferred plant species and trampling and feed refusal resulting from contamination by dung and urine. Both the quality and quantity of field-cured hay that can be harvested depends on such factors as (1) maturity when cut, (2) method of handling, (3) moisture content, and (4) weather conditions during harvest. For example, rain on freshly cut hay will cause little damage; however, when hay is partially dried, rain is very damaging. Early cut hay in many areas is often of low quality because of rain and resultant spoilage, leaching, and leaf loss. One report indicates dry matter losses ranging from about 6% for artificially dehydrated hay up to 33% or more for rain-damaged field-cured hay. Another report indicated the following losses: plant respiration losses (before plant is dry), 3.5% in 24 h; leaching by rain, 5 to 14%; and leaf shattering in legume hays, 3 to 35%. Thus, it is clear that very substantial amounts of dry matter may be lost in haymaking under adverse conditions; an average loss of 15 to 20% is not abnormal for legume hays.

Changes during Drying.   Rapid drying, provided it is not accomplished at excessively hot temperatures, results in the least changes in chemical components of forage. Machines such as crimpers (or conditioners) are used to crush the stems of plants such as alfalfa and speed up the drying process. If drying is slow in the field, stack, or bale, appreciable changes may occur as a result of activity by plant enzymes and microorganisms or because of oxidation. After the plant is cut, the cells continue to function for a time, with the result that soluble carbohydrates may be oxidized. Oxidative reactions may continue for some time, depending on the temperature and how the hay is stored. The most obvious change is a loss of pigmentation as plants lose carotenes by oxidation. Proteins may be modified as some hydrolysis occurs, resulting in relatively greater amounts of NPN, primarily amino acids. Slow drying almost always is accompanied by excessive mold growth, which reduces the palatability and nutritive value of hay, and in some cases the molds may be toxic.

Hay stored in the stack or bale while too wet to dry rapidly may undergo enough fermentation to result in marked temperature increases (largely because of growth of thermophilic molds), which may cause browning and sometimes spontaneous combustion. Newly baled alfalfa hay should have no more than 20 to 25% moisture to avoid these problems. Excess heating or molding results in a marked reduction in digestibility of protein and energy.

If drying is not complicated by weather factors, relatively little change in the composition takes place between green plant and hay, nor is there any pronounced effect on nutrient utilization (see Table 19.3). Animals generally ingest dry matter from fresh herbage at a slower rate than from hay, and some slight differences may occur in rumen fermentation, digestibility of the herbage, and site of digestion in ruminants, but the differences appear to be inconsequential.

Thus, we see that drying need not have any great negative effect on forage utilization. However, in practice, we must expect some loss of leaves in legumes and a reduction in soluble carbohydrates. Hay, if made from moderately mature plants, will have lower protein and digestible energy than that from young herbage. Nutritive properties of hays are, then, similar to those of forages, but with slightly to greatly reduced values depending on degree of weather damage and method of harvesting.

Artificially Dried Forage

Rapid drying is required for good haymaking. At one time in the United States and northern Europe, particularly in areas where weather was a problem, considerable interest was shown in barn drying. This is accomplished by circulating air through the hay after storage. Although a very good product can be produced, interest has declined with greater use of silage and developments in bale-handling machinery.

Dehydration of herbage with appropriate machinery is a viable industry in the United States

TABLE 19.3   Composition and digestibility of green ryegrass and material from the same field which was dried artificially or made into wilted or unwilted silage.[a]

ITEM	FRESH GRASS	DRIED GRASS	WILTED SILAGE	UNWILTED SILAGE
Composition of dry matter				
Organic matter, %	90.8	92.0	92.2	91.7
Total water-soluble carbohydrate, %	9.2	8.4	trace	trace
Cellulose, %	24.2	24.3	25.0	26.8
Hemicellulose, %	14.0	13.3	12.9	13.1
Total N, %	2.85	2.99	3.09	3.08
Gross energy, Kcal/g	4.59	4.55	4.46	4.89
Digestibility, %				
Energy	67.4	68.1	67.5	72.0
Cellulose	75.2	75.5	76.5	80.6
Hemicellulose	59.4	57.7	59.9	63.2
N	75.2	71.0	76.5	76.4

[a]From Beever et al. (1971). In this experiment, the fresh grass was quick frozen so that it could be fed at the same time as the other forms.

and some other countries. In the United States, alfalfa is the primary crop that is dehydrated, but in Europe, grasses or grass–clover mixtures are used more commonly. In making dehydrated alfalfa the herbage is cut at a prebloom stage, dried quickly at relatively high temperatures, ground, and sometimes pelleted. Because carotene and xanthophyll pigments are important for poultry feeds, the product often is stored using inert gases such as N_2 to reduce oxidation. Herbage so processed is high in crude protein, moderately low in fiber, high in carotene, and highly digestible. The relatively low fiber content makes such feedstuffs more suitable for species such as poultry and swine and, in the United States, a high proportion of dehydrated alfalfa goes into commercial diets for these species. For poultry, the carotenes and xanthophylls serve to increase pigmentation of the skin of broilers or the yolk of eggs. For swine, particularly sows, nutrients of interest are the vitamins, Ca, and trace minerals. These feeds serve very well for horses or ruminants, but generally are more expensive per unit of nutrients provided, when compared with other roughage sources.

Straws and Chaff

An appreciable amount of straw and chaff is available for animal feed in most farming areas. Straw consists mainly of the stems and variable amounts of leaves of plants that remain after the removal of the seeds. Chaff consists of the small particles removed from the seed head along with limited amounts of small or broken grains.

The primary supply of straw and chaff comes from the small cereal grains—wheat, barley, rye, rice, oats—but, in some areas, substantial amounts may be available from the grass or legume seed industry and from miscellaneous crops. As a whole, straws are very low in digestible protein; very high in fiber and, usually, lignin; and generally poor in feeding value. For example, values given from the National Research Council (NRC) (1982) for winter wheat straw (dry basis) are crude protein, 3.2%; crude fiber, 40.4%; and digestible energy (cattle), 1.92 Mcal/kg or 47% TDN. Of the various cereal straws, that from oats is regarded as the best, partly because the grain is often harvested before it is fully ripe. As a feedstuff, straws are best used as a diluent in high-concentrate rations or as the basal feed for wintering cattle when properly supplemented with deficient nutrients—protein, vitamin A, minerals. Even though straws are low in metabolizable energy, the energy derived from that which is digested and from the heat increment (see Chapter 10) provides energy that can be used by animals, such as pregnant cows, that have a low productive requirement.

Miscellaneous

In the United States nearly all of the grain from corn and sorghum now is field harvested. Often the cobs or threshold seed heads are left in the field, along with a substantial amount of uncaptured grain, in some cases. This material provides a useful resource that should be utilized, even though it is of low quality. The nutritive problems here are the same as with straws, and some supplementary feeding is required to supply needed minerals, carotene, and protein. Protein may be less critical than with straws.

In cotton-growing areas, cottonseed hulls and cotton gin trash are available in substantial amounts, and these are consumed readily by ruminants. In addition, roughage sources such as pineapple greenchop, pineapple juice presscake, pineapple stump meal, rice mill feed, sugarcane bagasse, and sugarcane strappings may be available where the base crops are grown (Bath et al., 1980).

Figure 19.4 *One type of forage harvester for making silage or green chop. Courtesy of Deere & Co.*

❏ HARVESTED HIGH-MOISTURE FORAGES

Green Chop (Soilage)

Green chop (or soilage) is herbage that has been cut and chopped in the field and then fed to livestock in confinement. Plants used in this manner include the forage grasses, legumes, sudan grass and other sorghum family species, the corn plant, and, at times, residues of food crops used for human consumption.

Although this is one of the simplest means of mechanically harvesting herbage (Fig. 19.4), it requires constant attention to animal needs as opposed to other methods of harvesting herbage. A major advantage of green chop is that more usable nutrients can be salvaged per unit of land than with other methods such as pasturing, haymaking, or ensiling. Thus, it is sometimes feasible to harvest in this manner rather than using such crops in other ways, provided land productivity is high (this reduces harvesting costs). When herbage growth outruns daily need, the excess can be made into hay or silage before it becomes too mature for efficient use. Weather, of course, is less of a factor than in haymaking.

Most data indicate that beef cattle fed soilage gain weight as well as when pastured using intensive systems such as short-term rotations or strip grazing and that dairy cows do equally well when fed soilage as when fed alfalfa preserved in other ways. Practical experience has indicated that optimal usage is obtained when green chop is fed along with hay or silage rather than by itself, partly because total dry matter consumption tends to be greater.

Silage

Silage has been used for feeding animals, primarily ruminants, for many years (Fig. 19.5). It is the material produced by controlled fermentation of high-moisture herbage. When such material is stored under anaerobic conditions and, if the supply of fermentable carbohydrates is adequate, sufficient lactic acid is produced to stabilize the mass so that fermentation stops. If undisturbed, silage will keep for an indefinite period. Alternate methods, primarily used in Europe, require the addition of strong acids that lower the pH, thus preventing fermentation, or the use of formaldehyde with or without organic acids such as propionic. Such combinations inhibit most fermentation that would otherwise occur.

Good silage is a very palatable product that is utilized well (see Table 19.3), and excellent results may be obtained with high-producing animals such

Figure 19.5 The face of a bunker (trench or pit) silo, a type used in many areas for storage of large amounts of silage. This type facilitates use of equipment for filling and unloading. Courtesy of Bill Fleming, Beef Magazine.

Most silage in the United States is made from the whole corn plant (*Zea mays*) or from any of a number of sorghum varieties in areas where rainfall is insufficient for growing corn. Grass, grass–legume, or legume silages are used extensively because technology is adequate to produce very good silage from these crops. For that matter, palatable and nutritious silage can be produced from a wide range of herbage, including weeds and low-moisture herbage. The chemical composition of alfalfa silage from early or late cut forage compared with that of orchard grass silage from early or late cut forage is shown in Table 19.4. Note the higher neutral detergent fiber (NDF) and lower soluble residues of orchard grass silage compared with alfalfa silage and the differences in composition of most constituents in early versus late cut forages. Growth and metabolism studies in yearling dairy animals fed these silages (Tyrrell et al., 1992) showed that the metabolizable energy available for growth from alfalfa silage resulted in 15% more energy retained than that from orchard grass silage. No differences between alfalfa and orchard grass were noted in metabolizable energy required for maintenance. The data illustrate the generally superior nutritive value of legumes compared with grasses for ruminants, whether the forages are fed in fresh, dry, or ensiled form.

as lactating cows. In addition to advantages in harvesting or in nutritive properties, the fermentation that occurs usually will reduce greatly the nitrate content, if nitrate is present, and the amounts of other toxic materials, such as hydrocyanic acid, will be reduced. However, it is important to note that fermentation requires energy; thus, these losses caused by ensiling reduce the overall nutrient content of material in the silo after ensiling compared with that of the starting material. This may or may not be reflected in normal analytical data.

TABLE 19.4 Chemical composition of silage dry matter from alfalfa or orchard grass fed to yearling dairy cattle.

	MEAN VALUES BY SPECIES AND MATURITY			
FRACTION	ALFALFA	ORCHARD GRASS	EARLY	LATE
DM, %	26.6	24.4	25.1	25.9
Organic matter, % of DM	90.9	92.9	92.4	92.4
Energy, Mcal/kg of DM	4.84	4.92	4.93	4.83
Nitrogen, % of DM	3.38	2.46	3.23	2.61
Ether extract, % of DM	3.77	4.33	4.54	3.57
Carbon, % of DM	47.7	48.1	48.5	47.4
NDF, % of DM	48.0	70.4	56.4	62.0
Hemicellulose, % DM	7.6	29.2	18.3	18.6
Cellulose, % of DM	28.4	32.5	29.0	31.9
Lignin, % of DM	11.0	8.0	8.6	10.4
ADF ash, % of DM	0.98	0.54	0.45	1.07
Soluble residue, % DM	18.1	2.9	11.4	9.6

Source: From Tyrrell et al. (1992).

Chemical Changes during Fermentation.
Chemical changes during silage fermentation are complex, and a complete discussion is beyond the scope of this book. Other sources (Heath et al., 1973; Watson and Nash, 1960; data generated and self-published by silage production and storage companies) contain additional information.

The chemical changes are the result of plant enzyme activity and action of microbes that are present on the herbage or that find their way into it from other routes. The plant enzymes continue to be active for the first few days after cutting while oxygen is available, resulting in some metabolism of soluble carbohydrates to CO_2 and water and the production of heat. Optimum temperatures during fermentation are said to be between 80 and 100°F. Excessive heat is objectionable and is unlikely if the silage is well packed to exclude most of the air.

Plant proteins are broken down partially by cellular enzymes, resulting in an increase in NPN compounds such as amino acids. Anaerobic microorganisms multiply rapidly, using sugars and starches as primary energy sources and producing mainly lactic acid with lesser amounts of acetic acid and small amounts of others such as formic, propionic, and butyric; little butyric acid is present in well-preserved silage. Continued action occurs on N-containing compounds with further solubilization and production of ammonia and other NPN compounds (Table 19.5). The level of lactic acid rises in well-preserved corn or sorghum silage, eventually reaching levels of 7 to 8%, and the pH drops to about 4.0, depending on buffering capacity and dry matter content of the crop in question. Grass silages will usually have a higher pH of about 4.5. If the silage is too wet or the supply of soluble carbohydrates too low, the pH will not go this low, allowing the development of clostridia bacterial species, relatively large amounts of butyric acid, and further fermentation of NPN compounds, resulting in production of amines such as tryptamine, histamine, and others. These amines have undesirable odors and tastes (i.e., rotten), and they may be toxic. On the other hand, if the mass is too dry or poorly packed, excess heating may occur, and molds may develop, producing unpalatable and, sometimes, toxic silage.

TABLE 19.5   Composition and nutritive value of third cutting alfalfa harvested as hay, low-moisture and high-moisture silage.

ITEM	HAY	WILTED FORAGE[a]	LOW-MOISTURE SILAGE	GREEN CHOP[b]	HIGH-MOISTURE SILAGE
Dry matter, %	92.6	58.1	59.0	28.8	28.1
Others on DM basis					
Crude protein, %	20.6	18.1	20.4	19.8	21.6
Cellulose, %	34.0	33.3	37.4	30.3	37.9
Soluble carbohydrates, %	7.8	5.2	2.5	3.8	3.7
Total N, %	3.3	2.9	3.3	3.2	3.4
Soluble N, % of total N	31.8	37.2	51.1	32.6	67.0
NPN, % of total N	26.0	28.4	44.6	22.6	62.0
Volatile fatty acids, %					
Acetic			0.4		4.2
Propionic			0.05		0.14
Butyric			0.002		0.11
pH			4.7		4.7
Dry matter intake, g/day/kg body wt.	25.3		21.8		25.8
Digestibility, %					
Dry matter	64.5		59.0		59.5
Nitrogen	77.5		63.0		72.8
Cellulose	58.8		60.4		64.3

[a,b]Wilted herbage was used to make low-moisture silage, and the green chop was used for the high-moisture silage.
Source: Taken from Sutton and Vetter (1971).

For most crops, dry matter contents of about 35% and a soluble carbohydrate content of 6 to 8% (dry basis) are near optimal for silage making. Consequently, if grass or legume silage is to be made, usually it is wilted some before ensiling. If direct-cut herbage is used, usually it will be too wet for good silage, allowing *Clostridia* bacteria to multiply. When herbage such as grass or legumes is direct-cut and ensiled, preservatives or sterilants may be added. Additional dry matter from sources such as ground corn cobs, straw, or hay, may be used to soak up some of the moisture, and soluble carbohydrates may be added in the form of grain or molasses; these provide useful insurance when making grass–legume silage. Silage sterilants such as formic acid, sulfur dioxide, or sodium metabisulfite may be advantageous when moisture content is high, but are of doubtful value in silage made from wilted grasses or legumes or from corn or sorghum plants.

Microbial Inoculation of Silage.

Exposure of silage to air before feeding can result in substantial nutritive losses. This is a practical problem in areas such as the southern United States where large dairies often contract for delivery of silage sufficient for several days of feeding and store it unprotected (Pitt et al., 1991). The aerobic deterioration of such silage is caused by metabolism of sugars and organic acids by yeasts and bacteria (Spoelstra et al., 1988). Speculation that such losses could be reduced by inoculation of the silage with lactic acid bacteria has led to many experiments with silage inoculants such as *Lactobacillus plantarum*, *Streptococcus faecium*, and other microorganisms in an effort to decrease aerobic losses and the nutritional value of the silage. In some cases, antifungal agents such as potassium sorbate added in small amounts, for example, 0.5 mg/g of wet herbage, are added in conjunction with the bacterial inoculant. Research by numerous investigators has produced inconsistent results. The variable results (Rust et al., 1989; Sanderson, 1993; Wohlt, 1989) of studies concerning the effects of bacterial inoculants on aerobic stability and fiber digestibility of silage suggest that such inoculants should be viewed primarily as fermentation aids (Sanderson, 1993).

Losses from Ensiling.

Variable losses occur during ensiling because of fermentative activity and resultant heat produced. Gaseous losses are said to range from 5 to 30% of original dry matter. Most of the losses originate from soluble and highly digestible nutrients, with the result that silage (Table 19.5) is almost certain to contain a higher percentage of fibrous and insoluble ingredients.

Very substantial amounts of seepage may occur in high-moisture silage, particularly when the dry matter content is less than 30%. Above this level such losses are moderate to low. Seepage moisture contains many soluble nutrients, and such losses must be avoided if silage feeding is to be highly efficient. Further losses occur from molding. Molding nearly always takes place around the perimeter and on top of the silage. Data on alfalfa silage indicate spoilage losses on the order of 4 to 12% of original dry matter ensiled. Such material is not only unpalatable but may also be toxic. If field losses are included, total losses may be expected to be about 20 to 25% of the herbage dry matter present in the field.

Nutritive Properties of Silages.

One of the nutritional problems associated with feeding silages is that consumption of dry matter in the form of silage nearly always is lower than when the same crop is fed as hay. This seems to apply whether the crop in question is legume, grass, or other herbage. Intake of silage usually is greater as the dry matter content increases. Data covering 70 different silages (Wilkins et al., 1971) indicated that ad libitum intake was correlated positively with the dry matter, N, and with lactic acid as a percentage of total acids. Intake was negatively correlated with acetic acid content and ammonia as a percentage of total N. Intake was related positively to digestibility for legumes but negatively for grasses other than ryegrass. This information would indicate that maximum intake should be achieved with silage containing just enough moisture to allow for preservation with minimum production of ammonia and acetic acid, which tend to increase at higher moisture contents (see Table 19.5).

Differences reported in digestibility and in animal performance between direct cut or wilted silage and hay have been inconsistent. Unfortunately, errors associated with silage analysis have been included in many experimental reports. If the test silage is dried in an oven, substantial amounts of volatile materials are lost, thus giving a low value for nutrient content of silage as fed and an underestimate of organic matter intake and digestibility.

Grass–Legume Silages.

Grass or legume silages are normally high in crude protein (20% or more) and carotene, but only moderate in digestible energy. This level of protein is too high for most ruminants; more nearly optimum results may be expected by supplementing grass silage with some form of energy (grains) or by diluting the protein by feeding some other form of low-protein roughage.

Digestibility (Table 19.5), retention of N, and efficiency of utilization of N for growth or lactation generally are lower with silage than with dry crops of comparable composition. This may be related to the presence of many different forms of nonprotein N found in silage.

Low-moisture silage, often called haylage, is a very palatable feed, probably because it tends to have relatively less acetic acid and less N in the form of ammonia and other NPN compounds (Table 19.5) than does high-moisture silage. It is used to a great extent in the United States for dairy cows. This type of silage is made best in air-tight silos now in common use in many areas (Fig. 19.6) or with reusable plastic bags (Fig. 19.7).

Figure 19.7 *A plastic bag used for silage making. Such equipment costs much less initially than other types, but it is a temporary solution if silage will be used continually. This type of silo does have the advantage that it can be located in different areas.*

Corn and Sorghum Silage

Corn silage is the most popular silage in the United States in areas where the corn plant grows well. It is becoming more popular in other areas of the world because, except for sugar cane, maximum yields of digestible nutrients per unit of land can be harvested from this crop. In addition, the corn plant can be handled mechanically at a convenient time of the year.

Well-made corn silage is very palatable and has a moderate to high content of digestible energy, but it usually is low to moderate in digestible protein. Although corn silage contains much grain (up to 50% in well-eared crops), maximum animal growth rates or milk yields cannot be obtained without energy and protein supplementation.

The effect of stage of maturity on nutritive value of corn silage is shown in Table 19.6. Note that the crude fiber declines and NFE increases as the corn becomes more mature; this is a reflection of the increased grain content. Digestibility of energy is quite constant because of the offsetting effect of increasing energy from grain versus the lower digestibility of the more mature stalk. The data show clearly that intake and digestibility were greater for the medium–hard dough stage.

Some additives may improve corn silage. Treatment at ensiling time with limestone (0.5–1%) or other Ca salts tends to buffer the acids produced during fermentation and results in a very substantial increase in lactic acid production and, usually, an improvement in intake and animal performance.

Figure 19.6 *An example of gas-tight silos that have been widely used for forage and high-moisture grain. Courtesy of the A. O. Smith Co.*

TABLE 19.6   Effect of maturity on nutritive value of corn silage as fed.[a]

	STAGE OF MATURITY			
ITEM	SOFT DOUGH	MED.-HARD DOUGH	EARLY DENT	GLAZED AND FROSTED
Dry matter, % as fed	24	27	32	39
Crude protein, % of DM	8.0	7.7	7.8	8.0
Crude fiber, % of DM	24.0	20.9	17.8	16.6
NFE, % of DM	58.4	62.9	65.9	68.3
Dry matter intake, % body wt./day	1.66	2.02	1.90	1.71
Gross energy digestibility, %	67.7	68.1	65.7	65.0
Protein digestibility, %	53.7	53.9	53.8	54.6
Energy balance, Kcal/day	962	3285	2870	2341

[a]Data from Colovos et al. (1974). Silage fed ad libitum along with a small amount of molasses–urea supplement.

Likewise, addition of urea (0.5–1%) or anhydrous ammonia to corn silage will increase the crude protein content and has worked out well for dairy cows in particular. The value of the added N will depend somewhat on fertilization practices used on the crop as heavy fertilization with N will increase the N content of the plant.

These same general comments apply to sorghum silage. As a rule, sorghum silage has a somewhat lower nutritive value than corn silage. This is a result of lower digestibility and lower sugar content of the stalk, and the passage through the GI tract of the small seeds if not broken up during ensiling. When the seeds have been partially broken, then sorghum silage is comparable to corn silage if the grain content is comparable.

Miscellaneous Silages

A wide variety of other herbaceous material has been used to make silage. Waste from canneries processing food crops such as sweet corn, green beans, green peas, and various root or vegetable residues have been used successfully. Residues of this type are difficult to use on a fresh basis because of a variable daily supply or because they are available for only short periods of time. Ensiling is advantageous in that it tends to result in a more uniform feed and a known supply provides for more efficient planning. Some low-quality roughages such as grass straws can be ensiled; when additives such as molasses, urea, and NaOH are added, a reasonably palatable and digestible product can be produced.

Silage Use in Tropical Climates

Limited use is made of silage in most tropical countries. However, there is adequate evidence that silage of satisfactory quality can be made from grasses such as elephant grass (*Pennisetum purpureum*) and several of the panic grasses and forage cereals. Wilkinson (1983) pointed out that tropical grasses generally contain lower levels of water-soluble carbohydrates than do temperate species. He suggested that at least 3% (fresh weight) of soluble carbohydrates should be present to ensure adequate production of lactic acid and to avoid the undesirable secondary fermentations resulting in high butyric acid production. Because of the low soluble carbohydrate content, grasses need to be wilted, but overwilting prior to ensiling can occur easily under tropical conditions. In one study with forage maize (corn), it was shown that losses associated with silage production (those measured) were mechanical losses in the field, 6%; respiration and fermentation during storage, 7%; surface waste, 7%; during removal from storage, 5%. Silos to be used in the tropics should be designed so that they are long and narrow, thereby minimizing the face area exposed between feeding times and thus reducing deterioration of the silage (Wilkinson, 1983).

❏ HIGH-ENERGY FEEDSTUFFS

Feedstuffs included in this class are those that are fed or added to a diet primarily to increase energy intake or to increase the energy density of the diet. High-energy feedstuffs include the various cereal grains and some of their milling byproducts. In addition, several liquid feeds—primarily molasses and mixtures in which molasses predominates—and fats and oils are included.

High-energy feedstuffs generally have low to moderate levels of protein. However, several of the high-protein meals could be included on the basis of available energy. Furthermore, on the basis of energy per unit of dry matter, some of the roots and tubers may be included; because of their high water content, they are usually considered separately. The same comments apply to fluid milk.

Energy from high-energy feedstuffs is supplied primarily either by readily available carbohydrates—sugars and/or starches—or by fat. Depending on the type of diet and the class of animal involved, such feedstuffs may make up a substantial percentage of an animal's total diet. Other nutrients provided by these feeds—amino acids, minerals, vitamins—must also be considered, but quantitatively these nutrients are of less importance than the available energy. Veum (2005) summarized the use of high-energy feedstuffs for livestock.

Cereal Grains

Cereal grains are produced by members of the grass family (Gramineae) grown primarily for their seeds. Vast quantities of cereal grains and their milling and distillery byproducts are produced annually in the United States primarily for use in animal feed. Information about the production and use of feedgrains and other feedstuffs fed to animals in the United States is available from the National Agricultural Statistics Service, U.S. Department of Agriculture at www.usda.gov/nass/pubs/agstats.htm. Of course, some of these grains go into human food. For example, corn may be consumed as popped corn, corn flakes, corn flour, corn starch, corn syrup, or corn oil, but the amount used in this manner is less than that used for animal feed. Wheat and rice are grown primarily for human consumption, although substantial amounts of wheat may go into animal feed in the United States when price and supply allow it. Barley and oats, though high-quality feeds, are relatively less

important because their yields per unit land area are less than those of corn and their demand for direct human consumption is high. Other grains such as millet and rye find only limited use as feeds. A wheat–rye cross, triticale, is produced in only small quantities for animal feed. Of the total grains (not counting byproducts) fed in 1985, corn accounted for 74.7%, and this was in a year when much more wheat was fed than is the usual practice.

Average values for some nutrients in cereal grains are given in Table 19.7. Note here that there are relatively small compositional differences among grains. Although grains tend to be less variable in composition than roughages, many factors influence nutrient composition and, thus the feeding value for a given grain. For example, factors such as soil fertility and fertilization, cultivar (genetic variety), weather and rainfall, and insects and disease may all affect plant growth and seed production so that average book values may not be meaningful. With wheat, for example, if a hot, dry period occurs while the grain is ripening, shriveled, small seeds may result. Although they may have a relatively high protein content, the starch content is apt to be much lower than usual, the weight per unit volume will be low, and the feeding value appreciably lower.

Although the crude protein content of feed grains is relatively low, ranging from 8 to 12% for most grains, we may find some lower than this and, particularly with wheat, some may be much higher, sometimes as high as 22% crude protein. Sorghum grains are also quite variable in protein content.

Of the nitrogenous compounds, 85 to 90% is in the form of proteins; most cereal grains are moderately low to deficient for nonruminant species in lysine and often in tryptophan (corn) and threonine (sorghum and rice) and in methionine for poultry, whose requirement is higher than that of pigs.

The fat content may vary greatly, ranging from less than 1% to more than 6%, with oats usually having the most and wheat the least. Most of the fat is found in the seed embryo. Seed oils are high in linoleic and oleic acids, both being unsaturated fatty acids that tend to become rancid, particularly after the grain is processed.

The carbohydrates in grains, with the exception of the hulls, are primarily starch with small amounts of sugars. The starch, which makes up most of the endosperm, is highly digestible, pro-

TABLE 19.7 Average composition of the major cereal grains, dry basis.

ITEM	CORN, DENT	OPAQUE-2 CORN	WHEAT HARD WINTER	WHEAT SOFT WHITE	RICE, WITH HULLS	RYE	BARLEY	OATS	SORGHUM (MILO)
Crude protein, %	10.4	12.6	14.2	11.7	8.0	13.4	13.3	12.8	12.4
Ether extract, %	4.6	5.4	1.7	1.8	1.7	1.8	2.0	4.7	3.2
Crude fiber, %	2.5	3.2	2.3	2.1	8.8	2.6	6.3	12.2	2.7
Ash, %	1.4	1.8	2.0	1.8	5.4	2.1	2.7	3.7	2.1
NFE, %	81.3	76.9	79.8	82.6	75.6	80.1	75.7	66.6	79.6
Total sugars, %	1.9		2.9	4.1		4.5	2.5	1.5	1.5
Starch, %	72.2		63.4	67.2		63.8	64.6	41.2	70.8
Essential amino acids, % of DM									
Arginine	0.45	0.86	0.76	0.64	0.63	0.6	0.6	0.8	0.4
Histidine	0.18	0.44	0.39	0.30	0.10	0.3	0.3	0.2	0.3
Isoleucine	0.45	0.40	0.67	0.44	0.35	0.6	0.6	0.6	0.6
Leucine	0.99	1.06	1.20	0.86	0.60	0.8	0.9	1.0	1.6
Lysine	0.18	0.53	0.43	0.37	0.31	0.5	0.6	0.4	0.3
Phenylalanine	0.45	0.56	0.92	0.57	0.35	0.7	0.7	0.7	0.5
Threonine	0.36	0.41	0.48	0.37	0.25	0.4	0.4	0.4	0.3
Tryptophan	0.09	0.16	0.20		0.12	0.1	0.2	0.2	0.1
Valine	0.36	0.62	0.79	0.56	0.50	0.7	0.7	0.7	0.6
Methionine	0.09	0.17	0.21	0.19	0.20	0.2	0.2	0.2	0.1
Cystine	0.09	0.22	0.29	0.34	0.11	0.2	0.2	0.2	0.2
Minerals, % of DM									
Calcium	0.02		0.06	0.09	0.06	0.07	0.06	0.07	0.04
Phosphorus	0.33		0.45	0.34	0.45	0.38	0.35	0.30	0.30
Potassium	0.33		0.57	0.44	0.25	0.52	0.63	0.42	0.39
Magnesium	0.12		0.11	0.11	0.11	0.13	0.14	0.19	0.22

[a]Analytical data taken from NRC publications.

viding hull permeability allows access of digestive juices. Starch from the different grains has specific physical characteristics that can be identified by microscopic examination, and the size of starch granules varies from grain to grain. Minor differences are noted in chemical characteristics, for example, ratio of amylin to amylopectin in the starch fraction (see Chapter 7), which may affect nutritional value.

Hulls of seeds have a substantial effect on feeding value. Most hulls (or seed coats) must be broken to some extent before feeding for efficient utilization, particularly for ruminant animals. Barley and oats sometimes are known as rough grains because of the very fibrous hulls that are relatively resistant to digestion. Rice hulls are almost totally indigestible. For more details on the cereal grains and their byproducts, see Church (1988), Pond et al. (1991), Seerley (1991), and Sauber and Owens (2001).

Corn (Maize) (Zea mays). In areas where it grows well, corn will produce more digestible nutrients per unit of land area than any other grain crop (Fig. 19.8). Yields of corn on small test acreages have approached 400 bushels/acre (more than 11 tons). In addition, corn is a very digestible and palatable feed, being relished by all domestic animals. Corn grain is harvested and shelled mechanically (Fig. 19.9).

The chemical composition of corn has been studied in detail. Zein, a protein in the endosperm, makes up about half the total protein in the kernel of most varieties. This protein is low in many of the essential amino acids but particularly so in lysine and tryptophan; the total protein of corn is

Figure 19.8   A field of corn (maize), a crop used to feed millions of animals when harvested in different ways. Photo courtesy of R. W. Henderson.

deficient in these amino acids for nonruminants. Therefore, supplementation of corn with protein sources providing a satisfactory balance of essential amino acids is required for adequate animal performance. N fertilization has been shown to increase protein content and decrease protein quality, primarily because of an increase in the zein fraction.

The energy value of corn is used as a standard of comparison for other grains (Table 19.8). Thus, if the relative energy value of corn is taken at 100, the value of other cereal grains is usually lower. This is partially accounted for by the low fiber content of the corn kernel and the high digestibility of its starch. Wheat compares favorably with corn, but diets composed of a high percentage of wheat often promote lower feed intake, particularly by ruminants.

White and yellow corn have similar compositions except that yellow corn has a much higher content of carotene and xanthophylls, vitamin A precursors. Both white and yellow corn are fair sources of vitamin E but low in the B vitamins and devoid of vitamin D. Corn is very deficient in Ca. Although the P content is relatively high, much of it is in the form of phytic acid P (phytin P), which has a low availability to most monogastric animals. Trace elements are relatively low to deficient in corn.

Several genetic mutants of corn have been identified. One of these, known as opaque-2, is of particular interest because it has a high level of lysine and increased levels of most other essential amino acids (see Table 19.7). This change results from a reduction in zein and an increase in glutelin, a protein found in both the endosperm and the germ. The result is an improvement in the quality of the corn protein. Performance of nonruminant animals may be improved considerably by the use of this mutant. Yields of opaque-2 are not equivalent to common corn, nor are prices proportionately higher, so corn producers have not had economic incentive to markedly increase production of this and other high-lysine mutants, including floury-2 and others. Additional experimental mutants include those designated as high-fat, high-amylose, and brown midrib.

Grain Sorghum (*Sorghum vulgare*).   Sorghum is a hardy plant that is able to withstand heat and drought better than most grain crops. In addition, it is resistant to pests, such as root worm and the corn borer, and is adaptable to a wide variety of soil types. Consequently, sorghum is grown in many areas where corn does not do well. Sorghum yields less than corn, where corn thrives. The seed from all varieties is small and relatively hard, and usually requires some processing for good animal utilization.

A wide range of sorghum varieties (all *Sorghum vulgare*) is used for seed production. These include milo, various kafirs, sorgo, sumac, millet, hegari, darso, feterita, and cane. Milo is a favorite in drier areas because it is a short plant adaptable to harvesting with grain combines. Development of hybrid varieties that have higher yields has increased production greatly in recent years.

Chemically, grain sorghums are similar to corn. Protein content averages about 11%, but the amount varies widely according to cultivar and growing conditions. Lysine and threonine are the most limiting amino acids. Content of other nutrients, except for carotenes, which are very low in sorghum, is similar to that of corn. Feeding trials indicate that sorghum grains are worth somewhat less than corn. Some bird-resistant varieties, whose seed coats are high in tannin, are not well liked by most animals, and performance is thereby depressed if the high-tannin sorghum is not diluted with low-tanning sorghum to improve palatability.

Wheat (*Triticum spp.*).   Wheat is not often grown intentionally for animal feed in North America, and almost all commonly grown varieties were

Figure 19.9 *A combine shells corn as it is harvested from the field. Courtesy of Massey Ferguson, N.A.*

developed with flour milling qualities in mind rather than feeding values. All harvesting is done mechanically in North America (Fig. 19.10). On a worldwide basis about 20% of the wheat produced is used as feed. The hard winter wheats are high in protein, averaging 13 to 15%, but the soft white wheats have less protein. The amino acid distribution is better than that of most cereal grains, and wheat is a very palatable and digestible feed, having a relative value equal to or better than corn for most animals. Feeding wheat to ruminants requires some caution because wheat is more apt

TABLE 19.8 Relative values (dry basis) of different cereal grains as given by NRC publications.

		ENERGY				
	DIGESTIBLE PROTEIN, CATTLE	TDN		ME CHICKENS	NE$_m$ CATTLE	NE$_g$ CATTLE
GRAIN		CATTLE	SWINE			
Corn, #2 dent	100	100	100	100	100	100
Barley	131	91	88	77	93	95
Milo	95	88	96	103	81	83
Oats	132	84	79	74	76	77
Wheat, hard red winter	152	97	99	90	95	96

Figure 19.10 *Grain combines being used to harvest wheat in the Pacific Northwest. Photo by R. W. Henderson.*

than other grains to cause acute indigestion in animals unadapted to it. Some processing (grinding, rolling) is required to optimal utilization. When available for feed, it can be substituted equally for corn on the basis of digestible energy.

Barley (Hordeum vulgare or H. distichon). Barley is grown widely in Europe and in the cooler climates of North America and Asia. Although a small amount goes into human food and a substantial amount is used in the brewing industry in the form of malt, a significant proportion of it is used for animal feeding.

Barley contains more total protein and higher levels of lysine, tryptophan, methionine, and cystine than corn, but its feeding value is appreciably less in most cases, because of the relatively high fiber content of the hull and the lower digestible energy. Barley is a very palatable feed for ruminants, particularly when rolled before feeding, and few digestive problems result from its use. Hullless varieties are roughly comparable to corn and, therefore, more suitable for swine and poultry feeding; however, not much hull-less barley is produced. A new high-lysine barley containing 25% more lysine than common barley shows promise as an energy source for nonruminants; this would allow a saving in supplemental protein.

Oats (Avena sativa). Oats represent only about 4.5% of the total world production of cereal grains, most of the production being concentrated in northern Europe and America. The protein con-

tent of oats is relatively high, and the amino acid distribution is more favorable than for corn, but oats are not used widely for feeding of swine or poultry because of the hull, which makes up about one-third of the seed. The hull is quite fibrous and poorly digested. Even when ground, the result is a very bulky feed. Including a high percentage in the diet does not allow for optimal feed intake for growing poultry or swine, although oats have some value in protecting young pigs from stomach ulcers. For ruminants and horses, oats is a favored feed for breeding stock, one well liked by animals. However, for feedlot cattle oats do not supply sufficient energy for maximal gains. Oat groats (whole seed minus the hull) have a feeding value comparable to corn, but the price usually is not favorable for animal use.

Triticale. Triticale is a hybrid cereal derived from a cross between wheat (*Triticum duriem*) and rye (*Secale cereale*). Its feeding value as an energy source is comparable to that of corn and other cereal grains for swine. Triticale can replace all of the grain sorthum and part of the soybean meal in diets for growing pigs. Total protein content tends to be higher than that of corn and grain sorghum and similar to that of wheat.

Buckwheat. Buckwheat (*Fagopyrum esculentum*) is not a true cereal (Gramineae). It belongs to the family Polygonaceae. Buckwheat is grown primarily for human consumption but it has potential in animal feeding. The average protein content of buckwheat is about 11% and is about 74% digestible. Whereas cereals are low in lysine content, buckwheat has a high content of this amino acid. The average lysine content of buckwheat on an as-fed basis is 0.61% compared with only 0.30 to 0.45% for either wheat or barley and 0.18% for maize.

Grain Amaranth. An ancient crop used as human food for centuries in some parts of the world has received recent attention as a potentially valuable feed for pigs. Several cultivars of *Amaranthus* species are currently under study to identify superior agronomic and nutritional sources for propagation. Feeding studies with laboratory animals indicate that some cultivars of grain amaranth have excellent protein quality and that grain from these cultivars can be used as the sole source of energy and protein for normal growth of nonruminant animals. *Amaranthus cruentus* and *Amaranthus hypochondriacus* cultivars have received the greatest attention as feed and food resources, and selected populations of these species show particular nutritional promise.

Milling and Distillery Byproducts of Cereal Grains

The milling of cereal grains for production of flour, starch, or other products results in production of a number of byproducts that are used primarily by the commercial feed trade. Details of milling procedures and seed composition and more complete descriptions of byproducts may be found elsewhere (Anonymous, 1980). In addition, a considerable tonnage of distillery and brewery products is available for use in animal feeds. To illustrate the quantities of these byproducts used in animal feeding in the United States in recent years, Holden and Zimmerman (1991) prepared tabular information, some of which is shown in Table 19.9.

Brewers' dried grains are "the dried extracted residue of barley malt alone or in a mixture with other cereal grains or grain products resulting from the manufacture of wort or beer" (American Association of Feed Control Officials, 1987). Distillers' dried grains are the product "obtained after removal of ethyl alcohol by distillation from yeast fermentation of a grain or a grain mixture by condensing and drying at least three-fourths of the solids of the resultant stillage" (American Association of Feed Control Officials, 1987). Corn gluten feed is "the part of the commercial shelled corn that remains after extraction of the larger part of the starch, gluten, and germ by the processes employed in the wet milling manufacture of corn starch or syrup" (American Association of Feed Control Officials, 1987). Corn gluten meal is "the

dried residue from corn after removal of the larger part of the starch and germ, and the separation of the bran by the process employed in the wet milling manufacture of corn starch or syrup or by the enzymatic treatment of the endosperm" (American Association of Feed Control Officials, 1987). Corn gluten feed contains 20% protein of poor quality. Corn gluten meal is sold as two products containing 40% or 60% protein of poor quality. Rice millfeed and wheat millfeed consist of the outer layers of the seed coat removed during the regular milling process of each. These millfeed products are sold as bran or, in the case of wheat, as wheat middlings (a mixture of bran, shorts, germ, flour). Wheat bran consists of the outer covering of the seed along with small amounts of flour and finely ground contaminating weed seeds. It contains 13.5 to 15% minimum protein, 2.5% minimum fat, and 12% maximum fiber. Wheat mill run is a blend of bran and middlings and contains 14 to 16% protein, 3% fat, and 9.5% fiber. Wheat shorts consists of primarily wheat middlings and flour (more flour than middlings); it has the same minimum protein content as mill run (14–16%) but 3.5% fat and 7% maximum fiber. Wheat red dog contains more flour than other mill feeds and a minimum protein of 13.5 to 15%, 2% minimum fat, and 4% maximum fiber (Holden and Zimmerman, 1991).

The production of grain byproducts in the United States exceeds the domestic demand; therefore, the export market is substantial. If the domestic demand should increase significantly as a result of increased use of corn for ethanol production to augment gasoline as a motor vehicle fuel, the amount of distillery byproducts available for animal feeds in the United States can be expected to increase.

Milling byproducts from wheat account for about 25% of the kernel. These are relatively bulky and laxative feeds, particularly bran, but are quite palatable to animals. The bran and middlings are from the outer layers of the seed and contain more protein than the grain. Protein quality is improved somewhat, although these products are deficient in lysine and methionine as well as some other essential amino acids. These outer layers of the seed are relatively good sources of most of the water-soluble vitamins, except for niacin, which is entirely unavailable. These feedstuffs are low in Ca and high in P and Mg. The bulk of the production of wheat byproducts is used in swine and poultry feeding, although when bran is available, it is a fa-

TABLE 19.9 Amounts of grain byproducts fed to animals in the United States in 1980, 1984, and 1988 (1000 metric tons).

FEEDSTUFF	1980	1984	1988
Brewers' dried grains	290	142	110
Distillers' dried grains	453	807	970
Corn gluten feed and meal	763	1876	1320
Rice millfeed	706	456	580
Wheat millfeed	4638	5084	5789
Total	6850	8365	8769

Source: From Holden and Zimmerman (1991).

vored feed in rations for dairy cows, all breeding classes of ruminants, and horses.

Corn milling byproducts are somewhat different than wheat byproducts because corn is often milled for purposes other than flour production. When milled for corn meal, hominy feed is produced, a product that consists of corn bran (hulls), corn germ, and part of the endosperm. It is similar to corn meal but higher in protein and fiber. Some wet-milling processes are used for various purposes, resulting in byproducts such as corn gluten meal, corn gluten feed, germ meal, corn solubles, and bran. The gluten meal is a high-protein product. Of these byproducts, hominy feed is most valuable as an energy feed, and it is used in a wide variety of diets. It is quite digestible but of variable energy value depending on its fat content.

Milling byproducts of barley, sorghum, rye, and oats are similar, more or less, to those of wheat or corn and of comparable nutritional value. With rice, the bran may be of good quality but is apt to be quite variable because of inclusion of hulls. The bran tends to become rancid rapidly because of its relatively high content of unsaturated fats.

❏ LIQUID ENERGY SOURCES

Molasses

Molasses is a major byproduct of sugar production, the bulk of it coming from sugar cane. About 25 to 50 kg of molasses results from production of 100 kg of refined sugar. Molasses also is produced from sugar beets and other products as shown in Table 19.10; hemicellulose extract (wood molasses, Masonex) is a somewhat similar product.

Molasses is essentially an energy source, and the main constituents are sugars. Cane molasses contains 25 to 40% sucrose and 12 to 25% reducing sugars, with total sugar content of 50 to 60% or more. Crude protein content normally is quite low (about 3%) and variable, and the ash content ranges from 8 to 10%, largely made up of K, Ca, Cl, and sulfate salts. Molasses is a good source of the trace elements but has only a moderate to low vitamin content. In commercial use molasses is adjusted to about 25% water content, but it may be dried for mixing into dry diets.

One of the problems in molasses feeding is that such products (except for corn molasses) are quite variable in composition. Age, type, and quality of the cane; soil fertility; and system of collection and processing have a bearing on composition of molasses. As an example, the Ca content of 11 samples of cane molasses ranged from 0.3 to 1.68% (Pond and Maner, 1984).

Molasses is utilized widely as a feedstuff, particularly for ruminants. In the United States alone, about 2.5 million tons are used annually and large amounts are used in Europe and in other areas where it is produced. The sweet taste makes it appealing to most species. In addition, molasses is of value in reducing dust, as a pellet binder, as a vehicle for feeding medicants or other additives, and as a liquid protein supplement when fortified with a N source. The cost often is attractive as compared to grains. Most molasses products are limited in use, however, because of milling problems (sticky consistency of molasses) or because levels exceeding 15 to 25% of the diet are apt to result in digestive disturbances, diarrhea, and inefficient animal performance. The diarrhea largely is a result of the high level of K and other mineral salts in

TABLE 19.10   Analytical and TDN values of different sources of molasses.[a]

ITEM	CANE[b]	BEET	CITRUS	CORN	SORGHUM	REFINERS
	SOURCE OF MOLASSES					
Total solids, %	75	76	65	73	73	73
Crude protein, %	3.0	6.0	7.0	0.5	0.3	3.0
Ash, %	8.1	9.0	6.0	8.0	4.0	8.2
Total sugars, %	48–54	48–52	41–43	50	50	48–50
TDN, %	72	61	54	63	63	72

[a]Also known as blackstrap molasses.
Source: Anonymous (1978).

most molasses products. The problem is not due to sugar content, because pigs or ruminants can utilize comparable amounts of sugar supplied in other forms. High-test molasses, which has a lower ash content, can be fed at very high levels to either pigs or cattle without any particular problem, but not much high-test molasses is available for animal feeding.

Other Liquid Feeds

Whey, a byproduct of cheese or casein production, is used similarly to dried skim milk. It may be purchased as is (no processing), condensed, or dried. The high lactose and mineral content may be laxative, so the amount that can be used is limited. Whey is relatively low in protein (13 to 17% dry basis), although the protein is of high biological value. It is an excellent source of B-vitamins. Delactosed whey is available in some areas.

Various other liquids may be used in feeds or liquid protein supplements. However, with the exception of propylene glycol, nearly all would be classed as protein sources because they contain greater than 20% crude protein (CP) on a dry basis.

❑ OTHER HIGH-CARBOHYDRATE FEEDSTUFFS

Root Crops

Root crops used in feeding animals, particularly in northern Europe, include turnips, swedes, fodder beets, carrots, and parsnips. These crops may be dug and left lying in the field to be consumed as desired when used as animal feed. The bulky nature of these feeds limits their use for swine and poultry, so most are fed to cattle and sheep.

Root crops are characterized by their high water (75–90% or higher), moderately low fiber (5–11% dry basis), and crude protein (4–12%) contents. These crops tend to be low in Ca and P and high in K. The carbohydrates range from 50 to 75% of dry matter and are mainly sucrose, which is highly digestible by ruminants and nonruminants. Animals (sheep, cattle) not adapted to beets (*Beta vulgaris*) tend to be subject to digestive upsets, probably because of the high-sucrose content.

A tropical root crop (*Manihot esculenta*) called cassava, yuca, manioc, tapioca, or mandioca is of great potential importance as a livestock feed. It is ninth in world production of all crops and fifth among tropical crops, and has been shown in experimental plots to be capable of yielding 75 to 80 tons/hectare/year, which is many times greater than can be produced by rice, corn, or other grains adapted to the tropics (Pond and Maner, 1984). Although it is strictly a tropical plant, significant amounts of cassava are used at present in the United States and Europe for livestock feeding. Cassava root contains approximately 65% water, 1 to 2% protein, 1.5% crude fiber, 0.3% fat, 1.4% ash, and 3% NFE. Thus, its dry matter largely is readily available carbohydrates. Dried cassava is equal in energy value to other root crops and tubers and can be used to replace all of the grain portion of the diet for growing–finishing pigs if the amount of supplemental protein is increased to compensate for the low-protein content of cassava. It can be used as the main energy source in diets of gestating and lactating swine. The stalk and leaf portion of the plant is utilized well by ruminants, but is too high in fiber for nonruminants.

Freshly harvested cassava roots and leaves may be high in prussic acid (hydrocyanic acid). Oven drying at 70 to 80°C, boiling in water, or sun drying are effective in reducing the HCN content of freshly harvested cassava. As improved harvesting and processing methods for cassava are developed, large-scale commercial production of cassava can be expected to make a significant contribution to the world energy feed supply for animals.

Tubers

Surplus or cull white potatoes (*Solanum tuberosum*) are used for feeding cattle or sheep in areas where commercial potato production occurs. Pigs and chickens do poorly on raw potatoes, but cooking improves digestibility of the starch so that it is comparable to corn starch. Potatoes are high in digestible energy (dry basis), which is derived almost entirely from starch. Water content is about 78 to 80%; crude protein content is low, and the quality of the protein is poor. The Ca content is usually low. Potatoes and particularly potato sprouts contain a toxic compound, solanin, which may cause problems if potatoes are fed raw or ensiled. In cattle finishing rations, cull potatoes frequently are fed at a level to provide up to half of the dry matter intake.

Various byproducts of potato processing are available in some areas as high percentages of white potatoes are processed to some extent before entering the retail trade in the United States. Potato meal is the dried raw meat of potato residue left from the processing plants. Potato flakes are residues remaining after cooking, mashing, and drying; potato slices are the residue after raw slices are dried with heat; and potato pulp is the byproduct remaining after extraction of starch with cold water. The raw meals have about the same relative nutrition values as raw cull potatoes. A product called dried potato byproduct meal is produced in some areas; it contains the residue of food production such as off-color french fries, whole potatoes, peelings, potato chips, and potato pulp. These are mixed, limestone is added, and the mixture dried with heat. Potato slurry is a high-moisture product remaining after processing for human food; it contains a high amount of peel. Generally, the value of these potato products is comparable roughly to that of raw or cooked cull potatoes, depending on how the byproduct is dried. However, residues of the potato chip processing may have much higher levels of fat, so the energy value may be increased accordingly. For swine, cooked potato products are usually restricted to 30% or less of the diet.

Dried Bakery Product

This is a product produced from reclaimed (unused) bakery products, candy, nuts, and other pastries. While relatively little is available, it is an excellent feed as most of the energy is derived from starch, sucrose and fat. It is utilized well by pigs and is a preferred ingredient in starter rations. A wide variety of unconsumed bakery products make very satisfactory feeds for swine or ruminants. Where available, it is often used in dairy feed mixes.

Dried Beet and Citrus Pulp

Dried beet pulp is the residue remaining after extraction of sugar from sugar beets. Molasses may be added before drying and the product may be sold in shredded or pelleted form. The physical nature and high palatability make it a favored feed for cattle or sheep, and it has a feeding value comparable to cereal grains. Unfortunately, the supply continues to decrease because fewer and fewer sugar beets are being produced.

Citrus pulp is a somewhat similar product (but with a lower energy value) remaining after the juice is extracted from citrus fruits; it includes peel,

pulp, and seeds. Citrus pulps are used primarily as feedstuffs for cattle or sheep.

❏ FATS AND OILS

Surplus animal fats and, occasionally, vegetable oils frequently are used in commercial feed formulas, depending on relative prices. A variety of different products are available, and these have been described as to composition and source by the American Association of Feed Control Officials. For the most part, feeding fats are animal fats derived by rendering of beef, swine, sheep, or poultry tissues. Vegetable oils generally command better prices for use in producing margarine, soap, paint, and other industrial products so that usually they are priced out of use in animal feeding.

Although most animals need a source of the essential fatty acids (see Chapter 8), these are supplied in sufficient amounts in natural feedstuffs, and supplementation is not required except when low-fat energy sources are used. Fats are added to rations for several reasons. As a source of energy, fats are highly digestible, and digestible fat supplies about 2.25 times that of digestible starch or sugar; therefore, they have a high caloric value and can be used to increase the energy density of a ration. Fats often tend to improve the diet by reducing dustiness and increasing palatability. There is some evidence that fats will reduce bloat in ruminants. In addition, the lubrication value on milling machinery is often of interest. Fats are subject to oxidation (see Chapter 8) with development of rancidity, which reduces palatability and may cause some digestive and nutritional problems. Feeding fats usually have antioxidants added, especially if the fats are not to be mixed into the diet and fed immediately.

Adding fat at low to moderate levels sometimes can increase total energy intake through improved palatability, although animals usually consume enough energy to meet their demand when it is physically possible. In swine and poultry diets, 5 to 10% fat is often added to creep diets for pigs or to broiler rations. Amounts above 10 to 12% usually will cause a sharp reduction in feed consumption, so concentration of other nutrients may need to be increased in order to obtain the desired consumption.

For ruminants, high levels of fats are used in milk replacers; depending on the purpose of the replacer, it may contain 10 to 30% added fat. Ruminants on dry feed, however, are less tolerant of high-fat levels than are nonruminants. Concentrations of more than 7 to 8% are apt to cause digestive disturbances, diarrhea, and greatly reduced feed intake. In practice, 2 to 4% fat may be added in finishing rations for cattle, and some fat occasionally is added to rations for lactating dairy cows.

❏ PROTEIN CONCENTRATES

Protein is one of the critical nutrients, particularly for young, rapidly growing animals and high-producing adults, although it may be secondary to energy or other nutrients under some conditions. In addition, protein supplements usually are more expensive than energy feeds, so optimal use is essential in any practical feeding system. Most energy sources, except for fat, starch, or refined sugar, supply some protein, but usually not enough to meet total needs except for adult animals. Fur-

thermore, for nonruminant animals the quality of protein from a given source rarely is adequate to sustain adequate production.

Protein supplements (greater than 20% crude protein) are available from a wide variety of animal and plant sources, and nonprotein N sources such as urea and ammonium phosphates are available from synthetic chemical manufacturing processes. It is beyond the scope of this discussion to detail all of the available sources; refer to Pond and Maner (1984) and Chiba (2001) for more detailed discussion and data on amino acid composition of some of the less well-known protein sources. Table 19.11 contains data on the composition of a few common protein supplements.

Protein Supplements of Animal Origin

Protein supplements derived from animal tissues originate primarily from inedible tissues from meat packing or rendering plants, from surplus milk or

TTABLE 19.11   Composition of several important protein supplements.

ITEM	ALFALFA, DEHY.	DRIED SKIM MILK	MEAT MEAL[a]	FISH MEAL, HERRING	FEATHER MEAL	COTTONSEED MEAL, SOL. EXT.[b]	SOYBEAN MEAL, SOL. EXT.[c]	BREWERS GRAINS, DEHY.[d]
Dry matter, %	93.1	94.3	88.5	93.0	91.0	91.0	89.1	92.0
Other components, %[a]								
Crude protein	22.1	36.0	55.0	77.4	93.9	45.5	52.4	28.1
Fat	3.9	1.1	8.0	13.6	2.6	1.0	1.3	6.7
Fiber	21.7	0.3	2.5	0.6	0.0	14.2	5.9	16.3
Ash	11.1	8.5	21.0	11.5	3.5	7.0	6.6	3.9
Ca	1.63	1.35	8.0	2.2	0.4	0.2	0.3	0.29
P	0.29	1.09	4.0	1.7	0.5	1.1	0.7	0.54
Essential amino acids								
Arginine	0.97	1.23	3.0	4.5	6.9	4.6	3.8	1.4
Cystine		0.48	0.4	0.9	4.1	0.7	0.8	
Histidine	0.43	0.96	0.9	1.6	0.6	1.1	1.4	0.5
Isoleucine	0.86	2.45	1.7	3.5	4.8	1.3	2.8	1.6
Leucine	1.61	3.51	3.2	5.7	8.7	2.4	4.3	2.5
Lysine	0.97	2.73	2.6	6.2	2.0	1.7	3.4	1.0
Methionine	0.32	0.96	0.8	2.3	0.6	0.5	0.7	0.4
Phenylalanine	1.18	1.60	1.8	3.1	4.8	2.2	2.8	1.4
Threonine	0.97	1.49	1.8	3.2	5.1	1.3	2.2	1.0
Tryptophan	0.54	0.45	0.5	0.9	0.7	0.5	0.7	0.4
Valine	0.11	2.34	2.2	4.1	7.8	1.9	2.8	1.7

[a]Banned from use in ruminant diets in the United States as a result of a threat of bovine spongioform encephalopathy (BSE).
[b]Solvent extracted.
[c]Dehydrated.
[d]Composition on dry basis. Data from NRC publications.

milk byproducts or from marine sources. Those of animal origin include meat meal, meat and bone meal, blood meal and feather meal; milk products include dried skim and whole milk and dried buttermilk; fish products include whole fish meal made from a variety of species, meals made from residues of fish or other seafood, and fish protein concentrate.

Meat Meal, Meat and Bone Meal, Blood Meal

Meat meal and meat and bone meal have been banned from use in ruminant diets in the United States because of the threat of bovine spongioform encephalopathy (BSE). These animal byproducts are used almost exclusively for swine, poultry, and pet diets in the United States, although they may be used for ruminants in some areas. Palatability is moderate to good. However, the quality of meat or meat and bone meal varies considerably depending upon dilution with bone and tendinous tissues and on methods and temperatures used in processing. The biological value of these proteins is lower, generally, than for fish or soybean meal. Growth studies with pigs indicate that these products are best used as a part rather than the total source of supplemental protein. Blood meal is a high-protein source (80–85% CP, dry basis), but it is quite deficient in isoleucine and is best used as a partial supplementary protein source, also. Dried blood plasma proteins are now routinely used in starter diets for early-weaned pigs to enhance growth and feed consumption during transition from sow milk. The mode of action of this beneficial response is believed to be related to the presence of peptide growth factors in the blood plasma that stimulate maturation and function of the intestinal mucosal cells and/or improve immune function.

Poultry Byproducts

The most important poultry byproduct protein source is feather meal. This meal is quite high in protein (85%) and, when diets are formulated to adjust for its amino acid deficiencies, it can be used sparingly in diets for nonruminants; it is a satisfactory protein for ruminants. Hair meals can also be produced in this manner. Both feathers and hair require more extensive cooking (to make their proteins digestible) than other byproduct animal meals.

Milk Products

Dried skin or whole milk, although excellent protein sources, usually are quite expensive compared to other protein supplements. These products are used primarily in milk replacers or in starter diets for young pigs or ruminants. The quality of dried milk can be impaired by overheating during the drying process (drum drying); therefore spray-dried milk is preferred. Poor quality milk, when used in milk replacers, is apt to lead to diarrhea and digestive disturbances.

Marine Protein Sources

Fish meals are primarily of two types—those from fish caught for making meal and those made from fish residues processed for human food or other industrial purposes. Herring or related species provide much of the raw material for fish meals. These fish have a high body oil content, much of which is removed in preparation of the finished fish meal.

Good quality fish meals are excellent sources of proteins and amino acids, ranking close to milk proteins. The protein content is high (see Table 19.11), and it is highly digestible. Fish meals are especially high in essential amino acids, including lysine, which is deficient in the cereal grains. In addition, fish meal normally is a good source of the B-vitamins and most of the mineral elements. As a result, fish meal is a highly favored ingredient for swine and poultry feeds, although the cost of such meal is considerably higher in most areas than for other protein sources except milk. Some fish meal is used for ruminants in Europe, but very little is used for this purpose in North America. Some use of fat-extracted meals appears feasible in milk replacers.

The quality of fish may be variable because of factors such as partial decomposition before processing or overheating during processing. Excess oil may lead to rancidity, and inadequate drying may allow mold growth. Even with good quality meal, high levels, when fed to swine, may result in fishy flavored pork if extraction of fat from the meal is incomplete.

❏ PLANT PROTEIN CONCENTRATES

The most important commercial sources of plant protein concentrates are derived from soybeans and cottonseed, with smaller amounts from peanuts, flax (linseed), sunflower, sesame, safflower, rapeseed, various legume seeds, and other miscellaneous sources. Meals made from the above seeds (except for some beans and peas) are called oilseed meals because these seeds are all high in oils that have a number of important commercial uses.

Three primary processes are used for removing the oil from these seed crops. These are known as expeller (screw press), prepress–solvent, and solvent extraction. In the expeller process the seed, after cracking and drying, is cooked for 15 to 20 minutes, then extruded through dies by means of a variable pitch screw. This results in rather high temperatures, which may cause reduced biological value of the proteins. The solvent-extracted meals are extracted with hexane or other solvents, usually at low temperatures. In the prepress–solvent process the seed oils are removed partially with a modified expeller process and then extracted with solvents; this usually is done with seeds containing more than 35% oil because they are not suitable commercially for direct solvent extraction.

As a group, the oilseed meals are high in crude protein, most being over 40%; the protein content is standardized before marketing by dilution with hulls or other material. A high percentage of the N is present as true protein (as high as 90%) that is highly digestible and of moderate to good biological value, although usually of lower value than many animal protein sources. Most meals are low in cystine and methionine and have a variable and usually low lysine content; soybean meal is an exception with its high lysine content. The energy content varies greatly, depending on processing methods; solvent extraction leaves less fat and thus reduces the energy value. Ca content is generally low, but most plant protein supplements are high in P content, although half or more is present as phytin P, a form poorly utilized by nonruminants. Oilseed meals contain low to moderate levels of the B-vitamins and are low in carotene and vitamin E.

Soybean Meal

Whole soybeans (*Glycine max*) contain 15 to 21% oil, which is removed by solvent extraction or by a combination of mechanical procedures and solvent extraction. In processing, the meal is toasted, a procedure that improves the biological value of its protein by destroying various inhibitors; the protein content is standardized at 44% or 50% by dilution with soybean hulls. Soybean meal is produced in large amounts in the United States and elsewhere and is a highly favored feed ingredient because it is palatable, highly digestible, and of high-energy value and results in excellent performance when used for different animal species. Methionine and lysine are the most limiting amino acids for swine and poultry, and the B vitamin content is marginal. In overall value, soybean meal is the best widely available plant protein source.

As with most other oilseeds, soybeans have a number of toxic, stimulatory, and inhibitory substances. For example, a goitrogenic material is found in the meal, and its long-term use may result in goiter in some animal species. It also may contain antigens that are especially toxic to young preruminants. Of major concern in nonruminant species is the presence of trypsin inhibitors that inhibit digestibility of protein. Fortunately, these inhibitors and other factors (saponins and a hemagglutinin) are inactivated by proper heat treatment during processing. Soybeans also contain genistin, a plant estrogen that may account, in some cases, for part of their growth-promoting properties.

Whole soybeans may be fed as a major source of energy and protein after appropriate heat processing (100°C for 3 min) to inactivate the trypsin inhibitors. This whole bean product is known in the feed trade as full fat soybean meal; it contains about 38% CP, 18% fat, and 5% CF. Heating-extruding equipment has been developed for on-the-farm processing and its use appears feasible in relatively small operations. Such meal has found some favor in dairy cow diets and, in moderate amounts, in diets for swine and poultry. Heat-treated soybeans can be used to replace all of the soybean meal in corn–soy diets from growing–finishing pigs.

Cottonseed Meal

Because of the growth habits of cotton (*Gossypium* spp.), cottonseed meal (CSM) is available in many areas where soybeans do not grow, particularly in some areas in South America, northern Africa, and Asia. CSM protein is of good, though variable, quality. Most meals are standardized (in the United States) at 41% crude protein. The protein is low in cystine, methionine, and lysine, and the meal is low in Ca and carotene, although palatable for ruminants. CSM is less well liked by swine and poultry. Nevertheless, it finds widespread use in animal feeds, although because of various problems, its use is more limited than that of soybean meal.

The cotton seed (and CSM) contains a yellow pigment, gossypol, which is relatively toxic for nonruminant species, particularly young pigs and chicks. In addition, feeding CSM results in poor egg quality because gossypol tends to produce green egg yolks. Sterculic acid, which is found in CSM, can cause egg whites to turn pink.

Gossypol is found bound to free amino groups in the seed protein or in a free form that can be extracted with solvents. The free gossypol is the toxic form. Gossypol toxicity can be prevented in most instances by addition of ferrous sulfate and other iron salts. Choice of an appropriate processing method such as prepress–solvent extraction or more complicated extraction methods can remove most of the free gossypol. Further improvement can be made by plant breeding. Gossypol is produced in glands that can be reduced in size or removed by genetic changes in the plant. Meals from glandless seeds have resulted in good performance in poultry, although not for young pigs when given most of their supplementary protein from CSM (LaRue et al., 1985). Biologically tested meals that are screened by feeding to hens are available in some areas, but at a higher price than normal meals.

Other Oilseed Meals

Linseed meal is made from flax seed (*Linum usitatissimum*), which is produced for the drying oils it contains. Linseed meal accounts for only a small part of the total plant proteins produced in the United States. The crude protein content is relatively low (35%) and is deficient in lysine. Although favored in ruminant diets, linseed meal is used sparingly for most species.

Substantial amounts of peanuts (*Arachis hypogaea*) are grown worldwide (second to soybeans among seed legumes), mostly for human consumption. Peanut meals are deficient in lysine, and the protein is low in digestibility. Peanuts contain a trypsin inhibitor, as do many beans, and they may be contaminated with molds such as *Aspergillus flavus,* which produce potent toxins (aflatoxins) that are detrimental, particularly to young animals.

Safflower (*Carthamus tinctorius*) is a plant grown in limited amounts for its oil. The meal is high in fiber and low in protein unless the hulls are removed. The protein is deficient in S-containing amino acids and lysine. This meal finds most use in ruminant rations because of the fiber content and inferior amino acid content.

Sunflowers (*Helianthus annuus*) are produced for oil and seeds, primarily in northern Europe and Russia because they grow in relatively cool climates, although production is increasing in North America. The meals, though high in protein, are high in fiber and have not produced performance in swine or dairy cows comparable to that obtained with other protein sources.

Rapeseed meal (from *Brassica napus* or *Brassica campestris*) is of interest because it grows in more northern climates than most oilseed plants. Most rapeseed meals are about 40% crude protein, with an amino acid pattern similar to that of other oilseed meals. Use of rapeseed meal from older varieties of rape is limited by the content of erucic acid. Plant breeders have developed low erucic acid cultivars called canola to distinguish them from the older cultivars. These newer varieties of canola and improved processing methods have resulted in meals of much better quality and improved palatability.

Crambe abyssinica, a member of the Cruciferae family, is increasingly investigated in many countries as a potential oilseed crop. Feeding studies in cattle suggest that crambe meal (the protein-rich byproduct resulting from the extraction of fat high in erucic acid) may be a useful protein supplement for ruminants. For nonruminant animals, the presence of glucosinolates in untreated crambe meal reduces protein utilization. Water-extracted crambe meal appears to be a satisfactory protein supplement for nonruminants. The meal contains about 40% protein.

Other Plant Protein Sources

Coconut or copra meal, the residue after extraction and drying the meat of the coconut, is available in many areas of the world. The crude protein content is low (20–26%) and of variable digestibility. It can be used to partially supply protein needs of most species.

Field beans and peas of a number of species are sometimes available for animal feeding, although they may be grown primarily for human food.

These seeds generally contain about 22 to 26% protein, and the proteins tend to be deficient in S-amino acids and tryptophan. Some seeds contain toxic factors, such as trypsin inhibitors, as well as other toxins. Because of the toxins and poor protein quality, use in feed for nonruminant species is limited unless they are processed to inactivate some of the inhibitors. Some of these grain legumes hold promise as complete energy–protein feeds for swine in tropical areas of the world. Their favorable chemical composition is illustrated by data in Table 19.12. Considerable research effort currently is being expended to identify high-yielding varieties and to develop economical methods of destroying inhibitors and toxins. Recent work suggests that autoclaving or extruding are satisfactory methods for some bean species.

New and Miscellaneous Protein Sources

As the world protein shortage becomes more acute, efforts continue to identify and develop additional sources for use in livestock feeding. Potential sources include animal wastes, seaweed, and single-cell protein products such as algae, bacteria, and yeasts.

Animal and poultry wastes (feces and urine) have potential value as animal feed. To date, it appears that ruminants are best able to utilize them and until technological advances provide the knowledge needed to convert these waste products to feed useful to other species, cattle and sheep probably will be the main avenue of utilization. Research efforts are aimed at possible utilization of microbial populations grown on animal wastes as a means of improving the nutritive value of these wastes for other animal species.

Algae is an attractive possibility as a protein source, except for the high moisture content. Results with *Spirulina* species, particularly *Arthrospira platensis*, a cultivated freshwater algae, indicate a potential for about 10 times as much protein per unit of land area as soybeans. Algae contains about 50% protein, 6 to 7% fiber, 4 to 6% fat, and 6% ash. A major problem, of course, is how to harvest and dry a product of this type. A particularly promising bluegreen algal feedstuff is *Arthrospiral platensis*, which contains more than 60% protein of high biological value.

Seaweed is plentiful in many areas of the world, but most research work indicates that it is not a good source of either energy or protein and should be used mainly as a mineral supplement that should not exceed 10% of the diet. It contains about 2% Ca and 0.4 to 0.5% P, is a good source of Fe, and is extremely high in I.

There is also interest in preparing leaf protein concentrates from crops such as alfalfa because of its high yield of protein per unit of land. Various trypsin inhibitors have been found in a variety of leaf extracts (Humpfries, 1980).

It seems likely that single-cell protein sources (yeasts, bacteria) may become much more important in the future as animal feeds. Yeasts and/or bacteria can be grown on many different kinds of wastes. This offers particular advantages if the wastes are quite dilute and if means of concentrating and harvesting the microorganisms can be developed. Some yeasts are used at this time in animal feeds, but the prices are relatively high. Bacterial proteins also are on the market, being produced from soluble wastes from breweries or wood pulping (paper) operations. Feeding values have not been well characterized for most animal

TABLE 19.12 Chemical composition of some grain legumes.

ITEM	MOISTURE, %	COMPOSITION, % OF DM				
		PROTEIN	ETHER EXTRACT	FIBER	ASH	NFE
Chickpeas (forage)	11.8	19.6	4.0	9.5	3.2	63.7
Cowpeas	11.0	26.4	2.0	4.8	3.9	62.9
Dry beans	11.0	26.8	1.5	4.7	3.8	63.2
Field beans	11.0	26.3	2.2	8.8	3.8	58.9
Field peas	10.1	28.8	1.7	6.7	2.8	60.0
Pigeon peas	11.0	23.5	1.9	9.0	3.9	61.7

species at this time, but preliminary data indicate that such products can be used satisfactorily for ruminant and nonruminant animals under appropriate conditions. Biotechnology using recombinant DNA methodologies offer great potential for mass production of microorganisms engineered for prescribed amino acid composition or for producing vast quantities of individual amino acids for use as crystalline amino acid supplements in animal diets. Such technologies are already in place for the production of lysine, threonine, tryptophan, and other amino acids.

❏ NONPROTEIN NITROGEN

Nonprotein nitrogen (NPN) includes any compounds that contain N but are not present in the polypeptide form of precipitable protein. Organic NPN compounds include ammonia, amides, amines, amino acids, and some peptides. Inorganic NPN compounds include a variety of salts such as ammonium chloride and ammonium sulfate. Although some feedstuffs, particularly some forages, contain substantial amounts of NPN, from a practical point of view NPN in formula feeds usually refers to urea or, to a lesser extent, such compounds as the ammonium phosphates, mainly because other NPN compounds often are too costly to use for feeding of animals at the present time, and those such as ammonium sulfate are not suitable sources of N.

NPN, especially urea, is primarily of interest for feeding of animals with a functioning rumen. The reason for this is that urea is hydrolyzed rapidly to ammonia, which is then incorporated into amino acids and microbial proteins by rumen bacteria that are utilized later by the host. Thus, the animal itself does not utilize urea directly.

A variety of factors must be considered in utilizing urea in feeds, but complete discussion of the subject is too complex for this chapter. Urea, if consumed too rapidly, may be toxic or lethal. Furthermore, urea must be fed with some readily available carbohydrate for good utilization and to prevent toxicity, and some adaptation time is required for the animal. Where livestock management is good and feed is formulated and mixed properly, urea can provide a substantial amount of the supplemental N required, even to the point where it supplies all of the N in purified diets. In practical rations current recommendations are that

not more than one-third of the total N be supplied by urea. Most states require labeling of feed tags with maximal amounts of urea as a protective measure for the buyer.

In nonruminants, the only microorganisms that can convert urea to protein are found in the lower intestinal tract at a point where absorption of amino acids, peptides, and proteins is believed to be rather low or nonexistent. Research with pigs, poultry, horses, and other species indicates that some ^{15}N from urea can be recovered in amino acids and proteins from body tissues, probably as a result of its incorporation into tissue amino acids by various known pathways of amination and transamination. However, little if any net benefit is to be expected from feeding urea to nonruminants. The most likely situation in which it might be useful would be when the supply of essential amino acids is adequate and the nonessentials in short supply; then supplementing with urea might be of some slight benefit.

❏ MINERAL SUPPLEMENTS

Although minerals make up only a relatively small amount of the diet of animals, they are vital to the animal, and, in most situations, some diet supplementation is required to meet needs. Of course, all of the required mineral elements are needed in an animal's diet, but needed supplementary minerals will vary according to the animal species, age, production, diet, and mineral content of soils and crops in the area where grown. Generally, those minerals of concern include common salt (NaCl), Ca, P, Mg, and, sometimes, S of the macrominerals and, of the trace elements, Cu, Fe, I, Mn, and Zn and, in some places, Co and Se. Most energy and N sources—fat and urea being two marked exceptions—provide minerals in addition to the basic organic nutrients, but the flexibility needed in formulating diets usually requires more concentrated sources of one or more mineral elements. Some of these sources are discussed briefly.

Mineral Sources

Common salt (NaCl) is required for good animal production except in areas where the soil and/or water are quite saline. It is a common practice to add 0.25 to 0.5% salt to most commercial feed formulas, although this probably is more than actually is required for most species. Salt often is fed ad libitum, particularly to ruminants and horses, because their requirement probably is higher than for swine or poultry, and different feeding methods lend themselves to this practice. Excess salt may be a problem for all species, but particularly for swine and poultry because they are much more susceptible to toxicity than other domestic species, especially if water consumption is restricted.

In many cases, salt is used as a carrier for some of the trace elements (Table 19.13). The composition of a typical trace mineralized salt premix is shown in Table 19.13. Additional sources of the trace elements can be added to feedstuffs as inorganic salts in areas where they are known to be needed, or feedstuffs high in a needed mineral can be used in the feed formula.

Ca and P are required in large amounts by the animal body, and many feedstuffs are deficient in one or both of these elements. Some of the common supplements are described in Table 19.14. Most sources of Ca are well utilized by different animal species. Although net digestibility may be low, particularly in older animals, there is little difference between Ca sources. This statement does not apply to P, however, because sources differ greatly in availability. In plants about half of the P is bound to phytic acid, and P in the product,

TABLE 19.13 Composition of a typical trace-mineralized salt and its contribution to the diet when included at a level of 0.40%.

MINERAL[a]	AMOUNT IN SALT MIXTURE, %	AMOUNT PROVIDED IN COMPLETE DIET WHEN 0.4% INCLUDED, PPM
NaCl	Not less than 97.000 or more than 99.000	4000.00
Co	Not less than 0.015	0.56
Cu	0.023	0.92
I	0.07	2.80
Fe	0.117	4.32
Mn	0.225	9.00
S	0.040	1.60
Zn	0.008	0.32[b]

[a]Ingredients are: salt, cobalt carbonate, copper oxide, calcium iodate, iron carbonate, manganese oxide, sodium sulfate, zinc oxide, yellow prussate of soda (sodium ferrocyanide) added as an anticaking agent.

[b]This amount supplies only about 1% of the needs of swine.

phytin, is utilized poorly by monogastric species. The usual recommendation is to consider only half of plant P available to monogastric species, although ruminants utilize it very well. Marked differences also exist in the biological availability of inorganic sources. Phosphoric acid and the mono-, di-, and tricalcium phosphates are utilized efficiently. Sources such as Curacao Island and colloidal (soft) phosphates are utilized less well by most animals. Some sources are high in fluoride and must be defluorinated to prevent toxicity when the supplement is fed for an extended period of time.

Of course, some differences exist in utilization of some of the inorganic salts and oxides of different elements. For example, Mg in magnesite is used very poorly; Fe from ferric oxide is almost completely unavailable, and I from some organic compounds such as diiodosalicylic acid, although it is absorbed, is also almost all excreted. Se from plant sources is more available biologically than that from inorganic sources. Legume forage and animal and fish proteins are excellent sources of the trace elements.

There is continuing interest in utilization of chelated trace minerals. Chelates are compounds in which the mineral atom is bound to an organic complex (hemoglobin, for example). Feeding of chelated minerals has been proposed on the basis that chelates will prevent the formation of insoluble complexes in the GI tract and reduce the amount of the particular mineral that will be required in the diet. Limited evidence indicates that this works sometimes (with Zn in poultry diets) but not in all, and the cost per unit of mineral is appreciably higher. Further data are required to know how useful chelated trace minerals will be for all species.

❑ VITAMIN SOURCES

Nearly all feedstuffs contain some vitamins, but vitamin concentration varies tremendously because it is affected by harvesting, processing, and storage conditions as well as plant species and plant part (seed, leaf, stalk). As a rule, vitamins are destroyed quickly by heat, sunlight, oxidizing conditions, or storage conditions that allow mold growth. Thus, if any question arises as to the adequacy of a diet, it is better to err on the positive side than to have a diet that is deficient.

TABLE 19.14   Some common supplements for Ca, P, and Mg.

MINERAL SOURCE	CA, %	P, %	MG, %
Bone meal, steamed	24–29	12–14	0.3
Bone black, spent	27	12	
Dicalcium phosphate	23–26	18–21	
Tricalcium phosphate	38–39	19–20	
Defluorinated rock phosphate[a]	31–34	13–17	
Raw rock phosphate[b]	24–29	13–15	
Sodium phosphate		22–23	
Soft phosphate	18	9	
Curacao phosphate	35	15	
Diammonium phosphate		20	
Oyster shell	38		
Limestone[c]	38		
Calcite, high grade	34		
Gypsum	22		
Magnesium oxide			60–61

[a]Usually less than 0.2% F.
[b]2–4% F.
[c]Dolomite limestone contains up to 10% or more Mg.

As a general rule, vitamins likely to be limiting in natural diets are vitamins A, D, E, riboflavin, pantothenic acid, niacin, choline, and B_{12}, depending on the species and class of animal with which we are concerned. For adult ruminants only A, D, and E are of concern; A is the most likely to be deficient. Vitamin K and biotin can be a problem for both swine and poultry under some conditions, in addition to those vitamins listed previously.

Fat-Soluble Vitamins

The best sources of carotene and vitamin A are green plants and organ tissues such as liver meal. Commercially, most vitamin A now is produced synthetically at such low prices that the cost of adding it to rations is negligible, and there is no economic reason not to add it where any likelihood of a need exists. High-quality forages are good sources of vitamin D; very high activity is obtained from fish liver oils or meals and irradiated yeast or animal sterols are common sources. Vitamin E is found in highest concentrations in the germ or germ oils of plants and in moderate concentrations in green plants or hay. Vitamin K is available in synthetic form (see Chapter 14), and, where gastrointestinal synthesis is not adequate, it can be added at reasonable costs.

Water-Soluble Vitamins

Animal and fish products, green forages, fermentation products, oil seed meals, and some seed parts are good sources of the water-soluble vitamins. The bran layers of cereal grains are fair to moderate sources, and roots and tubers are poor to fair sources. Commercially, some of the crystalline vitamins or mixtures used for humans are prepared from liver and yeast or other fermentation products, and some are produced synthetically. Thus, these various sources can be used to meet vitamin needs in animal diets. Vitamin B_{12} is produced only by microorganisms; it is stored in animal tissues and is found in fermentation products and in animal manures. Animal and fish products are good sources.

❏ NONNUTRITIVE FEED ADDITIVES

Feed additives may be defined as feed ingredients of a nonnutritive nature that stimulate growth or other types of performance (such as egg production), improve the efficiency of feed utilization, or are beneficial to the health or metabolism of the animal. Of the commonly used feed additives, many are antimicrobial agents—compounds that include antibiotics, antibacterial agents, and antifungal agents. Hormonelike substances may be used as growth stimulators. They may be feed additives or, in some cases, may be administered subcutaneously or intramuscularly rather than orally.

Feed additives are used extensively in the United States. The development of resistance to antibiotics by some microbes has caused concerns that the use of antibiotics in animal feeds may be detrimental to humans (White, 2004; Bischoff, 2005). The clear beneficial effects on animal performance as a result of adding subtherapeutic levels of antibiotics to feed are well documented (Hays, 2005; Cromwell, 2001). The ultimate legal restrictions on the use of antimicrobial feed additives in animal production in the United States are uncertain.

❑ FEED LABELING AND MEDICATED FEED

At least 80 to 85% of manufactured feeds are medicated, and more than 80% of the eggs and meat consumed in the United States come from animals fed medicated feed at one time or other during their lifetimes. In the United States, all commercially produced feeds must be registered at the state level and be labeled; unfortunately, labeling requirements are not standard throughout the country. In any case, if the feed contains any additive that is not on the GRAS (generally recognized as safe) list, the feed is under the control of the Food and Drug Administration (FDA). The FDA requires extensive documentation on safety, efficacy, and tissue residues before it approves use of a new drug. When approved and put into use, any feed containing such a drug must contain appropriate labels (Feed Additive Compendium, 2004) that contain the word "medicated" directly following and below the product name. The purpose of the medication must be stated, the names and amounts of the active ingredients must be listed, and a warning statement (when required by the FDA) for a withdrawal period before slaughter for human food, conditions against misuse, and appropriate directions must be given for use of the feed. Custom-mixed feeds are subject to the same requirements.

Problems do arise occasionally with the use of medicated feeds. In addition to outright mistakes that may be made in the manufacturing process, there may be inadvertent contamination in the feed

mill when a nonmedicated feed is mixed and processed in equipment following a batch of medicated feed. It is difficult to remove all traces of previous feed from the various pieces of equipment used in feed mills. Another common problem is the use of medicated feeds in combinations not approved for feeding to animals other than those on the approved list. These situations may develop as a result of requests by local veterinarians and/or nutritionists for a "special mix" of medicated feeds for their clients. If any of the parties knowingly produces and feeds an illegal mix, they are liable for legal action by the FDA.

❑ TYPES OF FEED ADDITIVES

Many different types of feed additives have been fed to domestic animals at one time or another, but many of them have not withstood the test of time and careful experimentation as well as of practical usage. Additives have come and gone as a result of such factors as cost, tissue residues, toxicity, or, more commonly, no beneficial response by the animal. The nonnutritive additives used as production improvers in the United States at the present time are the antibacterial agents—antibiotics, arsenicals, and nitrofurans—and hormones or hormonelike compounds intended to stimulate gain and improve feed efficiency in young, rapidly growing animals. With the primary exceptions of buffers, monensin, and lasalocid, mature animals do not respond with improved performance.

Additives listed in Table 19.15 include those for which the manufacturer claims improved growth or feed efficiency, and those of a more specific nature that are used for other purposes, such as control of internal and external parasites, insect pests, or a wide variety of infectious diseases. The various additives classed as drugs and controlled by the FDA are listed in an industry publication, *The Feed Additive Compendium*. Several of these drugs may be used in combination with other controlled products. The various listings are too detailed to give here, but the reader should be aware that it is illegal to use any combination of controlled drugs that has not been approved by the FDA.

Note (Table 19.15) that relatively few additives have been approved for horses, rabbits, sheep, and ducks. The primary reason for this is the cost in relation to potential sales volume. The approval procedure must be followed with each species. Thus, if projected sales do not indicate sufficient

TABLE 19.15   Feed additives approved by the FDA commonly used in animal feeds in the United States.

CATTLE

NUTRITIONAL USES

Feed Efficiency
Bacitracin Zinc ....
Bambermycins ....
Chlortetarcycline ....
Laidlomycin ....
Lasalocid ....
Melengestrol Acetate ....
Monensin ....
Oxytetracycline ....
Virginiamycin ....

Growth Promotion
Bacitracin Zinc ....
Bambermycins ....
Chlortetracycline ....
Laidlomycin ....
Lasalocid ....
Melengestrol Acetate ....
Monensin ....
Oxytetracycline ....
Virginiamycin ....

MEDICINAL USES

Anaplasmosis
Chlortetracycline ....

Bacterial Calf Diarrhea
Chlortetracycline ....
Oxytetracycline ....

Bacterial Pneumonia
Chlortetracycline ....

Bloat
Poloxalene ....

Cecal Worms (see Worms)

Coccidiosis
Amprolium ....
Decoquinate ....
Lasalocid ....
Monensin ....

Colibacillosis
Neomycin................................

Cooperia Worms (see Worms)

Face Flies
Rabon ....

Fecal Flies
Rabon ....

Foot Rot
Chlortetracycline ....

Hairworms (see Worms)

Hookworms (see Worms)

Horn Flies
(S)-Methoprene ....
Rabon ....

House Flies
Rabon ....

Liver Abscesses
Bacitracin Methylene Disalicylate ....
Chlortetracycline ....
Oxytetracycline ....
Tylosin ....
Virginiamycin ....

Nodular Worms (see Worms)

Respiratory Infection
Chlortetracycline ....

Scours (see Bacterial Calf Diarrhea)

Shipping Fever
Chlortetracycline ....
Oxytetracycline ....

Stable Flies
Rabon ....

Stomach Worms (see Worms)

Suppression of Estrus
Melengestrol Acetate ....

Worms
Coopers Worms—Cooperia
Fenbendazole ....
Levamisole Hydrochloride ....
Morantel Tartrate ....

Hair Worm—Trichostrongylus
Fenbendazole ....
Levamisole Hydrochloride ....
Morantel Tartrate ....

Hookworm—Bunostomum
Fenbendazole ....
Levamisole Hydrochloride ....

Large Intestinal Worms—Oesophagostomum —Nodular Worm
Fenbendazole ....
Levamisole Hydrochloride ....
Morantel Tartrate ....

Lungworms—Dictyocaulus viviparus
Levamisole Hydrochloride ....

Stomach Worm—large, medium and small Haemonchus, Ostertagia, Trichostrongylus
Fenbendazole ....
Levamisole Hydrochloride ....
Morantel Tartrate ....

Thread-Necked Strongylus—Nematodirus
Fenbendazole ....
Levamisole Hydrochloride
Morantel Tartrate

CHICKENS
NUTRITIONAL USES

Egg Hatchability
Bacitracin Zinc

TABLE 19.15—*Continued.*

Egg Production
  Bacitracin Methylene Disalicylate
  Bacitracin Zinc

Feed Efficiency
  Arsanilic Acid ....
  Bacitracin Methylene Disalicylate ....
  Bacitracin Zinc ....
  Bambermycins ....
  Chlortetracycline ....
  Lincomycin ....
  Oxytetracycline ....
  Penicillin ....
  Roxarsone ....
  Tylosin ....
  Virginiamycin ....

Growth Promotion
  Arsanilic Acid ....
  Bacitracin Methylene Disalicylate ....
  Bacitracin Zinc ....
  Bambermycins ....
  Chlortetracycline ....
  Lincomycin ....
  Oxytetracycline ....
  Penicillin ....
  Roxarsone ....
  Tylosin ....
  Virginiamycin ....

Pigmentation
  Arsanilic Acid ....
  Roxarsone ....

MEDICINAL USES

Blackhead
  Nitarsone ....

Breast Blisters
  Novobiocin ....

Capillary Worms (see Worms)

Cecal Worms (see Worms)

Cholera, Fowl
  Novobiocin ....
  Oxytetracycline ....
  Sulfadimethoxine and Ormetoprim 5:3 ....

Chronic Respiratory Disease (CRD)
  Erythromycin ....
  Chlortetracycline ....
  Oxytetracycline ....

Coccidiosis
  Amprolium ....
  Decoquinate ....
  Diclazuril ....
  Halofuginone Hydrobromide ....
  Lasalocid ....
  Maduramicin Ammonium ....

  Monensin ....
  Narasin ....
  Narasin/Nicarbazin ....
  Nicarbazin ....
  Robenidine Hydrochloride ....
  Salinomycin ....
  Semduramicin ....
  Sulfadimethoxine and Ormetoprim 5:3 ....
  Zoalene ....

Colibacillosis
  Sulfadimethoxine and Ormetoprim 5:3 ....

Coryza, Infectious
  Erythromycin ....
  Sulfadimethoxine and Ormetoprim 5:3 ....

Fly Control
  Cyromazine ....

Hepatitis, Infectious
  Oxytetracycline ....

Heterakis (see Worms, Cecal)

Histomaniasis (see Blackhead)

Large Roundworms (see Worms)

Mud Fever (see Bluecomb)

Necrotic Enteritis
  Bacitracin Methylene Disalicylate ....
  Lincomycin ....
  Virginiamycin ....

Non-Specific Enteritis (see Bluecomb)

Roundworms (see Worms)

Synovitis
  Chlortetracycline ....
  Novobiocin ....
  Oxytetracycline ....

Ulcerative Enteritis (see Quail Disease)

Worms
  Capillary Worms
    Hygromycin B ....
  Cecal Worms (Heterakis)
    Hygromycin B ....
  Large Roundworms (Ascaris)
    Hygromycin B ....
    Piperazine ....

DUCKS
MEDICINAL USES

Fowl Cholera
  Chlortetracycline ....
  Novobiocin ....
  Sulfadimethoxine and Ormetorpim 5:3 ....

Serositis, Infectious
  Novobiocin ....

continued

TABLE 19.15—*Continued.*

FISH

MEDICINAL USES
Furunculosis
Oxytetracycline . . . .
Sulfadimethoxine and Ormetoprim 5:1 . . . .
Hemorrhagic Septicemia
Oxytetracycline . . . .
Pseudomonas Disease
Oxytetracycline . . . .
Ulcer Disease
Oxytetracycline . . . .

GOATS

MEDICINAL USES
Coccidiosis
Decoquinate . . . .
Monensin . . . .
Colibacillosis
Neomycin . . . .
Gastrointestinal Roundworms
Morantel Tartrate . . . .

HORSES AND MULES

MEDICINAL USES
Fly Control
Rabon . . . .

MINK

NUTRITIONAL USES
Growth Promotion
Chlortetracycline . . . .
Pelt Promotion
Chlortetracycline . . . .

MEDICINAL USES
Fly Control
Rabon . . . .
Staphylococcal Infection
Novobiocin . . . .

PARTRIDGES

MEDICINAL USES
Coccidiosis
Lasalocid . . . .

PHEASANTS

NUTRITIONAL USES
Growth Promotion
Bacitracin Methylene Disalicylate . . . .
Bacitracin Zinc . . . .
Penicillin . . . .

MEDICINAL USES
Coccidiosis
Amprolium . . . .

PSITTACINE BIRDS

MEDICINAL USES
Psittacosis
Chlortetracycline . . . .

QUAIL

NUTRITIONAL USES
Growth Promotion
Bacitracin Methylene Disalicylate . . . .
Bacitracin Zinc . . . .
Penicillin . . . .

MEDICINAL USES
Coccidiosis
Monensin . . . .
Salinomycin . . . .
Ulcerative enteritis
Bacitracin Methylene Disalicylate . . . .

SHEEP

NUTRITIONAL USES
Growth Promotion
Chlortetracycline . . . .
Oxytetracycline . . . .

MEDICINAL USES
Bacterial Diarrhea
Oxytetracycline . . . .
Bacterial Pneumonia
Oxytetracycline . . . .
Bankrupt Worms (see Worms)
Coccidiosis
Lasalocid . . . .
Colibacillosis
Neomycin . . . .
Scours (see Bacterial Diarrhea)
Vibrionic Abortion
Chlortetracycline . . . .

SWINE

NUTRITIONAL USES
Growth Promotion
Arsanilic Acid . . . .
Bacitracin Methylene Disalicylate . . . .
Bacitracin Zinc . . . .
Bambermycins . . . .
Carbadox . . . .
Chlortetracycline . . . .
Chlortetracycline/Sulfamethazine/Penicillin . . . . . .

TABLE 19.15—*Continued.*

Chlortetracycline/Sulfathiazole/Penicillin ....
Lincomycin ....
Oxytetracycline ....
Penicillin ....
Ractopamine ....
Roxarsone ....
Tiamulin ....
Tylosin ....
Tylosin/Sulfamethazine ....
Virginiamycin

Feed Efficiency
Arsanilic Acid ....
Bacitracin Methylene Disalicylate ....
Bacitracin Zinc ....
Bambermycins ....
Carbadox ....
Chlortetracycline ....
Chlortetracycline/Sulfamethazine/Penicillin ......
Chlortetracycline/Sulfathiazole/Penicillin ....
Lincomycin ....
Oxytetracycline ....
Penicillin ....
Ractopamine ....
Roxarsone ....
Tiamulin ....
Tylosin ....
Tylosin/Sulfamethazine ....
Virginiamycin ....

Increased Leanness
Ractopamine ....

MEDICINAL USES

Atrophic Rhinitis
Chlortetracycline/Sulfamethazine/Penicillin ......
Chlortetracycline/Sulfathiazole/Penicillin ....
Tylosin ....
Tylosin/Sulfamethazine ....

Bacterial Swine Enteritis (Scours)
Apramycin ....
Carbadox ....
Chlortetracycline ....
Chlortetracycline/Sulfamethazine/Penicillin ......
Chlortetracycline/Sulfathiazole/Penicillin ....
Oxytetracycline

Cervical Abscesses
Chlortetracycline ....
Chlortetracycline/Sulfamethazine/Penicillin ......
Chlortetracycline/Sulfathiazole/Penicillin ....

Colibacillosis
Apramycin ....
Neomycin ....

Clostridial enteritis
Bacitracin Methylene Disalicylate ....

Dysentery
Arsanilic Acid ....
Bacitracin Methylene Disalicylate ....
Carbadox ....
Chlortetracycline/Sulfamethazine/Penicillin ......
Chlortetracycline/Sulfathiazole/Penicillin ....
Lincomycin ....
Roxarsone ....
Tiamulin ....
Virginiamycin ....

Dysentery, Vibrionic
Carbadox ....
Chlortetracycline/Sulfamethazine/Penicillin ......
Chlortetracycline/Sulfathiazole/Penicillin ....
Oxytetracycline ....
Tylosin ....
Tylosin/Sulfamethazine ....

Fly Control
Rabon ....

Large Roundworms (see Worms)

Leptospirosis
Chlortetracycline ....
Oxytetracycline ....

Mange Mites
Ivermectin ....

Mycoplasma Pneumonia
Lincomycin ....

Necrotic Enteritis
Carbadox ....
Chlortetracycline/Sulfamethazine/Penicillin ......
Chlortetracycline/Sulfathiazole/Penicillin ....
Lincomycin ....
Oxytetracycline ....
Virginiamycin ....

Respiratory Disease
Timicosin

Roundworms (see Worms)

Salmonellosis (see Necrotic Enteritis)

Scours (see Bacterial Swine Enteritis)

Stress
Chlortetracycline ....
Chlortetracycline/Sulfamethazine/Penicillin ......
Chlortetracycline/Sulfathiazole/Penicillin ....

Worms
  Kidney Worms
  Fenbendazole ....
  Ivermectin ....
  Levamisole Hydrochloride ....
  Large Roundworms
  Dichlorvos ....
  Fenbendazole ....
  Hygromycin B ....

continued

TABLE 19.15—*Continued.*

Ivermectin . . . .
Levamisole Hydrochloride . . . .
Piperazine . . . .
Pyrantel Tartrate . . . .

Lungworms
Fenbendazole . . . .
Ivermectin . . . .
Levamisole Hydrochloride . . . .

Nodular Worms
Dichlorvos . . . .
Hygromycin B . . . .
Ivermectin . . . .
Levamisole Hydrochloride . . . .
Piperazine . . . .
Pyrantel Tartrate . . . .

Small Stomach Worms/Red Stomach Worms
Fenbendazole . . . .
Ivermectin . . . .

Thick Stomach Worms
Dichlorvos . . . .
Ivermectin . . . .

Threadworms
Ivermectin . . . .
Levamisole Hydrochloride . . . .

Whipworms
Dichlorvos . . . .
Fenbendazole . . . .
Hygromycin B . . . .

TURKEYS
NUTRITIONAL USES

Bluecomb
Oxytetracycline . . . .

Feed Efficiency
Arsanilic Acid . . . .
Bacitracin Methylene Disalicylate . . . .
Bacitracin Zinc . . . .
Bambermycins . . . .
Chlortetracycline . . . .
Oxytetracycline . . . .
Penicillin . . . .
Roxarsone . . . .
Virginiamycin . . . .

Growth Promotion
Arsanillic Acid . . . .
Bacitracin Methylene Disalicylate . . . .
Bacitracin Zinc . . . .
Bambermycins . . . .
Chlortetracycline . . . .
Oxytetracycline . . . .
Penicillin . . . .
Roxarsone . . . .
Virginiamycin . . . .

Pigmentation
Arsanilic Acid . . . .
Roxarsone . . . .

MEDICINAL USES

Blackhead
Nitarsone

Bluecomb (Non-Specific Enteritis)
Chlortetracycline . . . .
Oxytetracycline . . . .
Penicillin . . . .

Breast Blisters
Novobiocin . . . .

Cholera, Fowl
Novobiocin . . . .
Sulfadimethoxine and Ormetoprim 5:3 . . . .

Chronic Respiratory Disease (CRD)
Erythromycin . . . .
Penicillin . . . .

Coccidiosis
Amprolium . . . .
Diclazuril . . . .
Halofuginone Hydrobromide . . . .
Lasalocid . . . .
Monensin . . . .
Sulfadimethoxine and Ormetoprim 5:3 . . . .
Zoalene . . . .

Coronaviral Enteritis
Oxytetracycline . . . .

Heterakis (see Worms, Cecal)

Hexamitiasis
Chlortetracycline . . . .
Oxytetracycline . . . .

Histomoniasis (see Blackhead)

Leucocytozoonosis

Mud Fever (see Bluecomb)

Non-Specific Enteritis (see Bluecomb)

Paratyphoid
Chlortetracycline . . . .

Sinusitis, Infectious
Chlortetracycline . . . .
Oxytetracycline . . . .

Synovitis
Chlortetracycline . . . .
Novobiocin . . . .
Oxytetracycline . . . .

Transmissible Enteritis
Bacitracin Methylene Disalicylate . . . .
Oxytetracycline . . . .

Worms
Large Roundworms
Piperazine . . . .

Source: Adapted from *2004 Feed Additive Compendium,* Miller Publishing Co., Minneapolis, MN.

volume to justify the cost, no approval will be requested. This is especially true for drugs that are approved for use with cattle but might be effective for sheep and goats. Quite often some of the research will be done with sheep and later with cattle, but very seldom is the drug put up for approval for sheep. Fortunately, rulings by the FDA on minor species may help to rectify this problem.

❑ ANTIMICROBIAL DRUGS

Antibiotics are compounds produced by one microorganism that inhibit the growth of another organism; they are the most widely used of the antimicrobial drugs. A list of those in common use in the United States as feed additives for domestic species is shown in Table 19.15. Chlortetracycline, oxytetracycline, and bacitracin zinc have enjoyed wider usage than most of the other antibiotics, although there are many others and new antibiotics are continually being isolated and studied. The approved use of various sulfa drugs (not antibiotics) is very limited. Arsenicals are used primarily for poultry and swine. When used as grown promotants, these various additives are added to feed at low levels. Therapeutic levels (for treating disease) normally are considerably higher.

Antibiotics are still in use, where allowed by law, because they usually give a response in terms of growth, improved feed efficiency, and, generally, improved health. The response in growth and improved feed efficiency may be rather variable. For example, there is little or no response in new animal facilities, in very clean surroundings, or in germ-free animals raised under aseptic conditions. The response that may be expected in growing pigs has been reviewed by Cromwell (1991) and is shown in Table 19.16.

Ionophore Antibiotics

Monensin and lasalocid, two of several antibiotics produced by various strains of *Streptomyces* bacteria, have been used widely in beef cattle since their approval by the FDA. The effects of monensin and lasalocid on the performance of growing and finishing cattle are shown in Table 19.17 (Berger et al., 1981; Goodrich, 1984). They are referred to as polyether ionophore antibiotics because these compounds interfere with passage of ions through various membranes. A symposium on monensin in

TABLE 19.16 Efficacy of antibiotics from 1950 to 1977 and from 1978 to 1985.

STAGE	IMPROVEMENT FROM ANTIBIOTICS (%)	
	1950–1977[a]	1978–1985[b]
Starting phase		
Daily gain	16.1	15.0
Feed/gain	6.9	6.5
Growing-finishing phase		
Daily gain	4.0	3.6
Feed/gain	2.1	2.4

[a]Data from 378 and 279 experiments, involving 10,023 and 5,666 pigs for the two phases, respectively; Hays (1977).
[b]Data from 75 and 164 experiments, involving 3609 and 7474 pigs for the two phases, respectively; Zimmerman (1986).
Source: Cromwell (2001).

cattle (Ramsey et al., 1984) contains reviews by several authors on mode of action, effects on productive efficiency of animals, and toxicity of monensin. Initially, ionophores were approved for use with poultry as cocciodiostats; later, it was learned that they were effective production improvers for cattle as well as cocciodiostats for cattle and sheep.

These ionophore antibiotics have a number of general effects. Although the effects of monensin and lasalocid are not identical, in general they both cause a shift in rumen volatile fatty acids to relatively more propionic acid accompanied by less loss of energy as methane, resulting in more efficient gain of growing animals or, what is unusual, increased weight gain of mature animals. The exact reason for the improved efficiency remains speculative, but it appears to be related to reduction in rumen degradation of protein and an increase in postruminal digestion of starch. Energy losses in feces and urine are reduced in most instances. Monensin, but apparently not lasalocid, has been shown to be effective as an inhibitor of rumen bloat, possibly because it acts to control protozoal growth. Monensin has also been shown to result in a reduced retention of Na and an increased retention of K in lambs (Kirk et al., 1985), and one paper suggests that it will prevent the development of ruminal parakeratosis, a problem that sometimes appears in animals fed high-grain or pelleted diets.

Monensin results in a reduction in feed consumption, little effect on gain in feedlot cattle, and

TABLE 19.17 Effect of monensin and lasalocid on performance of cattle.

| | MEASURE OF PERFORMANCE | | |
TREATMENT	FEED CONSUMED, KG/DAY	DAILY GAIN, KG/DAY	FEED TO GAIN RATIO
Feedlot cattle[a]			
Controls	8.27	1.09	8.09
Monensin	7.73	1.10	7.43
Controls	8.89	1.14	8.01
Monensin	8.65	1.18	7.43
Implant	9.61	1.28	7.63
Monensin + implant	8.97	1.32	6.85
Pasture cattle[a]			
Controls		0.61	
Monensin		0.69	
Feedlot cattle[b]			
Control	8.68	0.99	8.75
Lasalocid, 30 g/ton	8.27	1.02	8.09
Lasalocid, 45 g/ton	8.09	1.04	7.79
Monensin, 30 g/ton	8.09	1.03	7.93

[a]Data excerpted from a review (Goodrich et al., 1984) that is a compilation of many different experiments done in many different places.
[b]Data from Berber et al. (1981) from cattle (48 per treatment) fed a 60% high-moisture corn, 30% corn silage diet (dry basis).

an improvement in feed efficiency (feed to gain ratio). There also appears to be a slight additive effect when fed to animals given implants of growth promotants, such as (in this case) Ralgro and Synovex. On pasture, monensin generally results in increased liveweight gain. When research has been done with cattle on forage rations fed in confinement, feed intake may stay the same or be reduced and liveweight gain may usually be increased slightly.

Data shown for lasalocid indicate essentially the same response with feedlot cattle as for monensin. In most instances, it appears likely that either product could be substituted for the other.

General Comments on Antibiotics

Individuals and organizations that object to the continuous use of low levels in animal feeds do so on the basis that resistant pathogenic strains of microorganisms may develop that can be harmful to humans. In the United States, antibiotics have been widely used for about 40 years as feed additives. Although it is true that microbial resistance to antibiotics was shown from the first (indeed some microorganisms may come to require an antibiotic), no convincing, documented instance has been shown in which more virulent pathogens have been isolated as a result of feeding low levels of antibiotics to animals [American Society of Animal Science (ASAS) Symposium Proceedings, 1986]. In fact, the evidence shows that resistant strains are nearly always less virulent (Hirsch and Wiger, 1978; ASAS Symposium Proceedings, 1986). While overwhelming evidence exists on the favorable effects of subtherapeutic dietary levels of antibiotics on animal performance, it is clear that relatively satisfactory growth and health during the late stages of growth does not require dietary antibiotics. Although it is true that antibiotics and other drugs are sometimes used in lieu of good management practices, the beneficial response to antibiotics in livestock enterprises attests to the assertion that the cost in productivity in animal agriculture in the United States would be considerable if routine use of antibiotics were not allowed. The ratio of benefit to risk, a common measure used to evaluate drugs, greatly favors continued use of antibiotics.

❑ PROBIOTICS

Probiotics are live bacteria that may be added to the diet in an attempt to control intestinal infections in animals and enhance nutrient utilization. Recent interest in the use of probiotics in livestock feeds is the result of concern about the use of antibiotics in the feed to improve animal performance. Commonly studied probiotics in poultry, swine, and cattle diets include members of the *Lactobacillus* genus (*L. acidophilus*, *L. casei*, *L. fermentum*, and *L. reuteri*), and *Streptococcus faecalis* (Gilliland, 2005). Selected cultures of *Lactobacillus* species have been used to control salmonellosis in poultry. Some reports indicate reduced mortality and increased growth in swine fed a mixture of *Bacillus pseudolongum* and *L. acidophilus*. A significant reduction in frequency of occurrence of *Escherichia coli* 0157:H7 in feedlot cattle by feeding a particular strain of *L. acidophilus* has been reported. Such examples of observed beneficial effect on livestock and poultry health and performance encourage research to quantify the importance of dietary probiotics in animal nutrition and production.

❑ HORMONES

Many different hormones have been fed or injected into animals at one time or another with the intent of increasing growth or milk production or to modify the normal fattening processes. This list includes somatotropin (growth hormone), natural or synthetic adrenal cortical hormones, natural and synthetic estrogens, androgens, progestogens, androgen–estrogen combinations, thyroid, and antithyroid compounds.

At one time, diethylstilbestrol (DES) was widely used as either an oral feed additive or a subcutaneous implant for cattle or as an implant for sheep, but its use is no longer allowed in the United States because of potential residue problems. With swine, hormones have not shown any consistent benefit. For practical use with calves and feedlot cattle, compounds of interest include hexestrol, a synthetic estrogen used outside the United States; and melengestrol acetate, a synthetic progestogen used orally for heifers. Several products containing various combinations of estrogens and progesterone or estrogen and testosterone are used

as subcutaneous implants to promote growth of cattle. Most of these products also appear to work well for feedlot lambs, although much lower dosages are required.

In ruminants the various natural or synthetic hormones appear to produce a response that results from increased N retention accompanied by an increased intake of feed. The result is usually an increased growth rate, an improvement in feed efficiency, and, frequently, a reduced deposition of body fat, which may, at times, result in a lower carcass grade for animals fed to the same weight as nontreated animals.

Bovine somatotropin has had FDA approval as of 1994 for use in dairy cows to increase milk production. Its availability for commercial use has been made possible by biotechnology, which enables the production of mass quantities of the hormone by microbes into which the somatotrophin gene has been introduced.

Thyroid-active hormones have been used from time to time, particularly with cattle. Thyroprotein, or iodinated casein, which produces the same physiological response, may be used over short periods of time to stimulate milk production, particularly when cows are past their peak of production. Continued use during the remainder of a normal lactation is not apt to result in any improvement in milk production, and withdrawal will result in a prompt decrease in milk production. Thus, there is little current interest in the use of such products, and they are no longer approved for use with lactating dairy cows. No beneficial response is likely in growing and fattening animals. Antithyroid compounds also have been used on the assumption that a decrease in thyroid activity may be beneficial in the fattening process. Some studies have shown favorable response, particularly with respect to improved feed efficiency, but others have not and there is no current interest (or approved use) in feeding such products.

❑ OTHER FEED ADDITIVES

Beta-Adrenergic Agonists

There is active interest in the possibility of using orally effective synthetic beta-adrenergic agonists (beta-agonists) as a means of repartitioning nutrients to favor lean meat and reduce fat deposition. These compounds are catecholamine derivatives that have some pharmacological properties similar to those of epinephrine. The basis for the physiological effect appears to be the action of beta-agonists on beta-adrenergic receptors located on cell membranes to effect stimulation of cyclic adenosine monophosphate (AMP) production within the cell (Fain and Garcia-Sainz, 1983). The mechanism by which this response alters lipid metabolism in adipose tissue and decreases protein turnover in skeletal muscle is still unclear, but data reported from work with lambs, steers, chicks, and swine all show improvements in carcass leanness of animals fed the agents (Dalrymple et al., 1984). Three such agents are clenbuterol, raptopamine, and cimaterol, each similar in structure to epinephrine.

Raptopamine, approved by the FDA for use in beef cattle and swine feeds to improve carcass leanness, is widely used in both species. The total impact of this family of compounds on the livestock industry as a means of improving efficiency of lean meat production will depend on economic and biological factors not yet completely ascertained and on the regulatory obstacles to their clearance for use in food animal production. Cimaterol (Yen et al., 1990a) and raptopamine (Dunshea et al., 1993; Yen et al., 1990b) appear to increase leanness in both genetically obese and lean pigs and in rapidly growing pigs.

Zeolites

Zeolites are crystalline, hydrated aluminosilicates of alkali and alkaline earth cations that possess infinite three-dimensional structures. They are able to gain and lose water reversibly and to exchange some of their constituent cations without a major change in structure (Sheppard, 1984). Some 50 species of natural zeolites are present in sedimentary deposits of volcanic origin around the world. Clinoptilolite is probably most abundant in nature and has received attention in animal nutrition because of its cation-binding properties. Research from Japan, Russia, Yugoslavia, Czechoslovakia, the United States, and other countries provides evidence for improved body weight gain and efficiency of feed utilization of ruminant and nonruminant animals fed diets supplemented with 1 to 5% clinoptilolite. The physiological bases for the observed responses are unknown, but probably are related to the ammonia and other cation-binding capacities of clinoptilolite. Clinoptilolite protects rats, pigs, and sheep against acute ammonia

toxicity and against the toxic effects of Cd and Pb. The positive growth response is not obtained consistently; the variable response may be associated with differences in purity, particle size, or physicochemical variation in the raw ore.

A synthetic zeolite, Zeolite A, with a structure similar to that of natural zeolites, has been studied as a feed additive for swine, cattle, sheep, and poultry. It may have effects on rumen function, plasma Ca level, and egg shell quality. The ultimate usefulness of these natural mineral ores as feed additives for livestock production will be determined by the success of researchers and livestock producers in identifying the conditions under which positive responses to supplementation are obtained.

Miscellaneous Additives

A variety of feed additives is used from time to time for specific purposes that may or may not be related to stimulation of growth or other forms of production. For example, activated carbon has shown some promise for reducing absorption of certain pesticides that may be contaminants in the diet. A variety of anthelmintics are used routinely as medicinals for control of stomach and intestinal worms. Antioxidants have some value in reducing oxidation of nutrients such as vitamin E and unsaturated fats, with the result that dietary requirements of vitamin A and E may be reduced. Sodium bentonite, a clay, is used as a pellet binder and shows some promise of improving N utilization in ruminants. Zeolites and other aluminosilicate compounds are receiving wide attention as feed additives to counteract mycotoxins present in cereal grains, corn, and other feedstuffs infected with molds that are detrimental to livestock (Harvey, 1993). A variety of surface-active compounds are used for prevention and treatment of bloat in ruminants. Various buffers such as $NaHCO_3$ aid in the prevention of indigestion following sudden ration changes in ruminants and may help to relieve the low-fat milk problem in dairy cows. Other buffers are useful in the prevention and treatment of urinary calculi. Various feed flavors are used in different animal rations; although published research data are not overly encouraging, they suggest that the effect of flavors is mainly on preference when a choice of rations is given rather than on total feed consumption. Enzymes have been tried extensively for chicks, pigs, and ruminants, but results generally have not been encouraging.

Organic iodides have value for prevention of foot rot in cattle. Drugs that inhibit methane production may have some potential for ruminants if nontoxic ones can be developed. Live yeast cultures and dried rumen cultures sometimes are promoted, but with little factual evidence of their worth. Extracts from the plant *Yucca schidigera* contain urease inhibitors that appear to reduce tissue ammonia in animals and may protect against ammonia toxicity.

Further information on some of these miscellaneous feed additives appears in some of the later chapters dealing with specific animal species.

❏ SUMMARY

- A vast variety of plant, animal, microbial, mineral, and synthetic feed ingredients is available in our modern world for use in animal diets. Although many of these items are used only in small amounts or for specific situations, improved knowledge of animal nutrition has allowed more complete utilization of many products that were unmarketable at one time. Unfortunately, even with all of the sophisticated analytical equipment available, a limiting feature in utilization of feedstuffs is the dependence on the book analytical values of nonstandardized feed ingredients. Feeds actually used may be quite different than expected—partly because of natural variation when harvested or manufactured and partly because some organic components, particularly vitamins and fatty acids, may deteriorate with time in storage. Another important factor is that in many instances digestibility values are calculated values, which may not always be accurate. Last, but not least, animal utilization, particularly with ruminants, is highly dependent upon the level of feeding because digestibility and body metabolism will change as the quantity of a nutrient changes in the diet.

- Feed additives have important functions in modern animal diets. Antibiotics and other antimicrobial agents allow more rapid growth and improve efficiency of feed utilization. A family of beta-adrenergic agonists, similar in structure to the epinephrine, are of interest in improving the leanness of animals. Only raptopamine among these nutrient repartitioning agents has been approved by the FDA for use in animal feeds,

but they are potentially useful in improving lean meat production in food animals. Other additives are used for specific purposes, ranging from control of internal and external parasites (anthelmentics) in animals to improving pellet stability (pellet binders) in pelleted feeds.

❏ REFERENCES

American Association of Feed Control Officials. 1987. Official Publication. Association of American Feed Control Officials, West Virginia Department of Agriculture, Charleston, WV.

American Society of Animal Science Symposium Proceedings. 1986. Public health implications of the use of antibiotics in animal agriculture. *J. Anim. Sci.* **62** (Supplement 3): 1–106.

Anonymous. 1978. *Liquid Ingredients Handbook.* National Feed Ingredients Association, Des Moines, IA.

Anonymous. 1980. *Millfeed Manual.* Miller's National Federation, Chicago, IL.

Bath, D. L., et al. 1980. *By-product and Unusual Feedstuffs in Livestock Rations.* West Regional Extension. Publ. 39.

Beever, D. E., et al. 1971. *Brit. J. Nutr.* **26:** 123.

Berger, L. L., et al. 1981. *J. Animal Sci.* **53:** 1440.

Bischoff, K. M. 2005. Antibimicrobial use in food animals: potential alternatives. In *Encyclopedia of Animal Science,* W. G. Pond and A. W. Bell (eds., Marcel Dekker, New York (In Press).

Chiba, L. I. 2001. Protein supplements. In *Swine Nutrition* (2nd ed.), Chapter 36, A. J. Lewis and L. C. Southern (eds.). CRC Press, Boca Raton, FL.

Church, D. C. (ed.). 1984. *Livestock Feeds and Feeding,* 2nd ed. Prentice-Hall, Englewood Cliffs, NJ.

Church, D. C. (ed.). 1988. *The Ruminant Animal.* Prentice-Hall, Englewood Cliffs, NJ.

Colovos, N. F., et al. 1970. *J. Anim. Sci.* **30:** 819.

Cromwell G. L. 1991. In *Swine Nutrition* Chapter 17, p. 297. E. R. Miller et al. (eds.). Butterworth-Heinemann, Boston.

Cromwell, G. L. 2001. Antimicrobial and promicrobial agents. In *Swine Nutrition* (2nd ed.), Chapter 18, A. J. Lewis and L. L. Southern (eds.). CRC Press, Boca Raton, FL.

Dalyrmple, R. H., et al. 1984. *Proc. Georgia Nutr. Conf.,* p. 111. University of Georgia, Athens, GA.

Dunshea, F. R., et al. 1993. *J. Anim. Sci.* **71:** 2919.

Ely, R. E., et al. 1953. *J. Dairy Sci.* **36:** 334.

Fain, J. N. and J. A. Garcia-Sainz. 1983. *J. Lipid Res.* **24:** 945.

Feed Additive Compendium. 2004. *Feed Additive Compendium.* Miller Publ. Co., MN.

Gilliland, S. 2005. Probiotics. In *Encyclopedia of Animal Science,* W. G. Pond and A. W. Bell (eds.). Marcel Dekker, New York (In press).

Gohl, B. 1975. *Tropical Feeds.* FAO, Rome, Italy.

Goodrich, R. D., et al. 1984. *J. Anim. Sci.* **58:** 1484.

Harvey, R. B. 1993. *Zeolite '93, 4th International Conference on Occurrence, Properties, and Utilization of Natural Zeolites,* p. 118. International Commission on Natural Zeolites, SUNY-Brockport, NY.

Hays, V. W. 1977. *Effectiveness of Feed Additive Usage of Antibacterial Agents in Swine and Poultry Production.* Office of Technology Assessment, U.S. Congress, Washington, DC.

Hays, V. W. 1979. In *Drugs in Livestock Feed,* Vol. I, pp. 29–36. Technical Report, Office of Technology Assessment, Congress of the United States, Washington, DC.

Hays, V. W. 2005. Antibiotics: subtherapeutic levels. In *Encyclopedia of Animal Science,* W. G. Pond and A. W. Bell (eds.). Marcel Dekker, New York, (In Press).

Heath, M. E., D. S. Metcalf, and R. F. Barnes (eds.). 1973. *Forages,* 3rd ed. Iowa State University Press, Ames, IA.

Hirsch, D. C., and N. Wiger. 1978. *J. Anim. Sci.,* **31:** 1102.

Holden, P. J., and D. R. Zimmerman. 1991. In *Swine Nutrition,* Chapter 36, p. 585, E. R. Miller et al. (eds.), Butterworth-Heinemann, Boston.

Humpfries, C. 1980. *J. Sci. Food Agric.* **31:** 1225.

Huntington, G. B., and C. K. Reynolds. 1987. *J. Nutr.* **117:** 1167.

International Feed Institute. 1984. *IFI Tables of Feed Composition.* P. V. Fonnesbeck, H. Lloyd, R. Obray, and S. Romesburg (eds.). International Feedstuffs Institute, Utah State University, Logan, UT.

Jung, H. G., and G. C. Fahey, Jr. 1983. *J. Anim. Sci.* **57:** 206.

Kirk, D. J., et al. 1985. *J. Anim. Sci.* **60:** 1479.

Kiser, J. S. 1976. *J. Anim. Sci.* **42:** 1058.

LaRue, D. C., et al. 1985. *J. Anim. Sci.* **60:** 495.

Lewis, A. J., and L. L. Southern. 2001. *Swine Nutrition,* 2nd ed. CRC Press, Boca Raton, FL. 1008 pp.

Marn, N. J., et al. 1980. *J. Appl. Bacteriol.* **49:** 75.

McDowell, L. R., et al. 1974. *Latin American Tables of Feed Composition.* Department of Animal Science, University of Florida, Gainesville, FL.

Meyer, R. O. 2004. Feedstuff: nonconventional energy sources. In *Encyclopedia of Animal Science.* W. G. Pond and A. W. Bell (eds.). Marcel Dekker, New York. (In press).

Morley, F. H. W. (ed.). 1981. *Grazing Animals.* Elsevier Scientific Publishing Co., Amsterdcam and New York.

Morrison, F. F. 1956. *Feeds and Feeding,* 22nd ed. The Morrison Publishing Co., Clairmont, Alberta, Canada.

NRC-89 Committee on Confinement Management of Swine. 1984. *J. Anim. Sci.* **58:** 801.

National Research Council (NRC). 1982. *United States–Canadian Tables of Feed Composition,* 3rd revision. National Academy of Sciences, Washington, DC.

National Research Council. 2003. *Nutrient Requirements of Nonhuman Primates,* 2nd rev. ed. National Academy Press, Washington, DC.

Owens, F. N. 2005. Biotechnology: genetically modified feeds. In *Encyclopedia of Animal Science,* W. G. Pond and A. W. Bell (eds.). Marcel Dekker, New York (In Press).

Patience, J. F., and P. A. Thacker. 1989. *Swine Nutrition Guide.* Prairie View Swine Center, University of Saskatchewan, Saskatoon, Canada.

Pitt, R. E., et al. 1991. *Grass Forage Ser.* **46:** 301.

Pond, W. G. 1991. In *Swine Nutrition,* Chapter 1, p. 3, E. R. Miller et al. (eds.). Butterworth-Heinemann, Boston.

Pond, W. G., and J. H. Maner. 1984. *Swine Production and Nutrition.* AVI Publishing Co., Westport, CT.

Pond, W. G., J. H. Maner, and D. Harris. 1991. *Pork Production Systems.* Van Nostrand–Reinhold, New York, 439 pp.

Preston, R. L., et al. 1987. *J. Anim. Sci.* **65:** 481.

Ramsey, T. S. 1984. *J. Anim. Sci.* **58:** 1461.

Ramsey, T. S., W. G. Bergen and D. B. Bates; Goodrich, R. D. et al.; Potter, E. L., et al; Todd, G. C., et al.; Schelling, G. T.; Donoho, A. L. 1984. Symposium on Monensin in Cattle. *J. Anim. Sci.* **58:** 1461–1539.

Rayburn, E. B. 1986. *Seneca Trail RC&D* Technical Manual. Frankinville, NY.

Rust, S. R., et al. 1989. *J. Prod. Agric.* **2:** 235.

Sanderson, M. A. 1993. *J. Anim. Sci.* **71:** 505.

Sauber, T. E., and F. N. Owens. 2001. Cereal grains and by-products for swine. In *Swine Nutrition* (2nd ed.), Chapter 35. A. J. Lewis and L. L. Southern (eds.). CRC Press, Boca Raton, FL.

Seerley, R. W. 1991. In *Swine Nutrition,* Chapter 28, p. 451, E. R. Miller et al. (eds.). Butterworth-Heinemann, Boston.

Sheppard, R. A. 1984. In *Zeo Agriculture—Use of Natural Zeolites in Agriculture and Aquaculture,* p. 31. W. G. Pond and F. A. Mumpton (eds.). Westview Press, Boulder, CO.

Spoelstra, S. F., et al. 1988. *J. Agric. Sci.* (Cambridge). **111:** 127.

Sutton, A. L., and R. L. Vetter. 1971. *J. Anim. Sci.* **32:** 1256.

Thacker, P. A., and R. N. Kirkwood. 1992. *Non-Traditional Feeds for Use in Swine Production.* CRC Press, Boca Raton, FL.

Tyrrell, H. F., et al. 1992. *J. Anim. Sci.,* **70:** 3163.

Veum, T. L. 2005. Feedstuffs: High-Energy Sources. In *Encyclopedia of Animal Science*, W. G. Pond and A. W. Bell, eds., Marcel Dekker, New York (In Press).

Wallace, H. D. 1970. *J. Anim. Sci.* **31:** 1118.

Watson, S. J. and M. J. Nash. 1960. *The Conservation of Grass and Forage Crops.* Oliver and Boyd, Edinburgh, Scotland.

White, B. R. 2004. Antibiotics: Microbial Resistance. In *Encyclopedia of Animal Science,* W. G. Pond and A. W. Bell (eds.), Marcel Dekker, NY (In Press).

Wilkins, R. J., et al. 1971. *J. Agric. Sci.* **77:** 531.

Wilkinson, J. M. 1983. *World Animal Rev.* **45** (Jan–March): 37.

Wohlt, J. E. 1989. *J. Dairy Sci.* **72:** 545.

Yen, J. T., et al. 1989. *Proc. Soc. Exp. Biol. Med.* **190:** 393.

Yen, J. T., et al. 1990a. *J. Anim. Sci.* **68:** 2698.

Yen, J. T., et al. 1990b. *J. Anim. Sci.* **68:** 3705.

Zimmerman, D. R. 1986. *J. Anim. Sci.* **62** (Suppl. 3): 6.

20

Feed Preparation and Processing

FEED–FORAGES, GRAINS, VITAMIN and mineral supplements, and so on—represents the major cost in animal production. In ruminants, which typically consume more forage than other domestic species, feed may represent 50% or more of total production costs, and in nonruminants, the feed cost may be 80% or more of total production costs. Thus, it is imperative to supply a nutritionally adequate diet and to prepare the diet in a manner that encourages consumption without waste and allows high efficiency of feed utilization.

Feed processing may be accomplished by physical, chemical, thermal, bacterial, or other alterations of a feed ingredient before it is fed. Feeds may be processed to alter the physical form or particle size, to preserve, to isolate specific parts, to improve palatability or digestibility, to alter nutrient composition, or to detoxify.

The feed may be modified during the many steps involved between crop harvest and feeding. These processes may directly affect the feeding value as well as the ease of handling and mixing of the products treated. Therefore, it is important to be aware of the processes to which a feed has been subjected and the effect each process has on the nutritive value of the feed. The effects of processing on the efficiency of utilization and nutritive value of feeds were detailed by the National Academy of Sciences (1973) and swine feeds, by Hogberg et al. (1980) and Hancock and Behnke (2001). Generally, feed preparatory methods become more important as level of feeding increases and when maximum production is desired. This is so because animals fed for high

369

production become more selective and because, in ruminants, digestibility tends to decrease as level of feeding increases. The decrease in digestibility occurs primarily because feed does not remain in the GI tract long enough to allow maximal effect of the various digestive processes. Feed preparation is often more important for larger animal production operations with greater mechanization. This applies particularly to roughage because long or baled hay is less convenient to handle than chopped, pelleted, or cubed hay.

Feed preparatory methods that are used for swine and poultry are relatively simple and few as compared to the variety of methods available for ruminant feeds. Our discussion deals with methods used primarily with roughages, forages, and grains and not with methods used for processing oil-seed meals and grain byproducts.

❏ GRAIN PROCESSING METHODS

Grain processing methods can be divided conveniently into dry and wet processes or into cold and hot processing. Heat is an essential part of some methods, but it is not utilized at all in others. Similarly, added moisture is essential in some methods but may be detrimental in others. Examples of milo (grain sorghum) processed in different ways are shown in Fig. 20.1.

❏ COLD PROCESSING METHODS

Methods (or machinery) used for cold processing include rollermills, hammermills, soaking, reconstitution, ensiling at high-moisture content, and preservation with added chemicals.

Several types of mill designs are used to grind feedstuffs. The most common mills for grains are hammermills and rollermills. Their basic design and use are described and illustrated by Koch (1996).

Rollermill Grinding

Rollermills act on grain by compressing it between two smooth or corrugated rollers that can be screwed together to produce smaller and smaller particles. With grains such as corn, wheat, or sorghum (milo), the product can range in size from cracked grain to a rather fine powder. With the coarse grains—barley and oats—usually corrugated rollers are used, and the product may range in size from a flattened seed to a finely ground product, but the hulls will not be ground as well as with other types of grinding mills. Corrugated rollers create a product called crimped grain, for example, crimped oats. Rollermills produce a less dusty feed than that produced by hammermills. If the feed is not ground too finely, its physical texture is very acceptable to most species. Rollermills are not used with roughages.

Hammermills

A hammermill processes feed with the aid of rotating metal bars (hammers) that blow the ground product through a metal screen. The size of the product is controlled by changing the screen size. These mills can grind anything from a coarse roughage to any type of grain, and the product size will vary from particles of a size similar to cracked grain to that of a fine powder. A significant amount of dust may be lost in the process, and the finished product is usually dustier than that produced with a rollermill.

Soaked Grain

Grain soaked for 12 to 24 h in water has long been used by livestock feeders. The soaking, sometimes with heat, softens the grain, which swells during the process, making a palatable product that should be rolled before using in finishing rations. Research results do not show any marked improvement in animal performance when compared to other methods. Space requirements, problems in handling, and potential souring have discouraged large-scale use.

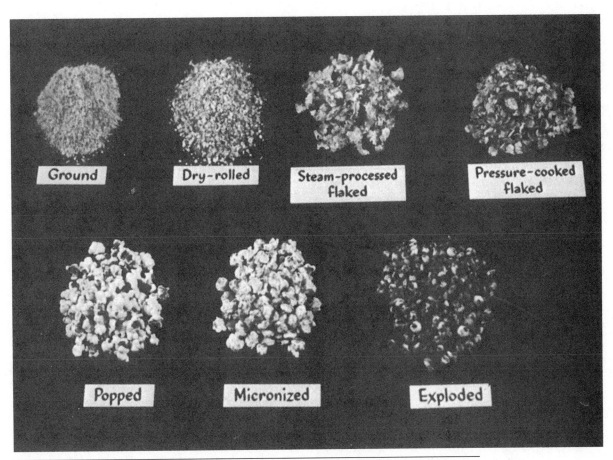

Figure 20.1 *Examples of sorghum grain processed by several different methods. Courtesy of W. H. Hale, University of Arizona.*

Reconstitution

Reconstitution is similar to soaking and involves adding water to mature dry grain to raise the moisture content to 25 to 30% and storage of the wet grain in an oxygen-limiting silo for 14 to 21 days prior to feeding. This procedure works well with sorghum in particular, resulting in improved gain and feed conversion (Table 20.1) compared with results obtained with cattle fed high-concentrate diets containing dry-rolled whole grain.

High-Moisture Grain

High-moisture grain is harvested at a high moisture content (20–35%) and stored in a silo to preserve the grain, which would spoil unless ensiled or treated with chemicals. The grain may be ground before ensiling or ground or rolled before feeding. This is a particularly useful procedure when weather con-

ditions do not allow normal drying in the field; the procedure obviates the need to dry the grain artificially. Storage costs may be relatively high, but high-moisture grain produces good feedlot results. Feed conversion, particularly, is improved. Ensiled wet grains are not as easily sold or transported as grains not preserved in this way.

Acid Preservation of High-Moisture Grains

With higher fuel costs, increased interest has developed in eliminating artificial drying of newly harvested cereal grains. Data with barley or corn for pigs and research with corn or sorghum for beef cattle show promise for the use of acids to preserve high-moisture grains. Thorough mixing of 1 to 1.5% propionic acid, mixtures of acetic–propionic acid or formic–propionic acid into high-moisture (20–30%) whole corn or other cereal

TABLE 20.1   Examples of effect of grain processing on steer performance.

ITEM[a]	AV. DAILY GAIN, KG	DAILY FEED INTAKE, KG	FEED TO GAIN RATIO, KG
Milo, dry-rolled	1.28	10.31	8.02
vs. steam-flaked	1.41	10.60	7.64
Sorghum grain, dry-rolled	1.00	8.26	8.21
vs. reconstituted	1.20	8.40	7.02
Barley, dry-rolled	1.31	9.44	7.22
vs. steam-flaked	1.41	10.31	7.32
Corn, dry-rolled	1.38	9.17	6.70
vs. pressure cooked	1.52	9.53	6.33
Whole corn, reconstituted	1.18	9.31	7.85
vs. steam-flaked	1.20	9.22	7.66
Whole corn	1.25	7.01	5.62
vs. whole steamed	1.31	7.57	5.79
vs. steam-flaked	1.33	6.71	5.06

[a]All except the last comparison excerpted from Church (1984); the last comparison is from Ramirez et al. (1985); in this last example, the feed intake is expressed on a dry matter basis. Comparisons among treatments should be made only as listed because the other treatments may have been done in a different location at a different time.

grains retards molding and spoilage without affecting animal performance appreciably, compared with that obtained with dried grains.

❑ HOT PROCESSING METHODS

Methods used for heat processing grains and other products (such as oilseed meals and pet food) include steam rolling, steam flaking, popping, micronizing, roasting, pelleting, and extruding. Pressure cooking and exploding are methods that have been tried at the feedlot level. However, cost of the equipment and maintenance problems have discouraged continued use.

Steam-Rolled and Steam-Flaked Grains

Steam-rolled grains have been used for many years, partly because the process kills weed seeds. The steaming is accomplished by passing steam up through a tower above the rollermill. Grains are subjected to steam for only a short time (3 to 5 min) prior to rolling. Most results indicate little, if any, improvement in animal performance as compared to dry rolling, but use of steam does allow production of larger particles and fewer fines.

Steam-flaked grains are prepared in a similar manner but with relatively rigid quality controls.

Grain is subjected to high-moisture steam for a sufficient time to raise the water content to 18–20%, and the grain is then rolled to produce a rather flat flake. Feedlot data with cattle indicate that best response is produced with thin flakes, which apparently allow more efficient rupture and exposure of starch granules and produce a more desirable physical texture. Steam flaking of corn, barley, and grain sorghum usually results in increased cattle weight gain; efficiency of feed utilization is improved with flaked corn and grain sorghum, but less so with flaked barley.

Feedlot performance of cattle is affected by degree of processing of corn (Brown et al., 2000). Increasing the degree of corn processing by steam flaking for varying times (40, 60, and 80 min, which increased bulk density) increased starch availability and improved the rate and efficiency of gain.

Pelleting

Pelleting is accomplished by grinding the feed ingredients or feedstuffs and then forcing them through a die. Feedstuffs usually, but not always, are steamed to some extent before pelleting. Pellets can be made in different shapes, diameters, lengths, and degree of hardness. All domestic animals generally like the physical nature of pellets, particularly as compared to meals, and a high proportion of poultry and swine feeds are pelleted. However, results with ruminants on high-grain diets have not been particularly favorable because of decreased feed intake, even though feed efficiency is usually improved over other methods. Pelleting the diet fines (finely ground portions of a diet) frequently is desirable because the fines often will be refused otherwise. Animals fed pellets cannot sort and select individual components of the diet, in contrast to the case with diets fed in meal form. Supplemental feeds such as protein concentrates are pelleted in many instances to prevent sorting; they can then be fed on the ground or in windy areas with little loss.

Popping and Micronizing

Popped corn is produced by the action of dry heat, causing a sudden expansion that ruptures the endosperm of the seed. This process may increase rumen and intestinal starch utilization but results in a low-density feed. Consequently, corn usually is rolled before feeding to reduce bulk. Micronizing is essentially the same as popping, except that heat is provided in the form of infrared energy.

Extruding

Feeds for pets generally are extruded. Extruded grains or grain mixtures are prepared by passing the grain or mixture through a machine with a spiral screw that forces the grain through a tapered head. In the process, the grain is ground and heated, and mixed with other ingredients, producing a ribbonlike product. Results with cattle fed diets containing extruded grain are similar to those obtained with other processing methods. The extrusion process is more widely used for pet diets than for diets of food-producing animals.

Spraying Feeds with Molasses, Fat, or Other Liquids

Molasses or fat often is sprayed onto the surface of feeds during the mixing process to improve palatability and reduce dustiness. Spraying the diet with fat also increases the energy density of the feed and improves the ease of mechanical handling and serves as a lubricant in augers, pipes, and storage bins. Other liquids often are added to feeds during mixing as a means of providing micronutrients, amino acids, flavor compounds, and mold inhibitors.

❏ FEED PROCESSING FOR NONRUMINANTS

Swine and Poultry

Grinding and pelleting are the most common means of preparing feed for swine and poultry. Generally, results show that grinding to a medium to moderately fine texture results in better performance than when grains are finely ground. Feed particles of different ingredients should be of a similar size so that animals do not sort out the coarse particles and leave the fines. Digestibility of grains by swine generally is improved by grinding, and finely ground (0.16-cm screen size) grain promotes improved efficiency, compared with coarse grinding (1.27-cm screen or larger) in a hammermill. However, finely ground feed is associated with an increased incidence of stomach ulcers in pigs.

As particle size of cereal grains is reduced, the incidence of esophagogastric ulcers tends to increase in nursery pigs, growing–finishing pigs, and lactating sows, possibly as a result of increased fluidity of stomach contents and/or increased pepsin activity in the stomach contents (reviewed by Hancock and Behnke, 2001). It has been suggested that genetic factors (Berruecos and Robison, 1972) and stressful housing and transport environments (Lawrence et al., 1998) may affect the incidence of ulcers. Despite producers' concerns about the high frequency of stomach ulcers associated with diets with small particle size, performance benefits of finely ground grains clearly exist. Figure 20.2 illustrates the benefit of corn of small particle size on daily feed intake and digestible energy of lactating sows and on litter weight gain during the nursing period (Hancock and Behnke, 2001; data adapted from Wondra et al., 1995).

The effects of pelleting and other feed processing methods on performance of growing swine have been reviewed (Danielson and Crenshaw, 1991; Hogberg et al., 1980; Holden and Zimmerman, 1991; Liptrap and Hogberg, 1991; Hancock and Behnke, 2001). Pelleting usually results in a 3 to 5% improvement in rate of weight gain partly because animals tend to eat more in a given period. Efficiency often is apparently improved 5 to 10%, sometimes because of less feed wastage with pellets. A common practice, particularly for poultry, is to pellet meal diets, then roll them, and screen out fines, producing a product called crumbles. The texture of crumbles is well liked, compared with that of pellets, particularly if the pellets are

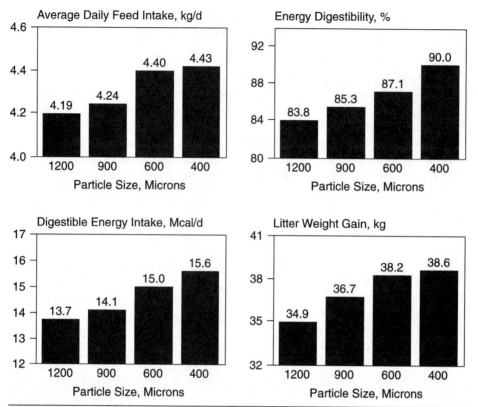

Figure 20.2   *Effects of corn particle size on lactation performance of primiparous sows and apparent digestibility and intake of energy. (Data from Wondra et al., J. Anim. Sci., 73: 421, 1995.)*

quite hard. It has been noted that birds fed pellets tend to exhibit more cannibalism than those given feed in mash form.

Pelleting generally improves swine gain and feed efficiency to a greater extent when applied to feed grains that are low in fiber (e.g., corn, wheat, sorghum), but the process has less effect on barley and oats. Although the efficiency of utilization and growth of swine is improved considerably by medium to fine grinding of oats, pelleting may improve efficiency with little or no effect on daily gain.

In sow feeding, particularly in outdoor systems where feed is dispensed on the ground or on a concrete pad, a complete feed mixture is provided in cube form (much larger size than pellets) to minimize wastage and feeding equipment costs.

Many seed legumes such as soybeans contain one or more heat-labile inhibitors of protein utilization and other heat-labile toxic factors. These inhibitors and toxins are of more practical concern for nonruminants than for ruminants because the rumen microflora generally destroy or inactivate

them. However, feeding some of these seed legumes to swine and poultry without heat treatment results in low-feed intake and poor growth. Commercial roasters are available for on-farm use in cases in which economic factors favor the direct feeding of these legumes to animals rather than marketing them for further processing as human foods or for other uses. Danielson and Crenshaw (1991) have reviewed the use of cooked soybeans in swine feeding.

Other feed preparatory methods have been used with varying degrees of success for both pigs and poultry. With barley, soaking and treatment with different enzymes may result in some improvement, but little use is currently being made of these methods, partly because little barley is fed to poultry, and the improvement for swine probably does not justify the expense under most conditions.

Recent research in swine fed cultivars of barley containing high levels of beta-glucans suggests that pretreatment of the barley with amylases may improve animal performance. This type of feed preparatory technology probably will continue to find application in animal feeding to improve the efficiency of nutrient and energy utilization.

Other Animals

Processing of horse feed is often a matter of convenience for the feeder or for purposes of avoiding dust and mold. Generally, efficiency is not an important consideration, in contrast to the case for a fast-growing chick or a high-producing cow. Horse feeds often contain rolled or coarsely cracked grains, with liberal amounts of molasses to avoid problems with dust. Some pelleted or cubed hays are fed, but only as a small percentage of the total forage fed to horses. Cubed hay has occasionally resulted in problems with choking.

Rabbits are usually fed pelleted, mixed feeds and/or long roughages. Commercial growers use pelleted feeds because of convenience and to reduce waste.

A substantial amount of pet food is either pelleted or processed through extruders that alter the physical form as well as the nutritive value of the feed by the heat generated in the process.

❑ GRAIN PROCESSING FOR RUMINANTS

Grain processing primarily improves digestibility and efficiency of utilization because grain is naturally in a physical form that can be handled mechanically with few problems. Improvement in animal utilization usually can be obtained by various means that break up the hull or waxy seed coat and improve digestibility of starch in the endosperm. Some methods may provide a more favorable particle size and/or density that facilitates more optimal passage through the rumen and may improve palatability. For a method to be effective, it must (a) reduce wastage, (b) increase consumption and rate of gain, or (c) be utilized more efficiently. Grain (or roughage) is expected to give greater returns per unit of cost when feed intake is high. Animals on a maintenance diet normally would not be fed much grain, and any improvement in efficiency normally would not return the added cost of processing.

In diets typically in use in the United States, grain and other concentrates in finishing diets for cattle may account for 60 to 95% of the total intake.

The benefits of grain processing can be shown clearly in most instances when higher levels of grain are fed ad libitum. Some research evidence indicates that whole corn grain may be an exception to this generally rule when corn makes up a high percentage of the diet.

Economics of Feed Processing

Escalating energy costs have created increased interest in minimizing feed processing costs or in obtaining the maximum return per unit of cost for processing. The fossil fuel energy used for the methods listed in Table 20.2 provides an appreciation of the comparative costs involved in feed processing. The actual costs of processing feed are not easy to compute and are apt to vary considerably from installation to installation.

The data in Table 20.3 suggest that grain processing (more than dry rolling) normally does not result in any marked increase in daily gain of cattle and that the primary benefit is some improvement in feed efficiency. Of course, good processing may stimulate feed consumption just as poor processing may depress feed consumption, and many reports show appreciable differences in gain. Wastage may be reduced by appropriate processing.

The relative value of different grain processing methods for feedlot cattle have been calculated by Schake and Bull (1980). Daily gains and efficiencies shown in Table 20.3 were compiled from a number of experimental trials done in Texas. Note that flaking, reconstitution, or early harvesting all resulted in some improvement in feed efficiency when compared to dry ground or rolled grain. However, because of higher costs for equipment,

TABLE 20.2 Fossil fuel energy use for some feed processing methods.

PROCESS	MCAL/TON
Drying corn	240
Propionic acid	200
Grinding corn	40
Pelleting corn	80
Steam flaking corn	120
Dehydrating and pelleting alfalfa	1800
Drying wet beet pulp	1750

Source: Data from G. W. Ward, unpublished data, Colorado State University, Fort Collins, CO, 1981.

TABLE 20.3 Effects of feed processing of corn and sorghum on daily gain and
feed conversion.

METHOD	CORN			SORGHUM		
	GAIN, LB	FC,[a] LB	FC, %[b]	GAIN, LB	FC,[a] LB	FC, %[b]
Dry ground or rolled	2.62	5.79	—	2.50	6.09	—
Steam flaked	2.68	5.36	7.46	2.65	5.60	8.11
Reconstituted	2.61	5.33	7.97	2.54	5.34	12.32
Early harvested						
Ground, ensiled	2.57	5.14	11.14	2.64	5.52	9.35
Acid treated	2.68	5.56	3.97	2.45	5.60	8.11

[a]FC = feed conversion (feed/gain).
[b]% improvement in FC of grain dry matter compared to dry processing.
Source: Data from Schake and Bull (1980).

power, storage (early harvesting requires that all the grain be purchased at one time), and other factors, the most efficient methods did not necessarily result in the greatest net value per ton of grain. Larger feedlots had some advantage in costs for flaking and reconstitution, but not with early-harvested grain whether ensiled or acid treated.

Feed Processing for Dairy Cows

Feed processing may result in somewhat different responses in dairy cows than for growing or finishing cattle or lambs. Generally, feeding lactating cows high-grain rations, particularly heat-treated grains, or all of their roughage or forage in ground or pelleted form results in reduced rumen acetate to propionate ratios and lower milkfat percentages. Total milkfat production may not be decreased because feedstuffs that are capable of causing low milkfat may also stimulate increased total milk production. Most commonly, lactating cows are fed concentrates in which the grains have been coarsely ground or steam-rolled. However, pelleted feeds are fed in milking parlors in many operations because cows will eat more pellets in a short period of time while being milked. Pelleting also avoids some of the dust otherwise encountered in such situations. Some dairy farmers utilize ground and ensiled high-moisture grains, particularly corn.

Grain Processing for Sheep

Very few benefits can be shown for processed grains fed to sheep. Diet sorting and wastage can be reduced by pelleting. Except for the advantages of pelleting, sheep chew grains (with the possible exception of sorghums) well enough that processing is not necessary. In addition, it is not a common practice to feed as much grain (as a percentage of the diet) to sheep as to cattle, so any possible benefits of processing are reduced. In some feedlots it is a common practice to pellet finishing rations. This allows greater use of forages or roughages, and such diets can be self-fed.

❑ FEED MIXING

Feed ingredients generally are mixed together in a commercially available mixer in the proper ratios to provide a nutritionally balanced mixture. There are two broad types of mixers: batch and continuous flow.

Batch Mixers

There are two configurations of batch mixers, vertical and horizontal. Vertical mixers are cylindrically shaped with an inverted cone-shaped bottom and one or more vertical augers to move feed from the bottom to the top of the mixer. A uniform mixture of all ingredients is attained within 15 to 20 min. Horizontal mixers are U shaped, with paddles or ribbonlike blades attached to a horizontal revolving shaft. Horizontal mixers can accommodate mixtures of chopped hay, silage, grains, molasses and other ingredients differing in particle size, density, and texture. They are often mounted

Figure 20.3 *Feed ingredients can be added to this mixer wagon to provide a total mixed ration. The mixer wagon can then deliver and discharge the feed directly into the feeder. Courtesy of John Deere & Co.*

on mobile wagons or trucks and powered by the transport vehicle (Fig. 20.3). Such portable mixers are widely used for preparing and dispensing complete growing and finishing diets for feedlot cattle and sheep. Mixing time is less than that of batch mixers (2–15 min), an asset that adds to their versatility. To produce a desired quantity of mixed feed [e.g., 250–6000 pounds (113.5–2724 kg)], individual feed ingredients are separately added to the mixer in the predetermined amounts. For most efficient mixing, usually the ingredient used in the largest amount is added first, and then ingredients that make up a smaller proportion of the final diet are added in descending order of their contribution to the total diet. When added fat is a constituent of the mixture, all other constituents are mixed first and the fat is added last and the final diet is mixed again.

Continuous Flow Mixers

Continuous flow mixers rely on continuously blending all the ingredients in the proper ratios to produce the desired diet. Feed ingredients are simultaneously augured and blended. Proper calibration is essential to ensure that the desired composition is produced. The system, once calibrated,

can be operated automatically without direct intervention. There is less flexibility in changing feed ingredients than with batch mixers, and most bulky ingredients such as hay and silage cannot be used. Continuous flow mixing systems with automated ingredient weighing and dispensing into the mixer are most commonly used in large swine and poultry production units in which diet composition and ingredients are relatively constant.

❏ DRY FORAGE AND ROUGHAGE PROCESSING FOR RUMINANTS

Baled Dry Forage and Roughage

Baling is still one of the most common methods of handling dry forage and roughage, particularly where it is apt to be sold or transported some distance. Baling has a considerable advantage over loose hay stacked in the field or roughage in other less dense forms. Although baled hay now can be handled mechanically, it still requires considerably more hand labor than many other feedstuffs. Furthermore, considerable waste may occur in feeding, depending on how it is fed (feed bunks, on the ground) and on the level of feeding. Animals fed for high levels of production, such as dairy cows, may be quite selective so that stems will not be consumed. Thus, a high loss of nutrients occurs in feeding baled hay. Consumption is not adequate for high levels of performance when baled hay provides the only feed for ruminants. Small rectangular bales [30–70 lb (13.6–31.8 kg)] are less common than large round bales [500–1000 lb (227–454 kg)] because less labor is generally required in feeding systems using large round bales (less frequent feeding). Dehydrated bagged alfalfa, though expensive, is increasingly popular for feeding horses as well as ruminants.

Chopped and Ground Forages and Roughages

Chopping or grinding puts forages and roughages into a physical form that allows easier handling by some mechanical equipment, tends to provide a more uniform product for consumption, and usually reduces feed refusals and waste. However, additional expense may be incurred by grinding and loss of part of the feed as dust may be appreciable from grinding in hammermills. This dust loss sometimes is reduced in commercial mills by spraying fat on bales before they are ground. Ground hays are, as a rule, dusty and may not be readily consumed. Adding molasses, fat, or water usually improves intake. Chopping produces a physical texture of a more desirable nature than grinding, but chopped hay does not lend itself as well to incorporation into mixed feeds as does ground hay.

Grinding as well as pelleting or cubing results in a feed with a less coarse physical texture. This may not be an advantage with herbivorous animals.

Pelleting

Pelleted forages are readily consumed by ruminants, provided they are not too dense to eat readily, or too soft, which results in too many fines. Forages such as long hay must be ground before pelleting; this is a slow, costly process compared to similar treatment of grains. Thus, cost of pelleting is of greater concern than for most other feed processing methods. Pelleting usually gives the greatest relative increase in performance for low-quality forages and roughages. This appears to result from an increase in diet density and greater feed intake associated with more rapid passage through the GI tract and not from any great improvement in digestibility. Pelleted forages and roughages are metabolized somewhat differently. As a result of more rapid passage out of the rumen, less cellulose is digested, and relatively less acetic acid is produced with relatively greater digestion in the intestines. Utilization of metabolizable energy usually is more efficient. Pelleted high-quality forages and roughages produce performance (gain in weight) in young cattle or lambs almost comparable to high-grain feeding. An example of the change in density achieved with grinding and pelleting is shown in Fig. 20.4.

Cubed Roughages

Cubing is a process in which dry hay is forced through dies that produce a square product (about 3 cm in diameter) of varying lengths. Grinding before cubing is not required, but usually water is sprayed on the dry hay as it is cubed. Field cubers have been developed, and stationary cubers are used to process hay from stacks or bales. Alfalfa hay produces the best cubes. Research data indicate that cubes produce satisfactory performance in cattle, provided the cubes are not too hard. Al-

Figure 20.4   *An example of the densification of alfalfa hay which can be achieved by baling, grinding and pelleting (left to right). Each sample contains 2.27 kg of hay.*

though cubes have some advantages, particularly that they can be handled with mechanized feeding equipment, they have never been as popular as expected. A partial answer may be that it is difficult to detect (visually) low-quality hay after it has been cubed, especially if green dyes have been added.

Chemical Treatments

Poor quality forage and other roughage materials (e.g., wood sawdust, sugar cane bagasse, pineapple waste) high in cellulose represent a substantial resource for animal feed. Such materials are not highly digestible; considerable research is underway on this subject in an attempt to improve energy utilization of such products by the animal. Methods have ranged from soaking straws in alkali solutions for up to a month to treatments involving heat, pressure, and chemicals.

Digestibility of a roughage such as cereal straw can be increased a moderate amount (i.e., dry matter digestibility from about 40 to 50%) by spraying with NaOH solutions or ensiling with NaOH, molasses, water, and so on, but costs in North America are not such that it is a feasible method at this time. In areas where roughage is in shorter supply, it may be a feasible method. In fact, several processing plants have been built in northern Europe.

Another method of improving energy utilization of fibrous feeds is by treatment with anhydrous ammonia. In this process, N is added to the straw to enhance the activity of rumen microorganisms, thereby improving breakdown and utilization of the fiber present in straw. In this instance sheep were fed a diet made up primarily of straw plus a supplement. Untreated straw was fed as a negative control and straw was treated by direct ammoniation of the bale (which was covered), or high-moisture straw was treated in a similar manner and either fed directly out of the sacks used for storage or exposed to air before being fed. The ammoniation of poor quality roughage generally increases consumption of the roughage, and digestibility of cellulose and hemicellulose is improved. Animal live-weight gain is usually improved to a moderate degree.

Because of its lower cost, urea is used in place of the more expensive and volatile anhydrous ammonia as a source of N in many cases. Urea is sprayed or poured onto the straw or roughage in a manner similar to that used for anhydrous ammonia, but its use is safer because (unlike the volatile ammonia) it is a stable crystal at room temperature and it is readily soluble in water.

Another possible treatment is the use of alkaline hydrogen peroxide (AHP). AHP causes some solubilization of lignin in highly lignified forages such as straw. In this case, treatment of straw with AHP markedly increased dry matter intake of the high-straw diet. In addition, treatment increased the digestibility of dry matter and cellulose markedly.

❏ SUMMARY

- Numerous feed processing methods are available that reduce waste, discourage selectivity of ingredient consumption, and usually increase feed consumption and efficiency of feed utilization. Feed processing is of particular importance for animals consuming feed at high levels and for animals expected to perform at high levels of production. However, higher relative fuel costs have resulted in more selectivity of processing methods because efficiency of fuel use varies among the different procedures. With forages and roughages, some degree of processing is required when they are incorporated into complete diets, to facilitate diet mixing and the use of feed distribution systems. Feed wastage

can be reduced considerably by processing baled or long hays, but the costs also will be increased substantially. Newer methods of chemically treating low-quality roughages may be feasible economically. The addition of anhydrous ammonia or urea to low-quality roughages (which are also low in crude protein) generally results in increased consumption of the roughage, improvement in digestibility of dry matter, and utilization of the ammonia as a N source for rumen microorganisms. Treatment of low-quality roughages with alkaline hydrogen peroxide (which solubilizes some of the lignin) appears to result in a marked increase in digestibility of the fibrous carbohydrates in roughages that are otherwise not well utilized without extensive chemical treatments.

❏ REFERENCES

Berruecos, J. M., and O. W. Robison. 1972. *J. Anim. Sci.* **35**: 20.

Brown, M. S., et al. 2000. *J. Anim. Sci.* **78**: 464.

Church, D. C. (ed.). 1984. *Livestock Feeds and Feeding.* 2nd ed. Prentice-Hall, Englewood Cliffs, NJ.

Danielson, D. M., and J. D. Crenshaw. 1991. Raw and processed soybeans in swine diets. In *Swine Nutrition,* Chapter 35, pp. 573–584, E. R. Miller, D. E. Ullrey, and A. J. Lewis (eds.). Butterworth-Heinemann, Boston, MA.

Hancock, J. D., and K. C. Behnke. 2001. Use of ingredient and diet processing technologies to produce quality feeds for pigs. In *Swine Nutrition* (2nd ed.), Chapter 21, pp. 469–497, A. J. Lewis and L. L. Southern (eds.). CRC Press, Boca Raton, FL.

Hogberg, M., et al. 1980. *Physical Forms of Feed-Feed Processing for Swine.* Pork Industry Handbook PIH 71, Cooperative Extension Service, Purdue University, West Lafayette, IN.

Holden, P. J., and D. R. Zimmerman. 1991. Utilization of Cereal Grain By-products in Swine Feeding. In *Swine Nutrition,* Chapter 36, pp. 585–593, E. R. Miller, D. E. Ullrey, and A. J. Lewis (eds.). Butterworth-Heinemann, Boston, MA.

Kerley, M. S., et al. 1986. *J. Anim. Sci.* **63**: 868.

Koch, K. 1996. *Coop. Ext. Bull.* MS-496, Kansas State University, Manhattan, KS.

Lawrence, B. V., et al. 1998. *J. Anim. Sci.* **76**: 788.

Liptrap, D. O., and M. G. Hogberg. 1991. Physical Forms of Feed: Feed Processing and Feeder Design and

Operation. In *Swine Nutrition,* Chapter 22, pp. 373–386, E. E. Miller, D. E. Ullrey, and A. J. Lewis (eds.). Butterworth-Heinemann, Boston, MA.

National Academy of Sciences. 1973. *Effect of Feed Processing on the Nutritional Value of Feeds.* NAS, Washington, DC, 494 pp.

Ramirez, R. G., et al. 1985. *J. Anim. Sci.* **61**: 1.

Schake, L. M., and K. L. Bull. 1980. *Result of Food Processing on the Nutrient Value of Food.* College Station, TX. Texas Agricultural Experiment. Station Technical Report, 81-1.

Streeter, C. L., and G. W. Horn. 1984. *J. Anim. Sci.* **59**: 559.

Wondra, K. J., et al. 1995. *J. Anim. Sci.* **73**: 421.

21

Diet Formulation

DIET FORMULATION IS A very important aspect of animal production. The success of any animal production enterprise depends, to a large extent, on proper nutrition and feeding based on economic diets. The owner or animal production practitioner should have a good knowledge of nutrition, feeding, physical and chemical characteristics of feedstuffs, and feedstuff interactions and limitations, as well as the economics of production. Diet formulation is a task that meets the nutrient requirements of the animal with combinations of various feed ingredients. Most of the mathematical techniques required in diet formulation are quite simple, although for complex diets the mathematics become complicated and are facilitated with the use of computers. Personal computers with appropriate spreadsheet or specialized software can be used by almost anyone to formulate diets. In this chapter we present some of the simple techniques, as well as the basics for complex diet formulation.

❏ INFORMATION NEEDED

Two pieces of information are needed to balance a diet for any animal at any physiological stage. First, the nutrient requirements of that animal need to be determined, and second, the chemical composition of the available feedstuffs must be known. With these pieces of information, the diet can be rapidly formulated.

Nutrient Requirements of the Animal

The first step is to know the nutrient requirements of the particular animal species for a particular physiologic function, for example, growth and lactation. Nutrient requirements for animals are expressed in terms of energy, fat, protein (and amino acids), mineral elements, and vitamins. In the United States, the National Research Council (NRC) recommendations are the most commonly used guides for nutrient requirements of farm animals, dogs, cats, and horses. The titles of these and other related publications and nutrient requirement for some animals are presented in Appendix Table 1.

These recommendations are subject to continuing modification as new information is acquired and when the formulator has knowledge indicating a change should be made. For example, high-producing dairy cows may need more liberal nutrient allowances than indicated in current requirement tables, or adjustments may be needed to compensate for weight loss following environmental stress or for exposure to a cold, hot, or humid climate.

Feedstuffs and Nutrient Composition

The next step is to list all the available feedstuffs (as presented in the Appendix, Table 2) to be considered in the diet for the particular animal. It is best to obtain the nutrient composition from a laboratory analysis of the feedstuffs. Most states have such services available at a nominal charge. Commercial laboratories specializing in such analyses are also common. Determining the composition of the feedstuff through analyses has some disadvantages: the extra cost of the analysis and the time delay in getting the results from the lab. The nutrient composition of the feed ingredients also can be estimated from published values. It must be recognized that published values usually represent the average analysis from a number of samples. In addition to the average, some tables will include the range of values that are typically observed. How satisfactory the diet will meet the needs of the animal will be dependent on the reliability of the feedstuff analysis. Updated analytical data on feedstuffs are preferred when available; if not, average composition data can be used from the NRC tables (National Research Council, 1982) or composition data from any other reliable source such as International Feed Institute (1984). Diets are only as good as the analytical information used to formulate them. In the case of feedstuffs utilized for nonruminants, it may be appropriate to include "available" nutrients as opposed to "total" nutrients (as in the case of P and amino acids).

Some feedstuffs may be limited in use due to digestibility or palatability. Therefore, both the feedstuff analysis and the maximum percentage allowances of a particular feedstuff for a diet should be considered in formulation. For example, in some cases urea could be used as a nitrogen source for adult ruminants instead of a protein feed, but not with young nonruminant species whose ability to utilize nonprotein nitrogen is limited. In areas where one of the trace minerals in native forages is low, a supplement may be appropriate to provide the deficient minerals, whereas normally it might be ignored in routine diet formulations.

Furthermore, one must consider whether the feedstuff should be processed and, if so, in what way and at what cost. Types of grain and hay processing (Chapter 20) vary as to effect and cost. Is it a palatable feedstuff, or will the mixture be palatable? Does the ingredient present a problem in handling, mixing, or storing? Many factors must be considered.

Obviously, it takes previous knowledge and experience to answer many of the questions correctly. The ability to answer these questions correctly is essential for a practicing nutritionist. For our purposes, the beginner need not address such questions; time and experience will help provide the answers that cannot be covered in this chapter.

❏ MECHANICS OF WORKING WITH PERCENTS

Feedstuff compositions and nutrient requirements are usually expressed as a percent (e.g., % crude protein, % Ca, % P). Often there is a need to convert from an as-fed basis (containing water) to a dry matter basis (without water). All of these conversions require the ability to work with percents. Remember that a percent is a fraction in which the denominator is equal to 100 (e.g., $15\% = \frac{15}{100} = 0.15$). Addition or subtraction of percents is just like whole numbers (e.g., 15% + 25% = 40%; 25% − 15% = 10%). Multiplication and division with percents is a little more challenging. Remember that percents are really fractions with the denominator of 100. When percents are multiplied, it is similar to multiplying fractions (e.g., $10\% \times 25\% = \frac{10}{100} \times \frac{25}{100} = \frac{250}{10,000} = 2.5\%$). A general rule for multiplication of percents is to convert one of the two percents to the decimal form and multiply it by the other expressed as a percent (e.g., 10% × 25% = 0.10 × 25% = 2.5%). Remember that when dividing percents, they need to be treated like fractions with the common denominator of 100. In dividing fractions, one of the fractions is inverted followed by multiplication (e.g., 25% divided by $10\% = \frac{25}{100} \times \frac{100}{10} = \frac{25}{10} = 2.5 = 250\%$). A general rule for division of percents is to convert the denominator to a decimal value (e.g., $\frac{25\%}{10\%} = \frac{25\%}{0.10} = 250\%$).

❏ MECHANICS OF DIET FORMULATION

When diets are formulated by hand, generally they are first formulated for one nutrient. Then the other nutrients are checked to determine whether the feedstuffs used will meet the requirements or whether alternative feeds need to be included in the diet. One method is to balance for protein first and then to check energy levels to see if they are met. Next the ration can be checked for other nutrients such as calcium and phosphorus.

When the number of nutrients specified is small, diet formulation can be adequately carried out through simple calculations. The calculations involved in this type of formulation are simple and few. However, as the number of nutrient specifications and/or the number of feedstuffs available increases, the mathematics can get quite involved and confusing. In the following sections, various techniques for diet formulation will be covered with the use of examples. The first examples will be easy and can be completed with a simple calculator. Later examples are appropriate for use with a computer spreadsheet or specialized program. Try to follow

the setup and rationale for what is being done. Try not to worry about the mathematics but understand how and what is being done.

There are two main methods (algebraic and Pearson's square) of balancing diets for one to two nutrients using two or more ingredients or using ingredients at fixed levels. Examples of both the algebraic and Pearson's square methods are included in Figs. 21.1 through 21.4.

This procedure works very well to produce a mix with exact specifications for two nutrients. If we want to add more nutrients, it will require more than three squares. Diets that have exact requirements for more than two nutrients are tedious to formulate by hand.

❏ SAME EXAMPLE USING THE ALGEBRAIC METHOD

Some individuals prefer to solve simple diet formulation using simultaneous equations. The amounts of the ingredients will be the unknowns. For the same problem as with the single Pearson's square, the approach would be:

$$X = \% \text{ of SBM in mix}$$
$$Y = \% \text{ of grain sorgum in mix}$$
$$X + Y = 100 \text{ (equation for total amount of each feedstuff)}$$
$$0.49X + 0.045Y = 16 \text{ (equation for CP)}$$

To solve this problem it is necessary to multiply the members of the first equation ($X + Y = 100$) by a unit that will allow one of the unknowns (X or Y) in the second equation to factor out. Thus, if we multiply it by 0.095 we have:

Amount equation times 0.095:
$0.095X + 0.095Y = 9.5$
CP equation $0.49X + 0.045Y = 16$
Subtract from it $0.095X + 0.095Y = 9.5$
The answer is $0.395X = 6.5$

Then $X = \frac{6.5}{0.395} = 16.46$, and substituting the value of X in the amount equation, we obtain

$$16.46 + Y = 100$$
$$Y = 100 - 16.46 = 83.54$$

Obviously, the answer is the same as with the Pearson's square method. The reader should keep in

Figure 21.1 Example—Balancing for one nutrient with two ingredients.
Balance a 100-kg diet so that it contains 16% crude protein (CP) using grain sorghum (9.5% CP) and soybean meal (49% CP).

METHOD 1: ALGEBRAIC
Let **X** be the kg of soybean meal (SBM) in 100 kg of feed.
Let **(100-X)** be the kg of grain sorghum in 100 kg of feed.

Set up the following equation:

$$
\underset{\substack{\uparrow \\ 100\text{ kg} \\ \text{feed}}}{100\text{ kg}} \times \underset{\substack{\uparrow \\ \text{desired} \\ \text{CP} \\ \text{level}}}{16\%} = \underset{\substack{\uparrow \\ \text{kg of} \\ \text{SBM}}}{[X\text{ kg}]} \times \underset{\substack{\uparrow \\ \%\text{ CP in} \\ \text{SBM}}}{(49\%)} + \underset{\substack{\uparrow \\ \text{kg of} \\ \text{grain} \\ \text{sorghum}}}{[(100-X)]} \times \underset{\substack{\uparrow \\ \%\text{ CP in} \\ \text{grain} \\ \text{sorghum}}}{(9.5)}
$$

$$
\underset{\substack{\uparrow \\ 100\text{ kg}}}{100\text{ kg}} \times 0.16 = \underset{\substack{\uparrow \\ [X\text{ kg}]}}{[X\text{ kg}]} \times \underset{\substack{\uparrow \\ \%\text{ CP in} \\ \text{SBM}}}{(0.49)} + [(100-X)] \times \underset{\substack{\uparrow \\ \%\text{ CP in} \\ \text{grain} \\ \text{sorghum}}}{(0.095)}
$$

$$16\text{ kg} = 0.49X\text{ kg}$$
$$16\text{ kg} - 9.5\text{ kg} = 0.49X\text{ kg} + 9.5\text{ kg} - 0.095X\text{ kg}$$
$$6.5\text{ kg} = 0.395X\text{ kg} \qquad - 0.095X\text{ kg}$$

$$\frac{6.5\text{ kg}}{0.395\text{ kg}} = X$$

$$\frac{X}{100\text{-}X} = 16.46\text{ kg SBM in 100 kg feed}$$

$$100 - 16.46 = 83.54\text{ kg grain sorghum in 100 kg feed}$$

CHECK: 16.46 kg SBM × 49% CP = 8.07 kg CP
83.54 kg sorghum × 9.5% CP = 7.93 kg CP
16.00 kg CP

METHOD 2: PEARSON'S SQUARE

To use the square, place the desired percent of protein in the center of the square (16% in this example). Place the percent of protein in the two feeds at the left-hand corners of the square (9.5% CP for grain sorghum and 49% for SBM). Then subtract (diagonally) the smaller percent from the larger percent and place the answers in the right-hand corners of the square (16 − 9.5 = **6.5** is placed in the bottom right and 49 − 16 = **33** is placed at the top right).

Grain sorghum 9.5 CP

33 parts

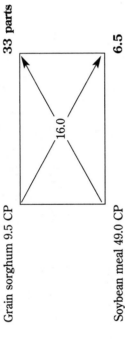

16.0

Soybean meal 49.0 CP

6.5

The figures at the right-hand corners are called parts (33 parts grain sorghum and 6.5 parts soybean meal). The total parts are determined by adding up the individual parts (33 + 6.5 = 39.5 parts). The individual parts are changed to percentages by dividing the individual parts by the total parts (33/39.5 = 83.54% for grain sorghum and 6.5/39.5 = 16.46% for SBM).

Grain sorghum 9.5 CP

33 parts

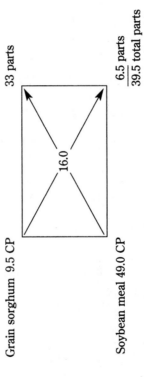

16.0

Soybean meal 49.0 CP

6.5 parts
————————
39.5 total parts

Percent grain sorghum in the diet → (33/39.5) × 100 = 83.54%
Percent SBM in the diet → (6.5/39.5) × 100 = 16.46%

Note that this is the same answer obtained by the algebraic method!

No check is needed.

385

Figure 21.2 *Example—Balancing a diet using two nonfixed and other fixed feed ingredients.*

A swine producer wants to balance a 100-kg diet for sows that contains 16% CP with the following ingredients: 20% ground hay (fixed), 10% oats (fixed), 5% vitamin and mineral premix (fixed), corn and soybean meal (SBM). The CP composition of the feed ingredients is as follows:

FEED INGREDIENT	% CRUDE PROTEIN
Hay	6.0
Oats	12.0
Premix	0.0
Corn	10.0
Soybean meal	44.0

METHOD 1: ALGEBRAIC

First determine the total kg of fixed feed ingredients: 20% ground hay + 10% oats + 5% vitamin and mineral premix = 35% (or 35 kg of a 100-kg ration). Next subtract the total kg of fixed ingredients from the 100-kg final ration: 100 kg − 35 kg = 65 kg. Therefore, the fixed ingredients will make up 35% of the ration, and the corn and SBM will make up the remaining 65%. Let X be the kg of corn and $65 − X$ be the kg of SBM. The formula should be set up like this:

$$
\begin{array}{c}
100\ \text{kg} \\ \downarrow \\ \text{Total} \\ \text{kg}
\end{array}
\times
\begin{array}{c}
16\% \\ \downarrow \\ \text{Desired} \\ \text{CP}
\end{array}
=
\begin{array}{c}
[20\ \text{kg} \times 6\%) \\ \downarrow \\ \text{Hay}
\end{array}
+
\begin{array}{c}
(10\ \text{kg} \times 12\%) \\ \downarrow \\ \text{Hay}
\end{array}
+
\begin{array}{c}
(5\ \text{kg} \times 0\%)] \\ \downarrow \\ \text{Premix}
\end{array}
+
\begin{array}{c}
(X\ \text{kg} \times 10\%) \\ \downarrow \\ \text{Corn}
\end{array}
+
\begin{array}{c}
((65 − X) \\ \downarrow \\ \text{SBM}
\end{array}
\times
44\%)
$$

$$16\ \text{kg} = 1.2\ \text{kg} + 1.2\ \text{kg} + 0\ \text{kg} + 0.1X + 28.6\ \text{kg} − 0.44X$$

$$
\begin{aligned}
16\ \text{kg} &= 31\ \text{kg} \\
0.34X &= 31\ \text{kg} \\
0.34X &= 15\ \text{kg} \\
X &= 44.12\ \text{kg of corn} \\
65 − X &= 65 − 44.12 \\
&= 20.88\ \text{kg of SBM}
\end{aligned}
$$

FINAL DIET AND CHECK:

FEED INGREDIENT	% IN DIET	% CRUDE PROTEIN	KG CP/100-KG FEED
Hay	20.0	6.0	1.2
Oats	10.0	12.0	1.2
Premix	5.0	0.0	0.0
Corn	44.12	10.0	4.41
Soybean meal	20.88	44.0	9.19
Totals	100.00		16.00

METHOD 2: PEARSON'S SQUARE

First determine the amount of crude protein contained in the fixed feed ingredients in a 100-kg diet:

FEED INGREDIENT	IN DIET		% CRUDE PROTEIN		KG CP IN FIXED FEED
Hay	20.0	×	6.0	=	1.2
Oats	10.0	×	12.0	=	1.2
Premix	5.0	×	0.0	=	0.0
	35.0				2.4 kg CP in 35 kg feed

The complete balanced diet should contain 16% CP or 16 kg of CP per 100 kg of diet. The fixed feed ingredients will provide 2.4 kg of CP, so the nonfixed feeds (corn and SBM) must provide the rest of the CP: 16 − 2.4 = 13.6 kg CP. The corn and SBM must provide 13.6 kg of CP in 65 kg of feed. Therefore, the percentage of protein in the 65 kg is (13.6/65) × 100 = 20.92% CP. Using the square to balance with corn and SBM to provide 20.92%, CP is done as follows. Place the desired CP percentage in the center (20.92%) and the CP for corn (10%) and SBM (44%) in the left-hand corners.

Corn 10.0 CP   23.08 parts

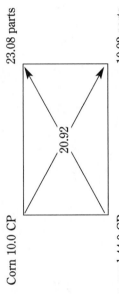

20.92

Soybean meal 44.0 CP   10.92 parts
34.00 total parts

Percent corn in 65 kg of the diet → (23.08/34) × 100 = 67.88%
Percent SBM in 65 kg of the diet → (10.92/34) × 100 = 32.12%

To compute the percentage of corn and SBM in the 100-kg total diet, the above percentages need to be multiplied by 65 kg:

Percent corn in the 100-kg diet → 67.88% × 65 kg = 44.12 kg
Percent SBM in the 100-kg diet → 32.12% × 65 kg = 20.88 kg

The answer obtained by the Pearson's square is the same as with the algebraic method, so no check is needed.

387

Figure 21.3 Example—*Balancing a diet using a fixed ratio of ingredients and other fixed ingredients*

A beef producer found a good buy on wheat and wants to balance a diet that will contain wheat and grain sorghum at a 60:40 ratio in the grain mix. The producer wants to use hay fixed at 20% of the diet and use cottonseed meal (CSM) as the other ingredient. The producer wants a diet balanced for 14% CP. The grain mix will be 60% wheat (13% CP) and 40% sorghum (9% CP). The hay will be fixed at 20% of the total diet and contains 8% CP and CSM contains 41% CP.

METHOD 1: ALGEBRAIC

$$100 \text{ kg} \times 14\% = (20 \text{ kg} \times 8\%) + \{[60\% \times (X \text{ kg} \times 13\%)] + [40\% \times (X \text{ kg} \times 9\%)]\} + [(80 - X) \times 41\%]$$

		Total Kg	Desired CP	Hay		Wheat	Sorghum		CSM	

$$100 \text{ kg} \times 0.14 = (20 \text{ kg} \times 0.08) + \{[0.6 \times (X \text{ kg} \times 0.13)] + [0.4 \times (X \text{ kg} \times 0.09)]\} + [(80 - X) \times 0.41]$$

$$14 \text{ kg} = 1.6 \text{ kg} + 0.078X + 0.036X + 32.8 \text{ kg} - 0.41X$$

$$14 \text{ kg} = 34.4 \text{ kg} + 0.114X - 0.41X$$

$$0.296X = 20.4 \text{ kg}$$

$$X = 20.4 \text{ kg}/0.296 = 68.92 \text{ kg of grain mix}$$

kg of wheat in final 100-kg diet = 68.92 kg × 0.6 (60% wheat) = 41.35 kg
kg of sorghum in final 100 kg diet = 68.92 kg × 0.4 (40% sorghum) = 27.57 kg
kg of CSM = 100 kg − [20 kg (hay) + 68.92 kg (wheat and sorghum)] = 11.08 kg

FINAL DIET AND CHECK:

FEED INGREDIENT	% IN DIET	% CRUDE PROTEIN	KG CP/100 KG FEED
Hay	20.0	8.0	1.6
Wheat	41.35	13.0	5.38
Sorghum	27.57	9.0	2.48
Cottonseed meal	11.08	41.0	4.54
Totals	**100.00**		**14.00**

METHOD 2: PEARSON'S SQUARE

First determine the amount of CP in the fixed feed ingredients: 20 kg hay × 8% CP = 1.6 kg CP in hay. The complete balanced diet should contain 14% CP or 14 kg of CP per 100 kg of diet. Next determine the amount of protein short that needs to be covered by the other ingredients: 14 kg (needed) − 1.6 kg (from hay) = 12.4 kg CP short. There needs to be 12.4 kg of CP in the 80 kg of nonfixed ingredients (100 kg (total ration) − 20 kg (hay) = 80 kg for nonfixed ingredients). This means that to balance the diet, the percentage of CP in the 80 kg is 15.5%.

(12.4 kg short/80 kg) × 100 = 15.5% CP

The protein content of 100 kg of the 60% wheat: 40% sorghum grain mix is:

60 kg wheat × 13% CP =   7.8 kg CP
40 kg sorghum × 9% CP =   3.6 kg CP
                        ──────────
                        11.4 kg CP/100 kg grain mix, or 11.4% CP

Using the Pearson's square to balance for 15.5% CP with the grain mix (11.4% CP) and cottonseed meal (CSM, 41% CP) is done as follows. Place the desired CP percentage in the center (15.5%) and the CP for the grain mix (11.4%) and for the CSM (41%) in the left-hand corners.

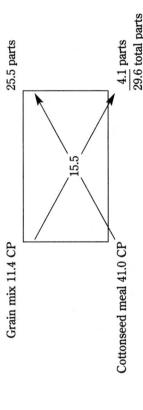

Grain mix 11.4 CP   25.5 parts

                              15.5

                                                     4.1 parts
Cottonseed meal 41.0 CP   ──────────
                                                     29.6 total parts

Percent grain mix = 25.5/29.6 = 86.15% of 80 kg → 68.92 kg grain mix
Percent CSM = 4.1/29.6 = 13.85% of 80 kg → 11.08 kg CSM

The total diet and check:

		AMOUNT	% CP	CHECK
Wheat	0.6 × 68.92 =	41.35 kg	13	5.38
Sorghum	0.4 × 68.92 =	27.57 kg	9	2.48
CSM	=	11.08 kg	41	4.54
Hay	=	20.00 kg	8	1.60
		100.00		14.00

389

Figure 21.4 Example—Balancing a diet for two nutrients and three Pearson's squares.

Sometimes it might be desired to have the exact amounts of two major nutrients, such as CP and energy. This can be accomplished by going through three Pearson's squares. Suppose a final mix with 12% CP and 74% TDN is desired. Corn has 8.8% CP and 81% TDN, CSM has 40.9% CP and 68.6% TDN, and oat hay has 8.1% CP and 53.8% TDN. For this diet, a minimum of three feedstuffs is required. First, utilize two squares and get a mix exact for each nutrient—in this example balance for CP first. One mix will have 12% CP and greater than 74% TDN, and one mix with 12% CP and less than 74% TDN.

Mix 1. 12% CP, >74% TDN.

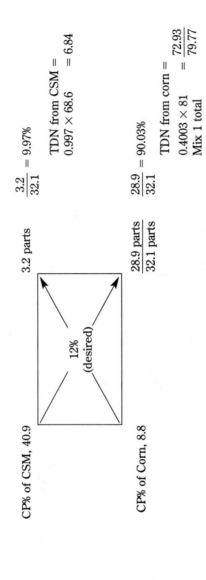

CP% of CSM, 40.9   3.2 parts $\dfrac{3.2}{32.1} = 9.97\%$

                                                         TDN from CSM =
                                                         $0.997 \times 68.6 = 6.84$

CP% of Corn, 8.8 $\dfrac{28.9 \text{ parts}}{32.1 \text{ parts}}$ $\dfrac{28.9}{32.1} = 90.03\%$

                                                         TDN from corn =
                                                         $0.4003 \times 81 = \dfrac{72.93}{79.77}$
                                                         Mix 1 total

Mix 2. 12% CP, <74% TDN.

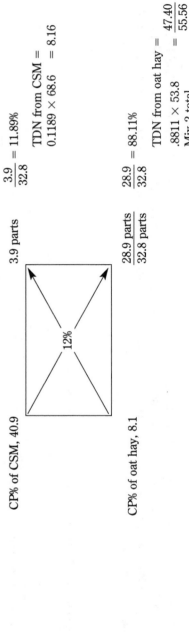

CP% of CSM, 40.9   3.9 parts $\dfrac{3.9}{32.8} = 11.89\%$

                                                         TDN from CSM =
                                                         $0.1189 \times 68.6 = 8.16$

CP% of oat hay, 8.1 $\dfrac{28.9 \text{ parts}}{32.8 \text{ parts}}$ $\dfrac{28.9}{32.8} = 88.11\%$

                                                         TDN from oat hay =
                                                         $0.8811 \times 53.8 = \dfrac{47.40}{55.56}$
                                                         Mix 2 total

Then solve for TDN.

Alignment of equal signs on right side of table OK? Followed old book for style.

390

Mix 3. 12% CP and 74% TDN.

TDN% in Mix 1, 79.77

18.44 parts = 76.17%

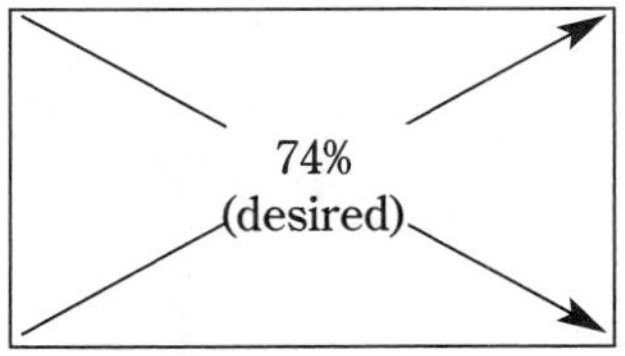

$$
\begin{array}{c}
\boxed{\begin{array}{c} 74\% \\ \text{(desired)} \end{array}}
\end{array}
$$

TDN% in Mix 2, 55.56

$\dfrac{5.77 \text{ parts}}{24.21 \text{ parts}}$ = 23.83%

Calculate ingredient composition:

CSM % in Mix 1, 9.97 × (76.17% of Mix 1 in Mix 3)	=	7.594	
CSM in Mix 2, 11.89 × (23.83% of Mix 2 in Mix 3)	=	2.833	
CSM in Mix 1 and Mix 2.		10.427	

Corn % in Mix 1, 90.03 × (76.17% of Mix 1 in Mix 3) = 68.57

Oat % in Mix 2, 88.11 × (23.83% of Mix 2 in Mix 3) = 21.00

Check CP = (10.427 × 0.409 + 68.57 × 0.088 + 21 × 0.081) = 4.27 + 6.03 + 1.70 = 12.00
Check TDN = (10.427 × 0.686 + 68.57 × 0.81 + 21 × 0.538) = 7.16 + 55.54 + 11.30 = 74.00

391

mind that for this system to work, the number of equations should be equal to or larger than the number of unknowns.

Work through the examples listed in Figs. 21.1 through 21.4 and find the procedure that is easiest for you to follow and utilize. Simple diet formulation can be completed without the use of a computer or specialized program.

Mathematical Programming

Producers, feed companies, consulting nutritionists, and public institutions widely use mathematical programming. The first large-scale public service to provide feed programming and forage testing in the United States was started at the Pennsylvania State University in January 1959. Most states offer feed formulation services through the cooperative extension service or through state departments of agriculture.

With the Pearson's square method and the simultaneous equations methods, the mix obtained is of a predetermined nutrient content (i.e., CP percentage, energy level, etc.). The amounts of the ingredients considered for the mix are therefore fixed. In reality, the percentage of protein required or the amount of energy in a diet is considered a minimum, and therefore the final mix should have "at least" the required amount of the nutrient but could have more than required. Sometimes the nutrient is specified within a range, that is, at least some minimum quantity but less than a maximum quantity. Unfortunately, the Pearson's square and the simultaneous equations methods cannot handle inequalities or ranges, and both methods are independent of price. When price is considered, developing a low-cost diet is done by trial and error.

Diet formulation becomes much more complicated when there are multiple nutrient requirements, multiple feed sources, price considerations, or requirements expressed as being greater than and/or less than some level. With these added complexities, the formulation should be completed by mathematical programming. This technique, when properly used, helps to achieve both a nutritionally balanced diet and economic optimization because it allows for simultaneous consideration of economical and nutritional parameters.

To formulate diets, nutritionists should have a good knowledge of diet specifications, be familiar with interpretation of results, and be concerned with the solution process, not the mechanics of the mathematical solution of the linear programming matrix. Diet formulation by mathematical programming should be treated as an interactive process in which the nutritionist should verify, interpret, and iterate, if necessary, all diet formulas.

With the availability of personal computers, mathematical programming can be performed quite easily if the programming is done properly. Spreadsheets can be used to formulate diets with nutrient and composition databases or with simple input of the requirements of the animal and input of the feedstuff chemical composition. Spreadsheets are available on the Web (www.asft.ttu.edu/ansc3307/ttu-diet.xls). We define the term *programming* as the planning of "economic" activities for the sake of optimization, subject to some constraints. A large number of computer programs exist for the solution of mathematical programming problems, so a detailed explanation of the mathematics involved is not given here.

Least-Cost Formulation

Feed costs represent 50 to 80% or more of the cost of production; therefore, diet formulation using least-cost techniques is used extensively to minimize the cost of the diet. Once the appropriate equations are set up properly, any linear programming software package can be used to solve the diet formulation problem. After a solution is obtained, it is up to the nutritionist to evaluate and verify that the solution conforms with nutritional knowledge (i.e., that the diet will be palatable to the animal, that it does not contain nutrients and other substances in toxic amounts, and that the effects of other factors related to the specific nutrient requirements and feed consumption patterns of the animals for which the diet has been formulated will be optimal).

Formulating Premixes and Supplements

Premixes are used in animal nutrition in small amounts to supplement vitamins, mineral elements, protein, amino acids, probiotics and other feed additives. As with the diet, premixes can be formulated using any of the methods described previously. Proper analysis is required to obtain the desired final premix results, and prices are necessary when premixes are to be least cost. When premixes have only a few constituents, the Pearson's square or simultaneous equations can be used.

❏ SUMMARY

- Feed represents a large percentage of the cost of production; therefore, diet formulation is a very important aspect of animal production. Diet formulation allows the nutritionist to develop a diet that can be eaten and utilized by the animal to provide an economically feasible level of production. For diets that contain only a few feedstuffs and in which the nutrient requirements are fixed, the techniques are quite simple. In these cases the Pearson's square or algebraic equations can be utilized. When diets become more complex, the use of the computer is desired to complete the diet formulation. The advent of the personal computer has allowed the nutritionist to formulate diets without an in-depth understanding of the mathematics involved. A key point to remember is that diet formulation by any technique or program requires interaction and revision. Just because a mathematical solution is reached doesn't mean the animal will eat the diet or that it will be a suitable feed. Once formulated, the diet should be reviewed, and if not satisfactory it should be reformulated until the diet has all the desired nutritional specifications.

❏ REFERENCES

International Feed Institute. 1984. IFI Tables of Feed Composition. P.V. Fonnesbeck, H. Lloyd, R. Obray, and J. Romesburg (eds.). International Feedstuffs Institute, Utah State University, Logan, UT.

National Research Council. 1982. *United States–Canadian Table of Feed Composition,* 3rd ed. National Academy Press, Washington, DC.

22

Photograph courtesy USDA, ARS, USMARC.

Beef Cattle

CALVIN L. FERRELL,

A PRIMARY GOAL IN beef cattle production is to maximize production efficiency; this goal generally involves efforts to optimize production versus production costs. It should be emphasized that maximum efficiency often does not occur at maximum production, and conversely maximum efficiency frequently does not occur at minimum costs. Moreover, maximum biological efficiency may not equate to maximum economic efficiency. Although many variables affect production costs and efficiency of production (Heitschmidt et al., 1996), the major cost of beef cattle production is incurred in providing feed or nutrients to the animals. Several estimates indicate that over 65% of the total production costs were associated with providing feed to the cattle. In this chapter, we provide an overview of current information on practical feeding of beef cattle during various phases of the life cycle.

All cattle require adequate supplies of good-quality water. Water is needed for regulation of body temperature and acts as a solvent necessary for transport of nutrients, metabolites, and waste products, joint lubrication, nervous system cushioning, sound transport, and eyesight. The water requirement reflects needs for accretion in body tissues (e.g., growth, pregnancy) and for milk production plus compensation for water lost through respiration, perspiration, urination, and fecal excretion. The requirements for water are influenced by environmental temperature and humidity, rate and composition of gain, pregnancy, lactation, activity, type of diet, feed intake, and breed of cattle. Because feeds contain water and oxidation of nutrients produces water, not all water needs must be provided by drinking. Water requirements are affected by many factors and thus are difficult to define. However, if an adequate supply of water is provided, animals will drink to fill their needs. NRC (1996) has provided guidelines indicating approximate water intake of cattle.

Restriction of water intake reduces feed intake and consequently cattle productivity. Cattle may consume water from surface water sources such as lakes, ponds, and streams and from ground water sources such as wells.

Both water availability and water quality are important in maintaining adequate water consumption. For example, several studies have shown that animal performance is improved when a clean source of water replaces a stagnant, muddy pond. In addition, grazing behavior or grazing pattern can be influenced substantially by location of water sources. Cattle tend to graze most heavily near a source of water and generally move as nearby food supplies diminish.

❑ COW-CALF

In a cow-calf enterprise, it is important to develop strategies for the feeding and management of the female to ensure the biological and economic sustainability of the enterprise. In general, one must ensure that adequate nutrients are available to replacement heifers so that they grow to adequate size and attain puberty at a desired age. Adequate nutrients should be available to cows so that they maintain themselves, produce healthy, viable offspring, produce adequate milk for those offspring, and rebreed within a desired interval. More specifically, feeding strategies for females should consider environmental influences such as temperature and humidity, as well as differences in metabolism and maintenance needs among animals. They must incorporate an understanding that cattle must cope with large changes in nutrient demand during the annual production cycle, and must accommodate large, asynchronous fluctuations in availability and quality of grazed forages. The variable ability of animals to select nutritious plant components should be considered.

Feeding strategies should also be based on an understanding of the underlying control of reproductive and other productive processes and the vulnerability of those processes to dietary deficiencies or excesses. That understanding provides the basis for making decisions regarding whether and when to supplement, composition of the supplement, and amount of supplement that is required, or whether supplementation is economically justified. Implicit in the formulation of feeding and management strategies is the need to understand the extent to which the maternal body can

serve as a reserve to buffer dietary imbalances, deficiencies, or excesses while still maintaining the integrity of the productive processes. Several reviews are available in the area of nutrition–reproduction of beef cattle for readers interested in greater detail (Dunn and Moss, 1992; Patterson et al., 1992; Robinson et al., 1999).

Replacement Heifers

Management of replacement heifers generally focuses on factors that enhance physiological processes that promote puberty and subsequent productivity. Age at puberty is an important production trait in most species of farm animals. In beef cattle, many currently used management systems require that heifers be bred at 14 to 16 months of age to calve at approximately 24 months of age. In addition, the breeding season is frequently restricted. It is important, especially in these types of management systems, that heifers attain puberty at a young age and conceive early in the breeding season. It is generally believed that maximum productive performance can be achieved in replacement heifers if they attain puberty by 12 to 13 months of age. Conception rate increases as the number of estrus cycles prior to breeding increases, and heifers that conceive early in the first breeding season have a greater probability of weaning more and heavier calves during their lifetimes. In addition, Vizcarra et al. (1995) observed that in Angus × Hereford heifers, those that reached puberty first were the last to return to anestrus following nutritional restriction. Those authors suggested that selection for heifers for earlier puberty may decrease feed costs for maintenance and increase overall reproductive performance when, as cows, they are exposed to environmental or nutritional stress.

In cattle, seasonal conditions of early and late postnatal periods influence the timing of puberty onset (Schillo et al., 1992). Autumn-born heifers attain puberty at younger ages than spring-born heifers, and exposure to spring–summer temperatures and photoperiods during the second six months of life reduces age at puberty, regardless of season of birth. Pre- and postweaning nutrition, however, is the major cause of variation in age at puberty. Rate of body weight gain is the primary indicator of nutritional status. The extent to which age at puberty may be influenced by nutritional status during the developmental period was documented by Rege et al. (1993), who reported age at first calving as great as 2527 days in a low-intensity production system. They suggested this could be reduced to 40 months by supplementation during the dry season. Age at first estrus in dairy heifers decreased from 498 to 252 days, as the daily postweaning growth rate increased from 0.45 to 0.85 kg (Sejrsen and Purup, 1997). Angus-Hereford crossbred heifers fed to gain 0.27, 0.45, or 0.68 kg/day after weaning reached puberty at average ages of 433, 411, and 388 days, respectively (Short and Bellows, 1971). Although these differences seem relatively small, pregnancy rates after a 60-day breeding season were 50, 86, and 87%, respectively.

Observations that weight is less variable than age at puberty has led several authors to conclude that size or weight is more important than age in determining time of onset of puberty. Several studies are available, however, indicating that weight, as well as age, at puberty is influenced by level of nutrition. Those studies indicate that heifers fed to gain less rapidly attain puberty at an older age, but lighter weight than those fed to gain more rapidly.

Both age and weight at puberty differ substantially among breeds of cattle, and both are affected somewhat by heterosis. Within beef breeds, those having larger mature size tend to attain puberty at an older age and heavier weight. However, *Bos indicus* heifers generally reach puberty at an older age than *Bos taurus* heifers. These differences in general reflect the rates at which different breeds mature physiologically. Heifers from higher milk-producing breeds are generally younger at puberty than those from breeds having lower milk production. Some of this difference is attributable to

greater preweaning rates of gain by calves from high milk-producing dams.

Numerous data indicate that neither age nor weight is a reliable indicator of reproductive development, but that threshold values for each of these traits must be attained before puberty can occur. This concept, in essence, states that heifers should be fed to attain a "target" weight at a given age. In general, heifers of typical beef breeds (e.g., Angus, Charolais, Hereford, Limousin) are expected to attain puberty at about 60% of mature weight. Heifers of dual purpose or dairy breeds (e.g., Braunvieh, Friesian, Gelbvieh, Holstein, Red Poll, Simmental) tend to attain puberty at a younger age and lower percentage of mature weight (55%) than heifers of the more typical beef breeds. Conversely, *Bos indicus* breeds (e.g., Brahman, Nellore, Sahiwal) tend to attain puberty at older ages and greater percentage of mature weight (65%) than European beef breeds. Generally, most *Bos taurus* breeds will attain puberty by 14 months of age if fed to achieve the above percentages of mature weight at or before that age. Threshold ages may be greater for some *Bos indicus* breeds. As an example, an Angus heifer is expected to have a mature weight of 600 kg. Puberty is expected at about (0.6 × 600) 360 kg. If weaning weight at 205 days is 230 kg, the heifer should be fed to gain (360 − 230) 130 kg or 0.74 kg/day for 175 days after weaning to attain puberty by 380 days (12.5 months) of age. It should be noted that the production environment may have a major effect on phenotypic expression of mature size. Environmental effects on mature size should be considered when estimating expected mature size and determining target weights and rates of growth.

It might be assumed from the preceding brief review that maximum growth rates during the development period are desirable. In addition to increasing feed costs, overfeeding heifers during development detrimentally affects expression of behavioral estrus (i.e., increased incidence of "silent heat"), decreases conception rate, decreases embryonic and neonatal survival, and increases calving difficulty. It has been clearly shown for both cattle and sheep that high planes of nutrition during the pre- and peripubertal period, when there is a spurt in mammary growth, can decrease mammary development and subsequent milk production.

The effects of early nutrition on milk production capability have important implications for beef production systems. High growth rates during the preweaning period are important to achieve high weaning weights, and weaning weight is often used as a breeding selection criterion. However, high rates of growth of heifers prior to weaning (resulting from high milk production of their dams) are associated with reduced milk production capability during their first as well as subsequent lactations (Baker, 1980). This effect results in intergenerational cycling of lactation ability.

This discussion assumes that heifers have access to diets that are balanced adequately to achieve the desired rates of growth. However, in addition to the general relationships among age, weight, growth rate, and plane of nutrition, several specific nutrients have been shown to influence onset of puberty. The source and amount of energy is generally the first consideration in determining a balanced ration for beef cattle. When energy is limiting, supplemental protein will, in part, be used to provide energy until energy needs are met, or energy and protein are equally limiting. Protein deficiency results in delayed puberty in most species. However, within ranges of protein in the diet compatible with an adequate growth rate, dietary protein does not appear to substantially affect the onset of puberty. Other observations suggest that the effects of protein on puberty are primarily related to its relationship to growth. In cattle, intake of amino acids is not generally of concern, provided protein intake is adequate. Amino acids supplied from ruminal microbes are generally adequate in terms of quantity and balance to meet the animals' needs for reproductive functions. Grossly excessive protein intake may be detrimental to reproduction, due to the toxic effects of high circulating levels of ammonia.

The roles of minerals and vitamins in reproduction have been reviewed by Hurley and Doane (1989) and NRC (1996). All essential vitamins and minerals are required for reproduction because of their roles in cellular metabolism. In addition, many of these nutrients have specific roles and requirements in reproductive tissues, and these requirements may change during puberty, estrus cycles, pregnancy, parturition, and lactation. The synthesis of B-vitamins and vitamin K by rumen microbes is generally adequate in forage-based diets to meet the animals' requirements. Vitamins of particular concern to beef cattle include vitamin A and its metabolites and precursors, vitamin D and its metabolites, and vitamin E. High-quality forages contain large quantities of vitamin A precursors and vitamin E, but supplements may be needed with prolonged feeding of harvested forages because vitamins tend to be lost with time in storage. Vitamin D is synthesized by animals exposed to direct sunlight and is found in large amounts in sun-cured forages, but may be of concern during the winter in extremely northern or southern climates, or if animals are confined indoors for extended periods of time. Several minerals, including Ca, P, Zn, Cu, Co, Mo, Se, and Mg, have essential cellular and subcellular roles in animal function and metabolism and are involved in numerous aspects of reproductive function. Deficiencies or toxicities may occur, depending on location and the minerals provided to plants by soil.

Once the female has attained puberty and has become pregnant, the primary goals are to feed the heifer to achieve desired rates of gain and to provide adequate nutrients for pregnancy. Heifers should be of adequate size to deliver a healthy, live calf and in adequate body condition to rebreed within a reasonable time period. As a general "rule of thumb," heifers should be fed to achieve about 80 to 85% of mature body weight at first parturition. Alternative nutrition and management strategies may be imposed (Clanton et al., 1983; Park et al., 1987; Freetly et al., 2001) to achieve appropriate end points.

Cow

An adequate assessment of the nutritional status of the cow is necessary so that optimal rates of reproduction and production may be maintained. Although numerous experimental approaches are available for this purpose, the options are rather limited for practical application. Visual evaluation of the cows' condition provides a useful, practical approach for general assessment of the cows' nutritional status. With experience, body condition (or fatness) can be visually evaluated with a fair degree of accuracy. A scoring system similar to those reported by Wagner et al. (1988) and NRC (1996) is illustrated in Table 22.1. Body condition is most generally associated with energy and protein status, but may reflect deficiencies of other essential nutrients as well.

Postpartum Reproduction.   In cattle, the interval from calving to first ovulation (postpartum interval) is of concern because if this interval exceeds 80 days, which frequently occurs, a 365-day

TABLE 22.1 System of body condition scoring of beef cattle.

CONDITION SCORE	PERCENTAGE IN BODY		APPEARANCE OF COW
	FAT	PROTEIN	
1	3.8	19.4	**Emaciated**—Cow is extremely thin with no fat detectable by sight or touch; all bone structure easily visible.
2	7.5	18.8	**Very thin**—Cow appears thin, but tail head and ribs are less prominent; individual spinous processes are sharp to touch, but some tissue cover exists along the spine.
3	11.3	18.1	**Thin**—Ribs are individually identifiable but not sharp to the touch; there is palpable fat along spine and over tail head with some tissue cover over dorsal portion of the ribs; no visible fat in brisket.
4	15.1	17.0	**Borderline**—Individual ribs are not visually obvious; spinous processes can be identified individually by palpation but feel rounded rather than sharp; some fat cover over ribs, transverse processes, and hip bones.
5	18.9	16.8	**Moderate**—Cow has generally good overall appearance; upon palpation, fat cover over ribs feels spongy and areas on either side of tail head now have palpable fat cover.
6	22.6	16.1	**Good**—Ribs fully covered and not noticeable to the eye; hindquarters plump and full; noticeable sponginess to covering of ribs and on each side of the tail head; firm pressure required to feel transverse processes; some fat in brisket.
7	26.4	15.4	**Fat**—Cow appears fleshy and obviously has considerable fat; very spongy fat cover over ribs and around tail head; some fat around vulva and crotch; brisket is full.
8	30.2	14.8	**Very fat**—Cow very fleshy and overconditioned; spinous processes almost impossible to palpate; large fat deposits over ribs, around tail head and below vulva; brisket is distended with fat; fat cover thick with patchiness likely.
9	33.9	14.1	**Obese**—Cow is obviously extremely fat and patchy; bone structure is not easily seen or felt; tail head and hips buried in fat; mobility may be impaired.

Source: Based on Wagner et al. (1988) and NRC (1996).

calving interval cannot be maintained. Calves born to cows with extended postpartum intervals of anestrus are younger and weigh less at weaning. In addition, cows with extended postpartum intervals become cyclic late in the breeding season, or fail to exhibit estrus at all, and thus have less opportunity to rebreed.

Duration of the postpartum anestrus interval in beef cows is affected by parity, season, calf presence and suckling, breed, dystocia, and nutritional status (Short et al., 1990). The interval from calving to first ovulation increases with parity and is longer in cows calving in the spring than in cows calving in the fall. Cows with suckling calves have longer intervals from parturition to ovulation than milked or nonlactating cows. This effect appears to result primarily from calf contact rather than from suckling per se. In addition, calf stimulus interacts with nutritional status of the cow such that thin cows or cows on low-energy diets are more sensitive to the calf stimulus than cows on high-energy diets, and consequently have longer intervals from calving to first ovulation. Early weaning calves, short-term weaning, and partial weaning, such as once per day suckling, have reduced the postpartum interval in beef cows. The successful uses of these approaches require high levels of management and other inputs.

Cows that are obese at parturition have greater incidence of metabolic, infectious, digestive, and reproductive disorders than cows that are in moderate body condition. However, the principal cause of poor reproductive performance or extended postpartum anestrus interval is general undernutrition due to feed shortage or poor quality feed. There is a general consensus that the effects of prepartum nutrition are more important than postpartum nutrition in influencing the length of the postpartum anestrus interval in cattle. In addition, body condition at calving appears to be more important than pattern of body condition or body weight change during the prepartum period. Cows that are in good body condition at calving (i.e., condition score 5 or above) are affected little by either pre- or postpartum weight changes. Postpartum interval is increased by prepartum weight loss in cows that are in thin to moderate body condition at calving. This problem is exacerbated by deficient levels of feed postpartum. The effects of poor body condition at calving can only be partly overcome by high levels of feeding postpartum. It is difficult for cows, especially first-calf heifers, to consume enough energy during early lactation to compensate for poor body condition at calving. Nonetheless, the importance of body condition at calving becomes less pronounced with time after calving, and postpartum dietary energy plays a more prominent role.

In species such as the cow, which usually give birth to a single progeny, nutrition has little, if any, effect on ovulation rate but certainly can influence whether an animal ovulates and expresses estrus. Short-term (1–3 weeks) increases in energy and/or protein intake (flushing) result in increased incidence of ovulation in thin cows, but generally do not totally overcome the adverse effects of chronic deficiencies. Flushing has little effect in cows with adequate body condition. Response to superovulation is less in heifers fed low levels of energy than in heifers fed adequately. Although nutritional influences on fertilization or embryonic mortality have not been well defined, it is evident that nutrition has important influences on conception rate, as indicated by services per conception. In cattle, as in other farm animal species, an intermediate, optimal level of energy and protein intake promotes maximum conception rate. Excessive under- or overfeeding tends to depress conception rate.

Gestation.   Once an animal becomes pregnant, the next goal is to ensure that pregnancy is maintained to term with the production of a live, healthy offspring. Meeting the nutrient requirement of the pregnant female is important to ensure adequate nutrient supply for normal growth and development of the fetus, to ensure the female is in adequate body condition to calve and lactate adequately, as well as to rebreed within 80 days after calving, and to insure, in the case of the 2- and 3-year-old heifer, adequate nutrients for continued growth. Although the nutrient demands of the early embryo are quantitatively minute, they are very specific in both qualitative and temporal expression. Failure to meet these needs can have detrimental effects on the establishment of pregnancy. At the other extreme, rapid growth of the fetus in late pregnancy can create demands for macronutrients in excess of that which can be supplied by forage diets. Mobilization of maternal tissues is frequently required to supplement dietary derived nutrients during late pregnancy and early lactation (Bell et al., 2004).

Estimates of the requirements of specific nutrients for pregnancy are based on their rates of accretion in gravid uterine tissues, which in turn are calculated from equations describing the temporal accumulation of nutrients in those tissues. The resulting values provide direct estimates of net requirements, which are then transformed into dietary amounts using appropriate factors for efficiency of absorption and utilization. It is generally assumed that the macronutrient needs for pregnancy are proportional to birth weight of the calf. Thus, it is assumed that factors that affect calf birth weight have a proportional effect on nutrient requirements for pregnancy. Factors that affect birth weight include fecundity, gender, parity, breed or breed cross, heat or cold stress, and maternal nutrition. In general, birth weight of each calf born as a twin is about 25% less than one born as a single, but total weight of twin fetuses is about 150% of that of a single fetus. Birth weight of males is about 7% greater than females, and weight of calves born to 2-, 3-, and 4-year-old cows, respectively average 8, 5, and 2% less than weight of calves born to cows 5 years of age and older. Birth weight is increased by cold, but decreased by heat or poor maternal nutrition; the magnitude of these effects depends upon the severity of the conditions. Of the factors affecting calf birth weight, breed or genotype of the sire, dam, and/or calf generally have the greatest influence. Mean birth weights of several breeds available in the United States are shown in Table 22.2.

Fetal growth follows an exponential pattern, as shown in Fig. 22.1. Fetal weight increases slowly during early gestation but quite rapidly during the latter stages. In cattle, about 90% of birth weight is achieved during the latter 40% of gestation. In addition to rapid changes in weight, percentage of dry matter, protein, and fat increases twofold or more during this interval. Mineral content of the

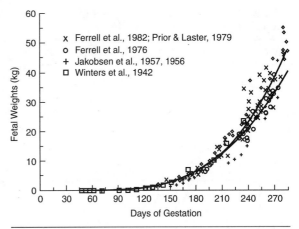

Figure 22.1 *Relationship between fetal weight and day of gestation in cattle. The lower curve is the curve describing the relationship obtained by Ferrell et al. (1976), whereas the upper curve is the best fit curve for the combined data set.*

TABLE 22.2 Typical birth weights (kg) of several breeds and breed crosses common in the United States.

BREED	BIF	AFRC	MARC
Angus	31	26	35
Brahman	31	—	41
Braford	36	—	—
Brangus	33	—	—
Braunvieh	—	—	39
Charolais	39	43	40
Chianina	—	—	41
Devon	32	34	—
Galloway	—	—	36
Gelbvieh	39	—	39
Hereford	36	35	37
Jersey	—	25	31
Limousin	37	38	39
Longhorn	—	—	33
Main Anjou	40	—	41
Nellore	—	—	40
Piedmontese	—	—	38
Pinzgauer	33	—	40
Polled Hereford	33	—	36
Red Poll	—	—	36
Sahiwal	—	—	38
Santa Gertrudis	33	—	—
Salers	35	—	38
Shorthorn	37	32	39
Simmental	39	43	38
South Devon	33	42	38
Tarentaise	33	—	38

Sources: Beef Improvement Federation (1990); Agricultural Food Research Council (1990); Meat Animal Research Center, from data reported by Cundiff et al. (1988) and Gregory et al. (1982) and are for progeny from sire breed shown bred to mature Angus or Hereford cows.

fetus and associated tissues follow similar patterns. The rapid changes in both fetal weight and composition during this interval are indicative of increased nutrient needs. Estimates of energy, protein, Ca, and P needs for pregnancy are shown in Table 22.3. Severe energy or protein malnutrition during late gestation may substantially affect fetal growth, and consequently birth weight in cattle.

An example of the influence of cow feeding level, as indicated by cow condition score, on calf birth weight is shown in Fig. 22.2. In general, calf birth weight is not greatly altered by maternal nutritional status within a broad range of condition scores from about 4 to 7. This reflects, in part, the tremendous capacity of the maternal system to buffer dietary nutritional inadequacies. Birth weight is reduced, however, if cows are overly fat (condition score > 7) or exceedingly thin (condition score < 4). In addition to negative influences on birth weight and neonatal survival, inadequate food intake during pregnancy is associated with weak labor, increased calving difficulty (dystocia), reduced milk production and growth of the calf, extended postpartum interval, and poorer rebreeding performance. Effects of undernutrition during pregnancy are more severe in first-calf heifers than in mature cows. Although not frequently reported in beef cattle, a serious consequence of overfeeding in dairy cows during gestation has become known as fat cow syndrome. It is characterized by increased occurrence of metabolic, infectious, digestive, and reproductive dis-

TABLE 22.3   Estimates of net energy (NE_m), net protein, Ca, and P requirements at different stages of gestation in cattle.[a]

DAY OF GESTATION	ENERGY, MCAL/DAY	PROTEIN, G/DAY	CA, G/DAY	P, G/DAY
130	0.33	13	0.56	0.47
160	0.63	25	1.62	1.18
190	1.17	46	3.76	2.42
220	2.03	80	6.62	4.04
250	3.33	136	8.42	5.20
280	5.17	223	6.84	4.64

[a]Birth weight of 38.5 kg was assumed.

orders. Reproductive disorders include retained fetal membranes, metritis, delayed uterine involution, and delayed rebreeding.

Lactation.   Milk production by the beef cow is difficult to assess. In contrast to the dairy cow, which is generally milked two or more times per day by machine, the beef cow is typically in a pasture or range environment, and milk is consumed by the suckling calf. Milk yield estimates vary depending on method used. In addition, milk yields differ based on the genetic potential and/or breed of the cow, age of the cow, capacity of the calf, nutritional status of the cow, thermal environment, and stage of lactation. Composition, as well as total yield, may vary. Typical patterns of milk production by beef cows, as determined by weigh-suckle-weigh procedures, are shown in Fig. 22.3. Peak milk yield of beef cows with suckling calves measured by these procedures typically range from about 5 kg per day to about 14 kg per day. This range encompasses most of the reported values for beef cows with suckling calves. Peak production generally occurs at about 8.5 weeks after parturition, which is about 2 to 4 weeks later than typically observed in milked cows. Limited capacity of the calf to consume milk during the first few weeks after birth is likely a major contributor to

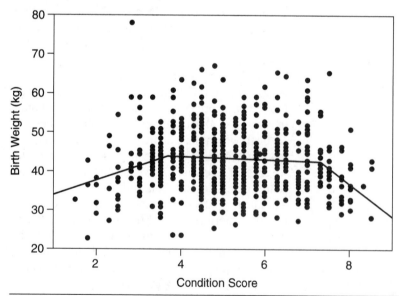

Figure 22.2   *Relationship between calf birth weight and cow condition score. The 383 observations shown are from nine breeds of mature cows that were fed at varying levels of feed intake to achieve a wide range in condition.*

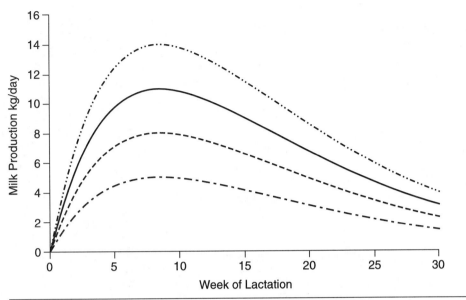

Figure 22.3 *Representative lactation curves for beef cattle.*

this difference. Nutrient requirements for lactation (Table 22.4) follow the pattern of milk production.

Estimates of the daily net energy requirements (NE_m) during a production year of a mature, 515-kg cow that gives birth to a 38.5-kg calf, and has a peak milk yield of 11 kg per day are shown in Fig. 22.4. The major portion (typically 70 to 75%) of the energy required by a cow during the year is used to maintain body weight and condition, that is, for maintenance. The annual requirement for maintenance (Fig. 22.4) is about 3040 Mcal NE_m as compared to 330 Mcal for pregnancy and 1100 Mcal for lactation. During the production year, minimal requirements for energy, as well as for other nutrients, occur around the time of weaning, increase rapidly during the latter stages of pregnancy, and peak at the time of peak milk production (about 2 months postcalving). Requirements at that time are, for the typical beef cow, about twofold higher than at the time of weaning. The first-calf heifer should also be growing at that time. In addition, the cow or heifer must rebreed while producing

TABLE 22.4 Pattern of nutrient requirements for milk production in beef cows.[a]

WEEK OF LACTATION	ENERGY, MCAL/D	PROTEIN, G/D	CA, G/D	P, G/D
3	5.32	252	8.15	6.79
6	7.48	354	11.46	9.55
9	7.88	373	12.07	10.06
12	7.39	350	11.32	9.43
15	6.49	307	9.94	8.28
18	5.47	259	8.38	6.98
21	4.48	212	6.86	5.72
24	3.60	170	5.52	4.60
27	2.85	135	4.37	3.64
30	2.19	105	3.36	2.80

[a]Assumed production at peak lactation is 11 kg/day of milk containing 4.0% fat, 3.4% protein, 8.3% solids not fat, 1.2% Ca, and 1.0% P. Energy is expressed as NE_m, and protein is expressed as net protein.

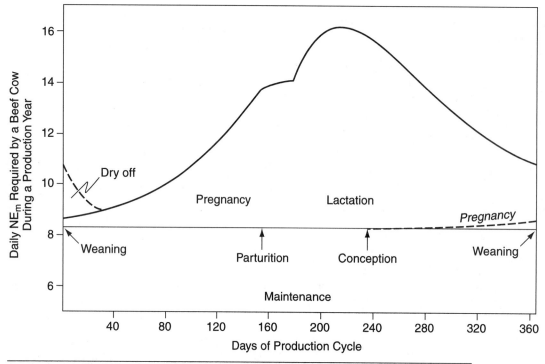

Figure 22.4   *Representation of the combined net energy requirements for a typical beef cow during a production year.*

nearly maximal amounts of milk—that is, while her nutrient requirements are nearly maximal. It is difficult to obtain required nutrients from forages to meet these requirements, much less achieve substantial body weight gains during this phase of the production cycle. This reemphasizes previous statements to the effect that it is usually advantageous to ensure the cow is in adequate body condition at calving. Heifers not only have higher nutrient requirements than mature cows, but are generally of lower rank in the herd and are less aggressive. As a result, it is often advisable to separate heifers or young cows from the cow herd and to provide additional nutrients as required.

Maintenance.   In the example provided in Fig. 22.4, the daily maintenance requirement was assumed to be constant at 77 Kcal NE_m/kg$^{0.75}$. Maintenance requirements, however, vary with weight, breed or genotype, age, season, temperature, physiological state, and previous nutritional status. In addition, variable amounts of activity associated with foraging or grazing may result in large differences in animal energy expenditures (NRC, 1996). A positive relationship exists between the

genetic potential for production (e.g., growth rate, milk production) and maintenance requirements. Commonly accepted values for daily NE_m requirements in thermoneutral environments, and with minimal activity, are 65 to 90 for nonlactating beef cows, 85 to 110 for lactating beef cows, and up to 140 Kcal/kg$^{0.75}$ for lactating dairy cows. Increases in NE_m allowances may be required to account for energy expenditures for walking and grazing activities (10 to 50% depending on terrain and forage quality and abundance) and in climatic conditions in which the effective ambient temperature lies above or below the thermoneutral zone (NRC, 1996). Adjustments needed for temperature are highly dependent on physiological state and level of feed consumption, for these directly affect the need for increased heat to maintain body temperature or, conversely, the amount of heat the animal must dissipate. Genotype differences in maintenance energy requirements are positively related to differences in production potential in terms of calf birth weight and milk production. Thus, breeds with genetic potential for high productivity in a good nutritional environment may have less advantage or be at a disadvantage in a nutritionally (forage quantity and quality) or environmentally (e.g., high heat and humidity) restrictive environment. This is often forgotten when animals with high genetic potential for production are introduced into adverse nutritional environments. It is also important to recognize, in this context, that *Bos taurus* cattle selected in temperate environments are less tolerant of high environmental temperatures and humidity than tropically adapted *Bos taurus* (e.g., Tuli) or *Bos indicus* (e.g., Brahman) cattle.

Body Weight and Composition Change. The ruminant animal typically undergoes a cyclical pattern of live weight and body composition change due to seasonal fluctuations in forage availability and quality, and due to the effects of photoperiod and temperature on voluntary food intake, nutrient partitioning, and basal metabolism. Consequently, the contribution of mobilized body tissue to the metabolic pool varies throughout the year, and by physiological state, energy balance, and N balance. Freetly and Nienaber (1998) demonstrated that dietary energy and nitrogen use efficiencies in cows that lost and then regained body weight was equivalent to, or greater than, efficiencies in cows fed to maintain weight. Those results were consistent with numerous reports demonstrating that energy expenditures (e.g., basal metabolism and maintenance) decline in animals with reduced feed intake. They are also consistent with numerous reports documenting increased efficiency of nutrient use upon increasing feed intake of previously restricted animals (e.g., compensatory gain). Those results demonstrated that weight cycling is not necessarily an inefficient process, as once believed, but reflects the animals' adaptability to constantly changing nutritional and environmental conditions. Those results suggest that opportunities may exist to better capitalize on the animals' adaptive capacity to improve efficiency of production without adversely affecting productivity.

There is ongoing controversy regarding composition of weight change in cows. The NRC (1996) concluded that body weight loss or gain of mature cows contains 5.82 Mcal. This value equates to 570 g fat and 81 g protein per kilogram of live weight change. Relationships (Fig. 22.5) of body fat or protein to shrunk live weight, calculated from NRC (1996), show that a mature cow weighing 600 kg at condition score 5 contains about 96 kg fat and 85 kg protein (1890 Mcal total energy), but only 14.7 kg fat and 75.6 kg protein (570 Mcal) at condition score 1. Conversely, the same cow at con-

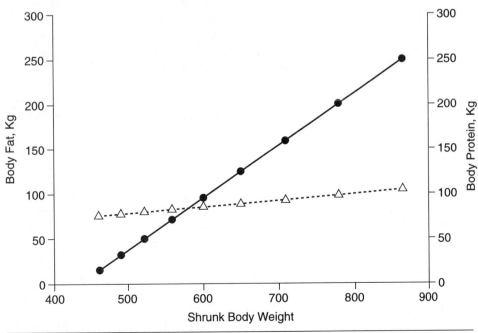

Figure 22.5 *Relation of body fat or body protein to shrunk body weight.*

dition score 9 contains about 249 kg fat and 104 kg protein (2937 Mcal). The NRC (1996) model indicated body weight change necessary to change body condition score increased as condition score increased. For example, for the cow described above, change from condition score 1 to 2 was associated with a shrunk body weight change from 459 to 488 kg (29 kg difference). In contrast, change from condition score 8 to 9 resulted in body weight change from 779 to 866 kg (87 kg difference). Others have concluded, as did Robinson et al. (1999), that the energy content of body weight change varies, depending on the condition or empty body weight of the animal and the rate at which body weight changes. Additional research efforts are needed to better describe the dynamics of body compositional changes in mature cows; there is virtually no information available for young (2- to 4-year-old) cows.

Regardless of which concept described above is correct, application of either results in the conclusion that cows in poor or thin body condition have less available energy and protein reserves than cows in good body condition. Differences in quantity of reserves may, in addition to the direct effects on amount of energy and amino acids available for mobilization, influence the pattern of nutrients mobilized, as described by Robinson et al. (1999). In addition, increased body condition, associated with increased subcutaneous fat deposits, is expected to provide increased insulation and, as a result, reduce susceptibility to cold stress (but increase susceptibility to heat stress). The animal in good body condition is expected to have greater capability to respond to nutritional and other environmental challenges.

Regardless of the physiological state discussed in the above sections, negative consequences have been associated with both inadequate and excessive body condition. Maximum productivity of beef cows, as indicated by weight of calf weaned per cow in the herd, is consistently associated with moderate body condition (typically, 4.5 to 6 body condition score). It is important that those levels of body condition are achieved at critical times during the production cycle, as previously described. However, it may not be biologically or economically justified to maintain those levels of body condition throughout the annual production cycle.

In meeting the nutrient requirements of grazing cattle, the condition of the pasture or range and the quality and amount of available forage must be considered. Forage intake is certainly reduced if the quantity of pasture is inadequate. If forage availability is limited, supplementation with forage or concentrates may be required, especially at critical times during the production cycle. When supplementation is needed, it is often more economical to provide small amounts of concentrates than supplemental forages. Intake is generally reduced on poor quality or mature forages. Poor quality forages typically provide inadequate amounts of energy, protein, or both energy and protein. Supplementation with the most limiting nutrients generally results in increased forage intake. These effects are most frequently seen with high protein, high energy, and protein supplements, but may be observed with supplementation of deficient minerals as well. In general, if protein or energy is limiting, supplementation with about 1 kg of a high-protein or high-energy supplement results in increased forage intake, but greater amounts of supplement may displace forage in the diet. Providing supplementation two or three times per week appears to produce animal responses similar to those of daily supplementation.

Preweaning Growth. The rumen of the calf is not fully developed for several weeks following birth. Prior to the development of the rumen, dietary and nutrient requirements of the calf parallel that of a nonruminant. During early growth, the calf consumes increasing amounts of forage or other available feedstuffs, in addition to milk. Feedstuffs consumed stimulate rumen development and provide an increasing proportion of the calf's nutrient needs. Depending on feeds available, energy, protein, minerals, or all may limit growth. Of primary concern are amount of milk available, forage quality and availability, supplementation, weaning, and marketing strategies.

Very young calves often cannot consume all of the milk that a cow is capable of producing. However, as the calf grows, its intake capacity and nutrient needs for maximum growth increase, but milk production begins to decline. In addition, in many management situations, the period of increasing calf requirements coincides with reduced forage availability and quality, in addition to the normal decrease in the dam's milk production. Where possible, choice of calving season, such that forage quality is highest when the young calf can use it and the cow's nutrient needs are the greatest (i.e., around peak milk production), provides an optimum match of forage resources to nutrient needs. In the young calf, forage quality may be more important than availability because of the calf's limited capacity to consume forage. However, as the calf develops, milk consumption decreases and forage consumption increases. Availability, as well as quality, becomes increasingly important.

In many production systems, the suckling calf's requirements for maximum growth are not met when the calf's growth potential is high, especially when nutrient requirements and forage production are not in synchrony. Several management options are available to ensure that adequate nutrients are available to allow the calf to express its genetic potential for growth. Calves may be weaned early and placed on high-quality grazed or harvested forage. Early weaned calves have relatively high protein requirements; therefore, protein supplementation may be needed. If weaning is not an economical or practical option, calves may be supplemented separately from the cow by creep feeding or creep grazing. With appropriate feeders or gates, an appropriate concentrate, or improved pasture may be provided when milk availability becomes limiting to calf growth. Advantages of creep feeding include greater efficiency from feeding the calf directly rather than feeding the cow, improved cow condition, heavier weaning weights, economical gains, reduced weaning stress, and capability to deliver feed additives. Conversely, creep feeding poses several disadvantages, notably, reduced use of the dam's milk production capabilities, management difficulties, particularly under many range conditions, and potentially excessive fat deposition. The benefits of increased weaning weights from creep feeding are diminished if ownership is retained, especially when calves are subsequently managed as stockers or are "backgrounded" on growing diets because growth rate will likely be reduced as compared to calves that were not creep fed. In addition, excessive fatness in heifers to be returned to the breeding herd results in diminished milk production and productivity as cows.

Bulls

Much less information is available on the nutrient needs of the bull during growth and development for breeding purposes than is available regarding the female. Sexual behavior, consisting of libido or sex drive and mating ability, and semen quantity and quality are the primary factors of concern in the bull. Malnutrition can adversely affect all of these components. Negative effects are more evident in young than in mature bulls and are more likely to be permanent. The mature bull is remarkably resilient to nutritional stress, and nutritionally related infertility problems are not often encountered. Severe overfeeding or underfeeding and deficiencies of specific nutrients, especially vitamin A, are the most common causes of impaired reproduction.

Puberty. In the male, as in the female, malnutrition, particularly low-energy intake, reduces rate of growth and delays puberty. If severe enough, sperm output may be permanently impaired. Nutrient deficiency is associated with reduction in testicular weight, secretory output of accessory sex glands, and sperm concentration and motility. Conversely, the reproductive potential of young bulls may be impaired by overfeeding. Scrotal circumference, epididymal sperm reserves, and semen quality may be reduced in bulls fed excessive energy.

Bulls fed vitamin A-deficient diets have delayed puberty, reduced libido, and reduced spermatogenesis. Vitamin A deficiency is associated with degeneration of the germinal epithelium, resulting in reduction or cessation of spermatogenesis. Vitamin E deficiency may result in testicular degeneration. Of the minerals, zinc appears to be of primary importance for reproductive function. Deficiency may result in retarded testicular growth, atrophy of tubular epithelium, reduced pituitary gonadotropin output, and reduced androgen production. Other essential minerals, including Cu, Co, and I, are associated with depressed libido and reduced semen quality.

Mating Behavior. Mating behavior is an important aspect of male reproduction, for it has a direct relationship to the number of females mated. In general, unless the male is severely deficient in energy or protein, there is little effect on sexual or mating responses. Other specific nutrient deficiencies may result in lowered physical ability to mate in addition to the specific effects noted above. Overly fat males may become less physically able to inseminate a female.

Nutrient Requirements. Considerably less attention has been given to the proper feeding of bulls than to that of females or steers, primarily because bulls typically constitute a low proportion (3 or 4%) of the cattle herd. Developing bulls are commonly fed to achieve nearly maximal rates of gain postweaning. However, this varies depending on specific production and marketing goals. Energy and protein requirements (NRC, 1996) are about 15% greater for growing bulls than for steers or heifers of the same size. This is primarily a result of the higher protein content of the body and of body weight gain. Conversely, bulls have more protein and water and less energy in body weight gain. Consequently, energy required per unit live weight gain is lower for bulls than for steers or heifers. Requirements for other nutrients change approximately in proportion to the needs of energy and protein. In addition to increased maintenance requirements per unit body size, mature bulls are approximately 60% larger in size than mature cows of the same breed. These effects combined result in mature bulls having daily maintenance requirements about twofold greater than mature cows of the same breed. During the breeding season, increased activity of breeding bulls may increase energy needs by an additional 50% or more, depending on the number of cows to be mated, terrain, and so on. As a consequence of mating activities, grazing time is reduced in bulls during the breeding season. During this time interval weight loss commonly occurs. Thus, it is important to ensure that bulls are in adequate body condition at the beginning of the breeding season.

❏ STOCKER, BACKGROUNDING, AND GROWING PROGRAMS

The primary goal of a stocker program is to utilize available forages for inexpensive growth of calves after weaning. Forage sources typically include vegetative/growing annuals (e.g., wheat pasture), growing perennial grasses (e.g., Bermuda grass), dormant perennial grasses (e.g., native grasses), and/or crop residues (e.g., corn residue). Expected rates of gain vary, depending on forage source and production goals, but they generally range from about 0.5 to 1.0 kg/day. Supplementation varies with forage source and production goals, but frequently involves only minerals, and perhaps protein, that are inadequate in the basal forage to support desired rates of growth. The optimum source and amount of supplemental protein differs depending on the nature and amount of protein as well as the amount of readily fermentable carbohydrate contained in the basal forage source. If sufficient readily fermentable carbohydrate is available, a nonprotein nitrogen source such as urea may provide a portion of the required protein. Conversely, if the protein available is rapidly fermentable in the rumen, a less rapidly degraded protein may provide a more effective source. The level and type of protein provided should reflect the feedstuff base and rumen fermentation conditions, such that adequate levels of amino acids are provided to the animal to allow maximal growth with the available energy.

Backgrounding programs typically are of relatively short duration and have the main objective of ensuring that calves are healthy, are eating feed from a bunk, and the like, before going to either the stocker or growing or the finishing program. Backgrounding programs typically depend on harvested feeds but usually have target gains similar to those of a stocker program. In the past, these programs have depended primarily on harvested forages as the primary feed source, but more recently producers have relied more heavily on grain diets to achieve the desired rates of gain. Inadequate feed consumption is a common concern. As a result, in order to get adequate protein intakes, preformed plant protein sources are preferred and levels of 16 to 18% are recommended. Vitamins A, D, and E are frequently given as intramuscular injections.

Growing programs are most commonly used in feedlots and are designed to provide sufficient nutrients to the animal for maximal protein and skeletal growth rates, while allowing minimal fat deposition. The duration of a growing program may range from less than 100 to more than 200 days, depending on the age and breed or type of calves. Yearling cattle, from stocker or similar types of programs, usually bypass the growing program and go directly to finishing programs. Energy may be provided as a harvested forage or forage-grain mixtures (generally equivalent to about a 50% concentrate diet) or from limit feeding a high-concentrate diet. The proportion of concentrate and/or level of feeding may vary depending on production goals, prices of diet ingredients, and type of cattle. Protein, making up 12 to 14% of the diet, is usually provided by a combination of preformed plant protein and urea, with urea providing up to one-third of the protein equivalent. Cattle from growing programs are commonly switched directly to finishing programs by reducing forage and increasing the grain component, or, if grown by limiting intake of a high-concentrate diet, by increasing the feed amounts provided.

❏ FINISHING PROGRAMS

The finishing phase may be as short as 60 days or may cover the entire period after weaning. For most purposes, animals are fed to achieve maximal rates of gain during this period and are usually fed to a predetermined target degree of fatness (e.g., 60% of the cattle in the pen grading low choice). Expected rates of gain are about 1.3 kg/day or greater, depending on cattle type, condition, weather conditions, and so on. Rates of gain and feed efficiency are generally improved by growth-promoting implants such as those containing estradiol or tremblelone acetate. Feeding programs have varied over time and may vary extensively among different localities, depending on availability and price of feedstuffs. In most large feedlots, complete, mixed diets are used. Feedlots in the Great Plains, which typically encompass about 85% of the cattle on feed, depend heavily on corn, corn byproducts, and, to a lesser extent, sorghum grains as primary energy sources in diets containing minimal forage. Supplemental protein is supplied primarily by urea, but some preformed plant protein such as soybean or cottonseed meal may also be provided. West Coast feedlots utilize byproduct and opportunistic feeds much more extensively, whereas Midwest feedlots depend more heavily on ensiled forages (e.g., corn silage).

In the growing or finishing animal, 30% or more of the feed consumed is used for maintenance (NRC, 1996). For typical crosses of British breeds of steers or heifers, daily maintenance requirements average 0.077 Mcal NE_m per kg body weight raised to the 0.75 power (metabolic body size). This value is less (10%) for *Bos indicus* cattle but is greater (up to 20%) for dairy or dual purpose breeds of cattle, is greater (15%) for intact males, and is influenced by season and weather conditions as well as by previous nutrition of the animal.

Growing–finishing animals typically consume the feed equivalent of 2.5 to 3.0 times maintenance when fed high-quality, balanced diets to appetite. On feedlot finishing diets, feed intake is thought to be limited by the animal's ability to utilize energy. As the forage concentration of diets increases, which is typically the case on growing or stocker programs, the ability of the animal to consume bulk or volume of feed increasingly limits intake of energy. Feed intake may differ substantially among breeds or genotypes of cattle. *Bos indicus* cattle typically have lower ad libitum intakes of high-quality diets than *Bos taurus* cattle. Conversely, dairy or dairy crosses generally consume greater amounts of feed than more typical beef breeds. Feed intake is reduced by high environmental temperature. This effect is accentuated in fat animals because of their reduced ability to dissipate heat. Feed intake is reduced for a short time after a sudden change in weather but is often increased for a day or two before the weather change. Feed intake is reduced by the addition of monensin, and perhaps other ionophores, to high-energy diets, but rates of gain are not adversely affected because efficiency of utilization is increased. An additional benefit of ionophore addition to high-concentrate diets is reduced digestive and metabolic problems such as acidosis.

The terms *compensatory gain* or *catch-up growth* are used to describe the greater than normal growth rate that occurs in animals in which growth has previously been limited by nutrition or other environmental factors, upon realimentation or relieving the factors that have limited growth. Maintenance is reduced in animals in which growth has been nutritionally limited. Feed intake, and perhaps efficiency of nutrient use above maintenance of animals, is greater for a period of time following growth restriction and accounts for a large part of the compensatory gain response. For example, yearling cattle typically consume more feed and gain weight more efficiently than calves of the same weight.

❏ SUMMARY

- Beef cattle primarily harvest and utilize feed resources that are not usable or poorly utilized by nonruminants and convert those feedstuffs to high value products. Less than 15% of the feed resources utilized for beef cattle production are of potential value for direct human consumption. Beef cattle are produced in a diversity of environments and management systems. Goals of beef cattle production and management, as a result, are generally directed toward optimization of production vs. input costs within a production environment rather than maximizing production. Production environment is, thus, a major consideration in developing management strategies and contributes to segmentation of the beef cattle industry into cow-calf, stocker/backgrounder, and feedlot segments. Each segment, to a large extent, has evolved to more effectively utilize available feed resources. Certain principles, some of which have been described in this chapter, are inherent in beef cattle production. However, an often underrated attribute of beef cattle is their ability to buffer nutritional or other environmental insults and rapidly adapt to changing environmental conditions. Thus, in many situations alternative strategies are available to achieve similar biological results, but may produce substantially different economical outcomes.

❏ REFERENCES

Agricultural Food Research Council (AFRC). 1990. Agricultural Food Research Council Technical Committee on Responses to Nutrients, Report No. 5. Nutrient Requirements of Ruminant Animals: Energy. *Nutr. Abstr. Rev.* (Ser. B) **60**: 729.

Baker, R. L. 1980. The role of maternal effects on the efficiency of selection in beef cattle—a review. *Proc. N. Z. Soc. Anim. Prod.* **40**: 285–303.

Beef Improvement Federation (BIF). 1990. *Guidelines for Uniform Beef Improvement Programs,* 6th ed. Stillwater, OK, Oklahoma State University.

Bell, A. W., C. L. Ferrell, and H. C. Freetly. 2004. Pregnancy and fetal metabolism. In *Quantitative Aspects of Digestion and Metabolism* (2nd ed.), J. Dijkstra (ed.). Marcel Dekker, New York.

Clanton, D. C., L. E. Jones, and M. E. England. 1983. Effect of rate and time of gain after weaning on the development of replacement beef heifers. *J. Anim. Sci.* **56**: 280–285.

Cundiff, L. V., R. M. Koch, and K. E. Gregory. 1988. Germ plasm evaluation in cattle. Beef Research Progress Report No. 3. USDA-ARS-ARS-71, pp. 3–4.

Dunn, T. G., and G. E. Moss. 1992. Effects of nutrient deficiencies and excesses on reproductive efficiency of livestock. *J. Anim. Sci.* **70**: 1580–1593.

Ferrell, C. L., W. N. Garrett, and N. Hinman. 1976. Growth, development and composition of the udder and gravid uterus of beef heifers during pregnancy. *J. Anim. Sci.* **42**: 1477–1489.

Ferrell, C. L., D. B. Laster, and R. L. Prior. 1982. Mineral accretion during prenatal growth of cattle. *J. Anim. Sci.* **54**: 618–624.

Freetly, H. C., C. L. Ferrell, and T. G. Jenkins. 2001. Production performance of beef cows raised on three different nutritionally controlled heifer development programs. *J. Anim. Sci.* **79**: 819–826.

Freetly, H. C., and J. A. Nienaber. 1998. Efficiency of energy and nitrogen loss and gain in mature cows. *J. Anim. Sci.* **76**: 896–905.

Gregory, K. E., L. V. Cundiff, and R. M. Koch. 1982. Characterization of breeds representing diverse biological types: preweaning traits. Beef Research Program Progress Report No. 1. USDA-ARS, ARM-NC-21, pp. 7–8.

Heitschmidt, R. K., R. E. Short, and E. E. Grings. 1996. Ecosystems, sustainability and animal agriculture. *J. Anim. Sci.* **74**: 1395–1405.

Hurley, W. L., and R. M. Doane. 1989. Recent developments in the roles of vitamins and minerals in reproduction. *J. Dairy Sci.* **72**: 784–804.

Jakobsen, P. E. 1956. Protein requirements for fetus-formation in cattle. Proc. 7th *Int. Congr. Anim. Husb.* **6**: 115–126.

Jakobsen, P. E., P. Havskov Sorensen, and H. Larsen. 1957. Energy investigations as related to fetus formation in cattle. *Acta. Agric. Scand.* **7**: 103–112.

National Research Council (NRC). 1996. *Nutrient Requirements of Beef Cattle.* National Academy Press, Washington, DC.

Park, C. S., G. M. Erickson, Y. J. Choi, and G. D. Marx. 1987. Effect of compensatory growth on regulation of growth and lactation: response of dairy heifers to a stair-step growth pattern. *J. Anim. Sci.* **64**: 1751–1758.

Patterson, D. J., R. C. Perry, G. H. Kiracofe, R. A. Bellows, R. B. Staigmiller, and L. R. Corah. 1992. Management considerations in heifer development and puberty. *J. Anim. Sci.* **70**: 4018–4035.

Prior, R. L. and D. B. Laster. 1979. Development of the bovine fetus. *J. Anim. Sci.* **48(6)**: 1546–1553.

Rege, J.E.O., R. R. von Kaufmann, W.M.M. Mwenya, E. O. Otchere, and R. I. Mani. 1993. On-farm performance of Bunaji (White Fulani) cattle. 2. Growth, reproductive performance, milk offtake and mortality. *Anim. Prod.* **57**: 211–220.

Robinson, J. J., K. D. Sinclair, R. D. Randel, and A. R. Sykes. 1999. Nutritional management of the female ruminant: mechanistic approaches and predictive models. In *Nutritional Ecology of Herbivores,* Proc. Vth International Symposium for Nutrition of Herbivores, pp. 550–608, H. G. Jung and G. C. Fahey, Jr. (eds.). American Society of Animal Science. Savoy, IL.

Schillo, K. K., J. B. Hall, and S. M. Hileman. 1992. Effects of nutrition and season on the onset of puberty in the beef heifer. *J. Anim. Sci.* **70**: 3994–4005.

Sejrsen, K., and S. Purup. 1997. Influence of puberty and feeding level on milk yield potential of dairy heifers: a review. *J. Anim. Sci.* **75:** 828–835.

Short, R. E., R. A. Bellows, R. B. Staigmiller, J. G. Berardinelli, and E. E. Custer. 1990. Physiological mechanisms controlling anestrus and infertility in postpartum beef cattle. *J. Anim. Sci.* **68:** 799–816.

Short, R. E., and R. A. Bellows. 1971. Relationships among weight gains, age at puberty and reproductive performance in beef heifers. *J. Anim. Sci.* **32:** 127–131.

Vizcarra, J. A., R. P. Wettemann, and D. K. Bishop. 1995. Relationship between puberty in heifers and the cessation of luteal activity after nutritional restriction. *Anim. Sci.* **61:** 507–510.

Wagner, J. J., K. S. Lusby, J. W. Oltjen, J. Rakestraw, R. P. Wettemann, and L. E. Walters. 1988. Carcass composition in mature Hereford cows: estimation and effect on daily metabolizable energy requirement during winter. *J. Anim. Sci.* **66:** 603–612.

Winters, L. M., W. W. Green, and R. E. Comstock. 1942. Prenatal development of the bovine. *Minn. Exp. Stat. Tech. B.* No. **151,** pp. 3–47.

23

Dairy Cattle

MICHAEL J. VANDEHAAR

Photograph courtesy North Carolina State University.

HUMANS HAVE USED THE capability of the ruminant digestive system to process fibrous plant material into the high-quality nutrients of milk for millennia. At one time, the cow obtained most of her nutrients from grass or stored forage and produced enough milk for perhaps one family. But with genetic selection and progressive management, a modern high-producing dairy cow will produce about 40 to 50 kg of milk per day in early lactation, and production as high as 60 kg is not uncommon. The current world record Holsteins produce over 30,000 kg of milk in a year—that's almost 100 kg per day averaged over a lactation.

The typical cow has a maintenance requirement of about 10 Mcal Net Energy for Lactation (NE_L) per day. Each kg of milk takes an additional 0.7 Mcal NE_L. Thus, the cow producing 45 kg of milk per day needs four times as much energy as she needs for her maintenance requirement alone. The elite cow producing 90 kg/day needs seven times as much energy as she needs for maintenance.

Not only must the cow receive enough energy, but the energy must come in the proper form. Energy is provided with cell-wall carbohydrate (fiber), non-fiber carbohydrates (starch and sugar), protein, and fat. The major challenge in feeding high-producing dairy cows is to find the right balance of these nutrients to promote rumen health and to maximize feed energy intake and nutrient flow to the mammary gland for milk synthesis. In addition, the cow needs minerals and vitamins and may benefit from one or more of several feed additives. The science of dairy cattle nutrition has grown tremendously in the past century, but dairy cattle feeding still involves considerable art. This chapter provides the basics of the science and art of feeding dairy cattle.

413

❑ THE DAIRY COW LIFE CYCLE

Ideally, a dairy cow calves for the first time at 2 years of age and then once every year thereafter. A dairy calf is usually separated from its mother soon after birth and is fed mostly milk until it is weaned between 5 and 8 weeks of age. During this time, the most important goal is to keep the calf healthy and vigorous. After weaning, the goal is to promote sound growth at a reasonable cost, with breeding occurring at 13 to 15 months. For Holsteins, the target body weight gain should be ~800 g/day until breeding and then 900 g/day once bred. For minimal calving problems and highest milk production, Holstein heifers should weigh ~630 kg just before calving (~570 kg after calving). Current data suggest that first-lactation milk yield will be reduced 70 kg for every 10 kg body weight below this optimum.

The lactation curve for a typical mature Holstein is shown in Fig. 23.1. Traditionally, the ideal lactation length was considered to be 305 days with a calf born once every 365 days. Most cows are "dried-off" (milking is stopped) ~60 days prepartum and will remain "dry" (nonlactating) until parturition. Daily milk yield peaks at 40 to 60 days postpartum and then gradually declines. Feed intake also increases as lactation progresses but usually peaks later than milk yield. Thus, most cows are in negative energy balance during the first 60 days postpartum. In other words, they cannot eat enough to meet the high-energy demands for milk production, and they consequently mobilize body tissues to make up for this deficit. During the first month of lactation, one-third of the cow's energy needs may be derived from body stores. Some loss of body stores in early lactation and replenishment in later lactation are desirable for maximization of lactation efficiency and profitability, but excessive loss of body condition will impair fertility and prolong the calving interval. With gestation being 280 days, the cow must conceive at 85 days postpartum to maintain a 365-day calving interval. Conception generally occurs much later than 85 days for most dairy cows, so the calving interval on most farms is 13 or 14 months, with lactation lengths of 320 to 360 days being more common. A high-producing cow with fertility problems may be milked for 400 or even 500 days after calving.

The idea that cows should calve once every 12 months is based partly on the desirable season of calving for pasture-based dairy farming in temperate climates. In addition, with a 12-month calving interval, the cow produces one calf per year and spends a greater proportion of her time at peak lactation than if she had a longer calving interval. Some have called this practice into question and have proposed that a longer calving interval (up to 18 months) might be more profitable for modern dairy farms using confined feeding systems and

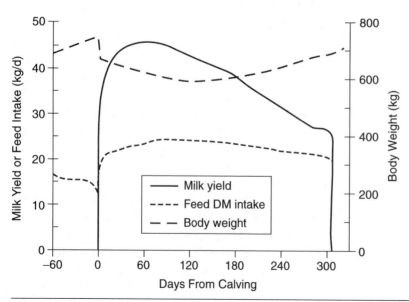

Figure 23.1 *Typical lactation curve for a Holstein cow.*

lactation enhancers such as bovine somatotropin (bST), which enhances the persistency of lactation. Because most health problems occur in the first two weeks after calving, the economic disadvantages of less milk per day during lactation and fewer calves per cow may be outweighed by fewer health problems, less difficulty getting cows bred in a timely manner, and fewer days dry relative to days being milked. Some also have proposed that a dry period less than 60 days might be just as cost-effective as the standard 60 days.

❏ NUTRITIONAL GOALS IN THE DAIRY INDUSTRY

The major goal for most dairy farms is to maximize profits. Because feed accounts for half of all costs on a dairy farm, many farmers are tempted to lower feed costs, especially when feed prices are high. However, feed for lactating cows is obviously not a frivolous expense but an investment. Good dairy farmers continually seek feed sources that cost less but yield the same returns. Often they are successful, but sometimes they are not, and without proper nutrition cows are unable to achieve their genetic potential for milk production. Milk yield is one of the most important factors in determining profitability of dairy cows, and high milk production is almost always more important for high profitability than is low feed cost (VandeHaar, 1998). A cow's level of milk production is determined by (1) the ability of the mammary gland to produce milk, (2) the ability of the cow to provide the mammary gland with nutrients, and (3) the ability of the farmer to manage and care for the cow.

Another important goal in farming is to practice good stewardship. There are four major areas to consider in agricultural stewardship. A good steward in dairy farming is one who (1) is environmentally friendly, (2) makes efficient use of the earth's natural resources, (3) produces quality milk and meat, and (4) practices good animal husbandry. Nutrition impacts each of these stewardship goals. Excess feeding of N and P contributes to air and water pollution as ammonia emissions, N contamination of ground water, and P-induced eutrophication and oxygen consumption of surface waters. Proper nutrition is also important in using the earth's resources efficiently; higher milk production is associated with a greater portion of feed nutrients being converted to milk (Fig. 23.2). Nu-

trition, though having a less important role in milk quality, can alter the fatty acid profile of milk fat. Finally, nutrition will impact animal health and well-being. Poor nutrition is a particular problem during stressful situations such as weaning and calving. Animals that are fed properly have fewer metabolic diseases and better immune function.

❏ THE BIOLOGY OF LACTATION

Mammary Biology

The udder of a cow contains two types of tissue, the parenchyma and the extraparenchymal fat pad. The parenchyma contains the epithelial cells that produce milk during lactation and is surrounded by the fat pad. The ability of the mammary gland to produce milk is largely dependent on the number of these epithelial cells (Tucker, 1987). The number of epithelial cells is determined both by genetics and by the environment during mammary development. Mammogenesis, or mammary development, occurs primarily in three phases of allometric growth (when the gland grows at a faster rate than does other body tissues). The first phase begins at about 2 months of age and ends around the time of puberty at 7 to 10 months of age. During this time, the parenchyma extends outward from the teat into the mammary fat pad in a "broccoli-like" fashion and forms the ductal foundation for later mammary development. Feeding excess energy before puberty will impair mammogenesis and decrease subsequent milk production. Exactly what constitutes "excess" energy is debatable, but most scientists agree that diets that promote growth rates up to 900 g/d are acceptable. In early pregnancy, the gland again undergoes allometric growth, and finally the amount of parenchymal tissue will double around the time of calving. Feeding excess energy does not seem to impact these later phases of mammogenesis. At the end of mammogenesis, the gland is structurally ready to produce milk. The alveoli are lined with epithelial cells and connected to ductules that join to form the ducts that will carry milk to the gland cistern and on out the teat (Fig. 23.3).

Lactogenesis is the initiation of lactation. During this transformation, major changes in gene expression and protein synthesis occur within the milk-secreting cells. Alpha-lactalbumin synthesis is turned on, which initiates lactose synthesis.

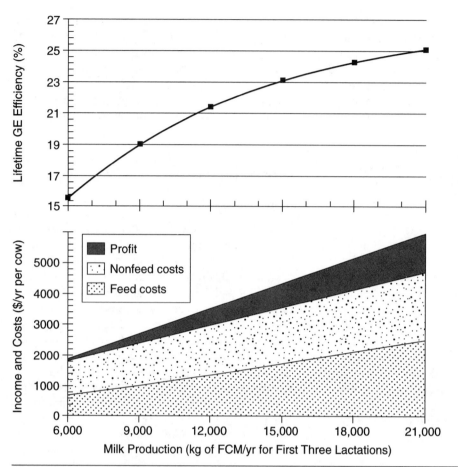

Figure 23.2   *Projected changes in the gross energetic (top panel) and profitability (bottom panel) with increasing production of 4%-fat-corrected milk (FCM).*

Casein synthesis is initiated, as well as several enzymes involved in fatty acid synthesis. Galactopoiesis is the synthesis of milk during an established lactation. Nutrition has little impact on lactogenesis but certainly impacts galactopoiesis. Milk synthesis is the predominant metabolic priority for a lactating cow, especially in early lactation (first 2 months). The gland will continue to take up metabolites from blood throughout the day and use them to synthesize milk components and secrete milk into the alveoli until the pressure within the alveoli builds up. Most cows are milked twice per day, which relieves the pressure and promotes continued milk synthesis. More frequent milking increases milk yield by 10 to 20%, and the decreased alveolar pressure is likely part of this response. Temporal changes in nutrient inflow into blood throughout the day have little impact on to-

tal milk production, but severe nutrient shortages will decrease milk production over time. Mammary involution is the gradual loss of lactational capacity and culminates in the transformation back to a nonlactating state. Inadequate energy or protein during mid- or late lactation (100 or more days after parturition) can advance the involution process and decrease milk production for the duration of the lactation. Involution usually will not be completed without cessation of milking for several days. A severe nutrient or water shortage will aid in this final step of involution at dry-off.

Metabolism during Lactation

The principal organic components of milk are lactose, triglycerides, and proteins. To make these components, the ruminant mammary gland uses

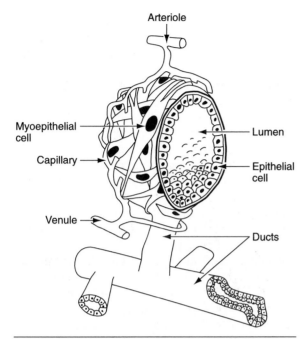

Figure 23.3 *The mammary alveoli is a microscopic saclike structure that is the basic unit for milk secretion. The epithelial cells that line it synthesize and secrete milk into the lumen of the sac. With proper stimulation, the myoepithelial cells that surround the sac contract and squeeze the sac so that milk travels down the ducts and eventually out the teat. Drawing courtesy of John VanderHear, Eaton Rapids, MI.*

glucose, acetate, ketones, fatty acids, and amino acids. Milk production will be greatest and most efficient when the optimal amounts of each metabolite are supplied to the mammary gland. These metabolites serve as the building blocks and the fuel for synthesis of milk components (Table 23.1).

A shortage of nutrients in the maternal diet can be overcome, at least in the short run, by mobilization of nutrients from maternal body stores. Milk production will be impaired if the shortage is severe enough or maternal supplies are depleted. Nutrient supply to the mammary gland seems quantitatively important only if demands of the mammary gland are not met. Infusing glucose into the mammary blood supply of a fasted cow will improve milk yield dramatically, but providing extra glucose, amino acids, or acetate to an already well-supplied mammary gland generally does not increase milk synthesis. The mechanisms by which inadequate maternal nutrition impairs lactation are likely twofold: direct effects of nutrient shortage on the mammary gland and indirect effects on the mammary gland through nutritional modulation of the endocrine system. As the milk-producing potential of the mammary gland is enhanced through normal breeding or through biotechnological approaches, the nutrition of the cow becomes a more important determinant of production.

Glucose Metobolism. Lactose is the predominant carbohydrate of milk, and lactose synthesis is the major osmotic regulator of total milk production. Lactose synthesis requires glucose, and the cow is adapted to preserve glucose for this function. In the mammary gland of ruminants, glucose is not converted to fatty acids, almost no glucose is oxidized in the citric acid cycle (TCA or Krebs cycle), and only half of the NADPH needed for fatty acid synthesis is produced by oxidation of glucose. Instead, acetate is the building block for fatty acids, the major fuel for ATP production, and supplies

TABLE 23.1 Milk components and their precursors in cattle.

MILK COMPONENT	CELLULAR METABOLITES	METABOLITES NEEDED FROM BLOOD	PRIMARY SOURCE	MAJOR FEED COMPONENT
Proteins	Amino acids	Amino acids	Absorbed AA, Muscle AA	Protein
Lactose	Glucose, Galactose	Glucose	Absorbed propionate	Starch, fiber
Triglycerides	Fatty acids	Triglycerides	Absorbed TG	Lipid
		NEFA	Adipose TG	
		Acetate	Absorbed acetate	Fiber, starch
		β-Hydroxybutyrate	Absorbed butyrate	Sugar, Fiber
	Glycerol	Glucose		
	Cellular fuels			
	ATP	Acetate		
	NADPH	Acetate, glucose		

half the NADPH via the isocitrate pathway. Some glucose is converted to glycerol for fatty acid esterification. In other body tissues, acetate and ketone bodies serve as the major fuel sources. Bovine milk is ~5% lactose, so a cow producing 50 kg of milk converts ~2.5 kg of glucose to lactose, uses ~1.1 kg of glucose for other mammary metabolic functions, and uses ~1.4 kg of glucose for other body functions. Thus, she needs ~5 kg of glucose daily, but because most carbohydrate is fermented by ruminal microbes, glucose absorption is low. In addition, much of the glucose that is absorbed in the small intestine is metabolized by gut tissues so that very little glucose enters the bloodstream from the gastrointestinal tract of a cow. Instead, the cow relies on gluconeogenesis to meet over 90% of her glucose needs. This major and continuous function of the liver is evidenced by the constantly high activity of liver gluconeogenic enzymes. Thus, in contrast to nonruminants, the rate of gluconeogenesis in ruminants is greatest after a meal and is directly proportional to absorption of glucogenic compounds. The major precursors for gluconeogenesis are propionate, lactate, glycerol, and amino acids, with estimates for gluconeogenic carbon at 50 to 60% propionate, ~10% gut lactate, ~10% endogenous lactate, 5 to 10% glycerol, and 5 to 20% amino acids. Of these, proprionate is the most important, so cows with high glucose demands benefit from diets that promote rumen proprionate production. The contribution of amino acids to gluconeogenesis is not clear; some studies have found as much as 40% of glucose comes from amino acids. Most gluconeogenesis occurs in the liver; hence, a healthy liver is essential for high milk production.

Protein Metabolism. The major milk proteins of ruminants are casein, β-lactoglobulin, and α-lactalbumin. The mammary gland takes up an excess of essential amino acids and insufficient nonessential amino acids. Some essential amino acids are deaminated and used for synthesis of nonessential amino acids. Amino acids are not major fuels for lactating mammary gland. Protein synthesis in mammary tissue is similar to that of other tissues. The amino acid profile of individual milk proteins is predetermined by the genetic code, not by dietary constraints. Thus, a deficiency of a single amino acid relative to its requirement by the gland will decrease total synthesis of protein. In ruminants, the amino acid that seems most limit-

ing for milk synthesis is methionine. Lysine may be limiting in some situations, but protein produced by ruminal microbes during fermentation is relatively high in lysine, and thus the gland typically receives enough lysine if total protein nutrition is adequate.

A lactating cow producing 50 kg of milk secretes 1600 g of milk protein per day; this is equivalent to ~8 kg of muscle tissue accretion per day. To produce this much milk, the cow must absorb 3 to 4 kg of amino acids daily. During most of lactation, the cow is in body protein equilibrium, and those amino acids that are not captured in milk are catabolized. Catabolism of amino acids is the result of (1) absorption of protein with an amino acid profile that does not match requirements, (2) the need for glucogenic compounds and metabolic fuels, and (3) the extent to which proteins are "turned over." For a cow producing 50 kg of milk per day, the amount of body protein that is broken down and resynthesized each day is likely 2 to 4 kg, and many of the amino acids from endogenous proteolysis are not recaptured in proteins but instead are deaminated and oxidized. During the first month of lactation, body protein catabolism is greater than resynthesis, and the cow typically loses 5 to 10 kg of body protein, which is equivalent to 25 to 50 kg of muscle because muscle is ~20% protein and ~80% water. If half of the protein lost were captured by milk protein synthesis, these endogenous amino acids would account for 5 to 10% of milk protein secretion in the first month of lactation. However, the carbon skeletons of many of these amino acids are used for gluconeogenesis or directly as fuel sources, and the resulting amino group is converted to urea and excreted. When cows are fed protein-deficient diets, up to 20 kg of body protein can be lost during early lactation. Any body protein lost during early lactation must be replenished during mid and late lactation.

Lipid Metabolism. The fatty acids of milk fat are derived from two sources: preformed lipids from blood and de novo synthesis of fatty acids within the gland, each accounting for about half of the milk fatty acids. Palmitic acid and the C18 fatty acids, stearic and oleic, are the principal fatty acids in plasma lipoproteins of ruminants. Lipoprotein lipase within the capillary wall hydrolyzes the triglycerides of plasma lipoproteins and thus enables transport of these fatty acids into mammary cells, whereupon some of the stearate is desaturated to oleate. These fatty acids from blood are used almost exclusively for milk fat synthesis and account for all of the C18 and half of the C16 milk fatty acids. Mammary tissue may also use nonesterified fatty acids (NEFA) from blood, and NEFA uptake by mammary tissue is directly related to the NEFA concentration of blood. Net mammary uptake of NEFA occurs when cows are in negative energy balance and NEFA release from adipose tissue is high so that blood NEFA concentrations are above 0.3 mM; such is the case during the first month or two of lactation. In ruminants, the primary substrates for de novo synthesis of fatty acids are acetate and β-hydroxybutyrate; of these, acetate accounts for ~80% of de novo fatty acid-carbon. Nearly all of the fatty acids with four to fourteen carbons (C4 to C14 fatty acids) and 60% of the palmitic acid (C16) are synthesized de novo in the gland. High concentrations of long-chain fatty acyl-CoA seem to inhibit de novo fatty acid synthesis within the mammary gland. Therefore, in early lactation, when much body lipid is being mobilized and long-chain fatty acids are more available to the mammary gland, milk fat will contain less short-chain and more C18 fatty acids.

During the first month of lactation, as much as one-third of the energy needed for milk production may be from body reserves, and during the first five or six weeks, the total amount of body fat mobilized is often 40 to 60 kg. Thus, during early lactation, mammary gland is favored at the expense of adipose. Ruminant liver takes up NEFA proportional to their concentration in blood; generally, this amounts to ~25% of whole-body NEFA flux during lactation. Most NEFA cleared by the liver are oxidized incompletely and released as ketones, with the rest released as acetate, completely oxidized to CO_2, or reesterified. Because lipoprotein synthesis in ruminant liver is very low, the refomed triglycerides are not quickly exported. Thus, in cases of high-lipid mobilization, a significant amount of triglyceride may accumulate in the liver, causing hepatic lipidosis or fatty liver.

As lactation progresses, the cow eats enough to meet her energy requirements, net lipid mobilization ceases, and repletion of adipose tissue begins to occur. Most lactation diets contain between 3 and 6% lipid, and for a Holstein producing 50 kg of milk per day, the 1.8 kg of milk fat would require ~0.9 kg of fatty acids from blood. If the diet contained more than ~4% fat, some dietary fat would be available for adipose repletion. However, most fatty acids for adipose triglycerides are from de novo synthesis using acetate and ketones as the major precursors.

Milk fat is one of the more saturated fats consumed by humans and is one of the major sources in the human diet of myristic acid (C14:0), an atherogenic fatty acid. However, milk fat is also one of major sources in the human diet of some good fatty acids, including various isoforms of conjugated linoleic acid (CLA) and of vaccenic acid (*trans-11* C18:1). The major fatty acids of forages and cereal grains are linolenic acid and linoleic acid, respectively. Most of these unsaturated acids are biohydrogenated to stearic acid in the rumen so that unsaturated fatty acids make up less than 20% of absorbed fatty acids. However, some of the linoleic acid (*cis-9, cis-12* C18:2) is isomerized to *cis-9, trans-11* C18:2 and to *trans-10, cis-12* C18:2. These fatty acids have been shown to be antitumerogenic and to decrease fat accretion in animal studies. For example, consumption of *cis-9, trans-11* CLA or of *trans-11* C18:1, which is metabolized by animal tissues to CLA, reduces cancer risk in rats (Corl et al., 2003). The production of these fatty acids in the rumen may also explain why some diets cause a depression in milk fat synthesis. Nutrition plays a major role in ruminal fatty acid metabolism, and scientists are now trying to devise strategies for controlling milk fat synthesis, energy balance, and fertility of cows and for enhancing the healthfulness of milk fat.

Acetate and Ketone Metabolism. Blood acetate in ruminants is derived from two sources: microbial fermentation and endogenous production. In a fed lactating cow in early lactation, the amount of acetate entering blood may exceed 5000 g/d, with >60% from ruminal fermentation, 8% from hindgut fermentation, and up to 30% from partial catabolism of lipids and amino acids. Acetate is a major metabolic fuel for gut tissues and skeletal muscle, and acetate uptake is directly proportional to its concentration in blood. Throughout lactation, most ketones, which include β-hydroxybutryate and acetoacetate, are synthesized from ruminally produced butyrate in the epithelium of the rumen and in the liver. However, during early lactation when cows are in negative energy balance, incomplete oxidation of NEFA in liver may account for half of ketogenesis. Occasionally, ketogenesis exceeds the body's ability to clear ketones from blood and the condition known as ketosis may occur.

Ketosis is characterized by high blood ketones and low blood glucose, and can be fatal if untreated.

❑ NUTRIENT REQUIREMENTS

Nutrient requirements and recommended nutrient densities for dairy cattle are shown in Tables 23.2 and 23.3, excerpted from the 2001 version of the NRC for Dairy Cattle (NRC, 2001). The concept of a nutritional requirement has limitations, as discussed in Chapter 18. Dairy cattle respond to increasing amounts of a nutrient in a curvilinear fashion such that each increment in intake of a nutrient results in a reduced increment of response. This is known as the law of diminishing returns and holds true for most nutrients. Indeed, for some nutrients, feeding more than is needed for maximal milk production will result in decreased milk production. Nutrient responses are complex and are governed by more than just the need for a nutrient to pass through its metabolic pathways. Nutrients also impact feed intake and nutrient partitioning through neural and endocrine mechanisms.

❑ NUTRITION FOR LACTATING COWS

The high-producing cow needs enough cell-wall carbohydrate (fiber) to keep her rumen healthy, enough nonfiber carbohydrate (starch and sugar) to provide the glucose precursors needed for making milk, enough rumen-degradable protein to enable optimal rumen fermentation, sufficient rumen-undegradable protein to supply the necessary amino acids to the udder and other tissues, some fat for essential fatty acids and extra energy substrate, as well as essential minerals and vitamins for support of metabolism and transfer to milk. In early lactation, particularly in the first month postpartum, it is nearly impossible to meet all of these nutrient needs at the same time. Thus, the cow will mobilize body tissues to help meet the requirements for maintenance and lactation. This nutrient mobilization and storage must be considered when we balance diets. Cows in early lactation may need extra protein to supplement the energy from body fat, and cows in late lactation need more energy than what their milk production alone would indicate.

Energy Nutrition for the Lactating Cow

Although protein, minerals, and vitamins are important, the overwhelming consideration when feeding cows in early lactation is energy intake. Energy intake is a function of the energy density of the diet and feed dry matter intake. Energy density is a function of diet composition. High-fiber diets are generally less digestible and thus less

TABLE 23.2   Formulas for energy and protein requirements from the 2001 Dairy NRC.

NET ENERGY FOR LACTATION REQUIREMENT

Milk	Kg milk $\times$ (0.0929 $\times$ %fat + 0.0547 $\times$ %CP + 0.0395 $\times$ %lactose) (%CP = true protein/0.93)
Work	0.00045 $\times$ kg BW $\times$ km/day + 0.0012 $\times$ kg BW + 0.006 $\times$ kg BW if pasture is "hilly"
Pregnancy	0.64 $\times$ (0.00318 $\times$ DaysPregnant $-$ 0.0352) $\times$ CalfBirthWt/0.14
Body reserves	(9.4 $\times$ kg fat gain + 5.6 $\times$ kg protein gain) $\times$ 0.85 if gain or 0.82 if loss
Growth	5.67 $\times$ kg BW gain$^{1.097}$ $\times$ (BW$^{0.75}$/matBW$^{0.75}$)/0.7

METABOLIZABLE PROTEIN REQUIREMENT

Maintenance (scurf + urinary)	(0.0002 $\times$ BW$^{0.6}$ + 0.00275 $\times$ BW$^{0.5}$)/0.67
Metabolic fecal	0.03 $\times$ DMI $-$ (0.125 $\times$ 0.64 $\times$ MCP)
Gut proteins	0.4 $\times$ 0.0119 $\times$ 6.25 $\times$ DMI/0.67
Milk	kg milk $\times$ %true protein/0.67 [TP = CP/0.93]
Pregnancy	0.00069 $\times$ DaysPregnant $-$ 0.0692) $\times$ CalfBirthWt/0.33
Body reserves	protein gain/0.492 if gain or 0.67 if loss (protein gain depends on BW and body condition)
Growth	Gain $\times$ (0.268 $-$ 0.0294 $\times$ retained energy/Gain)
Work	no

TABLE 23.3 Recommended nutrient densities for dairy cattle from 2001 dairy NRC.

	LACTATING			DRY COW	CLOSE-UP[b]	HEIFERS		
	LOW	HIGH[a]	FRESH			6 MOS	12 MOS	18 MOS
Days in milk	90	90	11					
Milk yield, kg/d	25	45	35					
Days pregnant				240	279			90
Body weight, kg	680	680	680	730	760	200	300	450
Weight gain, kg/d	0.5	0.1	−0.6	0.67	0.67	0.80	0.87	0.86
DM intake, kg/d	20.3	26.9	18.8	14.4	10.1	5.2	7.1	11.3
Minimum NDF, %	28	25	25	33	33	30	30	30
Undiscounted TDN, %	65	71	71	51	63			
NE_L, Mcal/kg	1.37	1.55	1.85	0.97	1.44			
ME, Mcal/kg						2.04	2.28	1.79
MP, %	9.2	11.0	12.0	6.0	8.0	8.0	7.7	5.6
RUP, %	4.6	6.2	5.6	2.2	2.8	3.4	2.9	0.8
RDP, %	9.5	9.8	10.3	7.7	9.6	9.3	9.4	8.6
CP, %[c]	16	18	18	12	14	15	14	13
Calcium, %	0.62	0.67	0.68	0.44	0.48	0.41	0.41	0.37
Phosphorus, %	0.32	0.36	0.37	0.22	0.26	0.28	0.23	0.18
Magnesium, %	0.18	0.20	0.24	0.11	0.16	0.11	0.11	0.08
Chloride, %	0.24	0.28	0.33	0.13	0.20	0.11	0.12	0.10
Potassium, %	1.0	1.1	1.1	0.51	0.62	0.47	0.48	0.46
Sodium, %	0.22	0.22	0.28	0.10	0.13	0.08	0.08	0.07
Sulfur, %	0.20	0.20	0.20	0.20	0.20	0.20	0.20	0.20
Cobalt, ppm	0.11	0.11	0.11	0.11	0.11	0.11	0.11	0.11
Copper, ppm	11	11	13	12	18	10	10	9
Iodine, ppm	0.60	0.40	0.64	0.40	0.50	0.27	0.30	0.30
Iron, ppm[d]	12	17	19	13	18	43	31	13
Manganese, ppm[e]	14	13	17	16	24	22	20	14
Selenium, ppm	0.3	0.3	0.3	0.3	0.3	0.3	0.3	0.3
Zinc, ppm	43	52	60	21	30	32	27	18
Vitamin A, IU/kg	3700	2800	4000	5600	8200	3100	3400	3200
Vitamin D, IU/kg	1000	760	1100	1500	2200	1200	1300	1200
Vitamin E, IU/kg	27	20	29	81	120	31	34	32

[a]The author recommends that densities of iodine and vitamins A, D, and E be at least as high as that of cows producing less milk.
[b]Data are inconclusive on the optimal nutrient concentrations for close-up dry cows, and NRC gives several options for close-up cows. The author's choice for feeding cows in the last two to three weeks before calving are those given in Table 14-9 of the Dairy NRC for cows at 279 days of gestation. The energy and protein densities before calving should be 10% higher for heifers than the close-up cows.
[c]Author's recommendations. NRC values are lower because NRC assumes perfect balance of RDP and RUP.
[d]Previous Dairy NRC recommended 50 ppm iron for most dairy cattle.
[e]Previous Dairy NRC recommended 40 ppm manganese for most dairy cattle.
Source: From 2001 Dairy NRC Tables 14-7, 14-8, 14-9, 14-10, and 14-16.

energy dense; in contrast, starches and sugars are more digestible and so are more energy dense. Fats also increase energy density. Just as importantly, however, diet composition affects feed intake, and this effect is difficult to predict. To further complicate matters, level of intake affects feed passage rate and thus the amount of time available for digestion and the energy available from a diet. And finally, diet composition and feeding management can alter nutrient partitioning and thus impact milk yield. An understanding of these considerations will enable one to better meet a high-producing cow's energy requirements.

Diet Composition and Voluntary Feed Intake. Feed intake is a major determinant of energy intake. Available data support the idea that there is an optimal level of fiber for which feed energy intake will be maximized, a concept recently reviewed by Allen (2000). This optimal level likely is between 25 and 28% neutral detergent fiber (NDF). If the diet contains more NDF, feed and energy intake will be lower because feed particles must be small and dense enough before they can leave the rumen and pass on down the gastrointestinal tract. If the diet has less NDF and more starch, the cow may eat less feed because of elevated propionate or lactate production, two potential metabolic signals of satiety. In addition, long particles of NDF are needed to stimulate rumination, which stimulates salivation, and thus helps to buffer the rumen. The NDF that stimulates rumination is known as *effective NDF*. If most of the particles of fiber in a feed are less than 1 cm in length, it likely is less effective at stimulating rumination. Some feeds, such as high-fiber byproduct, corn distiller's grains, or soybean hulls are high in NDF but have little effective NDF. If the diet contains inadequate effective NDF, the cow will be subject to problems associated with rumen acidosis. These include sudden depressions in feed intake, milk fat depression, perakeratosis of ruminal papillae, and laminitis of the feet. Of these, laminitis is the most detrimental to the cow's well-being but may take weeks to appear. So the optimal NDF for a high-producing cow is that at which energy intake is maximized while rumen pH is maintained at an acceptable level for most of the day. Feeding cows right at this edge of sufficient but not excess NDF generally maximizes milk yield and therefore profitability; however, feeding at the edge always involves risk. Furthermore, the optimal NDF will not be the same for all cows in a group, so the farmer must decide how much risk to take and how many cases of laminitis per year are acceptable.

Although the optimal level of NDF for a lactation ration is usually around 27%, this may be adjusted based on several factors. If the diet contains shorter fiber particles, such as occurs with many high-fiber byproduct feeds or finely chopped silage, the optimal NDF can be as high as 30 or even 32% NDF. If the diet contains more rapidly fermented starches, such as those in barley or high-moisture corn, the optimal NDF to maintain a healthy rumen pH will be even higher. In contrast, if the diet contains mostly slowly fermented grains, such as dry corn, the optimal NDF can be lower. In addition, if the cow is fed a total mixed ration (TMR, a blend of all the feedstuffs fed to the cow), the optimal NDF can perhaps be lower—as low as 25% NDF, whereas a cow that is fed grain separately will need a higher NDF diet. Finally, feeds generally vary in NDF content, so the actual NDF concentration of a diet often differs from that which is targeted. For example, a silo of alfalfa haylage often varies in quality and/or dry matter content over time; some farms are more prone to mixing errors. Excellent quality control that minimizes fluctuations in NDF level from day to day will enable a farmer to feed a lower NDF diet than when the NDF fluctuates widely.

Interestingly, finishing feedlot beef cattle are usually fed much lower NDF diets than are high-producing dairy cows, even though the cow has a much higher energy requirement relative to maintenance than does the feedlot steer. Adequate effective NDF is much more important for a dairy cow than for a beef steer because the cow has a higher level of intake and rate of passage, with more total fermentation acids produced per hour, and thus a higher need for the rumen buffering that comes with effective NDF.

Impact of Diet Composition and Intake on Energy Available from a Diet. As a cow eats more feed per day, the feed passes through the digestive tract faster. With faster passage rates, there is less time for the feed to be digested. Thus, even though a feed may have a lot of potentially digestible energy, some of it will not be digested at high intakes. This digestibility depression limits the accuracy of any energy system that uses a single energy value for a feed for dairy cattle. This was the case with older versions of the Dairy NRC, in which the same energy value was assigned to a feed whether it was fed for 30 or 120 pounds of milk per day. In reality, the amount of a feed that is digested depends on complex relationships among rate of passage, rate of digestion, and associative effects among feeds so that the digestive efficiency of a specific diet declines as intake increases (VandeHaar, 1998). And dairy nutritionists are not certain about the magnitude of this depression in digestibility, especially for cows producing more than 100 lb of milk.

In the 2001 Dairy NRC, the NE_L value of diet (the amount of energy available for milk and maintenance from a diet) is a function of the diet composition and the level of intake. Diet composition is used to calculate the DE (digestible energy) available at $1\times$ maintenance intake essentially as $4.2 \times (tdNFC + tdNDF) + 5.6 \times tdCP + 9.4 \times tdFA - 0.3 \times FFDMI$, where tdNFC is truly digestible nonfiber carbohydrate, tdNDF is truly digestible NDF, tdCP is truly digestible CP, tdFA is truly digestible fatty acid, and FFDMI is fat-free dry matter intake. This last term is used to correct for losses as metabolic fecal energy. One problem with the new NRC system is that the conversion of DE to ME is the same for high-protein feeds as for low-protein feeds, and thus the resulting NE_L value for high protein feeds is probably too high. (The conversion of DE to ME for high-protein feeds should be lower because some DE is lost as urinary energy.) Only about 20 to 30% of the N in a diet is actually captured even for a high producing cow, so the ME value of digested protein is closer to 4.5 kcal/g instead of 5.6 kcal/g, which would only be true if all the feed protein was captured in milk or meat. Another problem with the new energy system is that feeds with the highest 1X energy values are discounted the most for high-producing cows. This is an improvement over previous NRC versions when evaluating a diet that has already been fed because the actual feed intake can be used in the evaluation. However, the system is difficult to use as a ration formulator for high-producing cows because high-fiber feeds have essentially the same energy value as grains after the discounts are applied, and the NRC 2001 system does not alter predicted intake with diet composition.

Impact of Diet Composition and Intake on Nutrient Partitioning.

The metabolic priority of milk production is very high for a cow in early lactation, but changes in diet composition or intake can impact the partitioning of nutrients to the mammary gland relative to other body tissues. Thus, sometimes when a cow is switched to a diet high in fat or in rapidly fermentable starch, energy-corrected milk yield may not change or may decrease even though net energy intake increases. For example, in one recent study, feeding a high-grain diet increased energy intake, but most of the increased energy intake was partitioned toward body tissue gain. In contrast, feeding corn silage with greater fiber digestibility increased energy intake, with most of the increase partitioned to milk (Oba and Allen, 2000). Such changes in nutrient partitioning are regulated at least partly by the endocrine system. For example, insulin, which is needed for glucose uptake by fat cells but not mammary cells of cattle, is increased with higher grain feeding. Dietary protein also impacts partitioning; feeding a diet with inadequate protein may limit milk synthesis and thus result in more energy partitioned toward adipose tissue.

Feeding to Meet the Energy and Fiber Needs of Lactating Cows.

The goal for feeding energy to cows in early lactation are: (1) to meet the cow's high energy needs, (2) to provide enough fiber for optimal rumen function, and (3) to supply as much fermentable energy as possible to maximize microbial protein production.

Energy is provided with fiber, starches, sugars, proteins, and fats. The energy value and energy sources of selected feeds are given in Appendix Table 5. Fiber must be fermented in the rumen or hindgut to be useful to the cow, and its fermentation results in primarily acetic acid as well as some propionic acid. Most starch also is fermented in the rumen, with propionic acid as the primary end product along with acetate and butyrate. Forage to concentrate ratios in the diet will change the profile of acetate and propionate produced. When the rumen pH is low and fermentable starch is plentiful, some lactate may also be produced. Some starch, perhaps as much as 10 or 20%, passes down to the small intestine, where it is broken down to glucose for absorption. Sugars are rapidly fermented in the rumen, with butyrate being the major fermentation acid. Lipids may be hydrolyzed and biohydrogenated in the rumen, but most lipid digestion occurs in the small intestine and the resulting fatty acids are absorbed there. If protein is degraded in the rumen, the resulting carbon skeletons of amino acids may be fermented to unique short-chain fatty acids. If the protein escapes rumen degradation, most of it will be digested in the small intestine and absorbed as amino acids or di- or tri-peptides. Propionate, lactate, and glucose are considered glucogenic metabolites. Acetate and butyrate are lipogenic, and amino acids are aminogenic. Optimal blends of these metabolites have been proposed (i.e., 16% aminogenic, 29% glucogenic, 39% lipogenic, and 16% lipid as a percentage of total ME supply), but this is not clear.

Forage forms the base of the diet and typically makes up half the mass of the diet. Forages are often grown on or near the farm, so the rest of the diet should be formulated to match the forages. For high milk production, a cow should be fed a minimum amount of forage to meet her fiber requirements, and the forage should be of high quality. High-quality forages are palatable, free of mold, and well preserved, have long enough particles to stimulate rumination, and have highly digestible fiber. For silages, the fermentation acid profile is high in lactate with little butyrate. Corn silage provides a large amount of energy for the amount of effective fiber provided; hence, a blend of corn silage and early bloom alflalfa is a top choice of many nutritionists. Grass forages are also commonly used, but grasses typically ferment slowly, thus slowing the rate of passage and increasing gut fill; therefore, grasses are less useful for high-producing cows than are corn silage or alfalfa. Grasses may be the desired forage for low-producing cows because grasses often have as much fermentable energy as alfalfa if the passage rate is slower. Cows in late lactation can be fed lower quality forages; therefore within a farm it is beneficial to allocate the best forages for the cows in early lactation and to save poorer forages for the cows in late lactation.

High-fiber byproduct feeds may replace half of the forage, especially if forages are in short supply or expensive. The nutrient profile of byproduct feeds varies considerably, and these feeds are usually added to diets because they are relatively cheap on a nutrient basis. Some of the more common high-fiber byproduct feeds include soyhulls, beet pulp, almond hulls, citrus pulp, brewer's grains, and cottonseed hulls. Some high-fiber byproduct feeds are also good sources of other nutrients. These include cottonseeds (also a good fat source), distiller's grains (also a good source of rumen undegradable protein), and corn gluten feed (also high in protein). The fiber of these feeds is usually less effective than that of forages.

Cereal grains are fed to supply starch. A blend of rapidly and slowly fermentable grains is preferable. Those that ferment slowly are more likely to escape rumen fermentation and provide starch to the small intestine. Processing of feedstuffs can alter fermentation rate. For example, corn that is steam-flaked, high-moisture, or finely ground ferments quite rapidly, whereas dry cracked corn ferments slowly. Of the more commonly fed grains, barley ferments rapidly and milo ferments slowly, with corn being intermediate. Waste bakery products, which ferment rapidly, can be used as an alternative starch source. High-sugar feeds include molasses and whey.

Although a primary reason we feed protein is to provide the building blocks for making body or milk proteins, dietary protein also contributes to the energy value of a feed. About 15 to 25% of an animal's energy intake should come in the form of protein. About two-thirds of the amino acids consumed by a cow, even during peak milk production, will be deaminated and used as a fuel source by rumen bacteria or cow body tissues or as a glucogenic precursor by the cow. If the deamination occurred in the rumen, the resulting ammonia may be used to produce microbial protein. Excess ruminal ammonia and the nitrogen from body protein oxidation must be converted to urea for disposal. Urea is the major contributor to the energy lost in urine, and urinary energy (UE) losses are why the metabolizing energy (ME) value of protein is equivalent to that of starch even though protein has a higher gross energy (GE) and digestible energy (DE) value. The metabolic work of ureagenesis results in heat production. A common misconception in dairy cattle feeding is that excess dietary protein severely compromises the availability of energy to a cow. However, this cost is relatively low, and feeding as much as a 4% protein excess (feeding 20% crude protein when the requirement is 16%) will only increase NE_L requirements by ~1 Mcal (equal to ~1.2 kg of milk) (VandeHaar, 1998). The potential impact of feeding excess nitrogen on feed costs and on environmental stewardship are much more important issues.

Fat can be used to increase the dietary energy density. However, added fats often decrease feed intake, in which case very little gain in total energy intake might occur (Allen, 2000). In addition, added fat does not provide substrates for use by bacteria and thus a high-fat diet may decrease ruminal production of microbial protein. A general rule when supplementing with fat is first to use oilseeds, such as cottonseed, and then if more fat is desired, to use an animal fat or rumen-inert fat. One problem is that most fats are not completely rumen-inert and may impair rumen function. A typical base dairy diet is about 3% fat, and fat should not exceed 7% in the total diet.

Protein Nutrition for the Lactating Cow

The goals in feeding protein to lactating cows are (1) to provide enough metabolizable protein to meet the cow's needs for milk production, maintenance, and metabolic functions, and (2) to supply the microbes in the rumen with enough rumen-degraded protein (RDP) for optimal fiber fermentation. Metabolizable protein is defined as the protein that is absorbed and available for use by the cow's body tissues. Metabolizable protein has two sources: microbial protein (the protein in ruminal microbes that flush down into the small intestine), and rumen-undegraded protein or RUP (ingested protein that was not degraded in the rumen and passes to the small intestine). Dietary RDP provides a mixture of peptides, free amino acids, and ammonia for microbial growth and synthesis of microbial protein. Microbial protein typically makes up more than half of the protein that passes to the small intestine. Dietary RUP should be fed in sufficient quantity to make up for the difference between a cow's metabolizable protein requirement and the amount of that requirement met by microbial protein.

In considering the types of protein supplied by feeds, the 2001 Dairy NRC allocates protein to one of three fractions based on studies measuring the loss of feed protein from Dacron or nylon bags incubated in the rumen. Fraction A includes non-protein nitrogen (NPN) and a small amount of true protein that rapidly escapes from the bag because of high solubility or very small particle size. Fraction C is completely undegradable; it is the protein left in the bag after a long period of incubation. Fraction B is the rest of the protein, which is all potentially degradable, but the amount actually degraded depends on its rate of degradation relative to its retention time and rate of passage through the rumen. Rate of passage is higher (so retention time is lower) for cows with greater feed intakes; thus, the percentage of fraction B that is actually degraded will decline. Hence, the RUP percentage of a diet will increase as intake relative to body size increases. In Table 23.4, common dairy feeds are classified according to their rumen degradability, and in Table 23.5, the protein fractions of selected feeds are given.

The rate of degradation (k_d) of fraction B has a profound effect on the amount of a protein that is RUP or RDP. For example, solvent-extracted soybean meal is 15% fraction A and 84% fraction B with a k_d of 7.5%/hour; for a high-producing cow, 30 to 40% of this soybean meal is RUP. However, mechanically expelled soybean meal is 9% fraction A and 91% fraction B, with a k_d of 2.4%/hour; for a high-producing cow, 60 to 70% of this soybean meal is RUP.

Microbial protein is generally the cheapest source of metabolizable protein for a cow, and methods to maximize microbial protein synthesis are generally cost-effective. The most limiting factor for microbial growth is usually the amount of

TABLE 23.4 Rumen protein degradability of common feedstuffs.

ORIGIN	HIGH DEGRADABILITY[a]	MEDIUM DEGRADABILITY	LOW DEGRADABILITY
Legumes	Alfalfa haylage Alfalfa hay		
Grasses	Grass silage	Grass hay	
Cereal grains	Corn silage Corn gluten feed Oats	Corn distillers grains Corn grain	*Corn gluten meal*
Oil seeds	*Solvent-extracted SBM*[b] *Canola meal* *Soybeans*	*Roasted soybeans* Whole cottonseeds	*Expeller SBM*
Animal products			*Blood meal* *Fish meal*
Other	*Urea*		

[a]Italicized feeds are "protein feedstuffs."
[b]SMB = soybean meal.

TABLE 23.5   Protein fractions of alfalfa, corn grain, and common protein supplements from 2001 Dairy NRC.

		N FRACTIONS			K_d (%/H) OF B	LO K_p % RUP % CP[a]	HI K_p % RUP % CP[b]	DRUP % RUP[c]	EAA % CP[d]	LYS % EAA	MET % EAA
	% CP	% CP A	% CP B	% CP C							
Legume silage, mid-maturity	22	57	35	7	12.2	17	18	65	35.6	12.4	3.9
Corn grain, ground, dry	9	24	73	4	4.9	40	47	90	40.1	7.1	5.3
Blood meal, ring dried	96	10	61	29	1.9	73	77	80	56.4	15.9	2.1
Canola meal	38	23	70	6	10.4	29	35	75	42.6	13.2	4.4
Corn gluten meal, dried	65	4	91	5	2.3	68	74	92	45.2	3.7	5.2
Fish meal, menhaden	69	23	72	5	1.4	62	65	90	44.5	17.2	6.3
Soybean meal, expellers	46	9	91	0	2.4	62	68	93	45.5	13.8	3.2
Soybean meal, solvent 44%	50	23	77	1	9.4	28	34	93	45.4	13.8	3.2
Soybean meal, solvent 48%	54	15	84	1	7.5	35	42	93	45.3	13.9	3.2
Urea	280	100	0	0	—	0	0	—	0.0	0.0	0.0

[a]The RUP as a % of CP for cattle at low intake was calculated assuming dry matter intake at 2% of body weight and a diet containing 30% concentrates. Calculated passage rates were 5.1%/h for concentrates and 4.3%/h for wet forages.

[b]The RUP as a % of CP for cattle at high intake was calculated assuming dry matter intake at 4% of body weight and a diet containing 60% concentrates. Calculated passage rates were 7.2%/h for concentrates and 5.5%/h for wet forages.

[c]The expected digestibility (as a %) for the feed protein reaching the small intestine.

[d]Essential amino acids as a % of CP.

fermentable energy supplied. As fat is not fermentable, feeding high-fat diets may limit microbial growth. In addition, microbial protein production may be limited if the diet contains too much rapidly fermented starch, which increases the concentration of acidic fermentation products so that rumen pH is frequently below 5.5.

Occasionally, but not often, cows are fed inadequate RDP. Any diet containing at least 30% legume forage or high-quality grass silage likely will not be limiting in RDP. However, if corn silage is used as the sole forage in a diet for a high-producing cow and the protein supplements used are relatively high in RUP, the cow may benefit from special feeds that are high in RDP. The first choice in RDP supplements is usually urea (because it is cheap), but in most cases, simply using protein sources with a more balanced rumen degradability will solve the RDP shortage. For many of the most common protein supplements, including solvent-extracted soybean meal, canola meal, and cottonseed meal, more than 50% of their protein will be degraded in the rumen.

The RUP requirement for a cow is the difference between her metabolizable protein requirement and the amount of that requirement that is met by microbial protein. For a low-producing cow, microbial protein along with the normal amount of RUP in a diet will generally meet the cow's metabolizable protein requirement. High-producing cows are often supplemented with special RUP feeds, such as fish meal, expeller soybean meal, corn gluten meal, corn distillers grains, or roasted soybeans. However, the ability of any nutritional model to accurately predict microbial yield, and thus RUP requirement, is limited. Many times, lactating cows may not actually need the extra RUP that a computer model predicts is needed, and thus they do not respond to an RUP supplement. In a review of studies published over a 12-year span, Santos and co-workers (2000) found that response in milk yield to RUP supplements was highly variable, with increased milk production occurring in only about one of five cases; the RUP supplements that were most often beneficial were soy or fish based. Many researchers design experiments to maximize the possibility of seeing an effect. Thus, the response to feeding extra RUP cannot be predetermined with certainty by any model and must be monitored on farm.

Part of the reason that some feeds are high in RUP is that they were heated, which decreases rumen protein degradability. However, too much heating can decrease not only rumen degradability but also digestibility in the small intestine. For RUP to be of any value to the cow, it must of course be intestinally digested. If a feedstuff looks burned, there is a good chance that its intestinal digestibility is compromised. Because of concerns about the transmission of bovine spongiform encephalopathy, some feedstuffs, such as blood meal and meat meal, that were commonly used as RUP supplements are no longer legal for use in the U.S. dairy industry.

Dairy cows, like other animals, require not just a specific amount of protein but also a specific amount of each amino acid. Because much feed protein is degraded in the rumen, and because many of the amino acids absorbed by the cow come from microbial protein, research on the amino acid nutrition of cattle is considerably behind that in swine or poultry. Lysine and methionine have been identified most frequently as the most limiting amino acids in the metabolizable protein of dairy cattle. Microbial protein is relatively consistent in its amino acid profile and is relatively high in lysine. Thus, the most limiting amino acid for cattle depends on the amino acid profile of the RUP in the diet and the relative contribution of microbial protein to metabolizable protein. If the RUP is mostly from feeds of corn origin, lysine will likely be first-limiting. In contrast, if the RUP is mostly of soybean or animal origin, especially in conjunction with high-forage, low-grain diets, methionine will likely be first-limiting. In many cases, lysine and methionine may be co-limiting. Maximizing microbial protein yield and ensuring that RUP is from several types of feeds may be a reasonable approach to considering amino acid nutrition in dairy cows. In recent years, rumen-protected amino acid products have become available for use in feeding dairy cows. Cows frequently respond positively to rumen-protected methionine although the magnitude of the response does not always justify the added cost of the supplement. Moreover, the capture of the supplemented methionine into milk protein is often less than 10%, suggesting that the absorbed methionine may have biological effects in addition to its use as a protein building block.

Minerals and Vitamins for the Lactating Cow

Mineral and vitamin requirements can be found in Table 23.3, and concentrations in some common feeds are in Appendix Table 6. The macrominerals are generally supplemented as needed to meet any deficit provided by the base diet. Lactating cows require a relatively large amount of calcium. Unless legumes make up the majority of a cow's diet, she will need a Ca supplement, and limestone is the supplement of choice. Phosphorus is often overfed; when it is needed, dicalcium phosphate is generally the supplement. Some byproduct feeds are high in phosphorus, and proper nutrient management may preclude their use for some farms with limited land base per cow. Magnesium supplementation is often needed with corn silage-based diets, and the most common supplement is magnesium oxide. All dairy cows are supplemented with salt, usually at 0.25 to 0.5% of the diet DM to provide Na and Cl. If a diet contains a substantial amount of the rumen buffer sodium bicarbonate, less salt need be fed, but some will usually be needed to ensure Cl requirements are met. Potassium supplementation varies considerably with geographical region and forage type. In the northern United States, most legume and grass forages are quite high in K (as high as 4% K). In contrast, corn silage has much less K. There is some benefit to feeding extra K in hot weather, and potassium chloride is a common supplement. Sulfur supplementation is seldom needed, although some data support its supplementation to cows fed mostly corn silage-based diets, especially if the corn silage was ammoniated or the diets contain urea. The supplement of choice for adding S is gypsum (calcium sulfate). The seven trace minerals usually fed to dairy cows are copper, iron, manganese, zinc, cobalt, iodine, and selenium. Of these, the two that are almost always needed are I and Se. However, because trace minerals take very little space in a diet, and because trace mineral concentrations in feeds can vary considerably, all the trace minerals are often supplemented at close to 100% of their requirements.

In addition to these minerals, most cows are fed the three fat-soluble vitamins (A, D, and E). High concentrations of vitamins A and E may be beneficial to cows in late gestation or soon after calving to reduce the incidence of mastitis. Generally, the water-soluble vitamins are not supplemented to dairy cows. However, there is some evidence that cows in very early lactation or stressed cows may benefit from supplementation of niacin, choline, thiamin, and/or biotin.

Feed Additives and Metabolic Modifiers

Medicated feeds are not used in diets for lactating cows because residues may be carried into milk. The feed additives that are used are either natural compounds or have been extensively tested before receiving FDA approval. Rumen buffers, such as sodium bicarbonate and magnesium oxide, are often fed in conjunction with high-grain diets. Ionophores such as monensin are not currently legal for lactating cows in the United States, but their future approval seems likely. Monensin might be useful in early lactation to promote propionate production and thus enhance glucose homeostasis and feed efficiency and prevent ketosis. Feeding yeast or yeast extracts may improve rumen function and milk production efficiency in some cases. Many cows are given injections of bovine somatotropin (bST) starting at 70 days postpartum to lengthen the time of peak milk production; bST increases milk yield 10 to 15% (Fig. 16.7). Cows given bST should be fed and managed the same as cows that produce more milk because of superior genetics.

Formulating Diets for High-Milk Production

An example of diets for cows in early lactation is shown in Table 23.6.

1. Use high quality forage with high fiber digestibility as the base of the diet. Corn silage provides a great amount of energy relative to the amount of effective fiber provided. A blend of corn silage and early bloom alfalfa is the choice of many nutritionists.

2. Determine the amount of any byproduct feeds that will be included. These often comprise as much as 50% of the diet, especially if forages are in short supply or expensive on a nutrient basis.

3. Use grain to supply starch. A blend of grains to supply some that are fermented rapidly and some that are fermented more slowly will generally result in maximal energy intake and microbial protein yield. Ground dry corn alone also works well.

4. Balance the relative amounts of the forages, byproduct feeds, and grains to optimize NDF content and thus promote optimal rumen function and maximize energy intake. Guidelines for optimal NDF were given earlier.

5. Add fat sources to further increase energy concentration. However, fats may decrease feed intake. Many nutritionists prefer to first use oilseeds, such as cottonseed, and then if more fat is desired, to use a rumen-inert fat (most fats are not completely rumen-inert and may impair rumen function).

6. Replace some of the grain with protein supplements to provide about 18% crude protein.

7. Supplement with additional rumen-degraded protein if needed. Usually it is not.

8. Determine whether the estimated metabolizable protein supplied by the diet seems adequate. Replace some of the protein supple-

TABLE 23.6 Example of diets for lactating cows at low or high milk yield, fresh cows, dry cows, and heifers (% of dry matter intake).

	LOW	HIGH	FRESH	DRY COW	CLOSE-UP[a]	6 MOS	12 MOS	18 MOS
Corn silage	30%	20%	20%		35%	30%	49%	49%
Alfalfa haylage, early		20%	25%					
Alfalfa haylage, mid	30%				35%	53%	49%	49%
Mixed grass/alfalfa hay				98%				
Cottonseeds	0%	10%	10%					
Corn grain	32%	28%	25%		22%	10%		
Soybean meal, 48%	6%	8%	6%		5%	5%		
Soybean meal, expeller		6%	6%					
Distillers grains	0%	5%	5%					
Minerals and vitamins	2%	3%	3%	2%	3%	2%	2%	2%

ments with a feedstuff high in rumen-undegradable protein if necessary.

9. Supplement with macrominerals as needed.

10. Supplement with trace minerals and vitamins A, D, and E at 50 to 100% of recommended concentrations.

11. Consider additional supplements such as rumen buffers, rumen-protected amino acids, monensin (where legal), probiotics, and water-soluble vitamins. Responses of cows to these supplements are seldom predictable and must be monitored carefully.

12. Carefully monitor the cows when new diets are formulated and refine the diets as needed.

❏ NUTRITION OF DRY COWS

A cow undergoes major metabolic shifts around the time of calving and consequently is at risk for several metabolic disorders; these disorders have been reviewed (Drackley, 1999; Goff and Horst, 1997). A good dry-cow program results in a cow that has a body condition score of 3.5 to 4.0, is physically fit and healthy, and is ready to meet abrupt metabolic changes at calving. Cows that are fat at calving are more susceptible to metabolic diseases in the periparturient period. It also seems reasonable that cows that are too thin may have

less body reserves to support high-milk production in early lactation, but there is little published evidence to support this idea; thin cows also are generally more difficult to re-breed. In the last week before calving, dry matter intake often decreases dramatically, and the cow will start to mobilize body fat and accumulate triglycerides in the liver. Feeding a higher energy diet in the "close-up period" (the last three weeks before calving) reduces lipid mobilization and may help to prevent fatty liver and other metabolic diseases. In addition, feeding higher grain close-up diets may help to adapt the rumen microflora to the nutrient density of the lactation diet she will receive after calving. However, there is no solid evidence that high-grain close-up diets improve energy homeostasis, milk production, or health of cows after calving.

In contrast, calcium homeostasis at calving can be improved by special attention to close-up feeding (Fig. 23.4). In particular, close-up dry-cow diets should have a low value for Dietary Cation-Anion Difference (DCAD). The most commonly used method to calculate the DCAD value of a diet is to subtract the milliequivalents of Cl and S from the milliequivalents of Na and K in a diet. A DCAD value of -10 mEq per a 100 g DM or less will acidify the urine, slightly decrease blood pH, and prepare the bone for rapid mobilization of calcium to help prevent hypocalcemia at calving. The preferred method to lower the DCAD value of dry-cow

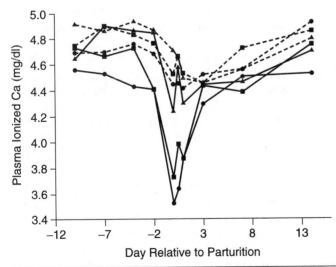

Figure 23.4 *Effect of low dietary cation-anion difference (DCAD) on blood calcium around the time of calving. Mean concentration of ionized Ca in plasma of cows (—) and heifers (- - -) fed a control diet (●) or a diet with DCAD of 0 (■) or −15 (▲) meq/100 g dietary DM. (From Moore et al., 2000.)*

diets is to avoid high potassium feeds so that the diet is not more than 1.0% K. Legumes and grasses often contain more than 3% K, especially if they were grown on soils where manure was frequently applied. If low K is not possible or does not result in the desired DCAD value, several supplements high in Cl or S are available to drop the DCAD value. These include hydrochloric acid-treated soy or canola products and "anionic salts" such as $CaCl_2$. Most anionic salts are not palatable and may decrease feed intake, which could be counterproductive to energy homeostasis; thus, these should be fed with care. If low DCAD diets are fed, the general recommendation is that the Ca percentage should be increased to 1.2 to 1.5%. Examples of dry-cow diets are presented in Table 23.6.

❑ NUTRITION FOR CALVES AND HEIFERS

Calf Nutrition

The most critical period for the heifer is the first three weeks of life, when mortality can be high. The goal for feeding calves is to keep them alive and healthy and start them growing well. The calf is born with an immature immune system and must receive colostral immunoglobulins via the mother's milk soon after birth to fight environmental pathogens. The immunoglobulin concentration of colostrums is variable but is generally highest in multiparous cows. Colostrum need not be from the calf's dam but should be from a cow exposed to the same environmental pathogens. Calves should ingest at least three liters of colostrum within an hour of birth. Colostrum also provides nutrients essential to the neonate, and the growth factors in colostrums may benefit the calf as well. Colostral substitutes are available commercially, but they are not as good as natural colostrum. Individual housing that allows for no physical contact between calves helps to minimize the transfer of infectious diseases.

The young calf is functionally a nonruminant because of its immature digestive system; thus, easily digested carbohydrates, proteins, and lipids should be fed to meet its nutrient requirements. Dairy calves are usually fed milk replacer instead of real milk. The young calf's ability to digest nonmilk proteins is limited. Milk replacers seldom contain caseins because dried skim milk is ex-

pensive. However, the young calf also efficiently uses whey proteins. The replacer proteins should be from whey products during the first 3 weeks, but half the protein can be from processed soy, gluten, or plasma products after this period. Milk replacers seldom contain real milk fat (which is expensive); because calves have limited ability to use vegetable fats, animal fats such as tallow are used. These fats should be blended with an emulsifier to enhance mixing characteristics and digestibility. Generally, calves are fed milk replacer at 1% of body weight per day on a dry basis. Replacers are normally reconstituted with water to 12% solids, and the mix is fed at ~10% of body weight per day over two feedings. Higher rates of milk feeding will enhance growth rates substantially but will also decrease grain intake, delay rumen development, and may delay the transition to a solid diet. Scours, or diarrhea, is not uncommon in young calves and causes loss of the electrolytes Na, K, and Cl. When severe, it should be treated with medication and electrolyte therapy. However, problems can usually be averted if the calf is fed electrolytes in addition to its normal diet of milk or milk replacer as soon as watery feces are noticed.

During the milk-fed stage, the calf should have free access to water and grain throughout most of the day. The rumen rapidly develops as the calf begins to eat dry feed, and hopefully by six weeks, the calf is ready to obtain all its nutrients without liquid feeds. At one time, it was common to feed hay to young calves for the purpose of stimulating rumen development. However, studies have shown that grain stimulates rumen development better than hay, and dairy calves need no hay until after weaning. Calves may be weaned after they are consistently consuming 700 grams of starter grain mix per day. Weaning usually occurs between 4 and 8 weeks of age. A coccidiostat, such as decoquinate, lasalocid, or monensin should be considered to help the calf thrive postweaning. After weaning, the calf is fed forage and grain, with the amount of each depending on the type and quality of the forage and the age of the calf. Recommended nutrient densities for milk replacer and starter grain are given in Table 23.7.

In the past 10 years, there has been much interest in accelerated calf feeding programs. These programs use milk replacers with higher protein concentrations, and calves are fed about twice as much replacer per day. The growth rates of these calves are substantially greater before weaning,

TABLE 23.7 Recommended nutrient densities of milk replacer and starter grain compared with whole milk (DM basis).

	MILK REPLACER	STARTER GRAIN	WHOLE MILK
ME, kcal/g	4.3 to 4.7	3.3 to 3.7	5.2 to 5.5
Protein, %[a]	20 to 30[a]	18 to 24[a]	25
Fat, %	15 to 20	4	28 to 35
NDF, %	0	17 to 20	
Nonfiber carbohydrate, %	40 to 50	45 to 50	35 to 40
Calcium, %	1.00	0.7	0.95
Phosphorus, %	0.70	0.45	0.76
Iron, ppm	100	50	3
Selenium, ppm	0.30	0.3	~0.1
Vitamin A, IU/kg	9000	4000	12,000
Vitamin D, IU/kg	600	600	300
Vitamin E, IU/kg	50	25	8

[a]The higher protein densities are recommended for accelerated rearing programs to encourage high lean to fat gain.
Source: Based partly on Tables 10-6 and 10-8 from 2001 NRC.

and the calves will likely reach breeding size earlier. Whether the increased cost associated with these feeding programs is economically beneficial in the long term (through changes in subsequent growth, reproduction, and milk production) remains to be determined.

Heifer Nutrition

A good heifer rearing program is critical to producing animals at first calving that have good body size and body condition capable of high-feed intake and delivery of nutrients to the mammary gland, and that have well-developed mammary glands capable of maximizing the animal's genetic potential. All available data suggest strongly that if heifers between 3 and 9 months of age are grown too fast or with too much body fat or if heifers at calving are too small or too thin, future productivity will be compromised. Most consultants would agree that the following targets for intensively managed Holstein heifers are consistent with optimal productivity:

- Age at first calving = 22 to 24 months
- Body weight after calving = 570 kg
- Height at calving = 140 cm at the withers
- Body condition score (BCS) at calving = 3.0 to 3.5
- Growth rate from 3 to 10 months of age = 750 to 850 g/day

Feeding heifers is not difficult, but too often inadequate attention is given to meeting these modest goals.

One of the most important determinants of age at calving is age at breeding, and breeding (for Holsteins) should occur when heifers weigh 360 to 390 kg and stand 130 cm at the withers. Ideally, this size will be attained at about 13 to 15 months of age, but heifers should be bred on size, not age. If several heifers in a group (for example, >20%) are already at the proper size for breeding at 10 months of age, they have grown too fast and may have impaired mammary development, but they should be bred anyway—the damage has already been done. However, a major adjustment to the young heifer program should be considered. Heifers that take several breedings before conception are more likely to gain excess fat during pregnancy because they are more mature. Thus, there is some value in feeding a low-energy diet during breeding. After breeding, heifers can be grown at a rapid rate without detriment to mammary development, as long as they are not fat (BCS > 3.5) at the time of calving.

Both before and after breeding, one of the most important aspects of heifer diet formulation is choosing the energy density of the diet, or on a practical basis, deciding how much grain, if any, to supplement available forages. Too often, dairy heifers are overfed energy. One reason for this is that current nutrition models are not very accurate at predicting if the supply of nutrients from a diet matches the heifer's nutrient requirements. Most nutrition models, including the 2001 Dairy NRC model, do a poor job of predicting intake and the efficiency of utilizing dietary energy for tissue gain. This problem is especially notable when heifers are fed high-forage TMRs. For example, observed gains in a study of 270 heifers fed diets with 75% forage were ~16% faster than expected based on actual energy intake (Van Amburgh et al., 1998). In three recent studies at Michigan State University, young Holstein heifers fed diets ad libitum of mostly poor-quality forage (oatlage, very mature alfalfa, and wheat straw) grew ~750 g/d but, based on NRC predictions, should have gained less than half as much. Modern Holstein heifers fed high-forage TMRs probably grow with more lean gain and less fat gain than expected and thus more live-weight gain per calorie of feed energy intake. In any case, the important point is that nutrition models are useful for balancing diets, but actual growth should be monitored.

Protein is needed for skeletal growth and maintenance in heifers, and feeding inadequate protein can partition more energy toward fat instead of lean gain. More importantly, inadequate protein may impair mammary development of heifers (Whitlock et al., 2002). As a heifer gets older, a larger proportion of her body weight gain each day is fat instead of lean tissue. Consequently, the requirement for protein relative to energy decreases as the animal ages. Composition of gain also changes with increasing rate of weight gain; as an animal is fed a higher energy diet to promote faster rates of gain, the ratio of fat to lean usually increases. In other words, if an animal is fed to grow slowly, it will gain mostly lean tissue, but if it is fed to grow rapidly, it will also gain a substantial amount of fat. At the least, if the energy concentration of a diet is increased, the protein concentration should be increased proportionally. Increasing the protein density relative to the energy density may minimize the addition of fat gain with rapid growth. However, a high protein to energy ratio will not prevent overfattening—even animals fed high-protein diets can gain too much body fat in confined feeding systems when they are fed high-energy diets.

Meeting the protein requirement of heifers is not difficult. Heifers seldom need special protein supplements high in RUP because microbial protein generally supplies a large portion of a heifer's metabolizable protein requirement. Thus, complex protein systems are hardly needed for heifers. Balancing diets on a crude protein basis works fine for growing heifers if most of the protein supplements are true protein sources (such as soybean meal). A little urea, or other source of NPN, is okay if enough fermentable starch is available to enable its use.

Recommended daily gains and dietary energy and protein concentrations are given in Table 23.8. The recommendations apply to healthy heifers fed a TMR in a comfortable, confined, group-housing environment with water and feed available most of the day. In group-feeding situations, ad libitum intake is desirable because it promotes more uniform growth rates. Restricted feeding is sometimes more cost-effective, but adequate bunk space for each heifer to eat at the same time is essential when restricting feed for groups. In some situations, higher energy diets may be needed to meet the target gains.

In developing a cost-effective heifer rearing program, one must weigh the costs of heifer rearing versus the potential impact on net income of the animal after calving. For most well-managed, intensive-feeding heifer operations, the most profitable age for first calving is likely 22 to 24 months. First calving at greater than 24 months will likely reduce profitability, unless feed or fixed costs are unusually low, as may be the case for heifers grown on pasture. Pasture generally has a very low cost per unit energy. Even in pasture systems, however, gains of 1.8 lb/day are attainable through intensive grazing or grain supplementation, and 22 to 24 months may be most profitable.

One key to success in heifer nutrition is a good monitoring program. Weight and height should be measured at weaning, at about 5 months, at breeding, and just after calving. If weights cannot be measured with a scale, a weigh tape is acceptable. The values obtained should be compared to standard weight and height curves (Table 23.8). There is currently no way to assess mammary development during the critical phase of development—teat length and udder volume are meaningless. Other factors to monitor include dry matter intake and estimated energy intake. If heifers are growing too fast, the diet should be changed so that estimated energy intake decreases. Example diets for heifers are presented in Table 23.6.

TABLE 23.8   Guidelines for feeding Holstein heifers ad libitum to calve at 23 months.[a]

AGE MONTHS	WEIGHT KG	GAIN G/DAY	HEIGHT CM[b]	ME[c] MCAL/KG	NE_M MCAL/KG	% CP[d]	CP:ME G/MCAL[e]
2	77	726	86	2.9	1.8	18.4	63
4	127	816	97	2.6	1.6	17.0	63
6	177	816	104	2.4	1.5	15.0	63
8	227	816	112	2.4	1.5	14.5	60
10	277	862	119	2.4	1.5	14.0	60
12	332	907	124	2.4	1.5	13.0	53
14	386	907	130	2.4	1.5	12.5	53
16	441	907	132	2.4	1.5	12.5	53
22	605	907	140	2.4	1.5	12.5	53
23 precalf	632	907	140	2.6	1.6	15.0	56
23 postcalf	564		142	2.9	1.7	18.0	63

[a]Targets for other breeds can be calculated. Goal is 90% of mature weight at 24 months.
[b]Height at the withers.
[c]Concentration of metabolizable energy. ME is approximately NE_m divided by 0.6.
[d]Special rumen-undegraded protein sources usually are not needed. If the %CP is at the suggested level, rumen-undegraded protein should be 25 to 35% of total CP.
[e]To calculate CP:ME, multiply %CP by 10, divide by Mcal ME/kg.

❏ FEEDING MANAGEMENT

On modern farms, dairy cattle are usually fed in groups. Because not all animals within a group have the same requirements, a decision must be made as to which animals will be the targets for balancing diets. The target animal should be in the top 50th to 90th percentile (relative to the range of nutrient requirements in the group), depending on economics and the risks of underfeeding versus overfeeding. For example, underfeeding cows in early lactation may limit milk production for the entire lactation, so they should be fed at the 90th percentile. Conversely, cows in later lactation may become too fat if overfed, so the target animal should be closer to the 50th percentile when considering requirements for milk, maintenance, frame growth, and replenishing body condition. Feeding grains or even forages at restricted intake, as is sometimes done for maintenance-type animals, can be a problem for group-fed animals because the animals with the greatest needs are often those that get the least. Whenever animals are limit-fed, feed bunk space must be adequate for all animals to eat at the same time. In addition, it is generally cost-effective to have several feeding groups on a farm. There is no point in feeding expensive protein supplements or buffers or the best forages on the farm to all cows; these feeds are most effectively used for cows in early lactation or primiparous cows. However, grouping strategies on farms should consider not only the nutrient needs of each animal but also animal interactions. Movement to a new group requires establishment of a new pecking order, which can be stressful and reduce productivity. Cow density is also an important consideration. Overcrowding is most detrimental to those cows at the end of the pecking order who have high nutrient needs (for example, primiparous cows in early lactation).

Average nutrient values for feeds can be found in tables; however, the actual nutrient composition of a feed can vary considerably from these book values. This is especially true for forages and byproduct feeds. To maximize production and to feed cows most efficiently and cost-effectively, the actual feeds used on a farm should be analyzed for nutrient content. Whether or not to test a feed would depend on the potential benefit to be gained by obtaining better information. For small farms, the benefit may never outweigh the cost of testing forages. For large farms, it may be beneficial to test all feedstuffs including the grains and protein supplements. Whenever feeds are to be tested, it is absolutely essential that the sample sent to the lab be representative of what the cows will be fed.

A high priority in feeding cows also should be placed on ensuring that the diet actually consumed by the cows is as close as possible to the diet formulated by the nutritionist. All too frequently, the diet on paper, the diet fed, and the diet consumed are not the same! To ensure continuity among these three diets, the mixing equipment and scales must be working properly, and, just as importantly, the nutritionist and the person mixing the diets must communicate with each other effectively. Feedstuffs should be checked visually for quality. The moisture content of wet feedstuffs should be checked weekly. This is best done on the farm with a microwave or portable tester. When mixing diets, it is essential that weights are accurate and that mixing is uniform and thorough. On-farm quality control becomes more important if a farm uses more raw ingredients when mixing diets compared to purchasing premixes and complete grain mixes.

Consistency within the rumen ecosystem improves the efficiency by which nutrients are used and enhances production. Total mixed rations (TMRs) have become very popular in the United States because the cow eats a little of every feed ingredient in every bite, and milk production is in-creased when cows are fed TMRs. Even with TMR feeding, however, rumen pH will vary throughout day. Cows eat 8 to 12 meals per day when feed is constantly available and rumen pH drops after each meal. Promoting more meals of smaller size would promote consistency in the rumen environment. One way to do this is to push the feed up to the feed bunk several times per day. Rumen pH may decrease considerably with slug grain-feeding—feeding more than 6 pounds of grain as a separate ingredient during the day, as is sometimes done in the milking parlor or with a computer feeder. In addition, whenever cows are switched to a high-grain diet from a low-grain diet, the change should be made over several days to give the rumen microflora time to adapt.

Maximizing dry matter intake is one of the most important ways to improve profitability on a dairy farm because increasing milk production without buying more expensive feeds will especially increase profits (Fig. 23.5). Feeding a well-balanced diet with minimum fiber is the first step in maximizing feed intake. However, the most important factors in maximizing feed intake have nothing to do with the diet on paper; in other words, an evaluation of the diet on paper will not reveal potential

Figure 23.5   *Maximizing dry matter intake is a key to high production and profitability in the dairy industry. Corbis Digital Stock/Corbis Images.*

problems. Along with feeding a well-balanced diet, the ingredients used should be of high quality. Forages should have high NDF digestibility. All feeds should be palatable and relatively free of mold and other contaminants. The feed should be kept fresh; old feed should be removed from the feed bunk before it spoils. In hot weather, this might require removal of all feed once a day. The feed should be available to the cows most of the day, and cows should not be without feed for more than two hours at a time. Ad libitum intake, however, requires feeding more than the cows will eat, which results in orts (uneaten feed). Many farmers will feed this refused material to other groups of animals. Bunk space should not be limiting. A general recommendation is that cows should have two feet of bunk space per cow, but some studies have shown that as little as 1 ft/cow may be compatible with maximum feed intake. Freshwater should always be available. Too few spaces for cows to drink, especially during hot weather, will decrease feed intake. Since high-producing cows will have virtually 100% turnover of water per day, water availability is a significant concern. The environment, especially near the feed bunk, should be comfortable. High-producing cows produce a considerable amount of body heat and have a much lower thermoneutral zone than humans and may be heat stressed at 25°C. Providing shade and a cool breeze near the bunk will encourage cows to eat. During cold weather, protection of the bunk area from wind and rain or snow may stimulate intake. In addition, stress, overcrowding, and muddy lots will decrease feed intake. Finally, using lights to simulate a long day photoperiod will enhance feed intake and milk production.

Nutritional Evaluation

Successful dairy farmers are good at monitoring their nutritional program. Metabolic problems are one of the first things managers notice (Table 23.9), and good records on their incidence are essential. Just as important, but more time consuming, are records for milk output, feed intake, and body condition. These records are especially helpful when making ration changes to determine whether the change was profitable. Many dairy farms measure the milk production of each cow once a month with the help of organizations like the Dairy Herd Improvement Association (DHIA). However, monthly milk yield is not frequent

enough for use in making day-to-day decisions. The preferred way to evaluate the success of a ration change is to randomly assign the high-producing cows to two groups that have similar production and stage of lactation. Half of the cows should be fed the new ration, while half stay on the old one. Milk yield and feed intake should be monitored over the next two weeks; if cows fed the new ration produce more and eat more, the switch was likely beneficial. However, most farms are not capable of conducting such an experiment. Instead, they must feed all the cows the new ration and rely on the farm's total daily milk production (the amount of milk shipped per day from the farm), which is affected by many factors other than diet, including cow groupings, milking procedures, and weather. Thus, it is difficult to use total milk yield per day in analyzing diets for one cow group out of several. Even if changes in milk yield per cow can be monitored, differences in milk composition and in body condition can make interpretation of milk yield data misleading. For example, an increase in milk yield may not be desirable if milk fat or protein concentrations or body condition were decreased at the same time. Although measuring body condition is helpful, it is also difficult to use in making day-to-day decisions.

Because changes in milk yield and composition can be difficult to detect or interpret, feed intake and energy intake by a group of cows should be closely monitored to determine whether a diet change resulted in increased energy intake. Energy intake is calculated by multiplying DM intake by the energy density of the ration. Every extra Mcal of NE_L intake is equivalent to an extra 1.3 kg of milk, provided the diet is not deficient in other nutrients. Increased energy intake usually will be beneficial for cows, especially in early lactation. However, using energy intake as the only evaluation criterion could cause cows in late lactation to become excessively fat. Because the NE_L concentration of forages is difficult to measure accurately, the effect of adding or subtracting purchased feeds from a diet is most easily detected when the forage base of the diet is consistent. This type of weekly nutritional assessment enables the nutritionist to fine-tune the diet for optimal health and production.

Farms can greatly benefit from a periodic and joint evaluation of the entire nutrition program by all those involved in managing and feeding the cows. This evaluation would include assessment

Table 23.9   Common problems related to dairy nutrition.

PROBLEM	COMMENT
Ketosis	Low-blood glucose along with high-blood ketones, usually within 6 wk after calving. Usually caused by low-feed intake related to a sudden change in feed or other illness. High starch and monensin might help. Treat with propylene glycol.
Milk fever	Low-blood Ca around the time of parturition. If blood Ca drops <6 mg/dl, muscle and nerve function are impaired. Prevent by feeding low K and Na relative to Cl and S along with high Ca during last 3 wks prepartum. Treat with intravenous and oral calcium.
Retained placenta	Placenta and fetal membranes are not expelled within 48 hours after calving. May be caused by poor muscle tone from hypocalcemia or poor immune function. For problem herds, try feeding low DCAD along with additional Se and vitamin E to dry cows.
Delayed conception	Failure to show estrus or conceive results in long calving interval. May be associated with excessive loss of body condition in early lactation. Supplemental fat may help.
Displaced abomasums	Twisted abomasum impairs flow of digesta. May be caused by inadequate effective fiber or secondary to poor appetite
Founder/laminitis	Inflammation of the lamina causes sore feet and lameness. Symptomatic of a herd fed too much grain.
Low milk fat content	High-grain diet can depress the fat concentration of milk. Low-rumen pH causes production of vaccenic acid, which impairs milk fat synthesis. Can be alleviated by feeding adequate effective fiber or a rumen buffer.
Underconditioned cows	Body condition scores typically drop to ~2.5 at 2–3 months postpartum, but lower scores can be detrimental to milk yield and fertility. May be caused by poor bunk management, internal parasites, mycotoxins, or poorly balanced diet.
Overconditioned cows	If body condition scores are >4 in late dry period, the incidence of dystocia and metabolic diseases will be increased. Grouping cows and feeding to prevent excess body condition in late lactation and the early dry period is essential.
"Poor production"	Nutritional causes include inadequate energy or protein in the diet, inadequate water, and poor feed intake due to insufficient access to feed, spoiled feed, or uncomfortable environment by the feed bunk.
"Poor production" in first lactation	Nutritional causes include inadequate body size due to poor heifer growth and too much competition with older cows at the feed bunk once lactation begins.

of feed intake, milk production, body condition, health, and fertility for both multiparous and primiparous cows, in addition to growth and health of heifers and calves. Although this type of analysis depends largely on computer-generated data, good quality control demands that managers pay attention to the cows themselves. Even the most sophisticated models of dairy nutrition are often inaccurate and unable to predict the effects of dietary changes before they are made. Although computer software programs for dairy nutrition such as the CPM Dairy, Brill, Dalex, Spartan Dairy, and Formulate2 Dairy, are essential for balancing and evaluating diets for high-producing dairy cows, no computer can replace a person who has good "cow sense" and can perceive the well-being of a cow by her behavior and appearance.

❏ REFERENCES

Allen, M. S. 2000. Effects of diet on short-term regulation of feed intake by lactating dairy cattle. *J. Dairy Sci.* **83:** 1598–1624.

Corl, B. A., D. M. Barbano, D. E. Bauman, and C. Ip. 2003. cis-9, trans-11 CLA derived endogenously from trans-11 18:1 reduces cancer risk in rats. *J. Nutr.* **133:** 2893–2900.

Drackley, J. K. 1999. ADSA Foundation Scholar Award. Biology of dairy cows during the transition period: the final frontier? *J. Dairy Sci.* **82:** 2259–2273.

Goff, J. P., and R. L. Horst. 1997. Physiological changes at parturition and their relationship to metabolic disorders. *J. Dairy Sci.* **80:** 1260–1268.

Moore, S. J., M. J. VandeHaar, B. K. Sharma, T. E. Pilbeam, D. K. Beede, H. F. Bucholtz, J. S. Liesman, R. L. Horst, and J. P. Goff. 2000. Effects of altering dietary cation-anion difference on calcium and energy metabolism in peripartum cows. *J. Dairy Sci.* **83:** 2095–2104.

National Research Council (NRC). 2001. *Nutrient Requirements of Dairy Cattle,* 7th ed. National Academy Press, Washington, DC.

Oba, M., and M. S. Allen. 2000. Effects of brown midrib 3 mutation in corn silage on productivity of dairy cows fed two concentrations of dietary neutral detergent fiber: 1. Feeding behavior and nutrient utilization. *J. Dairy Sci.* **83:** 1333–1341.

Santos, F. A., J. E. Santos, C. B. Theurer, and J. T. Huber. 1998. Effects of rumen-undegradable protein on dairy cow performance: a 12-year literature review. *J. Dairy Sci.* **81:** 3182–3213.

Tucker, H. A. 1987. Quantitative estimates of mammary growth during various physiological states: a review. *J. Dairy Sci.* **70:** 1958–1966.

Van Amburgh, M. E., D. G. Fox, D. M. Galton, D. E. Bauman, and L. E. Chase. 1998. Evaluation of National Research Council and Cornell Net Carbohydrate and Protein Systems for predicting requirements of Holstein heifers. *J. Dairy Sci.* **81:** 509–526.

VandeHaar, M. J. 1998. Efficiency of nutrient use and relationship to profitability on dairy farms *J. Dairy Sci.* **81:** 272–282.

Whitlock, B. K., M. J. VandeHaar, L. F. Silva, and H. A. Tucker. 2002. Effect of dietary protein on prepubertal mammary development in rapidly growing dairy heifers. *J. Dairy Sci.* **85:** 1516–1525.

24

Sheep and Goats

Photograph courtesy North Carolina State University.

SHEEP AND GOATS HAVE contributed greatly to food and fiber production worldwide for centuries and continue to do so. They are raised to produce meat, milk, and/or fiber (wool, mohair, and cashmere). Production systems include those involving large flocks and herds on the extensive range areas of the western and southwestern United States and in many regions throughout the world. Small flocks and herds are also used widely for intensive grazing of improved pastures. Some owners keep sheep or goats to utilize areas that are not suitable for cropping and to control unwanted vegetation.

In this chapter, we examine nutritional requirements and the feed resources appropriate to meet these requirements for the various physiological states and functions. Overviews of sheep and goat nutritional management are available (Dove, 2002, 2004; Freer, 2002, Sahlu and Goetsch, 2004, AFRC, 1998). Several features of pastures, mainly quantity and quality (including stage of maturity of the forage), influence intake. Management of sheep and goats emphasizes management of the amount and quality of forage resources. These resources are influenced by climate, seasonal weather patterns, and by the plant species adapted to the local conditions. Intake can be increased by ensuring that animals have access to large amounts of forage of high digestibility. Pasture management should not be focused on measuring intake of individual animals but on optimizing animal production per unit land area. In any pasture system, internal parasite control is required because infestation leads to dramatically reduced productivity and profit.

❏ GENERAL NUTRIENTS REQUIRED BY SHEEP AND GOATS

Nutrient requirements have been summarized for sheep (National Research Council, 1985; Agricultural Research Council, 1980; Bocquier and Theriez, 1989; Sheep Council of Australia, 1990) and for goats (National Research Council, 1981; Morand-Fehr and Sauvant, 1989; Sheep Council of Australia, 1990). These publications serve as guidelines for formulating diets to meet the nutrient requirements of sheep and goats. The general nutrients and feedstuffs needed to meet the requirements of sheep and goats are discussed in the following sections.

Energy

Generally, the most limiting nutrient in ewe and doe nutrition is energy. The major sources of energy are from pasture (forage, range, and browse), hays, silage, byproduct feeds, and grains. Energy deficiencies result in reduced growth or weight loss, reduced reproductive efficiency, reduced milk or fiber production, and increased death loss.

Good quality pasture, hay, silage, or byproducts can usually meet the energy requirements most economically. The high-moisture content of some silages or pastures (lush spring growth) may limit the animals' intake and thereby necessitate supplemental energy feeding in the form of grains. Meeting the high-energy demands of rapid growth, rapid fetal development, or lactation may also require supplementation with grain. In most cases, pasture, hay, silage, or byproduct feeds should be the major feedstuffs used to meet the energy requirements of sheep and goats.

Protein

As ruminants, sheep and goats rely on the microbial population in their rumens to manufacture many of the amino acids and vitamins required for desired production. Therefore, the quantity of protein in the diet is more important than the quality of the protein. However, young animals (milk fed) do not have a developed rumen or an active microbial population and require high-quality protein in their diet.

Rumen microbes utilize nitrogen from proteins of feed origin and nitrogen from nonprotein nitrogen (NPN) sources to manufacture amino acids. Feeds high in protein are usually the most expen-

sive, and therefore diets often contain urea, a cheap NPN source of nitrogen. Guidelines for using urea (Sheep Industry Development, 1992) in sheep diets are as follows:

1. Urea can be used up to 1% of total diet or 3% of the concentrate portion, but should not exceed one-third of the total N in the diet.
2. Urea should not be used in diets of young lambs or in creep feeder diets.
3. Urea should be introduced into the diet gradually to allow for adaptation by rumen microbes (full adaptation takes two to three weeks).
3. Urea should be thoroughly mixed into the diet to prevent high levels of intake.

Pastures that are in vegetative growth or regrowth and most browse generally meet the protein requirements of sheep and goats. Some hays and silages and many byproduct feeds are low in protein and must be supplemented with protein to achieve the desired performance. The oilseed meals (soybean meal, cottonseed meal, peanut meal, linseed meal, etc.) though expensive, are the most common protein sources used to supplement sheep and goats.

Protein blocks (20–200 kg) that contain natural and NPN sources of nitrogen are often provided to animals on pasture and range for protein supplementation and, by changing block location, as a tool to more effectively manage the forage resource.

Minerals

Mineral requirements are affected by several factors, including breed, age, sex, growth rate, physiological state, level and chemical form of ingested minerals, and interaction with other minerals in the diet (Sheep Industry Development, 1992). The mineral requirements and toxicities are fairly well established for sheep (National Research Council, 1985) but have not been definitively established for goats (National Research Council, 1981). The macromineral requirements of sheep and goats are for Na and Cl, K, Ca, P, S, and Mg. The micromineral requirements for sheep and goats include I, Cu, Co, Fe, Mn, Mo, Se, and Zn. Tabular descriptions and pictures of deficiencies are presented by Ensminger et al. (1990) in the chapters on feeding sheep and feeding goats.

Salt, the most common form of Na and Cl, is inexpensive and highly palatable and is usually provided free choice to sheep and goats. Providing it in loose form is preferred over blocks because sheep and goats tend to bite the block (wearing or damaging teeth) rather than licking the block. Potassium is generally high in most forages and is seldom deficient in the diet. Supplementation may be needed when animals are on high-grain diets or are grazing mature or drought-stressed pastures. Increased levels (up to 2% K) may also be appropriate when sheep and goats are stressed by transport and placed on new concentrate-based diets.

Calcium and P are closely related and important in the normal development of teeth and bone. Deficiency in either, or a Ca:P ratio of less than 1.2:1 may cause reduced growth and other possible metabolic problems. Lactating animals have a higher requirement for Ca and P than nonlactating animals, and milk production becomes limited if either Ca or P is deficient. Sheep and goats efficiently reuse P by recycling P through the saliva. In some cases, sheep recycle more P per day through the parotid salivary gland than is required in the diet (National Research Council, 1985). Calcium and P utilization are influenced by hormones and vitamin D.

Sulfur is an important mineral because it functions in the synthesis of methionine and cysteine (S-containing amino acids). Methionine and cysteine are important for live-weight growth but also for wool and hair growth. Methionine is usually the most limiting amino acid for fiber production. Most feeds contain adequate S, but some mature hays or drought-stressed pastures may be deficient. A ratio of dietary N to S should be 10:1 for both sheep and goats. This becomes particularly important when NPN is being fed because the rumen microbes need a source of S and N available at the same time in order to synthesize methionine and cysteine.

Magnesium deficiency is most commonly associated with sheep and goats grazing lush, fast-growing pastures, in the spring, when the grass is high in N and K and low in Mg. If blood levels of Mg are low, the animal may suffer from grass tetany. The animal may fall to the ground, froth at the mouth, and rigidly extend and relax the legs. Untreated animals can die. Prevention by feeding adequate levels of Mg is best, but animals can be treated by intravenous administration of Ca and Mg in the gluconate form.

Of the microminerals, Cu and Se have the narrowest range between what is required and what is potentially toxic. Sheep are more sensitive than goats to high levels of Cu. Copper is relatively cheap and is added at fairly high levels to beef, dairy, horse, and swine feeds. Feeding sheep these feeds over time can result in Cu toxicity and death; therefore, feeds and minerals formulated for any of these species should be checked for Cu level (needs to be <25 ppm) before they are fed to sheep.

There is also a relationship between Cu and Mo. In sheep or goats, if a feed has a normal or low level of Cu with a high level of Mo, then a Cu deficiency may occur. Adding Mo to the diet of sheep also can be effective in reducing the toxicity of Cu.

Selenium is regulated by the Food and Drug Administration because of its potential toxicity. Deficiency in Se results in white muscle disease (stiff lamb disease) in sheep or goats and can also reduce reproductive efficiency. States in the northwestern, northeastern, and southeastern United States can be Se deficient. To meet requirements, Se can be purchased as part of the mineral mix or as part of complete feeds. Selenium toxicity can occur if plants high in Se are consumed over a prolonged period. Selenium-accumulating plants are a problem in some range areas.

Iron, a major component of hemoglobin, is generally adequate in most feeds, and deficiency symptoms are rare in healthy animals; however, blood loss caused by internal parasites can cause anemia. Anemia can also develop in young lambs and kids because of low body stores of Fe and low Fe content of milk. Iron can be given intramuscularly to improve Fe status, but management or regular treatment to reduce internal parasites will reduce the chances of iron deficiencies developing in sheep and goats.

Iodine is generally adequate in most feeds except when grown in some areas of the northeastern, Great Lakes, and Rocky Mountain regions of the United States. Deficiency is generally exhibited in newborns by an enlarged thyroid gland clearly visible on the neck (goiter). Some severely deficient lambs are stillborn without any wool. Iodized salt is the most common supplement.

Zinc is not widely stored in the body, so the diet should have a continuous supply. Deficiencies are rare but can result in reduced fertility, stiffness of joints, reduced weight gain, and parakeratosis.

Manganese is required by both sheep and goats, but precise levels have not been established. Most forages have >50 ppm Mn and most grains have from 15 to 40 ppm Mn, so deficiencies are rare. Deficiency signs include difficulty in walking, skeletal abnormalities, and reduced reproductive efficiency.

Cobalt is needed by the microbes in the rumen to synthesize vitamin B_{12} but is usually adequate in most feedstuffs.

Vitamins

Sheep and goats require dietary sources of fat-soluble vitamins (A, D, E, and K), but adequate quantities of water-soluble vitamins are usually produced by the rumen microbes. Grazing animals generally obtain sufficient vitamin or vitamin precursor to meet requirements, but confinement-fed or high-producing dairy animals may have to be supplemented.

Vitamin A does not occur in forages, but ample amounts of carotene are contained in green vegetation. Carotene is converted to vitamin A. Approximately 1 mg of carotene is equivalent to 400 IU of vitamin A in sheep and goats. Vitamin A and carotene are stored in the body to meet requirements for three to six months after removal from pasture. Vitamin A deficiency is rare, but symptoms include night blindness, poor reproductive performance, keratinization of epithelial tissues, and lower resistance to infections. Vitamin A can be supplemented to the diet or injected, or feeds high in natural sources can be fed (i.e., green grass or legume hay).

Vitamin D is contained in green forages and sun-cured hays. Animals exposed to sunlight should obtain sufficient vitamin D to meet requirements. However, animals in confinement, animals with heavy fleeces, or dark-pigmented animals may need supplementation. Deficiency symptoms include development of rickets or osteomalacia. Vitamin D can be fed or injected to prevent deficiencies.

Vitamin E is a biological antioxidant. It is important in its role with Se in preventing white muscle disease (stiff lamb disease), and it aids in increasing the shelflife of milk. It is not stored in the body in large quantities and is often included in the supplement for fast-growing lambs and kids and for dairy sheep and goats.

Vitamin K is contained in green leafy forages and can also be synthesized by the rumen microbes. Deficiency is not generally seen.

The water-soluble vitamins (vitamin B complex) are not stored in the body for very long but are synthesized by the rumen microbes. Vitamin C is synthesized by the animals' tissues in sufficient quantities to meet requirements. Polioencephalomalacia (PEM) is a disorder in sheep in which thiamin (a B vitamin) is destroyed. Treatment with injections of thiamin reverses the symptoms. Under normal conditions, there is not a dietary requirement for water-soluble vitamins. Young milk-fed animals receive sufficient vitamins from the milk.

Water

In addition to drinking water, sheep and goats obtain water from their feed, snow, and dew. The total water required varies by size and physiological state of the animal, the environmental temperature, and the animal's level of intake. Table 24.1 presents ranges in water intake requirements for sheep and goats in various physiological states and at different temperatures. Water intake is expressed as kg of water needed per kg of dry matter intake.

❏ FEEDSTUFFS USED FOR FEEDING SHEEP AND GOATS

A variety of feeds are appropriate for feeding to sheep and goats. As ruminants, they consume forage, including hays, silage, and range. They utilize agroindustrial byproducts and feeds not appropriate for monogastrics. Chapter 19 covers many of the potential feeds used for sheep and goats, and Appendix Table 1 (National Research Council, 1985) lists the energy and protein composition of some of the common roughage and forage feeds for sheep and goats.

❏ SPECIFIC REQUIREMENTS AND FEEDING OF SHEEP

A successful sheep production system must effectively manage and provide the needed nutrients to ewes (at breeding, through gestation, and during lactation), to rams, and to lambs. If wool production or milk is a major contribution to net income, then requirements for wool growth and milk production need to be met.

Ewes are the backbone of any sheep production enterprise. They must remain fertile and produce lambs, wool, and milk. During the annual life cy-

TABLE 24.1 Desired water intake[a] (kg water/kg dry matter consumed) of sheep and goats in different physiological states and at different temperatures.

	TEMPERATURE (°C)			
	15	20	25	30
Sheep				
Lambs, growing or finishing	2.0	2.6	3	4
Ewes, nonpregnant or early pregnant	2.0–2.5	2.6–3.3	3–3.75	4–5
Ewes, late pregnancy				
With single	3.0–3.5	3.9–4.6	4.5–5.3	6–7
With twin	3.5–4.5	4.6–5.9	5.3–6.8	7–9
Ewes, lactating				
First month	4.0–4.5	5.2–5.9	6–6.8	8–9
Later months	3.0–4.0	3.9–5.2	4.5–6	6–8
Goats				
Does, early pregnancy	2.0–3.0	2.6–3.9	3–4.5	4–6
Does, late pregnancy	3.5–4.0	4.6–5.2	5.3–6	7–8
Does, lactating	4.0–5.0	5.2–6.5	6–7.5	8–10

[a] Water intake includes water from all sources (drinking, feed, snow, dew, etc.).
Source: Adapted from Jarrige (1989).

cle of a ewe, one of the easiest ways to determine nutritional status is by weighing the ewe and comparing her weight to desired standards. Figure 24.1 illustrates the body weight changes expected in a normal 70-kg ewe throughout the year.

In a one time per year lambing system, the ewe will be exposed to the ram once per year and the ewe will lamb once per year. In this system, the ewe will be bred during a period of weight gain (flushing) and then will receive slightly over maintenance requirements for the first 3 to $3\frac{1}{2}$ months of gestation. Approximately two-thirds of fetal growth occurs in the ewe during the last 6 to 8 weeks of gestation; therefore, nutrient requirements are higher, and level of feeding and/or quality of the feed needs to be increased. At lambing, weight loss is dramatic and nutrient demand is very high because of the onset of lactation. During lactation a gradual decrease in body weight is expected because nutrient demand for milk production usually is higher than nutrient intake. The female mobilizes and utilizes body stores of energy, protein, and minerals. At weaning, the demand and use of milk is drastically reduced, resulting in the reduction in nutrient requirements and the drying of the mammary glands. If the ewe is a dairy breed, then the offspring are usually re-

moved from the dam 4 to 10 days after birth and the dam is milked until production decreases or until approximately 1 month before breeding for the next lactation.

After weaning, the ewe replenishes body reserves and gains weight before rebreeding. Such dramatic weight changes are normal, but ewes need to be properly fed to avoid metabolic disorders and improve the likelihood of producing healthy, fast-growing lambs.

Weight is a simple method to monitor the changing nutritional status of ewes. However, measurements of weight do not indicate whether the ewe is in excellent or poor condition. A system of scoring the condition of the ewe has been developed (Meat Livestock Commission, 1983) and is an excellent complement to the weight of the ewe. In this system, ewes are "condition scored" by evaluating the fat cover over the back and are assigned a score ranging from 1 (thin) to 5 (very fat). The chart below describes the characteristics of animals representing each condition score.

With the combination of changing weight and changing condition, a more exact evaluation of nutrition status of the ewe is possible. A lactating ewe that loses 5 kg and goes from a condition score of 4 to 2 may be in poorer condition and require a

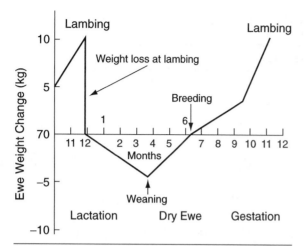

Figure 24.1 *Normally expected weight changes of a 70-kg ewe with twins during a 1-year production schedule. Weight changes for the same ewe carrying a single lamb would be approximately two thirds the above changes. Adapted from Sheep Industry Development (1992).*

higher quality diet than a ewe that loses 10 kg and goes from a condition score of 4 to 3. A combination of weight and condition score may be the best way to monitor nutritional status and the need for higher or lower quality diets.

The nutrient requirements (energy, protein, Ca, P, vitamins A and E) of sheep (National Research Council, 1985) are presented on a daily basis (Table 24.2). The requirements for various-size ewes from maintenance through lactation with twins and the requirements for growing lambs are included. Macromineral requirements (Table 24.3) and micromineral requirements and maximum tolerable levels (Table 24.4) give guidelines for dietary concentrations of mineral elements for sheep (National Research Council, 1985).

Feeding the Ewe

Flushing.   Flushing is the practice of improving the nutrient intake or nutrient utilization to increase body weight and/or condition before breeding, thereby improving the ovulation rate. This can be done by moving ewes to a better pasture, treating with an anthelmintic to reduce internal parasite burden, or feeding supplements. The period of flushing usually starts 3 to 4 weeks before breeding and continues 1 week into the breeding season. Results of flushing are greatest when breeding occurs outside of the normal peak in breeding season (when ovulation rates are already at a

high). The key to flushing is to have the ewes gaining weight and/or condition score. Ewes beginning flushing with condition scores of 2.5 or less will have the largest response and generally exhibit increased conception rate, ovulation rate, and therefore higher lambing percentages.

Gestation.   The gestation length of a ewe is approximately 5 months (147–150 days). The National Research Council (NRC) (1985; Tables 21.3 and 21.4) lists requirements for the first 15 weeks of gestation and then higher requirements for the last 4 weeks. Separate requirements are listed for ewe lambs that are bred to lamb at approximately 1 year of age. Their requirements are higher because not only are they providing nutrients for fetal development and lactation but they are also still growing.

From breeding through embryo implantation (approximately 40 days after fertilization), the embryo must survive in the uterine fluids. Nutrient needs to ensure survival of the embryo through implantation are higher than those of midgestation and are probably closer to the requirements that are listed for flushing (Table 24.2). From implantation until the last 4 to 6 weeks of gestation the ewe is essentially fed for requirements that are only slightly above maintenance. It is important to make sure the ewe does not become fat because excess weight can result in further complications at lambing.

Two-thirds of the fetal growth occurs during the last 6 weeks of gestation. The requirements in Table 24.2 are different for ewes expecting a 130 to 150% lambing rate as compared to a 180 to 225% lambing rate. These requirements are to provide for fetal growth and to provide for increasing the ewes condition so that future milk production needs can be met. Inadequate nutrition during this most critical stage of gestation can result in ketosis (threatening the lives of the lambs and ewes; see Avoiding Nutritional and Metabolic Diseases later in this chapter), low birth weights, low survival rates, low milk production, and low growth rates of the lambs. Adequate nutrient intake is critical during this period. It is interesting that the fetal growth that occurs during the last 4 weeks of gestation effectively reduces the capacity of the gastrointestinal tract right at the time of high nutrient demand. More energy and protein-dense feeding is required to meet this demand and smaller space in the gastrointestinal tract.

TABLE 24.2 Daily nutrient requirements of sheep[a]

BODY WEIGHT		WEIGHT CHANGE/DAY		DRY MATTER PER ANIMAL[a]			ENERGY[b]				NUTRIENTS PER ANIMAL					
							TDN		DE	ME	CRUDE PROTEIN		CA	P	VITAMIN A ACTIVITY	VITAMIN E ACTIVITY
KG	LB	G	LB	KG	LB	% BODY WEIGHT	KG	LB	(MCAL)	(MCAL)	G	LB	(G)	(G)	(IU)	(IU)
Ewes[c]																
Maintenance																
50	110	10	0.02	1.0	2.2	2.0	0.55	1.2	2.4	2.0	95	0.21	2.0	1.8	2350	15
60	132	10	0.02	1.1	2.4	18	0.61	1.3	2.7	2.2	104	0.23	2.3	2.1	2820	16
70	154	10	0.02	1.2	2.6	1.7	0.66	1.5	2.9	2.4	113	0.25	2.5	2.4	3290	18
80	176	10	0.02	1.3	2.9	1.6	0.72	1.6	3.2	2.6	122	0.27	2.7	2.8	3760	20
90	198	10	0.02	1.4	3.1	1.5	0.78	1.7	3.4	2.8	131	0.20	2.9	3.1	4230	21
Flushing—2 weeks prebreeding and first 3 weeks of breeding																
50	110	100	0.22	1.6	3.5	3.2	0.94	2.1	4.1	3.4	150	0.33	5.3	2.6	2350	24
60	132	100	0.22	1.7	3.7	2.8	1.00	2.2	4.4	3.6	157	0.34	5.5	2.9	2820	26
70	154	100	0.22	1.8	4.0	2.6	1.06	2.3	4.7	3.8	164	0.36	5.7	3.2	3290	27
80	176	100	0.22	1.9	4.2	2.4	1.12	2.5	4.9	4.0	171	0.38	5.9	3.6	3760	28
90	198	100	0.22	2.0	4.4	2.2	1.18	2.6	5.1	4.2	177	0.39	6.1	3.9	4230	30
Nonlactating—first 15 weeks gestation																
50	110	30	0.07	1.2	2.6	2.4	0.67	1.5	3.0	2.4	112	0.25	2.9	2.1	2350	18
60	132	30	0.07	1.3	2.9	2.2	0.72	1.6	3.2	2.6	121	0.27	3.2	2.5	2820	20
70	154	30	0.07	1.4	3.1	2.0	0.77	1.7	3.4	2.8	130	0.29	3.5	2.9	3290	21
80	176	30	0.07	1.5	3.3	1.9	0.82	1.8	3.6	3.0	139	0.31	3.8	3.3	3760	22
90	198	30	0.07	1.6	3.5	1.8	0.87	1.9	3.8	3.2	148	0.33	4.1	3.6	4230	24
Last 4 weeks gestation (130–150% lambing rate expected) or last 4–6 weeks lactation suckling singles[a]																
50	110	180 (45)	0.40 (0.10)	1.6	3.5	3.2	0.94	2.1	4.1	3.4	175	0.38	5.9	4.8	4250	24
60	132	180 (45)	0.40 (0.10)	1.7	3.7	2.8	1.00	2.2	4.4	3.6	184	0.40	6.0	5.2	5100	26
70	154	180 (45)	0.40 (0.10)	1.8	4.0	2.6	1.06	2.3	4.7	3.8	193	0.42	6.2	5.6	5950	27
80	176	180 (45)	0.40 (0.10)	1.9	4.2	2.4	1.12	2.4	4.9	4.0	202	0.44	6.3	6.1	6800	28
90	198	180 (45)	0.40 (0.10)	2.0	4.4	2.2	1.18	2.5	5.1	4.2	212	0.47	6.4	6.5	7650	30

TABLE 24.2 Daily nutrient requirements of sheep.—*Continued.*

									NUTRIENTS PER ANIMAL					

BODY WEIGHT		WEIGHT CHANGE/DAY		DRY MATTER PER ANIMAL[a]			ENERGY[b]				CRUDE PROTEIN		CA	P	VITAMIN A ACTIVITY	VITAMIN E ACTIVITY
							TDN		DE	ME						
KG	LB	G	LB	KG	LB	% BODY WEIGHT	KG	LB	(MCAL)	(MCAL)	G	LB	(G)	(G)	(IU)	(IU)
Last 4 weeks gestation (180–225% lambing rate expected)																
50	110	225	0.50	1.7	3.7	3.4	1.10	2.4	4.8	4.0	196	0.43	6.2	3.4	4250	26
60	132	225	0.50	1.8	4.0	3.0	1.17	2.6	5.1	4.2	205	0.45	6.9	4.0	5100	27
70	154	225	0.50	1.9	4.2	2.7	1.24	2.8	5.4	4.4	214	0.47	7.6	4.5	5950	28
80	176	225	0.50	2.0	4.4	2.5	1.30	2.9	5.7	4.7	223	0.49	8.3	5.1	6800	30
90	198	225	0.50	2.1	4.6	2.3	1.37	3.0	6.0	5.0	232	0.51	8.9	5.7	7650	32
First 6–8 weeks lactation suckling singles or last 4–6 weeks lactation suckling twins[d]																
50	110	−25 (90)	−0.06 (0.20)	2.1	4.6	4.2	1.36	3.0	6.0	4.9	304	0.67	8.9	6.1	4250	32
60	132	−25 (90)	−0.06 (0.20)	2.3	5.1	3.8	1.50	3.3	6.6	5.4	319	0.70	9.1	6.6	5100	34
70	154	−25 (90)	−0.06 (0.20)	2.5	5.5	3.6	1.63	3.6	7.2	5.9	334	0.73	9.3	7.0	5950	38
80	176	−25 (90)	−0.06 (0.20)	2.6	5.7	3.2	1.69	3.7	7.4	6.1	344	0.76	9.5	7.4	6800	39
90	198	−25 (90)	−0.06 (0.20)	2.7	5.9	3.0	1.75	3.8	7.6	6.3	353	0.78	9.6	7.8	7650	40
First 6–8 weeks lactation suckling twins																
50	110	−60	−0.13	2.4	5.3	4.8	1.56	3.4	6.9	5.6	389	0.86	10.5	7.3	5000	36
60	132	−60	−0.13	2.6	5.7	4.3	1.69	3.7	7.4	6.1	405	0.89	10.7	7.7	6000	39
70	154	−60	−0.13	2.8	6.2	4.0	1.82	4.0	8.0	6.6	420	0.92	11.0	8.1	7000	42
80	176	−60	−0.13	3.0	6.6	3.8	1.95	4.3	8.6	7.0	435	0.96	11.2	8.6	8000	45
90	198	−60	−0.13	3.2	7.0	3.6	2.08	4.6	9.2	7.5	450	0.99	11.4	9.0	9000	48
Ewe lambs																
Nonlactating—first 15 weeks gestation																
40	88	160	0.35	1.4	3.1	3.5	0.83	1.8	3.6	3.0	156	0.34	5.5	3.0	1880	21
50	110	135	0.30	1.5	3.3	3.0	0.88	1.9	3.9	3.2	159	0.35	5.2	3.1	2350	22
60	132	135	0.30	1.6	3.5	2.7	0.94	2.0	4.1	3.4	161	0.35	5.5	3.4	2820	24
70	154	125	0.28	1.7	3.7	2.4	1.00	2.2	4.4	3.6	164	0.36	5.5	3.7	3290	26

Last 4 weeks gestation (100–120% lambing rate expected)

40	88	180	0.40	3.8	3.3	1.5	0.94	2.1	4.1	3.4	187	0.41	6.4	3.1	3400	22
50	110	160	0.35	3.2	3.5	1.6	1.00	2.2	4.4	3.6	189	0.42	6.3	3.4	4250	24
60	132	160	0.35	2.8	3.7	1.7	1.07	2.4	4.7	3.9	192	0.42	6.6	3.8	5100	26
70	154	150	0.33	2.6	4.0	1.8	1.14	2.5	5.0	4.1	194	0.43	6.8	4.2	5950	27

Last 4 weeks gestation (130–175% lambing rate expected)

40	88	225	0.50	3.8	3.3	1.5	0.99	2.2	4.4	3.6	202	0.44	7.4	3.5	3400	22
50	110	225	0.50	3.2	3.5	1.6	1.06	2.3	4.7	3.8	204	0.45	7.8	3.9	4250	24
60	132	225	0.50	2.8	3.7	1.7	1.12	2.5	4.9	4.0	207	0.46	8.1	4.3	5100	26
70	154	215	0.47	2.6	4.0	1.8	1.14	2.5	5.0	4.1	210	0.46	8.2	4.7	5950	27

First 6–8 weeks lactation suckling singles (wean by 8 weeks)

40	88	−50	−0.11	4.2	3.7	1.7	1.12	2.5	4.9	4.0	257	0.56	6.0	4.3	3400	26
50	110	−50	−0.11	4.2	4.6	2.1	1.39	3.1	6.1	5.0	282	0.62	6.5	4.7	4250	32
60	132	−50	−0.11	3.8	5.1	2.3	1.52	3.4	6.7	5.5	295	0.65	6.8	5.1	5100	34
70	154	−50	−0.11	3.6	5.5	2.5	1.65	3.6	7.3	6.0	301	0.68	7.1	5.6	5450	38

First 6–8 weeks lactation suckling twins (wean by 8 weeks)

40	88	−100	−0.22	5.2	4.6	2.1	1.45	3.2	6.4	5.2	306	0.67	8.4	5.6	4000	32
50	110	−100	−0.22	4.6	5.1	2.3	1.59	3.5	7.0	5.7	321	0.71	8.7	6.0	5000	34
60	132	−100	−0.22	4.2	5.5	2.5	1.72	3.8	7.6	6.2	336	0.74	9.0	6.4	6000	38
70	154	−100	−0.22	3.9	6.0	2.7	1.85	4.1	8.1	6.6	351	0.77	9.3	6.9	7000	40

Replacement ewe lambs[e]

30	66	227	0.50	4.0	2.6	1.2	0.78	1.7	3.4	2.8	185	0.41	6.4	2.6	1410	18
40	88	182	0.40	3.5	3.1	1.4	0.91	2.0	4.0	3.3	176	0.39	5.9	2.6	1880	21
50	110	120	0.26	3.0	3.3	1.5	0.88	1.9	3.9	3.2	136	0.30	4.8	2.4	2350	22
60	132	100	0.22	2.5	3.3	1.5	0.88	1.9	3.9	3.2	134	0.30	4.5	2.5	2820	22
70	154	100	0.22	2.1	3.3	1.5	0.88	1.9	3.9	3.2	132	0.29	4.6	2.8	3290	22

Replacement ram lambs[e]

40	88	330	0.73	4.5	4.0	1.8	1.1	2.5	5.0	4.1	243	0.54	7.8	3.7	1880	24
60	132	320	0.70	4.0	5.3	2.4	1.5	3.4	6.7	5.5	263	0.58	8.4	4.2	2820	26
80	176	290	0.64	3.5	6.2	2.8	1.8	3.9	7.8	6.4	268	0.59	8.5	4.6	3760	28
100	220	250	0.55	3.0	6.6	3.0	1.9	4.2	8.4	6.9	264	0.58	8.2	4.8	4700	30

TABLE 24.2 Daily nutrient requirements of sheep.—*Continued.*

| BODY WEIGHT | | WEIGHT CHANGE/DAY | | DRY MATTER PER ANIMAL[a] | | | ENERGY[b] | | | | CRUDE PROTEIN | | CA | P | VITAMIN A ACTIVITY | VITAMIN E ACTIVITY |
| | | | | | | | TDN | | DE | ME | | | | | | |
KG	LB	G	LB	KG	LB	% BODY WEIGHT	KG	LB	(MCAL)	(MCAL)	G	LB	(G)	(G)	(IU)	(IU)
Lambs finishing—4 to 7 months olds[f]																
30	66	295	0.65	1.3	2.9	4.3	0.94	2.1	4.1	3.4	191	0.42	6.6	3.2	1410	20
40	88	275	0.60	1.6	3.5	4.0	1.22	2.7	5.4	4.4	185	0.41	6.6	3.3	1880	24
50	110	205	0.45	1.6	3.5	3.2	1.23	2.7	5.4	4.4	160	0.35	5.6	3.0	2350	24
Early weaned lambs—moderate growth potential[f]																
10	22	200	0.44	0.5	1.1	5.0	0.40	0.9	1.8	1.4	127	0.38	4.0	1.9	470	10
20	44	250	0.55	1.0	2.2	5.0	0.80	1.8	3.5	2.9	167	0.37	5.4	2.5	940	20
30	66	300	0.66	1.3	2.9	4.3	1.00	2.2	4.4	3.6	191	0.42	6.7	3.2	1410	20
40	88	345	0.76	1.5	3.3	3.8	1.16	2.6	5.1	4.2	202	0.44	7.7	3.9	1880	22
50	110	300	0.66	1.5	3.3	3.0	1.16	2.6	5.1	4.2	181	0.40	7.0	3.8	2350	22
Early weaned lambs—rapid growth potential[f]																
10	22	250	0.55	0.6	1.3	6.0	0.48	1.1	2.1	1.7	157	0.35	4.9	2.2	470	12
20	44	300	0.66	1.2	2.6	6.0	0.92	2.0	4.0	3.3	205	0.45	6.5	2.9	940	24
30	66	325	0.72	1.4	3.1	4.7	1.10	2.4	4.8	4.0	216	0.48	7.2	3.4	1410	21
40	88	400	0.88	1.5	3.3	3.8	1.14	2.5	5.0	4.1	234	0.51	8.6	4.3	1880	22
50	110	425	0.94	1.7	3.7	3.4	1.29	2.8	5.7	4.7	240	0.53	9.4	4.8	2350	25
60	132	350	0.77	1.7	3.7	2.8	1.29	2.8	5.7	4.7	240	0.53	8.2	4.5	2820	25

[a]To convert dry matter to an as-fed basis, divide dry matter values by the percentage of dry matter in the particular feed.

[b]One kilogram TDN (total digestible nutrients) = 4.4 Mcal DE (digestible energy); ME (metabolizable energy) = 82% of DE.

[c]Values are applicable for ewes in moderate condition. Fat ewes should be fed according to the next lower weight category and thin ewes at the next higher weight category.

[d]Values in parentheses are for ewes suckling lambs the last 4–6 weeks of lactation.

[e]Lambs intended for breeding, thus, maximum weight gains and finish are of secondary importance.

[f]Maximum weight gains expected.

Source: From National Research Council (1985).

TABLE 24.3 Macromineral requirements of sheep (percentage of diet dry matter).[a]

NUTRIENT	REQUIREMENT
Na	0.09–0.18
Cl	—
Ca	0.20–0.82
P	0.16–0.38
Mg	0.12–0.18
K	0.50–0.80
S	0.14–0.26

[a] Values are estimates based on experimental data.
Source: From National Research Council (1985).

Separating ewes that are in less than desired condition and ewes that most likely will have multiple births from ewes in good condition that are expected to have singles is helpful in the proper nutrition of each group.

Milk Production. Milk production from the ewe is directly related to the genetic potential, nutrition, and number of lambs suckling. The genetic potential for milk production varies between and within breeds. Some breeds have been selected for milk production for human use or for use in making cheese (an example is Roquefort cheese from France). A ewe suckling twins produces 25 to 50% more milk than if she were suckling a single.

Therefore, requirements (Table 24.2) are separated for ewes raising singles compared to twins and for the first 6 to 8 weeks of lactation compared to the last 4 to 6 weeks of lactation. Requirements for lactation are the highest of all production states. The best quality pasture/hay should be saved for feeding during early lactation. The lactation curve of sheep is such that peak milk yield is at about 3 to 4 weeks, and 70% of milk production will occur in 8 weeks of lactation.

To provide adequate desired nutrients, it is best to feed ewes with twins separately from ewes with singles. Condition and weight loss of ewes during lactation is normal and expected. In systems of once a year lambing, the ewe has time to regain weight after weaning and before rebreeding (see Fig. 24.1). In accelerated lambing production systems, it is desirable to produce a lamb every 6 to 9 months (Hogue, 1987) and level of nutrition is critical to maintaining higher production.

Feeding Lambs

Suckling Lambs. The most critical need for the newborn lamb is to receive colostrums (the ewe's first milk). Colostrum is high in protein, fat, minerals, vitamins, and immunoglobulins (important for disease resistance). Lambs should consume at least 50 to 100 mL of colostrum. Fresh colostrum can be provided from another ewe, or colostrum from ewes can be frozen for later thawing and

TABLE 24.4 Micromineral requirements of sheep and maximum tolerable levels (ppm, mg/kg of diet dry matter).[a]

NUTRIENT	REQUIREMENT	MAXIMUM TOLERABLE LEVEL[b]
I	0.10–0.80[c]	50
Fe	30–50	500
Cu	7–11[d]	25[e]
Mo	0.5	10[e]
Co	0.1–0.2	10
Mn	20–40	1000
Zn	20–33	750
Se	0.1–0.2	2
F	—	60–150

[a] From National Research Council (1985).
[b] Values are estimates on experimental data.
[c] High level for pregnancy and lactation in diets not containing goitrogens; should be increased if diets contain goitrogens.
[d] Requirement when dietary Mo concentrations are <1 mg/kg DM.
[e] Lower levels may be toxic under some circumstances.

bottle or stomach tube feeding. The lambs rely solely on the milk from their mothers for nutrients for the first 3 to 4 weeks. Lambs will begin to nibble at high-quality feeds at 3 to 4 weeks of age, so providing high-quality hay, water, and concentrate is desirable to ensure that lambs adapt to dry feed.

Milk Replacers. Feeding orphaned or excess lambs can be accomplished with milk replacers. The decision to remove the lamb from the ewe should be made early (24–48 h after birth), and the lamb should be placed in an area away from the sounds and smells of the ewe. Many commercial milk replacers are available that contain 25 to 30% fat, 20 to 25% milk protein (casein), and 30 to 35% milk sugar (lactose). Initially, warm milk replacer can be given by bottle or from multiple nipple pails, but eventually lambs can be self-fed cold milk. Lambs should have access to fresh dry feed so that consumption is encouraged. Since milk replacers are expensive, lambs may be weaned from milk replacer and consume dry feed at 3 weeks of age.

Creep Feeds. To supplement the ewes' milk and to shift the diet of lambs to dry feed, a creep feeder is often set up. A creep feeder is an area where lambs have access to feed without competition from the ewes. Creep feeders are usually set up for lambs at about 2 weeks of age, and expected consumption should average 200 to 250 g/day from age 20 days through weaning. Creep rations need to be palatable, should contain at least 15% crude protein, and may contain antibiotics. Feeds that are palatable to lambs include cracked shelled corn, oats (better if rolled), bran, molasses, soybean meal, and high-quality legume hays (especially alfalfa). Suggested lamb creep diets are shown in Table 24.5.

Growing Lambs. Requirements (Table 24.2) are different for replacement ewe and ram lambs

TABLE 24.5   Suggested lamb creep diets.

| | | | | | LAMB CREEP DIETS | | | |
| | | | | | DIET 5 | | DIET 6 | |
INGREDIENT	DIET 1 (%)	DIET 2 (%)	DIET 3 (%)	DIET 4 (%)	COMPLETE (%)	SUPPLEMENT (%)	COMPLETE (%)	SUPPLEMENT (%)
Corn grain (dent yellow)	51.5	61.0	—	—	85.0	—	—	—
Sorghum grain (milo)	—	—	53.0	65.0	—	—	80.0	—
Alfalfa hay (midbloom)	20.0	20.0	15.0	15.0	—	—	—	—
Soybean meal (solvent 44% CP)	22.0	12.5	—	—	13.0	86.7	—	—
Cottonseed meal (solvent 41% CP)	—	—	25.0	13.0	—	—	17.0	85.0
Molasses (cane)	5.0	5.0	5.0	5.0	—	—	—	—
Trace mineral salt (sheep)	0.5	0.5	0.5	0.5	—	—	—	—
Calcium carbonate	1.0	1.0	1.5	1.5	1.5	10.0	2.5	12.5
Ammonium chloride	—	—	—	—	0.5	3.3	0.2	2.5
Vitamin A	.... 500 IU/lb					5000 IU/lb	500 IU/lb	3750 IU/lb
Vitamin E	.... 10 IU/lb					100 IU/lb	10 IU/lb	75 IU/lb

Source: From Sheep Industry Development (1992), p. 539.

as compared to finishing lambs. Replacement lambs are intended for breeding, and, therefore, maximum weight gain and finish is of secondary importance. Finishing lambs are expected to gain weight rapidly and attain the desired finish. In addition, requirements are separate for lambs with moderate as compared to rapid growth potential. Crude protein requirements for small, medium, and large mature-weight genotypes at different rates of gain are presented in Table 24.6. Higher requirements for faster growing animals and for animals of higher mature body weight will aid in meeting optimum growth rates.

Lambs can be grown and finished on high-quality pasture, on pasture plus supplementation, or in dry lot feeding systems. The type of system used will depend on desired marketing time (faster gains can be made in a dry lot), type of lamb to be sold (lean versus more fat cover), available feed resources, and economics.

Pastures of cool season annuals (oats, wheat, barley, rye, ryegrass), legumes (alfalfa, clover), or cool season perennials mixed with a legume (i.e., fescue or bluegrass with clover) can provide nutrients for excellent growth. If lambs are used to graze lower quality pastures or to clean up unwanted vegetation, then additional energy supplementation may be warranted. Whole or ground gains can be used.

Lambs usually are placed on grain diets in dry lot for the last 30 to 40 days before marketing. Desire for leaner lamb carcasses may reduce the need for dry lot grain feeding in the future. Dry lot feeding systems are designed for rapid lamb growth and usually involve two to three diet changes through the finishing period (Ely, 1991). Lambs

TABLE 24.6 Crude protein requirements for lambs of small, medium, and large mature weight genotypes[a] (g/day).

DAILY GAIN (G)[b]	BODY WEIGHT (KG)[c]							
	10	20	25	30	35	40	45	50
Small mature weight lambs								
100	84	112	122	127	131	136	135	134
150	103	121	137	140	144	147	145	143
200	123	145	152	154	156	158	154	151
250	142	162	167	168	168	169	164	159
300	162	178	182	181	180	180	174	168
Medium mature weight lambs								
100	85	114	125	130	135	140	139	139
150	106	132	141	145	149	153	151	149
200	127	150	158	160	163	166	163	160
250	147	167	174	175	177	179	175	171
300	168	185	191	191	191	191	186	181
350	188	203	207	206	205	204	198	192
400	209	221	224	221	219	217	210	202
Large mature weight lambs								
100	94	128	134	139	145	144	150	156
150	115	147	152	156	160	159	164	169
200	136	166	170	173	176	174	178	182
250	157	186	188	190	192	189	192	195
300	179	205	206	207	208	204	206	208
350	200	224	224	224	224	219	220	221
400	221	243	242	241	240	234	234	234
450	242	262	260	256	256	249	248	248

[a]Approximate mature ram weights of 95 kg, 115 kg, and 135 kg, respectively.
[b]Weights and gains include fill.
Source: From National Research Council (1985).

are usually adjusted from pasture-based feeding to concentrates by feeding ground hay (30 to 50%) mixed with grains (corn, oats, barley; 50 to 60%), molasses (10%; to improve palatability and reduce dust), and possibly an antibiotic premix (1%).

Once adjusted to higher grain diets, the lambs will be fed diets that utilize available feed resources. The most common diets are based on corn and soybean meal, with molasses added to increase palatability and decrease dust and mineral mix and ammonium sulfate added to reduce problems associated with urinary calculi. Table 24.7 presents two different diets for growing lambs during each of three stages of growth from 15 kg to market weight. Some producers will choose one diet and keep feeding it throughout the feeding period until the lamb is marketed. This reduces problems associated with changing diets but will underfeed lambs when they are light and overfeed lambs when they are heavy. The low cost of grain relative to hay has encouraged high-grain feeding of lambs. With high-grain feeding, all lambs should be vaccinated to prevent enterotoxemia (overeating disease) (see Avoiding Nutritional and Metabolic Diseases later in this chapter).

Replacement Ewes. Most producers strive to have replacement ewes lamb at 1 year of age. To accomplish this requires that the ewe be bred at 7 months of age. Replacement ewes are usually identified at 3 to 4 months of age and removed from the market lambs that are being finished for market. Replacement ewe lambs should not get fat because this can reduce milk production potential.

Replacement Rams. Rams should be grown on pasture and supplemented as necessary to obtain at least 150 to 200 g/day growth. Once mature, pasture alone will usually meet maintenance requirements.

❑ SPECIFIC REQUIREMENTS AND FEEDING OF GOATS

There are over 400 million goats in the world that produce meat, milk, fiber (cashmere and mohair), and skins. Unfortunately much less is known about goats than about sheep or cattle. Much of the nutrition information is derived from sheep and cattle. Many textbooks even omit the coverage of

TABLE 24.7   Examples of diets for growing lambs.

	LAMB WEIGHT		
DIETS	TO 30 KG (%)	30 TO 40 KG (%)	40 KG TO MARKET (%)
Diet 1			
Cracked corn	48.0	58.0	68.0
Chopped hay	33.0	23.0	13.0
Soybean meal, 44% CP	11.5	11.5	11.5
Liquid molasses	5.0	5.0	5.0
Dicalcium phosphate	1.0	1.0	1.0
Trace mineral salt and Se	1.0	1.0	1.0
Ammonium sulfate	0.5	0.5	0.5
Diet 2			
Ground ear corn	60.0	30.0	—
Cracked corn	—	30.0	60.0
Chopped alfalfa hay	27.5	27.5	27.5
Soybean meal, 44% CP	6.0	6.0	6.0
Liquid molasses	5.0	5.0	5.0
Trace mineral salt and Se	1.0	1.0	1.0
Ammonium sulfate	0.5	0.5	0.5

Source: Adapted from Ely (1991).

goats. Information on the nutrition and feeding of sheep is still a valuable guide for the proper feeding of goats.

The feeding behavior of goats is much different from that of other ruminants. Goats have very mobile prehensile lips and an agile tongue that allows them to select specific plants and specific plant parts. Goats utilize a bipedal stance to obtain desired feed and even climb into shrubs and trees to obtain feed.

Goats are very effective at selecting diets that are higher in quality than the average of what is on offer and generally better than that selected by sheep and cattle. Goats will select from trees, shrubs, forbs, and grasses. Often a single plant species will be preferentially removed, or specific plant parts will be removed (usually tender, palatable, highly nutritious growth). Goats also perform well when fed supplementary feed grains. Three types of goats have been developed to meet the market needs; these are the meat goat, angora goat (production of mohair), and dairy goat.

Meat Goat Requirements

Meat goats include a variety of breeds. In the United States, meat goats are quite often called Spanish goats but still originate from several different breeds. The most famous meat goat is the Boer goat of Africa. Recently, Boer goat embryos have been brought into the United States (via New Zealand) and should make a large impact on the genetic quality of meat goat production. To date, little genetic selection of meat goats has been completed, so the meat goat is a combination of breed mixes, sizes, shapes, and general characteristics.

The NRC (1981) developed nutrition requirements for goats that are set up in quite a different format as compared to other NRC guidelines (Table 24.8). The requirements are different for goats at maintenance, low activity (intensive management), medium activity (semiarid rangelands or slight hilly pastures), and high activity (arid rangeland with sparse vegetation or mountainous pastures). Additional functions such as late gestation, growth (at different rates), or milk production (at different fat percentages) have additive requirements listed.

Reproduction rates are quite high, and twinning is frequently observed. An average target of kid survival to weaning is a 180% kid crop. Meat goat does usually produce 0.8 l/day at the peak of lac-

tation (3 weeks after birth) and produce 0.35 l/day at 120 day postkidding. Higher milk production is expected for does with higher amounts of dairy breeding. Kid growth of 150 g/day can be obtained from birth through weaning.

Since most goats are traditionally raised on pasture or range, they receive little routine supplementation or creep diet. Protein is usually the limiting nutrient of range goats, and supplementation of 300 to 400 g of a soybean meal or cottonseed meal improves production.

Replacement does, does in late gestation, or replacement billies may require up to 500 g/day of a 25% protein supplement. Small herds are often supplemented with commercial goat feed that also provides needed minerals and vitamins. Horse sweet feed is often substituted as a goat feed for lactating does and growing kids.

Fiber-Producing Goats

Meat goats produce two types of fiber: a coarse guard hair of little marketable value, and a very fine hair, cashmere, that has high market value. There is some interest in selecting high cashmere-producing goats, but little is known about the nutrient requirements for cashmere growth. Angora goats have been extensively selected for mohair production. The goats are smaller than most meat goats but give birth to similar size kids, which grow at 50 to 100 g/day. Because of a fairly low reproductive rate and the high value of mohair, angora goats are seldom slaughtered for meat. The does are kept as replacements and males not used for breeding are castrated and kept solely to produce mohair.

Nutrient requirements for angoras are similar to those of meat goats (Table 24.8) with additional requirements for mohair production. Reproductive rate can be improved by flushing as discussed for sheep. Supplementation of 100 to 150 g/day of an alfalfa-based range cube works well.

Angora goats are very sensitive to climatic changes, especially after removal of fleece. In addition, they are sensitive to diet changes, and diet should be maintained from day 80 to 120 of gestation. The last 3 to 4 weeks of gestation may require additional supplementation if the pasture or range is in poor condition or limited growth is available. Corn (150–300 g/day) or cottonseed cake (100–200 g/day) is often used as a supplement.

TABLE 24.8 Daily nutrient requirements of goats.

BODY WEIGHT (KG)	FEED ENERGY				CRUDE PROTEIN		CA (G)	P (G)	VITAMIN A (1000 IU)	VITAMIN D IU	DRY MATTER PER ANIMAL			
											1 KG = 2.0 MCAL ME		1 KG = 2.4 MCAL ME	
	TDN (G)	DE (MCAL)	ME (MCAL)	NE (MCAL)	TP (G)	DP (G)					TOTAL (KG)	% OF KG BW	TOTAL (KG)	% OF KG BW
Maintenance only (includes stable feeding condition, minimal activity, and early pregnancy)														
10	159	0.70	0.57	0.32	22	15	1	0.7	0.4	84	0.28	0.24	0.24	2.4
20	267	1.18	0.96	0.54	38	26	1	0.7	0.7	144	0.48	2.4	0.40	2.0
30	362	1.59	1.30	0.73	51	35	2	1.4	0.9	195	0.65	2.2	0.54	1.8
40	448	1.98	1.61	0.91	63	43	2	1.4	1.2	243	0.81	2.0	0.67	1.7
50	530	2.34	1.91	1.08	75	51	3	2.1	1.4	285	0.95	1.9	0.79	1.6
60	608	2.68	2.19	1.23	86	59	3	2.1	1.6	327	1.09	1.8	0.91	1.5
70	682	3.01	2.45	1.38	96	66	4	2.8	1.8	369	1.23	1.8	1.02	1.5
80	754	3.32	2.71	1.53	106	73	4	2.8	2.0	408	1.36	1.7	1.13	1.4
90	824	3.63	2.96	1.67	116	80	4	2.8	2.2	444	1.48	1.6	1.23	1.4
100	891	3.93	3.21	1.81	126	86	5	3.5	2.4	480	1.60	1.6	1.34	1.3
Maintenance plus low activity (= 25% increment, intensive management, tropical range, and early pregnancy)														
10	199	0.87	0.71	0.40	27	19	1	0.7	0.5	108	0.36	3.6	0.30	3.0
20	334	1.47	1.20	0.68	46	32	2	1.4	0.9	180	0.60	3.0	0.50	2.5
30	452	1.99	1.62	0.92	62	43	2	1.4	1.2	243	0.81	2.7	0.67	2.2
40	560	2.47	2.02	1.14	77	54	3	2.1	1.5	303	1.01	2.5	0.84	2.1
50	662	2.92	2.38	1.34	91	63	4	2.8	1.8	357	1.19	2.4	0.99	2.0
60	760	3.35	2.73	1.54	105	73	4	2.8	2.0	408	1.36	2.3	1.14	1.9
70	852	3.76	3.07	1.73	118	82	5	3.5	2.3	462	1.54	2.2	1.28	1.8
80	942	4.16	3.39	1.91	130	90	5	3.5	2.6	510	1.70	2.1	1.41	1.8
90	1030	4.54	3.70	2.09	142	99	6	4.2	2.8	555	1.85	2.1	1.54	1.7
100	1114	4.91	4.01	2.26	153	107	6	4.2	3.0	600	2.00	2.0	1.67	1.7

Maintenance plus medium activity (= 50% increment, semiarid rangeland, slightly hilly pastures, and early pregnancy)

Body weight														
10	239	1.05	0.86	0.48	33	23	1	0.7	0.6	129	0.43	4.3	0.36	3.6
20	400	1.77	1.44	0.81	55	38	2	1.4	1.1	216	0.72	3.6	0.60	3.0
30	543	2.38	1.95	1.10	74	52	3	2.1	1.5	294	0.98	3.3	0.81	2.7
40	672	2.97	2.42	1.36	93	64	4	2.8	1.8	363	1.21	3.0	1.01	2.5
50	795	3.51	2.86	1.62	110	76	4	2.8	2.1	429	1.43	2.9	1.19	2.4
60	912	4.02	3.28	1.84	126	87	5	3.5	2.5	492	1.64	2.7	1.37	2.3
70	1023	4.52	3.68	2.07	141	98	6	4.2	2.8	552	1.84	2.6	1.53	2.2
80	1131	4.98	4.06	2.30	156	108	6	4.2	3.0	609	2.03	2.5	1.69	2.1
90	1236	5.44	4.44	2.50	170	118	7	4.9	3.3	666	2.22	2.5	1.85	2.0
100	1336	5.90	4.82	2.72	184	128	7	4.9	3.6	723	2.41	2.4	2.01	2.0

Maintenance plus high activity (= 75% increment, arid rangeland, sparse vegetation, mountainous pastures, and early pregnancy)

Body weight														
10	278	1.22	1.00	0.56	38	26	2	1.4	0.8	150	0.50	5.0	0.42	4.2
20	467	2.06	1.68	0.94	64	45	2	1.4	1.3	252	0.84	4.2	0.70	3.5
30	634	2.78	2.28	1.28	87	60	3	2.1	1.7	342	1.14	3.8	0.95	3.2
40	784	3.46	2.82	1.59	108	75	4	2.8	2.1	423	1.41	3.5	1.18	3.0
50	928	4.10	3.34	1.89	128	89	5	3.5	2.5	501	1.67	3.3	1.39	2.7
60	1064	4.69	3.83	2.15	146	102	6	4.2	2.9	576	1.92	3.2	1.60	2.7
70	1194	5.27	4.29	2.42	165	114	6	4.2	3.2	642	2.14	3.0	1.79	2.6
80	1320	5.81	4.74	2.68	182	126	7	4.9	3.6	711	2.37	3.0	1.98	2.5
90	1442	6.35	5.18	2.92	198	138	8	5.6	3.9	777	2.59	2.9	2.16	2.4
100	1559	6.88	5.62	3.17	215	150	8	5.6	4.2	843	2.81	2.8	2.34	2.3

Additional requirements for late pregnancy (for all goat sizes)

	397	1.74	1.42	0.80	82	57	2	1.4	1.1	213	0.71		0.59	

Additional requirements for growth–weight gain at 50 g/day (for all goat sizes)

	100	0.44	0.36	0.20	14	10	1	0.7	0.3	54	0.18		0.15	

Additional requirements for growth–weight gain at 100 g/day (for all goat sizes)

	200	0.88	0.172	0.40	28	20	1	0.7	0.5	108	0.36		0.30	

Additional requirements for growth–weight gain at 150 g/day (for all goat sizes)

	300	1.32	1.08	0.60	42	30	2	1.4	0.8	162	0.54		0.45	

TABLE 24.8 Daily nutrient requirements of goats—*Continued*.

BODY WEIGHT (KG)	FEED ENERGY				CRUDE PROTEIN		CA (G)	P (G)	VITAMIN A (1000 IU)	VITAMIN D IU	DRY MATTER PER ANIMAL			
	TDN (G)	DE (MCAL)	ME (MCAL)	NE (MCAL)	TP (G)	DP (G)					1 KG = 2.0 MCAL ME		1 KG = 2.4 MCAL ME	
											TOTAL (KG)	% OF KG BW	TOTAL (KG)	% OF KG BW
Additional requirements for milk production/kg at different fat percentages (including requirements for nursing single, twin, or triplet kids at the respective milk production level)														
(fat %)														
2.5	333	1.47	1.20	0.68	59	42	2	1.4	3.8	760				
3.0	337	1.49	1.21	0.68	64	45	2	1.4	3.8	760				
3.5	342	1.51	1.23	0.69	68	48	2	1.4	3.8	760				
4.0	346	1.53	1.25	0.70	72	51	3	2.1	3.8	760				
4.5	351	1.55	1.26	0.71	77	54	3	2.1	3.8	760				
5.0	356	1.57	1.28	0.72	82	56	3	2.1	3.8	760				
5.5	360	1.59	1.29	0.73	86	60	3	2.1	3.8	760				
6.0	365	1.61	1.31	0.74	90	63	3	2.1	3.8	760				
Additional requirements for mohair production by angora at different production levels														
Annual Fleece Yield (kg)														
2	16	0.07	0.06	0.03	9	6								
4	34	0.15	0.12	0.07	17	12								
6	50	0.22	0.18	0.10	26	18								
8	66	0.29	0.24	0.14	34	24								

Source: Adapted from NRC (1981).

Kids should grow from the milk provided by the doe but will also obtain nutrients from the range or pasture. A suggested ration for angora and meat goats, kids, and yearlings (Ensminger et al., 1990) is 32% alfalfa, 28% cottonseed hulls, 18% grain sorghum, 8% barley, 6% molasses, 6% cottonseed meal, and 2% salt/mineral mix.

Dairy Goats

Dairy goats have been extensively selected for milk production. The general requirements for maintenance and growth (Table 24.8) are the same as for meat goats. Requirements for lactation are as high as for any productive function; therefore, feeding dairy goats to meet requirements is as much of a challenge as feeding a high-producing dairy cow.

To meet requirements, goats should be fed as much high-quality forage as they want and then supplemented with a limited amount of grain. The forage can be grazed or in the form of hay. The quantity of grain required will depend on the quality of hay and level of milk production of the doe.

Dry does should be in good body condition at kidding because peak nutrient demand (early and midlactation) will be higher than the intake capacity of the doe. The doe needs to go into lactation with enough body reserves to maximize milk production by mobilizing body reserves.

As with sheep, body weight and condition scores are important ways to judge the nutrient status of the doe. The doe should gradually gain weight throughout gestation and be at desired body condition 6 to 8 weeks before kidding. The last 6 to 8 weeks before kidding, nutrient demands (because of fetal growth and preparation for lactation) are higher, and quality of diet needs to be increased. Competition for space in the body cavity between the gastrointestinal tract and the rapidly growing fetus(es) often causes a reduction in intake during the last 1 to 3 weeks. To meet the nutrient demands requires that a higher quality diet be fed. Usually this is done with grain concentrate feeding, and concentrate feeding is continued during lactation. As milk production decreases, concentrate feeding decreases until the doe is at desired condition and weight.

The proper nutrition and feeding of dairy requires the watchful eye of the owner and individual feeding to meet production.

General Guidelines for Feeding Goats

There are many commercially available grain mixes formulated for goats. Dairy cattle grain mixes can be substituted if goat mixes are not available or if demand for goat mixes is low and freshness of feed is a problem.

Goats accept a large variety of feedstuffs in their diet, so there is not one feed that is most appropriate. Giving goats an opportunity for selection is beneficial but not absolutely necessary. Goats consume several feedstuffs that are not palatable to sheep or cattle, so mixed grazing of all these species is complementary.

When grain supplementation is necessary to meet the protein requirements, a good rule of thumb is the "total 28 method" (Whitlow, 1983). Subtract the percentage of protein in the forage dry matter from 28. The result is the protein percentage required from the grain supplement. If the goats are grazing fescue (12% CP), then 28 minus 12 equals 16, so a 16% CP grain supplement is needed. Urea can be used but should not exceed 2% of the diet and is generally not used for high-producing dairy goats.

Consumption of colostrums during the first few hours of life is extremely important to ensure survival. Frozen colostrums can be thawed and used. Goat colostrums is preferred to cow colostrums, but cow colostrums can be used.

Growing replacement animals is similar to sheep, in that it is undesirable to have fat replacement animals. Most nutrients should be obtained from forages, but limited supplementation may be required to reach desired target weights.

❏ AVOIDING NUTRITIONAL AND METABOLIC DISEASES

Several nutritional and metabolic diseases affect sheep and goats. Most of these can be prevented or avoided by proper feeding and nutrition. Their symptoms and ways to avoid occurrence are as follows.

Urinary calculi (water belly) is a problem associated with males and wethers. Mineral stones develop in the bladder and may become lodged in the urethra, causing blockage of urination and possible bladder rupture and death. Symptoms include difficulty in urination, straining, and kicking at the belly. Incidence is reduced by having dietary

Ca:P ratios of greater than 2:1 and inclusion of ammonium sulfate (0.5% diet) in the diet.

Enterotoxemia type D (overeating disease) is one of the most common nutritional problems occurring with animals on high-quality (usually grain) diets. The disease is caused by *Clostridium perfringens* type D and results in rapid death in many animals. Symptoms include depression, abdominal pain, and of grinding teeth. Prevention is through an inexpensive vaccination program.

White muscle disease (stiff limb diseases) is a nutritional muscular dystrophy caused by a deficiency of Se, vitamin E, or both. Usually seen in animals 6 to 8 weeks old, it results in poor growth, difficulty in movement, stiff legs, and possibly death. Prevention can be accomplished by providing adequate Se and vitamin E in diet or through injection.

Toxic or poisonous plants include those that are cyanogenetic, contain alkaloids, are photodynamic, or produce mechanical injury. Generally sheep and goats avoid such plants but during times of feed shortage or new exposure, some animals may consume poisonous plants. The poisonous plants have been described for sheep (Sheep Industry Development, 1991) and for goats (Ace and Hutchinson, 1984).

Ketosis (pregnancy toxemia, lambing paralysis), a nutritional disorder, occurs at the very end of gestation and is associated with the high demand [by the fetus(es)] for carbohydrates (especially glucose or glucose precursors). It most often occurs in animals that are in poor condition and in animals carrying twins or triplets. The high demand for carbohydrates results in the depletion of body reserves and metabolically results in high levels of ketone bodies in the blood and urine. Individuals affected will exhibit labored breathing, lack of energy, staggering, and impaired vision. The breath and urine will often smell sweet. Prevention, by feeding adequate carbohydrate sources during late gestation is best, but if detected early, solutions carbohydrates can be administered orally. Propylene glycol or corn syrup can be drenched to the affected ewe two to five times per day until the dam recovers. Increasing the carbohydrate level of the feed for the rest of the gestating animal is also recommended.

❏ SUMMARY

- Sheep and goats provide food and fiber for people throughout the world. General nutrient requirements of sheep and goats, feedstuffs commonly used, specific requirements of feeding animals during all phases of the life cycle, examples of suitable diets for animals during each stage of production, and nutritional and metabolic diseases are examined.

❏ REFERENCES

Ace, D. L., and L. J. Hutchison. 1984. *Poisonous Plants.* Goat Extension Handbook, Fact Sheet C-4, Pennsylvania State University, University Park, Pennsylvania.

Agricultural Food Research Council (AFRC). 1998. *The Nutrition of Goats.* CAB International, New York.

Agricultural Research Council. 1980. *The Nutrient Requirements of Farm Livestock, No. 2: Ruminants* (2nd ed.). Commonwealth Agriculture Bureaux, Slough, England.

Bocquier, F., and M. Theriez. 1989. In *Ruminant Nutrition: Recommended Allowances and Feed Tables,* pp. 153–168, R. Jarrige (ed.). Institut National de la Recherche Agronomique, Paris.

Dove, H. 2004. Sheep: nutrition management. In *Encyclopedia of Animal Science,* W. G. Pond and A. W. Bell (eds.). Marcel Dekker, New York (In Press).

Dove, H. 2002. Principles of supplemental feeding, In *Sheep Nutrition,* pp. 119–142, M. Freer and H. Dove (eds.). CABI Publishing/CSIRO Publishing, Wallingford, UK.

Ely, D. G. 1991. In *Livestock Feeds and Feeding* (3rd ed.), pp. 336–349, D. C. Church (ed.). Prentice-Hall, Englewood Cliffs, NJ.

Ensminger, M. E., J. E. Oldfield, and W. W. Heinemann. 1990. *Feeds and Nutrition.* Ensminger Publishing Co., Clovis, CA.

Freer, M. 2002. The nutritional management of grazing sheep. In *Sheep Nutrition,* pp. 357–375. M. Freer and H. Dove (eds.). CABI Publishing/CSIRO Publishing, Wallingford, UK.

Hogue, D. E. 1987. In *New Techniques in Sheep Production,* pp. 57–63, M. Marai, V. B. Owen, and I. Fayze, (eds.). Butterworth & Co., London.

Jarrige, R. (ed.). 1989. Appendix Table 18.1. In *Ruminant Nutrition: Recommended Allowances and Feed Tables.* Institut National de la Recherche Agronomique, Paris p. 365.

Meat Livestock Commission. 1983. *Feeding the Ewe,* p. 83. Meat and Livestock Commission, Milton Keynes, UK.

Morand-Fehr, P., and D. Sauvant. 1989. In *Ruminant Nutrition: Recommended Allowances and Feed Tables,* pp. 169–180, R. Jarrige (ed.). Institut National de la Recherche Agronomique, Paris.

National Research Council (NRC). 1981. *Nutrient Requirements of Goats: Angora, Dairy and Meat Goats in Temperate and Tropical Countries.* National Academy of Sciences, Washington, DC.

National Research Council (NRC). 1985. *Nutrient Requirements of Sheep.* National Academy of Sciences, Washington, DC.

Sahlu, T., and A. Goetsch. 2004. *Encyclopedia of Animal Science,* W. G. Pond and A. W. Bell (eds.). Marcel Dekker, New York (In Press).

Sheep Council of Australia. 1990. *Feeding Standards for Australian Livestock. Ruminants.* Commonwealth Scientific and Industrial Research Organization, East Melbourne, Australia, 266 pp.

Sheep Industry Development. 1992. In *Sheep Industry Development Handbook,* pp. 501–548. Sheep Industry Development, Denver, CO.

Whitlow, L. 1983. *Dairy Goat Feeding and Nutrition.* North Carolina Agricultural Extension Service Publication 6-83-IM, Raleigh, NC.

Swine

SEVERAL PUBLICATIONS PROVIDE DE-TAILED information about the nutrient requirements of swine during the life cycle (Burrin, 2001; Close and Cole, 2000; Cole et al., 2000; Kellems and Church, 2001; Lewis and Southern, 2001; McGlone and Pond, 2003; Miller et al., 1991; National Research Council, 1998; Pond et al., 1991; Yen, 2001). Major as well as unique feed resources available for swine and the applied aspects of swine nutrition also have received detailed attention (Lewis and Southern, 2001; McGlone and Pond, 2003; Miller et al., 1991; Pond and Maner, 1984; Pond et al., 1991). Therefore, this chapter addresses only the general principles of swine nutrition and applied swine feeding. The nutrient requirements of the pig resemble those of the human in more ways than any other nonprimate mammal. In addition, as noted earlier, the digestive physiology of pigs and people is similar, and, for that reason, the pig has become an important animal model for use in research directed toward nutrition of humans. Pigs and people have been unique partners throughout history. This partnership has thrived through the pig's ability to adapt favorably to varying environments and living conditions and to the changing needs and aspirations of people. The worldwide demand for pork continues to grow as a result of the adaptability and versatility of swine and the acceptance of and desire for pork throughout most areas of the world. Worldwide, the total yearly consumption of pork exceeds that of any other meat source. Thus, the economic value of the swine industry is significant in both industrialized and developing nations as an important source of income for those involved directly and indirectly with swine production and as a valuable source of food for human nutrition.

We will review the current knowledge of the nutrient requirements of swine and describe briefly

461

some of the common feedstuffs and feeding guidelines for optimum nutrition of swine during all phases of the life cycle. At the outset, it is important to recognize that all of the nutrients required by swine can be supplied by a simple mixture of an energy source (e.g., corn or cereal grain) and a protein supplement fortified with minerals and vitamins in amounts and proportions appropriate for swine in any discrete stage of the postnatal life cycle. An example of such a diet was presented by National Research Council (1998). It is shown in Table 25.1.

TABLE 25.1   Fortified swine diet.[a]

INGREDIENT	AMOUNT IN DIET (%)	DIGESTIBLE ENERGY (KCAL/KG)	DIGESTIBLE ENERGY CONTRIBUTION (KCAL/KG)
Corn	74.44	3525	2624
Soybean meal	23.40	3685	862
Dicalcium phosphate	0.71		
Ground limestone	0.90		
Sodium chloride	0.25		
Vitamin premix	0.10		
Trace mineral premix	0.10		
Antibiotic premix	0.10		
Total	100.00		3486

[a]Note that in this example corn and soybean meal are the only sources of energy. All other ingredients are important to provide a complete diet, but they contribute no energy.
Source: National Research Council (1998).

❑ NUTRIENT REQUIREMENTS

The quantitative nutrient requirements for swine during the various stages of the life cycle have been described by the National Research Council (1998) and are not reproduced here.[*] Clinical signs of nutrient defiencies in swine have been summarized by the National Research Council (1979). Some nutrients are toxic to swine when ingested in amounts exceeding the pysiological requirement. (The toxic levels of these nutrients and the signs of toxicity were discussed in earlier chapters.)

Nutrients of Special Concern in Swine Diet Formulation

Whereas all of the individual nutrient requirements of swine as listed by the National Research Council (1998) must be supplemented during one or more stages of life, many are provided in adequate amounts in most feed ingredients and are therefore of only minor concern under most conditons. The specific nutrients that require special attention in the formulation of diets using common feedstuffs are discussed in this section. Swine tend to eat to meet energy needs; therefore, concentrations of specific nutrients in the diet are generally expressed in units per kg of total dry diet consumed ad libitum. The energy component of the diet is the largest single constituent and is therefore of major concern in diet formulation. The im-

[*]The National Research Council regularly updates the current knowledge of the nutrient requirements of swine and other animals and publishes the information in a readily useful format (see list of available National Academy of Sciences/National Research Council publications on Nutrient Requirements of Animals in Appendix I). Other publications, *Vitamin Tolerance of Animals* (National Research Council, 1987) and *Mineral Tolerances of Domestic Animals* (National Research Council, 1980) contain valuable information on the upper and lower limits of nutrient tolerance in swine.

portant topic of energy nutrition is deferred to the next section to accommodate a brief description of specific nutrient requirements.

Protein. Inadequate dietary protein probably is the most frequently deficient class of nutrients, primarily because most of the common feedstuffs available as sources of energy are low in protein and protein supplements are expensive. The protein requirement of swine is, in reality, satisfied by an appropriate array of dispensable amino acids plus an adequate supply of nonspecific sources of nitrogen for use in dispensable amino acid synthesis. The amino acid needs of swine have been reviewed by Lewis (1991), and the bioavailability of amino acids in feedstuffs for swine has been summarized by Knabe (1991). Baker et al. (1993) described the "ideal patterns of amino acids" for swine, which they defined as those that contain neither deficiencies nor excesses of amino acids.

Values are expressed as amounts of each amino acid as a percentage of lysine in the diet; 100% bioavailability of each amino acid is assumed. Newborn pigs have the highest protein requirement; milk is high in protein, so protein deficiency during the suckling period is not a problem. However, early-weaned pigs are especially vulnerable to protein inadequacy because their total protein requirement for normal growth is exceedingly high and remains high well into the early postweaning period. A suboptimum level of total protein reduces rate of growth and efficiency of feed utilization. Severe deficiency results in complete growth failure and severely depresses blood serum albumin, increases liver fat, and produces edema (fluid accumulation) in the jowl and umbilical area. These signs of protein deficiency are associated with kwasiorkor, a human infant protein–calorie malnutrition syndrome affecting millions of children worldwide. Because of the similarity of the deficiency signs in pigs and humans, the pig is a useful animal model for studying the effects of early protein deficiency and strategies for minimizing the long-term effects in human populations.

Lysine. Lysine is the first or second limiting amino acid in maize, barley, wheat, oats, sorghum, millet, and most other energy sources fed to swine. A deficiency of lysine results in reduced feed intake, growth, and efficiency of feed utilization, in young pigs. In older pigs, a moderate deficiency may cause less protein deposition in the body and

lower efficiency of feed utilization, with no apparent effect of rate of weight gain. There may be no outward appearance of a nutrient deficiency; for this reason, marginal protein or lysine deficiency may go unnoticed and result in significant unrecognized losses in growth rate and feed utilization.

Tryptophan. Tryptophan is commonly second limiting in all-plant diets for swine. Trypotophan is first or second limiting in maize and must be provided in adequate amounts to allow normal performance of pigs fed a lysine-supplemented maize diet. Conversely, tryptophan added to a maize diet must be supplemented with lysine (the other limiting amino acid) in order to produce normal performance. A deficiency reduces growth and feed utilization and produces eye cataracts. Tryptophan can be converted in the tissues of the pig to the vitamin niacin, so it is important to provide adequate niacin to spare the tryptophan requirement. The hydroxy- and keto-analogs of tryptophan are utilized to some degree for growth, indicating that either the intestinal microflora or the pig tissues, or both, are capable of adding ammonia to form the amino acid. The cost of crystalline tryptophan has been reduced drastically through its mass production by monocultures of bacteria produced by recombinant DNA technology. This technology has an economic impact on the amino acid fortification of swine diets.

Threonine. Threonine is second-limiting in wheat, rice, and grain sorghum. A deficiency depresses growth and the efficiency of feed utilization, as with deficiencies of other amino acids and of protein. Threonine, like tryptophan, is now produced in volume by recombinant DNA technology, which will have an economic impact on its use in swine diet formulation.

Calcium and Phosphorus. It is inappropriate to discuss mineral element deficiencies independent of one another because of their interrelationships in metabolism. Phosphorus is related to Ca metabolism, and the level of phosphorus in the diet affects the Ca requirement. An optimum ratio of Ca to P in the diet is between 1:1 and 2:1, depending on the bioavailability of these elements in each in the feedstuffs used. When the ratio is less than 1:1, even a nearly adequate level of Ca in the diet may be associated with signs of Ca deficiency. Absorption of Ca from the intestinal tract is increased by vitamin D, decreased by high dietary fat, decreased by acid pH in the contents, and decreased by phytin-P. A deficiency in growing pigs causes rickets (crooked legs, bone fractures) due to the failure of the protein matrix of the bones to ossify. The corresponding condition in adult swine is osteoporosis. A dietary imbalance of Ca:P (ratio less than 1:2) results in excessive bone resorption (fibrous osteodystrophy) due to nutritional hyperparathyroidism. Calcium deficiency during gestation may result in spontaneous long bone fractures, parturient paresis, or lameness of the dam. The fetus is partially protected by transfer of labile maternal stores of Ca across the placenta. A deficiency of P results in classical rickets. Plant sources of P(phytin P) are poorly utilized by the pig. The recent development of microbially produced phytase, the enzyme that releases organic P from phytate, has provided a practical means of increasing P bioavailability in diets of pigs and poultry. Phytase supplementation of diets high in plant sources of P allows the formulation of diets containing lower levels of total P and reduces P pollution of the environment. Although the total P contributed by grain and protein supplement may seem appreciable, P deficiency may exist unless phytase or a P source of high bioavailability is added to the diet. Di- and mono-calcium phosphate and phosphoric acid are excellent in P bioavailability. Steamed bone meal, defluorinated rock phosphate, and other inorganic sources are lower in bioavailability. The bioavailability of Ca and P from an array of sources for swine has been summarized by Peo (1991). Note that P bioavailability is generally less than 50% in most plant sources of P. Supplemental dietary P levels should be calculated on the basis of biologically available P rather than total P in the diet.

Zinc. The role of supplemental Zn in preventing skin lesions (parakeratosis) in swine was recognized many years ago. Most feedstuffs contain marginal levels of Zn; zinc deficiency can be precipitated by diets high in Ca. Phytic acid (phytin P) present in plants depresses Zn absorption and increases the incidence of parakeratosis. For this reason, the dietary Zn requirement is higher in swine fed an all-plant diet than in those fed a diet containing animal sources of protein or energy. The addition of inorganic Zn salts, zinc carbonate, zinc sulfate, zinc chloride, to the diet is a routine practice in swine feed formulation.

Sodium and Chlorine. These two nutrients are considered together because the major source is common salt (sodium chloride). Whereas the required levels of Na and Cl have been reported to be approximately 0.10 and 0.13% of the diet, respectively, in practice it is common to add 0.40% NaCl to the diet. Common feedstuffs do not contain appreciable quantities of sodium or chloride, so virtually all swine diets are supplemented with salt. Signs of deficiency in swine include anorexia, muscle weakness, and a depraved appetite (tendency to ingest inappropriate substances or objects). Salt toxicity can be induced by a diet containing as little as 2% sodium chloride if water intake is restricted. There is little danger of salt toxicity if water is not restricted.

Iron. Iron deficiency anemia is a serious problem in suckling pigs because milk is very low in iron and placental transfer of iron to the pig fetus is limited. Therefore, iron stores at birth are insufficient to sustain normal hemoglobin synthesis during the first few days, and a supplemental source of iron is critical. Piglets deprived of supplemental iron develop severe anemia (microcytic hypochromic) within a few days and become pale, flabby, and weak and exhibit labored breathing (thumps). Oral iron can be provided by tablet or drench (the ferrous form is more efficiently absorbed than ferric form), but a more efficient way is by intramuscular injection of an iron-dextran complex. Administration of 150 mg of Fe (1 to 2 cc of iron-dextran complex) at 2–3 days of age, followed by a second injection at about 3 weeks, provides sufficient tissue iron storage to promote normal hemoglobin formation until dry feed consumption provides an adequate supply. Most common feedstuffs contain considerable amounts of iron, but it is customary to include a ferrous salt of iron in the trace element premix added to most swine diets. There is a wide margin of safety for oral intake of Fe, so toxicity is not a major concern.

Iodine. Iodine deficiency in swine is manifested mainly by the birth of hairless pigs, often weak or stillborn. The addition of 0.4% iodized salt containing 0.007% iodine (0.035 ppm in the diet) to the diet of growing and adult swine prevents all signs of iodine deficiency and allows normal thyroid function. Iodine in the form of NaI or KI is highly bioavailable, but is easily lost by volatilization (evaporation). Calcium iodate and potassium iodatate are more stable forms in trace element premixes and complete feeds (Miller, 1991).

Vitamin E and Selenium. These two nutrients are considered together here because they are both involved in normal cell membrane integrity and a deficiency of both may be associated with liver necrosis, mulberry heart disease, a yellowish discoloration of body fat, and sudden death. When both vitamin E and Se are marginal, one or more of the deficiency signs may occur. Apparently, either one present in an adequate amount prevents deficiency signs. Common feedstuffs contain several tocopherols, the most biologically active form of which is alpha tocopherol. Although the level of vitamin E in maize and cereal grains is normally abundant, frequent field reports of vitamin E or Se deficiency suggest that some methods of harvesting and storage, for example, heat-drying, may destroy vitamin E or reduce its bioavailability so as to produce deficiency signs when Se is also limiting. The Se content of plants is closely related to the Se content of the soils on which they are grown. Some areas of the United States and other parts of the world have soils low in Se, and crops grown on these soils are, in turn, low in Se. The U.S. Food and Drug Administration has approved the use of 0.30 ppm of inorganic Se to swine diets in the United States to prevent vitamin E–Se deficiency. Se toxicity is a practical concern in some areas in which Se content of soils is high and from which crops contain excessive levels of the element. Toxicity is manifested by poor growth and reproduction. The reproductive failure associated with Se toxicity can be prevented by feeding organic arsenic, which reduces Se absorption from the intestinal tract.

Vitamin A. Most feedstuffs for swine are low in carotene and vitamin A; exceptions are yellow maize and forages. Since vitamin A is stored in large amounts in the liver, the consumption of a diet deficient in vitamin A and carotene may not be associated immediately with deficiency signs. After tissue stores of vitamin A are depleted, growth is reduced, and there may be paralysis, xerophthalmia, and incoordination. In sows, there is reproductive failure, and the birth of piglets with various teratogenic abnormalities, including malformations of the brain, spinal cord, and palate, eyes, limbs, heart, lungs, diaphragm, liver, and kidneys. Swine can convert carotene to vitamin A; therefore, blood concentrations of carotene are normally low relative to those of vitamin A.

Toxicity of vitamin A (hypervitaminosis A) can be produced by excess intakes over extended periods since liver storage of the vitamin is efficient. Teratogenic effects of excess vitamin A include anomalies of the eyes and major blood vessels.

Vitamin D. Vitamin D (vitamins D2 and D3 are equally utilized by the pig) is closely related to dietary Ca and P metabolism through its effect of their absorption and utilization. Vitamin D deficiency reduces absorption of dietary Ca and P. As discussed in Chapter 12, the physiologically active forms of vitamin D, mainly 1-hydroxy- and 1,25-dihydroxy vitamin D, are produced in liver and kidney and exert their effects through changes in movement of Ca and P among and within body tissues. Dietary vitamin D is not required in swine allowed exposure to direct ultraviolet light from the sun, owing to the conversion of 7-dehydrocholesterol to vitamin D3 in the skin. Signs of deficeincy include stiffness and lameness, even in the presence of adequate intakes of Ca and P. Lameness is followed by reduced feed intake and growth depression; deficiency signs are quickly alleviated by parenteral injection or feeding of vitamin D. As the trend toward complete indoor conefinement of swine throughout the life cycle continues, the importance of adequate dietary vitamin D is intensified.

Water-soluble vitamins. The requirements and deficiency signs of each of the water-soluble vitamins have been reviewed by Cook and Easter (1991). The concentrations of vitamin B-12, pantothenic acid, riboflavin, thiamin, and choline, and the amount of biologically available niacin are low or marginal in all-plant diets composed largely of maize or cereal grains and soybean meal or other processed plant protein supplements. Therefore, it is customary to provide vitamin premixes containing each of these vitamins in commercial swine diets to avoid subclinical or clinical signs of deficiencies. In addition, many vitamin premixes contain folacin and biotin, along with the fat-soluble vitamins A, D, E, and K. Vitamin K and biotin are normally not of concern because microbial synthesis in the large intestine is usually sufficient to meet metabolic requirements for them. However, pigs housed in slatted-floor pens with minimum access to feces may become deficient in biotin or vitamin K. Thiamin and vitamin B-12 can be stored in relatively high concentrations in pig tissues for limited periods of time. Therefore, the day-to-day ingestion of these two vitamins by the pig is less critical than that of most of the other water-soluble vitamins. Tryptophan can spare part of the niacin requirement, and methionine can spare part of the choline requirement. Vitamin C, though generally considered to be synthesized in sufficient amounts by the pig to meet metabolic requirements, may improve weight gain of early-weaned pigs subjected to the stress of crowding (Cook and Easter, 1991).

Energy Requirements and Feed Energy Sources for Swine

Feed represents 60% or more of the total cost of pork production. Seventy to 90% of the weight of a typical swine diet is from high-energy feedstuffs such as maize, cereal grains, tubers, and other high-carbohydrate plants. Therefore, a consideration of energy requirements involves major attention to the degree of digestibility of the energy in that high-carbohydrate component. Energy nomenclature and concepts of energy utilization were described in detail in Chapter 10. Ewan (1991) summarized the concepts of energy utilization in swine and presented a table of energy values for many commonly used feedstuffs for swine. The nutritional and overall characteristics of these feedstuffs and their processing methods are described in Chapters 18 and 19. Fats such as lard, soybean oil, and tallow, are more than twice as high in metabolizable energy value as carbohydrate sources. The use of fats in swine diets is usually restricted to preweaning starter diets and, when economically favorable, to diets for sows during lactation. Swine of all ages can utilize fat efficiently, but the amount added to the diet is limited to 5 to 10% (weight basis) by physical problems of mixing and handling. Early dry feed intake of preweaning pigs may be improved by including fat in the starter feed; piglet survival and sow weight maintenance during lactation may be improved by fat in the lactation diet. The relative costs of fat versus carbohydrate sources will dictate the extent of use of fat in swine diets under most conditions. The use of fat in swine feeding was reviewed by Pettigrew and Moser (1991) and Azain (2001).

❏ LIFE-CYCLE FEEDING OF SWINE

Estimates of quantitative nutrient requirements for each phase of the life cycle have been tabulated by the National Research Council (1998). As with other species, the tabulated nutrient requirements must be considered only as guidelines because differences in the genetic background of the animal and environmental or climatic conditions may be expected to affect the requirement for each nutrient.

The following phases of swine life-cycle feeding are addressed in this section: Suckling, Nursery, Growing, Finishing, Breeding (Boars and Sows), Pregnancy, Postbreeding Sows, and Lactation.

Suckling

The suckling period includes birth through the nursing stage, typically 2 to 4 weeks. The neonatal pig has almost no immunity to infectious organisms because the placenta is nearly impervious to the transfer of immune antibodies to the fetus. Therefore, the ingestion of colostrum immediately after birth is critical not only to supply nutrients but also to supply passive immunity via its high content of immunoglobulins (IGs), including gamma globulin. These large globular proteins must be ingested within a few hours after birth because their effectiveness in providing immunity depends on their absorption intact. The intestine loses its ability to transport large proteins, including IGs, within 48 to 72 hours after birth. Therefore, it is important that neonates nurse as soon as possible to ensure adequate immune protection and nutrient intake. Colostrum contains higher percentages of total solids (dry matter) and protein than milk produced later in lactation. Colostrum collected immediately after parturition contains 25 to 30% total solids, 3 to 7% fat, 15 to 20% protein, 5 to 7% lactose, and 0.63% minerals (ash) (Klobassa et al., 1987). By day 2 of lactation, corresponding concentrations are slightly higher for fat, lactose, and ash and much lower for protein as a result of the rapid decline in IGs. Values at week 6 of lactation are similar to those at 2 days. Total daily milk production rises from parturition to about 3 weeks of lactation, then declines at 6 weeks and beyond. The shape of the lactation curve explains the trend toward earlier weaning than in the past. By 3 weeks of age, the total milk required by the litter surpasses the milk supply, requiring that supplemental feed be provided to sustain normal growth of the litter. Therefore, it is customary to provide supplemental feed (termed *creep* or *prestarter feed*) from 3 weeks until weaning. Weaning at less than 3 weeks has become a common practice for two reasons: (1) to shorten the period to rebreeding,

TABLE 25.2 Sequence of phase-feeding programs for early and conventionally weaned pigs.

EARLY WEANING 5- TO 21-D WEANING		CONVENTIONAL WEANING GREATER THAN 21-D WEANING	
5–11 lb	SEW		
11–15 lb	Transition	11–15 lb	Phase 1
15–25 lb	Phase 2	15–25 lb	Phase 2
25–50 lb	Phase 3	25–50 lb	Phase 3

Source: Adapted from Tokach et al. (1997b), *Breeding Herd Recommendations for Swine,* MF2302, Kansas State University Agricultural Experiment Station and Cooperative Extension Service. Reproduced with permission from S. S. Dritz, R. D. Goodband, J. L. Nelssen, and M. D. Tokach.

thereby improving reproductive rate and (2) to control the transfer of infectious disease from sow to litter by moving the pigs to a location remote from their dam. The practice of weaning at 2 to 3 weeks of age has become known in the swine industry as *segregated early weaning* (SEW). Table 25.2 (Tokach et al., 1997a) summarizes the dietary phases for pigs as they progress from early weaning (SEW, 5- to 21-day weaning; body weight of 5 to 11 lb) through phase 1 (transition phase, 11 to 15 lb), phase 2 (15 to 25 lb), and phase 3 (25 to 50 lb). An example of a suitable sarter diet for SEW and phase 1 pigs is shown in Table 25.3. Many variations exist.

Nursery

The nursery period is a critical stage in the life of the pig because it represents the transfer of responsibility for food supply from the sow to humans. Social structure and diet composition are changed drastically. It is important to provide a highly palatable and nutritious diet to minimize the "growth check" associated with this dramatic change to postweaning life. The use of spray-dried animal plasma, blood meal, whey, and other animal products has evolved as a means of encouraging rapid adaptation to the starter diet (see Table 25.3). The nursery period usually extends to about 10 weeks of age (35 to 50 lb [14 to 20 kg] per pig). Examples of acceptable phase 2 and phase 3 diets for nursery pigs weighing 15 to 25 lb and 25 to 50 lb, respectively, are shown in Tables 25.4 and 25.5.

Growing

Transfer of pigs from the nursery to a separate growing facility coincides with changes in the

TABLE 25.3 A suitable starter diet fi pigs immediately after weaning[a] (less t body weight).

INGREDIENT
Corn or grain sorghum (ground)
Soybean meal (46.5% CP)
Spray-dried animal plasma
Spray-dried blood meal
Select menhaden fish meal
Spray-dried whey
Lactose
Choice white grease
Lysine HCl
DL-methionine
Monocalcium phosphate (21% P)
Limestone
Salt (NaCl)
Vitamin premix[b]
Trace mineral premix[b]
Antibiotic premix[c] (optional)

[a]The combination of ingredients used in t be changed according to price and availab gredients. For example, barley or grain s place some or all of the corn as an energy meal or another oilseed meal might replac bean meal.

[b]Vitamin and mineral premixes contain amou trients to meet dietary requirements of we weight. These added nutrients are in additi present in the feed ingredients.

[c]Antibiotics may be supplemented at subthe promote health and growth rate in these yo *Source:* Adapted from the suggested SEW die Goodband, and Nelssen (1997). Use of this m to S. S. Dritz, R. D. Goodband, J. L. Nelssen, a *Swine Nutrition Guide,* Kansas State Unive KS.

TABLE 25.4   Acceptable Phase 2 diets for pigs weighing 15 to 25 lb.

INGREDIENT, LB/TON	NO FAT		ADDED FAT	
	FISH MEAL	BLOOD MEAL	FISH MEAL	BLOOD MEAL
Corn or milo (grain sorghum)	1102.8	1117	962.6	976.7
Soybean meal, 46.5% CP	540	540	580	580
Select menhaden fish meal	80	—	80	—
Spray-dried blood meal	—	50	—	50
Spray-dried whey	200	200	200	200
Choice white grease	—	—	100	100
Monocalcium phosphate, 21% P	27	37	27	37
Limestone	14	19	14	19
Salt	5	5	5	5
Vitamin premix	5	5	5	5
Trace mineral premix	3	3	3	3
Lysine HCl	3	3	3	3
DL-Methionine	0.2	1	0.4	1.3
Antibiotic premix	20	20	20	20
Total	2000	2000	2000	2000

Calculated Analysis

Lysine, %	1.35	1.35	1.40	1.40
Met:lysine ratio, %	28	28	28	28
Met & Cys:lysine ratio, %	56	56	55	55
Threonine:lysine ratio, %	66	66	65	65
Tryptophan:lysine ratio, %	20	20	19	20
ME, Kcal/lb	1476	1473	1574	1570
Protein, %	21.1	20.8	21.4	21.1
Calcium, %	0.90	0.89	0.90	0.89
Phosphorus, %	0.80	0.80	0.80	0.80
Available phosphorus, %	0.54	0.54	0.54	0.54
Lysine:calorie ratio, g/Mcal ME	4.15	4.16	4.03	4.04

Source: Tokach et al. (1997a), *Starter Pig Recommendations,* Table MF2300, Kansas State University Agricultural Experiment Station and Cooperative Extension. Reproduced with permission from S. S. Dritz, R. D. Goodband, J. L. Nelssen, and M. D. Tokach.

composition of the diet. Daily weight gain continues to increase beyond 100 lb (45 kg), and daily accretion of protein increases progrssively (maximum of up to a daily deposition of 140 g) to about 135 lb (54 kg) body weight, then gradually declines to about 110 g/d at 250 lb (100 kg). Daily feed consumption continues to increase as the pig grows, but the percentage of protein in the diet is reduced as the pig grows only enough to ensure adequate protein deposition while avoiding excess fat deposition as a result of deaminating protein in excess of requirement and converting

it to body fat. This reduction in the protein content of the diet is done in a stepwise sequence involving one or more diet changes during the growing period and additional reductions in protein during the finishing period.

The dietary protein requirement is related directly to the essential amino acid requirements. Lysine is the first limiting amino acid in most feedstuffs for swine. Therefore, it has become customary to express all other amino acid requirements on the basis of the lysine requirement set at 100. The calculation of amino acid requirements

TABLE 25.5 Acceptable Phase 3 diets for pigs weighing 25 to 50 lb.

INGREDIENT, LB/TON	NO FAT		ADDED FAT	
	1.25	1.30	1.30	1.35
Corn or milo	1276.6	1242.5	1131.3	1096.2
Soybean meal, 46.5% CP	645	680	690	725
Choice white grease	—	—	100	100
Monocalcium phosphate, 21% P	30	29	30	30
Limestone	20	20	20	20
Salt	7	7	7	7
Vitamin premix	5	5	5	5
Trace mineral premix	3	3	3	3
Lysine HCl	3	3	3	3
DL-Methionine	0.4	0.5	0.7	0.8
Antibiotic premix (optional)	10	10	10	10
Total	2000	2000	2000	2000
Calculated Analysis				
Lysine, %	1.25	1.30	1.30	1.35
Met:lysine ratio, %	28	28	28	28
Met & Cys:lysine ratio, %	58	58	57	57
Threonine:lysine ratio, %	67	67	66	65
Tryptophan:lysine ratio, %	21	21	21	21
Me, Kcal/lb	1485	1486	1583	1583
Protein, %	20.4	21.1	20.9	21.5
Calcium, %	0.76	0.76	0.76	0.77
Phosphorus, %	0.70	0.70	0.70	0.70
Available phosphorus, %	0.39	0.38	0.39	0.40
Lysine:calorie ratio, g/Mcal ME	3.82	3.97	3.73	3.87

Source: Tokach et al. (1997a), *Starter Pig Recommendations,* Table 7, MF 2300, Kansas State University Agricultural Experiment Station and Cooperative Extension. Reproduced with permission from S. S. Dritz, R. D. Goodband, J. L. Nelssen, and M. D. Tokach.

for growing pigs based on modeling procedures has been described by the National Research Council (1998). Amino acid requirements are usually expressed as g/megacalorie (Mcal) or kilocalorie (Kcal) of digestible energy (DE). The nutrient requirements, including amino acid requirements, of growing and finishing pigs are based on the assumption that the diet is full-fed (free access). Examples of acceptable diets for use during the growing period (phase 4, body weight of 50 to 150 lb) are shown in Table 25.6.

Finishing

The nutrient requirements during the finishing period are similar to those during the growing pe-riod, except that protein and amino acid concentrations and Ca and P concentrations are less than those for growing pigs because of the lower daily protein, accretion rate and mineral of finishing pigs.

The protein and amino acid requirements of gilts and intact males are higher than those of barrows. Therefore, split-sex feeding during the growing and finishing periods is used to match protein intake with protein requirements, thereby reducing feed costs. Lean genotypes need a higher percentage of dietary protein than fatter genotypes. As a consequence, pigs of differing genetic capacity for lean carcasses should be penned separately and be fed diets containing protein levels appro-

TABLE 25.6 Acceptable grower diets (Phase 4) for pigs weighing 50 to 150 lb.[a]

INGREDIENT, LB/TON	GROWER DIETS (WITHOUT FAT) LYSINE, %					
	0.80	0.90	1.00	1.10	1.20	1.30
Corn or milo	1620.0	1551.0	1476.0	1405.9	1336.7	1261.5
Soybean meal, 46.5%	320	390	465	535	605	680
Choice white grease	0	0	0	0	0	0
Monocalcium phosphate, 21% P	25	24	24	24	23	23
Limestone	19	19	19	19	19	19
Salt	7	7	7	7	7	7
Vitamin premix	3	3	3	3	3	3
Trace mineral premix	3	3	3	3	3	3
Lysine HCl	3	3	3	3	3	3
DL-Methionine	0	0	0	0.1	0.3	0.5
Total	2000.0	2000.0	2000.0	2000.0	2000.0	2000.0
Calculated Analysis						
Lysine, %	0.80	0.90	1.00	1.10	1.20	1.30
Methionine:lysine ratio, %	31	30	28	28	28	28
Met & Cys:lysine ratio, %	68	65	62	60	59	58
Threonine:lysine ratio, %	74	72	70	69	68	67
Tryptophan:lysine ratio, %	22	22	21	21	21	21
ME, Kcal/lb	1502	1502	1502	1501	1501	1501
Protein, %	14.3	15.7	17.1	18.4	19.7	21.2
Calcium, %	0.66	0.66	0.67	0.67	0.67	0.68
Phosphorus, %	0.59	0.59	0.61	0.62	0.62	0.64
Available phosphorus, %	0.32	0.32	0.32	0.32	0.32	0.32
Lysine:calorie ratio, g/Mcal ME	2.42	2.71	3.03	3.32	3.61	3.93

[a]Note the continuum in lysine level and changes in concentrations of corn or milo and soybean meal throughout the grower and finisher periods in Tokach et al. (1997a, 1997b) and Dritz et al. (1997).
Source: Dritz et al. (1997), *Growing-Finishing Pig Recommendations,* Table 6, MF2301, Kansas State University Agricultural Experiment Station and Cooperative Extension. Reproduced with permission from S. S. Dritz, R. D. Goodband, J. L. Nelssen, and M. D. Tokach.

priate for their protein requirement to avoid the economically expensive use of excess protein to produce body fat. Acceptable diets for the finishing period are similar to those used for the growing period, but are lower in Ca and P (by 0.2 to 0.3 percentage units) and lower in protein (by 1 to 2 percentage units) (see Table 25.6 and Dritz et al., 1997).

Breeding (Boars and Sows)

Gilts and boars selected for breeding stock are typically full-fed adequate diets throughout all stages of growth including the finishing period (250–270 lb [113.5–122.6 kg] live weight at 6 months of age) to maximize growth rate and provide for identifying individuals with superior lean growth potential. To prevent overfattening, boars and gilts selected for breeding are limit-fed from 6 months of age to breeding at about 8 months. Sows to be rebred postweaning are similarly limit-fed from weaning to re-breeding (usually 5 to 7 days following a 2- to 3-week lactation). The degree of feed restriction is determined by the degree of fatness of the sow at weaning.

Males and females used for breeding must be fed nutritionally adequate diets in amounts appropriate to maintain body weight at a level compatible with a high level of reproductive performance and longevity. The amount fed daily to boars will depend on body size and fatness, but, in general, mature boars and mature pregnant sows are fed similar amounts daily (6000 to 7000 Kcal of digestible energy).

Total or partial confinement in slatted- or grid-floor pens is common in commercial swine breeding facilities. Therefore careful attention to adequate intakes of Ca, P, and other inorganic elements, as well as other nutrients, is important to ensure sound bone health, mobility, and herd longevity. Environmental temperature may be a factor in breeding stock kept outdoors in cold climates. During extended exposure to ambient temperatures below the zone of thermal neutrality (comfort zone), daily rations should be increased 10 to 20% to offset the extra energy cost of body temperature maintenance. Sows and boars that are underfed may become too thin to reproduce normally; similarly, those that are overfed may lack libido or physical stamina due to obesity. Careful attention to body condition and appropriate adjustment of daily feed allowance of individual boars and open and pregnant sows within the broad limits of recommendations (National Research Council, 1998) serves to maximize reproductive efficiency and herd longevity of male and female breeding swine.

Pregnancy

The nutrient requirements for the sow during gestation are influenced by two separate productive functions—the need for maintaining the pregnant sow and the provision of an adequate nutrient supply for the developing fetuses. In general, nutrient requirements in early gestation are modest as compared to late gestation and, particularly, lactation. It is possible, at least with some nutrients, to observe normal reproduction through one reproductive cycle on a diet that is clearly inadequate for the dam. For example, results with sows fed protein-free diets during gestation show that the dam can draw on her own reserves to meet the needs of the fetuses for growth and survival. However, such a diet cannot be considered adequate because it does not satisfy the long-term requirements of both dam and fetuses.

The nutrition of the fetus depends on the transfer of nutrients across the placenta from the maternal blood. Adequate fetal nutrition therefore depends on adequate levels of nutrients and their precursors circulating in the maternal blood. In the case of some nutrients, the concentration in the blood of the dam is maintained in close homeostatic control. For example, serum Ca is maintained within a narrow range even when dietary intake of Ca is low, due to the high sensitivity of the endocrine system to very small changes in concentration of Ca in the blood, resulting in removal of Ca from bone when serum Ca is low and deposition of bone Ca when serum Ca is high. In contrast, a deficiency of the water-soluble vitamin, riboflavin, for example, in the diet of the pregnant sow, causes a drop in maternal blood riboflavin because of limited tissue storage, and the amount of the vitamin available for transfer to the fetus is reduced accordingly. Placental transfer of Fe is also limited in the pig, resulting in fetal iron stores that do not permit adequate hemoglobin synthesis in the newborn pig. Parenteral Fe administration to the dam in late gestation does not cause an appreciable increase in fetal Fe stores, so the only satisfactory means of providing adequate Fe to maintain normal hemoglobin concentration in the blood of newborn piglets is by administration of Fe directly to the pig at a few days of age. On the other hand, placental transfer of iodine, manganese, and several other trace elements is relatively efficient and responsive to maternal diet. Therefore, in discussing fetal pig nutrition, it is important to specify the nutrient of concern.

The nutrient requirements for fetal growth are hidden under the umbrella of fetal plus maternal requirements, so that the question of the nutritional requirements of the fetus per se is, in a sense, academic. Consequently, the nutrient requirements of the pregnant sow are really the combined requirements for herself and those of her fetuses and other products of conception, that is, placenta and other fetal and maternal membranes associated with the pregnancy. In addition, the increased size of the empty uterus and the other components of "pregnancy anabolism," including the accretion of mammary tissue in late gestation in preparation for lactation, contribute to the increased nutrient requirements of the pregnant sow. In any case, fetal growth cannot occur without transfer of nutrients across the placenta from the dam. Some nutrients are transferred efficiently, others very inefficiently. All of the fat-soluble and water-soluble vitamins readily reach the pig fetus, while maternal intact protein crosses the placenta in only negligible amounts, but individual amino acids are efficiently transferred. Amino acids reaching the fetus from the maternal circulation form the main source of tissue protein synthesis in the pig fetus. The fetus uses nonprotein nitrogen to synthesize dispensable amino acids. Individual fatty acids, but not triglycerides, cross the placenta, but the main source of energy for the pig fetus is glucose from the dam.

The pig fetus undergoes enormous growth during the latter half of gestation, increasing in weight from about 200 g at mid-gestation to about 1200 g at term (Fig. 25.1).

Therefore, the fetal requirement for individual nutrients is greatest in late gestation, coinciding with rapid fetal tissue accretion. Transfer of water and inorganic elements to the fetus increases severalfold during late gestation. The inorganic element composition of maternal blood plasma and placenta, placental fluids, and fetus in the pig at 110 days after conception was reported by Hansard (1966). Transfer of inorganic elements to the fetus involves several factors, including placental type and degree of permeability. The complex placenta of swine is termed epitheliochorial. Six tissue layers separate the maternal and fetal blood in the pig, three maternal and three fetal; the extraembryonic membranes can be separated from the uterus with no damage or destruction of the uterine epithelium and, unlike the case in other species, the blood must pass through the chorion epithelium to reach the fetus. This architecture results in some unique consequences to the fetus, including the nearly complete lack of immune globulins transferred to the fetus. Therefore, the pig has no passive immunity at birth and must receive protection via colostrum intake during the first days after birth. Pregnant sows will consume far more feed voluntarily than is required and will become obese if intake is not restricted to a daily intake of approximately 4 to 5 lb (1.8 to 2.3 kg) of a high-concentrate diet (6600 Kcal of DE) throughout pregnancy until about the last 2 weeks before parturition, when daily feed should be increased to 6 to 8 lb (2.4 to 3.2 kg) to accommodate the surge in weight of the fetus in late pregnancy The specific amount of restriction needed is dependent on the size and body fatness of the sow. Examples of acceptable diets for pregnant sows are shown in Table 25.7.

Lactation

The nutrient output in sow milk during a five-week lactation is much greater than the nutrient deposition in fetuses and placental membranes during a 114-day gestation period. Therefore, as with ruminants, nutrient requirements of the sow are far more demanding for lactation than for gestation. Although the sow can deplete her own body reserves for milk production, complete or partial lactation failure results if nutrient restriction is severe or prolonged. The degree to which the dam's reserves can be drawn upon for lactation when dietary intake is insufficient is related to the specific nutrient of concern. For example, transfer of Fe into the milk is inefficient, so that even a high level of Fe in the diet of the dam does not appreciably increase the Fe concentration of the milk. In general, a deficiency of a particular nutrient is manifested more by a reduction in total milk production than by a decreased concentration of that nutrient in the milk.

The most striking effect of level of intake on total milk production is energy intake. If the lactating sow is not allowed to eat at or near ad libitum, milk production declines. In the well-fed sow, milk production increases as litter size increases, up to the point at which her genetic capacity for milk production is reached. Mature sows raising large litters

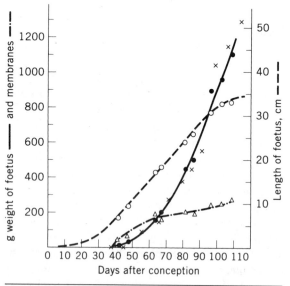

Figure 25.1   *Length (from end of snout to base of tail) and weight of foetuses and weight of membranes during pregnancy. X indicates American results; ●, ○, and △ indicate Danish results. From Pond, et al. (1991).*

TABLE 25.7 Acceptable gestation diets.

INGREDIENT, LB/TON	LYSINE, %		
	0.55	0.60	0.65
Corn or milo	1684	1665	1611
Soybean meal, 46.5% CP	225	260	300
Choice white grease	—	—	—
Monocalcium phosphate, 21% P	47	46	45
Limestone	21	21	21
Salt	10	10	10
Vitamin premix	5	5	5
Sow add pack	5	5	5
Trace mineral premix	3	3	3
Total	2000	2000	2000
Calculated Analysis			
Lysine, %	.55	.60	.65
Met:lysine ratio, %	41	39	37
Met & Cys:lysine ratio, %	90	85	81
Threonine:lysine ratio, %	94	91	88
Tryptophan:lysine ratio, %	27	26	26
ME, Kcal/lb	1480	1480	1480
Protein, %	12.8	13.5	14.2
Calcium, %	0.90	0.90	0.90
Phosphorus, %	0.80	0.80	0.80
Available phosphorus, %	0.55	0.54	0.53

Source: Tokach et al. (1997b), *Breeding Herd Recommendations for Swine,* MF2302, Kansas State University Agricultural Experiment Station and Cooperative Extension. Reproduced with permission from S. S. Dritz, R. D. Goodband, J. L. Nelssen, and M. D. Tokach.

often will consume 6.8 to 9.0 kg (15 to 20 lb) or more/day of a high-concentrate diet (3600 Kcal of ME/kg) at the peak of lactation (4–5 weeks postpartum). Daily yield of sow milk, which has a dry matter content of about 20% (compared to 13% for bovine milk), may reach a peak of 8 kg or more during the fourth or fifth week of lactation. The nutrient composition of sow milk and colostrum is summarized (Bowland, 1966). Since the caloric density of sow milk is considerably higher than that of cow milk, a large supply of energy is available to individual pigs during a period of very rapid growth. It is important to recognize that total sow feed requirements during lactation will vary depending on the energy density of the diet and the nutritional history of the sow. The energetic efficiency of lactation is higher when milk is produced by current energy intake than by dependence on body fat reserves. Therefore, in practical feeding, the highest efficiency of energy utilization is achieved by controlled feeding during gestation to minimize body fat mobilization during lactation for milk production.

Inadequate protein (indispensable amino acids) intake also results in reduced milk production, although the effect is less marked than for energy. As in other stages of the life cycle, dietary amino acids are met by formulating the lactation diet on the basis of the levels of amino acids and other nutrients in relation to the energy density of the diet. Deficiencies of other required nutrients may also reduce milk production, though to a lesser degree than for protein and energy.

Examples of acceptable diets to be fed ad libitum to lactating sows are shown in Table 25.8.

TABLE 25.8   Acceptable lactation diets.

INGREDIENT, LB/TON	LYSINE, %					
	0.70	0.80	0.90	1.0	1.1	1.2
Corn or milo	1581	1507	1438	1364	1295	1227
Soybean meal, 46.5% CP	330	405	475	550	620	690
Choice white grease[a]	0 to 5	0 to 5	0 to 5	0 to 5	0 to 5	0 to 5
Monocalcium phosphate, 21% P	45	44	42	41	40	38
Limestone	21	21	22	22	22	22
Salt	10	10	10	10	10	10
Vitamin premix	5	5	5	5	5	5
Sow add pack	5	5	5	5	5	5
Trace mineral premix	3	3	3	3	3	3
Total	2000	2000	2000	2000	2000	2000
Calculated Analysis						
Lysine, %	0.70	0.80	0.90	1.00	1.10	1.20
Met:lysine ratio, %	36	33	32	30	29	28
Met & Cys:lysine ratio, %	79	73	69	66	63	61
Threonine:lysine ratio, %	86	82	79	76	74	73
Tryptophan:lysine ratio, %	25	25	24	24	23	23
ME, Kcal/lb	1480	1480	1480	1480	1480	1480
Protein, %	14.8	16.5	17.5	18.9	20.2	21.6
Calcium, %	0.90	0.90	0.90	0.90	0.90	0.90
Phosphorus, %	0.80	0.80	0.80	0.80	0.80	0.80
Available phosphorus, %	0.53	0.53	0.51	0.50	0.50	0.48

[a]If adding fat, substitute for grain on an equal weight basis.
Source: Tokach et al. (1997b), *Breeding Herd Recommendations for Swine,* MF2302, Kansas State University Agricultural Experiment Station and Cooperative Extension. Reproduced with permission from S. S. Dritz, R. D. Goodband, J. L. Nelssen, and M. D. Tokach.

❏ FEEDSTUFFS USED IN SWINE FEED FORMULATION

An exhaustive list of potential feedstufs for swine is beyond the scope of this discussion. The classification and composition of feedstuffs is described in Chapter 19. Appendix Table 2 describes the composition of some common feed ingredients for swine (National Research Council, 1998). Most of the feedstuffs considered as protein or energy sources also supply significant amounts of minor constituents (vitamins and minerals). Therefore, when diets for swine are being formulated to meet all nutrient requirements, the amounts of supplemental vitamins and mineral elements needed to overcome the deficits existing in natural sources may vary from zero to considerable, depending on the nature of the feedstuffs used in the formula-tion. One must consider, therefore, the contributions of feedstuffs not only in terms of energy and protein but also in terms of their vitamin and inorganic element content.

Modern feed formulation laboratories use highly specialized equipment to determine the vitamin, mineral, amino acid, and fatty acid contents of feedstuffs. These analyses are expensive, but, with modern automated instrumentation, it is possible to obtain a complete nutrient analysis on a wide spectrum of feedstuffs, so that diets can be formulated within a few hours or days after the sample is obtained. An introduction to diet formulation for animals, including swine, was covered in Chapter 21. The success of diet formulation depends on accurate inputs: (1) tolerance (upper and lower) of specific ingredients, (2) feedstuff composition data, and (3) current ingredient costs.

A diet so formulated may meet all specifications, but the final measure of acceptability is the performance of the pig to which it is fed. Least-cost diet formulation does not account for factors, such as palatability or bioavailability of nutrients, that may have an important effect on animal response. In this connection, the processing of feedstuffs before diet mixing or the processing of the final mixed diet may have important effects on diet intake and animal performance. Chapter 20 addressed the overall subject of feed preparation and processing, based on reviews by Hancock and Behnke (2001), Hogberg et al. (1980), and Liptrap and Hogberg et al. (1991). The direction and magnitude of the response to a specific set of processing conditions is never completely predictable because of differences in pigs, plant growing and harvesting conditions, and plant variety (cultivar). Physiological, environmental, and nutritional factors affect feed intake in swine (National Research Council, 1987; Houpt and Houpt, 1991). Daily voluntary intake of growing swine can be predicted by estimating digestible energy consumption under thermoneutral conditions. This estimate can then be adjusted for physiological, environmental, and nutritional factors.

Common Feedstuffs for Swine

Feedstuffs used for swine are described in Chapter 19 of this work and in several other textbooks (Pond et al., 1991; Lewis and Southern, 2001; and McGlone and Pond, 2003).

Energy Sources. Grain and their byproducts are by far the most important sources of carbohydrates for swine. Corn (maize) (*Zea mays*) is the basis for most swine diets throughout the Americas, and in the United States alone, about half of all maize produced is fed to swine. In the temperate climates of Canada and Europe, maize is commonly replaced by barley, wheat, rye, oats, and triticale. Other seeds used as energy sources for swine include buckwheat and grain amaranth. The production of flour from maize and cereal grains for human consumption and of ethanol for use in motor vehicle fuels results in vast quantities of milling and distillery byproducts that are available at reasonable cost for use in swine diet formulation. These products are higher in fiber and lower in metabolizable energy than the grains from which they are derived. Other sources of energy for swine include roots and tubers, such as cassava, potatoes, sweet potatoes, bananas, cane sugar molasses, and fats and oils.

Protein Sources. Next to energy, protein is the class of nutrients swine need in the largest amount. Since protein sources are generally more expensive than energy sources, the best choice of protein to minimize total diet cost is that amount of one or more protein supplements in the diet that allow the formulation of a diet with the smallest amount of total protein and indispensable amino acids compatible with normal productive function (growth, gestation, lactation). A broad array of protein supplements is available. Only the most common ones are described here.

Animal Products. Animal proteins are generally more expensive than plant proteins, but they usually contain a better balance of amino acids of greater bioavailability and are also good sources of vitamins and trace mineral elements. The common animal protein supplements include meat meal, meat-and-bone meal, blood meal, and fish meal. In some parts of the world inedible or surplus byproducts of the dairy industry, such as milk whey, sometimes become available for swine feeding. Milk products are generally used only in diets of early-weaned pigs and in starter diets.

Plant Products. Plant protein supplements include primarily oilseed meals. Soybean meal is by far the most important plant protein supplement used in swine feeding in the United States as well as in many other countries. Oilseed meals are available for animal feeding as byproducts of the vegetable oil industry. The high-protein residue remaining after extraction of the oil from the seed provides an economical source of protein for swine. Most oilseed meals are deficient in one or more amino acids, requiring judicious selection of energy sources to complement the amino acids provided by the high-protein meal. A maize–soybean meal diet supplemented with vitamins and mineral elements provides a nutritionally well-balanced diet (see Table 25.1). Other common oilseed meals are coconut meal, cottonseed meal, linseed meal, peanut meal, canola meal, safflower meal, sesame meal, and sunflower meal. A second major group of plant protein supplements is the array of seed legumes. This list includes whole soybean (*Glycine max*), dry bean, also called kidney, snap, navy bean (*Phaseolus vulgaris*), mung bean (*P. mungo*), chick pea (*Cicer arietum*), cow pea

(*Vigna sinensis*), field bean (*Vicia fabia*), field pea (*Pisum sativum*), and others. Most of the seed legumes must be heated to remove an array of growth inhibitors. The seed legumes generally contain 20 to 30% protein, compared to 40 to 50% protein for oilseed meals and even higher levels of protein for some animal products. Other high-protein sources include lupin, alfalfa, and several single-cell proteins, for example, blue-green algae, fungi, bacteria, and yeast grown in continuous culture on a variety of substrates.

Mineral and Vitamin Supplements. Although the largest proportion of the diet is made up of energy and protein, the mineral elements and vitamins are vital to normal growth and reproduction. Most energy and protein sources provide some vitamins and minerals, but often it is necessary to add specific mineral and vitamin sources to balance the diet. Common sources of supplemental Ca and P were given in Chapter 19 (see Table 19.14). The composition of typical trace mineralized salt and its contribution to the diet when included at a level of 0.40% was given in Chapter 19 (see Table 19.13). The amount of Zn provided in the example in Table 19.14 is only about 1% of the amount required by swine. High-Zn trace-mineralized salt is available for swine. In large pork production enterprises, it is common to rely on commercially prepared trace element premixes and vitamin premixes as supplements to be added to the other feed ingredients during final diet mixing. The composition of a typical vitamin premix for swine is shown in Table 25.9.

TABLE 25.9 Composition of a typical vitamin premix for addition at 0.20% of the complete mixed diet.

VITAMIN	AMOUNT/KG OF PREMIX	AMOUNT/KG OF MIXED DIET WHEN ADDED AT 0.20%
Vitamin A	2,600,000 IU	5200 IU
Vitamin D	352,000 IU	704 IU
Vitamin E	17,600 IU	35.2 IU
Vitamin K	1760 mg	3.52 mg
Riboflavin	2640 mg	5.28 mg
d-Pantothenic acid	10,560 mg	21.12 mg
Niacin	14,080 mg	28.16 mg
Vitamin B_{12}	13.2 mg	26.40 mg
Thiamin	1100 mg	2.2 mg
Biotin	110 mg	0.88 mg
Carriers: Ca not more than 10%		
Roughage product not more than 40%		

pigs, boars and gilts to be kept for breeding, pregnant sows, and lactating sows; and common feedstuffs used in swine diet formulation.

❏ SUMMARY

- The pig has been an important source of human food for centuries. Worldwide, pork is consumed in larger quantities than meat from any other animal species. Feed represents 50 to 80% of the cost of swine production; therefore, high efficiency of growth and feed utilization through adequate nutrition is of high priority for a sustainable industry.

- The general principles of swine nutrition and applied swine feeding are addressed in this chapter. Major topical headings include quantitative nutrient requirements at each stage of the life cycle; applied feeding programs for suckling pigs, newly weaned pigs, growing and finishing

❏ REFERENCES

Azain, M. J. 2001. In *Swine Nutrition,* Chapter 6, p. 93, A. J. Lewis, and L. L. Southern (eds.). CRC Press, Boca Raton, FL.

Baker, D. H., et al. 1993. In *Growth of the Pig,* Chapter 8, p. 133 in G. R. Hollis (ed.). CAB International, Wallingford, UK.

Beermann, D. H. 1991. In *Growth of the Pig,* Chapter 11, p. 185, G. R. Hollis (ed.). CAB International, Wallingford, UK.

Bowland, J. P. 1966. In *Swine in Biomedical Research,* Table 23-8, L. K. Bustad and R. O. McClellan (eds.). Frayn Printing Co., Seattle, WA.

Burrin, D. G. 2001. In *Biology of the Domestic Pig,* Chapter 7, p. 309, W. G. Pond, and H. J. Mersmann (eds.). Cornell University Press, Ithaca, NY.

Close, W. H., and D. J. A. Cole. 2000. *Nutrition of Sows and Boars,* Nottingham University Press, Nottingham, UK.

Cole, D. J. A., et al. 2000. *Recent Developments in Pig Nutrition 2.* Nottingham University Press, UK.

Cook, D. A., and R. A. Easter. 1991. In *Swine Nutrition,*

Chapter 14, p. 235, E. R. Miller, et al. (eds.). Butterworth-Heinemann, Boston.

Dritz, S. S., et al. 1997. *Growing-Finishing Pig Recommendations.* MF 2301, Kansas State University, Manhattan, KS.

Ewan, R. C. 1991. In *Swine Nutrition,* Chapter 7, p. 121, E. R. Miller, et al. (eds.). Butterworth-Heinemann, Boston.

Hansard, S. L. 1966. In *Swine in Biomedical Research,* p. 79, L. K. Bustad and R. O. McClellan (eds.). Frayn Printing Co., Seattle, WA.

Garnsworthy, P. C., and J. Wiseman. 2001. *Recent Developments in Pig Nutrition 3.* Nottingham University Press, UK.

Hogberg, M. G., et al. 1980. *National Pork Industry Handbook,* No. 71. Purdue University,West Lafayette, IN.

Houpt, K. A., and T. R. Houpt. 1991. In *Swine Nutrition,* Chapter 21, p. 361, E. R. Miller, et al. (eds.). Butterworth-Heinemann, Boston.

Kellems, R. O., and D. C. Church. 2001. *Livestock Feeds and Feeding,* 5th ed. Prentice-Hall, New York.

Klobassa, F., et al. 1987. J. Anim. Sci. 64: 1458.

Knabe, D. A. 1991. In *Swine Nutrition,* Chapter 19, p. 327, E. R. Miller, et al. (eds.). Butterworth-Heinemann, Boston.

Lewis, A. J. 1991. In *Swine Nutrition,* Chapter 9, p. 147, E. R. Miller, et al. (eds.). Butterworth-Heinemann, Boston.

Lewis, A. J., and L. L. Southern. 2001. *Swine Nutrition,* 2nd ed. CRC Press, Boca Raton, FL, 1009 pp.

Liptrap, D. O., and M. G. Hogberg. 1991. In *Swine Nutrition,* Chapter 22, p. 373, E. R. Miller, et al. (eds.). Butterworth-Heinemann, Boston.

McGlone, J. J., and W. Pond. 2003. *Pig Production: Biological Principles and Applications.* Delmar Learning, a Division of Thomson Learning, Clifton Park, NY, 395 pp.

Miller, E. R. 1991. In *Swine Nurition,* Chapter 15, p. 267, E. R. Miller, et al. (eds.). Butterworth-Heinemann, Boston.

Miller, E. R., D. E. Ullrey, and A. J. Lewis. 1991. *Swine Nutrition.* Butterworth-Heinemann, Boston, MA, 673 pp.

National Research Council (NRC). 1980. *Mineral Tolerances of Domestic Animals.* National Academy of Science, Washington, DC.

National Research Council (NRC). 1987. *Predicting Feed Intake of Food-Producing Animals.* National Academy Press, Washington, DC.

National Research Council (NRC). 1998. *Nutrient Requirements of Swine,* 10th rev. ed. National Academy Press, Washington, DC.

National Research Council (NPC). 1979. *Nutrient Requirements of Swine, no. 2.* National Academy of Science, Washington, DC.

National Research Council (NRC). 1987. *Vitamin Tolerance of Animals.* National Academy Press, Washington, DC.

Patterson, B. W., et al. 1992. *J. Nutr.* **122:** 1830.

Peo, E. R., Jr. 1991. In *Swine Nutrition,* Chapter 10, p. 165, E. R. Miller, et al. (eds.). Butterworth-Heinemann, Boston.

Pettigrew, J. E., and R. L. Moser. 1991. In *Swine Nutrition,* Chapter 8, p. 133, in E. R. Miller, et al. (eds.). Butterworth-Heinemann, Boston.

Pond, W. G., and J. H. Maner. 1984. *Swine Production and Nutrition.* AVI Publishing Co., Westport, CT, 731 pp.

Pond, W. G., J. H. Maner, and D. L. Harris. 1991. *Pork Production Systems.* Van Nostrand Reinhold, New York, 439 pp.

Seerley, R. W. 1991. In *Swine Nutrition,* Chapter 28, p. 451, E. R. Miller, et al. (eds.). Butterworth-Heinemann, Boston.

Tokach, M. D., et al. 1997a. *Starter Pig Recommendations.* MF2300, Swine Nutrition Guide, Kansas State University, Manhattan, KS.

Tokach, M. D., et al. 1997b. *Breeding Herd Recommendations for Swine.* MF2302, Swine Nutrition Guide, Kansas State University, Manhattan, KS.

Veum, T. L. 1991. In *Swine Nutrition,* Chapter 29, p. 483, E. R. Miller, et al. (eds.). Butterworth-Heinemann, Boston.

Yen, J. T. 2001. *Digestive System in Biology of the Domestic Pig.* W. G. Pond and H. J. Mersmann (eds.)., p. 390. Cornell University Press, Ithaca, NY.

Photograph courtesy of ISA Babcock, Ithaca, NY.

26

Poultry

RICHARD E. AUSTIC

POULTRY IS A TERM that refers to domestic granivorous birds and waterfowl that are used for meat and egg production and are sometimes kept for pleasure by hobbyists, fanciers, and sportsman's organizations. Diverse as they appear to be, species of poultry have much in common in the kinds of feedstuffs that they consume and in the manner in which feeds are digested and nutrients are absorbed and metabolized.

Some 280 million chickens are used annually in the United States for the production of eggs for human consumption. Nearly 8.6 billion chickens, 270 million turkeys, and 24 million ducks are grown annually for the production of meat. Nutrient requirements vary among species, and also within species depending on the genetic potential for growth; and egg production, age, and physiological state of the bird. Requirements for egg production, for example, are different from the requirements for growth; requirements for growth of light breeds of pullets for replacement of egg production stocks are different from requirements of the heavy breeds of chickens that are grown as broilers, fryers and roasters. The poultry nutritionist, therefore, must combine a knowledge of avian biology with an understanding of the nutrient needs for growth, egg production, and maintenance of body weight in order to be successful in the formulation of diets for different kinds of poultry for a variety of productive purposes.

❏ ANATOMY AND PHYSIOLOGY OF AVIAN DIGESTION

The alimentary tract of poultry has specialized anatomical features, such as a crop and a gizzard, that are unique to avian species (Sturkie, 1986). The crop is an expansible pouch in the esophagus that stores food during a meal and meters its contents into the digestive tract. The stomach produces hydrochloric acid and pepsin and is specialized anatomically at its caudal end for grinding feed particles. This specialized structure, the ventriculus (gizzard), is composed of thick smooth muscle surrounding an epithelial layer that secretes a thick cornified epithelium. Feed particles are pressed against the epithelium by the muscle in a grinding action that reduces the particles to small size. The presence of grit in the gizzard facilitates this process, particularly when feed contains whole grains or seeds.

The acidic chyme from the gizzard passes into the duodenum, where the acid is neutralized by the secretions of the pancreas. Digestion in the small intestine is rapid; feed passes from the crop to the caudal end of the small intestine in a matter of a few hours. The absorption of nutrients occurs primarily in the small intestine. Some digesta passes into the paired cecum at the junction of the small and large intestine. Although the twin sacs of the cecum are sites of fermentation, they appear for most poultry not to generate sufficient quantities of absorbed fermentation products to add significantly to the energy-yielding nutrients absorbed in the small intestine of poultry (Moran, 1982).

The colon is very short in birds as compared to mammals. Despite its small size, it performs many important functions in poultry. It receives digesta from the small intestine and, intermittently, from the cecum. The posterior end of the large intestine contains expanded areas called the coprodeum and urodeum. The latter contains the distal openings of the ureters. Urine from both kidneys, products of the reproductive tract, and digesta are voided through a common anatomical chamber, the cloaca. The large intestine and cecum receive urinary excretions by retrograde movement of urine in the large intestine from the urodeum. The large intestine absorbs water and salts from digesta and from the portion of urine that undergoes retrograde movement in the alimentary tract (Skadhauge, 1967).

The major end products of digestion include amino acids, simple carbohydrates such as glucose and fructose, fatty acids, mono- and diglycerides, and other lipids, vitamins, minerals, and water. The requirements of poultry for the various classes of nutrients will be addressed in the following sections of this chapter.

❏ THE NUTRIENTS

Amino Acids

The proteins of body tissues, feathers, and eggs of poultry contain some 20 amino acids. Ten of these (listed below) are indispensable in the diet because poultry are unable to synthesize them or cannot synthesize them at a rate sufficient to meet their needs (National Research Council, 1994).

Arginine	Methionine
Histidine	Phenylalanine
Isoleucine	Threonine
Leucine	Tryptophan
Lysine	Valine

These are the same amino acids that are required by mammals. Mammals, however, can synthesize arginine using enzymes of the urea cycle, so their dietary requirements for this amino acid appear limited to periods of rapid growth, pregnancy, and perhaps lactation when the capacity for synthesis is insufficient to meet the need for the amino acids in protein synthesis. Birds do not have a full complement of urea cycle enzymes and, therefore, are unable to synthesize arginine from simple precursors. Consequently, their need for dietary arginine is absolute. Cysteine can be synthesized from methionine, and tyrosine can be derived from phenylalanine. In a sense, cysteine and tyrosine are essential amino acids: either they must be supplied in the diet, or the diet must contain sufficient methionine and phenylalanine to allow for their synthesis.

Several amino acids are dispensable; that is, they are not required in the diet by poultry. These include alanine, aspartic acid, asparagine, glutamic acid, and glutamine. As in mammals, these amino acids can be synthesized from carbohydrate precursors and nitrogen from other amino acids or from ammonium salts. The amino acid requirement of poultry, therefore, represents the requirement for the indispensable amino acids plus sufficient nitrogen in an appropriate chemical form for synthesis of the dispensable amino acids.

Three other amino acids—glycine, serine, and proline—have been reported to be essential for young growing poultry. Glycine and serine are metabolically interconvertible: the requirement for these amino acids can be satisfied by both or either one alone if provided in the diet in sufficient quantity to allow for the synthesis of the other (Akrabawi and Kratzer, 1968). A requirement for proline has been demonstrated in young chickens fed on diets lacking proline. Improvements in growth rate and efficiency of feed utilization were observed when the diet was supplemented with 0.2 to 0.4% L-proline (Graber and Baker, 1973). It is not clear which poultry require these amino acids and under what conditions the amino acids are needed. However, the responses of poultry to dietary supplements of glycine, serine, and proline indicate that amino acids synthesized by poultry may, under some physiological conditions, not be synthesized at sufficient quantities to meet the requirements.

Chickens are sensitive to the dietary balance of amino acids. Excesses of dietary amino acids increase the requirements of specific amino acids in a feed. Two major kinds of interactions among amino acids—imbalances and antagonisms—occur in poultry and other animals (Harper et al., 1970). In imbalances, excessive dietary levels of amino acids increase the requirement for the most limiting amino acid. Antagonisms are more specific, and an amino acid need not be first limiting to be affected by an antagonism. The lysine–arginine antagonism is an interaction in which excess dietary lysine increases the requirement for arginine. The branched-chain amino acid antagonism is another interaction, in which excesses of one or two branched-chain amino acids increases the requirement for the other branched-chain amino acid(s). Special consideration should be given to arginine or to isoleucine and valine in feed formulation if the dietary levels of these amino acids are marginal and dietary levels of lysine or leucine, respectively, appear excessive.

A relatively new concept of formulating feeds for amino acids is gaining acceptance in the feeding of poultry, particularly broilers. It is a procedure that partitions the amino acid requirements for maintenance of body weight and the accretion of protein during growth. The indispensable amino acid requirements for maintenance and growth are expressed in ratios to lysine, a convenient reference. The ratios differ substantially for maintenance and growth functions. By knowing the lysine requirement per day per kg body weight for maintenance and the lysine requirement per day for protein accretion, it is possible to estimate the lysine required per day for the combined functions. Similarly, the requirements of the other indispensable amino acids can be calculated from the ratios. The amino acid requirements, calculated in this manner, represent the amounts of amino acids required at the tissue level. Amino acids in feedstuffs are less than 100% digestible, and the digestibilities vary considerably among feedstuffs. Therefore, when using the ideal amino acid ratio concept, the digestible (rather than total) amino acid contents of feed ingredients must be used in formulating feeds. The reader is referred to Emmert and Baker (1997) and Baker et al. (2002) for more information on this topic.

Carbohydrates

Poultry are able to digest starches, glycogen, sucrose, maltose, and the simple sugars, glucose and fructose. Lactose, or milk sugar, is not well digested by birds because the intestine contains limited lactase activity and lactase is needed to hydrolyze lactose to its constituent monosaccharides, glucose and galactose. Pentoses, such as the ribose and deoxyribose of nucleic acids and xylose and arabinose are not well absorbed by poultry. Since they contain high concentrations of nucleic acids, yeasts or bacteria in excess of 10 to 15% of the diet can cause flatus and diarrhea because of the presence and fermentation of unabsorbed sugar in the lower gastrointestinal tract. Inclusion in the diet of similar levels of pentoses such as arabinose and xylose can yield the same undesirable consequences (Scott et al., 1982). Poultry have minimal capacity to digest crude fiber (Moran, 1982).

Lipids

Triglycerides are the major lipids in poultry diets. Fats usually are added to poultry diets as sources of energy and linoleic acid, a fatty acid that is required by poultry. The addition of fat to feeds also reduces the dustiness of the diet, an important practical consideration in mixing and handling feeds. The fats in poultry tissues contain higher quantities of unsaturated fatty acids than are found in most domestic animals. The fatty acid composition of poultry and eggs, moreover, can be influenced by the fatty acid composition of dietary lipids. This has encouraged researchers to investigate the possibility of enriching poultry products with polyunsaturated fatty acids of the omega-3 series (commonly found in fish) by use of oils rich in these fatty acids or their precursors (Suni and Sparks, 2001; Gonzalez-Esquerra and Leeson, 2001). These fatty acids may have health benefits for consumers.

Vitamins

Vitamins are a diverse category of nutrients that have been grouped together, historically, as organic micronutrients that are absolute dietary essentials. Poultry require 13 vitamins (National Research Council, 1994). All, except vitamins D and E, participate as cofactors in enzymatic reactions. Vitamin D is a precursor of 1,25-dihydroxycholecalciferol which is essential for calcium absorption and metabolism. Vitamin E serves antioxidant functions in tissues. Choline is an important source of methyl groups in metabolism; it is also a component of phospholipids and functions in neurotransmission.

Some vitamins can be synthesized by the bird, but usually not in sufficient quantity to meet the physiological demands of young growing poultry or laying hens. These include niacin and vitamin D_3, which are synthesized from tryptophan in liver, with 7-dehydrocholesterol in skin, respectively. Vitamin synthesis by microflora in the cecum and large intestine may contribute some B-vitamins (e.g., folic acid, vitamin B_{12}) for poultry. It is not known whether the vitamins are absorbed from the cecum. Poultry are coprophagic, however, and through digestion of ingested feces, these vitamins may contribute to the vitamin nutriture of the bird.

Some vitamins are susceptible to oxidative damage. The requirement for vitamin E, for example, is increased when diets contain high levels of polyunsaturated fatty acids, which tend to undergo oxidative rancidity. Vitamins A, D, and biotin are other vitamins that are adversely affected by oxidation.

Minerals

The approximately 13 inorganic elements required by poultry (National Research Council, 1994) perform a wide range of functions. In addition to having important roles in cellular metabolism, Ca and P are the major structural elements of bone, and

Ca is the major mineral of eggshell. Na, K, and Cl have physiological functions in acid-base balance, fluid balance, and membrane transport. The remaining minerals are cofactors in a wide variety of enzymatic reactions. Poultry, like other animals, require Cu, Fe, Mg, Mn, Zn, Mo, I, and Se. Some elements have been found to influence growing chicks under conditions in which chicks are grown in isolators and provided with highly purified diets, water, and filtered air. Si (Carlisle, 1980), Ni (Nielson and Ollerich, 1974), and V (Hopkins and Mohr, 1974) are three such elements. Whether the responses to these minerals reflect "requirements" is not clearly established at the present time.

The balance among selected minerals is an important consideration in poultry nutrition. A high-Ca diet increases the requirement of growing poultry for P. A Ca to available P ratio of 1:1 to 2:1 is acceptable for growing poultry. A very high ratio (e.g., 10:1 or more) is used in laying hen feeds because of the uniquely high Ca requirement for eggshell formation. Numerous studies have shown that the dietary balance of mineral cations and mineral anions (expressed in milliequivalents per unit weight of diet) can affect acid-base balance and some aspects of growth or egg production (Mongin, 1981). Diets rich in mineral cations tend to be alkalogenic, and those rich in mineral anions tend to be acidogenic.

Dietary sources of minerals can differ in biologically available mineral content. P in dicalcium phosphate has greater availability than P in raw rock phosphates (Scott et al., 1982; Sullivan et al., 1992). Phosphorus in the hydrated form of dicalcium phosphate ($CaHPO_4 \cdot 2H_2O$) may be more available to poultry than P in anhydrous dicalcium phosphate (Scott et al., 1982). Phytin phosphate, in contrast, a major form of P in plants, is poorly available to poultry. The phosphate in phytin can be released by intestinal phytase, but the activity of this enzyme is limited in nonruminant animals such as poultry. Phosphorus availability can be increased by the addition of phytase to diets (Simons et al., 1990). Although the actual availability can vary among feedstuffs, and also among birds, a useful rule of thumb in feed formulation is to assume that the P in feedstuffs of plant origin is 30% available (Scott et al., 1982).

Water

Poultry require an adequate supply of pure, clean water. If the supply of water is inadequate, birds become dehydrated, feed intake declines, and physiological functions become impaired. Water is eliminated from the body by respiration and excretion and in reproductive products such as eggs. Panting is a major mechanism for the regulation of body temperature in birds. Since respiration rate rises, and hence, evaporation of water from the respiratory tract increases, the requirement for water increases as the ambient temperature rises above the thermoneutral zone for poultry. Water intake may increase three- to fourfold as temperature rises to levels that cause severe heat stress in chickens (Wilson, 1948). If water supply is limited in hot environments, birds readily succumb to heat prostration because of loss of body fluids.

Excessive dietary levels of minerals increase water requirements because water is needed to clear the body of the minerals provided in excess of the requirement. There is a direct relationship between salt intake and the amount of water drunk and excreted. Excess K and Mg also increase water intake and excretion. The excretion of sulfates, phosphates, and nitrogenous products of protein metabolism contribute to increased water consumption and excretion in poultry fed excessive levels of dietary protein.

❑ DEFICIENCY SIGNS AND SYMPTOMS

The consequences of nutritional deficiencies vary greatly among the nutrients. Amino acid deficiencies suppress growth, rate of egg production, and egg size but result in few specific symptoms. Lysine deficiency inhibits melanin formation in feathers in bronze turkeys (Fritz et al., 1946), and arginine deficiency results in curled feathers (Klasing and Austic, 2003). Amino acid deficiencies increase uric acid excretion because of the degradation of amino acids that cannot be used for tissue or egg protein synthesis (Miles and Featherston, 1974). Deficiencies limit protein deposition and tend to increase the synthesis and deposition of body fat (Velu et al., 1972).

The gross signs of essential fatty acid deficiency also are nonspecific. Growth and egg production are depressed, as is the hatchability of fertile eggs (Hopkins et al., 1963; Menge, 1968). Essential fatty acid deficiency leads to decreased concentration of arachidonic acid and increased concentrations of eicosatrienoic fatty acid in tissues; consequently,

an increased ratio of C20:3 to C20:4 fatty acids can be used as a biochemical indicator of essential fatty acid deficiency (Holman, 1960).

The symptoms of vitamin deficiencies in the chicken are presented in Table 26.1. (See NRC, 1994; Klasing and Austic, 2003, for review.) They differ markedly among vitamins, reflecting the diverse biochemical functions of this class of nutrients. The chicken has been studied extensively, and, therefore, the gross symptoms of nutritional deficiencies are well known: less information is available on turkeys, ducks, geese, and other species.

Minerals, like vitamins, are a class of nutrients that have diverse metabolic functions. Deficiencies affect poultry in a variety of ways. Typical deficiency symptoms in the chicken are shown in Table 26.2.

Some deficiency symptoms can be caused by more than one nutrient. Anemia may be caused by copper, iron, folic acid, or vitamin B_{12} deficiency. Perosis, or slipped Achilles tendon, can result from deficiencies of Mn, choline, folic acid, and biotin. When these symptoms occur, analysis of feeds and tissues is required to determine whether a specific nutrient deficiency is the causative factor. Most of the vitamins and trace minerals participate as cofactors in enzymatic reactions. Therefore, enzyme assays have been used in research as a biochemical approach to assessing the adequacy of selected vitamins and minerals in humans. More research is needed to validate this approach for practical testing of nutrient adequacy in poultry.

❑ ENERGY

Poultry need energy for growth, maintenance of body tissue, production of eggs, regulation of body temperature, and activity. Energy is derived from the oxidation of carbohydrates, lipids, amino acids, and related organic compounds. The final steps in the overall metabolism of energy precursors involve the stepwise transfer of electrons to oxygen in mitochondria. Energy is sequestered in the form of high-energy phosphate bonds of ATP that are subsequently used in energy-requiring reactions in metabolism.

Energy that exists within the structures of the macronutrients in feeds (i.e., gross energy of feed) is not fully available to poultry because the protein, carbohydrates, and lipids in feedstuffs usually are

TABLE 26.1   Symptoms of vitamin deficiencies in poultry.

NUTRIENT	DEFICIENCY SYMPTOMS [a,b]
Vitamin A	Pustules in mucous membranes of esophagus and trachea; accumulation of exudate in eyes and nares; ataxia; visceral gout; increased cerebrospinal fluid pressure; slowed growth and maturation of epiphyseal cartilage; increased incidence of blood spots in eggs; decreased sperm motility, sperm counts; increased incidence of abnormal sperm
Vitamin B$_{12}$	Decreased egg size
Vitamin D	Low bone ash content; awkward gait; decreased eggshell thickness; increased incidence of soft-shelled eggs
Vitamin E	Encephalomalacia; exudative diathesis; muscular dystrophy; decreased fertility (males); erythrocyte fragility; decreased reticulocyte count; microcytic anemia; gizzard necrosis; and hock disorder (T)
Vitamin K	Subcutaneous intramuscular and internal hemorrhages; delayed blood clotting (increased prothrombin time)
Biotin	Crusty dermatitis of foot pad, eyes and mouth; perosis; poor feathering and broken feathers (T)
Choline	Perosis
Folic acid	Poor, ragged feathering; impaired feather pigmentation in colored breeds; cervical paralysis (T); depressed blood hemoglobin; decreased reticulocyte count
Niacin	Inflammation of oral cavity and tongue; enlarged hock; bowed legs (T,D)
Pantothenic acid	Dermatitis on ventral surface of feet, eyelids, and mouth; lymphoid cell necrosis in thymus and bursa of Fabricius
Pyridoxine	Depressed appetite hyperexcitability (jerky, convulsive movements when disturbed); weakness; perosis
Riboflavin	Anorexia; curled toe paralysis (toes curled medially); spraddled legs in severe deficiency; neuromalacia of sciatic and brachial nerves
Thiamine	Anorexia; polyneuritis (episthotonis); paralysis; irregular heart beat

[a]Symptoms pertain primarily to chickens. In turkeys and ducks, symptoms are often but not always similar. Those unique to turkeys are indicated by (T), to ducks by (D).
[b]See text for more information on protein and amino acid requirements.

not 100% digestible and because energy is lost in the digestion and assimilation of nutrients. Net energy, or the energy available to the bird for useful purposes is equal to the gross energy minus the losses of energy in feces and urine and the loss of energy as heat during the digestion and assimilation of nutrients (i.e., the heat increment of feeding).

Bird droppings contain urine and feces. Therefore, it is convenient to measure the energy losses of urine and feces together, and for this reason metabolizable energy has been used as a practical measure of energy for poultry (Hill and Anderson, 1958; Sibbald, 1982). Apparent metabolizable energy can be determined by measuring the combustible energy in feedstuffs and the combustible energy in excreta and then by subtracting the latter from the former. Some of the energy lost in excreta represents undigested endogenous secre-

tions of the digestive tract and endogenous urinary excretions. True metabolizable energy is the term applied to the energy value of feedstuffs when apparent metabolizable energy has been corrected for these endogenous losses. The latter are difficult to measure accurately. The energy content in excreta of birds from which feed has been withheld has been used as a crude measure of endogenous loss (Sibbald, 1982).

The energy requirements of poultry are determined mainly by maintenance functions, rate of growth, egg production, and physical activity. The energy requirement is markedly affected by ambient temperature. Low temperatures increase food intake to provide substrates for increased metabolic rate under these conditions. High temperature reduces food intake by mechanisms that are not fully understood.

TABLE 26.2 Symptoms of mineral deficiencies in poultry.

NUTRIENT	DEFICIENCY SYMPTOMS [a,b]
Ca	Rickets (see vitamin D_3); decreased eggshell thickness; depletion of medullary bone in laying hens; thinning of cortical bone; weakness; inability to stand; tetany and death in severe deficiency
Chloride	Nervous spasms in response to noise and handling; dehydration; increased hematocrit
Cu	Hypochromic microcytic anemia; impaired feather pigmentation; decreased elastin content of arteries and dissecting aneurysms; leg weakness and bone deformity; increased fragility of bones; thickened epiphyseal cartilage of long bones
I	Thyroid enlargement; delayed hatching time
Fe	Hypochromic microcytic anemia; impaired feather pigmentation
Mg	Anorexia; lethargy; nervous tremors; ataxia; tetanic convulsions; degenerative decreased eggshell weight and ash content
Mn	Perosis; shortening of long bones and spinal column; decreased bone ash, decreased eggshell strength; tetany and death in severe deficiency
P	Rickets; depletion of medullary bone in laying hens; cage layer fatigue (in conjunction with Ca and other factors)
K	Weakness; inability to stand; tetanic seizures; reduced bone ash; decreased eggshell thickness and eggshell weight
Se	Pancreatic fibrosis; exudative diathesis; muscular dystrophy
Na	Anorexia; increased hematocrit; softened bones
Zn	Poor feathering; shortening and thickening of long bones; decreased bone ash; impaired development of epiphyseal cartilage; enlarged hocks; perosis (T); scaly skin; foot dermatitis; increased hematocrit

[a]Symptoms pertain primarily to chickens. In turkeys and ducks, symptoms are often but not always similar. Those unique to turkeys are indicated by (T).
[b]See text for more information on protein and amino acid requirements.

The energy density of the diet is also an important practical consideration. Under normal environmental conditions, poultry may not be able to consume enough energy to meet their needs if the metabolizable energy content of the diet is much below 2500 Kcal/kg, because of the diet's bulkiness and the physical limitations of the gastrointestinal tract regarding the amount of feed that it can accommodate. Conversely, poultry tend to overconsume feed if the energy density of the diet is too high. This leads to excessive accumulation of body fat. Within the range of 2500 to 3200 Kcal metabolizable energy per kg of feed, feed intake tends to be inversely related to the metabolizable energy concentration of the feed. If the energy requirement of the bird and the energy density of the diet are known, the feed intake of the bird can be estimated (Scott et al., 1982). This has implications in the feeding of poultry as is discussed in the following section.

❏ ENERGY-NUTRIENT INTERRELATIONSHIPS

The requirements of poultry for protein, Ca, P, linoleic acid, and other nutrients are determined mainly by the rate of accretion of nutrients in body tissues or eggs and by the obligatory losses of nutrients that are associated with maintenance. Consequently, for any given age, body weight, and productive function (e.g., growth, egg production), the daily intake requirements for these nutrients are relatively constant.

The requirement for energy, in contrast to the need for most other nutrients, varies depending on the physical activity of birds and ambient temperature. Poultry nutritionists must take this into account when they formulate diets. The challenge is to formulate for dietary energy and nutrient levels in such a manner that the intakes of all nutrients are sufficient to satisfy requirements. Feed intake is determined primarily by the bird's need to obtain energy. If the metabolizable energy requirement of the bird and the metabolizable energy content of its feed are known, feed intake can be predicted. Consider, for example, a feed for laying hens that contains 3000 Kcal of metabolizable energy per kilogram. A hen requiring 300 Kcal of metabolizable energy per day will consume (300 Kcal ÷ 3000 Kcal/kg=) 0.100 kg (100 g) per day of this feed. If the daily amino acid requirements (i.e., milligrams of amino acids required per day) are known, then the amounts needed per kilogram of diet can be calculated. Let's assume that we wish to formulate for lysine, the requirement of which is 750 mg/hen/day. In our example of a hen consuming 100 g/day of a diet containing 3000 Kcal of metabolizable energy/kg, this represents a dietary lysine concentration of (0.75 g lysine/0.1 kg of diet=) 7.5 g lysine/kg of diet. If the energy density of the diet is altered, the amount of feed consumed each day will change, and this will affect the concentration of lysine that must be provided in the diet. If the energy density of the diet is decreased from 3000 Kcal to 2700 Kcal/kg, for example, food intake is expected to increase to (300 Kcal ÷ 2700 Kcal/kg=) 0.111 kg (111 g), and the lysine content of the diet can be safely decreased to (0.75 g ÷ 0.111 kg=) 6.76 g/kg of diet. Conversely, if the metabolizable energy density of the diet increases to 3300 Kcal/kg, then food intake is expected to decrease to (300 Kcal ÷ 3300 Kcal=) 0.09/kg/day, and the concentration of amino acid in the diet must be increased to (0.750 g ÷ 0.091 kg=) 8.24 g/kg of diet in order to provide for the same daily intake as is achieved when the hen receives the diet containing 3000 Kcal/kg of metabolizable energy. An example of the methionine and

lysine requirements of the layer at different dietary energy levels is provided in Table 26.3.

Similar interrelationships between energy and nutrient levels may be assumed to apply for protein and other macronutrients such as Ca, P, and linoleic acid. They also could be applied to vitamins and trace elements. However, these nutrients are added to the diet in small quantities and are relatively inexpensive. They routinely are provided in the diet with sufficient margin of safety so that requirements are satisfied in spite of variations in feed intake. Adjustments with regard to energy requirements usually are not made.

Maintenance requirements for energy are markedly affected by environmental temperature. Equations have been developed for calculation of energy requirements of poultry under a range of conditions of environmental temperature. Effectively used, such equations permit relatively accurate predictions of feed intake. Advances in the technology of monitoring the flock have led to the use of scales and flow meters to measure actual feed consumption. Under these conditions, it is possible to formulate diets on the basis of recent actual feed intake rather than on the basis of predicted feed intake.

❏ FEEDSTUFFS

Like other simple-stomached animals, poultry require feeds that can be digested by enzymes secreted by tissues and associated organs of the alimentary tract. These are feeds generally low in fiber content but rich in protein, starch, and lipids. The amino acids, monosaccharides, and fatty acids absorbed from these feeds are, with few exceptions, readily metabolized by poultry.

TABLE 26.3   Dietary methionine and lysine levels for layers in diets of varying energy density.[a]

METABOLIZABLE ENERGY, KCAL/KG	FEED INTAKE, G/HEN/DAY	METHIONINE, %	LYSINE, %
2600	115	0.30	0.65
2800	107	0.33	0.70
3000	100	0.35	0.75
3200	94	0.37	0.80

[a]Assumes an energy requirement of 300 Kcal metabolizable energy/hen/day and crude protein intake of approximately 17 g/hen/day. Requirements will vary depending on body size, rate of egg production, and ambient temperature.

Feedstuffs can be grouped according to their protein and energy contents and by other special nutritional attributes that make them useful in poultry feed formulations. Some of the ingredients that are used extensively in poultry feeds are described in Tables 26.4 and 26.5.

Poultry feeds typically comprise high-energy feed grains such as corn, milo, or wheat and often medium-energy ingredients such as wheat middlings, barley, and oats in combination with high-protein ingredients such as soybean oil meal, meat and bone meal, poultry byproduct meal, fish meal, or other protein-rich foodstuffs. Diets usually include supplements of fat or oil as well as sources of minerals and vitamins.

Carotenoid pigments are needed in feeds in order to satisfy consumer preference for yellow to orange color of skin and eggs in chickens, waterfowl, and other species (Janky et al., 1985; Karunajeewa et al., 1984). Natural sources of pigments are used in the United States. These include corn, corn gluten meal, and marigold petal meal among others. Synthetic carotenoids are added to feeds in other countries depending on local regulations concerning feed formulation.

Nonnutrient additives to poultry diets usually include antioxidants to help stabilize the vitamins and unsaturated fatty acids that are included in the diet. Antibiotics are often included, especially in diets of young poultry that are grown for meat. They tend to promote growth, possibly by suppressing the growth of undesirable intestinal microflora. Some antibiotics have specific antimicrobial functions. Anticoccidials, for example, control various species of *Eimeria*. These are organisms that infect the intestinal tract when birds are grown in floor pen environments where the *Eimeria* can be continuously reintroduced to the bird by consumption of droppings.

Plant feedstuffs may vary in composition depending on environmental factors during crop production and harvesting or the nature and extent of processing. Heat treatment is required to inactivate trypsin inhibitor activity and other toxic substances in beans. Underheated soybean meal, for example, can suppress the growth or egg production of poultry (Hill and Renner, 1960). Overheating may also suppress performance because of chemical damage to amino acids in the feedstuff (Nesheim and Carpenter, 1967). Lysine is particularly susceptible to destruction because of the reactivity of the ϵ-amino group with reducing sugars and certain amino acids. Nutritionists must be aware of the nutrient content and quality of ingredients when formulating feeds for poultry. Nutritional and toxicological problems can arise if ingredients are not tested for nutritional quality and presence of harmful constituents such as mycotoxins prior to their incorporation into poultry feeds.

TABLE 26.4 Grains used in the feeding of poultry.

GRAINS	PROTEIN, %	METABOLIZABLE ENERGY, KCAL/KG	REMARKS
Corn	8.5	3350	Excellent source of energy. High in linoleic acid and vitamin A precursors. Good source of xanthophyll pigments.
Milo	8.8	3288	Often used to replace corn. Lower than corn in linoleic acid and lacks xanthophyll pigments. High tannin content of some varieties may limit digestibility.
Wheat[a]	10.2	3120	Good source of energy. Lower than corn in linoleic acid and lacks xanthophyll pigments.
Barley	11.0	2640	Lower energy level than corn, wheat, or milo. Higher fiber content. Lower than corn in linoleic acid. Lacks xanthophyll pigments. Western U.S. varieties have reduced digestibility unless enzyme treated.
Oats	11.4	2550	Low-energy, high-fiber ingredient. Good source of linoleic acid. Lacks xanthophyll pigments.

[a]Soft white winter wheat.
Sources: Austic and Nesheim (1990) and National Research Council (1994).

TABLE 26.5   Selected byproducts commonly used in poultry feeds as sources of protein.

FEEDSTUFFS	PROTEIN, %	METABOLIZABLE ENERGY, KCAL/KG	LIMITING AMINO ACIDS	REMARKS
Soybean meal, dehulled	48.5	2440	Methionine	Excellent source of protein but must be properly heat treated to destroy antinutritional factors.
Soybean meal	44.0	2230	Methionine	Used less frequently than dehulled soybean meal. Must be properly heat treated to destroy antinutritional factors.
Corn gluten meal	62.0	3720	Lysine, tryptophan	Excellent source of xanthophyll pigments.
Canola meal	38.0	2000	Lysine	Made from a variety of rapeseed that has been genetically selected for low levels of antinutritional factors.
Cottonseed meal	44.7	1857	Lysine, methionine	Contains gossypol, which interferes with egg quality. Used sparingly in diets for laying hens.
Peanut meal	50.7	2200	Methionine, lysine	Can be contaminated with aflatoxin especially if produced in hot, humid regions of the world.
Meat and bone meal	50.4	2150	Methionine	Protein quality can be variable. Contains Ca and P.
Fish meal[a]	60.0	2820	None	Well-balanced amino acid profile. Source of Ca and P. Excessive (>10%) levels may lead to fish taint in eggs and meat.
Poultry byproduct meal	60.0	2950	None	Source of Ca and P. Protein quality varies depending on processing.
Hydrolyzed feather meal	81.0	2360	Methionine, lysine, histidine, tryptophan	Variable digestibility depending on processing. Limiting in several amino acids.
Blood meal	88.9	3420	Isoleucine	Good quality protein except for its unusually low isoleucine content.

[a] Menhaden meal.
Sources: Austic and Nesheim (1990) and National Research Council (1994).

❏ FEEDING POULTRY

Layers

The usual goal in the feeding of laying hens is to obtain the maximum production of eggs of large size and good internal and external quality with a least-cost feed formulation. If the layers constitute a breeding population, the goal is to obtain maximum production of eggs of an appropriate size for incubation with good potential for high yield of healthy, vigorous chicks. The nutritional requirements of breeder populations are not greatly different from those of commercial layers: generally they require higher levels of vitamins and minerals to ensure that the egg is sufficiently enriched in these nutrients to support optimal development and hatchability of the chick (NRC, 1994). For the sake of brevity, this chapter focuses on the nutrition of commercial laying hens.

Nutrition of Growing Pullets. Pullets reach sexual maturity at approximately 5 months of age, although the actual time is dependent on genetics, nutrition, and other environmental factors such as daylength. Nutrition is as important during the period of growth and development as it is after the pullet begins to lay eggs. Inadequate nutrition during the growing period may limit egg production throughout the laying period.

During the rearing period, the body weight of Leghorn pullets increases from approximately 40 g at hatching to nearly 1400 g at 20 weeks of age. The rate of growth, however, continually declines during this period: body weight may double in the first week of life; nearly double during the second week, etc.; less than double during the third week, etc., as the pullet grows to maturity. The higher rate of growth during the early weeks results in requirement for the highest nutrient density in feeds at this stage of life. Typical nutritional needs during the starting, growing, and developing period are shown in Table 26.6.

Nutrition of Laying Hens. Modern commercial laying hens are prolific egg layers. Egg production by individual hens may exceed 300 eggs per year (Heil and Hartmann, 1992); average flock production rates often exceed 275 eggs per hen per year. The latter represents, in weight, over 15 kg of eggs per year or nearly tenfold greater than the hen's body weight. The poultry nutritionist must formulate diets with appropriate nutrient levels to permit a high rate of production of eggs having appropriate internal quality and external quality. Eggshells, for example, must be strong enough to withstand the rigors of automatic collection, processing, packing, transport to retail outlets, and handling by consumers.

When a flock of pullets begins to lay, egg production increases rapidly to a maximum rate of lay within the first 2 months. Egg size increases at a rapid rate over this early period and continues to increase at a slower rate throughout the laying year. The average daily weight of eggs produced per day (i.e., % production × egg weight) per hen,

often termed *egg mass*, increases within a few weeks after peak rate of egg production occurs, remains relatively constant for a few months, and then slowly decreases during the remainder of the laying year (Leeson and Summers, 1991). The hen requires nutrients for the production of eggs, maintenance, growth (in the early stage of lay), and feather replacement. The requirements for feather growth are relatively insignificant after maximum body weight is achieved unless the hen undergoes molt.

Energy requirements of layers reflect the energy requirements of growth, maintenance, and egg production. Equations have been developed to predict energy requirements on the basis of ambient temperature, body weight, and egg production. One such equation (NRC, 1994) is as follows:

Metabolizable energy/hen/day
$$= W^{0.75} (173 - 1.95T) + 5.5\Delta W + 2.07EM$$

where W is body weight in kilograms, T is ambient temperature in °C, ΔW is the change in body weight in grams per day, and EM is egg mass. The energy requirements and feed intakes of typical 1.7-kg Single Comb White Leghorn hens housed at an ambient temperature of 22°C and laying at various rates from 25 weeks of age are shown in Table 26.7.

Laying hens need protein as a source of amino acids. As with all simple-stomached animals, there is no requirement for protein per se, only a requirement for indispensable amino acids and enough additional nitrogen to allow for the synthesis of the dispensable amino acids. Nevertheless, it is instructive to think of the amino acid requirement in terms of a protein requirement, with each indispensable amino acid representing a certain percentage of the total protein. These percentages can be calculated from the information presented in Table 26.6.

Let's consider the need for protein by a pullet in the early stages of production for the interval from 25 to 30 weeks of age. Such a pullet would weigh approximately 1600 g and, if a prolific layer, produce eggs nearly at the rate of an egg per day, say 0.97 egg/day on average. Egg size would be approximately 52 g. The endogenous nitrogen excretion is expressed by the equation, $N = (0.201 \times 1.5\ BW^{0.75})$, where N is nitrogen excretion in g/hen/day and $BW^{0.75}$ is body weight in kilograms to the 0.75 power (Scott et al., 1982). In our

TABLE 26.6   Nutrient requirements of Leghorn chickens.

NUTRIENT	STARTING CHICKS, 0–6 WK	GROWING CHICKENS, 6–12 WK	GROWING CHICKENS, 12–22 WK	LAYING HENS	BREEDING HENS
Protein, minimum %	21	16	14	17	17
Metabolizable energy, Kcal/kg	3000	3000	2850	3000	3000
Ca, %	1.0	0.8	0.8	3.5	3.3
P, available, %	0.45	0.4	0.4	0.4	0.4
Na, %	0.15	0.15	0.15	0.15	0.15
Chloride, %	0.15	0.15	0.15	0.15	0.15
K, %	0.40	0.40	0.40	0.40	0.40
Mg, %	0.06	0.05	0.05	0.05	0.05
Iodide, ppm	0.35	0.35	0.35	0.30	0.30
Mn, ppm	60	35	35	35	50
Fe, ppm	80	40	40	45	55
Cu, ppm	5	2	2	2	5
Zn, ppm	50	35	35	35	55
Se, ppm	0.15	0.10	0.10	0.10	0.10
Methionine, %	0.36	0.27	0.24	0.37	0.37
Methionine + cystine, %	0.71	0.54	0.48	0.65	0.65
Lysine, %	1.00	0.75	0.64	0.75	0.75
Arginine, %	1.18	0.90	0.78	0.80	0.80
Threonine, %	0.80	0.60	0.53	0.53	0.53
Tryptophan, %	0.20	0.15	0.13	0.16	0.16
Histidine, %	0.32	0.24	0.21	0.19	0.19
Leucine, %	1.18	0.90	0.78	0.85	0.85
Isoleucine, %	0.69	0.53	0.46	0.58	0.58
Valine, %	0.71	0.54	0.48	0.65	0.65
Phenylalanine, %	0.63	0.48	0.42	0.48	0.48
Phenylalanine + tyrosine, %	1.18	0.90	0.78	0.95	0.95
Vitamin A, IU/kg	6000	4500	4500	8800	11,000
Vitamin D_3, IU/kg	1000	650	650	1100	1100
Vitamin E, IU/kg	11	9	9	7	16
Vitamin K, mg/kg	2	2	2	2	2
Riboflavin, mg/kg	4.5	4.5	4.5	4.5	5.5
Nicotinic acid, mg/kg	35	25	25	25	35
Pantothenate, mg/kg	14	13	13	6	16
Folic acid, mg/kg	1.3	0.4	0.4	0.4	0.9
Choline, mg/kg	1300	1000	1000	1100	1100
Vitamin B_{12}, mg/kg	0.011	0.007	0.007	0.007	0.011
Thiamin, mg/kg	2	2	2	2	2
Pyridoxine, mg/kg	4.5	3.3	3.3	3.3	4.5
Biotin, mg/kg	0.13	0.11	0.11	0.11	0.18
Linoleic acid, %	1.2	0.8	0.8	1.2	1.2

Source: Adapted with modification from Scott et al. (1982), Austic and Scott (1984), and National Research Council (1994).

TABLE 26.7 Egg production, energy requirements, and feed intake of Leghorn hens at various times during the first year of egg production.[a]

AGE OF HEN WK	BODY WEIGHT[b] KG	RATE OF EGG PRODUCTION %	EGG WEIGHT G	EGG MASS G/HEN/DAY	METABOLIZABLE ENERGY REQUIREMENT[b] KCAL/HEN/DAY	FEED CONSUMED G/HEN/DAY
25	1.49	95	55	52.2	292	97
35	1.58	90	60	54.0	303	101
45	1.66	86	62	53.3	307	102
55	1.73	82	63	51.7	308	103
70	1.80	77	63	48.5	305	102

[a]Assuming that feed contains 3000 Kcal of metabolizable energy per kg and ambient temperature is 22°C.
[b]In the calculation of metabolizable energy requirement (see formula in the text) the body weight gains are considered to be 1.5, 1.3, 1.0, 0.8, and 0.5 g per hen per day for hens at 25, 35, 45, 55, and 70 weeks of age, respectively.

example, $N = 0.201 \times (1.5 \times 1.6^{0.75}) = 0.43$ g/hen/day. Since protein is approximately 16% nitrogen, a factor of 6.25 can be used to convert the endogenous nitrogen excretion to a protein equivalent ($6.25 \times 0.43 = 2.7$ g protein). Pullets typically gain weight at this age. This amounts to approximately 2.5 g/day, of which approximately 18% is protein. Therefore, the protein deposited in tissue (0.18×2.5) is nearly 0.5 g/day. Eggs contain approximately 12% protein. Therefore, the amount of protein used by this pullet for egg production = ($0.97 \times 52 \times 0.12$) = 6.0 g protein/day. The sum of protein for maintenance, growth, and egg production, 9.2 g/pullet/day, represents the net requirement for protein. Assuming that protein is used with 55% efficiency, the dietary need for protein for this pullet is ($9.2/0.55$) = 16.7 g/day, slightly less than the value listed in Table 26.6. If similar calculations are made for a 1.75 kg hen at 50 weeks of age, no longer growing but laying eggs weighing 61 g at a rate of 85%, the protein requirement for maintenance (2.9 g/day) and egg production (6.2 g/day) total 9.1 g/hen/day, and the dietary level of protein required is 16.5 g/hen/day.

The protein requirement was similar at both stages of production, although the relative contributions of maintenance, growth, and egg production differed. Similar approaches can be used to estimate protein, and thereby amino acid, requirements of other laying fowl.

The laying hen requires all of the minerals and vitamins that are essential for the growth of the young chick. The quantitative requirements differ, however, reflecting the use of nutrients in physio-logical processes that exist in hens in the reproductive state—namely, the maintenance of mature body size and the formation of eggs. The Ca requirement, for example, is markedly higher than that of the growing chick, reflecting the need of Ca for deposition of the eggshell. A typical large egg contains approximately 2 g of Ca, most of which is associated with the eggshell (Roland, 1982). If a layer forms an egg each day, 2 g of Ca is needed for egg formation in addition to a small amount of Ca that is needed for maintenance of body weight. Some Ca is required for other functions, but since the Ca requirement is primarily a function of egg formation, the requirement is proportional to the rate of egg production as indicated in Table 26.8. The requirement is based on an assumed 45% digestibility of Ca.

Broilers

Broiler is a term that defines a market category of poultry that can apply to all species. However, it is

TABLE 26.8 Calcium requirements of laying hens.

PRODUCTION %	AVERAGE DAILY CA DEPOSITED IN EGG G	CA % OF DIET[a]
70	1.4	3.1
80	1.6	3.6
90	1.8	4.0
100	2.0	4.4

[a]Assumes 45% apparent digestibility of Ca—and that feed contains 3000 Kcal of metabolizable energy per kg and ambient temperature of 22°C.

commonly taken, unless otherwise specified, to refer to young chickens. Broilers are young chicken that are grown to 5 to 7 weeks of age, at which time they are marketed for human consumption as whole birds, halves, parts, or further processed products. The commercial broiler was once a byproduct of the egg industry, in which males that would otherwise have been discarded at the hatchery were grown as broilers or fryers. As the broiler became popular, breeders began to specialize in the development of female and male broiler breeder lines. When crossed, these lines yielded commercial hybrid chicks that had improved potential for growth. The growth rates of broilers has continued to increase over the years as improvements have been made in genetics, nutrition, disease control, housing, and production management.

A distinguishing characteristic of modern broiler chickens is their rapid growth, deposition of a large proportion of breast and leg muscle, and their relative inactivity as compared to chickens of lighter breeds or hybrids that are used for egg production. The rate of growth decreases with age, resulting in progressively lower nutrient requirements (in amounts per kg of diet) as the broiler approaches market age. Typical body weight, rates of growth and feed consumption of broiler chicks are presented in Table 26.9.

The nutrient requirements of broilers are presented in Table 26.10. Amino acid requirements during the first 2 or 3 weeks of age (i.e., for starter diets) have been determined experimentally for most of the amino acids. Less research has been done on growing and finishing broilers, but the most limiting amino acids have received attention. Some of the requirements shown in Table 26.10 reflect extrapolations from the starter period on the assumption that the quantitative relationships among amino acids are constant throughout the growth of the broiler.

Food consumption is determined primarily by the energy requirement of the broiler and the energy concentration of the diet. The requirements for amino acids, and indeed the requirements for other nutrients, can be expressed in relation to energy requirements and energy density of the feed in order to ensure adequate intakes of these nutrients. Table 26.11 summarizes dietary protein levels for broilers at various dietary energy levels.

Although broilers reach market weight in a relatively short time (5 to 6 weeks, depending on market), there is sufficient change in rate of growth that three or more feed formulas may be used in broiler production. The starter diet, fed for 2 to 3 weeks, is higher in concentrations of amino acids and other nutrients than the grower diet and finisher diets, which are progressively lower in amino acids and certain other nutrients. The finisher diet has a higher energy to protein ratio than the starter and grower diets to increase the deposition of subcutaneous adipose tissue. The pigment in skin and

TABLE 26.9 Growth and feed consumption of broiler chickens.[a]

	MALES			FEMALES		
AGE, WK	BODY WEIGHT, G	CUMULATIVE FEED CONSUMPTION, G	FEED CONVERSION[b]	BODY WEIGHT, G	CUMULATIVE FEED CONSUMPTION, G	FEED CONVERSION[b]
1	175	150	1.07	165	135	1.07
2	450	475	1.16	410	430	1.16
3	885	1110	1.31	830	1040	1.32
4	1485	2035	1.41	1290	1825	1.46
5	2170	3195	1.50	1800	2820	1.60
6	2900	4645	1.62	2345	3995	1.73
7	3635	6200	1.73	2895	5335	1.87

[a]Assumes ambient temperature of 22°C after brooding and a diet containing approximately 3200 Kcal of metabolizable energy per kg of diet.
[b]Grams of feed consumed ÷ grams of body weight gain. Initial chick weight = 40 grams.

TABLE 26.10   Nutrient requirements of broiler chickens.

NUTRIENT	STARTING BROILERS, 0–2.5 WK	GROWING BROILERS, 2.5–5 WK	FINISHING BROILERS, 5–7 WK
Protein, minimum, %	23	20	18
Metabolizable energy, Kcal/kg	3200	3200	3200
Ca, %	1.00	0.90	0.80
P, available, %	0.45	0.40	0.40
Na, %	0.15	0.15	0.15
Chloride, %	0.18	0.15	0.15
K, %	0.40	0.40	0.40
Mg, %	0.06	0.05	0.05
Iodide, ppm	0.35	0.35	0.35
Mn, ppm	60	60	60
Fe, ppm	80	80	80
Cu, ppm	8	8	8
Zn, ppm	50	50	50
Se, ppm	0.15	0.15	0.15
Methionine, %	0.50	0.45	0.40
Methionine + cystine, %	0.90	0.80	0.70
Lysine, %	1.30	1.15	1.00
Arginine, %	1.25	1.10	1.00
Threonine, %	0.80	0.70	0.62
Tryptophan, %	0.24	0.21	0.19
Histidine, %	0.32	0.28	0.25
Leucine, %	1.20	1.05	0.95
Isoleucine, %	0.84	0.73	0.66
Valine, %	0.95	0.83	0.74
Phenylalanine, %	0.72	0.63	0.56
Phenylalanine + tyrosine, %	1.35	1.17	1.05
Vitamin A, IU/kg	5000	5000	5000
Vitamin D_3, IU/kg	1000	1000	1000
Vitamin E, IU/kg	10	10	10
Vitamin K, mg/kg	2	2	2
Riboflavin, mg/kg	4.5	4.5	4.5
Nicotinic acid, mg/kg	35	35	35
Pantothenate, mg/kg	14	14	14
Folic acid, mg/kg	1.3	1.3	1.3
Choline, mg/kg	1300	1300	1300
Vitamin B_{12}, mg/kg	0.011	0.011	0.011
Thiamin, mg/kg	2	2	2
Pyridoxine, mg/kg	4.5	4.5	4.5
Biotin, mg/kg	0.15	0.15	0.15
Linoleic acid, %	1.2	1.2	1.2

Source: Adapted with modification from Scott et al. (1982), Austic and Scott (1984), National Research Council (1994).

TABLE 26.11   Protein levels for broiler chickens.[a]

METABOLIZABLE ENERGY, KCAL/KG DIET	AGE WEEKS		
	0–3	3–6	6–8
		%	
2900	20.8	18.1	16.3
3000	21.6	18.8	16.8
3100	22.0	19.4	17.4
3200	23.0	20.0	18.0
3300	23.7	20.6	18.6

[a]Amino acid levels for diets containing 3200 Kcal of metabolizable energy per kg are shown in Table 26.10. Amino acids should be adjusted proportionately to protein.

subcutaneous fat gives the finished carcass the uniform yellow color that United States consumers prefer.

Broilers are prone to leg problems, which may be exacerbated if selected nutrients become marginal or limiting. Vitamin D, Ca, or P deficiency can result in weak or brittle bones. Tibial dyschondroplasia, a cartilage abnormality in the growing end of long bones, can cause distortions of the tibiotarsal–tarsometatarsal joint and the head of the tibia (Leach and Nesheim, 1972). This condition is exacerbated by diets rich in mineral anions such as phosphate, chloride, and sulfate and tends to be alleviated by increased dietary divalent cations, particularly Ca (Edwards, 1984; Nelson et al., 1981). Deficiency of salt suppresses water and feed intake, whereas excess salt increases water intake and excretion. Control of dietary salt is especially important in maintaining litter quality. Typically, Ca, P, salt, Mn, Zn, and Se must be added as mineral supplements to broiler feeds.

Supplements are needed to ensure adequate intakes of fat-soluble and water-soluble vitamins and choline. These usually include vitamin A, vitamin D_3, vitamin E, vitamin K, vitamin B_{12}, biotin, niacin, riboflavin, and pantothenic acid.

Turkeys

These birds usually are grown 14 weeks or more for marketing as roasters or as further processed products. Typical growth and feed consumption of female and male turkeys are presented in Table 26.12. Growth rates of turkeys are high early in the starting period and decline continuously throughout the long rearing period. Nutrient requirements decline over this interval, and consequently, several dietary formulas are used in the growing of turkeys for meat.

TABLE 26.12   Body weights and feed consumption of large-type turkeys.

AGE, WK	MALES			FEMALES		
	BODY WEIGHT, KG	CUMULATIVE FEED CONSUMPTION, KG	FEED CONVERSION[a]	BODY WEIGHT, KG	CUMULATIVE FEED CONSUMPTION, KG	FEED CONVERSION[a]
2	0.28	0.28	1.27	0.28	0.27	1.22
4	0.95	1.19	1.34	0.84	1.04	1.33
6	2.08	2.97	1.47	1.74	2.53	1.50
8	3.58	5.81	1.65	2.88	4.80	1.70
10	5.47	10.07	1.86	4.14	7.77	1.90
12	7.61	15.64	2.07	5.53	11.56	2.11
14	9.92	22.60	2.29	6.99	16.09	2.32
16	12.32	30.66	2.50	8.46	21.02	2.50
18	14.70	39.41	2.69	9.89	25.95	2.64
20	17.00	48.28	2.85	11.21	30.67	2.75
21	18.08	52.64	2.92	—	—	—

[a]Grams of feed consumed ÷ grams of body weight gain, assuming initial body weight = 59 grams.
Source: Calculated from the data of Ferket (2003).

The nutrient requirements of turkeys are presented in Table 26.13. Amino acid requirements of starting turkeys have been determined experimentally: however, requirements at later time periods are extrapolated except for total crude protein, sulfur amino acid, and lysine requirements which have been the subjects of investigation because of their first and second limiting status in practical diets. The specific amino acid needs of poultry can be estimated using the information presented in Tables 26.13 and 26.14.

Turkeys can accommodate a wide range of energy concentration in their feeds. Most commercial diets contain higher energy densities in starter diets, less energy in grower diets, and higher energy density in finisher diets. The finisher diet serves to increase subcutaneous fat deposition to improve the external appearance of roasters and

TABLE 26.13 Nutrient requirements of turkeys.

NUTRIENT	STARTING TURKEYS, 0–8 WK	GROWING TURKEYS, 8–16 WK	GROWING TURKEYS, 16–24 WK	BREEDING TURKEYS, LAYING
Protein, minimum, %	28	22	16	14
Metabolizable energy, Kcal/kg	2800	3000	3100	2850
Ca, %	1.40	1.00	0.60	2.00
P, available, %	0.70	0.60	0.50	0.60
Na, %	0.15	0.15	0.15	0.15
Chloride, %	0.15	0.15	0.15	0.15
K, %	0.60	0.60	0.60	0.60
Mg, %	0.06	0.06	0.06	0.06
Iodide, ppm	0.35	0.30	0.30	0.35
Mn, ppm	60	60	60	60
Fe, ppm	80	60	60	60
Cu, ppm	6	4	4	6
Zn, ppm	70	60	50	70
Se, ppm	0.20	0.15	0.10	0.20
Methionine, %	0.56	0.44	0.32	0.28
Methionine + cystine, %	1.05	0.84	0.60	0.50
Lysine, %	1.60	1.25	0.90	0.60
Vitamin A, IU/kg	11,000	11,000	11,000	11,000
Vitamin D_3, IU/kg	1500	1500	1500	1500
Vitamin E, IU/kg	15	13	7	30
Vitamin K, mg/kg	2.5	2.0	2.0	2.0
Riboflavin, mg/kg	5.5	4.5	4.5	5.5
Nicotinic acid, mg/kg	75	65	65	45
Pantothenate, mg/kg	15	11	11	20
Folic acid, mg/kg	1.3	0.9	0.9	1.1
Choline, mg/kg	2000	1750	1750	1300
Vitamin B_{12}, mg/kg	0.011	0.007	0.007	0.011
Thiamin, mg/kg	2.0	2.0	2.0	2.0
Pyridoxine, mg/kg	4.0	3.3	3.3	7.0
Biotin, mg/kg	0.26	0.22	0.22	0.26
Linoleic acid, %	1.1	0.8	0.8	1.0

Sources: Austic and Scott (1984), Scott (1987), National Research Council (1994).

TABLE 26.14 Amino acid requirements of starting and growing turkeys, expressed as percentage of dietary protein.[a]

AMINO ACID	REQUIREMENT
Arginine	5.7
Glycine + serine	3.6
Histidine	2.1
Isoleucine	3.9
Leucine	6.8
Lysine	5.7
Methionine	2.0
Methionine + cystine	3.8
Phenylalanine	3.6
Phenylalanine + tyrosine	6.4
Threonine	3.6
Tryptophan	0.93
Valine	4.3

[a]Based on amino acid requirements of the National Research Council (1994) for starting turkeys.

turkey parts. Turkeys are white skinned. Therefore, pigments that are used in chicken and duck production are not of concern in the formulation of diets for turkeys.

Turkeys, especially males, are grown to a large body size. This increases the tendency for leg problems if nutritional or environmental factors are marginal. Ca, P, vitamin D_3, biotin, choline, Mn, and Zn are nutrients that may affect leg development. Turkeys may develop tibial dyschondroplasia, usually at a later age than broiler chickens.

Ducks

There are many breeds of ducks, having different growth rates and different mature body sizes. This section concerns the nutrition of Pekin ducks, the breed traditionally used for meat production in the United States.

The growth of Pekin ducks is shown in Table 26.15. The birds grow rapidly, achieving 3.2 kg of body weight in just 7 weeks. The rate of growth declines during this time, permitting the use of diets of lower nutrient density as the duck gets older. Typical nutrient requirements of ducks are presented in Table 26.16. Amino acid requirements are not as well known as for other species of domestic fowl. Information is available on most limiting amino acids and crude protein levels that are consistent with good growth or egg production and efficiency of feed utilization when typical, practical diets based on corn, soybean, and meat meals are used.

A variety of minerals and vitamins are required, as for other poultry species. The same nutrients that are important in skeletal development and leg structure in other species also are important for ducks. Carotenoid pigments are a consideration in feed formulas for Pekin ducks as they are in broiler feeds.

TABLE 26.15 Body weights and feed consumption of Pekin ducks.

	MALES			FEMALES		
AGE, WK	BODY WEIGHT, KG	CUMULATIVE FEED CONSUMPTION,[a] KG	FEED CONVERSION[b]	BODY WEIGHT, KG	CUMULATIVE FEED CONSUMPTION,[a] KG	FEED CONVERSION[b]
1	0.27	0.25	0.92	0.28	0.25	0.89
2	0.80	1.10	1.38	0.81	1.04	1.28
3	1.40	2.34	1.67	1.38	2.31	1.67
4	1.92	3.62	1.88	1.93	3.71	1.92
5	2.46	5.13	2.08	2.45	5.32	2.17
6	2.95	6.74	2.28	2.85	7.04	2.47
7	3.28	8.43	2.57	3.11	8.66	2.78

[a]ME = approximately 3000 Kcal/kg feed.
[b]Feed consumed ÷ body weight.
Source: Adapted from data of Leeson et al. (1982).

TABLE 26.16 Nutrient requirements of ducks.

NUTRIENT	STARTING AND GROWING DUCKS, 0–7 WK	BREEDING DUCKS
Protein, minimum, %	16[a]	15
Metabolizable energy, Kcal/kg	2900	2900
Ca, %	0.6	2.8
P, available, %	0.4	0.4
Na, %	0.15	0.15
Chloride, %	0.12	0.12
K, %	0.60	0.60
Mg, %	0.05	0.05
Iodide, ppm	0.30	0.30
Mn, ppm	40	25
Fe, ppm	—[b]	—
Cu, ppm	—	—
Zn, ppm	60	60
Se, ppm	0.15	0.15
Methionine, %	0.32	0.30
Methionine + cystine, %	0.60	0.55
Lysine, %	0.90	0.70
Vitamin A, IU/kg	9000	9000
Vitamin D_3, IU/kg	1100	1100
Vitamin E, IU/kg	11	11
Vitamin K, mg/kg	2	2
Riboflavin, mg/kg	5.5	4.5
Nicotinic acid, mg/kg	55	45
Pantothenate, mg/kg	11	11
Folic acid, mg/kg	0.5	0.5
Choline, mg/kg	1750	1300
Vitamin B_{12}, mg/kg	0.009	0.005
Thiamin, mg/kg	2	2
Pyridoxine, mg/kg	3.5	3.5
Biotin, mg/kg	1.5	1.5
Linoleic acid, %	—	—

[a]Approximately 22% protein is required for maximum growth during the first 2 weeks.
[b]No information available.
Sources: Austic and Scott (1984), Scott and Dean (1991), National Research Council (1994).

Japanese Quail

The smallest of domestic fowl, *Coturnix coturnix Japonica*, the Japanese quail, has been used for egg and meat production in various countries, particularly in Asia. The growth and feed consumption of a strain selected for large body size is presented in Table 26.17.

Egg production commences at approximately 6 weeks and, under appropriate environmental conditions, persists for several months (Woodard et al., 1973). Egg size increases during the first several weeks of egg laying and then becomes constant during the remainder of the laying cycle. The egg production curve looks remarkably similar to that of laying chickens. The size of eggs produced, however, as percentage of body weight, is much larger than that of Leghorn hens (approximately 7% of body weight for quail as compared to about 4% of body weight for the Leghorn). This represents a large demand for nutrients in feeds that are used to support egg production in Japanese quail.

Although the nutrient requirements of growing and laying Japanese quail have not been explored as thoroughly as those of many other kinds of poultry, a substantial amount of information has been generated (Shim and Vohra, 1984). Nutrient requirements of growing and laying Japanese quail are shown in Table 26.18.

Other Species

Many species of birds are kept for food production. It is not possible to discuss all of them within the limited space available for this chapter. Geese, guinea fowl, pea fowl, bobwhite quail, pheasants, chuckar partridge, pigeons, ostriches, and emus are among the birds for which feeds must be

TABLE 26.17 Growth and feed consumption of Japanese quail.

AGE, WK	BODY WEIGHT, G	CUMULATIVE FEED CONSUMPTION, G	FEED CONVERSION[a]
1	27	32	1.18
2	57	95	1.67
3	91	185	2.03
4	123	295	2.40
5	149	437	2.93
6	172	592	3.44
7	182	734	4.03
8	180	844	4.69
9	181	958	5.29

[a]Feed consumed ÷ body weight.
Source: Adapted from Marks (1991). Sexes combined.

TABLE 26.18 Nutrient requirements of Japanese quail.

NUTRIENT	STARTING AND GROWING JAPANESE QUAIL, 0–6 WK	LAYING JAPANESE QUAIL
Protein, minimum, %	24	20
Metabolizable energy, Kcal/kg	2900	2900
Ca, %	1.0	2.5
P, available, %	0.4	0.4
Na, %	0.15	0.15
Chloride, %	0.15	0.15
K, %	0.40	0.40
Mg, %	0.05	0.05
Iodide, ppm	0.30	0.30
Mn, ppm	60	60
Fe, ppm	120	60
Cu, ppm	5	5
Zn, ppm	50	60
Se, ppm	—	—
Methionine, %	0.50	0.45
Methionine + cystine, %	0.75	0.70
Lysine, %	1.30	1.00
Vitamin A, IU/kg	5000	5000
Vitamin D$_3$, IU/kg	1100	1100
Vitamin E, IU/kg	12	25
Vitamin K, mg/kg	2	2
Riboflavin, mg/kg	4	4
Nicotinic acid, mg/kg	40	20
Pantothenate, mg/kg	10	15
Folic acid, mg/kg	1	1
Choline, mg/kg	2000	1500
Vitamin B$_{12}$, mg/kg	0.007	0.007
Thiamin, mg/kg	2	2
Pyridoxine, mg/kg	3	3
Biotin, mg/kg	0.3	0.15
Linoleic acid, %	1.0	1.0

Source: Modified from National Research Council (1994).

formulated. Pet birds and exotic birds also vie for the attention of practicing or research nutritionists. Experimental information is severely limited for many of these species. Nutritionists, however, can combine knowledge of traditional eating habits and limited experimental information with sound nutritional principles as a rational basis for the formulations of feeds. The extensive research literature on the nutrient needs of poultry serves as an excellent base from which to generate new knowledge on the nutritional requirements of diverse avian species that are held in captivity for food or for pleasure.

❏ REFERENCES

Akrabawi, S. S., and F. H. Kratzer. 1968. *J. Nutr.* **95:** 41.

Austic, R. E., and M. C. Nesheim. 1990. *Poultry Production*, 13th ed. Lea & Febiger, Philadelphia, PA.

Austic, R. E., and M. L. Scott. 1984. In *Diseases of Poultry* (8th ed.), pp. 38–64, M. S. Hofstad (ed.). Iowa State University Press, Ames, IA.

Baker, D. H., A. B. Batal, T. M. Parr, N. R. Augspurger, and C. M. Parsons. 2002. *Poult. Sci.* **81:** 485.

Carlisle, E. M. 1980. *J. Nutr.* **110:** 1046.

Edwards, H. M. 1984. *J. Nutr.* **114:** 1001.

Emmert, J. L., and D. H. Baker. 1997. *J. Appl. Poult. Res.* **6:** 462.

Ferket, P. R. 2003. "Turkey growth statistics." *Poultry USA.* **4**(7): 38.

Fritz, J. C., J. H. Hooper, J. L. Halpin, and H. P. Moore. 1946. *J. Nutr.* **31:** 387.

González-Esquerra, R., and S. Leeson. 2001. *Can. J. Anim. Sci.* **81:** 295.

Graber, G., and D. H. Baker. 1973. *Poult. Sci.* **52:** 892.

Harper, A. E., N. J. Benevenga, and R. M. Wohlheuter. 1970. *Physiol. Rev.* **50:** 428.

Heil, G., and W. Hartmann. 1992. *World's Poult. Sci. J.* **48:** 269.

Hill, F. W., and D. L. Anderson. 1958. *J. Nutr.* **64:** 587.

Hill, F. W., and R. Renner. 1960. *J. Nutr.* **80:** 375.

Holman, R. T. 1960. *J. Nutr.* **70:** 405.

Hopkins, D. T., R. L. Witter, and M. C. Nesheim. 1963. *Proc. Soc. Exp. Biol. Med.* **114:** 82.

Hopkins, L. L., Jr., and H. E. Mohr. 1974. *Fed. Proc.* **33:** 1773.

Janky, D. M., R. A. Voitle, and R. H. Harms. 1985. *Poult. Sci.* **64:** 925.

Karunajeewa, H., R. J. Hughes, M. W. McDonald, and F. S. Shenstone. 1984. *World's Poult. Sci. J.* **40:** 52.

Klasing, K. C., and R. E. Austic. 2003. *Diseases of Poultry*, 11th ed., pp. 1027–1053. H. J. Barnes, J. R. Glisson, A. M. Fadly, L. R. McDougald, and D. E. Swayne (eds). Iowa State University Press, Ames, IA.

Leach, R. M., Jr., and M. C. Nesheim. 1972. *J. Nutr.* **102:** 1673.

Leeson, S., and J. D. Summers. 1991. *Commercial Poultry Nutrition.* University Books, Guelph, Ontario, Canada.

Leeson, S., J. D. Summers, and J. Proulx. 1982. *Poult. Sci.* **61:** 2456.

Marks, H. L. 1991. *Poult. Sci.* **70:** 1047.

Menge, H. 1968. *J. Nutr.* **95:** 578.

Miles, R. D., and W. R. Featherston. 1974. *Proc. Soc. Exp. Biol. Med.* **145:** 686.

Mongin, P. 1981. *Proc. Nutr. Soc.* **40:** 285.

Moran, E. T., Jr. 1982. *Comparative Nutrition of Fowl and Swine. The Gastrointestinal Systems.* Office for Educational Practice, University of Guelph, Guelph, Ontario, Canada.

National Research Council (NRC). 1994. *Nutrient Requirements of Poultry,* 9th rev. ed. National Academy Press, Washington, DC.

Nelson, T. S., L. K. Kirby, Z. B. Johnson, and J. N. Beasely. 1981. *Poult Sci.* **60:** 1030.

Nesheim, M. C., and K. J. Carpenter. 1967. *Brit. J. Nutr.* **21:** 399.

Nielson, F. H., and D. A. Ollerich. 1974. *Fed. Proc.* **33:** 1767.

Roland, D. A. 1982. *Proc. Georgia Nutrition Conference* (Atlanta), February 17–19, p. 149.

Scott, M. L. 1987. *Nutrition of the Turkey.* M. L. Scott & Associates, Ithaca, NY.

Scott, M. L., and W. F. Dean. 1991. *Nutrition and Management of Ducks.* M. L. Scott & Associates, Ithaca, NY.

Scott, M. L., M. C. Nesheim, and R. J. Young. 1982. *Nutrition of the Chicken.* M. L. Scott & Associates, Ithaca, NY.

Shim, K. F., and P. Vohra. 1984. *World's Poult. Sci. J.* **40:** 261.

Sibbald, I. 1982. *Can. J. Anim. Sci.* **62:** 983.

Simons, P. C. M., H. A. J. Versteegh, A. W. Jongbloed, et al. 1990. *Br. J. Nutr.* **64:** 525.

Skadhauge, E. 1967. *Comp. Biochem. Physiol.* **23:** 483.

Sturkie, P. D. 1986. *Avian Physiology* (4th ed.) Springer-Verlag, New York.

Sullivan, T. W., J. H. Douglas, N. J. Gonzales, and P. L. Bond, Jr. 1992. *Poultry Sci.* **71:** 2065.

Surai, P. F., and N. H. C. Sparks. 2001. *Trends Food Sci. Technol.* **12:** 7.

Velu, J. G., D. H. Baker, and H. M. Scott. 1972. *Poult. Sci.* **51:** 938.

Wilson, W. O. 1948. *Poult. Sci.* **27:** 813.

Woodard, A. E., H. Abplanalp, W. O. Wilson, and P. Vohra. 1973. *Japanese Quail Husbandry in the Laboratory.* University of California Press, Davis, CA.

27

Horses

HAROLD F. HINTZ

Cornell University photograph by Russ Hamilton.

HORSES HAVE A LONG history of service to humans in terms of work, entertainment, and companionship. With the increased use of automobiles, tractors, and trucks in the early twentieth century, the number of horses in the United States decreased from more than 26 million in 1918 to about 3 million in 1960. No official federal government census has been taken since 1960, but the increased use of the horse for pleasure started about that time and the horse population increased to perhaps 10 million horses in the 1980s. Then, changes in the economy and in tax laws contributed to a decrease in the horse population. In 1984, there were about 324,000 new horses registered with the nine largest breed registries, but in 1992 only about 215,000 horses were registered in the nine breeds. But the horse has made another comeback. In 2003, over 300,000 horses were registered by the nine largest breeds. Thus, the horse still remains an important part of the U.S. culture.

The following is a brief review of the nutrition of the horse. More detailed information can be found in references such as NRC (1989), Pagan and Geor (2001), and Lewis (1995).

❑ DIGESTIVE PHYSIOLOGY

The horse is a nonruminant herbivore with significant microbial fermentation in the hind gut (cecum and colon) that permits the horse to utilize fibrous feeds (see Chapter 4).

The principal microbial inhabitants of the equine gut are bacteria, protozoa, and fungi (Julliand et al., 1992) and are similar to those found in the rumen (Argenzio, 1990). Fermentation in the hind gut results in the production of carbon dioxide, methane, and volatile fatty acids (VFA), primarily acetate, propionate, and butyrate. Fermentation also results in methane production, but it is equivalent to less than 3% of the total energy intake (Pagan and Hintz, 1986).

VFAs are readily absorbed from the hind gut and can be used as energy sources. In fact, 30% or more of the energy utilized by the horse may come from VFA (Bergman, 1991). Propionate can also result in significant glucose production (Ford and Simmons, 1985). Acetate and butyrate are not gluconeogenic.

The horse does not utilize fiber in feeds as efficiently as do cattle because the digesta in the horse has a faster rate of passage than in cattle. The microflora in the large intestine of the horse do not have as much time to ferment the material as do the microflora in the rumen.

Bacteria of the hind gut also produce protein and water-soluble vitamins, but the horse's utilization of these products is not great because absorption of them from the hind gut is not efficient (NRC, 1989).

Not only is the horse being fed but also the microflora of the digestive tract need to be fed and have a healthy environment. Proper feeding of horse and bacteria can greatly reduce the incidence of colic and other digestive problems. The horse evolved as a grazing animal and so it needs a slow and steady supply of fermentable material to the microflora. Clarke et al. (1990) demonstrated that frequent feeding is preferred to once or twice a day feeding if large amounts of grain are fed. They found that only twice a day feeding of a high-energy–low-forage diet adversely influenced microbial activity and fluid content of the large intestine and predisposed the horse to spontaneous digestive disturbances. Thus, when feeding large amounts of grain, three or more feedings per day is recommended and never more than 5 lb per feeding.

Dietary ingredients can greatly influence microbial activity. High-starch or -grain intakes can decrease the relative amount of acetate and increase the relative amount of propionate that is produced. Dietary changes should be made gradually because abrupt changes can lead to changes in microbial population, intestinal pH, and death of certain types of bacteria, which may lead to endotoxin release. It has been suggested that the release of endotoxins is one of the factors involved in the production of founder or laminitis. It is generally recommended that the horse be fed forage or forage equivalents at a rate of at least 1 lb of dry matter per 100 lb of body weight in order to maintain normal microbial function. Lower amounts of fiber can be fed under closely controlled conditions. But horses are usually fed more than 1% of body weight in forage.

Probiotics such as yeast cultures have been reported to enhance fiber digestion and phosphorus utilization (Pagan, 1990).

❑ GENERAL FEEDING

Forages

The feeding of horses can greatly be simplified by providing high-quality forage (pasture or hay). The forage, as mentioned above, can help to maintain normal microbial activity, and it provides energy, protein, vitamins, and minerals. Either legume or grass forage can be used. The most important characteristics to consider are the nutrient content and cost of the nutrients. Good-quality pasture can be an excellent basis for a feeding program. Proper use of pasture provides a much higher level of antioxidants such as vitamin E and carotene than are present in hay. Pasture can reduce the incidence of colic, ulcers, signs of respiratory diseases (due to decreased mold and dust), and abnormal behaviors.

Legumes usually contain a higher content of protein than do grasses, and thus they are of particular benefit to those classes of horses such as weanlings and lactating mares that require a high-protein diet. If the diet contains more protein than needed, the extra protein is catabolized. The carbon chain is utilized for energy, and the nitrogen is excreted in the urine. The excretion process requires water. Thus, water intake and urinary excretion are greater in horses fed legume hay than in those fed grass. One study reported that horses fed timothy hay produced 16 lb of urine daily, whereas those fed alfalfa hay produced 27 lb of urine (Henry, 1912). The additional excretion of urine led some horse owners to the erroneous conclusion that feeding alfalfa caused kidney damage.

The forage must be palatable and free of mold and toxins. Grazing on pasture containing fescue infected with the endophyte, *Acremonium coenophialium,* can result in prolonged gestation, thickened placentas, agalactia, dystocia in mares, and large, weak foals (Porter and Thompson, 1992). Removing the mares from infected pasture prior to the last third of gestation can prevent the problems.

Grazing on alsike clover has been reported to cause hepatic disease and photosensitization, but the association between alsike and these conditions still needs clarification (Nation, 1991).

Legumes, particularly red clover, infected with the fungus (*Rhizoctonia leguminicola*) can cause excessive slobbering and perhaps even abortion (Sockett et al., 1982). Kleingrass may cause chronic hepatic disease (Cornick et al., 1988). Sudangrass-sorghum hybrids may cause the cystitis syndrome (irritation of the bladder), which can lead to kidney infection and death (Morgan et al., 1990).

Ryegrass pasture or seed cleanings infected with the endophyte (*Acremonium lolii*) can cause rye grass staggers which is characterized by muscular tremors, incoordination, and tetany (Munday et al., 1985). The endophytes appears to be more common in New Zealand and Australia than in North America.

Pasture can be a rich source for parasitic infestation. Prompt removal of feces could greatly reduce the parasite load and improve pasture utilization.

Some forages, such as kikuyu and setaria, grown in subtropical and tropical areas may contain such a high concentration of oxalate that they can induce a calcium deficiency (Blaney et al., 1981).

Reed canary grass may contain an alkaloid, hordenine, that decreases palatability. In some states, the presence of hordenine in urine can cause disqualification of racehorses (Singh et al., 1992). Hordenine may also be found in the urine of horses fed sprouted barley.

Ingestion of alfalfa hay contaminated by blister beetles (*Epicauta sp.*) can cause severe irritation to tissues, heart failure, and death (Blodgett et al., 1991).

Forage Alternatives. Straw, cottonseed hulls, peanut hulls, soy hulls, rice mill feed, and paper with high-cellulose content have been used as roughage replacements when adequate amounts of high-quality forages were not available or not practical (NRC, 1989). The most commonly used alternative is probably beet pulp. It maintains normal intestinal activity of the microflora, but the fiber is highly digestible. In fact, the digestible energy content of a beet pulp is higher than that of most hays. Beet pulp is dust free and, therefore, excellent for rations of horses with respiratory problems. Beet pulp, however, has a low content of vitamins and certain minerals which must be considered when hay is replaced by beet pulp. When feeding large amounts of dried beet pulp, it should be soaked prior to feeding.

Grains

Oats are the traditional grain for horses probably because of their high palatability and relative safety. Oats contain more protein, more fiber, and less digestible energy than the other grains. Furthermore the starch of oats is readily digested in the small intestine and therefore less likely to cause digestive problems in the large intestine because of fermentation than the starch in corn or barley that is more resistant to digestion in small intestine. Oats can be fed whole. Crimping only increases digestibility of dry matter by 5 to 7%. Of course, for young foals and for old horses with poor teeth, crimping is of greater advantage (NRC, 1989).

Corn can be an excellent, economical source of energy when fed properly. It should be remembered that a volume of corn provides about the same energy as two volumes of oats. Therefore, feed should be measured by weight, not volume.

Barley has long been fed to horses. There has been a renewed interest in barley recently because it has an economical source of energy and is less likely to contain the mold *Fusarium moniliforme* than corn. The mold contains a toxin called fumonsin which causes atoxia, blindness, circling, hyperexcitability, and death due to degeneration of the subcortical white matter. The condition has been called leukoencephalomalcia or mycotoxic encephalomalacia (Ross et al., 1993). Corn screenings have been a particularly troublesome source of fumonsin.

Sorghum grains and wheat can be fed to horses but should be rolled, cracked, or steam flaked to increase utilization. Rye can be used if processed, but it may cause decreased palatability and should be limited to 20% of the grain mixture (NRC, 1989). Sorghum grains, wheat, and rye all contain low concentrations of fiber. Therefore, the same caution used with corn should be used with these grains.

Interest in the feeding of rice and rice byproducts has increased in recent years, particularly in Australia and to a lesser degree in the United States. The principal products are cracked or broken rice kernels and rice pollard (germ, rice polishings, and outer bran layer). The cracked rice was determined to contain 2.9 Mcal of DE per kg of dry matter (McMeniman et al., 1990). Rice pollard was valued at 2.8 Mcal. Use of rice bran has greatly increased. Rice bran is the germ and the bran layer that is removed to produce white rice. It may contain 15 to 22% fat and 12 to 15% protein.

Commercial Grains

Many horse owners use commercial mixtures. Such mixtures can have the advantages of containing added vitamins and minerals in balanced amounts. Furthermore, companies have the ability to use nutritional byproducts that are economical sources of nutrients but are not readily available for the horse owner to use.

There are three main types of commercial grains: (1) texturized, (2) pelleted, and (3) extruded. Combinations of the types can also be used. For example, a pelleted supplement containing protein, vitamins, and minerals may be added to a texturized feed. Sweet feed, a texturized grain with added molasses, is the most common type of commercial grain. Pelleting prevents sorting but may allow an increased rate of intake. Extruding, a process in which steam is added during the pelleting process, decreases the density. Extruded feed may weigh about one-half that of pelleted per unit of volume. The decreased density decreases rate of intake and may reduce the incidence of digestive disturbances. Extruding, however, increases the cost.

Commercial feeds designed by other species should be avoided. The nutrient balance might not be appropriate, and the feed may contain additives that are of no value or toxic to horses. Many horses have died owing to inadvertent feeding of ionophores such as monensin. Monensin enhances feed utilization in ruminants and is an excellent coccidiostat in chickens, but horses have a much lower tolerance for monensin than do cattle and chickens. Horses may experience colic, muscular weakness, and cardiac failure from ingestion of feed containing monensin (Whitlock, 1990). Other potentially dangerous ionophores include salinomycin, lasalocid, and narasin. Other antimicrobials such as lincomycin and clindamycin can also be dangerous (Whitlock, 1990).

❑ FEEDING THE HORSE AT MAINTENANCE

Nutrient requirements for all classes of horses have been summarized by the NRC (1989). Horses at maintenance have lower nutrient requirements than working horses or breeding animals. Good pasture or good-quality hay plus trace mineral salt may supply all the nutrients that are needed. NRC (1989) suggested that digestible energy (DE) requirements could be calculated by the equation DE (Mcal/day = $1.4 + 0.03W$ where W is weight in kg for horses weighing 600 kg or less and DE (Mcal/day) = $1.82 + 0.0383W - 0.000015W^2$ for horses over 600 kg. Horses at maintenance weighing 600 kg or less need about 1.5 to 2.0% of their body weight in hay to meet energy needs, depending on the quality of hay and the individuality of the horse. Larger horses need a smaller percentage of body weight, probably because their voluntary activity is less. Estimates of energy required and amounts of hay needed to meet the requirements are shown in Table 27.1. Cold weather would increase the amount of energy needed. Cymbaluk and Christison (1990) reported that the increase could be estimated by multiplying each degree below $-15°C$ by $0.00082 \times$ body weight in kg. Thus, a 500-kg horse at $-25°C$ would require 4.1 additional Mcal of DE, or 25% more than the maintenance requirement of 16.4 Mcal. The response would depend on body condition because thin horses have less insulation and require more energy than fat horses when exposed to cold environmental temperatures.

As mentioned above, the individuality of the horse also influences the amount of feed needed for maintenance. Unlike other species of livestock, horses have not been selected for feed efficiency. Thus, significant variation in energy requirements,

TABLE 27.1   Estimates of digestible energy requirements for maintenance calculated according to NRC equation and amount of hay needed to meet the requirements.

KG	BODY WT LB	DE (MCAL/DAY)	HAY[a] (LB/DAY)	PERCENT BODY WT	HAY[b] (LB/DAY)	PERCENT BODY WT
200	440	7.4	$9\frac{1}{4}$	2.1	$7\frac{1}{2}$	1.7
300	660	10.4	13	1.9	$10\frac{1}{2}$	1.6
400	880	13.4	$16\frac{3}{4}$	1.9	$13\frac{1}{2}$	1.5
500	1100	16.4	$20\frac{1}{2}$	1.9	$16\frac{1}{2}$	1.5
600	1320	19.4	$24\frac{1}{4}$	1.8	$19\frac{1}{2}$	1.5
700	1540	21.3	$26\frac{1}{2}$	1.7	21	1.4
800	1760	22.9	$28\frac{1}{2}$	1.6	23	1.3
900	1980	24.1	30	1.5	24	1.2

[a]Assuming 0.8 Mcal of DE/lb of hay (90% dry matter) such as timothy hay sun-cured.
[b]Assuming 1.0 Mcal of DE/lb of hay (90% dry matter) such as alfalfa hay, sun-cured, early bloom.

perhaps as much as 20%, can be found within a group of horses. Therefore, the best method to determine adequacy of energy intake is to weigh the horse and/or evaluate body condition. Several body score systems are available. Most systems evaluate body score by estimating condition over the back, ribs, and rump (Tables 27.2 and 27.3). If scales are not available, tapes calibrated to be used around the heart girth can be used to estimate weight.

Diets for horses at maintenance should contain at least 7.2% protein, 0.21% calcium, 0.15% phosphorus, and 0.3% potassium, plus adequate amounts of vitamins and trace minerals (NRC, 1989). Vitamin E has received a great amount of attention lately, both in human nutrition and equine nutrition. Studies showing that additional vitamin E could enhance the immune response and increase blood levels of vitamin E led NRC (1989) to increase the estimated requirement from 15 IU/kg to 50 IU/kg for horses at maintenance. The requirement of mares and growing and working horses was increased to 80 IU/kg.

There are also specific cases for additional vitamin E supplementation. Equine motor neuron disease (EMND), first reported by Cummings et al. (1990), can be prevented by vitamin E supplementation. Affected horses have generalized weakness, muscle fasciculations, muscle atrophy, and weight loss. There is degeneration of the motor neurons in the spinal cord and brain stem leading to axonal degeneration in the ventral roots and peripheral and cranial nerves. The changes are similar to those described in people with amyotrophic lateral sclerosis (ALS), also known as Lou Gehrig's disease. Almost all of the affected horses identified thus far have been housed in stables without access to pasture, and very low vitamin E plasma and adipose tissue levels have been reported in the affected horses (Divers, 1993).

Equine degenerative myeloencephalopathy (EDM) is a diffuse degenerative disease of the spinal cord and brain stem. It is most commonly found in young horses. Affected animals may show various signs such as clumsiness, inability to do complicated movements, malpositioning of limbs at rest, or obvious ataxia (Blythe and Craig, 1992a). Histologically, there can be neuroaxonal dystrophy in brain stem nuclei and throughout the spinal cord (Blythe and Craig, 1992a). It has been proposed that it is a familial disease and that a deficiency of vitamin E is involved (Blythe and Craig, 1992b). Supplementation with very high levels of vitamin E (6000 IU/day) may reduce the incidence of EDM in young horses from affected families or can cause improvement in some affected animals (Blythe and Craig, 1992b). The reason such high intakes of vitamin E are needed for certain families of horses is unknown, but it has been suggested that such horses are more susceptible to antioxidant deficiencies or have an increased antioxidant requirement during the first year of life (Blythe and Craig, 1992b).

Adequate amounts of clean water must be supplied. The amount of water needed will vary according to environmental temperature, dry matter

TABLE 27.2   Body condition score system.

SCORE	DESCRIPTION
1 Poor	Animal extremely emaciated. Spinous processes, ribs, tailhead, tuber coxae, and ischii projecting prominently. Bone structure of withers, shoulders, and neck easily noticeable. No fatty tissue can be felt.
2 Very thin	Animal emaciated. Slight fat covering over base of spinous processes, transverse processes of lumbar vertebrae feel rounded. Spinous processes, ribs, tailhead, tuber coxae, and ischii prominent. Withers, shoulders, and neck structures faintly discernible.
3 Thin	Fat buildup about halfway on spinous processes; transverse processes cannot be felt. Slight fat cover over ribs. Spinous processes and ribs easily discernible. Tailhead prominent, but individual vertebrae cannot be visually identified. Tuber coxae appear rounded but easily discernible. Tuber ischii not distinguishable. Withers, shoulders, and neck accentuated.
4 Moderately thin	Negative crease along back. Faint outline of ribs discernible. Tailhead prominence depends on conformation, fat can be felt around it. Tuber coxae not discernible. Withers, shoulders, and neck not obviously thin.
5 Moderate	Back level. Ribs cannot be visually distinguished but can be easily felt. Fat around tailhead beginning to feel spongy. Withers appear rounded over spinous processes. Shoulders and neck blend smoothly into body.
6 Moderately fleshy	May have slight crease down back. Fat over ribs feels spongy. Fat around tailhead feels soft. Fat beginning to be deposited along the side of the withers, behind the shoulders, and along the sides of the neck.
7 Fleshy	May have crease down back. Individual ribs can be felt, but noticeable filling between ribs with fat. Fat around tailhead is soft. Fat deposited along withers, behind shoulders, and along the neck.
8 Fat	Crease down back. Difficult to feel ribs. Fat around tailhead very soft. Area along withers filled with fat. Area behind shoulder filled with fat. Noticeable thickening of neck. Fat deposited along inner thighs.
9 Extremely fat	Obvious crease down back. Patchy fat appearing over ribs. Bulging fat around tailhead, along withers, behind shoulders and along neck. Fat along inner thighs may rub together. Flank filled with fat.

Source: Henneke et al. (1983).

intake, and nutrient content of the diet. Elevated temperatures can increase water by fourfold even in horses that are not working. The type of diet can influence water intake. As mentioned earlier, the ingestion of alfalfa hay causes greater water intake than the ingestion of timothy hay. Alfalfa contains higher concentrations of protein and minerals than does timothy. Water is needed for the excretion of the nitrogen and minerals that are not used. At comfortable environmental temperatures and when fed a grass hay ration, horses may need about 2 to 3 l of water per kg of dry matter intake (NRC, 1989).

If a horse weighs about 500 kg, the dry matter intake should be about 7 kg; therefore, the calculated water intake would be 14 to 21 l or about $3\frac{3}{4}$ to $5\frac{1}{2}$ gal.

Spier et al. (1990) suggested another method to estimate maintenance water needs. They suggested that horses require 30 ml of water per kg of body weight per day. Thus, a 500-kg horse would need 15 l. Therefore, the two methods are in reasonable agreement for horses at maintenance.

Even a slight deficiency of water could cause decreased feed intake and lead to impactive colic. A loss of body water equivalent to 5 to 7% of body weight might be considered mild dehydration (Spier et al., 1990). A loss of 8 to 10% body weight equivalent would be moderate. Signs would include sunken eyes and depression. A loss of over 10% body weight equivalent loss could be fatal according to Spier et al. (1990). Sneddon (1993), however, suggested that horses with Arabian-

TABLE 27.3 Body condition score system.

	NECK	BACK AND RIBS	PELVIS
0 Very Poor	Marked "ewe" neck Narrow and slack at base	Skin tight over ribs Spinous processes sharp and easily seen	Angular pelvis—skin tight Deep cavity under tail and either side of croup
1 Poor	"Ewe" neck Narrow and slack at base	Ribs easily visible Skin sunken either side of backbone Spinous processes well defined	Rump sunken but skin supple Pelvis and croup well defined Deep depression under tail
2 Moderate	Narrow but firm	Ribs just visible Backbone well covered Spinous processes felt	Rump flat either side of backbone Croup well defined, some fat Slight cavity under tail
3 Good	No crest (except stallions) Firm neck	Ribs just covered—easily felt No 'gutter' along back Spinous processes covered but can be felt	Covered by fat and rounded No "gutter" Pelvis easily felt
4 Fat	Slight crest Wide and firm	Ribs well covered—need firm pressure to feel "gutter" along backbone	"Gutter" to root of tail Pelvis covered by soft fat—felt only with firm pressure
5 Very Fat	Marked crest Very wide and firm. Folds of fat	Ribs buried—cannot feel Deep "gutter" Back broad and flat	Deep "gutter" to root of tail Skin distended Pelvis buried—cannot feel

Adjust the pelvis score by 0.5 point if it differs by 1 or more points from the back or neck scores to obtain the condition score.
Source: Carroll and Huntington, 1988.

based breeding could tolerate losses up to 12% of body weight.

Water temperature can influence water intake. Intake in the winter may decrease when the water temperature is less than 45 degrees.

❏ FEEDING THE BROOD MARE

Pre-Breeding

Energy intake and body condition appear to be the most important nutritional factors influencing conception rate. A few years ago, the consensus was that mares should be lean prior to breeding season and then fed to gain weight during breeding season. Henneke et al. (1984) demonstrated, however, that to improve chances of conception the mares should be in good condition at breeding time rather than lean. In fact, a little over-condition was preferred to undercondition. It was suggested that mares at breeding should have a body score of at least 5 (see Table 27.2). Swinkler et al. (1993) reported that a positive change in body weight increased the incidence of ovulation in mares.

Many nutrients can influence conception rate. For example, deficiencies of vitamin A and E can cause sterility. There is no evidence, however, that increasing the concentration of protein, minerals, or vitamins above that recommended for maintenance will significantly enhance rate of conception.

Gestation

The nutrient requirements depend on stage of production. The mare in early gestation does not have requirements much different from those of the open mare. Until recently, it was usually suggested that because the fetus does not increase greatly in size until the last third of gestation it was not necessary to increase feed intake above maintenance until that time (NRC, 1989). However, it has been suggested that it would be prudent to start to increase feed intake and weight gain as early as the sixth month of gestation. The total weight gained during gestation should be equivalent to about 18% of the mare's weight (Meyer, 1992). Some mares during gestation can attain reasonable weight gains without the need for grain and will need to be fed grain in order to obtain adequate energy for the development of the fetus. Some mares may voluntarily decrease hay intake during late gestation because of decreased abdominal space due to the growth of the fetus. And as mentioned earlier, some individuals appear to have higher energy requirements because of differences in behavior. Such mares, dubbed hard keepers, would need grain. NRC (1989) suggests that mares in late gestation may consume forage at a rate equivalent to 1 to $1\frac{1}{2}$% of body weight and grain at a rate of $\frac{1}{2}$ to 1% of body weight. Forages will usually be lacking in some trace minerals. The grain mixture should be fortified to balance the nutrients in the hay. Thus, those mares that require little grain may need a highly fortified concentrate to provide the daily requirement of minerals and vitamins. For example, most forages in the United States are likely to be lacking in copper and zinc. Forages may also be lacking in selenium, depending on the selenium content of the soil. The amount of protein needed in the concentrate is highly dependent on the type of hay. Average to good-quality legumes will usually provide more protein than is needed, whereas grass may require protein supplementation.

The total ration for a mare in late gestation should contain at least 10–11% protein, 0.45% calcium, 0.34% phosphorus, 10 mg copper, 40 mg zinc, and 0.1 mg selenium per kg of feed (dry matter basis). The protein, calcium, and phosphorus are necessary for the proper development of muscle and bone in the fetus. The selenium is needed to prevent white muscle disease in the foal. If the diet is lacking in selenium, the foal will be born weak, be unable to nurse, and will die. Selenium supplementation should be done carefully, however, for excessive levels (about 20 times the requirement) could be toxic. The selenium content of feeds varies greatly depending on the soil type in which the plants were grown. Iodine is also needed for the development of the fetus. Foals born to a mare fed inadequate levels of iodine may be stillborn or born weak, even though the mare appears normal. Affected foals usually have enlarged thyroid glands (goiters). Iodine intakes of 1 to 2 mg per day will meet the needs of the mare and fetus. As with selenium, excessive intakes must be avoided. Intakes as low as 40 mg of iodine per day have resulted in foals that were born with enlarged thyroids and immature skeletons. Thus, the signs of excessive and deficient iodine can be similar. Excessive intakes of iodine have resulted from overzealous use of supplements, particularly those containing seaweed.

Lactation

Lactation can increase energy demands by almost twice that of maintenance. NRC (1989) suggests that mares in early lactation (foaling to three months) may consume an amount of concentrate equivalent to 1 to 2% of body weight and forage in a similar amount. Thus, mares might have a total intake greater than 3% of body weight. In late lactation, intake of concentrate may be the equivalent of $\frac{1}{2}$ to $1\frac{1}{2}$% body weight, and forage intake may be 1 to 2% of body weight. Of course, the energy requirement depends on the amount of milk produced. The NRC estimates are based on an expected milk production equivalent to 3% of body weight during early lactation and 2% of body weight during late lactation. As with maintenance requirements, there can be significant variations among mares as to amount of milk production. Mares, at least in the United States, have not been selected for milk production. Thus, it is necessary to monitor the mare's body weight and body score to evaluate the adequacy of energy intake.

Milk production also greatly increases the requirement for other nutrients such as protein and calcium because milk is an excellent source of these nutrients. For example, the protein requirement (g/day) for a mare in early lactation is more than twice that for maintenance. The calcium requirement is almost three times maintenance. On a dry matter basis, the ration of a mare in early lactation should contain at least 13% protein, 0.5% calcium, and 0.34% phosphorus. As will be discussed in more detail later, the trace minerals, copper and zinc, are critical for normal skeletal development in the foal. Milk is not a particularly rich source of these trace minerals, but supplementing the lactating mare's diet with these minerals is not an effective method of increasing the milk content. If the foal needs additional copper and zinc, it is more effective to include them in a creep ration rather than increase the milk content by increasing the mare's intake of these nutrients.

❏ FEEDING THE FOAL

Suckling

Creep feeding is frequently used to increase the foal's rate of development, particularly when the mare is a poor milk producer and there is a need to decrease "postweaning slump." Many foals will lose weight the week or so after weaning because of the associated stress that can lead to decreased nutrient intakes. Foals that have been creep-fed and are accustomed to the grain mixture may have less difficulty adjusting to weaning and not decrease intake, thereby alleviating weight loss. Previously, it was recommended that a creep ration be specifically designed for creep feeding. The creep ration was higher in protein and mineral concentration than the typical weanling ration. But studies conducted at the University of Florida demonstrated that foals fed the weanling ration during the suckling period develop adequately and have less difficulty adjusting to weaning than when fed a different ration prior to weaning. Furthermore, studies have demonstrated that a creep ration containing 13% protein can be adequate. Use of only one ration, the weanling ration, also simplifies management. The grain in the weanling ration should be processed, such as crimping or rolling, because the teeth of the suckling are not effective for chewing whole grains. Corn, oats, and barley have been successfully used in weanling rations. The essential amino acids such as lysine must be supplied to enhance body development. Soybean meal is usually preferred over other plant protein sources because of its higher lysine content. Other plant protein sources such as cottonseed meal and lysine meal can be used, however, if the lysine content is adjusted correctly. Animal protein sources such as dried milk products or fish meal can also be used because they are excellent sources of the essential amino acids.

It is often recommended that foals be exposed to creep feed as early as one month of age, although some authorities recommend waiting until two months. The amount of feed is also debated. Feeding ad libitum is usually not recommended because of the concern that excessive rapid growth may lead to skeletal abnormalities such as epiphysitis and flexural deformities (formerly called contracted tendons). Lewis (1995) suggested that the intake of sucklings should be limited to 0.5–0.75% of body weight. A foal with an expected mature weight of 500 kg might weigh about 160 kg by 3 months and thus be fed 0.8 to 1.1 kg/day. Ott (1990) developed the thumb rule that foals should receive about one pound per day per month of age. Thus, a foal three months of age would be fed 3 lb or 1.35 kg daily. In any case, the foals should be observed closely and at the first indication of epiphysitis or flexural deformities, a deter-

mination should be made whether the grain intake should be reduced.

The Weanling

The ration for the weanling is perhaps the most critical of all the rations on the horse farm. It should contain at least 14% protein, 0.68% calcium, and 0.38% phosphorus. Lysine is the only amino acid for which a requirement has been established, that other amino acids are essential. Thus, the protein should provide a good array of amino acids. The diet should contain at least 0.6% lysine.

Considerable concern has been voiced about skeletal problems such as metabolic bone disease and developmental orthopedic disease in foals. For purposes of this chapter, we define metabolic bone disease as improper mineralization of bone. Nutritional secondary hyperparathyroidism is an example of metabolic bone disease. It is crucial that the body maintain a relatively consistent level of blood calcium because it is required for many functions such as normal muscle activity. When the blood level of calcium decreases, the parathyroid responds by releasing a hormone that removes calcium from the bone in order to maintain the critical levels in the blood. When the diet does not provide calcium, the parathyroid becomes very active and much calcium is removed from the bone—hence, the name, nutritional secondary hyperparathyroidism. Not only is the proper dietary level of calcium critical but also that of phosphorus. If phosphorus intake is significantly in excess of that of calcium, calcium–phosphorus complexes are formed and calcium utilization is greatly decreased. The Ca:P ratio of the diet should always be at least 1:1. Ratios as high as 3:1 are acceptable, however, providing phosphorus levels are adequate.

Developmental orthopedic disease (DOD) is a general term for several conditions such as osteochondrosis, flexural deformities, epiphysitis, and some forms of wobblers (stenosis of the vertebral canal causing ataxia). The basic lesion of DOD is improper maturation of cartilage into bone. DOD has probably been a problem for many years, but in the 1980s the number of cases appeared to be increasing greatly and it remains one of the most serious problems of foals. Jeffcott (1993) concluded that "osteochondrosis is present in the horse population at unacceptably high levels (i.e., 10–25%) across a range of different breeds, associated therefore with a corresponding inflated economic loss, on a global scale." Furthermore, he suggested that following the alert to the escalating problem in the 1980s little progress has been made in understanding the underlying cause.

DOD is a multifactorial problem; growth rate, energy intake, mineral nutrition, genetics, biomechanics, and endocrine elements are probably all involved. Not surprisingly, then, experiments have yielded conflicting results. Some studies have found that the incidence of DOD is greater in foals growing rapidly, but in other studies rapid growth was not found to consistently produce leg abnormalities. It seems prudent, however, to restrict growth rates to the moderate levels the NRC (1989) has listed for foals on farms where DOD is a problem. NRC points out that the optimum growth rate and energy for productivity and longevity have not been established, but that high intakes of soluble carbohydrates have been associated with DOD. NRC also stresses that any nutrient imbalance is exacerbated with high-energy intake. To illustrate the effect of energy intake and growth rate on requirements, NRC provides nutrient requirement values for foals with moderate growth and rapid growth. For example, moderate growth for foals six months of age (with an expected mature weight of 500 kg) is 0.65 kg per day compared to 0.85 kg for rapid growth. Foals with moderate growth were calculated to need 750 g of protein, 32 g of lysine, 29 g of calcium, and 16 g of phosphorus daily compared to 860 g of protein, 36 g of lysine, 36 g of calcium, and 20 g of phosphorus for foals with rapid growth.

As mentioned earlier, copper is a trace mineral of particular interest. Without doubt copper deficiency can cause DOD, but there is considerable debate as to the copper intake needed to prevent DOD. Several studies have found no lesions in foals fed diets containing a concentration of about 10 ppm (mg/kg) copper. Other studies have indicated that at least 30 to 50 ppm copper is needed. Fortunately, the higher levels of copper do not present any harmful side effects. Almost all feed companies have greatly increased the copper content of their feeds. Thus in recent years, when a commercial ration is fed, copper intake usually appears to be adequate even on those farms with complaints about DOD. A side issue is that the high copper content in the horse feeds are well above the level tolerated by sheep. Thus, many cases of copper toxicosis have been found in sheep with access to horse feed.

Frequency of feeding of weanlings may be important because of possible hormonal interactions. Frequent feeding causes less dramatic peaks in insulin secretion. High peaks of insulin secretion may decrease thyroxine production. A decrease in thyroxine production may lead to delayed maturity of the skeleton and, therefore, DOD. Thus, it seems reasonable to feed as often as practical.

The effect of exercise has also caused some controversy. Some believe that short, intense exercise is beneficial, whereas others suggest that forced exercise could be detrimental. Still others advocate severe restriction of exercise. Further studies on the effects of biomechanics are needed.

❑ FEEDING THE STALLION

Relatively few studies have been conducted on the nutrient requirements of the stallion. Energy is of considerable concern. Overfeeding can cause obesity and severe obesity has been shown to reduce sperm production in some species. Obesity can also predispose a horse to laminitis. Severe underfeeding may decrease libido. NRC (1989) suggested that stallions should be fed about 25% more DE during the breeding season than at maintenance. But as stressed earlier, the animal's body condition should be used to evaluate adequacy of energy intake. Protein requirements do not appear to be greatly increased during breeding season. NRC (1989) recommends 40 g crude protein/Mcal of DE, the same ratio as for maintenance. Mineral and vitamin requirements for stallions have not been determined, but it seems reasonable that the diet should contain at least the same ratios of these nutrients to energy as recommended for maintenance (NRC, 1989).

❑ FEEDING THE PERFORMANCE HORSE

Knowledge about the effect of nutrition on the performance of horses has greatly increased in recent years, with the greater use of the high-speed treadmill increasing opportunities to do meaningful studies. Nevertheless, many questions remain unanswered. Studies on performance horses are complex; many factors can influence performance. Training conditions, the individuality of the horse, the ability of the jockey or driver, and climate can all be factors. Thus, nutritional influences may be masked or obscured.

Energy is the dietary factor most influenced by exercise. NRC (1989) divides work into three categories—light, moderate, and intense. Examples of light work are Western and English pleasure, and bridle path hack. Moderate work includes ranch work, barrel racing, and roping. Examples of intense work are race training and polo. The categories are somewhat ambiguous, however; for example, they do not take in duration of exercise. However, because of the many factors that influence energy, it was thought that such a system would be practical. Horses are considered to require 25, 50, and 100% more energy at light, moderate, and intense work than at maintenance.

As discussed earlier, knowledge of the condition of the horse may be the best indicator of energy adequacy, but for the performance horse, visual appraisal such as body scoring is not adequate. Scales and perhaps even ultrasound to measure percent body weight are needed for such horses. Lim (1980) concluded that the optimal body weight of a race horse had a range of only plus or minus 1.5% of body weight. Thus, a 500-kg horse could be affected by a loss of 7.5 kg. Lawrence et al. (1992) reported that the top finishers in a 150-mile endurance ride had body fat of 6.5%, whereas nonfinishers had 11% body fat. Thus, scales and ultrasound could be useful in evaluating adequacy of energy intake and body condition.

The horse can use many dietary sources of energy. Fiber can be converted to volatile fatty acids by bacteria, but fibrous feeds are bulky and may limit energy intake. Furthermore, high intakes of fibrous feeds may increase the weight of gut fill. The effect of gut fill on the performance of horses has not been established, but it seems logical there must be some effect, particularly in horses such as sprinters also work at high rates. NRC (1989) suggested that a hay:grain ratio of around 1:1 to 1:2 would be reasonable. The hay:grain ratio for endurance horses could be higher in hay. Meyer (1987) suggested that endurance horses should be fed larger amounts of hay than that fed to horses racing at shorter distances. He also stated that the bulky rations dilate the intestine and increase the intestinal water and electrolyte reserve. Specifically, he reported that 4 to 6 kg of roughage per day would be adequate for race horses and that endurance horses should be given at least 6 to 8 kg of roughage daily.

Soluble carbohydrates such as starch are usually the greatest source of energy. Adequate levels of soluble carbohydrates are needed to maintain adequate storage of glycogen in the muscle. Several studies have shown that lack of adequate glycogen storage can lead to early fatigue. On the other hand, there is little evidence to suggest that significant dietary manipulation to promote glycogen loading such as is sometimes done by human athletes is needed in horses; in fact, it may be contraindicated.

Most conventional feedstuffs for horses do not contain high concentrations of fat. The addition of $1/2$ to 1 lb of vegetable oil or animal fat to the diets of hard-working horses increases the energy density of the diet and may enhance performance.

Horses can use protein for energy but do not utilize it efficiently. Diets containing 10% protein should provide adequate amounts of protein for working horses.

NRC (1989) suggested that calcium and phosphorus requirements are increased somewhat by exercise, but that further studies are needed to define exact amounts. In the absence of such studies, NRC suggested that if the calcium to energy ratio and phosphorus to energy ratio were maintained at the same rate as for maintenance, the intake of the minerals would be adequate.

Iron is necessary for hemoglobin formation. An iron-deficient horse becomes anemic and cannot then properly utilize energy. Iron deficiency, however, is probably unlikely unless the horse has experienced severe blood loss through parasites or trauma.

As mentioned earlier, interest in vitamin E has increased because of the vitamin's antioxidant activity and potential enhancement of the immune response. Vitamin E in the horse is also of interest because of reports in other species that exercise induces free-radical formation in muscle and liver, resulting in oxidative damage such as lipid peroxidation. It has been reported that the damage can be reduced by adding antioxidants such as vitamin E (Witt et al., 1992). Studies on the effect of vitamin E in working horses have not shown a dramatic increase in the vitamin E requirement. The

NRC (1989) value of 80 IU/kg of feed seems to be ample.

Ergogenic Aids

The horse owner is finding an ever increasing array of supplements at the marketplace. Owners of performance horses are particularly faced with many opportunities to buy products that are claimed to enhance athletic performance (ergogenic aids). Although it is impossible to evaluate the wide assortment of available products, some basic observations can be made. Sodium bicarbonate given in the correct dose and time prior to exercise can probably improve the performance of standardbreds and of thoroughbreds racing more than $1^1/_4$ miles. Theoretically, when sodium bicarbonate is given to horses, the blood is made more alkalotic, and, therefore, more lactic acid will be removed from the muscle, resulting in longer time to fatigue. However, the safety of the substance has been called into question, and as a result, many racing authorities have banned the use of sodium bicarbonate. Its use in endurance horses could be detrimental to performance because they are already alkalotic.

Carnosine (β)-alanyl-L-histidine) is found in high concentrations in equine muscle and is a major physiochemical buffer. Attempts to increase muscle carnosine by dietary methods have not been fruitful, and it is probable that the concentration of carnosine in muscle has a genetic basis.

Carnitine is a low-molecular-weight quaternary amine that may have an ergogenic activity because of its role in fatty acid metabolism. Again, dietary supplementation has not been an effective method to increase muscle carnitine, and, therefore, its value as an ergogenic agent in horses must be questioned.

❑ FEEDING THE OLDER HORSE

Just as the human population is growing older, so is the horse population. The older horse may have poor dentition such as missing or worn teeth, which may lead to decreased digestive efficiency and a higher incidence of colic. An older horse that begins to lose body condition that cannot be accounted for by another disease process may need a special diet. The older horse may respond to a diet that contains higher levels of protein and some

minerals and lower levels of fiber. Fortunately, many companies now produce diets designed for older horses. All the same, simply because a horse is 20 years or older does not mean that it must automatically be switched to a special diet.

❑ REFERENCES

Argenzio, R. A. 1990. Physiology of digestive secretory and absorptive processes. In *The Equine Acute Abdomen,* pp. 25–35, N.A. White (ed.). Lea and Febiger, Malvern, PA.

Bergman, E. N. 1990. Energy contributions of volatile fatty acids from the gastrointestinal tract in various species. *Physiol. Rev.* **70:** 567–590.

Blaney, B. J., R. J. W. Gartner, and R. A. Mckenzie. 1981. The inability of horses to absorb calcium from calcium oxalate (*Setaria sphacelata*). *J. Agric. Sci.* (Cambridge) **97:** 639–641.

Blodgett, S. L., J. E. Carrel, and R. A. Higgins. 1991. Cantharidin content of blister beetles (*Coleoptera meloidae*) collected from Kansas USA alfalfa and implications for inducing cantharidiasis. *Environ. Entomol.* **20:** 776–780.

Blythe, L. D., and A. M. Craig. 1992a. Equine degenerative myeloencephalopathy. 1. Clinical signs and pathogenesis. *Comp. Cont. Educ. Pract. Vet.* **14:** 1215–1221.

Blythe, L. D., and A. M. Craig. 1992b. Equine degenerative myeloencephalopathy. 2. Diagnosis and treatment. *Comp. Cont. Educ. Pract. Vet.* **14:** 1633–1637.

Carroll, C. D., and P. J. Huntington. 1988. Body condition scoring and weight estimation of horses. *Equine Vet. J.* **20:** 41–45.

Clarke L. L., M. C. Roberts, and R. A. Argenzio. 1990. Feeding and digestive problems in horses: physiologic responses to a concentrated meal. *Vet. Clinics N. Am., Equine Pract.* **6:** 433–450.

Cornick, J. L., G. K. Carter, and C. H. Bridges. 1988. Kleingrass-associated hepatotoxicosis in horses. *J. Am. Vet. Med. Assoc.* **193:** 932–935.

Cummings, J. F., A. DeLahunta, C. George, L. Fuhrer, B. A. Valentine, et al. 1990. Equine motor neuron disease: a preliminary report. *Cornell Vet.* **80:** 357–380.

Cunha, T. J. 1991. *Horse Feeding and Nutrition,* 2nd ed. Academic Press, San Diego, California.

Cymbaluk, N. F., and G. T. Christison. 1990. Environmental effects on thermoregulation and nutrition of horses. *Vet. Clin. North Am. Equine Pract.* **6:** 355–372.

Divers, T. 1993. Vitamin deficiency considered in equine form of Lou Gehrig's disease. *Equine Vet. Data* **141** (3): 341.

Ford, E. J. H., and H. A. Simmons. 1985. Gluconeogenesis for caecal propionate in the horse. *Br. J. Nutr.* **53:** 55–60.

Henneke, D. R., G. D. Potter, and J. L. Kreider. 1984. Body condition during pregnancy and lactation and reproductive efficiency of mares. *Theriogenology* **21:** 897–909.

Henneke, D. R., G. D. Potter, J. L. Kreider, and B. F. Yeates. 1983. Relationship between condition score, physical measurements and body fat percentage in mares. *Equine Vet. J.* **15:** 371–372.

Henry, W. A. 1912. *Feeds and Feeding,* 12th ed. Published by the author, University of Wisconsin, Madison.

Hintz, H. F. 1992. Dried beet pulp. *Equine Pract.* **14**(10): 5–7.

Hintz, H. F., and N. F. Cymbaluk. 1994. Nutrition of the horse. *Annl. Rev. Nutr.* **14:** 243.

INRA. 1992. *l'alimentation des chevaux.* W. Martin-Rosset (ed.). INRA, Paris.

Jeffcott, L. B. 1993. Problems and pointers in equine osteochondrosis. *Equine Vet. J.* (Suppl. 16): 1–3.

Julliand, V. 1992. Microbiology of the equine hindgut. Pferdeheilkunde 1. *Europaische Konferenz uber die Ernahrung des Pferdes,* pp. 42–47. Institut fur Tiernahrung Tierarztliche Hochsche Hannover.

Lawrence, L., D. Jackson, K. Kline, L. Moser, D. Powell, and M. Biel. 1992. Observations on body weight and condition of horse in a 150-mile endurance ride. *J. Equine Vet. Sci.* **12:** 320–324.

Lewis, L. D. 1995. *Feeding and Care of the Horse.* Lea & Febiger, Malvern, PA.

Lim, A. S. 1980. Body weight and performance in racehorses. *Proc. 4th International Conf. On control of the Use of Drugs in Racehorses,* pp. 93–94. Melbourne, Australia.

McMeniman, N. P., T. A. Porter, and K. Hutton. 1990. The digestibility of polished rice, rice pollard and lupin grains in horses. *Proc. Nutr. Soc. Aust.* **15:** 44–47.

Meyer, H. 1987. Nutrition of the equine athlete. In *Equine Exercise Physiology,* Part 2, pp. 644–773, J. Gillespie and N. Robinson (eds.). ICEEP Publication, Ann Arbor, MI.

Meyer, H. 1992. *Pferdefutterung.* Verlag Paul Parey, Berlin.

Morgan, S. E., B. Johnson, B. Brewer, and J. Walker. 1990. Sorghum cystitis ataxia syndrome in horses. *Vet. Hum. Toxicol.* **32:** 582.

Munday, B. L., I. M. Monkhouse, and R. T. Gallagher. 1985. Intoxication of horses by lolitrem B in ryegrass seed cleanings. *Aust. Vet. J.* **62:** 207.

Nation, P. N. 1991. Hepatic disease in Alberta horses: a retrospective study of "alsike clover poisoning," 1973–1988. *Can. Vet. J.* **32:** 602–607.

National Research Council (NRC). 1989. *Nutrient Requirements of Horses,* 5th ed. National Academy Press, Washington, DC.

Pagan, J. D. 1990. Effect of yeast culture supplementation on nutrient digestibility in mature horses. *J. Anim. Sci.* (Suppl. 1): 371 (Abstract).

Pagan, J. D., and R. J. Geor. 2001. *Advances in Equine Nutrition II.* Nottingham University Press, UK, 547 pp.

Pagan, J. D., and H. F. Hintz. 1986. Equine energetics. I. Relationship between body weight and energy requirements of horses. *J. Anim. Sci.* **63:** 815–821.

Porter, J. K., and F. N. Thompson, Jr. 1992. Effects of fescue toxicosis on reproduction in livestock. *J. Anim. Sci.* **70:** 1594–1603.

Ross, R. F., A. E. Ledet, D. L. Owens, L. G. Rice, H. A. Nelson, G. D. Osweiler, and T. M. Wilson. 1993. Experimental equine leukoencephalomalacia, toxic hepatosis, and encephalopathy caused by corn naturally contaminated with fumoninsins. *J. Vet. Diag. Invest.* **5:** 69–74.

Singh, A. K., K. Granley, Mishra, U., K. Naeem, T. White, and Y. Jiang. 1992. Screening and confirmation of drugs in urine interference of hordenine with the immunoassays and thin layer chromatography methods. *Forensic Sci. Int.* **54:** 9–22.

Sneddon, J. C. 1993. Physiological effects of hypertonic dehydration on body fluid pools in arid-adapted mammals. How do Arab-based horses compare? *Comp. Biochem. Physiol. A Comp. Physiol.* **104:** 201–213.

Sockett, D. C., J. C. Baker, and C. M. Stowe. 1982. Slaframine (*Rhizoctonia leguminicola*) intoxication in horses. *J. Am. Vet. Med. Assoc.* **181:** 606.

Spier, S. J., J. R. Snyder, and M. J. Murray. 1990. Fluid and electrolyte therapy for gastrointestinal disorders. In *Large Animal Internal Medicine,* B. P. Smith (ed.). C. V. Mosby Co., Baltimore, MD.

Swinkler, A. M., E. L. Squires, E. L. Mumford, J. E. Knowles, and D. M. Kniffen. 1993. Effect of body weight and body condition score on follicular development and ovulation in mares treated with GnRH analogue. *J. Equine Vet. Sci.* **13:** 519–520.

Whitlock, R. H. 1990. Feed additives and contaminants as a cause of equine disease. *Vet. Clinics N. Am. Equine Pract.* **6**(2): 467–478.

Witt, E. H., A. Z. Reznick, C. A. Viguie, P. Starke-Reed, and L. Packer. 1992. Exercise, oxidative damage and effects of antioxidant manipulation. *J. Nutr.* **122:** 766–773.

Photographs courtesy Duane E. Ullrey.

Dogs and Cats

DUANE E. ULLREY

DOGS WERE DOMESTICATED AND have been companions of humans for over 10,000 years and were even interred with humans about 12,500 years ago (Tchernov and Valla, 1997). The date of the domestication of cats is less certain, but estimates range from around 1600 to 7000 B.C. (Clutton-Brock, 1987). Recent discovery of a cat buried with a human about 9500 years ago suggests a long-standing close relationship (Vigne et al., 2004). Dogs and cats are the most popular pets in the United States, with 36% of U.S. households having one or more dogs and 34% having at least one cat; 15% have both a dog and cat. Dogs number about 60 million and cats about 75 million (Pet Food Institute, 2003). Most of these pets are fed manufactured foods, categorized as dry, canned, semimoist, or snacks/treats. The dry matter (DM) concentrations in these foods are about 90 to 94%, 22 to 26%, 70 to 85%, and 70 to 94%, respectively (Case et al., 2000). The dollar value of the various types of dog and cat food purchased by U.S. households in 2002 is shown in Table 28.1 and totals $12.35 billion.

TABLE 28.1 Dollar values of various types of dog and cat food purchased by U.S. households in 2002.

TYPE OF FOOD	DOLLAR VALUE (IN MILLIONS)	PERCENTAGE
Dog food		
Dry	5201.0	63.9
Canned	1379.7	17.0
Semimoist	80.3	1.0
Snacks/treats	1475.0	18.1
	8136.0	100.0
Cat food		
Dry	2334.0	55.6
Canned	1652.0	39.2
Semimoist	53.0	1.3
Snacks/treats	167.0	4.0
	4216.0	100.1
Dog and cat food		
Total	12,352.0	

Source: Pet Food Institute (2003).

515

❏ DIFFERENCES BETWEEN DOGS AND CATS

Although dogs and cats belong to the order Carnivora, there are important differences between them in dietary habits, digestive tract morphology, and nutritional requirements. Cats are almost totally carnivorous in the wild, whereas dogs tend to be more omnivorous. Both can use plant products in manufactured diets, but the metabolism of cats is rather specifically attuned to nutrients provided by animal prey. Even the shape and arrangement of the teeth are consistent with this difference. Dogs have more premolars and molars, with those in the rear designed for crushing, which is helpful in utilizing plant material. The upper P4 premolar (or carnassial) of the cat, when occluding with the lower molar, serves as a shear to cut animal flesh, and the cat does not have crushing molars. The cat has a relatively smaller cecum than the dog and a shorter intestine in proportion to body length, consistent with a carnivorous diet.

Cats have dietary requirements for certain nutrients that are found in animal tissues and rarely in plants. These include the essential fatty acid arachidonic acid, preformed vitamin A, and the aminosulfonic acid, taurine. Dogs, on the other hand, can generally synthesize sufficient arachidonic acid from linoleic acid, vitamin A from β-carotene, and taurine from sulfur-containing amino acids, nutrient precursors that are found in plants. Cats synthesize only nutritionally insignificant amounts of nicotinic acid from tryptophan, perhaps because both nutrients are generously supplied by animal tissue. Further evidence of carnivory can be seen in the high activity of gluconeogenic enzymes in the tissues of the cat and the limited ability to conserve nitrogen, which would not be limiting in the meat diet with which cats evolved. In addition, cats are sensitive to a number of the secondary compounds that occur in plants (e.g., benzoic acid) that can be detoxified by dogs.

❏ NUTRITION AND FEEDING OF DOGS

Ancestral dogs and their canid relatives are generally considered pack hunters, allowing them to kill prey as large, or larger, than themselves. Surplus food remaining after a meal is sometimes hidden for later recovery and consumption. Domestic dogs fed dry diets ad libitum eat 10 to 13 meals per day, consuming most of those meals during daylight (Mugford and Thorne, 1980). Preferences may be shown for meat-based canned diets, but properly formulated dry diets are nutritionally equivalent, less expensive, and less likely to result in accumulation of dental plaque and calculus. The palatability of dry diets can be improved slightly by adding small amounts of sucrose, but dry diets upon which fat and meat digests have been sprayed appeal particularly to dogs.

When not fed ad libitum, weaned puppies are commonly fed two to three times per day until 6 months of age and two times per day until one year. Idle adult dogs are sometimes fed only once per day, usually finishing their meal in 20 minutes or less. For those that are pregnant, lactating, or working, feeding ad libitum or two or more times per day may be desirable.

Water

Water is the most important nutrient for the dog, as it is for all living creatures. Dogs obtain water in liquid form, from food, and from oxidation of hydrogen during metabolism (metabolic water). About 10 to 16 g of metabolic water are produced for each 100 Kcal of energy metabolized. A dog consuming 2000 Kcal of metabolizable energy per day (e.g., a German Shepherd) will realize 200 to 320 g of metabolic water. Since semimoist and canned dog foods contain about 30% and 75% water, respectively, a dog eating these products would need to consume less liquid water than a dog eating a dry food containing 10% water. Growing puppies and idle adult dogs consume two to three times more water than they consume of DM. During lactation, hot weather, or severe exertion, water consumption may be four or more times DM intake.

Energy

The energy needs of dogs are commonly expressed in units of metabolizable energy (ME), and small dogs have higher ME requirements per unit of body weight than large dogs do. To account for this difference, the National Research Council (2004) has expressed the ME requirements of dogs of different size per unit of metabolic body size ($W_{kg}^{0.75}$). The exponent used in this expression is not universally agreed upon (Case et al.,

2000), but the NRC (2004) concluded that adult maintenance ME requirements for average laboratory kennel dogs or active pet dogs are about 130 Kcal/ $W_{kg}^{0.75}$/day. These requirements, along with the amount of dog food required to meet the ME needs for maintenance of different-size adult dogs, are shown in Table 28.2. Requirements may be higher for young adult laboratory dogs or young adult active pet dogs and for adult Great Danes and Terriers. Maintenance requirements may be lower for inactive pet dogs, older dogs, or laboratory Newfoundlands (NRC, 2004).

Although the NRC (2004) has published formulas for calculating ME requirements for other life stages, food requirements for growth of puppies of the weights shown in Table 28.2 may be estimated by multiplying maintenance requirements by two (Case et al., 2000). Working or lactating adult dogs may require two to three times more food than is required for maintenance. Dogs of different breeds, temperament, body condition, or subject to extremes of environment, exercise, or stress may vary from average, and adjustments in food offered should be made to accommodate individual needs.

Protein and Amino Acids

Dietary protein is required to provide essential amino acids that can't be synthesized rapidly enough by the dog's own tissues plus additional nitrogen for tissue synthesis of nonessential amino acids if preformed supplies are not adequate. Essential amino acids required by both puppies and adult dogs are shown in Table 28.3.

Earlier, Rose and Rice (1939) suggested that adult dogs do not require arginine, based on nitrogen balance. However, Burns et al. (1981) found that an arginine-free diet resulted in vomiting in adult dogs because of elevated levels of blood ammonia due to the need for arginine for ammonia disposal via the urea cycle.

Taurine is thought to be synthesized in adequate amounts in the liver of most dogs if sufficient sulfur amino acid precursors (methionine and/or cystine) are present in the diet. However, the feeding of commercial diets containing lamb meal and whole rice and/or rice bran has resulted in low plasma and whole blood taurine concentrations in some dogs unless the diets were supplemented with taurine (Delaney et al., 2003). This may be due in part to a reduced bioavailability of dietary sulfur amino acids, with potentially more serious consequences for large-breed dogs or breeds such as Newfoundlands, which have some individuals with a predilection for dilated cardiomyopathy (Tôrres et al., 2003).

Dietary requirements for protein depend upon protein quality (essential amino acid content), protein digestibility, energy intake per day, prior nutritional status, feeding pattern, age, growth rate, reproductive status, and dietary fat concen-

TABLE 28.2 Estimated daily ME and food allowances to meet the requirements for maintenance of adult dogs of various weights in a thermoneutral environment with moderate activity.

WEIGHT		ME	DRY[a]		SEMIMOIST[b]		CANNED[c]	
KG	LB	KCAL/DAY	G/KG BW	G/DOG	G/KG BW	G/DOG	G/KG BW	G/DOG
2.3	5	243	29	68	38	87	106	243
4.5	10	402	25	112	32	144	89	402
6.8	15	547	22	152	29	195	80	547
9.1	20	681	21	189	27	243	75	681
13.6	30	921	19	256	24	329	68	921
22.7	50	1352	17	476	21	483	60	1,352
31.8	70	1741	15	484	20	622	55	1,741
49.9	110	2441	14	678	17	872	49	2,441

[a]90% DM, 3.6 Kcal ME/g.
[b]75% DM, 2.8 Kcal ME/g.
[c]25% DM, 1.0 Kcal ME/g.
Source: NRC (2004).

TABLE 28.3 Essential amino acids required in the diet of puppies and adult dogs.

Arginine	Methionine[a]
Histidine	Phenylalanine[b]
Isoleucine	Threonine
Leucine	Tryptophan
Lysine	Valine

[a]Preformed cystine will meet up to 50% of the requirement for methionine.
[b]Preformed tyrosine will meet up to 50% of the requirement for phenylalanine.
Source: Schaeffer et al. (1989).

tration (protein–energy ratio). As little as 6% of a highly digestible, high-quality protein has been shown to support nitogen balance in a short-term study with adult dogs. However, protein requirements in practical diets are appreciably higher, and the NRC (2004) has proposed the recommended protein and essential amino acid allowances in the DM of dog food with 4 Kcal ME/g DM, as shown in Table 28.4.

ME densities of dog foods, as sold, can be estimated from label guarantees by use of the following equation:

$$ME(Kcal/g) = [(3.5 \times CP) + (8.5 \times EE) + (3.5 \times NFE)]/100$$

where CP = % crude protein, EE = % crude fat, and NFE = 100% − % crude protein, % crude fat, % crude fiber, % ash, and % moisture. The numerical coefficients represent typical ME values in Kcal/g for the indicated fractions.

Estimated ME values can be converted to a DM basis by dividing estimated ME on an as-sold basis by the % DM in the dog food (100% − % moisture) and multiplying by 100.

Correction of the protein and amino acid allowances in Table 28.4 to permit evaluation of an example dog food estimated to contain 4.5 Kcal ME/g DM requires that each value in Table 28.4 be multiplied by 4.5/4.0. Such a calculation results in protein minimums for growth in such a diet of 25.3% of DM and lysine minimums of 0.99% of DM.

High dietary protein concentrations have not been shown to induce renal pathology in normal

TABLE 28.4 Recommended allowances for protein and essential amino acids in DM of dog foods with 4 Kcal ME/g DM.

NUTRIENT	GROWTH[a]	LATE GEST/LACT[b]	ADULT MAINTENANCE[c]
Protein, %	22.5	20.0	10.0
Arginine, %	0.79	1.00	0.35
Histidine, %	0.39	0.44	0.19
Isoleucine, %	0.65	0.71	0.38
Methionine, %	0.35	0.31	0.33
Methionine + cystine, %	0.70	0.62	0.65
Leucine, %	1.29	2.00	0.68
Lysine, %	0.88	0.90	0.35
Phenylalanine, %	0.65	0.83	0.45
Phenylalanine + tyrosine, %	1.30	1.23	0.74
Threonine, %	0.81	1.04	0.43
Tryptophan, %	0.23	0.12	0.14
Valine, %	0.68	1.30	0.49

[a]For a 5.5-kg, 3-month-old puppy consuming 1000 Kcal ME/day.
[b]For a 22-kg bitch with eight puppies consuming 5000 Kcal ME/day.
[c]For a 15-kg adult dog consuming 1000 Kcal ME/day.
Source: NRC (2004).

dogs (Bovee, 1991; Case et al., 2000). However, dogs with advanced liver or kidney failure may have difficulty ridding the body of nitrogen in excess of need. Special diets have been formulated to assist in the medical management of such problems, and these diets are designed to provide minimum amounts of very high quality protein (Lewis et al., 1987).

Fat and Fatty Acids

Dogs do not require a certain amount of fat in the diet, although fat has a positive effect on diet palatability. Dry and semimoist dog foods commonly contain 8 to 12% fat on a DM basis. Water is sometimes added to a dry food formula and the product canned. The percent fat in such a canned dog food would also be 8 to 12% fat on a DM basis. Fat in meat-based canned dog food may constitute 40% of DM.

Dogs do have dietary requirements for essential fatty acids, including linoleic acid (18:2n-6) and α-linolenic acid (18:3n-3) or long-chain members of the n-6 and n-3 series, primarily arachidonic acid (20:4n-6), eicosapentaenoic acid (20:5n-3), and docosahexaenoic acid (22:6n-3). In the absence of n-6 essential fatty acids, the hair is coarse and dry, and skin lesions appear. Essential n-3 fatty acids are important for normal neural and retinal development. Hair and skin lesions can be prevented by linoleic acid, γ-linolenic acid (18:3n-6), or arachidonic acid. γ-linolenic and arachidonic acids are not major components of most food fats, but these essential fatty acids are metabolically interconvertible in the tissues of the dog, and n-6 essential fatty acid requirements are usually expressed in terms of linoleic acid. The Association of American Feed Control Officials (AAFCO) (2004) minimum for maintenance of adult dogs and for growth and reproduction is 1% linoleic acid in dietary DM (3.5 Kcal ME/g DM). The NRC (2004) has proposed a recommended allowance of 1.1% linoleic acid, 0.04% α-linolenic acid, and 0.01% eicosapentaenoic + docosahexaenoic acids in dietary DM (4 Kcal ME/g DM) for maintenance of adult dogs. For growth, late gestation, and lactation, NRC recommended allowances in dietary DM have been increased to 1.3% linoleic acid, 0.08% α-linolenic acid, and 0.05% eicosapentaenoic + docosahexaenoic acids. In addition, for growth, a recommended allowance of 0.03% arachidonic acid in dietary DM has been proposed.

Carbohydrates

Dogs, like other animals, have a metabolic requirement for glucose. Nearly all commercial dog foods contain starch, which when hydrolyzed during digestion will provide an ample glucose supply. However, if fed a meat-based diet without added carbohydrate, the dog must meet its needs from gluconeogenesis in the liver and kidneys, using precursors such as amino acids from protein or glycerol from fat to synthesize glucose. The weaned puppy is able to synthesize glucose at an adequate rate, but when carbohydrate-free diets, with 26% of ME supplied by protein, were fed to pregnant or lactating bitches, hypoglycemia developed the week before whelping, and fewer live pups were born (Romsos et al., 1981). In addition, the lethargy of the bitches induced by hypoglycemia impeded proper care of the pups. However, more recent studies suggested that these adverse effects can be prevented if reproducing bitches are fed a carbohydrate-free diet containing higher levels of protein (Blaza et al., 1989; Kienzle and Meyer, 1989). Thus, carbohydrate does not appear to be a dietary essential for the dog, as long as there are sufficient gluconeogenic precursors, such as alanine, glycine, or serine.

Puppies obviously digest lactose, a sugar that provides about 8% of the ME in bitch's milk. However, the concentration of lactose in bitch's milk is much lower than the 27% of ME provided by lactose in cow's milk. Thus, it is prudent to use a milk replacer for an orphaned puppy that simulates the composition of bitch's milk. By analogy with research on other mammals, sucrase activity may be limited early in life. Thus, use of table sugar (sucrose) in milk replacers for the puppy also should be limited.

Adult dogs that are unaccustomed to dietary lactose will exhibit severe diarrhea and watery feces when cow's milk is suddenly introduced. This can be prevented (Mundt and Meyer, 1989) by treating the milk with a commercial lactase preparation (Lactaid™, McNeil-PPC Inc., Fort Washington, PA 19034) or by using milk that has been previously treated (Lactaid™; Dairy Ease™, Sterling Winthrop Inc., New York, NY 10016).

The utilization of starch is improved by cooking, baking, or toasting. Starch granules in properly extruded dog food (dry or semimoist) are ruptured and partially dextrinized by heat and pressure, making them more digestible. The same is true for starches in baked or canned dog food. Thus, less starch escapes digestion in the small intestine, and problems of flatulence, colic, and soft stools associated with microbial fermentation of undigested raw starch that reaches the colon are minimized.

Structural carbohydrates, such as hemicellulose and cellulose, are not digested by endogenous enzymes and undergo only limited microbial fermentation in the cecum and colon. Optimal amounts of dietary fiber favor normal functioning of the gastrointestinal tract and proper stool formation. Insoluble fibers (primarily cellulose and hemicellulose) make diets less energy dense, promote a feeling of satiety, and maintain normal gastrointestinal motility and digesta transit time. Soluble fibers (fructans, mannans, galactans, pectic substances, gums, and mucilages) delay gastric emptying and may undergo colonic microbial fermentation, resulting in production of short-chain fatty acids—important energy sources for mucosal cells of the colon. Most dog foods have guaranteed maximum amounts of 3 to 5% crude fiber in diet DM, but higher concentrations may be found in diets for obese or inactive older dogs (Case et al., 2000).

Minerals

The recommended allowances (NRC, 2004) for minerals in the DM of dog foods with 4 Kcal ME/g DM are shown in Table 28.5. These values take into consideration the variable mineral availability in natural feedstuffs and, for most minerals, are higher than the minimum requirements. However, certain minerals in excess can cause harm comparable to the damage from deficiencies. This has been demonstrated in growing Great Danes fed diets containing 0.9% phosphorus and 0.55, 1.1, or 3.3% calcium on a DM basis (Hazewinkel, 1989). Young dogs fed 0.55% calcium ate more food and gained more height and weight than dogs fed 1.1% calcium but had bowed front legs, lordosis, exhibited pain when they walked, and developed osteoporosis and fractures. Young dogs fed 3.3% calcium grew slower than those fed lower concentrations of calcium, developed curvatures of the radius associated with osteochondrosis, and exhibited partial paralysis of the hind limbs, with neurologic signs of canine wobbler syndrome.

The last-named signs may have been caused by slowed bone turnover and reduced skeletal remodeling, limiting the expansion in volume of the

TABLE 28.5 Recommended allowances for minerals in DM of dog foods with 4 Kcal ME/g DM.

NUTRIENT	GROWTH[a]	LATE GEST/LACT[b]	ADULT MAINTENANCE[c]
Calcium, %	1.2	0.8	0.3
Phosphorus, %	1.0	0.5	0.3
Potassium, %	0.44	0.4	0.4
Sodium, %	0.22	0.2	0.04
Chloride, %	0.29	0.39	0.06
Magnesium, %	0.04	0.06	0.06
Iron, mg/kg	90	70	80
Copper, mg/kg	11	12.4	7.3
Manganese, mg/kg	5.6	7.2	5
Zinc, mg/kg	100	96	120
Iodine, mg/kg	0.9	0.9	1.5
Selenium, mg/kg	0.35	0.35	0.35
Chromium, mg/kg			0.70

[a]For a 5.5-kg, 3-month-old puppy consuming 1000 Kcal ME/day.
[b]For a 22-kg bitch with eight puppies consuming 5000 Kcal ME/day.
[c]For a 15-kg adult dog consuming 1000 Kcal ME/day.
Source: NRC (2004).

spinal canal that is necessary to accommodate the increasing size of the spinal cord. Similarly, the foramina through which nerves pass do not enlarge properly. The consequent pressure interferes with normal nerve function.

Vitamins

The recommended allowances (NRC, 2004) for vitamins in the DM of dog foods with 4 Kcal ME/g DM are shown in Table 28.6. No requirement has been given for biotin, although dogs are known to have a metabolic requirement for this vitamin. Microbial synthesis within the dog's digestive tract may supplement low biotin (and vitamin K) levels in natural dietary ingredients. However, to provide further assurance of adequacy, most high-quality dog foods include biotin (and vitamin K) in the vitamin premix.

Commercial Dog Foods

Many ingredients, of both animal and plant origin, are used in dog foods. Their suitability depends on nutrient and energy content, digestibility, palatability, and freedom from toxins. Ingredients that are deficient in one or more of these attributes (except for freedom from toxins) may still be made useful by incorporating them in mixtures with other ingredients that correct those deficiencies. It is important that these ingredients be free of mycotoxins or below limits designated as safe by the Food and Drug Administration. Animal products should also be free of pathogenic organisms such as *Salmonella*.

An ingredient list on a dog food label must conform to AAFCO (2004) definitions and might include some, but probably not all, of the following: corn, grain sorghum, wheat, rice, wheat middlings, corn gluten meal, soybean meal, meat, meat and bone meal, poultry, poultry meal, poultry byproduct meal, dried skimmed milk, dried whey, animal fat, L-lysine, DL-methionine, salt, calcium carbonate, dicalcium phosphate, trace minerals, and vitamins. Other ingredients are sometimes used. Appetite enhancers, such as an animal digest (produced by chemical and/or enzymatic hydrolysis of

TABLE 28.6 Recommended allowances for vitamins in DM of dog foods with 4 Kcal ME/g DM.

NUTRIENT	GROWTH[a]	LATE GEST/LACT[b]	ADULT MAINTENANCE[c]
Vitamin A, IU/kg	5050	5050	5050
Vitamin D, IU/kg	550	550	550
Vitamin E, mg/kg[d]	30	30	30
Vitamin K, mg/kg[e]	1	1	1
Thiamin, mg/kg	1.38	1.35	2.25
Riboflavin, mg/kg	10.5	10.5	5.25
Pyridoxine, mg/kg	1.5	1.5	1.5
Niacin, mg/kg	15	15	15
Pantothenic acid, mg/kg	15	15	15
Folic acid, mg/kg	0.27	0.27	0.27
Vitamin B$_{12}$, mg/kg	0.035	0.035	0.035
Biotin, mg/kg[f]			
Choline, mg/kg	1700	1700	1700

[a]For a 5.5-kg, 3-month-old puppy consuming 1000 Kcal ME/day.
[b]For a 22-kg bitch with eight puppies consuming 5000 Kcal ME/day.
[c]For a 15-kg adult dog consuming 1000 Kcal ME/day.
[d]As all-*rac*-α-tocopheryl acetate. Higher concentrations recommended for diets high in polyunsaturated fatty acids.
[e]As menadione.
[f]Diets containing raw egg white or intestinally active antibiotics may require supplementation.
Source: NRC (2004).

animal tissue), might be sprayed along with fat on the outside of dry dog foods.

Antioxidants, such as butylated hydroxy anisole, butylated hydroxy toluene, ethoxyquin, and propyl gallate, also may be present. Semimoist dog foods are protected against spoilage without refrigeration by sucrose, sorbates, and propylene glycol or by low pH (about 4.2, from use of phosphoric acid), plus mold inhibitors.

An example formula for a dry dog food is shown in Table 28.7. This product might be pelleted, baked and broken into kibbles, or water added and canned, but it typically would be extruded, and the fat, and possibly an animal digest, sprayed on the outside.

Feeding Guidelines

Dogs do not require variety in their diet, and abrupt food changes may induce digestive upsets, vomiting, and diarrhea. A high-quality, nutritionally complete commercial dog food plus fresh water will meet the dog's requirements without need for supplementation.

The amount to feed depends upon age, size, reproductive state, exercise, environmental temperature, genetic and individual differences between dogs, and the nutrient and energy density of the dog food. Hardworking dogs, pregnant or lactating bitches, and dogs that are injured or ill tend to have the highest energy requirements. Except for obese dogs, ad libitum feeding will ensure that energy intakes are adequate and appropriate. For

dogs that are obese, foods that are lower in energy density (less fat, more fiber) may be limit-fed to produce a gradual weight loss until an ideal weight is reached.

Growing puppies are often allowed all the food they want, but when Labrador Retrievers were fed a diet containing 3.94 Kcal ME/g DM ad libitum from weaning (at 8 weeks) to 2 years of age, there was an increased incidence of hip dysplasia (Kealy et al., 1992) compared to limit-fed (75% of ad libitum) controls. For breeds of dogs that are predisposed to canine hip dysplasia, it may be prudent to restrict food intakes somewhat during the postweaning period.

Because of the differences in nutrient and energy density between commercial dog foods, it is not possible to provide a universally applicable list of weights (oz, g) or volumes (cups) that should be fed. However, companies that have tested their products can provide useful feeding recommendations for the dog foods that they sell (Anonymous, 1990).

❑ NUTRITION AND FEEDING OF CATS

Most wild felids, except for the lion, are solitary hunters. The size and strength of lions and tigers is sufficient to allow them to kill prey larger than themselves. Thus, they are adapted to large but infrequent meals. Smaller wild felids, particularly those in the genus *Felis,* depend heavily upon small animals for food. An active adult domestic cat weighing 4 kg requires about 280 to 320 Kcal ME/day (Case et al., 2000; NRC, 1986) or the equivalent of nine to ten 26-g mice. Thus, a feral cat must practice frequent predation to meet this need. The meal patterns of cats offered commercial canned or dry extruded cat foods or purified diets have been investigated in two studies, and the mean number of meals in 24 hours ranged from about 10 to 17, with no apparent differences in frequency between light and dark periods (Morris and Rogers, 1989).

Cats have very particular tastes, favoring foods that were first presented when they were kittens. The proportion of canned food purchased in the United States for cats is slightly more than twice the proportion of canned food purchased for dogs. This difference may represent a decision made for kittens by their owners, and once established, a

TABLE 28.7   An example formula for a dry dog food.

INGREDIENT	PERCENTAGE
Ground yellow corn	56.0
Ground wheat	5.0
Corn gluten meal (60% CP)	5.0
Soybean meal (48% CP)	15.0
Meat and bone meal	10.0
Dicalcium phosphate	0.2
Salt	0.5
Trace mineral premix[a]	0.5
Vitamin premix[a]	0.8
Animal fat	7.0
Total	100.0

[a]Formulated to ensure that nutrient levels will meet or exceed AAFCO (2004) minimum limits without being excessive.

preference for canned food tends to persist into adulthood. The absolute cost of feeding a cat versus the cost of feeding a dog is sufficiently small that there may be inadequate economic incentive to limit purchase of the more expensive canned products. Nevertheless, in addition to being less expensive, dry cat foods have the benefit of promoting oral health.

Water

Water concentrations in cat food markedly influence the drinking pattern and amount of liquid water consumed. Cats given a canned food containing 78% water did not drink water after the first day, while those given a dry extruded diet drank an average of 16 times in a 24-hour period (Kane et al., 1981). The ratios of water to dry matter consumed by cats fed dry cat foods were 2 to 2.8:1, whereas water to dry matter ratios for cats fed canned cat foods were 3 to 5.7:1 (Anderson, 1983). These latter values reflect the obligatory water intake associated with consumption of a high-moisture food in amounts sufficient to meet energy needs. Although adult cats have been shown to remain healthy in the absence of drinking water when given high-moisture foods such as fish (Prentiss et al., 1959), provision of fresh water is recommended regardless of diet.

Energy

The energy needs of cats are commonly expressed in units of ME, although ME concentrations have been determined for few cat food ingredients. Most adult domestic cats weigh between about 2 and 6 kg (Case, 2000), a range that is relatively narrow compared to that of domestic dogs. As a consequence, the NRC (1986) chose to express ME requirements per unit of body mass, although the NRC (2004) expresses these requirements per unit of metabolic body size. Daily ME intakes observed for cats are shown in Table 28.8 (adapted from NRC, 1986).

Protein and Amino Acids

As is true for the dog, protein is required in the diet of the cat to provide essential amino acids as well as additional nitrogen for tissue synthesis of nonessential amino acids if preformed quantities are insufficient. Cats require the same dietary essential amino acids as the dog plus taurine. Taurine cannot be synthesized by the cat at a sufficient rate from methionine or cystine to meet tissue needs, and there is an obligatory loss of taurine in the bile as a bile salt conjugate. A deficiency of taurine results in central retinal degeneration, impaired vision, reproductive failure, cardiac abnor-

TABLE 28.8   Daily ME intakes observed for cats.

AGE (WEEKS)	PHYSIOLOGICAL STAGE	WEIGHT (KG)	ME INTAKE (KCAL/KG BW)
10	Kitten	0.9F, 1.1M	250
20	Kitten	1.9F, 2.5M	130
30	Kitten	2.7F, 3.5M	100
40	Kitten	3.0F, 4.0M	80
≥50	Adult, inactive		70
≥50	Adult, active		80
≥50	Adult, gestation		100
≥50	Adult, 1st wk lact, 4 kittens	3.8F	108
≥50	Adult, 2nd wk lact, 4 kittens		117
≥50	Adult, 3rd wk lact, 4 kittens		138
≥50	Adult, 4th wk lact, 4 kittens[a]		162
≥50	Adult, 5th wk lact, 4 kittens[a]		183
>50	Adult, 6th wk lact, 4 kittens[a]		228

[a]Values from 4th week onward include energy intake from queen's food by her kittens.
Source: NRC (1986).

malities, and impaired function of the immune system (Pion et al., 1990).

Minimum protein and essential amino acid requirements for the kitten have been proposed (NRC, 1986), but many of the essential amino acid requirements for adult maintenance and reproduction have not been defined. Taurine requirements have been shown to be affected by food processing methods (Douglass et al., 1991). Considering these uncertainties and the factors that influence nutrient requirements (described previously), the NRC (2004) has published the recommended allowances for cat food shown in Table 28.9.

Fats and Fatty Acids

Fat is a good energy source, favors absorption of fat-soluble vitamins, and usually enhances the palatability of cat food. Dry and semimoist cat foods commonly contain 8 to 13% fat, whereas canned cat foods contain 10 to 40% fat on a DM basis. However, a vital reason for including fat in the

diet is to provide sufficient amounts of the essential fatty acids. As in the dog, linoleic acid is important for good physical condition, particularly of the skin, but arachidonic acid also is required in the diet of the cat. Although conversion of linoleic to arachidonic acid at a rate sufficient to support spermatogenesis has been demonstrated in the testis of the adult male, dietary arachidonic acid must be provided to ensure normal pregnancy and lactation in the female (NRC, 1986). NRC (2004) has recommended an allowance of 0.55% linoleic acid and 0.02% of arachidonic acid in dietary DM (4.0 Kcal ME/g) for growth and reproduction. The recommended allowance of linoleic acid for adult maintenance also is 0.55% of dietary DM, but needs for arachidonic acid may be lower. NRC (2004) recommended allowances for n-3 essential fatty acids are 0.02% α-linolenic acid and 0.01% eicosapentaenoic + docosahexaenoic acid in dietary DM for growth and reproduction. Eicosapentaenoic + docosahexaenoic acid concentrations of 0.01% are recommended in dietary DM for adult maintenance. To provide these essential fatty

TABLE 28.9 Recommended allowances for protein and essential amino acids in DM of cat foods with 4 Kcal ME/g DM.

NUTRIENT	GROWTH[a]	LATE GEST[b]	LACT[b]	ADULT MAINTENANCE[c]
Protein, %	22.5	21.3	30.0	20.0
Arginine, %	0.96	1.5	1.5	0.77
Histidine, %	0.33	0.43	0.71	0.26
Isoleucine, %	0.54	0.77	1.20	0.43
Methionine, %	0.44	0.50	0.60	0.17
Methionine + cystine, %	0.88	0.90	1.04	0.34
Leucine, %	1.28	1.80	2.00	1.02
Lysine, %	0.85	1.10	1.40	0.34
Phenylalanine, %	0.50			0.50
Phenylalanine + tyrosine, %[d]	1.91	1.91	1.91	1.53
Threonine, %	0.65	0.89	1.08	0.52
Tryptophan, %	0.16	0.19	0.19	0.13
Valine, %	0.64	1.00	1.20	0.51
Taurine, %[e]	0.04	0.04	0.04	0.04

[a]For an 800-g kitten consuming 180 Kcal ME/day.

[b]For a 4-kg queen with four kittens in late gestation or lactation consuming 540 Kcal ME/day.

[c]For a 4-kg adult cat consuming 250 Kcal ME/day.

[d]Amount required to maximize black hair color; tyrosine should be half of total.

[e]Recommended taurine allowance in dry expanded or canned diet DM is 0.1% or 0.2%, respectively.

Source: NRC (2004).

acids, enhance palatability, support fat-soluble vitamin transport, and supply adequate caloric density, NRC (2004) has proposed a recommended minimum allowance of 9% fat from appropriate sources in dietary DM.

Carbohydrates

Cats do not have a dietary carbohydrate requirement but are able to meet tissue glucose needs by gluconeogenesis. However, dry commercial cat foods commonly contain 40% or more carbohydrate, and digestion of starch, dextrins, and sugars in these products generally exceeds 80% and may exceed 90%. Fine grinding or cooking increases the digestibility of starch in corn or wheat as compared to coarse grinding (Morris et al., 1977). High-temperature extrusion also increases starch digestibility. Inclusion of lactose at 15% of the ME in meat-based diets fed to adult cats leads to diarrhea, perhaps as a consequence of the decline in intestinal lactase activity with age. Cellulose is not digested.

Minerals

The recommended allowances (NRC, 2004) for minerals in cat food DM with 4 Kcal ME/g DM are shown in Table 28.10. These minerals are involved in tissue structure, acid-base balance, os-

motic regulation, and as components of or cofactors for enzymes. Both absolute dietary concentrations of minerals and interrelationships between them are important for the cat. For example, optimum absorption and utilization of calcium and phosphorus have been demonstrated at dietary ratios of 0.9:1 to 1.1:1 (Scott and Scott, 1967).

The most common nutritional bone disease is nutritional secondary hyperparathyroidism, seen in kittens or young adults as a consequence of feeding a meat-based diet inadequately supplemented with calcium. Muscle meat has a calcium:phosphorus ratio of 1:20 with a DM calcium concentration of 0.025%. Thus, bone formation does not proceed normally, bones become less dense, fine trabeculation is lost, and the cortices become thinner. Affected animals limp, show evidence of pain, and are reluctant to move. If stressed, the bones may fracture. The causes are inadequate calcium, interference in calcium absorption by the relative excess of phosphorus, hypocalcemia, and elevated secretion of parathormone. Elevated parathormone secretion promotes resorption of bone calcium in an attempt to restore serum calcium concentrations to normal, with a consequent loss of bone mass and development of pathology, variously called osteitis fibrosa, osteodystrophia fibrosa, osteogenesis imperfecta, and osteoporosis (NRC, 1986). This condition can be prevented by

TABLE 28.10 Recommended allowances for minerals in DM of cat foods with 4 Kcal ME/g DM).

NUTRIENT	GROWTH[a]	LATE GEST/LACT[b]	ADULT MAINTENANCE[c]
Calcium, %	0.80	1.08	0.29
Phosphorus, %	0.72	0.76	0.26
Potassium, %	0.40	0.52	0.52
Sodium, %	0.14	0.13	0.07
Chloride, %	0.19	0.20	0.10
Magnesium, %	0.04	0.06	0.04
Iron, mg/kg	80	80	80
Copper, mg/kg	8.4	8.8	5.0
Manganese, mg/kg	4.8	7.2	4.8
Zinc, mg/kg	75	60	75
Iodine, mg/kg	2.2	2.2	2.2
Selenium, mg/kg	0.40	0.40	0.40

[a]For an 800-g kitten consuming 180 Kcal ME/day.
[b]For a 4-kg queen with four kittens in late gestation or lactation consuming 540 Kcal ME/day.
[c]For a 4-kg adult cat consuming 250 Kcal ME/day.
Source: NRC (2004).

calcium supplementation and can be reversed, if not too severe, by raising dietary calcium:phosphorus ratios to 2:1 during recovery.

Mineral availability and form also must be considered. The cat readily digests and absorbs phosphorus in muscle, but a significant proportion of phosphorus in plant seeds is in the form of poorly available phytate. Furthermore, cat foods containing minimum protein levels and high proportions of vegetable matter can lead to excretion of alkaline urine and problems with struvite urolithiasis. Metabolism of high-protein, meat-based diets consumed by feral cats and other wild carnivores results in acidic urine (normal pH range 6 to 7). Struvite is a magnesium ammonium phosphate that can precipitate in alkaline urine and obstruct the urinary tract. It is largely soluble below pH 6.6 (Buffington et al., 1989).

Phosphoric acid can be used in the formula to provide a highly available form of phosphorus and to promote urine acidification. Urine pH has been successfully maintained at ≤6.4 in adult cats for one year by using a naturally acidifying diet without added acidifiers or by including 1.7% phosphoric acid in dietary DM (Fettman et al., 1992). Earlier conclusions that high dietary magnesium concentrations were principally responsible for struvite urolithiasis appear to have been biased by use of magnesium oxide (an alkalizing salt) as the magnesium source.

Vitamins

The recommended allowances (NRC, 2004) for vitamins in the dietary DM of cat foods with 4 Kcal ME/g DM are shown in Table 28.11. Little research has been conducted on vitamin requirements of the cat, and some of the levels represent extrapolation from other species. Because destruction of several of the vitamins may be promoted by light, heat, moisture, oxidation, rancidity, and reactive trace elements, supplementary levels should be sufficient to ensure that recommended concentrations are present when the food is consumed.

Conversely, excesses of any nutrient, if sufficiently large, have the potential to be harmful. Excesses of vitamin A have been shown to result in

TABLE 28.11   Recommended allowances for vitamins in DM of cat foods with 4 Kcal ME/g DM.

NUTRIENT	GROWTH[a]	LATE GEST/LACT[b]	ADULT MAINTENANCE[c]
Vitamin A, IU/kg	3550	7500	3550
Vitamin D, IU/kg	250	250	250
Vitamin E, mg/kg[d]	38	38	38
Vitamin K, mg/kg[e]	1	1	1
Thiamin, mg/kg	5.5	5.5	5.6
Riboflavin, mg/kg	4.25	4.25	4.25
Pyridoxine, mg/kg	2.5	2.5	2.5
Niacin, mg/kg	42.5	42.5	42.5
Pantothenic acid, mg/kg	6.25	6.25	6.25
Folic acid, mg/kg	0.75	0.75	0.75
Vitamin B_{12}, mg/kg	0.022	0.022	0.022
Biotin, mg/kg[f]	0.075	0.075	0.075
Choline, mg/kg	2550	2550	2550

[a]For an 800-g kitten consuming 180 Kcal ME/day.

[b]For a 4-kg queen with four kittens in late gestation or lactation consuming 540 Kcal ME/day.

[c]For a 4-kg adult cat consuming 250 Kcal ME/day.

[d]As all-*rac*-α-tocopheryl acetate. Higher concentrations recommended for diets high in polyunsaturated fatty acids.

[e]As menadione.

[f]Diets containing raw egg white or intestinally active antibiotics may require supplementation.

Source: NRC (2004).

abnormal bone metabolism, with damage to the cervical vertebrae (deforming cervical spondylosis) sufficient to impair head mobility and the ability to groom. Excesses of vitamin D result in calcification of soft tissue.

Commercial Cat Foods

Although cats and their wild relatives are carnivorous, entirely satisfactory cat foods can be prepared using significant proportions of ingredients of plant origin. However, an entirely vegetarian diet is not appropriate for cats because of their unique requirements for certain nutrients such as arachidonic acid and taurine.

Thus, the ingredient list on a cat food label is likely to include many of the items found in dog food, but with greater emphasis on meat and fish. Meat byproducts, animal liver, fish meal, condensed fish solubles, and fish oil are among the additional ingredients that may be listed. Phosphoric acid may be present as a source of phosphorus and as a urinary acidifier. An example formula for a dry cat food is shown in Table 28.12. It would typically be extruded and the fat and animal digest sprayed on the outside.

Feeding Guidelines

Cats, like dogs, do not require variety in their diet, but they are very individualistic in their feeding behavior and, as already noted, exhibit food preferences that have been conditioned by prior experience. A high-quality, nutritionally complete commercial cat food plus fresh water will meet their needs without supplementation. Avoiding the feeding of table scraps and treats will reduce problems with food rejection and consumption of nutritionally incomplete diets. Although nutritionally complete dry, semimoist, or canned commercial cat foods can all be used satisfactorily, dry foods are less expensive and favor oral health.

Kittens, active adults, and pregnant or lactating queens can be fed dry cat foods ad libitum. However, sedentary adults sometimes become obese when fed ad libitum and might best be limit-fed two times per day. Semimoist and canned foods are usually limit-fed twice a day to sedentary cats but may need to be fed more often to pregnant or lactating queens. Canned diets are sufficiently high in water that cats may have difficulty meeting their energy and nutrient needs if fed only twice a day. Estimated food al-

TABLE 28.12   An example formula for a dry cat food.

INGREDIENT	PERCENTAGE
Ground yellow corn	33.77
Poultry meal	17.4
Corn gluten meal (60% CP)	12.8
Soybean meal (48% CP)	10.0
Ground wheat	9.0
Meat and bone meal	3.0
Phosphoric acid solution (75% H_3PO_4)[a]	2.3
Calcium carbonate	1.16
Fish meal	1.0
Salt	0.7
Potassium chloride	0.48
Trace mineral premix[b]	0.5
Vitamin premix[b]	0.8
Taurine	0.08
Citric acid	0.01
Animal fat	6.0
Animal digest	1.0
Total	100.0

[a]Providing 1.7% phosphoric acid in the diet.
[b]Formulated to ensure that nutrient levels will meet or exceed AAFCO (2004) minimum limits without being excessive.
Source: Fettman et al. (1992).

lowances for cats are shown in Table 28.13. Because commercial cat foods differ in their nutrient and energy density, it may be preferable to follow the feeding recommendations of the manufacturer, assuming that those recommendations have been tested with cats.

Potential Toxins

Cats have limited ability to detoxify certain food constituents and are thus more susceptible to these compounds than some other mammals (NRC, 1986; NRC, 2004). Benzoic acid and benzoate salts are used as antifungal agents in human foods but should not be used in cat food because of the small margin of safety for the cat. Sorbic acid is an effective and safer alternative. Although evidence is sparse, it also may be prudent to avoid use of high levels of propylene glycol in cat foods because of the sensitivity of the cat erythrocyte to oxidative degradation of hemoglobin (Heinz body formation).

TABLE 28.13   Estimated food allowances for cats.

| WEIGHT | | DRY[a] | | SEMIMOIST[b] | | CANNED[c] | |
CAT	KG	G/KG BW	G/CAT	G/KG BW	G/CAT	G/KG BW	G/CAT
Kitten							
10 wk	0.9–1.1	78	70–86	83	75–91	227	204–250
20 wk	1.9–2.5	41	78–103	43	82–108	118	224–295
30 wk	2.5–3.8	31	78–118	33	83–125	91	228–346
40 wk	2.9–3.8	25	73–95	27	78–103	73	212–277
Adult							
Inact	2.2–4.5	22	48–90	23	51–99	64	141–288
Active	2.2–4.5	25	55–113	27	59–122	73	160–329
Gest	2.5–4.0	31	78–124	33	83–132	91	228–364
Lact[d]	2.2–4.0	78	172–312	83	182–332	227	499–908

[a]90% DM, 3.2 Kcal ME/g.
[b]70% DM, 3.0 Kcal ME/g.
[c]25% DM, 1.1 Kcal ME/g.
[d]Nursing 4–5 kittens in 6th week of lactation.
Source: NRC (1986).

❏ SUMMARY

• The qualitative and quantitative energy and nutrient requirements of the domestic dog and cat have been discussed in the context of natural dietary habits, digestive tract morphology and physiology, and intermediary metabolism. Though similar in many respects, important differences in nutrient needs have their origin in the carnivory of cats and the omnivory of dogs. These have been characterized through the efforts of numerous scientists, and minimum requirements and/or recommended allowances published by the National Research Council are presented in tabular form. Example formulas for dog and cat foods and feeding guidelines are also provided.

❏ REFERENCES

Anderson, R. S. 1983. Fluid balance and diet. In *Proceedings of the 7th Kal Kan Symposium,* Ohio State University, pp. 19–25, Columbus, OH.

Anonymous. 1990. *Canine Caloric Requirements: A Veterinarian's Guide to Feeding Recommendations.* Ralston Purina Co., Checkerboard Square, St. Louis, MO.

Association of American Feed Control Officials (AAFCO). 2004. *Official Publication.* R. J. Noel, Secretary-Treasurer. Office of the Indiana State Chemist, West Lafayette, Indiana.

Blaza, S. E., D. Booles, and I. H. Burger. 1989. Is carbohydrate essential for pregnancy and lactation in dogs? In *Nutrition of the Dog and Cat,* pp. 229–242, I. H. Burger and J.P.W. Rivers (eds.). Cambridge University Press, Cambridge, England.

Bovee, K. C. 1991. Influence of dietary protein on renal function in dogs. *J. Nutr.* **121:** S128–S139.

Buffington, C. A., N. E. Cook, Q. R. Rogers, and J. G. Morris. 1989. The role of diet in feline struvite urolithiasis syndrome. In *Nutrition of the Dog and Cat,* pp. 357–380, I. H. Burger and J. P. W. Rivers (eds.). Cambridge University Press, Cambridge, England.

Burns, R. A., J. A. Milner, and J. E. Corbin. 1989. Arginine: an indispensable amino acid for mature dogs. *J. Nutr.* **111:** 1020–1024.

Case, L. P., D. P. Carey, D. A. Hirakawa, and L. Daristotle. 2000. *Canine and Feline Nutrition, 2nd ed.* Mosby Inc., St Louis, MO.

Clutton-Brock, J. 1987. *A Natural History of Domestic Mammals.* University of Texas Press, Austin, TX.

Delaney, S. J., P. H. Kass, Q. R. Rogers, and A. J. Fascetti. 2003. Plasma and whole blood taurine in normal dogs of varying size fed commercially prepared food. *J. Anim. Physiol. Anim. Nutr.* **87:** 236–244.

Douglass, G. M., E. B. Fern, and R. C. Brown. 1991. Feline plasma and whole blood taurine levels as influenced by commercial dry and canned diets. *J. Nutr.* **121**: S179–S180.

Fettman, M. J., J. M. Coble, D. W. Hamar, R. W. Norrdin, H. B. Seim, R. D. Kealy, Q. R. Rogers, K. McCrea, and K. Moffat. 1992. Effect of dietary phosphoric acid supplementation on acid-base balance and mineral and bone metabolism in adult cats. *Am. J. Vet Res.* **53**: 2125–2135.

Hazewinkel, H. A. 1989. Calcium metabolism and skeletal development in dogs. In *Nutrition of the Dog and Cat,* pp. 293–302, I. H. Burger and J.P.W. Rivers (eds.). Cambridge University Press, Cambridge, England.

Kane, E., Q. R. Rogers, J. G. Morris, and P.M.B. Leung. 1981. Feeding behavior of the cat fed laboratory and commercial diets. *Nutr. Res.* **1**: 499–507.

Kealy, R. D., S. E. Olsson, K. L. Monti, D. F. Lawler, D. N. Biery, R. W. Helms, G. Lust, and G. K. Smith. 1992. Effects of limited food consumption on the incidence of hip dysplasia in growing dogs. *J. Am. Vet. Med. Assoc.* **201**: 857–863.

Kienzle, E., and H. Meyer. 1989. The effects of carbohydrate-free diets containing different levels of protein on reproduction in the bitch. In *Nutrition of the Dog and Cat,* pp. 243–257, I. H. Burger and J.P.W. Rivers (eds.). Cambridge University Press, Cambridge, England.

Lewis, L. D., M. L. Morris, Jr., and M. S. Hand. 1987. *Small Animal Clinical Nutrition III*. Mark Morris Associates, Topeka, KS.

Morris, J. G., and Q. R. Rogers. 1989. Comparative aspects of nutrition and metabolism of dogs and cats. In *Nutrition of the Dog and Cat,* pp. 35–36, I. H. Burger and J.P.W. Rivers (eds.). Cambridge University Press, Cambridge, England.

Morris, J. G., J. Trudell, and T. Pencovic. 1977. Carbohydrate digestion by the domestic cat *(Felis catus)*. *Br. J. Nutr.* **37**: 365–373.

Mugford, R. A., and C. J. Thorne. 1980. Comparative studies of meal patterns in pet and laboratory housed dogs and cats. In *Nutrition of the Dog and Cat,* pp. 3–14, R. S. Anderson (ed.). Pergamon Press, Oxford, England.

Mundt, H.-C., and H. Meyer. 1989. Pathogenesis of lactose-induced diarrhoea and its prevention by enzymatic splitting of lactose. In *Nutrition of the Dog and Cat,* pp. 267–274, I. H. Burger and J.P.W. Rivers (eds.). Cambridge University Press, Cambridge, England.

National Research Council (NRC). 1974. *Nutrient Requirements of Dogs*. National Academy of Sciences, Washington, DC.

National Research Council (NRC). 1985. *Nutrient Requirements of Dogs*. National Academy of Sciences, Washington, DC.

National Research Council (NRC). 1986. *Nutrient Requirements of Cats*. National Academy of Sciences, Washington, DC.

National Research Council (NRC). 2004. *Nutrient Requirements of Dogs and Cats*. National Academies Press, Washington, DC.

Pet Food Institute. 2003. *Pet Food Institute Fact Sheet*. Pet Food Institute, Washington, DC.

Phillips, T. 1993. Market trends. *Petfood Industry* **35**(6): 36–38 (November/December).

Pion, P. D., M. D. Kittleson, Q. R. Rogers, and J. G. Morris. 1990. Taurine deficiency myocardial failure in the domestic cat. In *Functional Neurochemistry of Taurine,* pp. 423–430, H. Pasante-Morales (ed.). Alan R. Liss, New York.

Prentiss, P. G., A. V. Wolf, and H. E. Eddy. 1959. Hydropenia in cat and dog. Ability of the cat to meet its water requirements solely from a diet of fish or meat. *Am. J. Physiol.* **196**: 625–632.

Rivers, J.P.W., and I. H. Burger. 1989. Allometry considerations in the nutrition of dogs. In *Nutrition of the Dog and Cat,* pp. 67–112, I. H. Burger and J.P.W. Rivers (eds.). Cambridge University Press, Cambridge, England.

Romsos, D. R., H. J. Palmer, K. L. Muiruri, and M. R. Bennink. 1981. Influence of a low carbohydrate diet on performance of pregnant and lactating dogs. *J. Nutr.* **111**: 678–689.

Rose, W. C., and E. E. Rice. 1939. The significance of the amino acids in canine nutrition. *Science* **90**: 186–187.

Schaeffer, M. C., Q. R. Rogers, and J. G. Morris. 1989. Protein in the nutrition of dogs and cats. In *Nutrition of the Dog and Cat,* pp. 159–205, I. H. Burger and J.P.W. Rivers (eds.). Cambridge University Press, Cambridge, England.

Scott, P. P., and M. G. Scott. 1967. Nutritive requirements for carnivora. In *Husbandry of Laboratory Animals,* pp. 163–186. Academic Press, London, England.

Tchernov, E., and F. F. Valla. 1997. Two new dogs, and other Natufian dogs, from the Southern Levant. *J. Archaeol. Sci.* **24**: 65–95.

Tôrres, C. L., R. C. Backus, A. J. Fascetti, and Q. R. Rogers. 2003. Taurine status in normal dogs fed a commercial diet associated with taurine deficiency and dilated cardiomyopathy. *J. Anim. Physiol. Anim. Nutr.* **87**: 359–372.

Vigne, J.-D., J. Guilaine, K. Debue, L. Haye, and P. Gerard. 2004. Early taming of the cat in Cyprus. *Science* **304**: 259.

29

Photograph courtesy Duane E. Ullrey.

Fish

DUANE E. ULLREY

AQUACULTURE IS BELIEVED TO have begun in China about 4000 years ago (Lovell, 1998). Descriptions of the culture of common carp were written in 475 B.C., the Romans built fishponds during the first century A.D., and fish culture spread throughout Eastern Europe during the Middle Ages. Interest in this subject spread from England to the United States before 1800, but much of the technical progress in this country and elsewhere has occurred during the latter part of the twentieth century.

Although the production of fish for human food is a major part of the industry, aquaculture in its broadest sense includes production of shrimp, crawfish, lobsters, oysters, clams, alligators, frogs, turtles, and eels. In addition, aquaculturists produce ornamental aquarium fish and aquatic plants, with the plants being used for aquariums, food, ornamental gardens, and wetland restoration.

531

❑ DIFFERENCES BETWEEN NUTRITION AND FEEDING OF FISH AND LAND ANIMALS

To the extent that they have been studied, the qualitative nutrient requirements of fish (finfish and crustaceans) are similar to those of land animals. Fish need protein (nitrogen and essential amino acids), essential fatty acids, minerals, vitamins, and energy sources. Energy requirements are relatively lower than for most land animals, and fish diets generally have a higher protein to energy ratio. Although it was once thought that fish require certain lipids that some land animals do not, specifically the omega-3 (n-3) fatty acids, further research has established that n-3 fatty acids are required by all land animals that have been studied. The ability of fish to absorb certain minerals, such as calcium, from the water surrounding them reduces the dietary need for these minerals. Some fish have a limited ability to synthesize vitamin C and must depend on dietary sources.

Fish food must have specific physical properties that facilitate feeding in water. It may be desirable that the food either float or sink, but it should not disintegrate before being consumed. Nutrients should be stable and not easily leached away. Particle size must be appropriate for the species and life stage, and may be particularly critical for larval forms. There should be minimum food waste so as not to adversely affect dissolved oxygen concentration and other features of water quality. Nutrient concentrations in the food should consider contributions of food organisms and nutrients in the water in which the fish live.

❑ NATURAL DIETARY HABITS AND DIGESTIVE TRACT MORPHOLOGY

Fish, like most animal groups, include species that are carnivorous, omnivorous, or herbivorous. Some consume food that is living or dead or both. Some are very opportunistic and can be carnivorous or herbivorous, according to need (Barrington, 1957). A few vary their diet seasonally, consuming plankton in the summer and fish in the winter.

Because more than 18,000 species of fish and lower chordates have been identified (Wilson, 1992) and only a few have been studied in detail, it is not appropriate to generalize about gastrointestinal structure and function. Analogies have been made between the digestive tracts of fish and those of mammals, and terms such as stomach, duodenum, ileum, colon, and rectum are found in the fish literature. However, these structures are not always present, or the function of structures in their respective locations may be different than in mammals (Smith, L. S., 1989; Rust, 2002).

Carnivorous fish commonly have short guts in relation to their body length, and herbivores have the longest. Omnivore guts are not necessarily intermediate in length, and the correlation of diet with gut length is confused by the influence of structures such as loops, folds, pyloric ceca, and spiral valves that may contribute surface area comparable to a much longer gut. Digestive tract configurations are shown in Figure 29.1 (from Smith, 1980) for rainbow trout (carnivore), catfish (omnivore with emphasis on animal food), carp (omnivore with emphasis on plant food), and milkfish (microphagous planktivore).

The digestive system of fish is typically simpler during the larval stage than in the adult. The gut of larval tilapia begins as a straight tube but develops loops and pouches by the third day after hatching (Ishibashi, 1974). Metamorphosis to adult tilapia with a longer, more complex gut accompanies a change in dietary habits from zooplankton to plants.

The means by which fish capture, sort, and reduce particle sizes of food varies immensely (Stevens and Hume, 1995). Filter feeders commonly capture plankton by swimming forward at appropriate speeds and using mouth suction. Many bony fish have teeth, but they differ in structure and may not be located on the jaws. In some species, they are found on the tongue, pharynx, or other orobranchial surfaces. They are often not used for chewing but are a means by which food is kept from escaping and moved from the mouth to the esophagus. Because continuous water flow through the mouth and gills is required for oxygen exchange and osmoregulation, in some species food is prevented from escaping through the gills by sievelike projections called gill rakers. For fish such as tuna that feed on large prey, these projections may be spaced quite far apart. For microphagous planktivore feeders like menhaden, the gill rakers form a fine meshwork.

Most fish species have a short, wide esophagus with a nondigestive lining that gives way to the secretory cells of the stomach or, in species

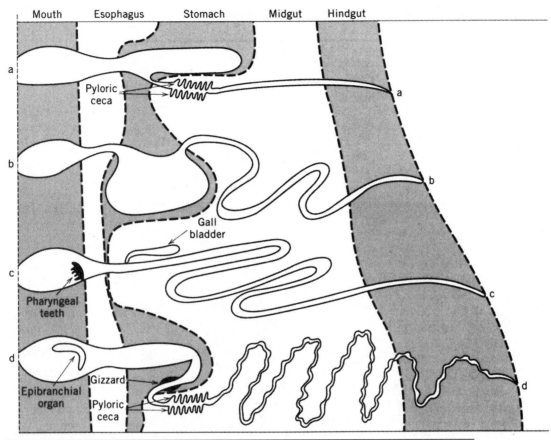

Figure 29.1 *Digestive tract configurations for: (a) rainbow trout (carnivore, Y-shaped stomach); (b) catfish (omnivore with preference for animal foods, pouched stomach); (c) carp (omnivore with preference for plant foods, no stomach); (d) milkfish (microphagous planktivore, tubular stomach with muscular gizzard). Adapted from L. S. Smith (1989).*

without stomachs, to the intestine. Taste buds are often present in the anterior esophagus. In species that have stomachs, the lining has both secretory and mucous cells, and there is an anterior cardiac and a posterior pyloric sphincter (Smith, L. S., 1989). Sometimes the stomach is followed by a muscular gizzard in which particle sizes of food are reduced by maceration with sand and digestive juices.

The midgut begins immediately posterior to the pylorus, and if pyloric ceca are present, they are located near the beginning of the midgut. Fish without a stomach have neither a pylorus nor pyloric ceca. The function of the pyloric ceca appears to be to continue digestive processes begun in the stomach and to absorb liberated nutrients. The pyloric ceca add significantly to intestinal surface area and have been shown in rainbow trout to take up amino acids and sugars (Diamond and Buddington, 1985).

The hind gut is an extension of the midgut with diminished digestive and absorptive functions but with a major role in osmoregulation.

❑ ENERGY

Fish are among the most efficient of animals in converting food energy to body tissue (Table 29.1), partly because they require less than 10% of the energy for maintenance required by birds or mammals of the same size. Upon first examination, fish foods seem high in protein, particularly when expressed as a percentage of dry matter. However, the energy requirements per unit of tissue gain are lower for fish than they are for land animals. Thus,

TABLE 29.1 Dietary protein and digestible energy (DE) requirements per unit of gain in fish and terrestrial animals.

| | FEED/GAIN | PROTEIN | | DIGESTIBLE ENERGY | | DE/PROTEIN |
| | | G/KG | G/KG | KCAL/KG | KCAL/KG | |
SPECIES	G/G[a]	DIET	GAIN	DIET	GAIN	KCAL/G
Rainbow trout	1.5	350	525	3000	4500	8.6
Catfish	1.8	300	480	3420	5472	11.4
Broilers	2.5	200	500	2950	7400	14.8
Swine	4.0	160	640	3300	13,200	20.6
Beef cattle	8.0	100	800	2500	20,000	25.0

[a]Air-dry feed/wet tissue gain.
Source: Adapted from R. R. Smith (1989).

what appears to be a high requirement for protein is really a low requirement for energy. As a consequence, protein efficiencies (protein gain/protein fed) of 45% have been seen in catfish fed diets containing 11.4 Kcal of digestible energy (DE) per gram of protein (Gatlin et al., 1986). Optimum DE:protein ratios that have been determined for fish are about 8 to 10 Kcal DE/g dietary protein compared to 14 to 20 Kcal DE/g dietary protein for terrestrial farm animals (Smith, R. R., 1989; Lovell, 1998).

The National Research Council (NRC, 1993) has chosen to relate digestible protein to digestible energy and has published optimum ratios for weight gain of several species of fish. These are shown in Table 29.2.

Proteins and lipids in food are important energy sources for many species of commercially raised fish, but the value of dietary carbohydrates varies appreciably among species. Warm-water fish such as Nile tilapia and channel catfish will digest over 70% of the gross energy in uncooked starch, whereas a cold-water species such as rainbow trout will digest less than 50% (NRC, 1993). When the heat and pressures of extrusion are applied to raw starch, it will be partially dextrinized, and digestibility for the rainbow trout may increase by 75%. Facultative anaerobic bacteria have been isolated from the digestive tract of some species of fish, and cellulase digestion has been demonstrated in the carp intestine (Stevens, 1988). The quantitative contribution of microbial fermentation to the energy budgets of fish has not been established but is generally considered to be small.

Demands for energy are influenced by physical activity, water temperature, body size, and stress.

TABLE 29.2 Optimum digestible protein: digestible energy ratios in air-dry diet for weight gain of several species of fish.

SPECIES	DIGESTIBLE PROTEIN %	DIGESTIBLE ENERGY KCAL/G	DIGESTIBLE PROTEIN/ DIGESTIBLE ENERGY MG/KCAL
Channel catfish	22.2–28.8	2.33–3.14	81–97
Red drum	31.5	3.20	98
Hybrid bass	31.5	2.80	112
Nile tilapia	30.0	2.90	103
Common carp	31.5	2.90	108
Rainbow trout	33.0–42.0	3.60–4.10	92–105

Source: Adapted from NRC (1993).

Although the bodies of fish are efficiently supported by the water in which they live, if fish are forced to swim actively against a strong current, appreciable energy will be lost as heat. When water temperatures decline, the metabolic rates and body temperatures of fish also decline. All fish have a lower lethal temperature limit, but some species have adapted to life in waters that are covered by ice for much of the year. Food and oxygen are in short supply, but metabolic rate is sufficiently reduced that little energy is required. There is no universal agreement on an equation relating body weight to metabolic rate in fish, but clearly small fish have higher energy requirements per unit of body weight than do large fish of the same species. Stresses such as poor water quality, low oxygen levels, crowding, accumulations of waste, and disturbance will increase energy requirements and adversely affect growth and reproduction of fish.

❑ PROTEIN AND AMINO ACIDS

Protein and essential amino acid requirements have been studied primarily in juvenile fish, using maximal weight gain as the principal criterion. These requirements, expressed as a percentage of air-dry diet, decline as fish approach maturity. Channel catfish weighing 14 to 100 g require 35%

protein, while channel catfish weighing 114 to 500 g require 25% protein (Page and Andrews, 1973). Similar differences with age have been described for salmonids, common carp, and tilapia (Wilson and Halver, 1986). Estimated dietary protein requirements (as fed [air-dry] basis) for 24 species of juvenile fish range from 30 to 55% (Table 29.3). These estimates have been determined with natural feed ingredients or with isolated protein sources, such as casein, egg, or soy protein, sometimes supplemented with amino acids.

Estimated essential amino acid requirements for six species of fish are shown in Table 29.4. These estimates were made mostly with chemically defined or purified diets in which the amino acid being studied was added in crystalline form. Since the availability of amino acids in natural ingredients may be different from those in crystalline form, there is a measure of uncertainty concerning the applicability of these values to practical diets.

Whereas amino acid deficiencies generally impair growth, a deficiency of methionine also leads to cataracts in Atlantic salmon, rainbow trout, and lake trout (Poston et al., 1977). Tryptophan deficiency, too, has been associated with cataracts in rainbow trout (Poston and Rumsey, 1983; Walton et al., 1984) and leads to scoliosis (lateral curvature of the spine) in rainbow trout (Walton et al., 1984), sockeye salmon (Halver and Shanks, 1960), and chum salmon (Akiyama et al., 1986).

TABLE 29.3 Estimated dietary protein requirements in air-dry diet for maximal growth of juvenile fish.

SPECIES	PROTEIN %	SPECIES	PROTEIN %
Atlantic salmon	45	Puffer fish	50
Channel catfish	32–36	Rainbow trout	40
Chinook salmon	40	Red sea bream	55
Coho salmon	40	Smallmouth bass	45
Common carp	31–38	Snakehead	52
Estuary grouper	40–50	Sockeye salmon	45
Gilthead sea bream	40	Striped bass	47
Grass carp	41–43	Blue tilapia	34
Japanese eel	44	Mossambique tilapia	40
Largemouth bass	40	Nile tilapia	30
Milkfish	40	Zillii's tilapia	35
Plaice	50	Yellowtail	55

Source: Adapted from NRC (1993).

TABLE 29.4 Amino acid requirements of juvenile fish, expressed as a percentage of dietary protein.

	FISH SPECIES					
NUTRIENT	CHINOOK SALMON	COHO SALMON	RAINBOW TROUT	COMMON CARP	CHANNEL CATFISH	NILE TILAPIA
Dietary protein	40–41	40	42–46	39	24	28
Arginine	6.0	5.8	3.6	4.3	4.3	4.2
Histidine	1.8	1.8	—	2.1	1.5	1.7
Isoleucine	2.2	—	—	2.5	2.6	3.1
Leucine	3.9	—	—	3.3	3.5	3.4
Lysine	5.0	—	4.2	5.7	5.1	5.1
Methionine	4.0[a]	—	2.2[b]	3.1[b]	2.3[b]	2.7[c]
Phenylalanine	5.1[d]	—	—	6.5[e]	5.0[f]	3.8[g]
Threonine	2.2	—	—	3.9	2.0	3.8
Tryptophan	0.5	0.5	1.4	0.8	0.5	1.0
Valine	3.2	—	—	3.6	3.0	2.8

[a]Diet contained 1.0% cystine.
[b]Diet contained 0.0% cystine.
[c]Diet contained 0.2% cystine.
[d]Diet contained 0.4% tyrosine.
[e]Diet contained 0.0% tyrosine. When diet contained 1.0% tyrosine, phenylalanine requirement was 3.4% of protein.
[f]Diet contained 0.3% tyrosine. When diet contained 0.6% tyrosine, phenylalanine requirement was 2.0% of protein.
[g]Diet contained 0.5% tyrosine.
Source: Adapted from NRC (1993).

❏ FATS AND FATTY ACIDS

Dietary fats are important sources of energy and essential fatty acids and assist in the absorption of fat-soluble vitamins. Like other animals that have been studied, most fish are unable to synthesize linoleic acid (18:2, n-6) or α-linolenic acid (18:3, n-3). Thus, one or both of these fatty acids must be supplied in the diet. In addition, fish species vary in their ability to convert 18-carbon unsaturated fatty acids to longer-chain, more highly unsaturated fatty acids in the same series (Owen et al., 1975). Freshwater fish generally require dietary linoleic acid or α-linolenic acid, or both, whereas stenohaline marine fish (fish that don't tolerate wide variations in water salinity) require dietary eicosapentaenoic acid (20:5, n-3) and/or docosahexaenoic acid (20:6, n-3).

The essential fatty acids are components of cell membranes, influence their fluidity, and serve as precursors of eicosanoids with a variety of metabolic functions. Changes in proportions of the two major membrane phospholipids, phosphatidylcholine and phosphatidylethanolamine, are observed during adaptation to changes in environmental temperature (Hazel, 1979; Hazel and Carpenter, 1985).

The percentage of fat to include in a practical fish diet involves consideration of the type of fat and the dietary protein and energy content. Excessive amounts of fat can produce an imbalance in the digestible energy:crude protein ratio and cause excessive fat deposition in the visceral cavity and tissues.

Essential fatty acid deficiencies retard growth, increase mortality, and induce abnormalities, such as underdeveloped swim bladders and scoliosis in red sea bream larvae (Watanabe et al., 1983).

Essential fatty acid requirements of several species of fish are shown in Table 29.5.

❏ CARBOHYDRATES

Fish do not appear to have a dietary requirement for carbohydrate but do have enzymes that permit carbohydrate digestion. However, the hormonal and metabolic regulation of carbohydrate metabolism varies among fish species and may be somewhat different from that in mammals (Shimeno,

TABLE 29.5 Essential fatty acid requirements in air-dry diet of fish.

SPECIES	FATTY ACID REQUIREMENTS[a]
Freshwater fish	
Ayu	1% α-linolenic acid or 1% eicosapentaenoic acid
Channel catfish	1–2% α-linolenic acid or 0.5–0.75% eicosapentaenoic and docosahexaenoic acids
Chum salmon	1% linoleic acid and 1% α-linolenic acid
Coho salmon	1–2.5% α-linolenic acid
Common carp	1% linoleic acid and 1% α-linolenic acid
Japanese eel	0.5% linoleic acid and 0.5% α-linolenic acid
Rainbow trout	0.8–1% α-linolenic acid
Nile tilapia	0.5% linoleic acid
Zillii's tilapia	1% linoleic acid or 1% arachidonic acid
Striped bass	0.5% eicosapentaenoic and docosahexaenoic acids
Marine fish	
Red sea bream	0.5% eicosapentaenoic acid or 0.5% eicosapentaenoic and docosahexaenoic acids
Giant sea perch	1% eicosapentaenoic and docosahexaenoic acids
Striped jack	1.7% docosahexaenoic acid or 1.7% docosahexaenoic and eicosapentaenoic acids
Turbot	0.8% eicosapentaenoic and docosapentaenoic acids
Yellowtail	2% eicosapentaenoic and docosapentaenoic acids

[a]Linoleic acid (18:2, n-6), α-linolenic acid (18:3, n-3), arachidonic acid (20:4, n-6), eicosapentaenoic acid (20:5, n-3), docosahexaenoic acid (22:6, n-3).
Source: Adapted from NRC (1993).

1974; Cowey and Walton, 1989). In general, rainbow trout, plaice, and yellowtail effectively use less than 25% dietary dextrin or gelatinized starch as an energy source, whereas channel catfish and common carp can use appreciably more (NRC, 1993). Because cereal grains are an inexpensive source of carbohydrate energy, they are commonly used in diets for farmed warm-water fish. Their use in diets for cold-water fish is much more limited.

❏ MINERALS

Assuming that these elements are present, fish can meet some of their requirements for calcium, magnesium, sodium, potassium, iron, zinc, copper, and selenium from the water in which they live. Depending upon the element, uptake occurs via gills, skin, and oral epithelium (NRC, 1993). In marine fish, which drink copiously, some elements may

be absorbed from water via the gut mucosa as they would be from food.

The availability of mineral elements in food varies with their form, with amounts of other interacting elements, and with species of fish. For example, the availability of phosphorus from monocalcium or dicalcium phosphate is higher than from tricalcium phosphate or calcium-magnesium phytates (phytin), the latter found in seeds and seed byproducts of plants like wheat and soybeans. Although dietary calcium concentration may not affect phosphorus availability for channel catfish, common carp, and rainbow trout, optimum dietary calcium:phosphorus ratios are important for red sea bream (1:2) and eels (1:1) (NRC, 1993). The availability of phosphorus in fish meal is higher for rainbow trout and chum salmon than for tilapia and common carp, perhaps because of the limited secretion of gastric juices by the latter warm-water species (Ogino et al., 1979; Yone and Toshima, 1979).

Calcium

Calcium requirements are met largely by absorption through gills in freshwater, and to a lesser extent through skin, and by drinking seawater in a marine environment. When reared in calcium-free water, growing catfish and tilapia require 0.45 and 0.7% calcium in the air-dry diet, respectively (Robinson et al., 1986, 1987). Uptake of calcium from seawater is sufficient for Atlantic salmon (Lall and Bishop, 1977) but insufficient for red sea bream, which still require 0.34% calcium in the air-dry diet (Sakamoto and Yone, 1973, 1976).

Bone is not as functional in calcium homeostasis in fish as in mammals because fish bone is acellular (without osteocytes) in most species. There is little calcium exchange between bones and body fluids in marine fish, but in a low-calcium, freshwater environment, mobilization of calcium from bones and scales may be necessary during ovarian maturation, starvation, and spawning migration (NRC, 1993). The rate of calcium exchange between scales and body fluids is three times that of bones (Berg, 1968).

Phosphorus

Total phosphorus requirements reported for finfish range from 0.5 to 0.8% in the air-dry diet, whereas available phosphorus requirements are about 0.42% (NRC, 1993). The availability of phosphorus in most plant products is low, estimated at 29 to 54% in soybean meal (Lovell, 1978; Wilson et al., 1982).

Deficiency signs include poor growth and feed efficiency, impaired bone mineralization, deformed bones, reduced serum inorganic phosphorus concentration, increased serum alkaline phosphatase activity, reduced liver glycogen, and increased carcass fat (NRC, 1993).

Magnesium

Magnesium requirements of fish can be met from diet or from water if the water contains sufficient amounts. Generally, marine fish in seawater do not require dietary supplementation. Magnesium requirements of rainbow trout fed a magnesium-free diet can be met if living in fresh water containing 46 mg magnesium per L. Magnesium requirements of fish species that have been studied range from 0.04 to 0.06% in the air-dry diet.

Deficiency signs include anorexia, poor growth, lethargy, convulsions, degeneration of muscle fibers and the epithelium of the pyloric ceca and gill filaments, cataracts, deformity of vertebrae, and calcinosis (accumulation of calcium) of soft tissues, particularly of the kidneys (NRC, 1993).

Sodium, Potassium, and Chlorine

Concentrations of these elements in seawater are sufficient to meet the needs of marine fish. Sodium and chlorine requirements appear to be met from natural levels in the ingredients commonly used in foods for freshwater fish. Juvenile chinook salmon reared in freshwater require 0.8% potassium in the air-dry diet (Shearer, 1988). Signs of potassium deficiency in chinook salmon include anorexia, convulsions, tetany, and death.

Iron

Food is the principal source of iron because water concentrations of soluble iron are generally low. If present, soluble iron can be absorbed through the gills. Iron requirements reported for channel catfish, Atlantic salmon, and eel are 30, 60, and 170 mg/kg of air-dry diet, respectively (NRC, 1993). Iron deficiency results in microcytic anemia, including reduced hemoglobin, hematocrit, plasma iron, and transferrin saturation.

Copper

Copper requirements have been reported to be 3 to 5 mg/kg of air-dry diet (NRC, 1993). Copper deficiency results in reduced activities of copper-containing enzymes such as heart cytochrome c oxidase and liver copper-zinc superoxide dismutase. Fish are very sensitive to dissolved copper in water, with 0.8 to 1.0 mg copper per L (as copper sulfate) producing mortality. Rainbow trout have tolerated copper concentrations of 150 mg/kg of air-dry diet for 20 weeks without deleterious effects (Knox et al., 1982), but 32 mg/kg caused growth depression and anemia in channel catfish (Lovell, 1998).

Zinc

Zinc requirements have been reported to be 15 to 30 mg/kg of air-dry diet when purified ingredients were used (NRC, 1993). However, the bioavailability of zinc is reduced by complexing with calcium phytates or tricalcium phosphate in diets containing natural ingredients. Thus, higher zinc levels must be included in natural diets (approximately 80 to 100 mg/kg) to compensate for lower zinc availability (Lovell, 1998). Zinc deficiency reduces appetite, growth, egg production, and hatchability, increases mortality, and results in cataracts, erosions of fins and skin, and dwarfism. Zinc concentrations of 1000 mg/kg of air-dry diet interfere with copper metabolism and reduce hemoglobin, hematocrit, and liver copper concentrations (Knox et al., 1982).

Manganese

Although manganese can be absorbed from water, and seawater can be a significant source, soluble manganese concentrations in freshwater are commonly low. Dietary manganese requirements have been reported to be 2.4 mg/kg air-dry matter for channel catfish and 13 mg/kg air-dry matter for common carp and rainbow trout (NRC, 1993). Manganese deficiency reduces growth and produces skeletal abnormalities. The activity of manganese-superoxide dismutase in cardiac muscle and liver also may be reduced.

Iodine

Iodine is obtained from water via branchial pumps and from food (Leloup, 1970). Dietary requirements that have been determined suggest a need of 0.6 to 1.1 mg/kg air-dry matter (NRC, 1983), although Lall et al. (1985) proposed that 4.5 mg/kg air-dry matter were necessary to protect Atlantic salmon from bacterial kidney disease. Iodine deficiency results in thyroid hyperplasia and a deficiency of thyroid hormone.

Selenium

Selenium requirements vary with form of selenium, polyunsaturated fatty acid and vitamin E concentrations in the diet, and concentrations of selenium in water. The bioavailability of selenium in fish meal is low but is higher in the form of sodium selenite, sodium selenate, or selenomethionine. Since selenium, in phospholipid hydroperoxide glutathione peroxidase, serves with vitamin E as an antioxidant, potential oxidative stress from polyunsaturated fats and the supply of vitamin E have opposite effects on the need for selenium. Although uptake by the gills of soluble selenium in water is very efficient (Hodson and Hilton, 1983), waterborne selenium concentrations are generally quite low.

When water selenium concentrations were 0.4 ug/L, rainbow trout required dietary selenium concentrations of 0.15 to 0.38 mg/kg air-dry matter (Hilton et al., 1980). Selenium toxicity has been seen in rainbow trout and channel catfish at concentrations of 10 to 13 and 15 mg/kg air-dry diet, respectively (Hilton et al., 1980; Hilton and Hodson, 1983; Gatlin and Wilson, 1984).

❑ VITAMINS

Vitamin needs of a few species of fish have been determined by feeding chemically defined diets that are deficient in one vitamin. In this way, quantitative requirements for most of the vitamins have been established for chinook salmon, rainbow trout, common carp, channel catfish, and yellowtail. Quantitative requirements for a few vitamins have been established for red sea bream and tilapia.

Vitamin A

Vitamin A_1 (retinol) is found in mammals and marine fish, whereas both vitamin A_1 and vitamin A_2 (3-dehydroretinol) are found in freshwater fish, which can oxidatively convert retinol to 3-dehydroretinol. Beta-carotene and canthaxanthin can be transformed in the liver of tilapia to vitamin A_1, whereas astaxanthin, zeaxanthin, lutein, and tunaxanthin can be converted to vitamin A_2 without vitamin A_1 as an intermediate (Katsuyama and Matsuno, 1988). Vitamin A deficiency results in poor growth, anorexia, high mortality, twisted gill opercula, hemorrhages in the skin, fins, eyes, liver, and kidneys, exophthalmia, displaced ocular lens, and retinal degeneration. Vitamin A requirements for fish that have been studied are 1000 to 4000 IU/kg of air-dry diet (NRC, 1993).

Vitamin D

Both dietary vitamin D_2 (ergocalciferol) and vitamin D_3 (cholecalciferol) are effective in meeting the requirements of rainbow trout and channel catfish, although vitamin D_3 seems most active. Published descriptions of vitamin D deficiency in fish are somewhat conflicting, but poor growth, impaired calcium homeostasis, lower body calcium and phosphorus concentrations, lower total body ash, tetany, and ultrastructural changes in white skeletal muscle fibers and epaxial musculature have been described (NRC, 1993). Weakness of the epaxial musculature may be related to a lordosis-like, droopy tail syndrome seen in vitamin D-deficient trout (Barnett et al., 1982). Vitamin D requirements appear to be in the range of 500 to 2400 IU/kg of air-dry diet (NRC, 1993). Dietary vitamin D_3 concentrations of 20,000 to 50,000 IU/kg are excessive for catfish and depress weight gain (Andrews et al., 1980).

Vitamin E

Vitamin E, along with selenium and other antioxidant compounds, protects polyunsaturated membrane phospholipids against the damage caused by reactive oxygen and other free-radical initiators of oxidation. Deficiency signs in fish include muscular dystrophy, edema, anemia, depigmentation, and reduced immune competency. Vitamin E requirements have been set at 50 to 100 IU/kg of air-dry diet (NRC, 1993), although requirements are increased when dietary polyunsaturated fatty acid levels are high (Watanabe et al., 1981; Cowey et al., 1983). Excessive vitamin E (1000 IU/kg of air-dry diet or higher) may increase clotting times and induce anemia.

Vitamin K

Vitamin K is important for normal blood coagulation. When fish consume natural diets, they derive their needs from plant or animal foods or from bacterial synthesis in the gut. The inclusion of sulfaguanidine (a bacterial growth inhibitor) in a vitamin K-deficient diet, and low water temperature, prolonged blood coagulation time and decreased hematocrit in brook trout (Poston, 1964). Estimates of vitamin K requirements (as menadione) for lake trout are 0.5 to 1 mg/kg of air-dry diet (Poston, 1976). The menadione derivatives commonly added to fish foods include menadione sodium bisulfite (50% menadione), menadione sodium bisulfite complex (33% menadione), or menadione dimethylpyrimidinyl bisulfite (45.5% menadione).

Thiamin

Thiamin, as thiamin pyrophosphate, functions in oxidative decarboxylation of pyruvate and alpha-ketoglutarate and in transketolase reactions in the pentose shunt. Thiamin deficiency results in anorexia, neurological disorders, including hyper-irritability, and declines in erythrocye transketolase activity and blood thiamin concentration,

which can be detected before neurological signs appear. Thiamin requirements are 0.5 to 1 mg/kg of air-dry diet (NRC, 1993).

Riboflavin

Riboflavin, as a component of flavin mononucleotide and flavin adenine dinucleotide, functions in the transfer of electrons during intermediary metabolic oxidation-reduction reactions. Riboflavin deficiency results in anorexia and poor growth in all fish studied and in photophobia, cataracts, corneal vascularization, and hemorrhages in salmonids. Some of these signs also have been seen in common carp and channel catfish. Riboflavin requirements are 4 to 9 mg/kg of air-dry diet (NRC, 1993).

Niacin

Nicotinic acid and nicotinamide have similar niacin activity for fish. Niacin, as a component of nicotinamide adenine dinucleotide and nicotinamide adenine dinucleotide phosphate, is required in several oxidation–reduction reactions involving transfer of hydrogen and electrons during carbohydrate, fat, and protein metabolism. Niacin is found in both animals and plants but is of limited bioavailability in corn, grain sorghum, wheat, barley, and their byproducts (unless fermented or produced by wet-milling) for fish. Niacin deficiency results in anorexia, reduced growth, photosensitivity, lesions of skin, fins, and intestine, hemorrhage, anemia, and ataxia. Niacin requirements are 10 to 28 mg/kg of air-dry diet (NRC, 1993).

Pantothenic Acid

Pantothenic acid, as a component of coenzyme A, acyl CoA synthetase, and acyl carrier protein, is involved in fatty acid synthesis and in acyl group transfers, resulting in energy production by incorporation of the carbon skeletons of glucose, fatty acids, and amino acids into the tricarboxylic acid cycle. Pantothenic acid deficiency results in anorexia, slow growth, lethargy, anemia, gill lamellar hyperplasia (clubbed gills), and high mortality. Pantothenic acid requirements are 10 to 30 mg/kg of air-dry diet (NRC, 1993).

Vitamin B_6

Pyridoxine is the principal vitamin B_6 form found in plants, whereas pyridoxal and pyridoxamine are the principal forms found in animals. All three forms can be converted by animals to the coenzymes, pyridoxal phosphate and pyridoxamine phosphate. These coenzymes are important during transamination, decarboxylation, and dehydration of amino acids, in catabolism of glycogen, in porphyrin synthesis, and in synthesis of the neurotransmitters, 5-hydroxytryptamine and serotonin, from tryptophan. Vitamin B_6 deficiency results in anorexia, poor growth, lesions in the intestine, liver, and kidneys, and neurological disorders, such as erratic swimming, hyperirritability, and convulsions. The activity of serum alanine and/or aspartate aminotransferase has been used to assess the vitamin B_6 status of fish. Vitamin B_6 requirements are 3 to 6 mg/kg of air-dry diet (NRC, 1993).

Biotin

Biotin functions in carbon dioxide transfer during carboxylation and decarboxylation reactions, and as part of the coenzymes, acetyl-CoA cocarboxylase, pyruvate cocarboxylase, and propionyl-CoA cocarboxylase. Biotin deficiency results in anorexia, poor growth, histopathology of gills, liver, and kidneys, and changes in skin pigmentation. Hepatic pyruvate carboxylase activity is reduced. Biotin requirements are 0.15 to 1 mg/kg of air-dry diet (NRC, 1993).

Folacin

Folic acid (pteroylmonoglutamic acid) and related compounds derive their biological activity through conversion to the coenzyme, tetrahydrofolic acid. This coenzyme functions as an intermediate carrier of methyl, methylene, and other one-carbon groups during the metabolism of certain amino acids and the synthesis of purines, pyrimidines, and the nucleotides found in DNA and RNA. Folacin deficiency results in anorexia, poor growth, macrocytic normochromic anemia (with megaloblastic proerythrocytes in the erythropoietic tissue of the anterior kidney), congestion in the fins and bronchial mantle, and dark skin coloration. There is some evidence of bacterial synthesis of folacin in the intestine of common carp (Aoe et al., 1967; Kashiwada et al., 1971). Folacin requirements are 1 to 2 mg/kg of air-dry diet (NRC, 1993).

Vitamin B_{12}

Cyanocobalamin and related corrinoids with vitamin B_{12} activity are required for normal develop-

ment and maturation of erythrocytes, metabolism of fatty acids, methylation of homocysteine to methionine, and recycling of tetrahydrofolic acid. Some vitamin B_{12} deficiency signs are similar to those of folacin deficiency because of the metabolic relationship between them. There is considerable evidence of vitamin B_{12} synthesis by intestinal microflora, and the concentrations of vitamin B_{12} and viable counts of *Bacteroides* type A in intestinal contents of fish are positively related (Sugita et al., 1991). Fish that have this bacterium in their gut appear not to have a dietary requirement for vitamin B_{12}. Vitamin B_{12} requirements of fish that do not show synthesis by intestinal bacteria are estimated to be 0.01 mg/kg of air-dry diet (NRC, 1993).

Choline

Choline is part of phospatidylcholine (a structural component of biological membranes), a precursor of acetylcholine (a neurotransmitter), and a labile methyl donor (for methylation reactions, such as formation of methionine from homocysteine). Choline deficiency results in anorexia, slow growth, anemia, and fatty livers. Certain fish species are able to meet part of their choline requirements from excess methionine. Choline requirements are 400 to 1000 mg/kg of air-dry diet (NRC, 1993).

Myoinositol

Myoinositol is a structural component of phosphatidylinositol in biological membranes. A hormonal stimulus of cell membrane receptors results in generation of secondary chemical messengers from phosphatidylinositol that modulate amylase secretion, insulin release, smooth muscle contraction, liver glycogenolysis, histamine secretion, platelet aggregation, and DNA synthesis in fibroblasts and lymphoblasts. Myoinositol deficiency results in anorexia, poor growth, anemia, fatty liver, fin erosion, dark skin coloration, delayed gastric emptying, and decreased cholinesterase and aminotransferase activities in several fish species. Myoinositol appears to be synthesized by intestinal bacteria in the common carp (Aoe and Masuda, 1967) and by intestinal bacteria and in the liver of channel catfish (Burtle and Lovell, 1989), although amounts synthesized in young carp were insufficient to meet need. High dietary glucose concentrations may increase the dietary myoinositol requirement (Yone et al., 1971). Myoinositol requirements are 300 to 440 mg/kg of air-dry diet (NRC, 1993).

Vitamin C

Many fish species cannot synthesize L-ascorbic acid (vitamin C) from D-glucose; thus, a dietary supply is required. Ascorbic acid, because it is a strong reducing agent, protects phagocytic cells and surrounding tissues from reactive oxygen radicals that are released during microbicidal activity. In turn, ascorbic acid is oxidized to dehydroascorbic acid. Dehydroascorbic acid can be reduced back to ascorbic acid in animal tissues, using reduced glutathione and NADPH for reducing equivalents. Ascorbic acid also facilitates absorption of iron and is involved in the hydroxylation of proline and lysine, resulting in hydroxyproline and hydroxylysine found in procollagen, the precursor of collagen.

Vitamin C deficiency results in defective connective tissue synthesis and structural deformities, such as scoliosis, lordosis, and abnormalities of cartilage supporting the eyes, gills, and fins. Ascites, hemorrhage, anorexia, slow growth, elevated plasma triglycerides and cholesterol, reduced reproductive performance, and impaired resistance to environmental contaminants have also been seen. Vitamin C requirements are 25 to 50 mg/kg of air-dry diet (NRC, 1993). Because ascorbic acid is so labile and easily destroyed during the manufacture of fish food, the more stable ethylcellulose-coated or fat-coated L-ascorbic acid have been used. Nevertheless, 50% of the added ascorbic acid may be destroyed during extrusion (Lovell and Lim, 1978). L-ascorbyl-2-monophosphate or L-ascorbyl-2-polyphosphate are biologically active and stable to extrusion and are now favored vitamin C supplements.

❑ NRC NUTRIENT REQUIREMENTS

Minimum nutrient requirements for maximum performance of five species of fish under experimental conditions are shown in Table 29.6 and have been cited in the previous discussions of individual nutrients. Requirements are expressed on an as fed or air-dry basis rather than a dry-matter basis. Air-dry feeds generally contain about 8 to 10% moisture. The values should meet needs of small fish fed diets composed of purified or chemically defined ingredients. However, no excesses are included to compensate for nutrient losses during food manufacture and storage, for the lower nutrient bioavailability expected when natural dietary ingredients are used, or for increased nutrient requirements associated with less than optimum environmental conditions. Thus, it may be necessary to include higher levels of some nutrients in practical fish diets.

❑ PIGMENTS

Free-living fish have characteristic colors of skin and flesh that are derived from natural pigments. Carotenoids are the most important of these, and since fish cannot synthesize them, they must be present in the diet if characteristic colors are sought. The red to orange color of the flesh, skin, and fins of salmonids comes from ingestion of two oxycarotenoids found in zooplankton, astaxanthin, and canthaxanthin. When these two pigments are not present and feedstuffs of plant origin containing lutein and zeaxanthin (such as yellow corn, dehydrated alfalfa meal, or algae) are used, the desired color of salmon flesh is not produced. Natural products containing the desired pigments, carotenoid extracts of natural products, or chemically synthesized pigments are commonly added to farmed salmonid diets to ensure appropriate flesh color (Halver and Hardy, 2002). Diets of channel catfish containing appreciable zeaxanthin and lutein result in flesh that has an undesirable yellow color. Thus, dietary ingredients low in these pigments must be used (Lee, 1987).

Differences between fish species in the accumulation of carotenoids may be due to differences in their absorption or metabolism. The absorption of astaxanthin and canthaxanthin by salmonids is 10 to 20 times more efficient than absorption of lutein and zeaxanthin (Schiedt et al., 1985). However, goldfish and fancy red carp absorb carotenoids in the order zeaxanthin-astaxanthin-lutein. Zeaxanthin, a yellow pigment, is readily metabolized to astaxanthin, a red pigment, by goldfish and fancy red carp. Goldfish metabolize no lutein and little beta-carotene to astaxanthin (Hata and Hata, 1972, 1973, 1976).

❑ DIETS AND FEEDING PRACTICES

Diet characteristics requiring consideration in the feeding of different species and sizes of fish include

TABLE 29.6 Nutrient requirements in air-dry diet for channel catfish, rainbow trout, Pacific salmon, common carp, and tilapia under experimental conditions (as fed basis).[a]

NUTRIENT	CHANNEL CATFISH	RAINBOW TROUT	PACIFIC SALMON	COMMON CARP	TILAPIA
DE, Kcal/kg[b]	3000	3600	3600	3200	3000
Crude protein, %	32	38	38	35	32
Digestible protein, %	28	34	34	30.5	28
Arginine, %	1.20	1.50	2.04	1.31	1.18
Histidine, %	0.42	0.70	0.61	0.64	0.48
Isoleucine, %	0.73	0.90	0.75	0.76	0.87
Leucine, %	0.98	1.40	1.33	1.00	0.95
Lysine, %	1.43	1.80	1.70	1.74	1.43
Meth. + cyst., %	0.64	1.00	1.36	0.94	0.90
Phenylal. + tyr., %	1.40	1.80	1.73	1.98	1.55
Threonine, %	0.56	0.80	0.75	1.19	1.05
Tryptophan, %	0.14	0.20	0.17	0.24	0.28
Valine, %	0.84	1.20	1.09	1.10	0.78
n-3 fatty acids, %	0.5–1	1	1–2	1	0
n-6 fatty acids, %	0	1	0	1	0.5–1
Calcium, %	+[c]	1E[d]	−[e]	−	+
Phosphorus, %	0.45	0.6	0.6	0.6	0.5
Magnesium, %	0.04	0.05	−	0.05	0.06
Sodium, %	+	0.6E	−	−	−
Potassium, %	+	0.7	0.8	−	−
Chlorine, %	+	0.9E	−	−	−
Iron, mg/kg	30	60	−	150	−
Copper, mg/kg	5	3	−	3	+
Manganese, mg/kg	2.4	13	+	13	+
Zinc, mg/kg	20	30	+	30	20
Iodine, mg/kg	1.1E	1.1	0.6–1.1	−	−
Selenium, mg/kg	0.25	0.3	+	−	−
Vitamin A, IU/kg	1000–2000	2500	2500	4000	−
Vitamin D, IU/kg	500	2400	−	−	−
Vitamin E, IU/kg	50	50	50	100	50
Vitamin K, mg/kg	+	+	+	−	−
Thiamin, mg/kg	1	1	+	0.5	−
Riboflavin, mg/kg	9	4	7	7	6
Niacin, mg/kg	14	10	+	28	−
Pantoth. acid, mg/kg	15	20	20	30	10
Vitamin B_6, mg/kg	3	3	6	6	−
Biotin, mg/kg	+	0.15	+	1	−
Folacin, mg/kg	1.5	1.0	2	0	−
Vitamin B_{12}, mg/kg	+	0.01E	+	0	0
Choline, mg/kg	400	1000	800	500	−
Myoinositol, mg/kg	0	300	300	440	−
Vitamin C, mg/kg	25–50	50	50	+	50

[a]Concentrations of some nutrients may need to be higher in practical diets.
[b]Typical digestible energy concentrations in commercial diets.
[c]Required but concentration unknown.
[d]Requirement estimated.
[e]Requirement not determined.
Source: NRC (1993).

(in addition to nutritional adequacy) palatability, particle size, color, texture, density, and whether the food needs to be living or can be nonliving.

Larval Fish

The larval stage is a period of metamorphosis extending from hatching until the juvenile fish assume the external characteristics and major organ functions of adults. There are three principal types of larval fish based on digestive tract morphology and the presence or absence of certain digestive enzymes (Dabrowski, 1984). The first type includes salmonids and channel catfish that have a relatively well-developed digestive system with a functional stomach, and when yolk sac reserves are depleted, feeding on prepared diets begins. The second type includes striped bass and many marine species that initially have a very rudimentary digestive system with no functional stomach or gastric glands but undergo a complex metamorphosis as they become juveniles. Feeding on prepared foods is prefaced by feeding on live foods. The third type includes carp that develop a functional digestive tract, allowing consumption of prepared foods quite early, but remain stomachless throughout life.

Larval forms of fish species with immature digestive systems initially have difficulty using prepared foods and may require some live foods in early feedings. This difficulty may be due to the lack of enzymes, hormones, or growth factors that are found in live foods (Lauff and Hofer, 1984; Baragi and Lovell, 1986). If available, the preferred live foods for larval fish would be those found in their natural environment. However, the only zooplankton that are available commercially in large quantities are rotifers (*Brachionus plicatilis*) and brine shrimp (*Artemia* spp.). The composition of zooplankton differs by geographical origin and culture system, and some enrichment with n-3 essential fatty acids may be necessary before feeding (Watanabe et al., 1983).

Successful prepared diets for larval salmonids and catfish contained 70 to 80% good-quality fish meal, although several possibilities exist (NRC, 1993). Diet particle size for larval fish should be no greater than 20% of the mouth opening (Dabrowski and Bardega, 1984). Frequent feeding is important (as often as 10 to 24 times per day) to ensure that their high metabolic needs are met. Larval fish consume 50 to 300% of their body weight per day compared to 2 to 10% for subadult fish that are being fed to market weight (Bryant and Matty, 1980, 1981).

Juvenile to Adult Fish

Channel catfish begin feeding on prepared diets 0.5 mm in diameter, can change to 1- to 3-mm particles at fingerling size of 10 g, and will feed on 3- to 5-mm pelleted or extruded particles at later stages. Pellets are less expensive than extruded foods, but because extruded foods float, feeding behavior is easier to observe, and there may be less waste. Initially, larval catfish may be fed hourly at 25% of body weight, with reductions to two to four times per day at 5 to 10% of body weight while in the hatchery. In rearing ponds, they may be fed one to two times per day, at 1 to 2% of body weight, six or seven days per week (NRC, 1993).

Striped bass and striped bass $\times$ white bass hybrids have rudimentary digestive systems and are usually fed zooplankton, such as rotifers or small brine shrimp nauplii, from 4 to 5 days posthatching, with prepared foods gradually replacing live foods by 14 to 28 days. Diets designed for trout or salmon can be successfully used for feeding these fish from juveniles to market size.

Rainbow trout and Pacific and Atlantic salmon begin feeding on prepared foods <0.6 mm in diameter when their yolk sac reserves are exhausted. As they grow, the particle sizes they will accept and prefer also increase. Feeding guides for rainbow trout, based on fish weight and water temperature, have been published by Cho (1992). Moist (28 to 32% H_2O) and semimoist (18 to 22% H_2O) diets have been used for Pacific salmon, but pelleted or extruded diets are generally less expensive to make and store. Juvenile Pacific and Atlantic salmon are reared in fresh water and subsequently raised to marketable size in saltwater. The saltwater stage may involve release for migration to the ocean, followed by return to near-shore capture fisheries, or the feeding of captive fish in near-shore net pens.

Further details on feeding practice and on other species have been published (Brandt, 1991; Cho, 1990, 1992; Dupree and Huner, 1984; Halver, 1989; Halver and Hardy, 2002; Hardy, 1989, 1991; Helland et al., 1991; Hepher, 1988; Hodson et al., 1987; Jauncey and Ross, 1982; Lim, 1989; Lovell, 1989, 1998; Robinson, 1991; Tucker and Robinson, 1991).

Natural Ingredient Diets

Examples of diets that have been successfuly used for Atlantic salmon, rainbow, brook, and brown trout, Pacific salmon, and channel catfish are shown in Table 29.7. Additional typical diet formulas may be found in Lovell (1998) and Halver and Hardy (2002).

❑ SUMMARY

- Aquaculture has a long history, but major advancements in nutrition and feeding of fish (fin-

fish and shellfish) are relatively recent. The aquatic environment in which fish live affects energy expenditure and provides some nutrients, thus influencing energy and nutrient need in formulated diets. Furthermore, the composition, physical form, and particle size of fish diets must consider natural dietary habits and gut morphology of the various species and life stages. Nutrient and energy requirements for many of these species have been identified by the NRC and are presented in this chapter. Also included are example formulas for natural ingredient diets and details of associated feeding practices.

TABLE 29.7 Examples of natural ingredient diets (%, as fed [air-dry] basis) for salmonids and channel catfish.

INGREDIENTS	ATLANTIC SALMON AND TROUT[a]	PACIFIC SALMON[b]	CHANNEL CATFISH[c]
Herring fishmeal	30	50	—
Menhaden fishmeal	—	—	8
Soybean meal (48% CP)	13	—	50
Corn gluten meal (60% CP)	17	—	—
Corn	—	—	34.1
Wheat middlings	16.5	12.2	5
Dried whey	10	5	—
Blood meal	—	10	—
Condensed milk solubles	—	3	—
Poultry byproduct meal	—	1.5	—
Wheat germ meal	—	5	—
Dicalcium phosphate	—	—	1
Marine fish oil	11.5	9	—
Catfish oil	—	—	1.5
Vitamin premix[d]	1	2.2	0.2
Trace mineral premix	1[e]	0.1[f]	0.2[g]
Pellet binder	—	2	—
Total	100.0	100.0	100.0

[a]Choy (1990).

[b]Hardy (1991).

[c]Robinson (1991).

[d]To meet NRC (1993) requirements plus allowance for processing and storage losses.

[e]To provide per kg air-dry diet: 6.25 mg copper (from copper sulfate), 13.3 mg iron (from ferrous sulfate), 21.5 mg manganese (from manganese sulfate), 6 mg iodine (from potassium iodide), 52 mg zinc (from zinc sulfate), and 3 g salt.

[f]To provide per kg air-dry diet: 75 mg zinc (from zinc sulfate), 20 mg manganese (from manganese sulfate), 1.5 mg copper (from cuprous sulfate), and 10 mg iodine (from potassium iodide).

[g]To provide per kg air-dry diet: 100 mg zinc, 30 mg iron, 5 mg copper, 5 mg iodine, 2.5 mg manganese, 0.3 mg selenium, and 0.05 mg cobalt.

Source: Adapted from NRC (1993).

❏ REFERENCES

Akiyama, T., T. Murai, and T. Nose. 1986. Oral administration of serotonin against spinal deformity of chum salmon fry induced by tryptophan deficiency. *Bull. Jpn. Soc. Sci. Fish.* **52:** 1249–1254.

Andrews, J. W., T. Murai, and J. W. Page. 1980. Effects of dietary cholecalciferol and ergocalciferol on catfish. *Aquaculture* **19:** 49–54.

Aoe, H., I. Masuda, T. Saito, and T. Takada. 1967. Water-soluble vitamin requirements of carp. 5. Requirement for folic acid. *Bull. Jpn. Soc. Sci. Fish.* **33:** 1068–1071.

Baragi, V., and R. T. Lovell. 1986. Digestive enzyme activities in striped bass from first feeding through larva development. *Trans. Am. Fish. Soc.* **115:** 478–484.

Barnett, B. J., G. Jones, C. Y. Cho, and S. J. Slinger. 1982. The biological activity of 25-hydroxycholecalciferol and 1,25-dihydroxycholecalciferol for rainbow trout (*Salmo gairdneri*). *J. Nutr.* **112:** 2020–2026.

Barrington, E.J.W. 1957. The alimentary canal and digestion. In *The Physiology of Fishes, vol. 1, Metabolism,* M. E. Brown (ed.). Academic Press, New York.

Berg, A. 1968. Studies on the metabolism of calcium and strontium in freshwater fish. 1. Relative contribution of direct and intestinal absorption. *Mem. Ist. Ital. Idrobiol. Dott. Maroco Marchi* **23:** 161–196.

Brandt, T. M. 1991. Temperate basses, *Morone* spp., and black basses, *Micropterus* spp. In *Handbook of Nutrient Requirements of Finfish,* pp. 161–168, R. P. Wilson (ed.). CRC Press, Boca Raton, FL.

Bryant, P. L., and A. J. Matty. 1980. Optimization of *Artemia* feeding rate for carp (*Cyprinus carpio* L.) larvae. *Aquaculture* **21:** 203–212.

Bryant, P. L., and A. J. Matty. 1981. Adaptation of carp (*Cyprinus carpio* L.) larvae to artificial diets. Optimum feeding rate and adaptation age for commercial diet. *Aquaculture* **23:** 275–286.

Burtle, G. J., and R. T. Lovell. 1988. Lack of response of channel catfish (*Ictalurus punctatus*) to dietary myo-inositol. *Can. J. Fish. Aquat. Sci.* **46:** 218–222.

Cho, C. Y. 1990. Fish nutrition, feeds, and feeding: with special emphasis on salmonid aquaculture. *Food Rev. Int.* **6:** 333–357.

Cho, C. Y. 1992. Feeding systems for rainbow trout and other salmonids with reference to current estimates of energy and protein requirements. *Aquaculture* **100:** 107–123.

Cowey, C. B., J. W. Adron, and A. Youngson. 1983. The vitamin E requirement of rainbow trout (*Salmo gairdneri*) given diets containing polyunsaturated fatty acids derived from fish oil. *Aquaculture* **30:** 85–93.

Cowey, C. B., and M. J. Walton. 1989. Intermediary metabolism. In *Fish Nutrition* (2nd ed.), pp. 259–329, J. E. Halver (ed.). Academic Press, San Diego, CA.

Dabrowski, K. 1984. The feeding of fish larvae. Present state of the art and perspectives. *Reprod. Nutr. Develop.* **24:** 807–833.

Dabrowski, K., and R. Bardega. 1984. Mouth-size and recommendation of feed size preferences in three cyprinid fish. *Aquaculture* **40:** 41–46.

Diamond, J. M., and R. K. Buddington. 1985. Pyloric ceca: a "new" absorptive organ. *Fed. Proc.* **44:** 811 (Abstr. 2338).

Dupree, H. K., and J. V. Huner. 1984. *Third Report to the Fish Farmers.* U.S. Fish and Wildlife Service, Washington, DC.

Gatlin, D. M., III, and R. P. Wilson. 1984. Dietary selenium requirement of fingerling channel catfish. *J. Nutr.* **114:** 627–633.

Gatlin, D. M., III, W. E. Poe, and R. P. Wilson. 1986. Protein and energy requirements of fingerling channel catfish for maintenance and growth. *J. Nutr.* **116:** 2121–2131.

Halver, J. E. (ed.). 1989. *Fish Nutrition,* 2nd ed. Academic Press, San Diego, CA.

Halver, J. E., and R. W. Hardy (eds.). 2002. *Fish Nutrition,* 3rd ed. Academic Press, San Diego, CA.

Halver, J. E., and W. E. Shanks. 1960. Nutrition of salmonid fishes. 8. Indispensable amino acids for sockeye salmon. *J. Nutr.* **72:** 340–346.

Hardy, R. W. 1989. Practical feeding—salmon and trout. In *Nutrition and Feeding of Fish,* pp. 185–203, T. Lovell (ed.). Van Nostrand Reinhold, New York.

Hardy, R. W. 1991. Pacific salmon, *Oncorhynchus* spp. In *Handbook of Nutrient Requirements of Finfish.* pp. 105–121, R. P. Wilson (ed.). CRC Press, Boca Raton, FL.

Hata, M., and M. Hata. 1972. Carotenoid pigments in goldfish. 4. Carotenoid metabolism. *Bull. Jpn. Soc. Sci. Fish.* **38:** 331–338.

Hata, M., and M. Hata. 1973. Carotenoid pigments in goldfish. 5. Conversion of zeaxanthin to astaxanthin. *Bull. Jpn. Soc. Sci. Fish.* **38:** 339–343.

Hata, M., and M. Hata. 1976. Carotenoid metabolism in fancy red carp, *Cyprinus carpio.* 2. Metabolism of ^{14}C-zeaxanthin. *Bull. Jpn. Soc. Sci. Fish.* **42:** 203–205.

Hazel, J. R. 1979. Influence of thermal acclimation on membrane lipid composition of rainbow trout liver. *Am. J. Physiol.* **236:** R91–R101.

Hazel, J.R., and R. Carpenter. 1985. Rapid changes in the phospholipid composition of gill membranes during thermal acclimation of the rainbow trout, *Salmo gairdneri. J. Comp. Physiol.* **155:** 597–602.

Helland, S., T. Storebakken, and B. Grisdale-Helland. 1991. Atlantic salmon (*Salmo solar*). In *Handbook of Nutrient Requirements of Finfish,* pp. 13–22, R. P. Wilson (ed.), CRC Press, Boca Raton, FL.

Hepher, B. 1988. *Nutrition of Pond Fishes*. Cambridge University Press, Cambridge, England.

Hilton, J. W., and P. V. Hodson. 1983. Effect of increased dietary carbohydrate on selenium metabolism and toxicity in rainbow trout (*Salmo gairdneri*). *J. Nutr.* **113**: 1241–1248.

Hilton, J. W., P. V. Hodson, and S. J. Slinger. 1980. The requirement and toxicity of selenium in rainbow trout (*Salmo gairdneri*). *J. Nutr.* **110**: 2527–2535.

Hodson, P. V., and J. W. Hilton. 1983. The nutritional requirements and toxicity to fish of dietary and waterborne selenium. *Environ. Biogeochem. Ecol.* (Stockholm) **35**: 335–340.

Hodson, R., T. Smith, J. McVey, R. Hawell, and N. Davis. 1987. *Hybrid Striped Bass Culture: Status and Perspective*. University of North Carolina Sea Grant College Publ. UNC-SG-87-03. University of North Carolina Press, Chapel Hill.

Ishibashi, N. 1974. In *The Early Life History of Fish*, pp. 339–344, J.H.S. Blaxter (ed.), Springer-Verlag, Berlin.

Jauncey, K., and B. Ross. 1982. *A Guide to Tilapia Feeds and Feeding*. Institute of Aquaculture, University of Stirling, Scotland.

Kashiwada, K., A. Kanazawa, and S. Teshima. 1971. Studies on the production of B vitamins by intestinal bacteria. 6. Production of folic acid by intestinal bacteria of carp. *Mem. Fac. Fish.* Kagoshima University **20**: 185–189.

Katsuyama, M., and T. Matsuno. 1988. Carotenoid and vitamin A and metabolism of carotenoids, beta-carotene, canthaxanthin, astaxanthin, zeaxanthin, lutein, and tunaxanthin in tilapia *Tilapia nilotica*. *Comp. Biochem. Physiol.* **90B**: 134–139.

Knox, D., C. B. Cowey, and J. W. Adron. 1982. Effects of dietary copper and copper:zinc ratio on rainbow trout, *Salmo gairdneri*. *Aquaculture* **27**: 111–119.

Lall, S. P., and F. J. Bishop. 1977. Studies on mineral and protein utilization by Atlantic salmon (*Salmo salar*) grown in sea water. *Fish. Mar. Serv. Tech. Rep.* **688**: 1–16.

Lall, S. P., W. D. Paterson, J. A. Hines, and N. J. Adams. 1985. Control of bacterial kidney disease in Atlantic salmon *Salmo salar* L. by dietary modification. *J. Fish. Dis.* **8**: 113–124.

Lauff, M., and R. Hofer. 1984. Development of proteolytic enzymes in fish and the importance of dietary enzymes. *Aquaculture* **37**: 335–346.

Lee, P. H. 1987. "Carotenoids in Cultured Channel Catfish." Ph.D. Thesis. Auburn University, Auburn, Alabama.

Leloup, J. 1970. Le mecanismes de regulation de l'iodouremie et leur controle endocrinien chez les teleosteens en eau douce. *Mem. Mus. Natl. Hist. Nat. Ser. A. Zool.* **62**: 1–108.

Lim, C. 1989. Practical feeding—tilapias. In *Nutrition and Feeding of Fish,* pp. 163–167, T. Lovell (ed.), Van Nostrand Reinhold, New York.

Lovell, R. T. 1978. Dietary phosphorus requirement of channel catfish (*Ictalurus punctatus*). *Trans. Am. Fish. Soc.* **107**: 617–621.

Lovell, T. (ed.). 1989. *Nutrition and Feeding of Fish*. Van Nostrand Reinhold, New York.

Lovell, T. (ed.). 1998. *Nutrition and Feeding of Fish,* 2nd ed. Kluwer Academic Publishers, Boston, MA.

Lovell, R. T., and C. Lim. 1978. Vitamin C in pond diets for channel catfish. *Trans. Am. Fish. Soc.* **107**: 321–325.

National Research Council (NRC). 1993. *Nutrient Requirements of Fish*. National Academy Press, Washington, DC.

Ogino, C., L. Takeuchi, H. Takeda, and T. Watanabe. 1979. Availability of dietary phosphorus in carp and rainbow trout. *Bull. Jpn. Soc. Sci. Fish.* **45**: 1527–1532.

Owen, J. M., J. W. Aldron, C. Middleton, and C. B. Covey. 1975. Elongation and desaturation of dietary fatty acids in turbot, *Scophthalmus maximus* L., and rainbow trout, *Salmo gairdnerri* Rich. *Lipids* **10**: 528–531.

Page, J. W., and J. W. Andrews. 1973. Interactions of dietary levels of protein and energy on channel catfish (*Ictalurus punctatus*). *J. Nutr.* **103**: 1339–1346.

Poston, H. A. 1964. Effect of dietary vitamin K and sulfaguanidine on blood coagulation time, microhematocrit, and growth of immature brook trout. *Prog. Fish-Cult.* **26**: 59–64.

Poston, H. A. 1976. Relative effect of two dietary water-soluble analogues of menaquinone on coagulation and packed cell volume of blood of lake trout, *Salvelinus namaycush. J. Fish Res. Board Can.* **33**: 1791–1793.

Poston, H. A., R. C. Riis, G. L. Rumsey, and H. G. Ketola. 1977. The effect of supplemental dietary amino acids, minerals and vitamins on salmonids fed cataractogenic diets. *Cornell Vet.* **67**: 472–509.

Poston, H. A., and G. L. Rumsey. 1983. Factors affecting dietary requirement and deficiency signs of L-tryptophan in rainbow trout. *J. Nutr.* **113**: 2568–2577.

Robinson, E. H. 1991. A practical guide to nutrition, feeds, and feeding of channel catfish. *Miss. Agric. For. Exp. Sta. Bull.* **979**.

Robinson, E. H., S. D. Rawles, P. B. Brown, H. E. Yette, and L. W. Greene. 1986. Dietary calcium requirement of channel catfish (*Ictalurus punctatus*) reared in calcium-free water. *Aquaculture* **53**: 263–270.

Robinson, E. H., D. LaBomascus, P. B. Brown, and T. L. Linton. 1987. Dietary calcium and phosphorus requirements of *Oreochromis aureus* reared in calcium-free water. *Aquaculture* **64**: 267–276.

Rust, M. B. 2002. Nutritional physiology. In *Fish Nutrition* (3rd ed.), pp. 367–452, J. E. Halver and R. W. Hardy (eds.). Academic Press, San Diego, CA.

Sakamoto, S., and Y. Yone. 1973. Effect of dietary calcium/phosphorus ratio upon growth, feed efficiency and blood serum Ca and P level in red sea bream. *Bull. Jpn. Soc. Sci. Fish.* **39:** 343–348.

Sakamoto, S., and Y. Yone. 1976. Requirement of red sea bream for dietary Ca. *Rep. Fish. Res. Lab. Kyushu Univ.* **3:** 59–64.

Schiedt. K., F. J. Leuenberger, M. Vecchi, and E. Glinz. 1985. Absorption, retention and metabolic transformations of carotenoids in rainbow trout, salmon and chicken. *Pure Appl. Chem.* **57:** 685–692.

Shearer, K. D. 1988. Dietary potassium requirements of juvenile chinook salmon. *Aquaculture* **73:** 119–130.

Shimeno, S. 1974. Studies on carbohydrate metabolism in fishes. *Rep. Fish Lab. Kochi Univ.* **2:** 1–107.

Smith, L. S. 1980. In *Fish Feed Technology,* pp. 3–18, T. V. R. Pillay (ed.). UNDP-FAO, Rome.

Smith, L. S., J.E. Halver and R.W. Hardy (eds.). 1989. Digestive functions in teleost fishes. In *Fish Nutrition* (2d ed.), pp. 331–421. Academic Press, San Diego, CA.

Smith, R. R., J.E. Halver and R.W. Hardy (eds.). 1989. Nutritional energetics. In *Fish Nutrition* (2nd ed.), pp. 1–29. Academic Press, San Diego, CA.

Stevens, C. E. 1988. *Comparative Physiology of the Vertebrate Digestive System.* Cambridge University Press, New York.

Stevens, C. E., and I. D. Hume. 1995. *Comparative Physiology of the Vertebrate Digestive System,* 2nd ed. Cambridge University Press, New York.

Sugita, H., C. Miyajima, and Y. Deguchi. 1991. The vitamin B_{12}-producing ability of the intestinal microflora of freshwater fish. *Aquaculture* **92:** 267–276.

Tucker, C., and E. H. Robinson. 1991. *Channel Catfish Farming Handbook.* Van Nostrand Reinhold, New York.

Walton, M. J., R. M. Coloso, C. B. Cowey, J. W. Adron, and D. Knox. 1984. The effects of dietary tryptophan levels on growth and metabolism of rainbow trout (*Salmo gairdneri*). *Br. J. Nutr.* **51:** 279–287.

Watanabe, T., C. Kitajima, and S. Fujita. 1983. Nutritional value of live organisms used in Japan for mass propagation of fish. A review. *Aquaculture* **34:** 115–143.

Watanabe, T., T. Takeuchi, and M. Wada. 1981. Dietary lipid levels and alpha-tocopherol requirements of carp. *Bull. Jpn. Soc. Sci. Fish.* **47:** 1585–1590.

Wilson, E. O. 1992. *The Diversity of Life.* Harvard University Press, Cambridge, MA.

Wilson, R. P., and J. E. Halver. 1986. Protein and amino acid requirements of fishes. *Annu. Rev. Nutr.* **6:** 225–244.

Wilson, R. P., E. H. Robinson, D. M. Gatlin, III, and W. E. Poe. 1982. Dietary phosphorus requirement of channel catfish *Ictalurus punctatus. J. Nutr.* **112:** 1197–1202.

Yone, Y., and N. Toshima. 1979. The utilization of phosphorus in fish meal by carp and black sea bream. *Bull. Jpn. Soc. Sci. Fish.* **45:** 753–756.

Yone, Y., M. Furuichi, and K. Shitanda. 1971. Vitamin requirements of red sea bream, I. Relationship between inositol requirements and glucose levels in the diet. *Bull. Jpn. Soc. Sci. Fish.* **37:** 149–155.

30

Zoo Animals

DUANE E. ULLREY

Photograph courtesy Duane E. Ullrey.

WILD ANIMALS MAY BE free-living or captive, but nowhere are they absent the influence of man. In the most isolated desert, aircraft can be heard and vapor trails seen in the skies above. On the Arctic tundra, the detritus of civilization is preserved in ice. In tropical rain forests, the sound of chain saws stills the birds, while the biodiversity of complex ecosystems is transformed into simpler, and perhaps less sustainable, interrelationships. Everywhere on the earth's surface there is evidence of the presence of 6.35 billion humans (World POPClock, 2004), and in practical terms, there no longer is any "wild."

Although we may claim this earth as our own, we share it with myriad other creatures, many of which have not been identified. Wilson (1992) has estimated that 1,143,000 species of living organisms are known. About 248,000 are higher plants, and of the remainder, 1,032,000 are animals, including 4000 mammals, 9000 birds, 6300 reptiles, 4200 amphibians, and 18,800 fish and lower chordates. Although the world's zoos contain thousands of individuals, only somewhat more than 3000 species of mammals, birds, reptiles, and amphibians are represented. This is about 13% of the known species of terrestrial vertebrates—not a very large number but still daunting when it comes to developing programs of dietary husbandry that will nourish these species appropriately.

Considering that most of our knowledge of nutrient needs has been established for just 11 terrestrial species (rats, mice, dogs, cats, chickens, turkeys, pigs, cattle, sheep, horses, and humans), it is apparent that the formulation of diets for zoo animals requires a great deal of extrapolation and considerable application of faith. Fortunately, the similarities among animal species are greater than their differ-

549

ences. The qualitative nutrient needs of animal tissues are nearly the same, and for only a few nutrients, like vitamin C, taurine, and arachidonic acid, are there species differences in metabolic synthetic ability. Thus, relatively few diets are required to successfully nourish a large number of apparently different creatures. In formulating diets for these creatures, it is helpful to understand the environment in which they evolved, the natural dietary items that they prefer, and the morphology and function of their digestive systems.

❏ HABITAT AND FEEDING STRATEGY

A useful aid in developing appropriate diets for captive wild animals is a matrix of feeding strategies and habitat use that was devised for mammals by Eisenberg (1981; Table 30.1). These habitat and food categories also apply well to birds, reptiles, and amphibians. The habitat categories represent the environments in which species evolved and spend the majority of their time, including foraging, resting, mating, and rearing young. The food categories represent the principal types of food that are consumed to meet energy and nutrient needs.

Fossorial animals (Habitat Category 1; HC1) are adapted to underground living, and those that eat earthworms and some vegetable matter, such as the eastern American mole (*Scalopus aquaticus*), are insectivore/omnivores (Food Category 09; FC09), occupying Habitat/Food niche 109.

Semifossorial animals (HC2) are adapted for burrowing, may have an underground refuge, may dig for food, but move quite freely on the surface. The American badger (*Taxidea taxus*) is such an animal, is a carnivore, eats small mammals, birds, reptiles, and arthropods, and fits in Habitat/Food niche 202.

Aquatic animals (HC3) live in freshwater, estuarine, or marine environments and have evolved to use the foods found there. The blue whale (*Balaenoptera musculus*), a planktonivore, feeds on small shrimplike crustaceans called krill (zooplankton) and fits in Habitat/Food niche 315.

Semiaquatic animals (HC4) spend part of each day out of water, some such as the Canadian river otter (*Lutra canadensis*) to groom its fur or to rest in its den. This species, a piscivore but an opportunist, eats fish and fits Habitat/Food niche 401 but also eats frogs, crayfish, crabs, other aquatic invertebrates, and some land mammals and birds.

Volant animals (HC5) can fly, and some such as horseshoe bats (*Hipposideros* spp.), aerial insectivores, catch most of their prey on the wing. They fit in Habitat/Food niche 507. Other volant ani-

TABLE 30.1   Principal habitats and food habits of different animal species, arranged in a matrix.[a]

| | PRINCIPAL FOOD OR FOOD HABIT | | | | | | | |
HABITAT	01 FISH & SQUID	02 CARNIVORE	03 NECTAR	04 GUMS	05 CRUSTACEANS & CLAMS	06 ANTS	07 FLYING INSECTS	08 FOLIAGE INSECTS
1 Fossorial	—	102	—	—	—	106	—	—
2 Semifossorial	—	202	—	—	—	206	—	—
3 Aquatic	301	302	—	—	305	—	—	—
4 Semiaquatic	401	402	—	—	405	—	—	—
5 Volant	501	502	503	—	—	—	507	508
6 Terrestrial	—	602	—	—	—	606	—	—
7 Scansorial	—	702	703	704	—	706	—	—
8 Arboreal	—	802	803	804	—	806	—	—

[a]If a matrix block contains a dash, no species are known to occupy that niche. If a matrix block contains a number (combination of habitat and food numbers), one or more species are represented.
Source: Adapted from Eisenberg (1981).

mals, such as fruit bats (e.g., flying fox [*Pteropus neohibernicus*]), frugivores, use flight to travel to a stationary food source. They fit in Habitat/Food niche 510.

Terrestrial animals (HC6) forage on the surface and do little digging and little climbing. Many herbivores are found in this category, with selective foragers (herbivore/browsers) like white-tailed deer (*Odocoileus virginianus*) fitting in Habitat/Food niche 613, while plains grazers (herbivore/grazers) like American bison (*Bison bison*) fit in Habitat/Food niche 614.

Scansorial animals (HC7) are well adapted for climbing but spend about equal time in trees and on the ground. Wood rats (*Neotoma fuscipes*) regularly nest and move about in trees, eat mostly plant tissues (herbivore/browsers) such as leaves, stems, roots, and seeds, with some invertebrates, and fit in Habitat/Food niche 713.

Arboreal animals (HC8) spend most of their lives in trees. Included in this category are many primates and the sloths. Orangutans (*Pongo pygmaeus*) eat principally fruit and vegetation (frugivore/herbivores), but also consume insects, small vertebrates, and eggs. They fit best in Habitat/Food category 812. Three-toed sloths (*Bradypus variegatus*) eat plant leaves, tender twigs, and buds (herbivore/browsers) and fit in Habitat/Food niche 813.

These examples illustrate how particular species can be grouped into categories that provide a useful perspective for diet development. However, as is common in nature, some species defy neat categorization and, depending on contemporary experience, may be found in Habitat/Food niches that are different than predicted.

❏ MORPHOLOGY AND FUNCTION OF THE DIGESTIVE SYSTEM

Oral anatomy and gastrointestinal tract morphology and physiology usually correlate well with the natural diet (Stevens and Hume, 1995). For example, the prehensile upper lip of the black rhinoceros (*Diceros bicornis*) is well adapted for browsing, whereas the lack of incisors makes grazing difficult. The white rhinoceros (*Ceratotherium simum*) also has no incisors (Nowak, 1999), but its broad, flexible lips are used to crop the leaves and stems of short grasses, and it is an efficient grazer (Clemens and Maloiy, 1982).

The presence of a ruminoreticulum in the common eland (*Taurotragus oryx*) or a large cecum plus a large, sacculated colon in the African elephant (*Loxodonta africana*) implies adaptation to fiber digestion like that occurring in the domestic cow or domestic horse. The ruminoreticulum and cecum and colon accommodate symbiotic bacteria that anaerobically ferment cellulose and hemicellulose to volatile fatty acids that can be absorbed and metabolized for energy. These bacteria also

TABLE 30.1 Principal habitats and food habits of different animal species, arranged in a matrix.[a]—*Continued*

| | PRINCIPAL FOOD OR FOOD HABIT | | | | | | | |
HABITAT	01 INSECTIVORE OMNIVORE	02 FRUGIVORE OMNIVORE	03 FRUGIVORE GRANIVORE	04 FRUGIVORE HERBIVORE	05 HERBIVORE BROWSER	06 HERBIVORE GRAZER	07 PLANKTON	08 BLOOD
1 Fossorial	109	110	111	112	113	114	—	—
2 Semifossorial	209	210	211	212	213	214	—	—
3 Aquatic	—	—	—	—	—	314	315	—
4 Semiaquatic	409	410	411	412	413	414	—	—
5 Volant	—	510	—	—	—	—	—	516
6 Terrestrial	609	610	611	612	613	614	—	—
7 Scansorial	709	710	711	712	713	714	—	—
8 Arboreal	809	810	811	812	813	—	—	—

[a]If a matrix block contains a dash, no species are known to occupy that niche. If a matrix block contains a number (combination of habitat and food numbers), one or more species are represented.
Source: Adapted from Eisenberg (1981).

synthesize B-complex vitamins, vitamin K, and microbial protein, with the potential to improve protein quality if certain essential amino acids are deficient in the diet. However, the latter improvement may be of little use if it occurs within fermentation compartments in the hind gut since proteolysis is limited as digesta moves beyond the gastric stomach and small intestine. Colobus and langur monkeys are arboreal folivores (leaf-eating herbivores) that have a four-compartment stomach but do not ruminate.

Some herbivores have fermentation vats within their digestive tracts that differ significantly in location or appearance from those typical of domesticated cattle, sheep, and horses. Foregut modifications are particularly noteworthy in kangaroos, wallabies, sloths, hippopotamuses (*Hippopotamus amphibius*), camels, llamas (*Lama glama*), and hoatzins (*Opisthocomus hoazin*). The hoatzin is a bird, larger than the American crow (*Corvus brachyrhynchos*), and a clumsy flier, with a crop modified for fiber digestion. The koala (*Phascolarctos cinereus*) has a cecum that is about three times its body length.

Carnivorous mammals and birds, like mink (*Mustela vison*) and red-tailed hawks (*Buteo jamaicensis*), usually have no or vestigial ceca and a short colon. They are totally dependent on their own digestive enzymes, and the diet must supply all nutrients that cannot be synthesized in their tissues.

Thus, information on the structure of the digestive system is helpful in devising diets for use in captivity. However, as noted for habitat and food preference, some of nature's creatures defy simple categorization. The giant panda (*Ailuropoda melanoleuca*) belongs to the order Carnivora, has a simple digestive tract with no cecum, but feeds almost exclusively on bamboo. The great horned owl (*Bubo virginianus*), like the red-tailed hawk, consumes small mammals and birds, but has paired, relatively long ceca that are greatly expanded in their distal third.

❑ EXTRAPOLATION FROM SPECIES WHOSE REQUIREMENTS ARE KNOWN

Because so few controlled studies of the nutrient requirements of wild animals have been conducted, there are few authoritative references for wild animals comparable to the nutrient require-

ment publications for domestic animals prepared by the Committee on Animal Nutrition of the National Research Council, National Academy of Sciences, Washington, D.C. This information deficit is a handicap, yet extrapolations can be made from domestic to wild species with similar dietary preferences and comparable gastrointestinal morphology and physiology.

Nutrient requirements of domestic cattle, sheep, and goats have application to the formulation of diets for wild ruminants and near-ruminants, such as deer, giraffe, pronghorn, antelope, cattle, goats, sheep, llamas, camels, and hippopotamuses. The nutrient requirements of domestic horses have application to the formulation of diets for elephants, rhinoceroses, horses, zebra, asses, and tapirs. Nutrient requirements of domestic swine have application to formulation of diets for wild swine and peccaries, and nutrient requirements of domestic poultry have application to formulation of diets for a wide variety of precocial and altricial birds.

A major difference between the formulation of diets for wild versus domestic species is that diets for the latter are generally designed for maximum weight gain or maximum productivity, while a long, healthy, reproductively successful life is a principal concern for captive wild animals. To avoid excessive accumulation of body fat, it may be important to control digestible energy supply in relation to need by adjusting dietary concentrations of fat or fiber. The amount of food offered also might be controlled, but this may be difficult in exhibits containing animals of mixed age, sex, or species. There frequently is a dominance hierarchy that puts submissive animals at a disadvantage when living space and feeding space and time are limited.

Some wild animals exhibit a seasonal fluctuation in body fat as an adaptation for survival during periods of food scarcity (Abbott et al., 1984; Watkins et al., 1991). For example, white-tailed deer that live in northern temperate regions accumulate body fat as days grow shorter after the summer solstice. This is a result of increased food intake and an increase in the activity of lipogenic enzymes, under regulation by the pineal gland. When winter comes, food is scarce in the wild, and a search for food may result in greater energy expenditure than the energy acquired. Deer become less active, seek shelter under evergreens to minimize heat (energy) loss from their warm bodies to the cold sky, and use body fat to meet their energy needs. This evolutionary adaptation is so strong that in captivity, even when given food ad libitum, deer will eat less in winter than in summer. If food offered during summer is excessively limited, accumulations of body fat will be insufficient to meet winter energy needs.

Obviously, where information is available specific to the requirements of a wild species, its use is preferable to extrapolation. Some data for primates have been published by the NRC (2003), and limited data for other species are available in specialized research journals. Robbins (1993), Kleiman et al. (1996), and Klasing (1998) also have assembled much useful information.

❏ DIETARY HUSBANDRY OF TERRESTRIAL HERBIVORES

Because many captive wild animals are terrestrial herbivores, it is instructive to consider how these species can be appropriately nourished. These animals have evolved strategies for making effective use of plant material, have no enzymes of their own that can degrade cellulose or hemicellulose, but usually have gastrointestinal compartments with an environment suitable for symbiotic microorganisms that can ferment plant cell walls. Hofmann (1973) found not only that African ruminants have such compartments but that the structure of these compartments can be used as an indicator for the genetically fixed, adaptive feeding habits that guide free-ranging ruminants in their natural habitat. Based on the studies of others and on his own field and laboratory research, Hofmann grouped East African ruminants into three main categories: concentrate selectors, bulk and roughage eaters, and intermediate feeders (seasonally and regionally adaptable). Hofmann (1989) and others have extended this classification system to cervids and to a variety of other ruminant and nonruminant species.

Concentrate selectors, sometimes called browsers, feed on trees, shrubs, forbs, and fruit. They rarely feed intensively on one plant but tend to wander about picking a variety of plants and plant parts. Those species that might be considered concentrate selectors include giraffe (*Giraffa camelopardalis*), lesser kudu (*Tragelaphus imberbis*), greater kudu (*Tragelaphus strepsiceros*), gerenuk (*Litocranius walleri*), Gunther's dikdik (*Madoqua guentheri*), Kirk's dikdik (*Madoqua kirki*), suni (*Nesotragus moschatus*), klipspringer (*Oretragus oreotragus*), gray duiker (*Sylvicapra grimmia*), Harvey's duiker (*Cephalophorus harveyi*), blue duiker (*Cephalophorus monticola*), bushbuck (*Tragelaphus scriptus*), and sitatunga (*Tragelaphus spekei*).

Bulk and roughage eaters feed primarily on grass and are sometimes termed grazers. Some are highly water dependent, preferentially feed on fresh grass, and may migrate to find it. Included in this subgroup are African buffalo (*Syncerus caffer*), oribi (*Ourebia ourebi*), bohor reedbuck (*Redunca arundinum*), Uganda kob (*Kobus kob*), waterbuck (*Kobus ellipsiprymnus*), and blue wildebeest (*Connachaetes taurinus*). Others are less dependent on water and show little inclination to migrate in search of water and fresh grass unless there is a severe drought. This subgroup includes hartebeest (*Alcelaphus buselaphus*), topi (*Damaliscus lunatus*), mountain reedbuck (*Redunca fulvorufula*), roan antelope (*Hippotragus equinus*), and sable antelope (*Hippotragus niger*). A few species are well adapted to dry, arid regions. This subgroup includes fringe-eared oryx (*Oryx gazella callotis*) and beisa oryx (*Oryx gazella gallarum*).

Intermediate feeders are more adaptable to changing habitats and vegetation than most of the concentrate selectors and bulk and roughage eaters. This adaptability is reflected in their food selection behavior and in apparently reversible morphologic adjustments of the rumen papillae and omasal laminae and, to a lesser extent, of the mucosa of the reticulum and abomasum. Impala (*Aepyceros melampus*) and Thomson's gazelle (*Gazella thomsoni*) are intermediate feeders that prefer grass. Eland (*Taurotragus oryx*), Grant's gazelle (*Gazella granti*), and steenbok (*Raphicerus campestris*) are intermediate feeders that prefer forbs and shrub and tree foliage.

The probable qualitative and quantitative nutrient and energy requirements of the above animal types have been described (Ullrey and Allen, 1986), and food habits and nutrient intakes by representative species in the wild have been summarized (Ullrey, 1989). Over 25 years of controlled and uncontrolled studies have established that the majority of these terrestrial herbivores can be fed successfully in captivity using just three commercially prepared pelleted feeds plus hay when appropriate.

TABLE 30.2 Nutrient specifications (air-dry basis) for a complete pellet (ADF16) designed for herbivorous concentrate selectors/browsers.[a]

NUTRIENT	CONCENTRATION
Crude protein, %	17 minimum
Lysine, %	0.8 minimum
Acid detergent fiber, %	13 minimum
Acid detergent fiber, %	17 maximum
Ether extract, %	3 minimum
Linoleic acid, %	1 minimum
Ash, %	8 maximum
Calcium, %	0.65 minimum
Calcium, %	1.00 maximum
Phosphorus, total, %	0.65 minimum
Sodium, %	0.25 minimum
Potassium, %	1.2 minimum
Magnesium, %	0.2 minimum
Sulfur, %	0.2 minimum
Iron, mg/kg	150 minimum
Copper, mg/kg	20 minimum
Copper, mg/kg	30 maximum
Manganese, mg/kg	90 minimum
Zinc, mg/kg	120 minimum
Cobalt, mg/kg	0.3 minimum
Iodine, mg/kg	0.8 minimum
Selenium, mg/kg	0.4 minimum
Thiamin, mg/kg	5 minimum
Riboflavin, mg/kg	9 minimum
Pantothenic acid, mg/kg	30 minimum
Niacin, mg/kg	50 minimum
Biotin, μg/kg	250 minimum
Vitamin B_{12}, μg/kg	20 minimum
Choline, mg/kg	1500 minimum
Beta-carotene, mg/kg	30 minimum
Vitamin A, IU/kg	5000 minimum
Vitamin D_3, IU/kg	1200 minimum
Vitamin E, IU/kg	300 minimum

[a]Expected nutrient concentrations at delivery.

Nutrient specifications and a formula for a complete pellet (ADF16) designed for concentrate selectors/browsers are shown in Tables 30.2 and 30.3. When formed into $\frac{3}{16}$ to $\frac{1}{4}''$ pellets, it is suitable for animals ranging widely in size. Larger pellets can be readily consumed by small antelope, although the danger of choking is increased somewhat. However, larger pellets may reduce unintended use by feral pigeons.

Although animals with a functional rumen might not require that the specified water-soluble vitamin concentrations be present, this diet may be used for a variety of nonruminant herbivores that are less likely than adult ruminants to derive their vitamin requirements from microbial synthesis in the digestive tract. Such animals include young ruminants in transition from milk to solid food; nonruminants such as horses, zebras, asses, elephants, rhinoceroses, and tapirs; and the herbivorous giant (Aldabra and Galapagos) tortoises. Nonprotein nitrogen sources such as urea should not be used, and proteins should be of high quality to ensure an appropriate supply and balance of essential amino acids.

Since ADF16 has a relatively high digestible energy and nutrient density, it is particularly suitable (with or without hay, depending upon species) for smaller, browsing ruminants or nonruminants that have evolved with only moderate capacity to digest plant fiber. Nevertheless, fiber concentrations are sufficient to inhibit development of the lactic acidosis sometimes seen with

TABLE 30.3 Formula (air-dry basis) for a complete pellet (ADF16) designed for herbivorous concentrate selectors/browsers.[a]

INGREDIENT	PERCENTAGE
Alfalfa meal, dehydrated (17% CP)	31.9
Wheat middlings	29.8
Corn grain	19.0
Soybean meal (48% CP)	11.0
Cane molasses	5.0
Soybean oil	1.0
Mono-dicalcium phosphate (16% Ca, 21% P)	1.0
Salt	0.6
Trace mineral premix	0.1
Vitamin premix	0.4
Choline chloride premix (60% choline)	0.1
Mold inhibitor	0.1
	100.0

[a]This product is commonly designated ADF16 for its typical acid detergent fiber concentration (dry basis). It can be formed into pellets $\frac{3}{16}$ to $1''$ in diameter. Usually fed with hay but may be the sole diet for highly selective browsers like white-tailed deer.

TABLE 30.4 Nutrient specifications (air-dry basis) for a complete pellet (ADF25) designed for bulk or roughage eaters/grazers.[a]

NUTRIENT	CONCENTRATION
Crude protein, %	15 minimum
Lysine, %	0.7 minimum
Acid detergent fiber, %	21 minimum
Acid detergent fiber, %	26 maximum
Ether extract, %	3 minimum
Linoleic acid, %	1 minimum
Ash, %	10.5 maximum
Calcium, %	0.85 minimum
Calcium, %	1.20 maximum
Phosphorus, total, %	0.75 minimum
Sodium, %	0.5 minimum
Potassium, %	1.5 minimum
Magnesium, %	0.25 minimum
Sulfur, %	0.20 minimum
Iron, mg/kg	200 minimum
Copper, mg/kg	20 minimum
Copper, mg/kg	30 maximum
Manganese, mg/kg	90 minimum
Zinc, mg/kg	120 minimum
Cobalt, mg/kg	0.3 minimum
Iodine, mg/kg	0.8 minimum
Selenium, mg/kg	0.4 minimum
Thiamin, mg/kg	5 minimum
Riboflavin, mg/kg	9 minimum
Pantothenic acid, mg/kg	30 minimum
Niacin, mg/kg	50 minimum
Biotin, μg/kg	250 minimum
Vitamin B_{12}, μg/kg	20 minimum
Choline, mg/kg	1500 minimum
Beta-carotene, mg/kg	50 minimum
Vitamin A, IU/kg	5000 minimum
Vitamin D_3, IU/kg	1200 minimum
Vitamin E, IU/kg	300 minimum

[a]Expected nutrient concentrations at delivery.

excessive intakes of high-concentrate feeds. This formulation is also useful, along with hay, for growing young, or pregnant or lactating adult bulk or roughage eaters/grazers. A slightly different version of this diet was fed ad libitum as the only food source to white-tailed deer at the Michigan Department of Natural Resources Houghton Lake Wildlife Research Station for over

20 years. Average annual fawn production during that period was 1.8 fawns per doe.

Nutrient specifications and a formula for a complete pellet (ADF25) designed for bulk or roughage eaters/grazers are shown in Tables 30.4 and 30.5. As in ADF16, water-soluble vitamins have been added, nonprotein nitrogen compounds have not been used, and protein is of high quality to ensure that animals with either pregastric or postgastric fermentation will be properly nourished. Pellets can be $\frac{3}{16}$ to 1″ in diameter but are usually produced in the larger diameters because most herbivore species in this category tend to be large. This pellet is usually fed with hay.

Some adult herbivores derive sufficient, or nearly sufficient, digestible energy from hay or pasture to maintain appropriate body condition but may require supplemental protein, minerals, or vitamins to ensure optimum health. Thus, a supplement pellet, designed for herbivores, with the nutrient specifications shown in Table 30.6, and the formula shown in Table 30.7, may be useful. Fresh forage or hay should be available ad libitum when this pellet is fed, and this pellet should constitute no more than 25% of total dietary dry matter intake. If more digestible energy is required than would be supplied by this level of feeding, use of ADF16 is recommended.

TABLE 30.5 Formula (air-dry basis) for a complete pellet (ADF25) designed for bulk or roughage eaters/grazers.[a]

INGREDIENT	PERCENTAGE
Alfalfa meal, dehydrated (17% CP)	59.7
Wheat middlings	31.6
Cane molasses	5.0
Sodium tripolyphosphate	1.5
Soybean oil	1.0
Salt	0.5
Trace mineral premix	0.1
Vitamin premix	0.4
Choline chloride premix (60% choline)	0.1
Mold inhibitor	0.1
	100.0

[a]This product is commonly designated ADF25 for the typical acid detergent fiber that it contains on a dry matter basis. It can be formed into pellets ranging from $\frac{3}{16}$ to 1″ in diameter, although the larger diameters are most appropriate for the larger herbivore species commonly fed this pellet. It is usually fed with hay.

TABLE 30.6   Nutrient specifications (air-dry basis) for a supplement pellet designed for herbivores.[a]

NUTRIENT	CONCENTRATION
Crude protein, %	23 minimum
Lysine, %	1.2 minimum
Acid detergent fiber, %	10 minimum
Acid detergent fiber, %	14 maximum
Ether extract, %	3 minimum
Linoleic acid, %	1.5 minimum
Ash, %	12 maximum
Calcium, %	1.2 minimum
Calcium, %	1.5 maximum
Phosphorus, total, %	0.9 minimum
Sodium, %	0.8 minimum
Potassium, %	1.5 minimum
Magnesium, %	0.25 minimum
Sulfur, %	0.25 minimum
Iron, mg/kg	300 minimum
Copper, mg/kg	35 minimum
Copper, mg/kg	50 maximum
Manganese, mg/kg	150 minimum
Zinc, mg/kg	300 minimum
Cobalt, mg/kg	0.4 minimum
Iodine, mg/kg	2.0 minimum
Selenium, mg/kg	0.8 minimum
Thiamin, mg/kg	7 minimum
Riboflavin, mg/kg	15 minimum
Pantothenic acid, mg/kg	75 minimum
Niacin, mg/kg	180 minimum
Biotin, μg/kg	300 minimum
Vitamin B_{12}, μg/kg	60 minimum
Choline, mg/kg	3500 minimum
Beta-carotene, mg/kg	20 minimum
Vitamin A, IU/kg	15,000 minimum
Vitamin D_3, IU/kg	3600 minimum
Vitamin E, IU/kg	900 minimum

[a]Expected nutrient concentrations at delivery.

❑ DIETARY HUSBANDRY OF OTHER SPECIES

Successful diets have been devised to nourish many species in addition to the terrestrial herbivores discussed here. However, only a few general comments can be made in the space remaining for this chapter. Ostrich, rheas, and emus are successfully reproducing, and young are being raised

on moderately high-fiber diets (Ullrey and Allen, 1996), providing the young are kept on nonslip floors and are provided exercise. Psittacine husbandry has been much improved by replacing traditional seed mixes with nutritionally complete extruded diets (Howard et al., 1991; Ullrey et al., 1991). Problems associated with feeding carnivorous mammals and birds have been described (Barbiers et al., 1982; Edfors et al., 1989, 1990; Haberstroh et al., 1984; Tabaka et al., 1996; Ullrey and Bernard, 1989; Vosburgh et al., 1982), and it is apparent that low-calcium intakes, when skeletal muscle is the principal food, and accumulation of calculus and oral health problems, when ground meat diets are fed to carnivorous mammals, are major issues.

Insectivorous animals fed crickets, meal worms, or wax moth larvae may suffer calcium deficiencies that can be prevented by feeding the insects calcium-fortified diets before they themselves become food for insectivores (Allen et al., 1993). Some nursing primates may require ultraviolet (UV) light exposure to support skin photobiogenesis of vitamin D because their mother's milk is low in this nutrient and consumption of vitamin D-containing solid food by infants may be delayed until 8 to 12 months of age (Ullrey, 1986). Certain

TABLE 30.7   Formula (air-dry basis) for a supplement pellet designed for herbivores.[a]

INGREDIENT	PERCENTAGE
Alfalfa meal, dehydrated (17% CP)	20.0
Wheat middlings	36.3
Soybean meal (48% CP)	30.0
Cane molasses	5.0
Soybean oil	2.0
Mono-dicalcium phosphate (16% Ca, 21% P)	2.0
Calcium carbonate (38% Ca)	0.8
Salt	2.0
Trace mineral premix	0.3
Vitamin premix	1.2
Choline chloride premix (60% choline)	0.3
Mold inhibitor	0.1
	100.0

[a]This product can be formed into pellets $\frac{3}{16}$ to 1″ in diameter. It should always be fed with hay or fresh forage at no more than 25% of total dietary dry matter intake.

basking lizards may be unable to use dietary sources of vitamin D and must have sunlight or artificial UV exposure (Bernard et al., 1991; Ullrey et al., 1986).

These and other interesting problems associated with feeding the diverse animal species found in nature provide stimulating opportunities for the intellectually curious. Research possibilities abound, and studies of wild animals offer the chance to make new discoveries that will elevate our understanding of biology both for our benefit and for the benefit of the other creatures with which we share the earth.

❏ SUMMARY

- Somewhat over 3000 species of higher animals are found in the world's zoos, providing a major challenge to those responsible for their care. Information on natural dietary habits, gastrointestinal morphology and physiology, and nutrient requirements of related species is helpful in formulating diets for animals whose needs are otherwise undefined. Organization of these species by habitat and food niches supports development of relatively few diets, each of which will adequately nourish several species found in the combined habitat/food niche. Examples of species found in these niches are provided, and nutrient specifications and example formulations of diets successfully used for terrestrial herbivores are presented.

❏ REFERENCES

Abbott, M. J., D. E. Ullrey, P. K. Ku, S. M. Schmitt, D. R. Romsos, and H. A Tucker. 1984. Effect of photoperiod on growth and fat accretion in white-tailed doe fawns. *J. Wildl. Manage.* **48:** 776–787.

Allen, M. E., O. T. Oftedal, and D. E. Ullrey. 1993. Effect of dietary calcium concentration on mineral composition of fox geckos (*Hemidactylus garnoti*) and Cuban tree frogs (*Osteopilus septentrionalis*). *J. Zoo Wildl. Med.* **24:** 118–128.

Barbiers, R. B., K. M. Vosburgh, P. K. Ku, and D. E. Ullrey. 1982. Digestive efficiencies and maintenance energy requirements of captive wild felidae: cougar (*Felis concolor*); leopard (*Panthera pardus*); lion (*Panthera leo*); and tiger (*Panthera tigris*). *J. Zoo Anim. Med.* **13:** 32–37.

Bernard, J., O. Oftedal, P. Barbosa, C. Mathias, M. Allen,

S. Citino, D. Ullrey, and R. Montali. 1991. The response of vitamin D-deficient iguanas (*Iguana iguana*) to artificial ultraviolet light. *Proc. Annu. Meet. Am. Assoc. Zoo Vet.,* Calgary, pp. 147–150.

Clemens, E. T., and G. M. O. Maloiy. 1982. The digestive physiology of three East African herbivores: the elephant, rhinoceros and hippopotamus. *J. Zool. Lond.* **198:** 141–156.

Edfors, C. H., D. E. Ullrey, and R. J. Aulerich. 1989. Prevention of urolithiasis in the ferret (*Mustela putorius furo*) with phosphoric acid. *J. Zoo Wildl. Med.* **20:** 12–19.

Edfors, C. H., D. E. Ullrey, and R. J. Aulerich. 1990. Effects of dietary calcium concentration and calcium-phosphorus ratio on growth and selected bone measures in young ferrets (*Mustela putorius furo*). *J. Zoo Wildl. Med.* **21:** 185–191.

Eisenberg, J. F. 1981. *The Mammalian Radiations.* University of Chicago Press, Chicago, IL.

Haberstroh, L. I., D. E. Ullrey, J. G. Sikarskie, N. A. Richter, B. H. Colmery, and T. D. Myers. 1984. Diet and oral health in captive amur tigers. *J. Zoo Anim. Med.* **15:** 142–146.

Hofmann, R. R. 1973. *The Ruminant Stomach: Stomach Structure and Feeding Habits of East African Game Ruminants.* East African Monographs in Biology, Vol. 2. East African Literature Bureau, Nairobi, Kenya.

Hofmann, R. R. 1989. Evolutionary steps of ecophysiological adaptation and diversification of ruminants: A comparative view of their digestive system. *Oecologia* **78:** 443–457.

Howard, K., D. Ullrey, and R. Howard. 1991. Dietary husbandry of twelve species of psittacines. *Proc. Annu. Meet. Am. Assoc. Zoo Vet.,* Calgary, pp. 261–265.

Klasing, K. C. 1998. Comparative Avian Nutrition. CAB International, New York, NY.

Kleiman, D. G., M. E. Allen, K. V. Thompson, and S. Lumpkin (eds.). 1996. *Wild Mammals in Captivity: Principles and Techniques.* University of Chicago Press, Chicago, IL.

National Research Council (NRC). 2003. *Nutrient Requirements of Nonhuman Primates,* 2nd rev. ed. National Academies Press, Washington, DC.

Nowak, R. M. 1999. *Walker's Mammals of the World,* Vol. I & II, 6th ed. Johns Hopkins University Press, Baltimore, MD.

Robbins, C. T. 1993. *Wildlife Feeding and Nutrition,* 2nd ed. Academic Press, San Diego, CA.

Stevens, C. E., and I. D. Hume. 1995. *Comparative Physiology of the Vertebrate Digestive System,* 2nd ed. Cambridge University Press, New York.

Tabaka, C. S., D. E. Ullrey, J. G. Sikarskie, S. R. DeBar, and P. K. Ku. 1996. Diet, cast composition, and energy and nutrient intake of red-tailed hawks (*Buteo jamaicensis*), great horned owls (*Bubo virginianus*), and turkey vultures (*Cathartes aura*). *J. Zoo Wildl. Med.* **27:** 187–196.

Ullrey, D. E. 1986. Nutrition of primates in captivity. In *Primates: The Road to Self-Sustaining Populations,* pp. 823–835, K. Benirschke (ed.). Springer-Verlag, New York.

Ullrey, D. E. 1989. Wild mammal nutrition research, 1968–1988. *Proc. 37th Annu. Pfizer Res. Conf.,* Kansas City, MO, pp. 198–212.

Ullrey, D. E., and M. E. Allen. 1986. Principles of zoo mammal nutrition. In *Zoo and Wild Animal Medicine,* (2nd ed.), pp. 515–532, M. E. Fowler (ed.). W. G. Saunders (ed.), Philadelphia, PA.

Ullrey, D. E., and M. E. Allen. 1996. Nutrition and feeding of ostriches. *Anim. Feed Sci. Technol.* **59:** 27–36.

Ullrey, D. E., M. E. Allen, and D. J. Baer. 1991. Formulated diets versus seed mixtures for psittacines. *J. Nutr.* **121:** S193–S205.

Ullrey, D. E., and J. B. Bernard. 1989. Meat diets for performing exotic cats. *J. Zoo Wildl. Med.* **20:** 20–25.

Ullrey, D. E., M. A. Strzelewicz, S. E. Kaupp, P. T. Robinson, J. Davis, and E. Mandela. 1986. Skylights and artificial lights for photobiogenesis of vitamin D_3. *Proc. Annu. Meet. Am. Assoc. Zoo Vet.,* Chicago, pp. 66–67.

Vosburgh, K. M., R. B. Barbiers, J. G. Sikarskie, and D. E. Ullrey. 1982. A soft versus hard diet and oral health in captive timber wolves (*Canis lupus*). *J. Zoo Anim. Med.* **13:** 104–107.

Watkins, B. E., J. H. Witham, D. E. Ullrey, D. J. Watkins, and J. M. Jones. 1991. Body composition and condition evaluation of white-tailed deer fawns. *J. Wildl. Manage.* **55:** 39–51.

Wilson, E. O. 1992. *The Diversity of Life.* Harvard University Press, Cambridge, MA.

World POPClock. 2004. *World Population Information for February 11, 2004.* International Program Center, U.S. Census Bureau, Washington, DC.

Photo Credits

Chapter 3
Fig. 3.1 (left): Courtesy of Enercorp Instruments, Ltd. Fig. 3.1 (right): Courtesy of Photovolt Instrument Inc. Fig. 3.5 (left): Beckman Instruments, Inc., Fullerton, CA. Fig. 3.5 (right): Courtesy of Waters Chromatography Division, Millipore Corporation, Milford, MA. Fig. 3.6: Courtesy of the Instrumentation Laboratory. Fig. 3.7: Rainin Instrumental Company, Inc. Emeryville, CA. Fig. 3.8: Courtesy of Ciba-Corning Diagnostics Corporation, Norwood, MA. Fig. 3.9: Photo courtesy of Beckman Coulter, Inc.

Chapter 4
Fig. 4.2: Photos by D. C. Church. Fig. 4.4: Courtesy of Jeanne M. Riddle, Wayne State University School of Medicine. Fig. 4.5: Photo by Don Heifer, Oregon State University. Fig. 4.6: Courtesy of CSIRO. Fig. 4.8: From Broughton and Lecce (1070). Courtesy of J. C. Lecce.

Chapter 5
Fig. 5.1: Courtesy of American Calan, Inc. Fig. 5.2: Courtesy of G. Fishwick, Glasgow University Vet School. Fig. 5.4: Photo courtesy of North Car-olina State University. Fig. 5.5: Photo by W. Pond. Fig. 5.6: Courtesy of C. Noel, University of Laval, Quebec, Canada.

Chapter 10
Fig. 10.4: Courtesy of P. W. Moe. Fig. 10.5: Courtesy of P. W. Moe, USDA, ARS, Beltsville, MD.

Chapter 11
Fig. 11.2: Courtesy J. W. Thomas, Michigan State University. Fig. 11.3: Courtesy of L. Krook, Cornell University. Fig. 11.5: Courtesy L. Krook and Cornell Veterinarian. Fig. 11.7: Courtesy R. B. Becker, Florida Agricultural Experiment Station. Fig. 11.8: Courtesy R. B. Becker, Florida Agricultural Experiment Station.

Chapter 12
Fig. 12.2: Photo courtesy of W. M. Hawkins, Montana State University. Fig. 12.5: Courtesy of W. M. Beeson. Fig. 12.8: By permission of G. L. McClymont.

Chapter 13
Fig. 13.1: University of Tennessee photograph.

Chapter 15
Fig. 15.1: Courtesy of Poultry Science Department Cornell University. Fig. 15.2: Courtesy of Chas Pfizer Co. Fig. 15.3: Courtesy of W. M. Beeson.

Chapter 16
Fig. 16.6: Photo courtesy of Elanco Animal Health.

Chapter 17
Fig. 17.1: Courtesy of P.J. Wangsness, Pennsylvania State University.

Chapter 19
Fig. 19.2: Courtesy of John Deere. Fig. 19.3: Courtesy of J.I. Case Implement Co. Fig. 19.4: Courtesy of John Deere. Fig. 19.5: Courtesy of BEEF Magazine. Fig. 19.6: Courtesy of Engineered Storage Products Company. Fig. 19.8: Photo courtesy of R. W. Henderson. Fig. 19.9: Courtesy of Massey Ferguson, N.A. Fig. 19.10: Photo by R.W. Henderson.

Chapter 20
Fig. 20.1: Courtesy of W.H. Hale, University of Arizona. Fig. 20.3: Courtesy of John Deere.

Chapter 23
Fig. 23.5: Corbis Digital Stock.

Index

Abomasum, 32, 37, 44, 436
Absorption, 25, 58
Acclimatization, 159
Acetate metabolism, 419
Acetyl-CoA carboxylase, 281, 282
Achromotrichia, 204
Acid-base balance, 178
Acid detergent extraction, 20
Acid detergent fiber (ADF), 20
Acidic amino acids, 116
Acid preservation of high-moisture grain, 371, 372
ACTH, *see* Adrenocorticotrophic hormone
Actin, 121
Activated carbon, 364
Active transport, 41, 78, 79
Additives:
    in feedstuffs, 355–365
    and lactation, 428
    nonnutritive, 354–355
    in poultry diet, 487
Adenine, 128
Adenosine diphosphate (ADP), 81, 82
Adenosine triphosphate (ATP), 75, 81–83, 85
ADF16 pellet, 554
ADF25 pellet, 555
ADF (acid detergent fiber), 20
Adipose tissue, 101, 106, 107, 281, 282, 287
ADP, *see* Adenosine diphosphate
Adrenocorticotrophic hormone (ACTH), 85, 283
Aflatoxins, 225
Ag, *see* Silver
Age:
    and basal metabolism, 156
    and body composition, 10, 11
    at calving, 431
    and mineral content, 164
    at onset of puberty, 396–397
Agricultural Research Council (ARC), 307
AHP (alkaline hydrogen peroxide), 379

AIN (American Institute of Nutrition), 229
Air-dried feedstuffs, 8
Al (aluminum), 212
Alanine, 115, 119
Albumin, 119, 121, 138, 192
Alfalfa, 325–328, 331, 333–335, 351, 378, 503, 506
Alfalfa hay, 8, 9, 37, 46
Algae, 351
Algebraic method, 383, 384, 386, 388, 392
Aliphatic amino acids, 115
Alkali disease, 210
Alkaline hydrogen peroxide (AHP), 379
Alopecia, 204
Alpha 1-4 linkage, 76
α-globulin, 121
α-receptors, 283
Alpha-lactalbumin synthesis, 415
ALS (amyotrophic lateral schlerosis), 505
Alsike clover, 503
Aluminum (Al), 212, 218, 223–224
Alveoli, mammary, 415, 417
Amaranth, 342–343
American Association of Feed Control Officials, 346
American Institute of Nutrition (AIN), 229
American Society of Animal Science (ASAS), 362
Amino acids, 113–142
    absorption of, 124
    after absorption, 126–129
    amino acid antagonism/toxicity/ imbalance, 136–137
    biological availability of, 139–141
    in dairy-cow diet, 427
    D-amino acids used by nonruminants, 133, 134
    deamination/transamination, 130
    degradation sites, 125
    in dog diet, 518

essential/nonessential, 117–120
fate of, after absorption, 126–129
in fish diet, 535, 536
functions of, 120–123
and gluconeogenesis, 82, 83
and lactation, 418
metabolism of, 123–132
and nitric oxide role, 126
nitrogen cycling in intestine, 125–126
nonprotein nitrogen, 135–136
in poultry diet, 480–481
requirements for, 132, 134, 135
roles of, in animals, 120
ruminant nitrogen metabolism, 126
synthesis/degradation of, 127–129
synthesis sites, 124–125
in turkey diet, 495, 496
urea cycle, 130–132
Amino acid analysis, 21
Amino acid antagonism, 136–137
Amino acid degradation, 125
Amino acid synthesis, 124–125
Amino acid toxicity, 136–137
Ammonia, 36, 130, 134, 135, 175
Ammonia toxicity, 137
Amylase, 38, 39
Amylopectin, 77–78
Amylose, 77–78
Amyotrophic lateral schlerosis (ALS), 505
Anabolism, protein, 123–124
Anaerobes, obligate, 33
Analytical methods, 15–23
    amino acid analysis, 21
    atomic absorption spectrophotometry, 21–22
    automated equipment for, 22–23
    bomb calorimetry, 20–21
    chemical, 16–17
    detergent extraction, 20
    dry matter determination, 17–18
    gas-liquid chromatography, 22
    microbiological, 17
    on-farm feed analysis, 23

Analytical methods (*Continued*)
  pH of feedstuffs, 20
  proximate analysis, 18–19
  sampling for analysis, 16
Analytical Ultracentrifuge, 21
Androgen, 278, 284, 285
Anemia, 4, 45, 46, 200, 203, 204, 207, 441, 483
Angora goats, 453, 456, 457
Animal body composition, 9–11
Animal nutrition, 5–13
  and animal products in human nutrition, 5
  carbohydrates, 8
  and chemical composition of whole plants, 8, 9
  and composition of animal bodies, 9–11
  and composition of animal feed, 6–9
  and crop production, 11–12
  lipids, 7, 8
  minerals, 8
  organic compounds, 7, 8
  proteins, 7, 8
  specialties in, 3
  vitamins, 8
  water, 6
Animal production, 11–12
Animal productivity, 317–318
Animal products (in feedstuffs), 425, 475, 521
Animal protein factor (APF), 262
Anorexia, 233, 253
Antagonism, amino acid, 136–137
Anthelmintics, 364
Antibiotics, 361–362, 487
Antibodies, immune, 122, 241, 466
Antimetabolites, 253*n*.
Antimicrobial drugs, 361–362
Antimony (Sb), 218, 224
Antioxidants, 364
APF (animal protein factor), 262
Apparent digestibility, 52, 54–56
Apparent digestible energy, 148
Appetite, 287, 292, 294, 296–300
Aquaculture, 11, 531. *See also* Fish
Aquatic animals, 550, 551
Arachidonic acid, 97, 524
AR (atrophic rhinitis), 168
Arboreal animals, 550, 551
ARC (Agricultural Research Council), 307
Arginase, 130–132, 518
Arginine, 116–119, 123, 136, 517
Argininosuccinate lyase (AS), 130
Argininosuccinate synthetase (ASS), 131, 132

Aromatic amino acids, 115
Arsenic (As), 209, 211, 212, 218, 224
Artificially dried forage, 330, 331
As, *see* Arsenic
AS (argininosuccinate lyase), 130
ASAS (American Society of Animal Science), 362
Ascorbic acid, 268–270. *See also* Vitamin C
Ash, 19, 53, 169
Asparagine, 116, 118
Aspartic acid, 116, 117, 119
Aspartic transaminase, 261
ASS, *see* Argininosuccinate synthetase
Associative effect, 54
Ataxia, 203, 206, 232
Atherosclerosis, 108–109
Atlantic salmon, 535, 538, 539, 544, 545
Atomic absorption spectrophotometry, 21–22
ATP, *see* Adenosine triphosphate
Atrophic rhinitis (AR), 168
Au (gold), 224
Autocrine factors, 279
Automated analytical equipment, 22–23
Available nutrients, 382
Avian species, 31, 480. *See also* Poultry
Ayu (fish), 537

B, *see* Boron
Ba, *see* Barium
Backgrounding programs, 408–409
Bacteria, 33–34, 351, 551, 552
Bakery product, 346
Balance trials, 55, 56
Baled dry forage and roughage, 377
Barium (Ba), 213, 218, 224
Barley, 339, 341, 342, 374, 503
Barley straw, 298, 299
Basal metabolic rate (BMR), 191
Basal metabolism, 155–157
Basic amino acids, 116
Bass (fish), 534, 535, 537, 544
Batch mixers, 376–377
Batch trials, 57
BCAA (branched-chain amino acids), 125
Beans, 350–351
Be (beryllium), 224
Beef cattle, 395–410
  body conditioning scoring of, 398, 399
  cow-calf enterprise, *see* Cow-calf enterprise

finishing programs, 409
stocker/backgrounding/growing programs, 408–409
Beet, 346
Beet pulp, 503
Beriberi, 253
Beryllium (Be), 224
β-adrenergic agonists, 283, 284, 363–364
β-carotene, 230, 234
β-globulin, 121
β-receptors, 283
Bi, *see* Bismuth
"Big head" disease, 168
Bile, 31, 39–40, 43, 98, 105
Bilirubin, 40
Bioenergetics, *see* Energy metabolism
Biological value (BV), 137–138
Biological variation in digestion/absorption, 58
Biotin, 266–267, 316, 541
Birds, 31, 480. *See also* Poultry
Birth weights, 400–402
Bismuth (Bi), 218, 224
Bladder wall, thickened, 233
Blind staggers, 210
Bloat, 36, 37, 328
Blood:
  composition of, 44–45
  glucose in the, 81
  and iron, 196, 197
  and nutrition, 44–46
  proteins in the, 121
  role of, in nutrient transport, 42–43
  and water absorption, 64
Blood clotting, 170, 245–247
Blood meal, 348, 426, 427
BMR (basal metabolic rate), 191
Boars, 470–471
Body condition score:
  for beef cattle, 398, 399, 401, 402, 405, 406
  for horses, 506, 507
  for sheep, 443–444
Body fat, 107, 405, 406
Body size:
  of breeding heifers, 431
  of cattle, 10
  and GI tract capacity, 33
  and heat production, 154–155
Body temperature, 63, 154, 158–159
Body tissues, 125
Body water, 63–64
Body weight (BW), 155
  of beef cows, 405–406
  and feed intake, 301, 304–305

of horses, 504, 505
and metabolizable energy, 157, 158
at onset of puberty, 397
of pigs, 10
of turkeys, 494
Bomb calorimetry, 20–21
Bone analyses, 318
Bone ash, 169
Bone density scan, 169
Bone formation:
 calcium in, 167
 in cats, 525, 526
 and copper, 201, 203–204
 and estrogen, 284
 and fluoride, 222
 in horses, 509–510
 and manganese, 206
 manganese in, 205
 and regulation of nutrient
  partitioning, 280, 281
 vitamin A in, 232
 vitamin D in, 238–240
 zinc in, 195
Bone meal, 348
Bone remodeling, 165–166
Boron (B), 186, 218, 224
Bos indicus cattle, 156, 314, 397, 408
Bos taurus cattle, 156, 314, 397, 408
Bovine social psychology, 427
Bovine somatotropin (bST), 415, 428
Bovine spongioform encephalopathy
 (BSE), 123, 348
Br, *see* Bromine
Bradycardia, 253
Bran, 343, 344
Branched-chain amino acids
 (BCAA), 125
Breed differences, 314
Breeding period, swine, 470–471
Breeding season, 396
British Thermal Units (BTUs), 141
Broilers:
 dietary formulation for, 491–494
 effect of energy level on
  performance of, 298
 feed intake expectations for, 305
 water consumption in, 69
Brome, 325
Bromine (Br), 213, 218, 224
Brood mares, 507–508
Brook trout, 545
Brown trout, 545
Browsers, 553
BSE, *see* Bovine spongioform
 encephalopathy
BST, *see* Bovine somatotropin

BTUs (British Thermal Units), 141
Buckwheat, 342
Buffer systems, 178
Bulk and roughage feeders, 553
Bulls, 407–408
Bunker silo, 333
Bunk space, cow, 435
BV, *see* Biological value
BW, *see* Body weight
Bypass proteins, 135
Byproducts:
 high-fiber feeds from, 424
 milling/distillery, 343–344
 from poultry, 348
 in poultry diets, 488

Ca, *see* Calcium
Cabbage family, 328
Ca-binding protein (CaBP), 165
Cadmium (Cd), 186, 212–213, 218,
 220–221, 224
Cage fatigue, 171
Cal, *see* Calories
Calcitonin, 169, 170, 239, 240
Calcium (Ca), 165–171
 in animals, 164–165
 deficiency signs for, 167–169, 316
 in feedstuffs, 8, 9
 in fish diet, 538
 functions of, 165–167
 in laying-hen diet, 491
 and lead absorption, 220
 maximum tolerable levels of, 218
 in sheep diet, 441
 sources of, 354
 in swine diet, 463–464
 tissue distribution of, 165
 toxicity, 170–171
 and vitamin D, 236, 238, 240
Calcium tetany, 169–170
California Experiment Station, 161
Calories (cal), 132, 141, 297–300
Calorimetry, 20–21, 152–153
Calves. *See also* Cow-calf enterprise
 body composition of, 9
 iron in, 197
 and nonprotein nitrogen, 136
 nutrition requirements for,
  430–431
 veal, 200
 water consumption in, 68, 70
Calving intervals, 414
Camels, 69, 70
CAMP, *see* Cyclic adenosine
 monophosphate
Cannulas, 57–59
Cannulated trachea, 153

Canola meal, 426
Carbohydrates, 8, 73–90
 abnormal metabolism of, 85–86
 absorption/transport of
  monosaccharides, 78–81
 in cat diet, 525
 classification and structure of,
  74–75
 deficiencies in, 316
 and diabetes mellitus, 85–87
 digestive enzymes and utilization
  of, 81
 in dog diet, 519–520
 effects of, on lipid metabolism, 107
 energetics of glucose catabolism,
  84–85
 in fish diet, 536, 537
 in food and feed, 88, 89
 functions of, 75–77
 glucogenesis, 82–84
 metabolic conversions, 81–82
 metabolism of, 76–85
 microbial fermentation of
  cellulose/plant fibers, 78–80
 plant cell walls and digestion of, 78
 and plant fibers as energy sources,
  87–88
 in poultry diet, 481
 preparation for absorption, 76–78
 in rumen fermentation, 34–36
 in urine, 47
Carbon—nitrogen balance, 153–154
Carboxypeptidase, 38, 39
Carnitine, 512
Carnivores:
 digestive tract of, 27
 fish as, 532
 tooth function in, 29
 in zoos, 556–567
Carnosine, 512
Carotenoids, 230–235
Carp, 532–535, 537, 540–544
Cartilage, 117
Caseins, 416, 430
Cashmere, 453
Cassava, 345
Cats, 515–516, 522–528
 carbohydrates, 525
 commercial foods for, 527
 dietary habits of, 516
 digestive tract of, 33
 energy, 523
 essential amino acids for, 118
 fat and fatty acids, 524, 525
 and fiber, 27
 guidelines for feeding, 527, 528
 heat production in, 155

Cats (*Continued*)
  mineral content in body of, 164
  mineral requirements for, 525–526
  potential toxins, 527
  protein and amino acids, 523, 524
  spending on food for, 515
  sugar absorption by, 80
  vitamins, 526–527
  water requirements for, 523
Catabolism, 84–85, 123–124
Cataracts, 254
Catch-up growth, 408
Catecholamines, 278, 280, 283
Catfish, 532–535, 537–545
Cattle. *See also* Cows
  beef, *see* Beef cattle
  body composition of, 10
  calcium excretion in, 167
  dairy, *see* Dairy cattle
  digestibility of alfalfa hay by, 46
  effect of monensin/lasalocid on, 362
  feed additives approved for, 356
  feed and growth relationship in, 50
  GI tract capacity of, 33
  heat increment of feeding for, 151
  ketosis in, 85
  magnesium in, 176
  mineral tolerances for, 218–219
  relative value of grains for, 341
  roughage consumption expectations for, 303–304
  water consumption in, 66–69
CCK, *see* Cholecystokinin
Cd, *see* Cadmium
Cecal digesters, 27
Cecum, 26–33, 44
Cell contents, 87
Cell membranes, 114, 117
Cell permeability, 242
Cellular metabolism, 166
Cellulose, 76–80, 87, 295–296
Cell walls, 78, 324
Cell-wall carbohydrate, 420
Cell wall contents (CWC), 20, 87
Central nervous system (CNS), 295, 296
Cereal grains, 338–344
  amaranth, 342–343
  barley, 342
  buckwheat, 342
  composition of, 338–339
  corn, 339–341
  as feed, 11–12
  and lactating cows, 424
  milling/distillery byproducts of, 343–344

oats, 342
  relative value of, 341
  rumen protein degradability of, 425
  sorghum, 340
  triticale, 342
  wheat, 340–342
Ceruloplasmin, 201
Cesium (Cs), 213, 224
CETP (cholesterol transport proteins), 116
Chaff, 331
Channel catfish, 534, 535, 537, 539–545
Chastek paralysis, 253
Chemical analysis, 16–17
Chemical treatments (of feedstuffs), 378–379
Chernobyl, 224
Chicks:
  on carbohydrate-free diets, 76
  curled-toe paralysis in, 254, 255
  in feed analysis, 16
  and selenium, 210
Chickens:
  body composition of, 9
  broilers, *see* Broilers
  digestive tract of, 31
  feed additives approved for, 356–357
  knowledge of nutrient requirements for, 4
  laying hens, *see* Laying hens
  relative value of grains for, 341
Chinese panda, 28, 29
Chinook salmon, 535, 538, 539
Chlorine (Cl), 176–181, 317, 464, 538
Cholecystokinin (CCK), 287, 297
Cholesterol, 96, 98–101, 104, 105
Cholesterol transport proteins (CETP), 116
Choline, 206, 244, 267–268, 316, 541
Chondroitin sulfate, 117
Chopped forages and roughages, 377–378
Chromic oxide, 53
Chromium (Cr), 186, 211, 219, 224, 317
Chronic wasting disease (CWD), 123
Chum salmon, 537
Chylomicrons, 99, 234
Chymotrypsin, 38, 39
Ciliate protozoa, 33–35
CIT (conjunctival impressin cytology) test, 231
Citric acid cycle, 82, 83, 176
Citrulline, 118

Citrus pulp, 346
CJD (Creutzfeldt-Jakob disease), 123
Cl, *see* Chlorine
CLA, *see* Conjugated linoleic acid
Clinoptilolite, 364
Closed-circuit calorimeters, 153
Clover, 328, 503
CNS, *see* Central nervous system
Co, *see* Cobalt
Cobalamin, 37, 316
Cobalt (Co):
  deficiency signs for, 317
  maximum tolerable levels of, 219
  requirements of, 186–189
  in ruminant diet, 37
  in sheep diet, 442
  and vitamin B$_{12}$, 263–265
Coccidiostat, 430
Coconut meal, 350
Coenzyme A (CoA), 258–259
Coenzyme R, *see* Biotin
Coenzymes, 229, 251
Coho salmon, 535, 537
Cold processing, 370–372
Collagen, 120, 204
Colon, 27, 29, 31–33, 480
Colonic digesters, 27
Colostrum, 39, 42, 311, 430, 449, 450, 457, 466
Commercial pet foods, 521, 522, 527
Common carp, 534, 535, 537, 540, 541, 543
Comparative slaughter, 154
Compensatory gain, 408
Compensatory growth, 312
Compound lipids, 91
Computer software programs, 436
Concentrate selectors, 553
Conception, delayed, 436
Conjugated linoleic acid (CLA), 92–94, 419
Conjunctival impressin cytology (CIT) test, 231
Continuous fermenters, 57
Continuous flow mixers, 377
Contractile proteins, 120, 121
Control mechanisms (of feed intake), 294–300
Cool-season grasses, 327, 328
Copper (Cu), 186, 201–205
  deficiency signs for, 203–204, 317
  in fish diet, 538
  functions of, 201–203
  in horse diet, 510
  and iron, 196
  maximum tolerable levels of, 219
  metabolism of, 202, 203

and molybdenum, 222
pollution potential of, 224
in sheep diet, 441
tissue distribution of, 201
toxicity, 204–205
and zinc, 196
Copra meal, 350
Coprophagy, 29, 125, 267
Corn, 8, 9
 amino acid composition of, 119
 amino acid deficiencies in, 133, 134
 composition of, 339
 effects of, particle size on lactation performance, 374
 as high-energy feedstuff, 339–341, 344
 as horse feed, 503
 molasses from, 344
 relative value of, 341
 and rumen protein, 426
 and soybean mixture, 137
Corn gluten meal, 343, 344
Corn picker, 341
Corn silage, 333, 336–337
Cortisol, 278, 283, 284, 286
Cottonseed meal (CSM), 349–350
Cows, 398–407. See also Calves; Heifers
 body weight and composition change, 405–406
 gestation, 400–402
 heat production in, 155
 lactation, see Cow lactation
 magnesium in, 175
 maintenance, 404–405
 and nonprotein nitrogen, 136
 postpartum reproduction, 398–400
 preweaning growth, 406–407
Cow-calf enterprise, 396–408
 bulls, 407–408
 cow, 398–407
 replacement heifers, 396–398
Cow lactation, 415–429
 in beef cows, 402–404
 diets for high-milk production, 428–429
 energy nutrition, 420–425
 feed additives/metabolic modifiers, 428
 in life cycle, 414–415
 mammary biology, 415–417
 metabolism during, 416–420
 mineral/vitamin nutrition, 427
 nutrient requirements, 420, 421
 protein nutrition, 425–427
Cr, see Chromium

CRABP (cytosolic retinoic acid-binding protein), 231
Crambe meal, 350
CRBP (cytosolic retinol-binding protein), 231
Creep feeding, 407, 450, 466, 509
Cretinism, 191
Creutzfeldt-Jakob disease (CJD), 123
Crops (bird anatomy), 31, 43
Crop production, 11–12
Crude fiber, 19, 323
Cs, see Cesium
CSM, see Cottonseed meal
Cu, see Copper
Cubed roughages, 378
Curled-toe paralysis, 254, 255
Cushing's Syndrome, 283
CWC, see Cell wall contents
CWD (chronic wasting disease), 123
Cyanocobalamin, 262, 263
Cyclic adenosine monophosphate (cAMP), 104, 364
Cysteine, 115, 441, 480
Cystine, 115, 119
Cystitis, 233, 503
Cytosine, 128
Cytosolic retinoic acid-binding protein (CRABP), 231
Cytosolic retinol-binding protein (CRBP), 231

Dairy cattle, 413–436
 calf nutrition, 430–431
 dry-cow nutrition, 429–430
 feeding management, 433–436
 feed processing for, 376
 heifer nutrition, 431–433
 lactation, see Cow lactation
 life cycle, 414–415
 nutritional evaluation, 435–436
 nutritional goals, 415, 416
 problems related to dairy nutrition, 436
 rumen fermentation in, 36
 water restriction in, 70
Dairy goats, 457
Dairy Herd Improvement Association (DHIA), 435
D-amino acids, 118, 133, 134, 136
DCAD, see Dietary cation—anion difference
DE, see Digestible energy
Deamination, 130
Deficiency signs/symptoms:
 biotin, 266–267
 calcium, 167–170
 choline, 268

cobalt, 189
copper, 203–204
folacin, 266
iron, 199, 200
magnesium, 175–176
manganese, 206–207
niacin, 256–257
and nutritional evaluation, 315–319
pantothenic acid, 259–260
phosphorous, 173–174
potassium/sodium/chlorine, 180
in poultry, 483–485
riboflavin, 254–255
sulfur, 181–182
thiamin, 253
vitamin A, 231–233
vitamin B_6, 261–262
vitamin B_{12}, 264–265
vitamin C, 269–270
vitamin D, 238–240
vitamin E, 242
vitamin K, 246
zinc, 195
Dehydration:
 animal, 69, 70
 of herbage, 330, 331
Delayed conception, 436
Deoxyribonucleic acid (DNA), 127, 128
Deposition, lipid, 99–101
Depot fat, 100, 101
Derangement of cell permeability, 242
Derived lipids, 91
Dermatitis, 97
DES (diethylstilbestrol), 363
Detergent extraction, 20
Developing countries, 4
Developmental orthopedic disease (DOD), 509–510
DHIA (Dairy Herd Improvement Association), 435
Diabetes mellitus, 85–87
Dicoumarol, 247
Diet(s):
 defined, 6
 for fish, 542, 544–545
 purified, 56, 57
 as water consumption factor, 67–68
Dietary cation—anion difference (DCAD), 178–179, 429–430
Dietary habits:
 of cats, 516
 of dogs, 516
 of fish, 532–533
Dietary intake protein (DIP), 308

Diet formulation, 2–3, 381–393
    algebraic method of, 383, 392
    and energy availability, 422–423
    and feed intake, 422
    feedstuffs/nutrient composition,
        382
    for high-milk production, 428–429
    mechanics of, 383–391
    mechanics of working with
        percents, 382–383
    and nutrient partitioning, 423
    nutrient requirements of animal,
        382
    for swine, 462
Diethylstilbestrol (DES), 363
Diffusion, 41
Digestibility, 46–47
    apparent vs. true, 54–56
    associative effects of, 54
    by difference, 53–54
    and energy utilization efficiency,
        160–161
    of lipids, 97
    variations in, by species, 58
Digestible energy (DE), 147–149,
    161, 298
Digestible protein (DP), 308
Digestion, 25, 58
Digestion trials, 52–53
Digestive enzymes, 81
Digestive tract, see Gastrointestinal
    tract
Diglycerides, 94–96
Dilutent, water as, 62
DIP (dietary intake protein), 308
Disaccharidases, 74, 75
Disease. See also specific diseases
    and nutrition, 315
    prion, 122–123
    and productivity, 315
    of sheep/goats, 457–458
    urinary indicators of, 47
Dispensible amino acids, see
    Nonessential amino acids
Displaced abomasums, 436
Distillery byproducts, 343–344
DM, see Dry matter
DNA, see Deoxyribonucleic acid
DOD, see Developmental orthopedic
    disease
Dogs, 515–522
    carbohydrates, 519–520
    commercial foods for, 521, 522
    dietary habits of, 516
    digestive tract of, 26, 30, 33
    energy, 516–517
    essential amino acids for, 118

fat and fatty acids, 519
and fiber, 27
guidelines for feeding, 522
heat production in, 155
iron in, 197
magnesium in, 175
mineral content in body of, 164
mineral requirements for, 520–521
in nutritional science, 2
osteoporosis in, 170
protein and amino acids, 517–519
regulation of drinking in, 66
spending on food for, 515
vitamins, 521
water requirements for, 516
Dominance hierarchy, 552
DP (digestible protein), 308
Dried bakery product, 346
Dried beet, 346
"Dried-off," 414
Drinking water, 211
Dry-cow nutrition, 429–430
Dry forages, 322, 329–332, 377
Dry matter determination, 17–18
Dry matter (DM), 53, 298, 308, 434
Dry roughage, 377–379
Ducks, 357, 496–497
Duodenum, 28, 30, 32, 78, 98
Dwarfism, 195

EBDC (ethylene carbamates), 192
ED, see Exudative diathesis
EDM (equine degenerative
    myeloencephalopathy), 505
EDTA, see Ethylene diamine
    tetraacetic acid
Eels, 535, 537, 538
EFA (essential fatty acids), 97
Effective NDF, 422
Efficiency, feed, 50
Efficiency of energy utilization,
    159–161
EGF (epidermal growth factor), 129
Egg albumin, 138
Egg mass, 489
Egg protein, 137
EGR (erythrocyte glutathione
    reductase), 254
Elastin, 120
Electrolytes, 177, 430
Electronic feeding devices, 51
Elephants, 155
ELISA (enzyme-linked
    immunosorbent assay), 23
Embden-Meyerhof pathway, 83
EMND (equine motor neuron
    disease), 505

Encephalomalacia, 242
Endocrine system, 277, 280
Endopeptidases, 122
Energetics of glucose catabolism,
    84–85
Energy:
    in cat diet, 523
    deficiencies in, 316
    defined, 146
    diet composition and availability
        of, 422–423
    in dog diet, 516–517
    environment and expenditure of,
        158–159
    fatty acids as source of, 97
    and feed consumption, 292
    in fish diet, 533–535
    in goat diet, 440
    in poultry diet, 483–486
    as requirement in diet, 2
    in sheep diet, 440
    in swine diet, 466, 475
Energy concentrates, 323
Energy metabolism, 145–162
    basal metabolism, 155–157
    digestible energy, 147–149
    and efficiency of energy utilization,
        159–161
    and environment, 158–159
    gross energy, 146–148
    heat production measurement,
        152–154
    lactation, 420–425
    maintenance, 157–158
    metabolizable energy, 149–150
    net energy, 150–152
    relationship of heat production to
        body size/weight, 154–155
    terminology used in ration
        formulation and feeding
        standards, 161
    total digestible nutrients, 149
Enteritis, 233
Enterokinase, 39
Enterotoxemia type D (overeating
    disease), 458
Environment:
    and beef cow maintenance,
        404–405
    and birth weight, 400
    and dairy cows, 435
    and energy expenditure, 158–159
    and mineral toxicities, 224
    as nutritional factor, 314–315
    trace minerals in the, 186–188
    as water consumption factor, 68
Environmental Protection Agency, 70

Enzymes, 121, 122
  and carbohydrate absorption,
    77–78
  and copper, 201
  digestive, 37–40
  and iron, 199, 200
  and selenium, 207, 208
  and utilization of carbohydrates,
    81
  and vitamin E, 241
  and zinc, 192
Enzyme-linked immunosorbent
  assay (ELISA), 23
Epidermal growth factor (EGF), 129
Epinephrine, 283
Epithelium, 41–43, 124
Equine degenerative
  myeloencephalopathy (EDM), 505
Equine motor neuron disease
  (EMND), 505
Ergogenic aids, 512
Ergosterol, 235–236
Eruction, 37
Erythrocyte glutathione reductase
  (EGR), 254
Erythropoietin, 120
Escape proteins, 126, 135
Esophagus, 28, 37, 43
Essential amino acids, 117–120
Essential fatty acids (EFA), 97, 316,
  536, 537
Esters, 94
Estradiol, 284
Estrogen, 278, 284, 285
Estuary grouper, 535
Ether extract, 19
Ethylene carbamates (EBDC), 192
Ethylene diamine tetraacetic acid
  (EDTA), 170, 193–195
Evaluating nutritional status of
  animals, 315–319
Ewes, 442–449
  flushing, 444, 445
  gestation, 444–449
  ketosis in, 85
  maintenance, 445
  milk production, 449
  replacement, 452
Examinations, physical, 318
Excretion, 43, 47
  calcium, 167
  copper, 203
  iron, 199
  manganese, 206
  phosphorus, 173
  potassium/sodium/chlorine,
    178–179

selenium, 209
Exercise, 136, 157, 169
Exogenous N, 55
Exopeptidases, 122
Exophthalmia, 232
External indicators, 53
Extruded feed, 504
Extruding grains, 373
Exudative diathesis (ED), 207, 209,
  210, 242, 243

F, see Fluorine
FA, see Fatty acids
Factorial method, 309
FAD (flavin adenine dinucleotide),
  254
Falling disease, 204
Fast, see Feast and fast cycle
Fat(s) and fatty acids:
  analysis of, 19
  in beef cows, 405, 406
  bile action on, 40
  in cat diet, 524, 525
  in dog diet, 519
  as feedstuff, 346–347
  in fish diet, 536, 537
  in human diets, 76
  and lactating cows, 424–425
  and metabolic water, 62
  in rumen fermentation, 36
  storage of, 76
  and water consumption, 68
Fat-corrected milk (FCM), 416
Fat cow syndrome, 401, 402
Fat-soluble vitamins, 229–248
  deficiencies in, 316
  and fatty acids, 98
  sources of, 354
  vitamin A, 230–235
  vitamin D, 235–240
  vitamin E, 241–245
  vitamin K, 245–247
Fattening function, 280–285, 311–312
Fatty acids (FA). See also Fat(s) and
  fatty acids
  absorption of, 98–99
  biosynthesis of, 102, 103
  catabolism of, 104, 105
  in chick diets, 76
  deficiencies in, 316
  and lactation, 418–419
  structure of, 92, 93
  transport/deposition of, 99–101
  and triglyceride metabolism,
    101–106
Fatty acid synthase, 281, 282
FCM (fat-corrected milk), 416

FDA, see Food and Drug
  Administration
FDM (fecal dry matter), 56
FE, see Feces energy
Fe, see Iron
Feast and fast cycle, 275–276
Feather meal, 348
Fecal analysis, 140
Fecal collection, 54
Fecal dry matter (FDM), 56
Fecal excretion, 47
Fecal metabolic N, 55
Fecal N (FN), 55, 56
Feces, 29, 66
Feces energy (FE), 147–149
Feed:
  carbohydrates in, 88, 89
  composition of animal, 6–9
  defined, 6
  percentage composition of, 9
  trends in use of cereal as, 12
The Feed Additive Compendium, 355
Feed additives, see Additives
Feed consumption, 291–306
  and appetite, 292, 294
  and chemostatic regulation of
    appetite, 296–297
  and control mechanisms of feed
    intake, 294–301
  and energy required to produce
    gain, 292
  and inhibition of feed intake,
    300–305
  long-term/short-term control of,
    296
  measurement apparatus for, 295
  and odor, 294
  and palatability, 292
  patterns of, 295–296
  and sight of food, 294
  and taste, 292–294
  and texture-physical factors, 294
Feed efficiency, 50
Feeding management:
  for dairy cattle, 433–436
  for fish, 542, 544–545
  for zoo animals, 550–551
Feeding standards, 161, 307–310
Feeding trials, 50–57, 309
  apparent vs. true digestibility,
    54–56
  associative effects, 54
  balance, 55, 56
  digestibility by difference, 53–54
  digestion, 52–53
  growth, 50–52
  purified diet, 56, 57

Feed intake:
  and body weight, 301
  for cattle/sheep/swine/poultry/
    fishes, 303–305
  control mechanisms of, 294–300
  and dairy cattle, 422
  inhibition of, 300–305
  and milk production, 303
  predicting, 301, 302
  regulation of, 286–287
Feed mixing, 376–377
Feed preparation and processing,
  369–379
  cold processing, 370–372
  dry roughage and roughage
    processing for ruminants,
    377–379
  economics of, 375–376
  feed mixing, 376–377
  grain processing, 370, 371
  hot processing, 372–373
  for nonruminants, 373–375
  for ruminants, 375–376
Feedstuffs, 321–365
  additives in, 355–365
  antimicrobial drugs in, 361–362
  β-adrenergic agonists in, 363–364
  for cats, 527
  classification of, 322–323
  comparison of, 8, 9
  composition of, 6–9
  defined, 6, 322
  and diet formulation, 382
  for dogs, 521, 522
  dried bakery product, 346
  dried beet and citrus pulp, 346
  fats and oils, 346–347
  forages and roughages, 323–326
  for goats, 442
  harvested dry forages, 329–332
  harvested high-moisture forages,
    332–337
  high-energy, 338–344
  hormones in, 363
  labeling of, 355
  liquid energy sources, 344–345
  medicated feeds, 355
  mineral supplements, 352–353
  nonnutritive feed additives,
    354–355
  nonprotein nitrogen, 352
  pasture and grazed forages,
    326–329
  pH of, 20
  plant protein concentrates,
    348–352
  for poultry, 486–487

probiotics in, 363
protein concentrates, 347–348
quality of, 434–435
root crops, 345
for sheep, 442
for swine, 474–476
tubers, 345–346
vitamin sources, 353–354
water content of, 65
zeolites in, 364
Feed utilization measurement, see
  Measurement of feed and nutrient
  utilization
Fermentation:
  cellulose used in, 78
  in gastrointestinal tract, 27, 31
  microbial, of cellulose/plant
    fibers, 78–80
  rumen, 34–37, 57–58
  in silage, 334–335
Ferritin, 45, 198, 199
Fertilization, 325
Fertilizer, 11, 176
FFA, see Free fatty acids
FGF (fibroblast growth factor), 122
Fiber, plant:
  as energy sources, 87–88
  microbial fermentation of, 78–80
  and water absorption, 63
Fiber (in feedstuffs), 8, 9
  analysis of crude, 19
  and birds, 31
  dietary function of, 27
  for lactating cows, 423–425
  in ruminant diet, 37
Fiber-producing goats, 453–457
Fibrin, 45
Fibroblast growth factor (FGF), 122
Finishing programs, 409, 448,
  469–470
Fish, 531–545
  carbohydrates, 536, 537
  dietary habits/morphology of,
    532–533
  diets and feeding practices for,
    542, 544–545
  energy, 533–535
  and fats/fatty acids, 536, 537
  fatty acids in, 92, 109
  feed additives approved for, 358
  feed intake expectations for, 305
  juvenile to adult fish, 544
  land animals vs., 532
  larval fish, 544
  minerals, 537–539
  natural ingredient diets, 545
  nutrient requirements for, 542, 543

pigments of, 542
protein and amino acids, 535–536
sugar absorption by, 80
vitamins, 539–542
Fish meal, 348, 426
Fistulation, 58
Flagellated protozoa, 34
Flavin adenine dinucleotide (FAD),
  254
Flavin mononucleotide (FMN), 254
Flavoproteins, 251
Flavoring agents, 293, 294
Fluorine (F), 186, 211–212, 219,
  221–222, 317
Flushing, 400, 444, 445
FMN (flavin mononucleotide), 254
FN, see Fecal N
Foals, 508–510
Foam, 328
Folacin, 263, 265–266, 316, 541
Folic acid, 252, 265–266
Food, 6, 88, 89
Food and Drug Administration
  (FDA), 355–361, 363, 364, 441, 521
Foodstuff, see Feedstuffs
Forages, 323–326
  baled dry, 322
  chopped/ground, 377–378
  dry, 322
  high-moisture, 322
  for horses, 502–503
  for lactating cows, 424
  pelleted, 378
  and water intake, 65
Fossorial animals, 550, 551
Founder, 436
Free fatty acids (FFA), 98, 100
Frogs, 80
Froth, 37
Fructose, 80
Fumonsin, 503
Fungi, 34, 173, 225

GAG (glycosaminoglycan), 232
Galactopoiesis, 285, 286, 416
Galactose, 36, 78, 80
γ-globulin, 121
Gas-liquid chromatography (GLC), 22
Gas production, 36
Gastric inhibitory peptide (GIP), 287
Gastrin, 287
Gastrointestinal tract (GI tract),
  25–48
  anatomy and function of, 26–32
  of avian species, 31
  blood and lymph in nutrient
    transport, 42–43

comparative capacities of, 32–33
control of activity in, 40, 41
digestibility and partial digestion, 46–47
fecal and urinary excretion, 47
of fish, 532–533
of horses, 502
major actions in, 43–44
in nonruminant species, 29–31
of poultry, 480
and relationship of blood to nutrition, 44–46
rumen metabolism, 33–37
of ruminants, 31–32
secretions of, 37–40
and transport of nutrients, 40–42
variations in types of, 26–29
of zoo animals, 551, 552
GE, *see* Gross energy
Gels, 63
Generally recognized as safe (GRAS), 355
Gene-splicing techniques, 88
Genetically modified (GM) crops, 11
Genetics, 314
Gestation. *See also* Pregnancy period
and beef cow nutrition, 400–402
and dairy cow nutrition, 414
and diet, 76
and horse nutrition, 507–508
and iron transfer, 199
and sheep nutrition, 443–449
and vitamin A, 235
Gestational diabetes, 86
GH (growth hormone), *see* Somatotropin
Giant sea perch, 537
Gigantism, 281
Gilthead sea bream, 535
Gilts, 470
GIP (gastric inhibitory peptide), 287
GI tract, *see* Gastrointestinal tract
Gizzard, 31, 44, 480, 533
GLC (gas-liquid chromatography), 22
Globulin, 119, 121
Glucagon, 86, 132, 278–280
Glucocorticoids, 132, 283–286
Gluconeogenesis, 82–84, 232, 418
Glucose, 8, 76, 78–85
catabolism of, 84–85
and chemostatic regulation of appetite, 296–297
and manganese, 206
metabolism of, 279–280, 417, 418
and vanadium, 213
and zinc, 195

Glucose-4-β-glucoside, 74, 78
Glucose-6-phosphate, 83
Glucose-6-phosphate dehydrogenase, 282
Glucose tolerance, 86, 211
Glutamic acid, 116, 119
Glutamine, 116, 118, 123, 125
Glutathione peroxidase (GSH-Px), 207, 208, 210
Glutelin, 119
Glycans, 75, 76
Glycerol, 75, 93, 94
Glycine, 115, 119, 481
Glycogen, 81, 83
Glycogenesis, 81–82
Glycogenolysis, 81
Glycolytic pathway, 83, 85
Glycoproteins, 116, 117
Glycosaminoglycan (GAG), 232
GM (genetically modified) crops, 11
Goats, 439–442, 452–458
daily nutrient requirements of, 454–456
dairy goats, 457
digestibility of alfalfa hay by, 46
digestive tract of, 33
diseases of, 457–458
energy, 440
feed additives approved for, 358
feedstuffs used for, 442
fiber-producing, 453–457
guidelines for feeding, 457
magnesium in, 175
meat goats, 453–456
minerals, 440–442
nutrients required for, 440–443
protein, 440
vitamins, 442
water, 442, 443
Goiter, 189, 191, 192
Gold (Au), 224
Goldfish, 542
Gonads, 284
"Goose-stepping," 259
Gossypol, 350
Grab sampling, 53
Grains:
and calves, 430, 431
commercial, 504
for horses, 503–504
in poultry diet, 487
Grain amaranth, 342–343
Grain processing, 370, 371
Grain sorghum, 340
GRAS (generally recognized as safe), 355
Grasses, 326–328, 424, 425

Grass carp, 535
Grass—legume silages, 335–336
Grass staggers, 176
Grass tetany, 176
Grazed forages, 326–329
Grazers, 553
Green chop (soilage), 332, 334
"Green revolution," 11
GRF (growth hormone releasing factor), 281
Gross efficiency, 160
Gross energy (GE), 146–148, 151
Ground forages and roughages, 377–378
Grouper, 535
Growing period/programs:
for beef cattle, 408–409
feed intake required for, 311–312
metabolic priorities during, 276
regulation of nutrient partitioning during, 280–285
and sex steroids, 284–285
for swine, 467–470
Growth, 50, 406–407
Growth assay, 140, 141
Growth failure, 233
Growth hormone (GH), 285–286, 297. *See also* Somatotropin
Growth hormone releasing factor (GRF), 281
Growth rate, 254
Growth trials, 50–52
GSH-Px, *see* Glutathione peroxidase
Guanine, 128
Guinea pigs, 2, 46, 155, 164

Habitat, zoo, 550–551
Hair analyses, 318
Hair meal, 348
Hammermills, 370
Hamsters, 80
Harvested dry forages, 329–332
Harvested high-moisture forages, 332–337
Harvester equipment, 332, 342
Harvest methods, 325–326
Hay, 329–331, 334. *See also* Alfalfa hay
Hay balers, 329
Haylage, 336
HCl (hydrochloric acid), 39
HDL, *see* High-density lipoproteins
HE, *see* Heat production
Heat increment (HI), 150–152
Heat of combustion, 146
Heat of fermentation (HF), 150, 151
Heat production (HE), 152–155, 158–159

Heifers:
    growth promotants for, 285
    nutrition of dairy, 431–433
    replacement, 396–398
    water consumption in, 69
Hematopoiesis, 201
Heme Fe, 198–199
Hemicellulose, 78
Hemoglobin, 196–198, 200
Hemoglobinuria, 204
Hemosiderin, 197, 199, 200
Hepatic disease, 503
Hepatocyte, 194
Herbivores, 27–29, 551–556
Heterocyclic amino acids, 116
HF, see Heat of fermentation
Hg, see Mercury
HI, see Heat increment
High-density lipoproteins (HDL), 45,
    92, 100, 114
High-energy feedstuffs, 343–344. See
    also Cereal grains
High-fiber byproduct feeds, 424
High-moisture forages, 322, 332–337
High-moisture grain, 371
High-moisture silage, 334
High-performance liquid
    chromatography (HPLC), 21, 22
Hind gut fermentors, 27, 28
Hippuric acid, 132
Histidine, 116, 118, 119, 136, 195, 518
Histimine, 195
Holstein cattle, 414, 431–433
Homeorhesis, 276–277
Homeostasis, 276, 277
Homeotherms, 154
Hominy feed, 344
Hoof-and-mouth disease, 315
Hormones, 40, 121, 122, 278–279,
    363
Horses, 501–512
    amino acid synthesis in, 125
    body condition score system for,
        506, 507
    brood mares, 507–508
    calcium deficiency in, 168
    composition of, 9
    digestibility of alfalfa hay by, 46
    digestive physiology of, 502
    digestive tract of, 26, 31–33
    feed additives approved for, 358
    feed processing for, 375
    and fiber, 28
    foals, 508–510
    forages, 502–503
    grains, 503–504
    heat production in, 155

iron in, 197
magnesium in, 175
maintenance-level feeding,
    504–507
mineral tolerances for, 218–219
older horses, 512
performance horses, 510–512
stallions, 510
tooth function in, 29
water consumption in, 69
Hot processing, 372–373
HPLC, see High-performance liquid
    chromatography
Humans:
    calcium deficiency in, 169
    calcium excretion in, 167
    composition of, 9
    copper metabolism in, 202
    heat production in, 155
    and iron, 200
    iron in, 197
    magnesium in, 175
    mineral content in body of, 164,
        165
    nutrients required by, 7
    obesity in, 106–107
    phosphorus in, 171
    size of GI tract in, 32, 33
    stomach lining of, 30
    sugar absorption by, 80
    tooth function in, 29
Human nutrition, 4, 5
Hunger, 294, 296
Hybrid bass, 534
Hydrochloric acid (HCl), 39
Hydrolysis, 62
Hydroxy proline, 116, 119
Hypercalcemia, 170
Hypercholesterolemia, 257
Hyperlipidemia, 108
Hyperparathyroidism, 168–169
Hypertension, 180
Hypocalcemia, 169
Hypogonadism, 195
Hypomagnesemia, 176
Hypothalamus, 283, 286, 296

I, see Iodine
IDDM (insulin-dependent diabetes
    mellitus), 86
IDL (intermediate-density
    lipoproteins), 100
IE, see Intake energy (IE)
IF (intrinsic factor), 264
IGF, see Insulinlike growth factors
IGF binding proteins (IGFBP), 282
IGFBP (IGF binding proteins), 282

Immune antibodies, 122, 241, 466
Immune function, 123, 315
Immune system, 244, 430
Immunity, 236
Immunoglobulins, 430
Indirect calorimetry, 152–153
Indispensible amino acids, see
    Essential amino acids
Infections, 233
Infrared analysis, 22
Inhibition of feed intake, 300–305
Inorganic elements, see Minerals
Inositol (myoinositol), 270
Insectivores, 556–567
Insulin, 81, 86, 132, 211, 278–280,
    285, 286
Insulin-dependent diabetes mellitus
    (IDDM), 86
Insulinlike growth factors (IGF),
    86–87, 101, 122, 129, 130, 278,
    281–286, 288
Intake energy (IE), 147–149
Intermediate-density lipoproteins
    (IDL), 100
Intermediate feeders, 553
Internal indicators, 52–53
International Committee on
    Nomenclature, 229
International Feed Institute, 382
International Union of Nutritional
    Science Committee, 252
International units (IU), 308
Intestine, 125–126
Intrinsic factor (IF), 264
In vitro techniques, 57–58
In vivo procedures, 59
Iodine (I), 186, 189–192, 219, 441,
    464, 539
Iodine number, 96
Iodothyronine deiodinases, 208
Iodothyronines, 191
Ionophore antibiotics, 361–362, 428,
    504
Iron (Fe), 186, 196–201, 219
    deficiency signs for, 199, 200, 317
    in fish diet, 538
    metabolism of, 197–199
    in sheep diet, 441
    swine, 464
    tissue distribution/function of,
        196–197
    toxicity, 200–201
    and vitamin E, 244
    and zinc, 196
Iron regulatory proteins (IRP), 197
IRP (iron regulatory proteins), 197
Irrigation, 11

Isoalloxazine, 255
Isoleucine, 115, 119, 518
IU (international units), 308

Japanese eel, 535, 537
Japanese quail, 497, 498
Jejunum, 30, 78, 98
Joule, 141
Journal of Nutrition, 229, 252

K, *see* Potassium
Kangaroo, 26, 28
Kcal, *see* Kilocalorie
Keratinization, 232
Keratins, 121
Keratomalacia, 233
Ketones, 47, 85, 106
Ketone metabolism, 419–420
Ketosis, 85, 86, 419–420, 436, 458
Kidney, 43, 64, 66, 81, 82, 175
Kidney excretion systems, 178–179
Kidney stones, 171
Kilocalorie (Kcal), 141, 149
Kjeldahl N analysis, 16, 18
Krebs cycle, 82, 83, 131
Kwashiorkor, 133
Kwasiorkior, 4

Labeling of feedstuffs, 355, 521
Laboratory animals as models, 59
Lactase, 38, 39, 81
Lactate, 83
Lactation function. *See also* Cow
  lactation
    feed intake required for, 313–314
    and horse nutrition, 508
    and iron transfer, 199
    ketosis during, 85
    metabolic priorities during, 276–277
    swine, 472–474
    water consumption during, 68, 69
Lactic acid, 80
Lactoferrin, 120, 124
Lactogenesis, 285–286, 415, 416
Lactose synthesis, 417, 418
Lake trout, 535
Lambs, 449–452
    copper deficiency in, 203
    effect of caloric density on
      feed/energy intake of, 298–300
    growing, 450–452
    iodine deficiency in, 191
    metabolism affected by age of, 156
    and nonprotein nitrogen, 136
    nutrient requirements for, 446–452
    suckling, 449, 450
Laminitis, 422, 436, 510

L-amino acids, 118, 136
Large intestine, 26, 28, 30–32, 44
Largemouth bass, 535
Larval fish, 544
Lasalocid, 361–362
L-ascorbic acid, 269
Lavoisier, Antoine, 3
Laying hens, 117, 488–491
    calcium requirement for, 171
    feed intake expectations for, 305
    iodine in, 190
    magnesium in, 175
    water consumption in, 69
    and zinc, 195
LDL, *see* Low-density lipoproteins
Lead (Pb), 186, 219, 220, 223, 224
Least-cost formulation, 392
Lecithin, 105–106
Leghorn chickens, 490, 491
Legumes, 8, 9, 324, 328, 350–351,
    425, 426, 502–503
Legume silage, 333
Lens, loss of, 233
Leptin, 107, 166, 278, 279, 284, 287,
    295
Leucine, 115, 119, 124, 518
Libitum, 284
Lignin, 52–53, 78, 87, 324
Liming, 176
Linoleic acid, 92–94, 97, 419, 519, 524
Linolenic acid, 97
Linseed meal, 350
Lips, 453
Lipids, 7, 8, 75, 91–111
    abnormalities in metabolism of,
      107–110
    absorption of, 98–99
    adipose tissue, 101
    in animal diets, 76
    composition of selected foods, 110
    conjugated linoleic acid, 92–94
    effect of dietary carbohydrate
      source on metabolism of, 107
    and effects of frequency of feeding
      on metabolism, 106
    fatty acid and triglyceride
      metabolism, 101–106
    fatty acids, 92, 93
    functions of, 96–98
    glycerol, 93, 94
    and growth hormones, 281
    and lactation, 418–419
    and manganese, 206
    monoglycerides/diglycerides/trigl
      ycerides, 94–96
    and obesity as related to lipid
      metabolism, 106–107

omega-3/omega-6 fatty acids, 92
    phospholipids, 96
    in poultry diet, 481–482
    in rumen fermentation, 36
    sterols, 96
    structure of, 92–96
    transport and deposition of, 45,
      99–101
Lipogenic enzymes, 281, 282
Lipoprotein lipase (LPL), 104
Lipoproteins, 114, 116, 117, 234
Lipothiamide (LTPP), 252
Liquid energy feedstuff sources,
    344–345
Liquid spraying on feeds, 373
Liver, 26, 30, 31, 43, 81, 82, 130, 132
    fatty, 108
    and regulation of nutrient
      partitioning, 283
    triglyceride metabolism in the, 101
    and vitamin A, 231–235
Liver fibrosis, 200–201
Liver necrosis, 242
Lizards, 556–567
Low-density lipoproteins (LDL), 45,
    92, 100, 108, 109
Lower critical temperature (T_{lc}), 159
Low milk fat content, 436
Low-moisture silage, 334
LPL (lipoprotein lipase), 104
LTPP (lipothiamide), 252
Lung abscesses, 233
Lymph, 42–43
Lysine, 116, 119, 133, 136, 316, 463,
    486, 518

Macrominerals, 163–183
    calcium, 165–171
    calcium content of animals,
      164–165
    deficiencies in, 316–317
    magnesium, 174–176
    magnesium content of animals,
      164–165
    microminerals vs., 185
    phosporous, 171–174
    phosporous content of animals,
      164–165
    potassium/sodium/chlorine,
      176–181
    sulfur, 181–182
Mad cow disease, 123
Magnesium (Mg), 174–176, 187, 188,
    219, 224
    in animals, 164–165
    deficiency signs for, 175–176, 316
    in fish diet, 538

Magnesium (Mg) (*Continued*)
functions of, 174, 175
metabolism of, 175
in sheep diet, 441
sources of, 354
toxicity, 176
Maintenance function:
beef cow, 404–405
and energy metabolism, 157–158
feed intake required for, 311
horses, 504–507
and metabolic partitioning, 277
sheep, 445
Maize, 78. *See also* Corn
Mammary biology, 415–417, 431
Mammary transfer, 199, 235, 244
Mammogenesis, 285, 415
Mandioca, 345
Manganese (Mn), 186, 205–207, 219, 317, 441–442, 539
Mannose, 78
Marasmus, 4
Marine protein sources, 348
Mass spectrometry, 3, 125–126
Mathematical programming, 392
Mating behavior, 407
Maturity, stage of, 325
Mcal (megacalorie), 146
ME, *see* Metabolizable energy
Measurement of feed and nutrient utilization, 49–60
apparent vs. true digestibility, 54–56
associative effects, 54
balance trials, 55, 56
and biological variation in digestion/absorption, 58
digestibility by difference, 53–54
digestion trials, 52–53
growth trials, 50–52
and laboratory animals as models, 59
purified diets, 56, 57
rumen digestion techniques, 57–58
surgical procedures, 58–59
in vitro techniques, 57–58
Meat and bone meal, 348
Meat goats, 453–456
Meat meal, 8–9, 348, 427
Medicated feeds, 355
Megacalorie (Mcal), 146
Melengesterol acetate (MGA), 285
Menadione, 245–247
Menaquinone, 245, 247
Menstruation, 200
Mercury (Hg), 186, 219, 221

Messenger RNA (mRNA), 127, 128, 231
Metabolic conversions, 81–82
Metabolic modifiers, 428
Metabolic rate, regulation of, 286–287
Metabolic regulation, 275–280
Metabolic size, 155
Metabolic water, 62–63
Metabolic weight, 155
Metabolism:
abnormal carbohydrate, 85–86
abnormalities in lipid, 107–110
basal, 155–157
cellular, 166
effect of dietary carbohydrate source on lipid, 107
effects of frequency of feeding on, 106
energy, *see* Energy metabolism
factors affecting, 156–157
during lactation, 416–420
of proteins, 123–132
rate of, 40, 41
regulation of nutrient partitioning for, 275–280
water and body, 62
Metabolism crate, 56
Metabolizable energy (ME), 149–152, 157, 158, 161, 308
Metabolizable protein (MP), 308
Metalloflavin enzymes, 200
Metalloporphyrin enzymes, 200, 202
Metalloproteins, 200, 202
Metaplasia, 233
Methane, 36
Methionine, 115, 119, 123, 136, 263, 268, 316, 441, 486
Methyltetrahydrofolate (MTHF), 266
Mg, *see* Magnesium
MGA (melengesterol acetate), 285
Mg tetany, 176
Mice, 155, 164, 175
Michigan Department of Natural Resources Houghton Lake Wildlife Research Station, 555
Microbes, 26, 27
Microbial fermentation, 34–37, 78–80
Microbiological analysis, 17
Microbiological assay, 139
Microminerals, 185–214
aluminum, 212
analytical detection of, 187
animal requirements for, 186–188
arsenic, 212
barium, 213
bioavailability of, 213
bromine, 213

cadmium, 212–213
chromium, 211
cobalt, 188–189
copper, 196, 201–205
deficiencies in, 317
fluorine, 211–212
functions of, 187, 188
iodine, 189–192
iron, 196–201
macrominerals vs., 185
manganese, 205–207
molybdenum, 212
nickel, 213
selenium, 207–211
silicon, 212
tin, 213
toxicities of, 223–224
vanadium, 213
zinc, 192–196
Micronizing, 373
Microorganisms, 33–34, 124–125
Middlings, 343
Milk components, 417
Milk fat, 419, 436
Milk fever, 170, 179, 436
Milkfish, 532, 533, 535
Milk production. *See also* Lactation function
effects of nutrition on, for beef cattle, 397–398
ewe, 449
and feed intake, 303
and feed sources, 415, 416
and nonprotein nitrogen, 136
and preweaning growth, 406–407
regulation of nutrient partitioning for, 285–286
and water consumption, 68
Milk products, 348
Milk replacers, 430, 431, 450
Milk yield, 435
Millet, 338
Milling byproducts, 343–344
Milo, 341
Minerals:
in beef cattle diet, 398
in cat diet, 525–526
in dog diet, 520–521
in feedstuffs, 8, 9
in fish diet, 537–539
in goat diet, 440–442
for lactating cows, 427
macro, *see* Macrominerals
in poultry diet, 482, 485
in sheep diet, 440–442, 449
in swine diet, 476
trace, *see* Microminerals

Mineral supplements, 323, 352–354
Mineral toxicities in food chain, 217–225
  cadmium, 220–221
  fluorine, 221–222
  lead, 220
  maximum tolerable levels of dietary minerals for domestic animals, 218–219
  mercury, 221
  molds, 224–225
  molybdenum, 222–223
  potential pollution of environment by, 224
  radioisotopes, 224
  trace minerals, 223–224
Mink, 358
Mitochondria, 83
Mn, *see* Manganese
Mo, *see* Molybdenum
Moisture meters, 17
Molasses, 344–345, 373
Molds, 225, 350, 503
Molecular nutrition, 3
Molybdenum (Mo), 186, 188, 212, 219, 222–223, 317, 441
Monensin, 361–362, 428, 504
Monitoring programs, 432, 435–436
Monogastric species, 3, 27*n*.
Monoglycerides, 94–96
Monosaccharides, 74–76, 78–81
Mouth, 43
MP (metabolizable protein), 308
MRNA, *see* Messenger RNA
MTHF (methyltetrahydrofolate), 266
Mucoproteins, 117
Mucosal block, 198
Mucosal cells, 30
Mulberry heard disease, 209
Mules, 358
Muscle growth, 280–282
Muscular dystrophy, 207
Muscular lesions, 242
Mycoplasmas, 34
Mycotoxins, 225
Myelin, 114
Myofibrilar proteins, 120
Myoglobin, 196, 198
Myoinositol (inositol), 252, 270, 541–542
Myopathies, 242
Myosin, 120, 121
Myristic acid, 419
Myxedema, 191

N-15 isotope, 125
Na, *see* Sodium

NaCl, *see* Salt; Sodium chloride
NAD (nicotinamide adenine dinucleotide), 256
NADP, *see* Nicotinamide adenine dinucleotide phosphate
National Academy of Sciences, 369
National Agricultural Statistics Service, 338
National Research Council (NRC), 59, 163–164, 301, 307, 310, 325, 331, 382
NDF, *see* Neutral detergent fiber
NE, *see* Net energy
NEFA (nonessential fatty acids), 419
NE$_L$, *see* Net energy for lactation
NEm (net energy of maintenance), 152, 161
Nephrosis, 233
NEp (net energy for production), 152
Nerve growth factor (NGF), 122
Net efficiency, 160
Net energy for lactation (NE$_L$), 152, 161, 408, 423, 435
Net energy for production (NEp), 152
Net energy (NE), 150–152, 308
Net energy of maintenance (NEm), 152
Net protein utilization (NPU), 138
Net protein value (NPV), 138
Neuroendocrine factors, 156
Neutral detergent extraction, 20
Neutral detergent fiber (NDF), 20, 87, 333, 422
NFE (nitrogen-free extract), 19
NGF (nerve growth factor), 122
Ni, *see* Nickel
Niacin (nicotinic acid), 85, 256–257, 316, 540
Niacin—tryptophan interrelationships, 257
Nickel (Ni), 213, 219, 223, 317
Nicotinamide, 256
Nicotinamide adenine dinucleotide phosphate (NADP), 256
Nicotine adenine dinucleotide phosphate (NADP), 83
Nicotinic acid, *see* Niacin
NIDDM (noninsulin-dependent diabetes mellitus), 86
Night blindness, 231, 233
Nile tilapia, 534, 537
Nitrates, 224
Nitric oxide (NO), 126, 132
Nitric oxide synthase (NOS), 126
Nitrites, 71

Nitrogen (N):
  analysis of, 18–19
  and carbon, 153–154
  and digestibility, 55–57
  nonprotein, 134–136, 308, 352
  plant vs. animal requirements of, 6
  in rumen fermentation, 36
Nitrogen cycling, 125–126
Nitrogen-free extract (NFE), 19
NMD, *see* Nutritional muscular dystrophy
NMR (nuclear magnetic resonance), 23
NO, *see* Nitric oxide
Nonessential amino acids, 117–120
Nonessential fatty acids (NEFA), 419
Noninsulin-dependent diabetes mellitus (NIDDM), 86
Nonnutritive feed additives, 323, 354–355
Nonprotein nitrogen (NPN), 134–136, 308, 352
Nonruminants:
  carbohydrate metabolism in, 78, 79
  and crude protein, 18
  D-amino acids used by, 133, 134
  defined, 3
  feed processing for, 373–375
  fermentation in, 27
  gastrointestinal tract in, 29–33
  gluconeogenesis in, 83
  nitrogen needs in, 126
  nonprotein nitrogen used by, 134, 135
Norepinephrine, 283
NOS (nitric oxide synthase), 126
NPN, *see* Nonprotein nitrogen
NPU (net protein utilization), 138
NPV (net protein value), 138
NRC, *see* National Research Council
NSHP, *see* Nutritional secondary hyperparathyroidism
Nuclear magnetic resonance (NMR), 23
Nucleic acids, 83
Nursery period, 467–469
Nutrient(s):
  classes of, 2
  defined, 2, 6
  human requirements for, 7
  plant/animal requirements for, 6, 7
  transport of, 40–42
Nutrient composition, 382
Nutrient partitioning, *see* Regulation of nutrient partitioning

Nutrient requirements. *See also specific animals*
  and diet formulation, 382
  factors affecting, 314–315
  during lactation, 420, 421
  methods of estimating, 309
  for productive functions, *see* Productive functions, nutrient requirements for
  variations in knowledge of species, 4
Nutrient utilization measurement, *see* Measurement of feed and nutrient utilization
Nutrition, 1–4, 44–46, 315. *See also* Animal nutrition
Nutritional evaluation, 315–319, 435–436
Nutritional goals in dairy industry, 415, 416
Nutritional history (of animal), 317
Nutritional individuality, 314
Nutritional muscular dystrophy (NMD), 209–210, 242
Nutritional science, 2–3
Nutritional secondary hyperparathyroidism (NSHP), 168–169
Nutritionists, 3
Nylon bag technique, 57–58

Oats, 339, 341, 342, 503
Obesity:
  and bone density, 166
  in bulls, 407
  in cows, 400
  in horses, 510
  and lipid metabolism, 106–107
Obligate anaerobes, 33
Odor, 294
Oils, 346–347
Oilseed meal, 349–350, 424, 425
Omasum, 32, 37, 44
Omega-3 fatty acids, 92, 109
Omega-6 fatty acids, 92
Omnivores, 27–28
On-farm feed analysis, 23
Opacity of cornea, 233
Open-circuit calorimeters, 153
Optic nerve constriction, 233
Orchard grass, 324, 325, 333
Organic compounds, 7, 8
Organic toxins in food chain, 217, 224–225
Ornithine cycle, *see* Urea cycle
Osmosis, 63
Ossification, 284

Osteocytic osteolysis, 168
Osteomalacia, 169, 238
Osteoporosis, 169, 170, 175, 211–212
Ovalbumin, 117
Ovary, 190
Overconditioned cows, 436
Overeating disease (enterotoxemia type D), 458
Ovomucoid, 117
Oxidation, 62
Oxygen bomb calorimeters, 20–21

P, *see* Phosphorous
PABA, *see* Para-aminobenzoic acid
Pacific salmon, 543–545
Pancreas, 26, 28, 30, 31, 37, 81
Pancreatic amylase, 38, 39
Pancreatic polypeptide, 40, 41
Panda, 28, 29, 552
Panting, 483
Pantothenic acid, 258–260, 316, 540
Papilledema, 233
Para-aminobenzoic acid (PABA), 252, 271
Paracrine factors, 279
Parakeratosis, 171, 192, 194, 195
Parasites, 200, 503
Parathyroid hormone (PTH), 165, 169, 236, 238, 239
Partial digestion, 46–47
Partridges, 358
Parts per million (ppm), 308
Pasture, 8, 9, 53, 326–329
Pb, *see* Lead
PBBs (polybrominated biphenyls), 192
PBI (protein-bound I), 191
PCBS (polychlorinated biphenyls), 192
Peanut meal, 350
Peanuts, 225
Pearson's square method, 383, 385, 387, 389–392
Peas, 350–351
Peat scours, 222
Pectins, 76, 77
Pekin ducks, 496
Pellagra, 256, 257
Pelleted food, 294, 378, 504, 544, 553–556
Pelleting, 372–374
PEM, *see* Polioencephalomalacia
Pennsylvania State University, 23, 392
Pentagastrin, 297
Pepsin, 38, 39
Peptide bond, 118

Peptides, 39–41, 122, 295
Percents, working with, 382–383
Performance horses, 510–512
Perosis, 206
Peroxides, 207
PER (protein efficiency ratio), 138
Pet foods, 375, 515, 521–523, 527
PGFs, *see* Polypeptide growth factors
pH, 20, 31
Phase 2 diets for pigs, 468
Phase 3 diets for pigs, 469
Phase 4 diets for pigs, 470
Phase-feeding programs, 467
Pheasants, 358
Phenylalanine, 115, 119
Phenylketonuria, 136, 315
Phosphate, 194
Phosphatides, 96
Phospholipids, 96, 98, 105–106, 171, 267
Phosphorous (P), 171–174, 219
  absorption of, 172
  in animals, 164–165
  deficiency signs for, 173–174, 316
  excretion of, 173
  in feedstuffs, 9
  in fish diet, 538
  functions of, 171
  metabolism of, 171
  and microbial phytases, 172–173
  in sheep diet, 441
  sources of, 354
  swine, 463–464
  tissue distribution of, 171
  toxicity, 174
  and vitamin D, 236
Photosensitization, 503
Phylloquinone, 245, 247
Physical examinations, 318
Phytases, 172–173
Phytate P, 172, 173
Phytic acid, 173
Pica, 174
Pigs. *See also* Swine
  amino acid deficiency in, 132–133
  amino acid requirements of, 133–135
  body composition of, 9, 10
  calcium deficiency in, 168, 169
  carbohydrate metabolism in, 80, 81
  digestibility in, 55
  digestibility of alfalfa hay by, 46
  effects of protein quality/quantity in growing, 129
  energy utilization efficiency of, 160
  excessive protein intake in, 136–137

feed and growth relationship in, 50
feed intake expectations for, 304–305
gastrointestinal tract of, 27–30, 32, 33, 42
as laboratory animals, 59
magnesium in, 175
mineral content in body of, 164
phosphorus deficiency/toxicity in, 174
and selenium, 210
selenium content of, 207, 208
and zinc, 195
Pigeons, 80
Pigments:
fish, 542
in poultry diet, 487, 496
and vitamin D deficiency, 240
wool, 204
Pinocytosis, 41
PL, see Pyridoxal
Placenta, retained, 436
Placental transfer, 199, 209, 244
Plaice, 535
Plants:
amino acids in, 118–119
chemical composition of whole, 8, 9
nutrients required by, 6, 7
Plant cell walls, 78
Plant fibers, 78–80, 87–88
Plant products, 475–476
Plant protein concentrates, 348–352
Plasma, 45, 139
and iron, 197, 199
and selenium, 209–210
and zinc, 194
Plasma lipid transfer proteins (PLTP), 116
PLTP (plasma lipid transfer proteins), 116
PM, see Pyridoxamine
PN, see Pyridoxine
Pneumonia, 233
Poisonous plants, 458
Polioencephalomalacia (PEM), 182, 442
Pollution of environment, 224
Polybrominated biphenyls (PBBs), 192
Polychlorinated biphenyls (PCBS), 192
Polycythemia, 189
Polyethylene glycol, 53
Polyneuritis, 253
Polypeptide growth factors (PGFs), 129, 130

Polypeptides, 122
Polysaccharides, 63, 75, 76
Polyunsaturated fatty acids (PUFA), 241, 243–244
Pony, 26
"Poor production" in first lactation, 436
Popped corn, 373
Population, human, 4
Postpartum reproduction, 398–400
Postweaning slump, 508–509
Potassium (K), 176–181, 219, 316, 538
Potatoes, 78, 345–346
Poultry, 479–498
broilers, 491–494
deficiency signs/symptoms in, 483–485
digestive anatomy and physiology of, 480
ducks, 496–497
energy, 483–485
energy-nutrient interrelationships in, 485–486
feed processing for, 373–374
feedstuffs used for, 486–487
Japanese quail, 497, 498
laying hens, 488–491
mineral tolerances for, 218–219
nutrient requirements, 480–483
turkeys, 494–496
water consumption in, 68, 69
Poultry byproducts, 348
Ppm (parts per million), 308
Pregastric fermentation, 27, 28
Pregnancy period. See also Gestation
ketosis during, 85
metabolic priorities during, 276
swine, 471–473
water consumption during, 68
water consumption in, 69
Premixes, 392
Preweaning growth, 406–407
Prions, 122–123
Prion protein (PrP), 122
Probiotics, 363
Processing methods, feed, 141
Productive functions, nutrient requirements for, 310–319
and disease, 315
evaluating nutritional status of animals, 315–319
factors affecting, 314–315
growth and fattening function, 311–312
lactation function, 313–314
maintenance function, 311

reproduction function, 312–313
work function, 312
Proenzymes, 39
Progesterone, 285
Programming, 392
Prolactin, 285
Proline, 116, 117, 119, 481
Prostaglandin synthesis, 241
Proteins, 7, 8, 113–142. See also Amino acids
analysis of crude, 18, 19
in beef cattle diet, 398
in beef cows, 405
for broiler chickens, 492, 494
in cat diet, 523, 524
and copper, 204–205
deficiencies in, 132–133, 316
digestibility of, 55
in dog diet, 517–519
excessive intake of, 136–137
and exercise, 136
in fish diet, 535–536
functions of, 120–123
in goat diet, 440
in heifer diet, 432
and lactation, 418, 420, 425–427
for lambs, 451
measures of nutritive value of, 137–141
metabolism of, 123–132
nonprotein nitrogen, 135–136
requirements for, 132, 134, 135
roles of, in animals, 120
in rumen fermentation, 36
in sheep diet, 440
structure of, 114–120
in swine diet, 463, 475
synthesis/degradation of, 127–129
urea cycle, 130–132
in urine, 47
and vitamin A, 234
whole-body protein turnover, 129, 130
Proteinases, 122
Protein blocks, 440
Protein-bound I (PBI), 191
Protein—carbohydrate complexes, see Glycoproteins
Protein concentrates, 323, 347–352
Protein degradation, 127–129
Protein efficiency ratio (PER), 138
Protein—lipid complexes, see Lipoproteins
Protein synthesis, 127–129
Protein turnover, 129, 130
Proteolytic enzymes, 38, 39
Proteome, 23

Protozoa, 33–35
Proventriculus, 31, 44
Proximate analysis, 18–19
PrP (prion protein), 122
Prussic acid, 224
Psittacine birds, 358, 556
PTH, see Parathyroid hormone
Puberty, 284–285, 396–398, 407
PUFA, see Polyunsaturated fatty
  acids
Pullets, 489, 490
Purified diets, 56, 57
Pylorus, 32
Pyridoxal (PL), 260, 261
Pyridoxamine (PM), 260, 261
Pyridoxine (PN), 260, 261, 316
Pyrithiamin, 253
Pyruvate, 83
Pyruvic acid, 253

Quail, 358, 497, 498

Rabbits:
  amino acid synthesis in, 124–125
  composition of, 9
  digestibility of alfalfa hay by, 46
  feed processing for, 375
  gastrointestinal tract of, 28, 29, 31
  heat increment of feeding for, 151
  heat production in, 155
  magnesium in, 175
  mineral content in body of, 164
  mineral tolerances for, 218–219
  sugar absorption by, 80
RACs, see Readily available
  carbohydrates
Ractopamine, 283, 284
Radiation, 21
Radioimmmune assays (RIA), 23
Radioisotopes, 224
Rainbow trout, 532–535, 537–539,
  543–545
Rams, 447, 450–452
Rapeseed meal, 350
Raptopamine, 364
Rare earth markers, 53
Rats:
  biological value of proteins for, 138
  calcium excretion in, 167
  digestive tract of, 26, 28
  essential/nonessential amino
    acids for, 117
  in feed analysis, 16
  heat increment of feeding for, 151
  heat production in, 155
  iron in, 197
  magnesium in, 175

mineral content in body of, 164
and nutritional science, 2
sugar absorption by, 80
Ration (term), 6
Ration formulation, 161
Rat poison, 247
Rb, see Rubidium
RBC, see Red blood cells
RBP, see Retinol-binding protein
RDP, see Rumen-degraded protein
RDR (relative dose response) test,
  231
Readily available carbohydrates
  (RACs), 35, 37
Recombinant DNA, 88
Reconstitution (of grain), 371
Rectum, 28, 32, 33
Red blood cells (RBC), 197, 198, 201,
  203, 204
Red carp, 542
Red drum (fish), 534
Red sea bream, 536–539
Reed canary, 325
Regulation:
  of body temperature, 63
  of drinking, 66–67
Regulation of nutrient partitioning,
  275–288
  and diet formulation, 423
  feast and fast cycle, 275–276
  for growth and fattening, 280–285
  for metabolic rate and feed intake,
    286–287
  for metabolism, 275–280
  for milk production, 285–286
Reichert-Meissl (RM) number, 95,
  96
Relative dose response (RDR) test,
  231
Relative nutritive value (RNV), 138
Rennin, 38, 39
Replacement ewes, 447, 450–452
Replacement heifers, 396–398
Replacement rams, 447
Reproduction function, 242, 284–285,
  312–313
Respiration systems, 178
Respiratory quotient (RQ), 152
Retained placenta, 436
Reticular groove, 32
Reticulorumen, 33, 43–44
Reticulum, 32
Retinal, 230, 232, 233
Retinoic acid, 230, 233, 235
Retinol, 230–235
Retinol-binding protein (RBP), 231,
  232, 234

RIA (radioimmmune assays), 23
Riboflavin, 253–256, 316, 540
Ribonucleic acid (RNA), 127, 128,
  245
Rice, 78, 339, 504
Rickets, 167, 168, 207, 238
RM number, see Reichert-Meissl
  number
RNA, see Ribonucleic acid
RNV (relative nutritive value), 138
Rodents, 29. See also Mice; Rats
Rollermill grinding, 370
Root crops, 345
Roughages, 299, 322–326
  cubed, 378
  defined, 323
  dry processing of, 377–379
  expected consumption of, 303–304
Roughened hair or feathers, 233
Rubidium (Rb), 213, 223, 224
Rumen, 32–37, 57–58, 78–80
Rumen buffers, 428
Rumen-degraded protein (RDP),
  425–427
Rumen fermentation, 34–37, 57–58
Rumen-undegraded protein (RUP),
  126, 425–427
Ruminants:
  calcium toxicity in, 171
  cellulose used by, 78
  defined, 3
  digestibility in, 55
  dry roughage and roughage
    processing for, 377–379
  energy costs and physical activity
    in, 157
  energy utilization efficiency of, 160
  feed processing for, 375–376
  fermentation in, 27, 34–37
  and fiber, 28
  gastrointestinal tract of, 31–37
  gluconeogenesis in, 83
  ketosis in, 85
  nitrogen metabolism in, 126
  nonprotein nitrogen used by,
    135–136
  phosphorus in, 172
Rumination, 37
Ruminoreticulum, 551
RUP, see Rumen-undegraded protein
Rye, 338, 339
Rye grass staggers, 503

S, see Sulfur
Sacculated colon, 27, 28
Safflower meal, 350
Saliva, 29, 32, 37, 66, 190

Salivary amylase, 37, 38
Salmon, 535, 537–540, 542, 544, 545
Salt (NaCl), 218, 219
    deficiency of, 180
    excess ingestion of, 180
    and feed intake inhibition, 300, 301
    in sheep diet, 440–441
    sources of, 352–353
    and taste, 293
    and water consumption, 68
    and water quality, 70–71
S-amino acids, 244
Sampling for analysis, 16
Saponification number, 95, 96
Satiety, 294, 296
Saturated fats, 108
Sb, see Antimony
Scandinavian Food Unit System, 152
Scansorial animals, 550, 551
Schizophrenia, 257
Scours, 430
Scrapie, 122, 123
Scurvy, 269, 270
Se, see Selenium
Sea perch, 537
Seasonal fluctuations, 552–553
Seaweed, 351
Secretin, 287
Secretions, 29, 30, 37–40
Segregated early weaning (SEW), 467
Selenium (Se), 186, 187, 207–211
    deficiency signs for, 209–210, 317
    in fish diet, 539
    functions of, 207, 208
    maximum tolerable levels of, 219
    metabolism of, 208–209
    pollution potential of, 224
    in sheep diet, 441
    swine, 465
    tissue distribution of, 207
    toxicity, 210–211
    toxicity of, 223
    and vitamin E, 241–243
Selenophosphate synthetase, 208
Selenoproteins, 208
Semiaquatic animals, 550, 551
Semifossorial animals, 550, 551
Serine, 115, 119, 481
Serum, 45, 174, 175
SE (Starch Equivalents), 152
SEW (segregated early weaning), 467
Sex steroids, 284–285
Sham-drinking, 66
Sheep, 439–452, 457–458. See also
    Ewes; Lambs
    and alfalfa hay, 46

body composition of, 10
copper deficiency in, 204
daily nutrient requirements of, 445–448
digestibility in, 58
digestive tract of, 26, 32, 33
diseases of, 457–458
energy, 440
ewes, 444–449
feed additives approved for, 358
feed processing for, 376
feedstuffs used for, 442
and fiber, 28
heat increment of feeding for, 151
heat production in, 155
ketosis in, 85
magnesium in, 175, 176
metabolism affected by age of, 156
and minerals, 218–219, 440–442
nutrients required for, 440–443
protein, 440
roughage consumption
    expectations for, 304
taste responses in, 293
vitamins, 442
and water, 65, 69, 442, 443
Si, see Silicon
Sight of food, 294
Silages, 322, 331, 332–337
    chemical composition of, 333
    corn/sorghum, 336–337
    equipment for storing, 336
    fermentation of, 334–335
    for lactating cows, 424
    and losses from ensiling, 335
    microbial inoculation of, 335
    nutritive properties of, 335
    in tropical climates, 337
Silica, 53
Silicon (Si), 186, 212, 219, 224
Silver (Ag), 219, 224
Simple lipids, 91
Skeletal structure, 165–166, 171
Skin, 117
Slaughter, comparative, 154
Small intestine, 27, 28, 30–33, 44, 76–78, 98
Sn, see Tin
Soaked grain, 370
Sockeye salmon, 535
Sodium (Na), 64, 78, 79, 176–181, 316, 464, 538
Sodium bentonite, 364
Sodium bicarbonate, 512
Sodium chloride (NaCl), 218, 219
Sodium—glucose cotransporters, 78
Soilage, see Green chop

Solvent, water as, 62
Somatostatin, 86, 287
Somatotropin, 278, 280–287, 363
Sorghum, 326, 328, 333, 339, 340, 344, 504
Sorghum silage, 337
Sows, 470–471
Soybean meal, 8, 9, 133, 137, 349, 426
Species differences, 314
Spectrometry, 3
Spectrophotometry. atomic absorption, 21–22
Spongioform encephalopathies, 122–123
Sr, see Strontium
Stallions, 510
Standards, feeding, 161, 307–310
Starch, 77–78, 81, 519–520
Starch Equivalents (SE), 152
"Stargazing," 182, 206
Steam-flaked grains, 372
Steam-rolled grain, 372
Steely wool, 204
Steers, 9, 69, 285, 298–300, 372
Steroid metabolism, 104, 105
Sterols, 91, 96
Stewardship, 415
Stiff lamb disease, 209, 458
Stocker programs, 408
Stomach, 29–33, 44
Storage methods, 325–326
Straws, 331
Striped bass, 537, 544
Striped jack, 537
Strontium (Sr), 219, 224
Suckling period, 32, 449, 450, 466–467, 508–509
Sucrase, 38, 39
Sucrose, 38, 39
Sugars, 80
Sugar beet, 344
Sugar cane, 344
Sulfates, 70, 71
Sulfur-containing amino acids, 115
Sulfur (S), 181–182, 219, 317, 441
Sunflower meal, 350
Supplements, premix, 392
Supplement pellet, 555, 556
Surgical procedures, 58–59
Sweat, 66, 312
Swine, 461–476. See also Pigs
    breeding period, 470–471
    calcium/phosphorus, 463–464
    energy, 466
    fatty acids in, 101
    feed additives approved for, 358–360

Swine (*Continued*)
feed processing for, 373–375
feedstuffs used for, 474–476
finishing period, 469–470
fortified diet for, 462
growing period, 467–470
heat increment of feeding for, 151
heat production in, 155
iodine, 464
iron, 464
ketosis in, 85
lactation period, 472–474
life-cycle feeding of, 466–474
lysine, 463
mineral tolerances for, 218–219
nursery period, 467–469
nutrient requirements, 462–466
pregnancy period, 471–473
protein, 463
relative value of grains for, 341
selenium, 465
sodium/chlorine, 464
suckling period, 466–467
threonine, 463
tryptophan, 463
vitamins, 465
water consumption in, 69
water-soluble vitamins, 465–466
zinc, 464
Symbiotic relationships, 87

TAG, *see* Triacylglycerol.
Tannins, 87
Tapioca, 345
Taste, 292–294
Taurine, 118, 517, 524
TCA (tricarboxylic acid) cycle, 104
TDN, *see* Total digestible nutrients
TDS (total dissolved solids), 71
Teart, 222
Teeth, 29, 31, 32, 211, 222, 516
Temperature, 63, 154, 158–159
Tendon, 117
Termites, 87
Terpenes, 91
Terrestrial animals, 550, 551
Terrestrial herbivores, 553–556
Testosterone, 284
Tetany, 176, 236
Texture of food, 294
Texturized feed, 504
TFAA (total fecal amino acids), 140
TGB (thyroxin-binding globulin), 191
TGF-β, *see* Transforming growth factor-β
Thallium (TI), 224

Thiamin, 37, 252–253, 316, 540
Thiaminases, 253
Thiamin pyrophosphate (TPP), 252
Thioredoxin reductases, 208
Three Mile Island, 224
Threonine, 115, 119, 135, 463
Thymine, 128
Thyroglobulin, 190–191
Thyroid, 156, 189–190, 234
Thyroid hormones, 278, 286, 287
Thyroid-stimulating hormone (TSH), 190, 191, 286
Thyroprotein, 363
Thyrotropic hormone-releasing factor (TRF), 191
Thyrotropin releasing hormone (TRH), 286
Thyroxin, 189, 234
Thyroxin-binding globulin (TBG), 191
Thyroxine, 286
Ti, *see* Titanium
Tilapia, 532, 534, 535, 537–539, 543
Timothy, 325
Tin (Sn), 213, 219, 224
Tissue analyses, 318–319
Tissue proteins, 120
Titanium (Ti), 219
TI (thallium), 224
T_{lc} (lower critical temperature), 159
TMR, *see* Total mixed ration
Tongue, 29, 31, 32, 292–293, 453
Total digestible nutrients (TDN), 149, 161, 308, 325
Total dissolved solids (TDS), 71
Total fecal amino acids (TFAA), 140
Total mixed ration (TMR), 422, 432, 434
Total nutrients, 382
Toxicity:
amino acid, 136–137
cadmium, 220–221
calcium, 170–171
and cats, 527
cobalt, 189
copper, 204–205
fluorine, 221–222
iodine, 192
iron, 200–201
lead, 220
magnesium, 176
manganese, 207
mercury, 221
mineral, *see* Mineral toxicities in food chain
molybdenum, 222–223
organic, 217, 224–225

potassium/sodium/chlorine, 180–181
riboflavin, 255
sulfur, 182
vitamin A, 235
vitamin D, 240
vitamin E, 245
vitamin K, 247
zinc, 196
Toxic plants, 458
TPP (thiamin pyrophosphate), 252
Trace minerals, *see* Microminerals
Trans-aconitate, 176
Transaminases, 261
Transamination, 130
Trans-fatty acids, 92
Transferrin, 197, 199, 200
Transfer RNA (tRNA), 127, 128
Transforming growth factor-β (TGF-β), 122, 285
Transport:
active, 41, 78, 79
of copper, 202, 203
of lipids, 99–101
of manganese, 205–206
of monosaccharides, 78–81
of nutrients, 40–42
of selenium, 209
water as medium for, 62
TRF (thyrotropic hormone-releasing factor), 191
TRH (thyrotropin releasing hormone), 286
Triacylglycerol (TAG), 95*n.*, 101
Tricarboxylic acid (TCA) cycle, 104
Trichothecenes, 225
Triglycerides, 76, 94–96, 98–100, 481
Triglyceride biosynthesis, 102, 104
Triglyceride catabolism, 102, 104
Triglyceride metabolism, 101–106
Triiodothryronine, 286
Triticale, 338, 342
TRNA, *see* Transfer RNA
Trout, 532–535, 537–540, 543–545
True digestibility, 55, 97
True digestible energy, 148, 149
Trypsin, 38, 39
Tryptophan, 116, 119, 133, 135, 136, 139, 261, 463
Tryptophan—niacin interrelationships, 257
TSH, *see* Thyroid-stimulating hormone
Tubers, 345–346
T_{uc} (upper critical temperature), 159
Tungsten (W), 219, 224
Turbot, 537

Turkeys, 69, 360, 494–496
Twins, 400, 444, 446, 447, 449
Tyrosine, 115, 119, 480
Tyrosinemia II, 136

U, *see* Uranium
UIP (undegraded intake protein), 308
Undegraded intake protein (UIP), 308
Underconditioned cows, 436
Undernourishment, 4
United Kingdom, 307
University of Florida, 509
Unsacculated colon, 27
Unsaturated fatty acids, 92, 93
Upper critical temperature (T_{uc}), 159
Uranium (U), 219, 224
Urea, 66, 124, 132, 135, 352, 379, 424, 426, 440
Urea cycle, 130–132, 137
Urease, 134
Uric acid, 66, 124, 132
Uridine triphosphate (UTP), 81
Urinary calculi, 171, 212, 457–458
Urinary excretion, 47
Urine, 47, 65–66
Urine analyses, 318–319
Urochrome, 47
Urohthiasis, 233
UTP (uridine triphosphate), 81

V, *see* Vanadium
Valine, 115, 119
Vanadium (V), 213, 219, 224
Vasoactive intestinal peptide (VIP), 40, 41
Veal calves, 200
Very low-density lipoproteins (VLDL), 100, 108
VFA, *see* Volatile fatty acids
Villi, 30
VIP, *see* Vasoactive intestinal peptide
Vitamins, 8
    ascorbic acid, 268–270
    in beef cattle diet, 398
    biotin, 266–267
    in cat diet, 526–527
    choline, 267–268
    in dog diet, 521
    fat-soluble, *see* Fat-soluble vitamins
    in fish diet, 539–542
    folacin, 265–266
    in goat diet, 442
    for lactating cows, 427
    during lactation, 427

myoinositol, 270
niacin, 256–257
pantothenic acid, 258–260
para-aminobenzoic acid, 271
    in poultry diet, 482, 484
riboflavin, 253–256
    in rumen fermentation, 36–37
    in sheep diet, 442
    sources of, 353–354
    as supplements, 323, 476
    in swine diet, 476
thiamin, 252–253
vitamin B_6, 260–262
vitamin B_{12}, 262–265
water-soluble, *see* Water-soluble vitamins
Vitamin A, 230–235
    absorption/transport/storage of, 231
    deficiency of, 4, 231–233, 316
    in fish diet, 539
    functions of, 231
    metabolism of, 232–235
    and ruminants, 37
    in sheep diet, 442
    structure of, 230
    in swine diet, 465
    toxicity, 235
    and vitamin E, 244
Vitamin B_6, 260–262, 540–541
Vitamin B_{12}, 37, 189, 244, 262–265, 541
Vitamin C, 244, 268–270, 316, 542
Vitamin D, 235–240
    deficiency signs for, 238–240, 316
    in fish diet, 539–540
    functions of, 236
    metabolism of, 236–239
    in sheep diet, 442
    structure of, 235–236
    in swine diet, 465
    toxicity, 240
Vitamin E, 207, 210, 235, 241–245
    deficiency signs for, 242, 316
    in fish diet, 540
    functions of, 241–242
    in horse diet, 505, 511–512
    and mercury toxicity, 221
    metabolism of, 242–244
    placental/mammary transfer of, 244
    in sheep diet, 442
    structure of, 241
    in swine diet, 465
    toxicity, 245
Vitamin H, *see* Biotin
Vitamin K, 245–247, 316, 442, 540

VLDL, *see* Very low-density lipoproteins
Volant animals, 550, 551
Volatile fatty acids (VFA), 78, 85, 87, 102, 502

W, *see* Tungsten
Warfarin, 247
Warm-season grasses, 327, 328
Wasting disease, 123, 188
Water, 61–72
    absorption of, 63
    in beef cattle diet, 395–396
    body water, 63–64
    in cat diet, 523
    in dairy cow diet, 435
    dietary factors in consumption of, 67–68
    in dog diet, 516
    environmental factors in consumption of, 68
    in feedstuffs, 6
    functions of, 62–63
    in goat diet, 442, 443
    in horse diet, 505–507
    for irrigation, 11
    and lead poisoning, 220
    loss of, from animal body, 65–66
    in poultry diet, 482–483
    quality of, 70–71
    and regulation of drinking, 66–67
    restrictions on supplies of, 69–70
    in sheep diet, 442, 443
    shortage of, 12
    sources of, 64–65
    volume of required, 68–69
Water belly, *see* Urinary calculi
Water-free feedstuffs, 8
Water intoxication, 70
Water loss (from animal body), 65–66
Water quality, 70–71
Waters Amino Acid Analysis System, 21
Water-soluble markers, 53
Water-soluble vitamins, 251–272
    ascorbic acid, 268–270
    biotin, 266–267
    choline, 267–268
    deficiencies in, 316
    folacin, 265–266
    myoinositol, 270
    niacin, 256–257
    pantothenic acid, 258–260
    para-aminobenzoic acid, 271
    riboflavin, 253–256
    sources of, 354

Water-soluble vitamins (*Continued*)
  in swine diet, 465–466
    thiamin, 252–253
    vitamin B_6, 260–262
    vitamin B_{12}, 262–265
Water turnover, 64
Weaning, 407, 430, 509–510
Weight, 50, 154–155. *See also* Body
  weight
Weight cycling, 405
Weight loss, 233
Wernicke's encephalopathy, 253
Wheat, 8, 9, 35–36, 78, 338–343
Wheat-pasture poisoning, 176
Wheat straw, 8, 9
Whey, 345, 430
White muscle disease, 209, 441,
  458
Whole-animal nutrition, 3

Whole-animal nutrition, molecular
  vs., 3
Whole body, 51–52, 129, 130
Wilted forage, 334
Wool, 201, 203, 204
Work function, 312
Wound healing, 195

Xerophthalmia, 232, 233
Xerosis of membranes, 233
Xylose, 81

Yeasts, 34, 351, 428
Yellowtail (fish), 537, 539
Yuca, 345

Zein, 119
Zenker's degeneration, 209
Zeolites, 364

Zillii's tilapia, 537
Zinc (Zn), 186, 192–196, 218
  deficiency of, 171, 317
  in fish diet, 538–539
  maximum tolerable levels of,
    219
  pollution potential of, 224
  in sheep diet, 441
  swine, 464
  and vitamin E, 244
Zoo animals, 549–557
  carnivores/insectivores/lizards,
    556–567
  digestive function and morphology
    of, 551, 552
  habitat and feeding strategy,
    550–551
  and known species, 552–553
  terrestrial herbivores, 553–556

BASIC ANIMAL

NUTRITION

AND FEEDING

APPENDIX

Fifth Edition

Wilson G. Pond / **David C. Church**

Kevin R. Pond / **Patricia A. Schoknecht**

Appendix

- List of Pertinent Publications of the National Research Council
- Tables of Energy, Amino Acid, Vitamin, and Mineral Element Content and Proximate Analysis of Selected Food Ingredients
- Conversion Tables for Weight, Temperature, Area, and Volume

NOTE: The Appendix Tables and Glossary can be accessed online, along with images and animations that supplement the bound 5th Edition of Basic Animal Nutrition and Feeding (www.wiley.com/college/pond).

The National Research Council (NRC) continually updates new knowledge of the nutrient requirements of individual species of the domestic animals. These publications are printed by the National Academy Press, 2101 Constitution Avenue, NW, Washington, D.C. The information contained in each publication is derived from research published in the scientific literature. The individual documents for each species of animal are written by subcommittees of scientists who are authorities on nutrition of the animals species in their respective fields of specialization and appointed by the National Research Council of the National Academy of Science and the Committee on Animal Nutrition of the National Research Council.

A comprehensive compilation of the chemical and nutrient composition of a broad array of feedstuffs used in animal production was published in 1982 by the National Academy Press. The document was prepared by the Subcommittee on Feed Composition of the Committee on Animal Nutrition, National Research Council, under the title, *United States—Canadian Tables of Feed Composi-* *tion*. This 148-page booklet is valuable as a reference for nutritionists formulating diets for all species of domestic animals. Most of the chapters in this book that describe the nutrition and feeding of individual species of animals (Chapters 22 through 30) include feed composition data derived from that publication.

Appendix Table 1 lists all of the current National Research Council publications pertaining to the nutrient requirements of the domestic animal species. Appendix Tables 2 and 3 list the amino acid and vitamin contents, respectively, of common feedstuffs. Appendix Table 4 lists the mineral element composition of mineral supplements. Appendix Tables 5 and 6 list the energy values and mineral concentrations, respectively, of some common feeds. Appendix Table 7 lists area conversions and Appendix Table 8, volume conversions, between metric and Imperial units. Appendix Table 9 lists weight-unit conversion factors used commonly in the English and metric systems of measurement. Appendix Table 10 lists temperature conversions between Celsius (centigrade) and Fahrenheit units.

APPENDIX TABLE 1   National Research Council publications on nutrient requirements of domestic animals.

Nutrient Requirements of Beef Cattle, seventh revised edition, 1996
Nutrient Requirements of Dairy Cattle, seventh revised edition, 2001
Nutrient Requirements of Goats; Angora, Dairy, and Meat Goats in Temperate and Tropical Countries, 1981
Nutrient Requirements of Mink and Foxes, second revised edition, 1982
Nutrient Requirements of Rabbits, revised edition, 1985
Nutrient Requirements of Swine, tenth revised edition, 1998
Nutrient Requirements of Cats, revised edition, 1986
Nutrient Requirements of Dogs, revised edition, 1985
Nutrient Requirements of Horses, fifth revised edition, 1989
Nutrient Requirements of Nonhuman Primates, second revised edition, 2003
Nutrient Requirements of Sheep, sixth revised edition, 1985
Nutrient Requirements of Poultry, ninth edition, 1994
Nutrient Requirements of Fish, 1993
Nutritional Energetics of Domestic Animals and Glossary of Energy Terms, 1981
Predicting Feed Intake of Food-Producing Animals, 1987
Mineral Tolerance of Domestic Animals, 1980
Vitamin Tolerance of Animals, 1987
Rangeland Health: New Methods to Classify, Inventory, and Monitor Rangelands, 1994
Selenium in Nutrition, revised edition, 1983
Designing Foods: Animal Product Options in the Marketplace, 1988
Metabolic Modifiers: Effects on the Nutrient Requirements of Food-Producing Animals, 1994

APPENDIX TABLE 2 Amino acid composition of selected feedstuffs.

	CRUDE PROTEIN, %	ARGININE	CYSTINE	GLYCINE	HISTIDINE	ISOLEUCINE	LEUCINE	LYSINE	METHIONINE	PHENYLALANINE	THREONINE	TRYPTOPHAN	TYROSINE	VALINE
Forage-Roughage														
Alfalfa, dehy, 15% CP	15.2	0.60	0.17	0.70	0.30	0.68	1.10	0.60	0.20	0.80	0.60	0.40	0.40	0.70
Alfalfa, dehy, 20% CP	20.6	0.90	—	1.00	0.40	0.80	1.50	0.90	0.30	1.10	0.90	0.50	0.70	1.19
Alfalfa leaf meal	21.3	0.90	0.34	0.90	0.33	0.90	1.25	0.95	0.30	0.80	0.70	0.25	0.60	0.90
Grass, dehy.	14.8	0.99	0.19	0.72	0.46	1.38	1.98	1.06	0.31	1.30	0.89	0.31	0.46	1.57
Energy Feeds														
Barley grain	11.6	0.53	0.18	0.36	0.27	0.53	0.80	0.53	0.18	0.62	0.36	0.18	0.36	0.62
Corn hominy feed	10.7	0.50	0.18	0.50	0.20	0.40	0.80	0.40	0.18	0.30	0.40	0.10	0.50	0.50
Corn germ meal	18.0	1.20	0.32	—	—	—	1.70	0.90	0.35	0.80	0.90	0.30	1.50	1.30
Corn grain	8.8	0.50	0.09	0.43	0.20	0.40	1.10	0.20	0.17	0.50	0.40	0.10	—	0.40
Millet grain	12.0	0.35	0.08	—	0.23	1.23	0.49	0.25	0.30	0.59	0.44	0.17	—	0.62
Oats grain	11.8	0.71	0.18	—	0.18	0.53	0.89	0.36	0.18	0.62	0.36	0.18	0.53	0.62
Potato meal	8.2	0.43	—	—	0.11	0.48	0.30	0.47	0.07	0.29	0.21	0.15	—	0.39
Rice grain w hulls	7.3	0.53	0.10	0.80	0.09	0.27	0.53	0.27	0.17	0.27	0.18	0.10	0.60	0.51
Rye grain	11.9	0.53	0.18	—	0.27	0.53	0.71	0.45	0.18	0.62	0.36	0.09	0.27	0.62
Sorghum grain, milo	11.0	0.36	0.18	0.40	0.27	0.53	1.42	0.27	0.09	0.45	0.27	0.09	0.36	0.53
Wheatgrain	12.7	0.71	0.18	0.89	0.27	0.53	0.89	0.45	0.18	0.62	0.36	0.18	0.45	0.53
Wheat shorts	18.4	0.95	0.20	0.40	0.32	0.70	1.20	0.70	0.18	0.70	0.50	0.20	0.40	0.77
Whey, dried	13.8	0.40	0.30	0.30	0.20	0.90	1.40	1.10	0.20	0.40	0.80	0.20	0.30	0.70
Plant Protein Sources														
Brewer's dried grains	25.9	1.30	—	—	0.50	1.50	2.30	0.90	0.40	1.30	0.90	0.40	1.20	1.60
Corn dist, solv., dehy.	26.9	1.00	0.60	1.10	0.70	1.50	2.10	0.90	0.60	1.50	1.00	0.20	0.70	1.50
Corn gluten meal	42.9	1.40	0.60	1.50	1.00	2.30	7.60	0.80	1.00	2.90	1.40	0.20	1.00	2.20
Cottonseed meal, solv.	50.0	4.75	1.00	2.35	1.25	1.85	2.80	2.10	0.80	2.75	1.70	0.70	0.80	2.05
Peanut meal, solv.	47.4	4.69	—	—	1.00	2.00	3.10	1.30	0.60	2.30	1.40	0.50	—	2.20
Rapeseed meal, solv.	39.4	2.16	—	1.88	1.05	1.43	2.63	2.09	0.76	1.49	1.65	0.48	0.83	1.90
Soybean meal, solv.	45.8	3.20	0.67	2.10	1.10	2.50	3.40	2.90	0.60	2.20	1.70	0.60	1.40	2.40
Sunflower meal, solv.	46.8	3.50	0.70	2.70	1.10	2.10	2.60	1.70	1.50	2.20	1.50	0.50	—	2.30
Yeast, brewer's, dried	44.6	2.20	0.50	1.70	1.10	2.10	3.20	3.00	0.70	1.80	2.10	0.50	1.50	2.30

AMINO ACIDS, AS-FED BASIS, %

APPENDIX TABLE 2 Amino acid composition of selected feedstuffs—*Continued.*

	CRUDE PROTEIN, %	AMINO ACIDS, AS-FED BASIS, %												
		ARGININE	CYSTINE	GLYCINE	HISTIDINE	ISO-LEUCINE	LEUCINE	LYSINE	METHIONINE	PHENYL-ALANINE	THREONINE	TRYP-TOPHAN	TYRO-SINE	VALINE
Animal and Fish Protein Sources														
Blood meal	79.9	3.50	1.40	3.40	4.20	1.00	10.30	6.90	0.90	6.10	3.70	1.10	1.80	6.50
Buttermilk, dried	32.0	1.10	0.40	0.60	0.90	2.70	3.40	2.40	0.70	1.50	1.60	0.50	1.00	2.80
Casein, dried	81.8	3.40	0.30	1.50	2.50	5.70	8.60	7.00	2.70	4.60	3.80	1.00	4.70	6.80
Fish meal, anchovy	66.0	4.46	1.00	5.10	1.84	3.40	7.01	5.40	2.19	2.48	3.04	0.80	1.77	3.54
Fish meal, herring	70.6	4.00	1.60	5.00	1.30	3.20	5.10	7.30	2.00	2.60	2.60	0.90	2.10	3.20
Fish meal, menhaden	61.3	4.00	0.94	4.40	1.60	4.10	5.00	5.30	1.80	2.70	2.90	0.60	1.60	3.60
Liver meal	66.5	4.10	0.90	5.60	1.50	3.40	5.40	4.80	1.30	2.90	2.60	0.60	1.70	4.20
Meat meal	53.4	3.70	0.60	2.20	1.10	1.90	3.50	3.80	0.80	1.90	1.80	0.30	0.90	2.60
Meat and bone meal	50.6	4.00	0.60	6.60	0.90	1.70	3.10	3.50	0.70	1.80	1.80	0.20	0.80	2.40
Meat meal tankage	59.8	3.60	—	—	1.90	1.90	5.10	4.00	0.80	2.70	2.40	0.70	—	4.20
Milk, dried, skim	33.5	1.20	0.50	0.20	0.90	2.30	3.30	2.80	0.80	1.50	1.40	0.40	1.30	2.20

APPENDIX TABLE 3 Vitamin content (ppm) of selected feedstuffs, fresh basis.

	CAROTENE	VITAMIN E	CHOLINE	NIACIN	PANTOTHENIC ACID	RIBOFLAVIN	THIAMIN	VITAMIN B6	VITAMIN B12
Plant Sources									
Alfalfa, dehy., 15% CP	102	98	1550	42	21	11	3.0	6.5	—
Alfalfa leaf meal	62	—	1600	55	33	15	—	11	—
Barley grain	—	11	1030	57	6.5	2.0	5.1	2.9	—
Brewer's dried grains	—	—	1587	43	8.6	1.5	0.7	0.7	—
Corn dist. solubles, dehy.	1	55	4818	115	21	17	6.8	10	—
Corn grain	4	22	537	23	5	1.1	4.0	7.2	—
Cottonseed meal, solve, 41% CP	—	15	2860	40	14	5.0	6.5	6.4	—
Oats grain	—	36	1073	16	13	1.6	6.2	1.2	—
Peanut meal, solv.	—	3	2000	170	53	11	7.3	10	—
Rice grown w hulls	—	14	800	30	3.3	1.1	2.8	—	—
Rye grain	—	15	—	1.2	6.9	1.6	3.9	—	—
Sorghum grain, milo	—	12	678	43	11	1.2	3.9	4.1	—
Soybean meal, solv., 45% CP	—	3	2743	27	14	3.3	6.6	8.0	—
Wheat grain	—	34	830	57	12	1.2	4.9	—	—
Wheat middlings	—	58	1100	53	14	1.5	19	11	—
Yeast, brewer's dried	—	—	3885	447	110	35	92	43	—
Animal Sources									
Buttermilk, dried	—	6	1808	9	30	31	3.5	2.4	0.02
Fish meal, herring	—	27	4004	89	11	9.0	—	3.7	219
Fish meal, menhaden	—	9	3080	56	9	4.8	0.7	—	0.1
Meal meal	—	1	1955	57	4.8	5.3	0.2	3.0	51
Meat and bone meal	—	1	2189	48	3.7	4.4	1.1	2.5	45
Liver meal	—	—	—	204	45	46	0.2	—	501
Milk, cow—s, dried skim	—	9	1426	11	34	20	3.5	3.9	42
Whey, dried	—	—	20	11	48	30	3.7	2.5	0.03

APPENDIX TABLE 4 Mineral element composition of salts and mineral supplements used in livestock diets, as-fed basis, feed grade purity.[a]

NAME	BASE CHEMICAL FORMULA	ELEMENTAL COMPOSITION, % FEED GRADE PRODUCTS										ACID-BASE REACTION
		CA	CL	F	K	MG	MN	N	NA	P	S	
Ammonium chloride	NH_4Cl		66.28					26.10				Acid pH 4.5-5.5
Ammonium phosphate												
Monobasic	$(NH_4)H_2PO_4$			0.0025[b]				12.18		26.93		Acid pH 4.5
Dibasic	$(NH_4)_2HPO_4$			0.0025[b]				21.21		23.48		Basic pH 8.0
Animal bone charcoal		27.1				0.5		1.4		12.7		
Animal bone meal, steamed		29.0				0.6	trace	1.9	0.5	13.6		
Calcium												
Calcium carbonate	$CaCO_3$	40.04										Basic
Limestone	$CaCO_3$	35.8				2.0	trace			trace		Basic
Limestone, dolomitic	$CaCO_3$	22.3			0.4	10.0	trace					Basic
Oyster shell	$CaCO_3$	38.0			0.1	0.3	trace		0.21	0.07		Basic
Calcium phosphates												
Monocalcium phosphate	$CaH_4(PO_4)_2$	15.9		0.0025[b]						24.6		Acid pH 3.9
Dicalcium phosphate	$CaHPO_4$	23.35		0.02						18.21		Acid pH 5.5
Tricalcium phosphate	$Ca_3(PO_4)_2$, CaO	38.76		0.18[b]						19.97		sl.[c] Acid pH 6.9
Defluorinated rock phosphates		32.0		0.18[b]	0.09				4.0	18.0		
Gypsum	$CaSO_4$	29.44									23.55	Acid
Magnesium carbonate	$MgCo_3$	0.02				25.2						sl. Basic
Magnesium oxide	MgO					60.3			0.5			sl. Basic
Manganese sulfate	$MnSO_4 \cdot H_2O$, MnO						28.7				14.7	
Potassium chloride	KCl		47.6		52.4							Neutral pH 7.00
Sodium chloride	$NaCl$		60.66						39.34			Neutral
Sodium phosphate												
Monobasic	$NaH_2PO_4 \cdot H_2O$			0.0025[b]					16.6	22.40		Acid pH 4.4
Dibasic	$NaHPO_4$			0.0025[b]					32.39	21.80		Basic pH 9.0
Tripoly	$Na_5P_3O_{10}$			0.0025[b]					24.94	30.85		Basic pH 9.9

[a]Elemental content varies according to source; use guaranteed analysis where available.
[b]Maximum.
[c]Slightly

APPENDIX TABLE 5 Energy values and energy sources of selected feeds from the 2001 Dairy NRC.

DIET INGREDIENTS	DM %	ME-3X[1] MCAL/ KG	NEL-3X MCAL/ KG	NEL-4X MCAL/ KG	LIPID %	NDF %	NFC[2] %	LIGNIN %	CP %	MAX RUP[3] %
Almond hulls	87	1.89	1.14	1.07	3	37	48	15	7	56
Bakery byproduct	89	3.07	1.97	1.88	4	10	74	3	9	21
Barley grain, rolled	91	2.92	1.86	1.76	2	21	62	2	12	24
Beet pulp, dried	88	2.36	1.47	1.38	1	46	36	2	10	76
Bermudagrass hay early head	87	1.73	1.02	0.96	3	73	6	7	10	30
Blood meal, ring dried[4]	90	3.58	2.33	2.21	1	0	1	0	96	78
Brewer's grains, wet	22	2.69	1.71	1.62	5	47	14	5	28	35
Canola seed	90	4.92	3.52	3.36	41	18	17	3	21	21
Canola meal, mech. extracted	90	2.75	1.76	1.66	5	30	20	10	38	36
Citrus pulp dried	86	3.44	1.76	1.66	5	24	57	1	7	32
Corn distiller's grains w/solubles, dried	90	3.72	1.97	1.87	10	39	16	4	30	51
Corn gluten feed, dried	89	3.43	1.73	1.64	4	36	30	2	24	30
Corn gluten meal, dried	86	4.43	2.38	2.25	3	11	18	2	65	75
Corn grain, cracked, dry	88	3.69	1.91	1.80	4	10	75	1	9	47
Corn grain, ground, dry	88	3.85	2.01	1.90	4	10	75	1	9	47
Corn grain, stem-flaked	88	3.97	2.09	1.98	4	10	75	1	9	75
Corn grain, ground, high moisture	72	3.96	2.09	1.97	4	10	75	1	9	43
Corn grain and cob, high moisture, ground	67	3.74	1.94	1.83	4	21	65	1	8	39
Corn hominy	89	3.64	1.88	1.78	4	21	60	2	12	31
Corn silage, immature	24	2.87	1.36	1.28	3	54	29	4	10	32
Corn silage, normal	35	2.99	1.45	1.38	3	45	39	3	9	35
Cottonseed, whole seed with lint	90	3.55	1.94	1.83	19	50	3	13	24	23
Cottonseed hulls	89	0.95	0.48	0.44	3	85	4	23	6	56
Fats - Calcium soaps of fatty acids	95	6.27	5.02	4.80	85	0	0	0	0	0
Fats - Hydrolyzed tallow fatty acids	100	6.76	5.41	5.17	99	0	1	0	0	0
Fats - Partially hydrogenated tallow	100	3.72	2.97	2.84	100	0	1	0	0	0
Fats - Tallow	100	5.66	4.53	4.33	100	0	0	0	0	0
Fishmeal, menhaden	91	3.52	2.33	2.20	10	0	1	0	69	66
Grasses, Cool Season, intensive pasture	20	2.46	1.54	1.45	3	46	15	2	27	26
Grasses, Cool Season, hay, immature	84	2.22	1.37	1.29	3	50	20	4	18	21
Grasses, Cool Season, hay, mid maturity	84	2.02	1.23	1.16	3	58	18	4	13	31
Grasses, Cool Season, hay, mature	84	1.85	1.11	1.04	2	69	11	6	11	44
Grasses, Cool Season, silage, immature	36	2.10	1.29	1.21	3	51	20	5	17	21
Grasses, Cool Season, silage, mid maturity	42	1.92	1.16	1.09	2	58	14	7	17	25
Mixed hay, mid maturity	85	2.05	1.25	1.17	2	51	19	6	18	24
Mixed hay, mature	90	1.86	1.12	1.05	2	56	14	7	18	32
Mixed silage, immature	46	21.00	1.29	1.21	2	45	22	6	20	19
Mixed silage, mid maturity	44	2.01	1.23	1.15	3	50	18	6	19	21
Legume hay immature	84	2.23	1.38	1.30	2	36	29	6	23	18
Legume hay, mid maturity	84	2.09	1.28	1.20	2	43	25	6	21	19
Legume hay, mature	84	1.88	1.13	1.06	2	51	21	7	18	24
Legume silage, immature	41	2.18	1.34	1.26	2	37	27	6	23	18
Legume silage, mid maturity	43	2.01	1.22	1.15	2	43	22	7	22	18

APPENDIX TABLE 5   Energy values and energy sources of selected feeds from the 2001 Dairy NRC—*Continued.*

DIET INGREDIENTS	DM %	ME-3X[1] MCAL/ KG	NEL-3X MCAL/ KG	NEL-4X MCAL/ KG	LIPID %	NDF %	NFC[2] %	LIGNIN %	CP %	MAX RUP[3] %
Legume silage, mature	43	1.84	1.10	1.03	2	50	17	8	20	27
Molasses, sugarcane	74	2.78	1.76	1.66	0	0	80	0	6	18
Oats grain, rolled	90	2.78	1.77	1.67	5	30	48	5	13	15
Oats silage, headed	35	1.91	1.15	1.08	3	61	13	6	13	39
Rye silage, vegatative	30	2.08	1.28	1.20	4	58	13	5	16	26
Sorghum grain, dry rolled	89	2.83	1.80	1.70	3	11	72	1	12	47
Sorghum grain, steam-flaked	89	3.17	2.04	1.93	3	11	72	1	12	61
Sorghum, sudan type, silage	29	1.79	1.07	1.00	4	63	11	6	11	51
Soybean hulls	91	2.34	1.46	1.37	3	60	18	3	14	45
Soybean meal, expellers	90	3.61	2.38	2.25	8	22	18	2	46	69
Soybean meal, solvent, 44% CP	89	3.31	2.13	2.02	2	15	27	1	50	35
Soybean meal, solvent, 49% CP	90	3.41	2.21	2.09	1	10	29	1	54	43
Soybean seeds, whole	90	4.05	2.75	2.62	19	20	16	1	39	30
Soybean seeds, whole roasted	91	4.00	2.72	2.58	19	22	11	3	43	39
Sunflower meal, solvent	92	2.24	1.38	1.30	1	40	22	10	28	16
Tomato pomace	25	2.37	1.52	1.43	13	60	2	13	19	32
Wheat, bran	89	2.55	1.61	1.52	4	43	30	3	17	21
Wheat grain, rolled	89	3.10	1.99	1.88	2	13	68	2	14	26
Wheat middlings	90	2.64	1.67	1.58	5	37	35	4	19	24
Wheat silage, early head	33	1.91	1.16	1.08	3	60	16	6	12	23
Wheat straw	93	1.44	0.82	0.76	2	73	13	9	5	78
Whey, wet, cattle	21	2.92	1.86	1.76	1	0	75	0	15	6

[1] The ME-3X energy value is the estimated metabolizable energy available for an animal consuming at 3 times its maintenance requirement.

[2] The NFC (Non fiber carbohydrate) value is an estimate of the content of starches and sugars in a feed and is calculated as 100 - %ash - %lipid - %protein - %NDF.

[3] The maximum RUP value is the estimated percentage of CP that will not be degraded in the rumen for a cow consuming a diet that is 60% concentrate at 4% of body weight per day.

[4] Restrictions may apply on feeding blood meal to dairy cattle due to concerns regarding transmission of bovine spongiform encephalopathy.

APPENDIX TABLE 6   Mineral concentrations (dry matter basis) of some common feeds from 2001 Dairy NRC.

	CA%	P%	MG%	K%	NA%	CL%	S%
Almond hulls	0.28	0.13	0.13	2.62	0.02	0.03	0.04
Bakery byproduct	0.17	0.29	0.10	0.33	0.59	0.69	0.10
Barley grain, rolled	0.06	0.39	0.14	0.56	0.02	0.13	0.12
Beet pulp, dried	0.91	0.09	0.23	0.96	0.31	0.18	0.30
Bermudagrass hay early head	0.49	0.27	0.19	1.80	0.47	0.67	0.48
Blood meal, ring dried	0.30	0.30	0.03	0.33	0.40	0.33	0.77
Brewer's grains, wet	0.35	0.59	0.21	0.47	0.01	0.12	0.33
Canola seed	0.44	0.68	0.21	0.91	0.03	0.00	0.42
Canola meal, mech. extracted	0.75	1.10	0.53	1.41	0.07	0.04	0.73
Citrus pulp dried	1.92	0.12	0.12	1.10	0.06	0.08	0.10
Corn distiller's grains w/solubles, dried	0.22	0.83	0.33	1.10	0.30	0.26	0.44
Corn gluten feed, dried	0.07	1.00	0.42	1.46	0.13	0.20	0.44

APPENDIX TABLE 6 Mineral concentrations (dry matter basis) of some common feeds from 2001 Dairy NRC—*Continued.*

	CA%	P%	MG%	K%	NA%	CL%	S%
Corn gluten meal, dried	0.06	0.60	0.14	0.46	0.05	0.11	0.86
Corn grain, ground, dry	0.04	0.30	0.12	0.42	0.02	0.08	0.10
Corn grain and cob, high moisture, ground	0.05	0.28	0.12	0.48	0.01	0.07	0.90
Corn hominy	0.03	0.65	0.26	0.82	0.01	0.10	0.12
Corn silage, normal	0.28	0.26	0.17	1.20	0.01	0.29	0.14
Cottonseed, whole seed with lint	0.17	0.60	0.37	1.13	0.02	0.06	0.23
Cottonseed hulls	0.18	0.12	0.17	1.16	0.02	0.06	0.07
Fats - Calcium soaps of fatty acids	12.00	0.00	0.00	0.00	0.00	0.00	0.00
Fishmeal, menhaden	5.34	3.05	0.20	0.74	0.68	0.80	1.16
Grasses, Cool Season, intensive pasture	0.56	0.44	0.20	3.36	0.02	0.56	0.20
Grasses, Cool Season, hay, immature	0.72	0.34	0.23	2.57	0.03	0.42	0.24
Grasses, Cool Season, hay, mid maturity	0.66	0.29	0.23	2.13	0.08	0.92	0.24
Grasses, Cool Season, hay, mature	0.47	0.26	0.18	1.97	0.02	0.66	0.17
Grasses, Cool Season, silage, immature	0.57	0.36	0.22	3.11	0.05	0.67	0.21
Grasses, Cool Season, silage, mid maturity	0.60	0.36	0.21	2.78	0.05	0.67	0.21
Mixed hay, mid maturity	1.04	0.32	0.25	2.59	0.03	0.80	0.24
Mixed hay, mature	0.97	0.37	0.26	2.24	0.01	0.93	0.28
Mixed silage, immature	1.08	0.35	0.28	2.89	0.01	1.77	0.16
Mixed silage, mid maturity	1.09	0.35	0.27	2.80	0.01	1.10	0.26
Legume hay immature	1.56	0.31	0.33	2.56	0.03	0.55	0.33
Legume hay, mid maturity	1.37	0.30	0.30	2.45	0.02	0.61	0.31
Legume hay, mature	1.22	0.28	0.27	2.38	0.02	0.48	0.23
Legume silage, immature	1.39	0.36	0.30	3.03	0.03	0.55	0.30
Legume silage, mid maturity	1.36	0.35	0.28	3.00	0.02	0.61	0.28
Legume silage, mature	1.30	0.33	0.26	2.87	0.02	0.48	0.28
Meat meal, rendered	8.86	4.20	0.26	0.49	0.78	0.44	0.51
Molasses, sugarcane	1.00	0.10	0.42	4.01	0.22	0.00	0.47
Oats grain, rolled	0.11	0.40	0.16	0.52	0.03	0.00	0.19
Oats silage, headed	0.52	0.31	0.20	2.89	0.24	1.34	0.19
Rye silage, vegetative	0.43	0.42	0.16	3.34	0.05	0.90	0.20
Sorghum grain, dry rolled	0.07	0.35	0.17	0.47	0.01	0.06	0.11
Sorghum, sudan type, silage	0.64	0.24	0.31	2.57	0.03	0.56	0.15
Soybean hulls	0.63	0.17	0.25	1.51	0.01	0.05	0.12
Soybean meal, expellers	0.36	0.66	0.30	2.12	0.04	0.10	0.34
Soybean meal, solvent, 44% CP	0.40	0.71	0.31	2.22	0.04	0.13	0.46
Soybean meal, solvent, 49% CP	0.35	0.70	0.29	2.41	0.03	0.13	0.39
Soybean seeds, whole	0.32	0.60	0.25	1.99	0.01	0.04	0.31
Sunflower meal, solvent	0.48	1.00	0.63	1.50	0.04	0.12	0.39
Tomato pomace	0.22	0.47	0.28	0.98	0.12	0.00	0.15
Wheat, bran	0.13	1.18	0.53	1.32	0.04	0.16	0.21
Wheat grain, rolled	0.05	0.43	0.15	0.50	0.01	0.11	0.15
Wheat middlings	0.16	1.02	0.42	1.38	0.03	0.10	0.18
Wheat silage, early head	0.38	0.29	0.16	2.28	0.07	0.83	0.17
Wheat straw	0.31	0.10	0.14	1.55	0.12	0.60	0.11
Whey, wet, cattle	1.37	1.04	0.22	3.32	1.40	2.41	1.15

APPENDIX TABLE 7 Area conversions.

Metric

1 square centimeter	=	0.55 square inch
	=	100 square millimeters
1 square meter	=	1,550 square inches
	=	10,764 square feet
	=	1,196 square yards
	=	10,000 square centimeters
1 square kilometer	=	0.3861 square mile
	=	1,000 square meters
	=	2,471 acres
	=	10,000 square meter

Imperial

1 square inch	=	6.452 square centimeters
	=	1/144 square foot
	=	1/1296 square yard
1 square foot	=	929,088 square centimeters
	=	0.0929 square meter
1 square yard	=	8,361.3 square centimeters
	=	0.8361 square meter
	=	1.296 square inches
	=	9 square feet
1 square mile	=	2.59 square kilometers
	=	640 acres
1 acre	=	0.4047 hectare
	=	43,560 square feet
	=	4,840 square yards
	=	4,046.87 square meters
1 kilometer	=	3,281 feet
	=	1.094 yards
	=	0.621 mile
	=	1,000 meters
1 inch	=	25.4 millimeters
	=	2.54 centimeters
1 foot	=	30.48 centimeters
	=	0.3048 meter
	=	12 inches
1 yard	=	0.9144 meter
	=	91.44 centimeters
	=	3 feet
1 mile	=	1,609 meters
	=	1.609 kilometers
	=	5,280 feet
	=	1,760 yards

APPENDIX TABLE 8 Volume conversions.

Metric

1 milliliter	=	1 cubic centimeter (cc)
1 liter	=	1.057 U.S. quarts, liquid
	=	0.9081 quart, dry
	=	0.2642 U.S. gallon
	=	0.221 Imperial gallon
	=	1,000 milliliters or cc

Imperial

1 fluid ounce	=	1/128 gallon
	=	29.57 cubic centimeters
	=	29.562 milliliters
	=	1,805 cubic inches
	=	0.0625 U.S. pint (liquid)
1 U.S. quart, liquid	=	946.3 milliliters
	=	57.75 cubic inches
	=	32 fluid ounces
	=	4 cups
	=	1/4 gallon
	=	2 U.S. pints (liquid)
	=	0.946 liter
1 quart, dry	=	1.1012 liters
	=	67.20 cubic inches
	=	2 pints (dry)
	=	0.125 peck
	=	1/32 bushel
1 cubic inch	=	16.387 cubic centimeters
1 cubic foot	=	28.317 cubic centimeters
	=	0.0283 cubic meter
	=	28.316 liters
	=	7,481 U.S. gallons
	=	1,728 cubic inches
1 U.S. gallon	=	16 cups
	=	3.785 liters
	=	231 cubic inches
	=	4 U.S. quarts, liquid
	=	8 U.S. pints, liquid
	=	8.3453 pounds of water
	=	128 fluid ounces
	=	0.8327 British Imperial gallon
1 British Imperial gallon	=	4.546 liters
	=	1.201 U.S. gallons
	=	277.42 cubic inches
1 U.S. bushel	=	35.24 liters
	=	2,150.42 cubic inches
	=	1.2444 cubic feet
	=	0.03524 cubic meter
	=	2 pecks
	=	32 quarts, dry
	=	64 pints, dry

APPENDIX TABLE 9 Weight-unit conversion factors.

UNITS GIVEN	UNITS WANTED	FOR CONVERSION MULTIPLY BY
lb	g	453.6
lb	kg	0.4536
oz	g	28.35
kg	lb	2.2046
kg	mg	1,000,000.0
kg	g	1,000.0
g	mg	1,000.0
g	μg	1,000,000.0
mg	μg	1,000.0
mg/g	mg/lb	453.6
mg/kg	mg/lb	0.4536
μg/kg	μg/lb	0.4536
Mcal	kcal	1,000.0
kcal/kg	kcal/lb	0.4536
kcal/lb	kcal/kg	2.2046
ppm	μg/g	1.0
ppm	mg/kg	1.0
ppm	mg/lb	0.4536
mg/kg	%	0.0001
ppm	%	0.0001
mg/g	%	0.1
g/kg	%	0.1

APPENDIX TABLE 10 Temperature conversions.

To convert a temperature, in either Celsius (centigrade) or Fahrenheit, to the other scale, find that temperature in the center column, and then find the equivalent temperature in the other scale, either in the Celsius column to the left or in the Fahrenheit column to the right. For example, if a given temperature is 40°F, the equivalent temperature on the Celsius scale will be 4.44°C (shown in the left-hand column); if the given temperature is 40°C, the corresponding Fahrenheit reading will be 104°F (shown in the right-hand column).

On the Celsius scale the temperature of melting ice is 0° and that of boiling water is 100° at normal atmospheric pressure. On the Fahrenheit scale, the equivalent temperatures are 32° and 212°, respectively. The formula for converting Fahrenheit to Celsius is $C = 5/9(F - 32)$, and the formula for converting Celsius to Fahrenheit is $F = (9/5C) + 32$.

°C	°F OR °C READING	°F	°C	°F OR °C READING	°F
−73.33	−100	−148.0	−53.89	−65	−85.0
−70.56	−95	−139.0	−51.11	−60	−76.0
−67.78	−90	−130.0	−48.34	−55	−67.0
−65.00	−85	−121.0	−45.56	−50	−58.0
−62.22	−80	−112.0	−42.78	−45	−49.0
−59.45	−75	−103.0	−40.0	−40	−40.0
−56.67	−70	−94.0	−37.23	−35	−31.0
−34.44	−30	−22.0	3.33	38	100.4

PPENDIX TABLE 10 Temperature conversions—*Continued.*

°C	°F OR °C READING	°F	°C	°F OR °C READING	°F
−31.67	−25	−13.0	3.89	39	102.2
−28.89	−20	−4.0	4.44	40	104.0
−26.12	−15	5.0	5.00	41	105.8
−23.33	−10	14.0	5.56	42	107.6
−20.56	−5	23.0	6.11	43	109.4
−17.8	0	32.0	6.67	44	111.2
−17.2	1	33.8	7.22	45	113.0
−16.7	2	35.6	7.78	46	114.8
−16.1	3	37.4	8.33	47	116.6
−15.6	4	39.2	8.89	48	118.4
−15.0	5	41.0	9.44	49	120.2
−14.4	6	42.8	10.0	50	122.0
−13.9	7	44.6	10.6	51	123.8
−13.3	8	46.4	11.1	52	125.6
−12.8	9	48.2	11.7	53	127.4
−12.2	10	50.0	12.2	54	129.2
−11.7	11	51.8	12.8	55	131.0
−11.1	12	53.6	13.3	56	132.8
−10.6	13	55.4	13.9	57	134.6
−10.0	14	57.2	14.4	58	136.4
−9.44	15	59.0	15.0	59	138.2
−8.89	16	60.8	15.6	60	140.0
−8.33	17	62.6	16.1	61	141.8
−7.78	18	64.4	16.7	62	143.6
−7.22	19	66.2	17.2	63	145.4
−6.67	20	98.0	17.8	64	147.2
−6.11	21	69.8	18.3	65	149.0
−5.56	22	71.6	18.9	66	150.8
−5.00	23	73.4	19.4	67	152.6
−4.44	24	75.2	20.0	68	154.4
−3.89	25	77.0	20.6	69	156.2
−3.33	26	78.8	21.1	70	158.0
−2.78	27	80.6	21.7	71	159.8
−2.22	28	82.4	22.2	72	161.6
−1.67	29	84.2	22.8	73	163.4
−1.11	30	86.0	23.2	74	165.2
−0.56	31	87.8	23.9	75	167.0
0	32	89.6	24.4	76	168.8
0.56	33	91.4	25.0	77	170.6
1.11	34	93.2	25.6	78	172.4
1.67	35	95.0	26.1	79	174.2
2.22	36	96.8	26.7	80	176.0
2.78	37	98.6	27.2	81	177.8
27.8	82	179.6	100	212	414
28.3	83	181.4	104	220	428
28.9	84	183.2	110	230	446
29.4	85	185.0	116	240	464
30.0	86	186.8	121	250	482
30.6	87	188.6	127	260	500
31.1	88	190.4	132	270	518
31.7	89	192.2	138	280	536
32.2	90	194.0	143	290	554

APPENDIX TABLE 10 Temperature conversions—*Continued.*

°C	°F OR °C READING	°F	°C	°F OR °C READING	°F
32.8	91	195.8	149	300	572
33.3	92	197.6	154	310	590
33.9	93	199.4	160	320	608
34.4	94	201.2	166	330	626
35.0	95	203.0	171	340	644
35.6	96	204.8	177	350	662
36.1	97	206.6	182	360	680
36.7	98	208.4	188	370	698
37.2	99	210.2	193	380	716
37.8	100	212.0	199	390	734
43	110	230	204	400	752
49	120	248	210	410	770
54	130	266	216	420	788
60	140	284	221	430	806
66	150	302	227	440	824
71	160	320	232	450	842
77	170	338	238	460	860
82	180	356	243	470	878
88	190	374	249	480	896
93	200	392	254	490	914
99	210	410	260	500	932

Figure 4.5 on page 30 of the 5th edition of Basic Animal Nutrition and Feeding should be replaced by the accompanying Figure.

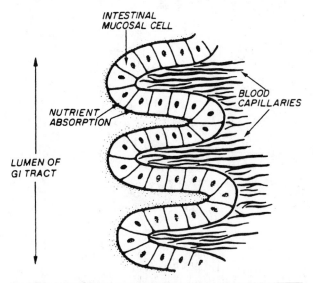

Figure 4.5 *Diagram of epithelial (mucosal) cells lining the small intestinal tract. Note the fingerlike projections, the villi, which greatly increase the absorptive surface area.*

You can find helpful animations of concepts covered in this text, along with text images and a downloadable copy of this appendix, on this text's website at www.wiley.com/college/pond.

www.wiley.com/college/pond

ISBN 0-471-73874-3

90000